PHYSICAL GEOGRAPHY

THE GLOBAL ENVIRONMENT

JOSEPH A. MASON
UNIVERSITY OF WISCONSIN–MADISON

PETER O. MULLER
UNIVERSITY OF MIAMI

JAMES E. BURT
UNIVERSITY OF WISCONSIN–MADISON

H. J. de BLIJ
MICHIGAN STATE UNIVERSITY

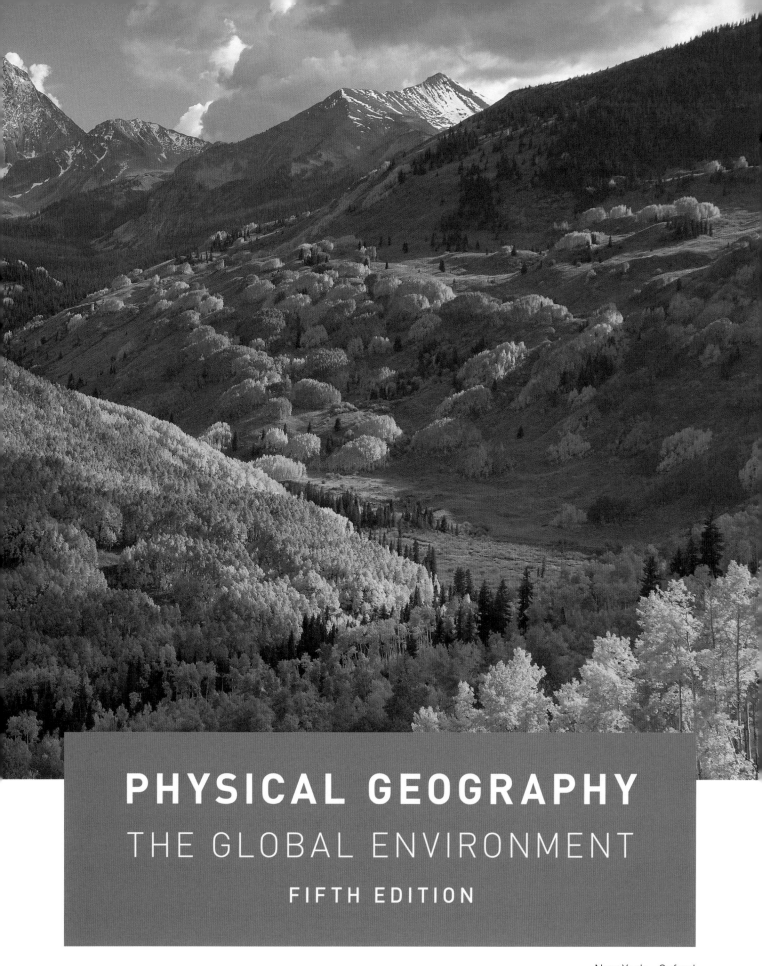

PHYSICAL GEOGRAPHY

THE GLOBAL ENVIRONMENT

FIFTH EDITION

New York   Oxford
OXFORD UNIVERSITY PRESS

Oxford University Press is a department of the University of Oxford.
It furthers the University's objective of excellence in research,
scholarship, and education by publishing worldwide.

Oxford   New York
Auckland   Cape Town   Dar es Salaam   Hong Kong   Karachi
Kuala Lumpur   Madrid   Melbourne   Mexico City   Nairobi
New Delhi   Shanghai   Taipei   Toronto

With offices in
Argentina   Austria   Brazil   Chile   Czech Republic   France   Greece
Guatemala   Hungary   Italy   Japan   Poland   Portugal   Singapore
South Korea   Switzerland   Thailand   Turkey   Ukraine   Vietnam

For titles covered by Section 112 of the US Higher Education
Opportunity Act, please visit www.oup.com/us/he for the
latest information about pricing and alternate formats.

Published by Oxford University Press
198 Madison Avenue, New York, New York 10016
http://www.oup.com

Library of Congress Cataloging-in-Publication Data

Mason, Joseph A.
   Physical geography : the global environment / Joseph A. Mason, University of Wisconsin-Madison, James E. Burt,
University of Wisconsin-Madison, Peter O. Muller, University of Miami, H.J. de Blij, Michigan State University. -- Fifth
edition.
      pages cm
   H. J. de Blij is listed as the first author of the fourth edition.
   ISBN 978-0-19-024686-0 (acid-free paper) 1. Physical geography. I. Title.
   GB55.D34 2016
   910'.02--dc23
                                                          2015030031

Printing number: 9 8 7 6 5

Printed in the United States of America
on acid-free paper

With great admiration for his remarkable and wide-ranging contributions to American geography, we proudly dedicate this edition to the memory of our founding co-author

HARM J. DE BLIJ (1935-2014)

Harm will long be remembered for his accomplishments during his highly successful career as a scholar, educator, and communicator—all of which were harnessed in his passionate, lifelong crusade to advance the cause of geography in the public arena.

BRIEF TABLE OF CONTENTS

TABLE OF CONTENTS

PREFACE

On March 25, 2014, American geography lost one its most distinguished and best-known practitioners of the past half-century, Dr. Harm J. de Blij, the John A. Hannah Professor of Geography at Michigan State University. For the authors and editors of this book, the loss was far more personal: Professor de Blij was the dynamic senior author of the four previous editions, our constant inspiration, and a cherished friend of many years. His training as a professional geographer, all the way through earning his Ph.D. at Northwestern University, focused on the geomorphology of the lithosphere (both terms that readers will shortly become familiar with); more specifically, he analyzed landform patterns in the southeastern African country of Swaziland, particularly rift valleys that would shortly become defined as pieces of the gigantic jigsaw puzzle that constitute global plate tectonics—a central component of the contemporary Earth sciences.

Perhaps the most important contribution of Professor de Blij, during the six-plus decades of his hugely successful career, was to almost single-handedly advance the cause of geography in the U.S. public arena. His talents and skills as a communicator and passionate ambassador for our scholarly discipline were truly legendary—a rich legacy that survives in a range of media, from YouTube videos of his lectures to his remarkable, user-friendly writings for college-level student readers (including many that still grace the pages of this book). No wonder he was so greatly admired for his work as Geography Editor in a seven-year stint on ABC's *Good Morning America*. He subsequently parlayed that platform into several books for general readers, all beautifully written, every page sparkling with insights and boundless enthusiasm for the subject he loved and literally pursued in his travels to the ends of the Earth. His three most successful forays into this realm were proudly published by Oxford University Press: *Why Geography Matters* (2005), *The Power of Place* (2009), and *Why Geography Matters . . . More Than Ever* (2012).

It is indeed an honor for his coauthors of this book to dedicate the Fifth Edition of *Physical Geography* to the memory of Professor Harm de Blij. In fact, his contribution to this book was so fundamental and enduring, we are able to reprint the immediately following opening section of this Preface as he revised it for the previous edition in 2013. As an introductory overview of the scope, importance, and current scientific context of physical geography with respect to our changing global environment—he absolutely nailed it!

Why spend a semester (or more) studying physical geography? You will find the answer in this course, but you can also find it elsewhere: in your daily check on the weather before you set out; in news reports about natural events such as earthquakes, floods, and volcanic eruptions; in sometimes-vociferous public debates about climate change;

in scientists' pleas for sustainable lifestyles. You will find that the components of nature that form our small planet's diverse environments are intricately interconnected not only with each other but also with you. A volcanic eruption on the other side of Earth can kill the crops on which people and animals depend half a world away—we know, because it happened as recently as the early 1800s. An earthquake near Indonesia can take lives thousands of kilometers away: it happened to more than a quarter of a million people in December 2004. When a combination of offshore earthquakes and resulting sea waves struck nuclear plants in northeast Japan in March 2011, radiation exposure threatened those who did not perish, and in the aftermath huge "rafts" of resulting debris, floating in the North Pacific Ocean, reached beaches in California and Oregon.

We live on a planet whose crust is still unstable, and we will show you maps that reveal where the risk is greater, and where it is less. But while the crust, after another few billion years, may finally come to rest, planetary environments are likely to continue changing. Again, these changes affect you directly: the continued thinning of the protective ozone layer makes sunburn ever more dangerous to your health, and the environmental swings associated with climatic reversals are related to severe droughts in some areas and, perversely, floods in others, resulting in crop failures and rising food prices (and hunger among the poorest of the world's poor). In August 2012 India was struck by a series of power outages called the worst in human history, the second one affecting 670 million people across the country's north. Explanations ranged from overuse of electricity to lack of maintenance of India's electrical "grid," but here is the real reason: India's annual monsoon had failed, hydroelectric dams stood silent at the outlets of nearly empty reservoirs, and the country's coal-fired power plants could not cope with demand. Thus a climatic event had consequences felt in cities, at airports, on trains, and along traffic-stalled highways.

In case you have not noticed, our society has its own shortages, and one of them relates to the way we look at the wider world. We tend to localize and specialize, focusing on detail and, as many writers have found, on the small and the familiar rather than the bigger picture and the wider world. That is not what you will find in this book. This is the story of the way the planet works, of its place in our solar system, of the amazing range of its natural environments, of the powerful forces that shape its landscapes, climates, ecologies, and—yes—geographies, because humans have become part of the planetary equation. Long before humans emerged on Earth, the planet's surface was continuously shaped and reshaped by nature's systems and cycles, whose properties we will examine in the pages that follow. It is a riveting story of ice ages and intervening warmth, of sudden death from space, of rising and falling sea levels, of shifting landmasses and shocking earthquakes, of choking eruptions and giant floods, of flowering life and devastating extinctions. Less than 200,000 years ago—a matter of minutes in your lifetime when compared to the planet's 4.6 billion—modern humans appeared in Africa, and now, more than 7 billion strong, we are transforming Earth as no creature has ever done. That is why some scientists are proposing that the current "geologic" epoch be given a human name: the *anthropocene*. Not only are we paving over entire urban regions, cropping vast expanses, and damming major rivers, we are also filling the atmosphere with pollutants that are, as we will see, affecting the course of climate change. This is why the human factor has some prominence in this book.

Interesting as these topics are, they do require careful study. Not everyone who participates loudly in the climate change debate is sufficiently well informed on the basics, nor is physical geography an easy field. But if you stay the course, you will be among the most geographically literate of citizens . . . a worthy objective, to be sure. And beyond that, what you learn will be of personal, practical value and use, not only making life more interesting but, quite possibly, giving you a new outlook on it.

Hallmarks of *Physical Geography: The Global Environment*

Studying a demanding field such as physical geography is a tall order, and the process is easier if strong educational materials aid one's efforts. This book and its ancillary materials have been specifically designed to support professors and students in this endeavor. The book employs a unique approach characterized by several important attributes.

- *Structure.* *Physical Geography* organizes the material into 49 short units instead of the 20 or so long chapters found in most physical geography books. These shorter and more self-contained units are a good fit for students who increasingly seem to prefer brief study periods sandwiched between other activities. (Of course, they can be read in succession and thus accommodate extended interludes.) These short units also provide instructors with much more flexibility in assigning readings or rearranging the order in which materials are taught. For those professors who wish to customize the book, the units offer an easy way to get just the right materials to the student audience.
- *Keeping humans in the picture.* Understanding the role of humans and how humans interact with Earth is critical to building a good geographic understanding. This book carefully includes the human role in the Earth system.
- *Writing with clarity and impact.* The book is known for clear, compelling prose that is free from extraneous detail and unimportant diversions.
- *Qualitative precision.* Descriptions and explanations have been methodically evaluated for accuracy, completeness, and currency. Thus, for example, there are no

references to "air holding water vapor," and the classic depiction of cold fronts as snowplows is offered as a historic (and convenient) conceptual model rather than a physical process rooted in theory. Instead, the book focuses on tying fundamental physical principles to observable phenomena, such as applying a knowledge of *how* heat is converted to kinetic energy to the energetics of weather systems and discussions of how climate change is manifested in extreme events, including recent incidences of extreme heat; these discussions are supported by examples such as specific coverage of California's recent and ongoing drought.

- *Strong reliance on a systems perspective.* For us that means an emphasis on process and interaction rather than obsession over the arcane terminology of formal systems theory. Throughout the book we appeal to conservation principles presented in Unit 1 for explanation of phenomena from molecular to global scales. The book pays particular attention to feedbacks and the timescales on which they operate. For example, students reading Unit 18 will understand why the carbon cycle acts as a brake on climate change over millions of years and yet can amplify change over shorter time periods. In every unit the authors have made a conscious attempt to explain *why* things are as they are, and to point out where the knowledge gaps are when that isn't possible.
- *Meticulously chosen examples illustrate the topic at hand clearly.* Clear, illustrative examples that vividly illustrate the concept at hand are critically important to learning geography. *Physical Geography* features many such highly illustrative examples.
- *Emphasis on current* science *rather than current* events. Because current events pass quickly, this book emphasizes basic geographic science and leaves it to the professor to contextualize that material with examples from the daily newspaper. Minimizing attention to the ephemeral allows the book to stay focused on underlying concepts and on the clearest, most illustrative examples.
- *Financial discipline and nonprofit status result in lower prices to students.* Oxford University Press, a department of the University of Oxford, is a nonprofit publisher devoted to furthering the University's objective of excellence in research, scholarship, and education. Since accessible materials clearly support this mission, the Press uses a combination of nonprofit status and financial discipline to offer course materials that generally cost students significantly less money than those offered by commercial publishers.

New to the Fifth Edition

The Fifth Edition has been extensively revised, updated, and improved. Some of the most important changes include:

- *Extensive new diagrams.* New figures were developed to guide students through complex concepts like biogeochemical cycles, the rock cycle, and feedbacks in soil formation. Instead of abstract boxes and arrows, these diagrams connect cycles and systems to the real world: the rock cycle is shown as taking place in a cutaway view of the lithosphere, and feedbacks in soil development are illustrated with a series of realistic soil profiles, for example. A new figure on the mechanics of slope failure makes an important broader point by showing how an abstract view of the balance of forces on a block of soil can be related to the potential failure of a realistic hillslope. In the units on tectonics and volcanism, many of the figures have been completely revised for up-to-date scientific accuracy and to illustrate features more effectively in three dimensions. Climagraphs in all units are plotted on the same scale for easier comparison.

- New climate maps based on state-of-the-art datasets. The authors constructed new global fields of temperature, precipitation, and evaporation by merging terrestrial and satellite-derived ocean data from the University of Delaware's Center for Climatic Research and the University of Hamburg Ocean-Atmosphere project, respectively. The resulting climate maps have details not seen in traditional textbook figures (e.g., summer-dry continental climates) and are far more defensible scientifically. New figures showing the latitudinal distribution of climate types were also computed from the merged data.

- *In-depth treatment of biogeochemical cycles.* The Fourth Edition was distinguished by its comprehensive, up-to-date coverage of climate change and its environmental impacts. The new edition adds a key element to this focus: a unit devoted to biogeochemical cycles, emphasizing cycles of carbon and nitrogen. For both elements, diagrams and text illustrate the important pools and fluxes and current estimates of their magnitude, as well as recent human impacts. Major uncertainties are noted while emphasizing the rapid progress toward understanding these cycles that has been made in recent years. As in the rest of the book, the process of scientific discovery is highlighted with well-chosen examples.

- *More effective, up-to-date presentation of the science.* The Fourth Edition was also marked by an effort to move away from well-worn but often simplistic or even misleading explanations of key processes toward an approach that clearly and accurately presents up-to-date science. The Fifth Edition continues and extends this effort in a number of areas, while updating topics that are the focus of active research; for example:
  - The many connections made between ecosystems, soils, and the global carbon cycle throughout Part Four (The Biosphere) have been updated to reflect recent progress in understanding these links. For example, the "missing sink" of carbon was highlighted as an intriguing problem in the Fourth Edition; this sink is now more definitely attributed to carbon uptake by terrestrial ecosystems, especially tropical forests, in line with recent assessments. The greatly enhanced coverage of the nitrogen cycle incorporates advances in understanding the complex

pathways of nitrogen (including emissions of the greenhouse gas N_2O) that have resulted from intensive study in recent years.
  - More systematic discussion of streams and glaciers as open systems, including forms of energy and work within these systems, has been included. At the same time, as in earlier editions, these somewhat-abstract discussions are balanced by vivid examples of the work of streams and glaciers, and by occasional historical background where it provides especially good illustrations of the scientific process. Explanations of other geomorphic and tectonic processes (e.g., wind erosion, mantle convection, volcanism associated with subduction) have also been carefully scrutinized and revised where needed to make them more effective, more consistent with current research, and more anchored in basic physical science
  - A substantially new discussion of evolution and its importance for biogeography has been added, including more up-to-date and effective explanations of key concepts such as natural selection, gene flow, and speciation, along with compelling examples. Here and in several other sections, the new edition retains sections on the historical background of important concepts, but in revised form, explaining how the science evolved but also leading more directly to our current understanding of it.

- Climatic time series have been updated with information through 2014, including the latest National Oceanic and Atmospheric Administration global mean temperature estimates (*Science*, June 2015) that debunked the suggestion of a warming pause or hiatus. The discussion and figures related to anthropogenic climate change have been revised based on the Fifth Assessment Report of the Intergovernmental Panel on Climate Change, and there is new coverage of hydrological impacts of greenhouse gas emissions.
  - A new section on atmospheric rivers has been added in accord with the growing realization of their importance in global energy/mass transport and flood events.

- *Reorganized for better flow and clarity.* In response to reviewer feedback, Part Five, on the lithosphere, has been reorganized to place the units on minerals and rocks before those that describe the broader structure of the lithosphere. In other words, students working through the units of this part in order will proceed from mineral structures through rock types to Earth's internal structure and tectonics. The number of units has been reduced from 50 to 49.

- Presentation enlivened and economized with digitally enhanced figures. Sometimes a difficult concept or important graph comes more quickly into focus if supported by a brief explanation, or connections between landscapes and processes become more vital when motion enhances the presentation. In this edition many figures have been enhanced with live content that runs for less than a minute per figure and can be easily played on a smartphone. These resources are designed to help users learn concepts and processes more quickly and efficiently, and thus do not burden the user with tedious re-recitations of vocabulary or foundational material already explicated in the book.

- *Digitally enhanced "Quick Review" end-of-unit materials.* In this edition students can use their smartphone to immediately check their understanding of the material by taking quizzes about the vocabulary and concepts presented in every unit.

Teaching and Learning Package

This book is supported by an extensive and carefully developed set of ancillary materials designed to support both professors' and students' efforts in the course. The depth, breadth, and quality of these materials was dramatically expanded between the Third and Fourth editions of the book, and they have been further refined and expanded for the Fifth Edition. We have been fortunate to work with a dedicated and creative group of authors when assembling this package. Steve LaDochy and Angel Hamane of California State University, Los Angeles, have been instrumental in the creation of all of the supplements and have made great contributions to the augmented figures and end-of-chapter materials. We also continue to benefit from the efforts of Pedro Ramirez, who worked with Drs. Ladochy and Hamane on the supplements for the Fourth Edition. Much of his work has been carried forward in revised form to the Fifth Edition. The result of this combined effort is the following:

- *Oxford University Press Animation Series.* Animation and visualization are very helpful when studying physical geography. Recognizing this, our authors have worked with leading animators to produce clear, dramatic, and illustrative animations and visualizations of some of the most important concepts in physical geography. Animations are available to adopting instructors at no charge.
- *Digital files and PowerPoint presentations.* Instructors will find all the animations—the Oxford University Press Animation Series, all of the images from the text, and some animations and visualizations from other sources—available to them, pre-inserted into PowerPoint. In addition, our ancillary author team has created suggested lecture outlines, arranged by chapter, in PowerPoint. These materials are free to adopting professors.
- *Interactive animation and visualization exercises.* Our animation and visualization exercises begin with the Oxford University Press Animation Series and incorporate interesting animations and visualizations from other sources. The exercises then guide the students through a series of activities based on the visualization at hand. Responses are automatically graded by the computer.

- *Review questions for students.* Carefully crafted to highlight the most important concepts in each chapter, these computer-graded review questions accompany each chapter in the textbook. Professors can assign them for homework, or students can use them independently to check their understanding of the topics presented in the book. Responses are automatically graded by the computer.
- *Test questions and testing software.* Answerable directly from the text, these questions provide professors with a useful tool for creating and administering tests.
- *Dashboard.* A text-specific, integrated learning system designed with clear and consistent navigation, the dashboard delivers quality content and tools to track student progress in an intuitive, Web-based learning environment. This tool features a streamlined interface that connects instructors and students with the functions they perform most, simplifying the learning experience to save time and put student progress first.
- *Course cartridges.* Instructors may order selected digital supplements free of charge in ready-to-upload form for the most popular course management systems, including Blackboard, D2L, Moodle, Canvas, and Angel, by contacting their Oxford University Press representative.
- *Open-access quizzing for students.* In keeping with Oxford University Press's mission to disseminate educational materials as widely as possible at the lowest cost, we have posted free review quizzes that offer immediate feedback at www.oup.com/us/mason.
- *Lab manual.* Dalton Miller and Andrew Mercer, both of Mississippi State University, have prepared a lab manual. While this manual can be employed by anyone, it has been specifically designed to accompany the base textbook. If the professor so chooses, the lab manual can be bundled together with the text for a significant discount.

Acknowledgments

Writing a modern textbook and its ancillary package is a team effort, and during the preparation of this new edition of *Physical Geography: The Global Environment*, we received advice, guidance, and input from many quarters. First and foremost, we thank our reviewers and colleagues who supplied valuable feedback on the previous edition and valuable advice about the revised manuscript. These are:

Edward Aguado, San Diego State University
Kent B. Barnes, Towson University
Jacob Bendix, Syracuse University
David Butler, Texas State University
Maria Caffrey, University of Tennessee
David Cairns, Texas A&M University
Mary Costello, San Diego State University
Richard Crooker, Kutztown University
Melinda Daniels, Kansas State University
Caroline P. Davies, University of Missouri, Kansas City

Thomas C. Davinroy, Metropolitan State University
of Denver
Carol Ann Delong, Victor Valley College
Jeremy E. Diem, Georgia State University
John D. Frye, University of Wisconsin, Whitewater
Sarah Gagne, University of North Carolina, Charlotte
Paul Geale, Glendale Community College
William L. Graf, University of South Carolina
Christine Hansell, Skyline College and Laney College
William Hansen, Worcester State University
James J. Hayes, California State University, Northridge
April Hiscox, University of South Carolina
Chris Houser, Texas A&M University
Daniel M. Johnson, Portland State University
Lawrence Kiage, Georgia State University
Stephen E. G. LaDochy, California State University,
Los Angeles
Phillip H. Larson, Minnesota State University
Kevin Law, Marshall University
Jeffrey Lee, Texas Tech University
Denyse Lemaire, Rowan University
Laszlo MariaHazy, Saddleback College
Blake Mayberry, Red Rocks Community College
Elliot G. McIntire, California State University, Northridge
Kendra McLauchlan, Kansas State University
William Monfredo, University of Oklahoma
Donald L. Morgan, Brigham Young University
Steve Namikas, Louisiana State University
Sarah Null, Utah State University
Narcisa Pricope, University of North Carolina, Wilmington
Steven Quiring, Texas A&M University
Anne Saxe, Saddleback College
Jacqueline Shinker, University of Wyoming
John F. Shroder, Jr., University of Nebraska at Omaha
Erica Smithwick, Penn State University
Jane Thorngren, San Diego State University
Paul Todhunter, University of North Dakota
Mark Welford, Georgia Southern University
Thomas A. Wikle, Oklahoma State University
Forrest D. Wilkerson, Minnesota State University, Mankato
Harry F. L. Williams, University of North Texas
Hengchun Ye, California State University, Los Angeles
Stephen Yool, University of Arizona
Craig ZumBrunnen, University of Washington
In addition to the many reviewers listed above who offered
comments both on paper and in focus groups, we'd like to
thank the hundreds of colleagues who responded to surveys
conducted by Oxford University Press. This broad-based
feedback provided invaluable information we used to make
a number of decisions about how the material should be
presented and arranged.

Several valued colleagues engaged in a technical review of
some of the units. These experts combed through those
units to make sure that the science was completely accurate
and current. For this we thank:

Ed Aguado, San Diego State University
Michael E. Mann, Penn State University
William Monfredo, University of Oklahoma

Paul Todhunter of the University of North Dakota contributed enormously to the project by writing initial drafts of
several units. We're not sure what we would have done
without Paul's graceful research and writing, and appreciate his attention to detail and timeliness more than we can
say here. This book is much better as a result of his efforts.

Perhaps nowhere in the textbook process is the notion
of a team more apparent than in the ancillary program. We
are deeply indebted to the individuals who have spent
many hours crafting ancillaries for the professors and students who will ultimately use this book for their classes.
Steve LaDochy, Pedro Ramirez, and Angel Hamane, all of
California State University, Los Angeles, have worked together and with our authors to create a superb selection of
instructor and student supplements.

Last, we are indebted to our colleagues at Oxford University Press, who superbly managed the preparation of
this Fifth Edition. Executive Editor Dan Kaveney set the
revision process into motion and insisted on the highest
standards as he oversaw the entire project. We are grateful
for his thoughtful reflection about how to improve the
book, for the many concrete ideas he contributed, and for
his ability to magically bring them about. Development
Editor Erin Mulligan and Director of Development Thom
Holmes admirably handled the developmental editing,
constantly exuding high energy and infectious enthusiasm. Sarah Vogelsong provided superb copyediting and
remedied many errors that had gone unnoticed by authors
and editors alike. Production Editor David Bradley coordi-
nated myriad details as he smoothed us through the production process, and the fantastic staff at the University of
Wisconsin Cartography Lab, ably led by Tanya Bucking-
ham, drafted the art program. Associate Editor Christine
Mahon and Editorial Assistant John Appeldorn gracefully
handled the logistics of reviewing and spent many hours
providing meticulous help in preparing the manuscript for
production. David Jurman, our indefatigable marketing
manager, and the Oxford University Press sales staff have
put forward tireless efforts on behalf of the book. Our sin-
cerest thanks to you all.

Joseph A. Mason
James E. Burt
Peter O. Muller

Introducing Physical Geography

Planet Earth and the Moon in the northern summer—clear U.S. skies, a hurricane off Mexico, snow-bearing clouds over Antarctica.

OBJECTIVES

- Identify the contemporary focus of physical geography
- Describe the relationship of physical geography to the other natural and physical sciences
- Explain the scientific method as a way of investigating the natural world
- Summarize the systems and modeling approaches of physical geography

This is a book about Earth's natural environments. Its title, *Physical Geography: The Global Environment*, suggests the unifying perspective. In this book, we survey Earth's human habitat and focus on the fragile layer of life that sustains us along with millions of other species of animals and plants. We examine features of the natural world such as erupting volcanoes, winding rivers, winter blizzards, advancing deserts, and changing climates not only as physical phenomena but also in terms of their relationships with human societies and communities.

Along the way, we will do more than study the physical world. We will learn how broad a field physical geography is, and we will discover many of the topics on which physical geographers do scientific research. To proceed, we need to view our planet from various vantage points. We will study Earth in space and from space, from mountaintops and in underground shafts. Our survey will take us from clouds and ocean waves to rocks and minerals, from fertile soils and verdant forests to arid deserts and icy wastelands.

Geography

Geographers are not the only researchers studying Earth's surface. Geologists, meteorologists, biologists, hydrologists, and scientists from many other disciplines also study aspects of the planetary surface and what lies above and below it. But only one scholarly discipline—geography—combines, integrates, and, at its best, *synthesizes* knowledge from all these other fields as it makes its own research contributions. Time and again in this book, you will become aware of connections among physical phenomena and between natural phenomena and human activities. Although you might be familiar with many of these phenomena from everyday life or perhaps have studied them in another course, you might not have encountered these connections.

Although geography is a modern discipline and its scientists increasingly use high-technology research equipment, the roots of geography extend to the very dawn of scholarly inquiry. When the ancient Greeks recognized the need to organize the knowledge they were gathering, they divided it all into two areas: geography (the study of the terrestrial world) and cosmography (the study of the skies, stars, and the universe beyond). A follower of Aristotle, a scholar named Eratosthenes (ca. 275–ca. 195 BCE), coined the term *geography* in the third century BCE. To him, geography was the accurate description of Earth (*geo*, meaning Earth; *graphia*, meaning description). During Eratosthenes's lifetime, volumes were written about rocks, soils, and plants. A magnificent library in Alexandria (in what is now Egypt) came to contain the greatest collection of existing geographical studies.

Soon the mass of information (i.e., the database) concerning terrestrial geography became so large that the rubric lost its usefulness. Scientific specialization began.

Some scholars concentrated on the rocks that make up the hard surface of Earth, and geology emerged. Others studied living organisms, and biology grew into a separate discipline. Eventually, even these specializations became too comprehensive. Some biologists, for example, began to exclusively focus on plants (botany) or animals (zoology). The range of scientific disciplines expanded—and continues to expand to this day.

This, however, did not mean that geography itself lost its identity or relevance. As science became more compartmentalized, geographers realized that they could contribute in several ways, not only by conducting specialized research but also by maintaining that connective, integrative perspective that links knowledge from different disciplines. One aspect of this perspective relates to the "where" with which geography is popularly associated. The location or position of features on the surface of Earth (or above or below it) may well be one of the most significant elements about these features. Thus, geographers seek to learn not only about the features themselves but also about their spatial relationships. The word *spatial* comes from the noun *space*—not the outer space surrounding our planet, but Earthly, terrestrial space.

The question is not only *where* things are located but also *why* they are positioned where they are and *how* they came to occupy those positions. To use more technical language: What is the cause of the variations in the distribution of phenomena we observe to exist in geographic space? What are the dynamics that shape the spatial organization of each part of Earth's surface? These are among the central questions that geographers have asked for centuries and continue to pose today.

GEOGRAPHY AND PHYSICAL GEOGRAPHY

Specialization has also developed within geography. Although all geographers share an interest in spatial arrangements, distribution, and organization, some geographers focus on physical features or natural phenomena, whereas others concentrate on people and their activities. This results in two very broad divisions of the discipline, *physical* and *human* geography. Other geographers work in *geographic information science*, developing new methods for the display, analysis, and management of spatial information. Still others emphasize two-way interactions between people and their physical environment in a subarea known as *nature-society relations*. And even within these four broad areas, there are subdivisions. For example, a physical geographer may work on shorelines and beaches, on soil erosion, or on climate change. A human geographer may be interested in urban problems, in geopolitical trends, or in health issues. As a result, geography today consists of an ever-evolving cluster of fields and subfields.

Our book's concern is with just one subfield—namely, the geography of the physical world. Much of the planet remains unaffected by humans and thus

FIGURE 1.1 This view of Phoenix, Arizona, points to many questions addressed by physical geography, such as: Why are arid climates found in this corner of the United States? What plant forms and plant communities develop in such locales? In what ways and why are the landforms and soils so different than those found in more humid landscapes? How do concrete, glass, and tall buildings affect the local climate in the city core?

remains in its "natural" state. Physical geography certainly studies the natural world, but this is not to say it concerns itself only with natural landscapes. For example, a geographer examining the effects of dam removal on flood events is doing physical geography. Similarly, determining the hydrologic response to the invasion of farmland by suburbs is a topic for physical geography. As suggested by Figure 1.1, studies of how cities affect their climate are physical geography, despite the obvious human role in creating the physical environment. The distinguishing feature of physical geography is that where human activity is involved, the human element is largely a given rather than the object of study. Untangling the economic and cultural forces that produce changes in land use, for example, is not considered a physical geography topic.

SUBFIELDS OF PHYSICAL GEOGRAPHY

Over the past century, physical geography has evolved into a cluster of subfields, the most important of which are diagrammed in the inner circle of Figure 1.2. To make things easier, we have numbered the eight subfields. Notice that each is connected to a related discipline. Physical geographers specializing in a subarea typically have substantial training in the associated field.

The geography of landscape, **geomorphology** (1 in Fig. 1.2), remains one of the most productive subfields of physical geography. As the term suggests (*geo*, meaning Earth; *morph*, meaning shape or form), this area of research focuses on the structuring of Earth's surface. Geomorphologists seek to understand the evolution of slopes, the development of plains and plateaus, and the processes shaping dunes and caves and cliffs—the elements of the

physical landscape. Often, geomorphology has far-reaching implications. From the study of landscape, it is possible to prove the former presence of ice sheets and mountain glaciers, rivers, and deserts. *Geology* (the study of Earth's physical structure) is geomorphology's closest ally, and there is enough overlap between the two that geomorphology is sometimes taught as a subfield in geology departments. But regardless of where it is located academically, within geomorphology the processes of running water, moving ice, surging waves, and restless air are the primary concerns because of their role in landscape genesis. Proceeding clockwise from the top of Figure 1.2, we observe that *meteorology* (the branch of physics that deals with atmospheric phenomena) and physical geography combine to form **climatology** (2), the study of climates and their spatial distribution. Climatology involves not only the classification of climates and the analysis of their distribution but also broader environmental questions, including those concerning climatic change, vegetation patterns, soil formation, and the relationships between human societies and climate.

As Figure 1.2 indicates, the next three subfields relate physical geography to aspects of *biology*. Where biology and physical geography overlap is the broad subfield of **biogeography** (3–5), and there are specializations within biogeography itself. Physical geography combined with botany is *plant geography* (also called phytogeography) (3); combined with zoology, it becomes *zoogeography* (5). Note that biogeography (4), itself linked to *ecology*, lies between these two subfields. Both zoogeography and plant geography are parts of biogeography. The next subfield of physical geography is related to soil science, or *pedology*. Pedologists' research tends to focus on the internal

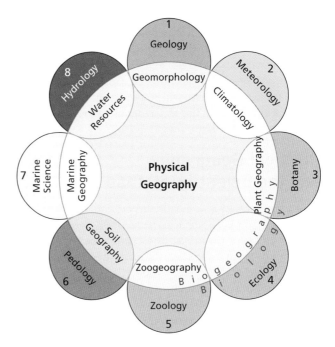

FIGURE 1.2 Important subfields of physical geography and related disciplines.

properties of soils and the processes of soil development. In **soil geography** (6), research centers on the spatial patterns of soils, their distribution, and their relationships to climate, vegetation, and humankind.

Other subfields of physical geography are **marine geography** (7) and the study of **water resources** (8). Marine geography, which is related to the discipline of *marine science*, also has human as well as physical components. The human side of marine geography has to do with maritime boundaries, the competition for marine resources, and the law of the sea; therefore, this subfield is closely allied with the geographic subdiscipline of political geography. The physical side of marine geography deals with coastlines and shores, beaches, river mouths, and other landscape features associated with the oceanic margins of the continents. The subfield of water resources (where *hydrology* and physical geography intersect) also has human as well as physical elements. The landmasses contain fresh water on their surfaces (in the form of lakes and rivers) and below (as groundwater). As we will see later in this book, the study of these waters may be approached from geomorphic as well as economic standpoints.

Figure 1.2 is a simplified view of physical geography, and the subareas themselves comprise more specialized fields. These days, when you ask a physical geographer what his or her specialty is, you frequently hear such answers as hydroclimatology, periglacial processes, paleogeography, or wetland ecosystems. All this helps explain the wide range of material you will encounter in this book, which is an overview of a broad field encompassing many topics. In the 49 units that follow, these various subfields of physical geography are examined in some detail, and the connections between physical and human geography are revealed. As you will see, we must go beyond the confines of Figure 1.2 to put our work in proper perspective. To understand the basics of physical geography, we must comprehend the general properties of our planet, not only deep below its surface but also far beyond as it orbits the Sun as part of the solar system in our tiny corner of the vast universe. Comprehending the general properties of Earth will be among our first tasks and will constitute much of the remainder of Part One.

Physical Geography and the Scientific Method

Physical geography is an Earth science, and like its related fields (and physical sciences generally), it relies on a framework called the **scientific method**. The term is a bit misleading, because the scientific method is more a way of knowing or learning about the world than a strict method or process. Humans have developed a variety of ways to "know" or claim something is one way and not another. We sometimes rely on common sense or intuition in asserting something as fact. Or we grant some persons the right to speak with certainty about some topics. This authority might be an expert with impressive scientific credentials, or it might be a religious leader who claims to have a divine source for his or her pronouncements. Of course, many oral and written traditions offer explanations for elements of the physical world, including Earth's formation. Any of these might yield a fact no reasonable person would dispute, but none is scientific.

Unlike other ways of knowing, science places demands on statements of fact. For example, intuition fails as a scientific approach because it is not reproducible. Studying the same phenomenon, two people might come up with completely different explanations. Likewise, there is no way to be sure a particular scholarly prophet has truly heard the voice of an all-knowing deity, and therefore science grants no speaker a special position. The scientific method consists of certain elements that together constitute a convention for obtaining facts or learning the "truth" about phenomena of interest. Although obviously not the only approach, this convention has been in use for centuries, and there is no denying its remarkable success in a multitude of areas. For example, we depend on vaccines to protect us from horrific diseases, we travel without qualms in metal contraptions miles above the ground, and we talk to friends on other continents using devices no larger than a child's hand. None of these would be possible or even imaginable without discoveries achieved through the scientific method.

The scientific method is typically presented as a series of steps, as shown in Figure 1.3, several of which deserve comment. First, *information gathering* is not just measurement but also includes subjective observation and the precise definition of terms. **Hypotheses** are tentative explanations of observations and measurements. For example, we might have observed stunted growth in a forest, defined as more than 50 percent of tree stems being smaller than 2 inches in diameter. As a hypothesis, we propose that unfavorable chemical conditions in the soil prevent the trees from obtaining essential nutrients. Like people, plants need calcium, iron, and other "nutrient" chemicals to grow. For a plant these must be present in the soil, and in addition, the soil must have the right level of acidity. If soil acidity is too low, nutrient uptake is hampered and plants suffer. For this example, we hypothesize that the soil acidity is too low for nutrient uptake.

A hypothesis helps us to generate *predictions*, which are not typically or even mainly forecasts. Rather, they are deductions that follow from the hypothesis. In our tree example, a prediction might be that although a certain soil contains abundant nutrients, the acidity is too low to permit nutrient uptake. The prediction leads to an **experiment**, which is a very general term. Growing plants in a greenhouse is an experiment, but so are performing calculations using a computer program or collecting soil samples and measuring acidity and nutrient content. Data resulting from the experiment are analyzed and compared with the predictions to rule on the hypothesis.

There is an important subtlety of the scientific method not always appreciated by the general public: the process can only disprove hypotheses. To see this, consider the

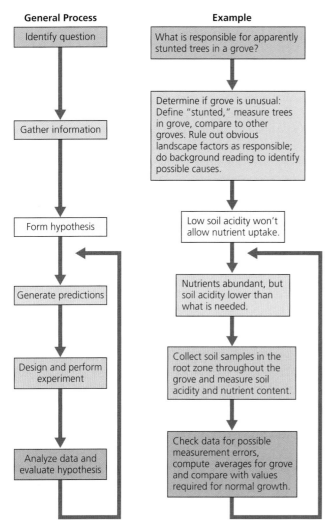

General Process

Identify question

↓

Gather information

↓

Form hypothesis

↓

Generate predictions

↓

Design and perform experiment

↓

Analyze data and evaluate hypothesis

Example

What is responsible for apparently stunted trees in a grove?

↓

Determine if grove is unusual: Define "stunted," measure trees in grove, compare to other groves. Rule out obvious landscape factors as responsible; do background reading to identify possible causes.

↓

Low soil acidity won't allow nutrient uptake.

↓

Nutrients abundant, but soil acidity lower than what is needed.

↓

Collect soil samples in the root zone throughout the grove and measure soil acidity and nutrient content.

↓

Check data for possible measurement errors, compute averages for grove and compare with values required for normal growth.

FIGURE 1.3 The scientific method in general terms (left) and by example (right).

possible outcomes of the example experiment, keeping in mind the hypothesis in Figure 1.3 (i.e., Low soil acidity won't allow nutrient uptake):

1. Soil acidity is suitable for growth. Whether nutrients are present or not, the hypothesized explanation is false because acidity can't be the explanation for the poor growth.
2. Soil acidity is unsuitable for growth, and soil nutrients are insufficient. We can't tell which (if either) aspect of the soil is responsible for low growth, and the hypothesis is neither confirmed nor rejected.
3. Soil acidity is unsuitable for growth, and soil nutrients are abundant. This is consistent with the hypothesis, but it does not prove the hypothesis. Possibly other, even more critical, resources for normal plant growth are lacking, such as water or sufficiently warm temperatures.

Notice that only outcome (1) leads to a firm decision about the hypothesis, and that outcome is rejection rather than

proof of the hypothesis. In both of the other cases, more research is needed. That is, new predictions and experiments are required. This example illustrates a very important point about scientific hypotheses: only after a hypothesis survives repeated attempts to be proven false can it be accepted as correct.

Notice also that if a hypothesis is to be useful scientifically, it must be possible to prove it false. Hypotheses for which there is no conceivable disproving experiment are unhelpful and in that sense not scientific. For example, one might hypothesize that all of life is an illusion contained in the dream of a sleeping turtle. Whether this hypothesis is true or not, there is no way to prove it false, and thus, this hypothesis about a dreaming turtle has no place in the scientific method. As a less fanciful example, in the last few decades theoretical physicists, in proposing ideas about the universe, have put forth predictions that are impossible to verify experimentally. Some scientists within the discipline have labeled such theories as unscientific.

Finally, we should say a word about what is meant by **theory** in science. Rather than mere speculation or conjecture, the term is reserved for a comprehensive body of explanatory knowledge that is both widely supported by experiment and generally accepted among scientists. For example, heat-transfer theory is used to design air conditioners, and gravitational theory underlies the calculation of satellite orbits. The scientific method cannot prove a theory is true, and theories are not considered immutable truth. Indeed, established theories are constantly refined as new evidence accumulates.

Scientists often speak of **scientific laws** in connection with theories. Laws are fundamental principles believed to hold without exception. One famous example is Newton's law, which states that the velocity of an object is constant unless acted on by an outside force. Like good hypotheses and theories, laws cannot be proven true, but there is so much confirming evidence for them that they are taken as absolute within the limits of their application. For instance, Newton's law applies to objects moving much slower than the speed of light. At higher speeds, the law needs modification, but for everyday phenomena like water plunging down a mountain stream, landslides, erupting volcanoes, and even tornado winds, we can be sure Newton's laws are unfailing. Theories are often constructed by combining and applying fundamental laws to particular problems.

Systems and Models in Physical Geography

Physical geography is a multifaceted science that seeks to understand major elements of our complex world. To deal with this complexity, physical geographers employ numerous concepts and specialized methods. We will discuss many of these in the units that follow. This section provides an introduction to that analysis by considering two general approaches to the subject: systems and models.

SYSTEMS AND FEEDBACKS

For the past few decades, many physical geographers have found it convenient to organize their approach to the field within a systems framework of thinking. For our purposes here, we define a **system** as any set of related events or objects and their interactions. For example, a city could be described as a large and elaborate system. Each day, the system receives an inflow of energy, food, water, and vast quantities of consumer goods. The various populations that reside in the urban center consume this energy and matter and change its form. At the same time, huge amounts of energy, manufactured goods, and services, along with sewage and other waste products, are produced in and exported from the city. As a physical geography example, consider the reservoir depicted in Figure 1.4. To explain even basic aspects of the reservoir, such as changes in water level, variations in water temperature, or why fish are near the bottom in winter but not summer, a systems approach is extremely helpful.

A system consists of interacting components, but what does "interacting" mean? In the physical world, it means there is an exchange of energy, mass (i.e., matter), or momentum. A precise definition of these terms isn't needed here; simply think of some familiar examples, such as the heat energy given off by hot coils in a hair drier, the transfer of paint (matter) from a paintbrush to a wall, or the momentum of a bowling ball rolling down an alley. If two physical systems are connected to each other or if there is interaction between parts of a system, there must be a transfer of one or more of energy, mass, or momentum. In the reservoir example, solar energy enters the system at the water surface, and most of it is absorbed in the upper part of the water. Some of that heat is carried downward by swirling eddies and other motions. Matter (water) enters the reservoir as rainfall and a stream inflow at the upper end, and there are losses as a result of evaporation, overflow through the spillway outlet, and downward percolation to deeper groundwater. Finally, winds blowing across the reservoir transfer momentum from the atmosphere to the water, creating small waves on the surface.

The power of the systems approach comes in large part because nature obeys certain broad principles known as the **conservation laws**. These laws are nothing more than accounting rules stating that energy, mass, and momentum cannot be created or destroyed. In other words, these do not simply appear and disappear.* Just as an accountant can track the flow of money through a corporation, a scientist can track the flow of energy, mass, and momentum through a system, knowing that ultimately everything must add up. For example, if the reservoir water inflow increases, those additional water molecules do not disappear into nothingness. Either they are stored and the reservoir level rises or the water outflow increases or perhaps both. The point is that all of the additional water can be accounted for because mass is conserved. Similarly, if the water level were to change, mass conservation demands that there be some reason for that change.

The conservation laws provide explanations for phenomena discussed throughout the book, and you will see them applied over a huge range of scales, from the molecular to the global. Energy conservation helps explain why air temperature in a parking lot often peaks after the Sun has reached its highest point in the sky and why continents get so much colder in winter than the surrounding ocean. Momentum conservation explains not only wave action in a small cove but also wind belts on the grandest scale.

Notice that nothing in the definition of a system says exactly what constitutes a system. Is a city a system and a neighborhood within a city not a system? Or are both systems? Or is the neighborhood a **subsystem** (a system within a system) of a city? These decisions are up to the researcher. For some purposes, neighborhood differences might not matter, whereas for others they might. The researcher also decides what components of the system need to be included. For example, a water supply specialist might be willing to ignore wave action, whereas someone interested in beach erosion would find wave action essential to the phenomenon. There are also no set rules about the boundaries of a system. Does the reservoir stop at the water's edge? Is it necessary to consider the movement of water through the muck and rock below as part of the reservoir

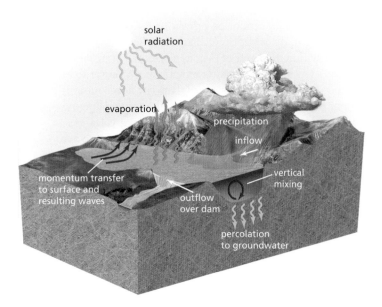

FIGURE 1.4 A reservoir can be considered as a system. This diagram shows some of the energy, mass, and momentum exchanges within the system and between the system and its surroundings.

---

* As with all laws, there are limits to the application of these conservation principles. In nuclear reactions, mass is converted into energy, and it becomes necessary to talk in terms of mass-energy conservation as captured by Einstein's law.

system? Again, it depends on the goal of the analysis, and decisions like these are made by the analyst.

In the reservoir system, energy, mass, and momentum move freely across its boundaries, making it an **open system** and underscoring its relationships with surrounding systems (the atmosphere and landscape within which it is embedded). There are many examples of open systems throughout this book, such as a weather system or a river drainage basin. In terms of energy flows, Earth itself is an enormous open system that comprises several interconnected lesser systems. Although it is difficult to find one on Earth's surface, we should also know what is meant by the term **closed system**: a self-contained system exhibiting no exchange of energy or matter across its boundaries.

System boundaries are called **interfaces**. The transfer or exchange of energy and matter takes place at these interfaces. Sometimes interfaces are visible: we can see where sunlight strikes the roof of a building. But often they are not visible: we cannot see the movement of groundwater, a part of the global water system, as it flows through the subterranean rocks of the geologic system. Many geographers focus their attention on these interfaces, visible or invisible, particularly when they coincide with Earth's surface. It is here that the greatest activity of our dynamic world occurs.

FROM THE FIELD NOTES

FIGURE 1.5 "South Florida's Atlantic coast, looking northward beyond Miami Beach. At present the beach seen in this photograph represents a system in dynamic equilibrium, with currents flowing parallel to the coast, continuously removing sand and at the same time depositing a replacement supply."

Two other ideas commonly referred to in systems approaches are **equilibrium** and **feedback**. A system is in equilibrium with respect to some property if that property is unchanging over time. Thermal equilibrium means temperature is constant through time; mass equilibrium means there is no change in the number of molecules; charge equilibrium means the electric charge is unvarying; and so forth. True stasis, meaning no movement of energy, matter, or momentum, is very rare in the natural world. Most of the time, there is continuous transfer; thus, if an equilibrium exists, any gains must be balanced by losses. There may be a continuous movement of something through the system, but as long as supply and removal are equal, the conservation laws state that the amount stored is unchanging. When supply and removal are equal, the system is in **dynamic equilibrium**. On a global scale, water falls from the atmosphere continuously as rain and snow, yet the atmosphere never dries out thanks to constant replenishment from the planet's surface. We might take this grand example of dynamic equilibrium for granted, but our very existence depends on it.

Feedbacks are important in understanding how a system responds to some disturbing force or impetus for change. A feedback process is a chain of connections that affects (that is, feeds back on) some initial disturbance. This is best explained by an example. Siberian winters are murderously cold. Meager amounts of solar radiation are the ultimate cause for this cold, but two important feedbacks affect the temperature response. The first of these is a **positive feedback**, which acts to accentuate the change. Thus, as temperatures fall, there is a feedback that makes temperatures colder than they would be otherwise. The sequence can be traced out by following Figure 1.6A. As incoming sunlight declines, less heat is supplied to the ground, and temperatures fall accordingly. This means more snow on the ground and trees. As alpine skiers know, snow is highly reflective, and therefore snow cover reduces the amount of sunlight absorbed. With less energy supplied, still lower temperatures result. The feedback is positive because the feedback process amplifies the initial change.

A **negative feedback** also operates in the Siberian example (Fig. 1.6B). Negative feedbacks act to resist, or counteract, change. In this case, cold Siberian temperatures contribute to a large temperature difference between the warm tropics and the much-colder middle latitudes. As we will see in later units, storms and other atmospheric disturbances are for the most part driven by this temperature imbalance, and these atmospheric disturbances are more vigorous in the winter when the temperature gradient is large. The net effect of these enhanced atmospheric motions is to move additional heat out of lower latitudes

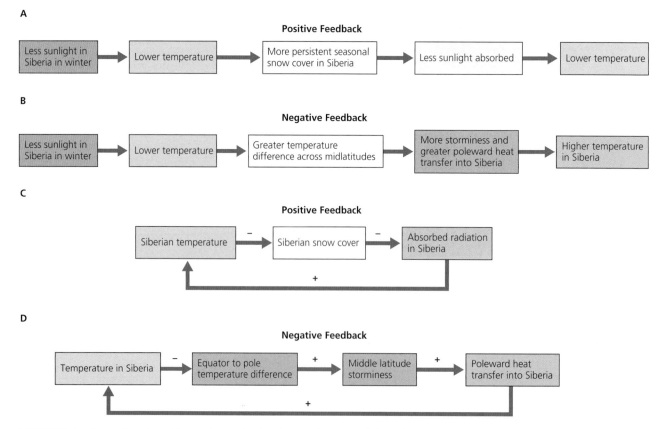

FIGURE 1.6 Examples of positive and negative feedbacks. Diagrams (C) and (D) represent the same feedback processes as (A) and (B), but as loops rather than as linear sequences. The two negative links in (C) combine to form a positive overall relationship. The single negative link in (D) means the overall loop comprises a negative feedback.

toward the poles, which benefits Siberians (human and otherwise) struggling against the cold. Nobody would say Siberia is warm in winter, but it would surely be colder without this feedback.

The two Siberian feedbacks are more usefully illustrated in parts C and D of Figure 1.6. As in parts A and B, arrows again indicate linkages, but in C and D, the arrows are labeled with plus and minus signs. A plus sign means that as one variable increases, the linked variable also increases. For example, as absorbed radiation in Siberia increases, there is a corresponding increase in Siberian temperature. Negative signs denote just the reverse. As Siberian temperature rises, the amount of snow cover falls. The negative feedback loop in Figure 1.6D has just one negative link. Thus, overall, the feedback loop is negative. In contrast, the positive loop in Figure 1.6C has two negative signs, yet the loop overall is positive. In effect, the two negative links cancel each other out, resulting in a positive relationship across them. If a loop has zero or an even number of negative links, the overall loop is a positive feedback. If a loop has an odd number of negative links, it constitutes a negative feedback. The somewhat abstract way of presenting feedbacks illustrated in Figures 1.6C and 1.6D is succinct and has other advantages. For example, we can see that the two lower diagrams work for summer as well as winter. If we wanted to draw something similar to Figures 1.6A and 1.6B for summer, the box contents would need to change ("less sunlight" would become "more sunlight," etc.). We use this more abstract presentation where feedbacks are discussed in the book.

Positive and negative feedbacks are present throughout the Earth system, operating at many spatial and temporal scales. The Siberian example has a fast-acting, molecular-scale positive feedback (individual snow crystals reflecting sunlight) and a slower large-scale feedback (atmospheric circulation changes). With multiple feedbacks, some amplifying change and others being restorative, many Earth systems can exhibit spontaneous change over time without the involvement of any external agent. That is, feedbacks give rise to changes in the system that are driven internally rather than by some outside process. This might seem strange, but pouring milk from a jug held at a steep angle (Fig. 1.7) provides an everyday example. The milk gurgles out in what is obviously highly variable flow, yet there are no changes in external forces. Instead, internal processes are to blame for the erratic flow. A little thought explains why. As milk flows out a partial vacuum forms in the jug that is able to temporarily hold the remaining milk against the force of gravity. Very quickly enough air enters the jug to destroy the vacuum, and milk again begins to flow, starting another cycle of a developing vacuum and another gurgle. In later units, you will encounter Earth-system examples of this type of behavior. For now, it is sufficient to know that in addition to external processes, internal feedbacks can generate variability in a system.

FIGURE 1.7 If a jug of milk is held at a high angle, the flow starts and stops in fits of gurgling motion. This is an example of system variation arising internally rather than in response to changes in external factors.

MODELS

Another way that physical geographers approach the study of Earth's phenomena is to make models of them. **Models** are simplified representations containing processes and features that are of interest. A wall map is one example of a model. Obviously, only certain phenomena are represented in the map, and the map departs greatly from the reality it represents. This is not a defect, provided the excluded phenomena are not important to the map's purpose.

Physical geographers today rely heavily on mathematical models in which equations represent processes and properties of the physical world. These equations can be developed from observations, or they may follow directly from physical theory. In either case, they provide a way for geographers to simulate real-world processes and understand relationships among variables of interest. Model building is complementary to systems analysis as a way of thinking about the world. It entails the controlled simplification of a complex reality, filtering out the essential forces and patterns from the myriad details in a complicated world. Such abstractions,

Sliding Spatial Scales

Imagine a couple sunbathing on Miami Beach. We can photograph them occupying a square of sand about 1 m (3.3 ft) on a side. If we move the camera higher, a square 10 (10^1) m wide reveals their companions. When we focus on a 100 (10^2) m area, we can see a crowd of people on the beach. A picture of 1,000 (10^3) m across includes the beach, some sea, and some land (as in Fig. 1.5). One with a perimeter of 10,000 (10^4) m captures most of the city of Miami Beach and parts of neighboring Miami across Biscayne Bay to the west.

Moving the camera still farther, we shoot a picture of a 100,000 (10^5) m square. It encompasses most of the South Florida region. The next step is 1,000,000 (10^6) m. This snapshot takes in the entire state of Florida, some neighboring states and Caribbean islands, and parts of the Atlantic Ocean, the Gulf of Mexico,

and the Caribbean Sea. A photo at the next level of generalization, showing a square of 10,000,000 (10^7) m, covers most of the visible Earth. And if the camera is far enough away in outer space to focus on a square of 100,000,000 (10^8) m, we see planet Earth as a small globe. Somewhere on it is that couple lying on a square of Miami Beach's seaside sand. A digital camera capable of resolving those beachgoers would need about 10 million megapixels. Cameras used by professional photographers these days have about 10 megapixels, a factor of 1 million smaller. Clearly, the phenomena of interest in physical geography span an enormous range of scales. Indeed, one of the discipline's great challenges is working out how small-scale processes manifest themselves at global scales.

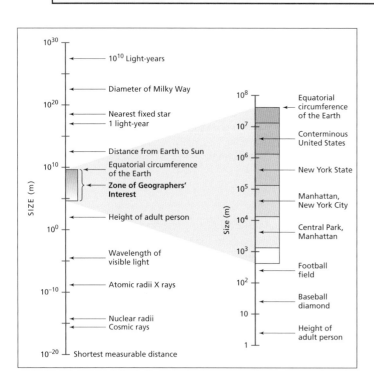

FIGURE 1.8 Orders of geographic magnitude. Geographers mostly operate in the context of the scales shown on the right, but sometimes they must think in much smaller or larger terms, as indicated on the left.

FIGURE 1.9 Central Park really does occupy a central location on the totally built-up island of Manhattan. This north-looking view of the 340-hectare (840-acre) green space shows the park, first laid out in 1856, flanked by tall buildings in all directions. Central Park for many years has been the subject of geographic studies, ranging from *biogeographic* research (the survival of plants and animals in this unique setting) to *human geography* research (the behavior of people as they use the park for purposes ranging from open-air symphony concerts to skateboarding).

which convey not the entire truth but a valid and reasonable part of it, are highly useful because they allow us to make generalizations. Much of what we know about Earth processes comes from modeling (especially mathematical

modeling), and therefore modeling is implicit in much of what you will read in the following units. In addition, we frequently make explicit use of models to penetrate complex subjects quickly and highlight their most essential aspects.

Geographic Magnitude

In approaching the real world, we must also consider the size of the subjects and phenomena that interest us.

Let us consider different sizes, or **orders of magnitude**. Figure 1.8 shows the various orders of magnitude with which we must become familiar. The scale in this figure is written in *exponential* notation. This means that 100 is written as 10^2 (10×10); 1,000 is 10^3 ($10 \times 10 \times 10$); 0.01 is 10^{-2} (1/100); and so forth. This notation saves us from writing numerous zeros. The scales geographers use most often—which are shown on the right side of Figure 1.8—span from about 10^3 or 10^4 m (11,000 yd), the size of Central Park in New York City (Fig. 1.9), up to about 10^8 m (62,500 mi), which is beyond the order of magnitude of Earth's circumference (see Focus on the Science: Sliding Spatial Scales). Physical geographers sometimes have to expand their minds even further. Cosmic rays with wavelengths of 10^{-18} m may affect our climate. The nearest fixed star is approximately 10^{16} m (25 trillion mi) away. Brighter stars much farther away sometimes help when navigating a path across Earth's surface. Occasionally (as in Unit 4), we must perform mental gymnastics to conceive of such distances, but one of the beauties of physical geography is that it helps us see the world in a different way.

We are now ready to embark on our detailed study of Earth. Central to this effort is the attempt to understand the environment at Earth's surface, a habitat we must all live within in a one-to-one relationship. Nobody who has become acquainted with physical geography is ever likely to forget that.

Key Terms

biogeography *page 3*

climatology *page 3*

closed system *page 7*

conservation laws *page 6*

dynamic equilibrium *page 8*

equilibrium *page 8*

experiment *page 4*

feedback *page 8*

geomorphology *page 3*

hypothesis *page 4*

interface *page 7*

marine geography *page 4*

model *page 9*

negative feedback *page 8*

open system *page 7*

orders of magnitude *page 11*

positive feedback *page 8*

scientific laws *page 5*

scientific method *page 4*

soil geography *page 4*

subsystem *page 6*

system *page 6*

theory *page 5*

water resources *page 4*

Scan Here for a quick vocabulary review

Review Questions

1. Define the term *spatial* and show how it is central to the study of geography.

2. What contributions did the Greeks and Romans make to the early evolution of physical geography?

3. What is physical geography? How does it differ from other sciences?

4. Define the eight major subfields of physical geography. What are their major foci of study? How do these differ from other fields in the natural and social sciences?

5. The in-laws of one of the authors, Bev and Lois, are sleeping under a heated blanket with dual controls. When either person is too hot or cold, he or she wakes up and adjusts his or her control. Is equilibrium possible? That is, can there be any setting of the two controls that makes both people comfortable? Now suppose (as actually occurred) the controls are crossed so that each person is controlling heat for the other. What happens if Lois feels the need for more heat and adjusts her control? Draw the situation as a set of linked boxes and explain how the process happens in terms of feedback loops. Would you describe this as a positive or negative feedback loop?

6. Describe how models can help us understand our complex physical world.

Scan Here for a quick concept review

www.oup.com/us/mason

A Planetary Perspective

Relationships Among the Five Spheres of the Earth System

Our planet is the manifestation of a set of interconnected, interactive, interlocking systems that generate the forces, processes, and landscapes with which we are all familiar. These geochemical, geophysical, and biological systems forge Earth as we know it. We can view the planet as five gigantic open (interacting) systems represented as *spheres*. These spheres, which we describe and discuss in detail in this part of the book, are the *atmosphere*, the *lithosphere*, the *hydrosphere*, the *cryosphere*, and the *biosphere*. The oldest is the lithosphere, the sphere of rocks (and earthquakes and volcanoes), and the youngest is the biosphere, the realm of plants and animals. We experience the atmosphere and its weather subsystems on a daily basis, and we know Earth as the "blue planet" because the waters of the hydrosphere cover about 70 percent of its surface. We are reminded of the cryosphere, which includes all forms of frozen water, in polar and high-altitude areas where Ice Age conditions, once much more extensive, still prevail. Within each sphere dominant processes explain many aspects of what we observe, but anything more than superficial understanding requires that we consider interactions between spheres. These interactive processes vary in both space and time.

UNIT 2

The Planet Earth

Despite their grandeur on human scales, not even the Andes Mountains of South America amount to a significant departure from Earth's almost perfectly spherical shape.

OBJECTIVES

- Recount Earth's shape and size
- Name the five spheres of the Earth system
- Discuss Earth's hemispheres
- Describe Earth's landmasses
- Describe the world's ocean basins and the topographic characteristics of the seafloor

When U.S. astronauts left Earth's orbit and reached the Moon for the first time, they were able to look back at our planet and see it as no one had ever seen it before. The television cameras aboard their spacecraft beamed spectacular pictures of Earth to television sets around the globe. Against the dark sky, Earth displayed a range of vivid colors, from the blue of the oceans to the green of forests to the brown of sparsely vegetated land to the great white swirls of weather systems in the atmosphere (see photo, p. 1). The astronauts also saw things no camera could adequately transmit. Most of all, they were struck by the smallness of our world, what architect Buckminster Fuller famously termed "Spaceship Earth," in the vastness of the universe. It was difficult to conceive at the time that more than 6 billion humans and their works were confined to, and dependent on, so tiny a planet. Every participant in those Moon missions returned with a sense of awe—and a heightened concern for the fragility of our terrestrial life-support systems.

In the units that follow, we examine these systems and learn, among other things, about both their resilience and their vulnerabilities to human activities. It is important to remember that all the systems we study—weather and climate, oceanic circulation, soil formation, vegetation growth, landform development, and erosion—ultimately are parts of one great Earth system. Even though this total Earth system is an open system with respect to energy flows, the amount of matter on and in Earth is pretty much fixed. Little new matter is being created, and so far, we know of nothing of consequence that permanently leaves Earth. This means that our material resource base is finite; in other words, parts of it can be used up. It also means that any hazardous products we create remain a part of our environment if they are not converted to more benign forms.

Earth's Shape and Size

If you look out your window, there is no immediate reason for you to suppose that Earth's surface is anything but flat. It takes a considerable amount of traveling and observation to reach any other conclusion. Yet the notion of Earth as a sphere is a long-standing one and was accepted by several Greek philosophers as far back as 350 BCE. By 200 BCE, Earth's circumference (approximately 40,000 km [25,000 mi]) had been accurately estimated by the Greek mathematician Eratosthenes (ca. 275–ca. 194 BCE) to within a few percentage points of its actual size. The idea of a spherical Earth continued to be challenged, however, and was not universally accepted until Portuguese explorer Ferdinand Magellan (1480–1521) led an expedition that successfully circumnavigated the globe in the early sixteenth century.

THE SPHERICAL EARTH

Why should Earth be spherical rather than some other shape such as cylindrical, cubical, or even a highly irregular blob? The answer is gravity, as well as the somewhat plastic nature of Earth. Although Earth's outer shell of rock is

presently quite rigid, this was not true early in its formation, and even now, most of the planet's interior is molten and able to flow. Gravity attracts all parts of Earth toward the center of its mass. On Earth, some points are higher than others. Just as water poured on a table becomes a flat layer, gravity attempts to pull the higher points of Earth down. A spherical shape is unique in that all points are as low as they can be without making any other point higher.

THE NONSPHERICAL EARTH

Our planet is not a perfect sphere. One important departure from this shape arises from the spin of Earth on its axis. In the same way a disc of pizza dough thrown in the air and spun expands outward, latitudes near Earth's equator bulge out from a purely spherical shape. The outward force is called a **centrifugal force**, because it pushes outward from a center. The entire mass of the planet has adjusted to the combination of gravity and centrifugal force, with the result that the spherical Earth is squashed into an **ellipsoid**, the term for a three-dimensional object that is elliptical in cross-section. As seen in Figure 2.1A, the equatorial radius is a little larger than the polar radius, and in cross-section, Earth is elliptical rather than circular (Fig. 2.1B).

The drawings in Figure 2.1 greatly exaggerate the flattening; the lunar astronauts saw no departure from a sphere, and none is visibly detectable from any image you might find on the Internet. The difference between the polar and equatorial radii is only about 20 km (12.4 mi), which amounts to just a few tenths of 1 percent.

Another departure from a perfectly spherical shape arises from Earth's rugged terrain—that is, the many mountains, valleys, and deep chasms that cover the continents and ocean floor. Terrain is part of the **topography** of an area, which includes both natural and artificial features found at the surface. Over the large distances we are considering here, buildings, highways, and other human artifices are obviously insignificant. Even the tallest natural features have little impact on the overall shape of Earth. True enough, a climb to Mt. Everest's top is a great achievement in human

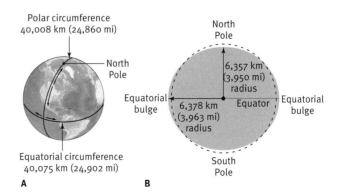

FIGURE 2.1 (A) A three-dimensional view of an ellipsoid showing the axis of rotation, the equator, and one polar geodesic. (B) Ellipse with equatorial and polar radii labeled 6,378 km and 6,357 km.

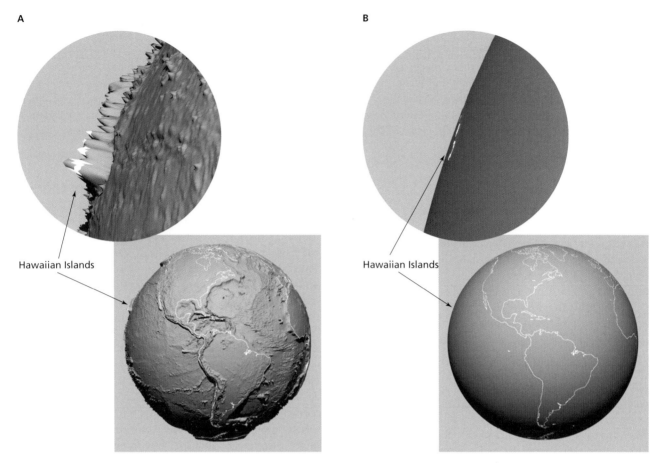

FIGURE 2.2 Earth's topography is impressive when drawn with the vertical dimension exaggerated by a factor of 50 (A). When drawn at true scale, the same data used in (A) reveal that Earth is extremely smooth (B).

terms. But from sea level, the summit of Everest is only 8.8 km vertically. At a depth of about 10.9 km below sea level, the deepest ocean trench is slightly more impressive. This means there is less than 20 km of vertical separation between the highest and lowest points of Earth, once again just a tiny fraction of Earth's radius. If you have wondered why mountains are often hard to recognize on photographs taken from space, this is the answer. At global scales, our planet is incredibly smooth (Fig. 2.2).

So, neither flattening caused by spin nor surface sculpting creates much in the way of departures from a purely spherical form. This is highly advantageous for introductory physical geography because a sphere is a wonderfully regular and simple shape. It can be completely described by a single number. Take your pick: radius, diameter, circumference, or other possibilities. Knowing any one enables us to calculate the others. Hence, as a model, the spherical Earth is very attractive. It is simple, and it closely agrees with reality. An ellipsoidal Earth is a more accurate model, but it comes with some disadvantages. Two pieces of information are needed for its description, and calculations of distance, area, and other quantities are vastly more complicated. An even more accurate model that included topographic departures would entail considerably more complexities. Luckily a spherical model serves admirably for most of this book's purposes.

Spheres of the Earth System

Our planet consists of a set of interacting shells or "spheres"; some extend over the entire globe, and others cover it partially. The **atmosphere** is the blanket of air that adheres to Earth's surface. It is our life layer, the mixture of gases we breathe. The atmosphere begins a few meters under the soil or on the water's surface and extends to a height of about 60,000 km (37,000 mi) above Earth. Heat energy from the Sun keeps the atmosphere in motion, causing weather systems to form and travel across land and sea. The atmosphere is densest at sea level and thins out with increasing altitude.

Below the atmosphere lies the outermost shell of the solid Earth, the **lithosphere** (*lithos* means rock). The lithosphere's upper surface is sculpted into the almost endless variety of landforms and physical landscapes that form Earth's scenery. The lithosphere continues under the oceans, as the surface of the seafloor.

Covering about 71 percent of Earth's surface, the oceans make up the largest segment of the **hydrosphere**, which contains all the water that exists on and within the solid surface of Earth and in the atmosphere above. The oceans are the primary moisture source for the precipitation that falls on the land areas, carried there in the constantly moving atmosphere.

Earth as an Ellipsoid: Who Cares?

We've discussed how flattening by centrifugal forces has caused a departure from a truly spherical form, but the degree of flattening is too small to be seen at global scales. Does that mean flattening is unimportant? Hardly!

To see why, consider the diagram on the left in Figure 2.3, which shows a flattened planet with an equatorial bulge that is completely covered by an ocean and an atmosphere. Centrifugal forces pull the ocean and atmosphere equatorward just as they pull the whole mass of the planet equatorward, so ocean and an atmosphere cover the solid planet as a uniform layer. On the planet shown in the figure, just like on Earth, ocean and atmosphere are everywhere.

Imagine another planet whose interior is perfectly rigid and spherical, like the one on the right in Figure 2.3. The fluid ocean and atmosphere are drawn equatorward and bulge outward just as they do on the planet on the left (and on Earth). Both the depth of water and the atmospheric thickness follow the shape of Earth. However, because ocean and atmosphere on the planet on the right overlay a solid sphere that is not similarly deformed, the depth of ocean and atmosphere decrease poleward. The result is no ocean or atmosphere covering for a sizable part of the planet, drawn in the buff color. If the solid and molten parts of Earth were truly spherical, as they are in the model on the right, only lower latitudes would be covered by life-sustaining fluids. With no air to breathe and no water in the Arctic Ocean or anywhere else outside the tropics, a round planet would be vastly different from our nearly ellipsoidal home.

Knowing Earth is close to an ellipsoid is also important in understanding how locations on Earth are specified. Unit 3 will explain why; for now, it is enough to note that there is good reason for bringing up what might seem like a trivial issue.

FIGURE 2.3 Two imaginary planets, one that deforms like Earth (left) and another whose solid part remains rigidly spherical as it spins (right). On the right, the higher latitudes of both hemispheres have exposed surfaces covered by neither ocean nor atmosphere (buff-colored areas).

The **cryosphere** includes all forms of frozen water, including glaciers, floating ice, snow cover, and *permafrost* (permanently frozen ground below the surface subsoil). Although it could logically be regarded as a component of the hydrosphere, the cryosphere has its own distinct properties and processes, and many scientists in recent years have come to treat it as a full-fledged Earth sphere.

The **biosphere** is the zone of life, the home of all living things. This includes Earth's vegetation and animals (including human beings) and the dead and decaying organic matter they produce. Since there are living organisms in the soil and plants are rooted in soil, the biosphere extends into the soil. Similarly, part of the biosphere is aquatic, existing within and beneath both freshwater lakes and oceans.

These five spheres—atmosphere, lithosphere, hydrosphere, cryosphere, and biosphere—are the key Earth layers we are concerned with in our study of physical geography. But other shells of Earth also play their roles. Not only are there outer layers atop the effective atmosphere, but there also are spheres inside Earth, beneath the lithosphere. In turn, the lithosphere is affected by forces and processes from above as well as below. Thus, it is important to keep in mind the interactions among the five spheres; they are not separate and independent segments of our planet but constantly interacting subsystems of the total **Earth system** (Fig. 2.4). It is interesting that as understanding of individual spheres has advanced, the need to understand interactions between the spheres has become more important. In the last twenty years, this has led to recognition of a new discipline, Earth system science. A large number of physical geographers take a holistic,

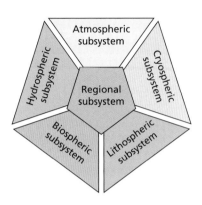

FIGURE 2.4   The five spheres of the Earth system and their interrelationships.

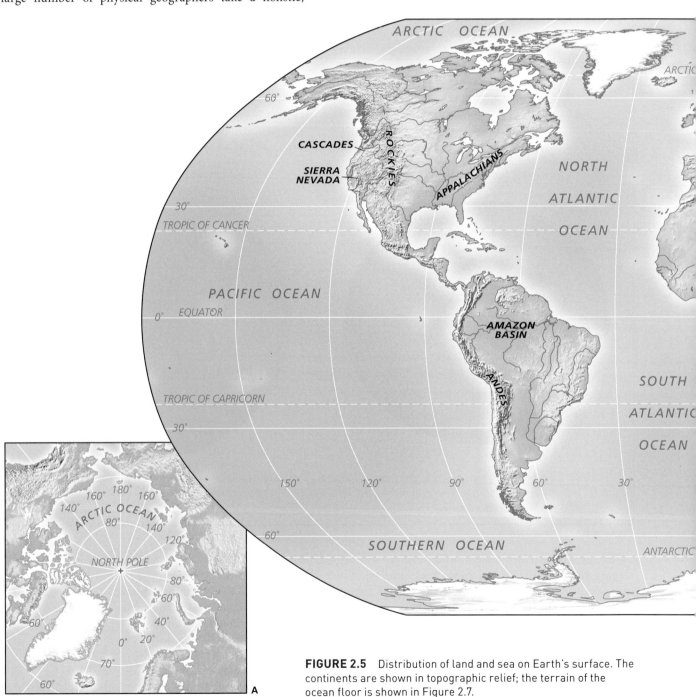

FIGURE 2.5   Distribution of land and sea on Earth's surface. The continents are shown in topographic relief; the terrain of the ocean floor is shown in Figure 2.7.

integrating approach to their research and in doing so contribute to this relatively new field.

Earth's Hemispheres

In addition to the five layered spheres (the atmosphere, lithosphere, hydrosphere, cryosphere, and biosphere), Earth can be divided into **hemispheres** (from the ancient Greek *hemi*, meaning half; *sphaira*, meaning sphere). The northern half of the globe, from the equator to the North Pole, is the Northern Hemisphere; the southern half is the Southern Hemisphere. There are many differences between the two hemispheres, as we shall see in later units that address patterns of seasonality, planetary rotational effects, and the regionalization of climates. Perhaps the most obvious difference concerns the distribution of land and sea (Fig. 2.5). Earth's continental landmasses are more heavily concentrated in the Northern Hemisphere (which contains about 70 percent of the total land area); the Southern Hemisphere has much less land and much more water than the Northern Hemisphere. The polar areas of each hemisphere also differ considerably. The Northern Hemisphere polar zone—the *Arctic*—consists of a central mass of sea ice floating atop the Arctic Ocean (Fig. 2.5, inset map A), surrounding seawater that freezes in winter, and islands partly covered by ice. The Southern Hemisphere polar zone—the *Antarctic*—is dominated by a large continental landmass

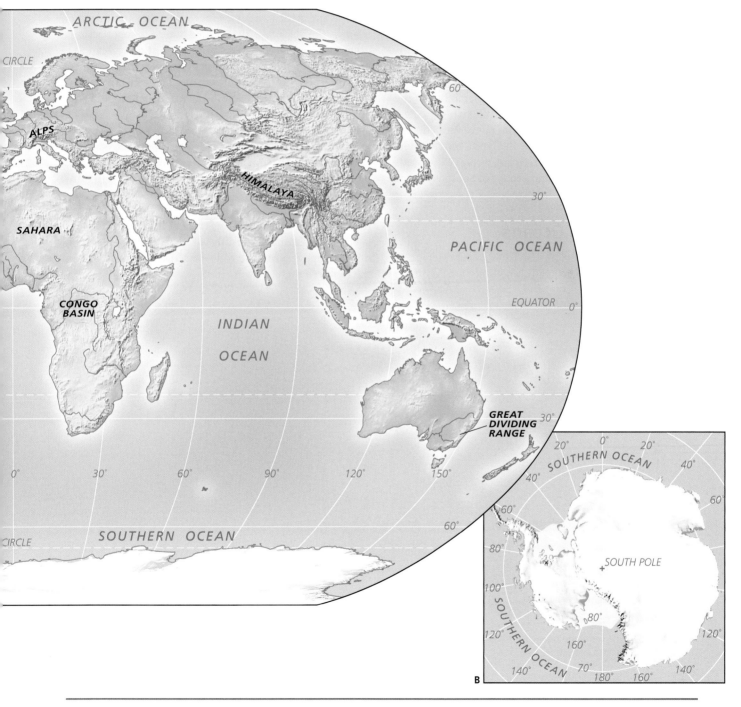

covered with the world's largest continuous ice sheet (Fig. 2.5, inset map B).

We can also divide Earth into Eastern and Western Hemispheres. Technically, the Eastern Hemisphere lies to the east of an imaginary line drawn from pole to pole through the Royal Observatory in Greenwich, England (part of the city of London), and through the middle of the Pacific Ocean on the opposite side of the world. In practice, however, the Western Hemisphere consists of the half of Earth centered on the Americas, and the Eastern Hemisphere contains all of Eurasia and Africa.

A look at any globe representing Earth suggests yet another pair of hemispheres: a **land hemisphere** and a **water** (or oceanic) **hemisphere** (Fig. 2.6). The landmasses are concentrated on one side of Earth to such a degree that it makes sense to refer to that half of Earth as the land hemisphere. This land hemisphere is centered on Western Europe, which is surrounded by the other continents: the Americas to the west, Eurasia to the east and northeast, and Africa to the south. The opposite hemisphere, the water hemisphere, has only one-eighth of the world's land area and is dominated by Earth's greatest ocean, the Pacific. When viewed from above the center of this water hemisphere (near New Zealand), only moderately sized Australia and Antarctica as well as the islands off Southeast Asia and the southern tip of South America are visible.

Continents and Oceans

There is an old saying that "the Earth has six continents and seven seas." In fact, that generalization is not too far off the mark (see Fig. 2.5). Earth has six continental landmasses: Africa, South America, North America, Eurasia (Europe and Asia occupy a single large landmass), Australia, and Antarctica. As for the seven seas, there are five great oceanic bodies of water and several smaller seas. The Pacific Ocean is the largest body of water. The Indian Ocean lies between Africa and Australia. The North Atlantic Ocean and the South Atlantic Ocean may be regarded as two discrete oceans that are dissimilar in a number of ways. Encircling Antarctica is the Southern Ocean. The sixth-largest body of water, and the largest of the smaller seas, is the Arctic Ocean, which lies beneath and around the floating Arctic icecap.

The seventh body of water often identified with these oceans is the Mediterranean Sea, which lies between Europe and Africa and is connected to the interior sea of Eurasia, the Black Sea. The Mediterranean is not of oceanic dimensions, but unlike the Caribbean or the Arabian Sea, it also is not merely an extension of an ocean. The Mediterranean is very nearly landlocked and has only one narrow natural outlet through the Strait of Gibraltar, between Spain and Morocco.

In our study of weather and climate, the relative location, general dimensions, and topography of the landmasses are important because these influence the movement of moisture-carrying air so important to rainfall and snowfall patterns. The topography of the ocean floor will feature prominently in later units describing the largest of Earth-shaping forces. Readers therefore need a basic understanding of both continental and marine topography. The following two sections contain the required material.

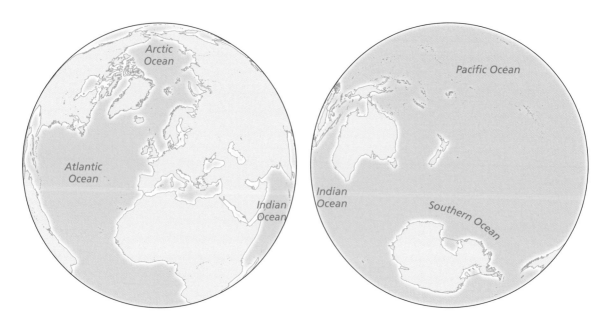

FIGURE 2.6 The land and ocean hemispheres. The land hemisphere cuts the planet in half in such a way that it contains the largest possible land area (left). The ocean hemisphere is the remaining half.

THE LANDMASSES

Only about 29 percent of Earth's surface consists of land; the remaining 71 percent is water or ice. Thus, less than one-third of our planet is habitable by human beings, and much of this area is too dry, too cold, or too rugged to allow large concentrations of settlement. Our livable world—where permanent settlement is possible—is small indeed (see Perspectives on the Human Environment: Human Population and Natural Processes).

Each of the six continental landmasses possesses unique physical properties, some of which are listed in Table 2.1. *Eurasia* (covering 36.5 percent of the land surface) is by far the largest landmass on Earth, and all of it lies in the Northern Hemisphere. The topography of Eurasia is dominated by a huge mountain chain that extends from west to east across the giant landmass. The mountains in the chain have many names in various countries, the most familiar of which are the Alps in Central Europe and the Himalayas in South Asia. In Europe, the Alps lie between the densely populated North European Lowland to the north and the subtropical Mediterranean lands to the south. In Asia, the Himalayas form but one of many great mountain ranges that extend from central Asia northeastward into Russia's Siberia, eastward into China, and southeastward into South and Southeast Asia. Between and below these ranges lie several of the world's most densely populated river plains.

Africa, which accounts for just over 20 percent of the Earth's total land area, is at the heart of the land hemisphere. Of all the landmasses, Africa alone lies astride the equator in such a way that large segments of it occupy the Northern as well as the Southern Hemisphere. Africa is called the plateau continent because much of its landmass lies above 1,000 m (3,300 ft) in elevation and coastal plains are relatively narrow. (The word *plateau* refers to any area of relatively flat high ground.) As Figure 2.5 reveals, a long steep slope, or *escarpment*, rises near the coast in many parts of Africa, leading rapidly up to the plateau surface of the interior. African rivers that rise in the interior plunge over falls and rapids before reaching the coast, limiting their navigability. Africa lacks one physical feature seen on all the other landmasses: a linear mountain range comparable to South America's Andes, North America's Rocky Mountains, Eurasia's Himalayas, or Australia's Great Dividing Range. The reason for this will become clear when the geomorphic history of that continent is discussed in Unit 30.

North America comprises one-sixth of the total land area on Earth and is substantially larger than South America. This landmass extends from Arctic to tropical environments. Western North America is mainly mountainous; the great Rocky Mountains stretch from Alaska to Mexico. West of the Rockies lie other major mountain ranges, such as the Sierra Nevada and the Cascades. East of the Rocky Mountains lie extensive plains covering a vast area from Hudson Bay south to the Gulf of Mexico and curving up along the Atlantic seaboard as far north as New York City. Another north–south-trending mountain range, the Appalachians, rises between the coastal and interior lowlands of the East Coast. Thus, the continental topography of North America is somewhat funnel-shaped. As a result, air from both polar and tropical areas can penetrate the heart of the continent, without topographic obstruction, from north and south. As a result, summer weather in the middle of North America can be tropical, whereas winter weather can exhibit Arctic-like extremes.

South America, occupying 12 percent of the world's land, is much smaller than Africa (Table 2.1). The topography of this landmass is dominated in the west by the gigantic Andes Mountains, which exceed 6,000 m (20,000 ft) in height in many places. East of the mountains, the surface of the continent becomes a plateau interrupted by the basins of major rivers, among which the Amazon is by far the largest. The Andes constitute a formidable barrier to the cross-continental movement of air, which has a major impact on the distribution of South America's climates.

Antarctica, the "frozen continent," lies almost entirely buried by the world's largest and thickest ice sheet. Beneath the ice, Antarctica (constituting the remaining 9.3 percent of the world's land area) has a varied topography that includes the southernmost link in the Andean mountain chain (the backbone of the Antarctic Peninsula). Currently, very little of this underlying landscape is exposed, but

TABLE 2.1 Dimensions of the Landmasses

	Area			Elevation		
Landmass	km²	mi²	Percentage of Land Surface	Highest Mountain	m	ft
Eurasia	54,650	21,100	36.5	Everest	8,850	29,035
Africa	30,300	11,700	20.2	Kilimanjaro	5,861	19,340
North America	24,350	9,400	16.3	McKinley	6,158	20,320
South America	17,870	6,900	11.9	Aconcagua	6,919	22,834
Australia	8,290	3,200	5.6	Kosciusko	2,217	7,316
Antarctica	13,990	5,400	9.3	Vinson Massif	5,110	16,864

Antarctica was not always a frigid polar landmass. As we will discuss in more detail in Parts 3 and 4 the Antarctic ice, the air above it, and the waters around it are critically important in the global functioning of the atmosphere, hydrosphere, and biosphere.

Australia is the world's smallest continent (constituting less than 6 percent of the total land area) and topographically its lowest. The Great Dividing Range lies near the continent's eastern coast, and its highest peak reaches a mere 2,217 m (7,316 ft). Australia's northern areas lie in the tropics, but its southern coasts are washed by the outer fringes of Antarctic waters.

THE OCEAN BASINS

Before the twentieth century, Earth's ocean basins were unknown territory. Only the tidal fringes of the continents and the shores of deep-sea islands revealed glimpses of what lay below the vast world ocean. Then, sounding devices were developed that beamed sound waves downward and measured reflections from the ocean floor. With this some of the ocean-floor topography became apparent, at least in cross-sectional profile. Later, equipment was built that permitted the collection of rock samples from the seabed along with small submarines that allowed marine scientists to venture down to deep areas of the ocean floor and watch volcanic eruptions in progress. More recently still, satellites measuring small variations in Earth's gravity have been used to obtain a global picture of the ocean floor. Undulations in underwater topography give rise to variations in the gravitational field, and with enough measurements and complex calculations the shape of the seafloor can be determined, a small part of which is shown in Figure 2.7. Thanks to all these technologies, the secrets of the submerged 71 percent of the lithosphere are finally being revealed.

When the seafloors were mapped and the composition and age of rock samples were determined, a remarkable discovery was made: the deep ocean floors are geologically different from the continental landmasses. The ocean floors have a varied topography, but this topography is not simply an extension of what we see on land. There are ridges and valleys and mountains and plains, but these are not comparable to, say, the Appalachians or the Amazon Basin.

We return to this topic in Part Five. For the present, we will acquaint ourselves with the main features of the ocean basins. If all the water were removed from the ocean basins, they would reveal the topography shown in Figure 2.8. The map reveals that the ocean basins can be divided into three regions: (1) the *margins* of the continents; (2) the *abyssal zones*—extensive, mound-studded plains at great depth; and (3) a system of *midoceanic ridges* flanked by elaborate fractures and associated reliefs.

The continental margin consists of the *continental shelf, continental slope,* and *continental rise.* The **continental shelf** is the very gently sloping, relatively shallow, submerged

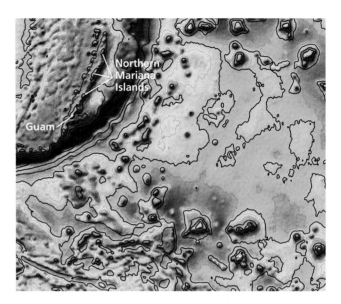

FIGURE 2.7 Seafloor topography inferred from gravity measurements. This small area in the Pacific near Guam has mountains and valleys dwarfing those found on land.

plain at the edge of the continent. The map shows the continental shelves to be continuations of the continental landmasses (see especially the areas off eastern North America, southeastern South America, and northwestern Europe). Generally, these shelves extend no deeper than 180 m (600 ft). Their average width is about 80 km (50 mi), but some are as wide as 1,000 km (625 mi). Because the continental shelves are extensions of the landmasses, the continental geologic structure continues to their edges. During the depth of the most recent Ice Age, when much ocean water was taken up by the ice sheets, sea levels dropped enough to expose most of these continental shelves. Rivers flowed across them and carved valleys that can still be seen on detailed maps. The oceans today are fuller than in the past: they have flooded extensive plains at the margins of the continents.

At a depth of about 180 m (600 ft), the continental shelf ends with a break in slope that is quite marked in some places and less steep in others. Here the **continental slope** begins and plunges steeply downward. Often, at the foot of the continental slope, there is a transitional **continental rise** of gently down-sloping material carried downward from the shelf and slope above (See Fig. 2.8). The continental rise leads into the abyssal zone. This zone consists mainly of the **abyssal plains**, large expanses of lower-relief ocean floor. The abyssal plains form the floors of the deepest areas of each ocean, except for even deeper *trenches*, which occur at the foot of some continental slopes (as along the Pacific margin of Asia). The abyssal zone is not featureless, however, and the extensive plains are diversified by numerous hills and *seamounts* (volcanic mountains reaching over 1,000 m [3,300 ft] above the seafloor), all of which are volcanic in origin. Lengthy valleys also exist, as though

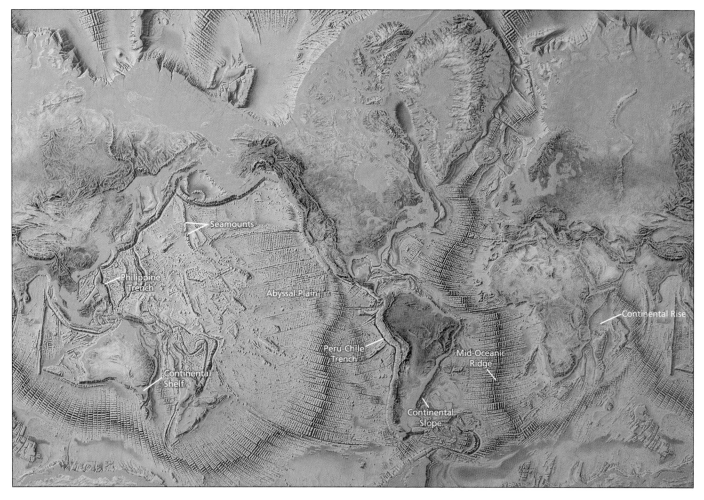

FIGURE 2.8 World ocean floor. Examples of the major features highlighted in the text on pp. 22–23 are labeled.

rivers had carved them there more than 1,800 m (6,000 ft) below the water surface. The origin of these valleys remains uncertain.

The third major ocean-floor feature is a global system of **midoceanic ridges**. These ridges are high, submarine, volcanic, and structural mountain ranges. The existence of one such ridge, the Mid-Atlantic Ridge, was known to scientists as early as the mid-1800s. But its properties were not understood until comparatively recently, when surveys that began in the 1950s made it clear that mid-oceanic ridges also extend across the Indian, Pacific, and Southern Ocean floors (see Fig. 2.8). As we will discuss in detail in Part Five, these midoceanic ridges are the scenes of active submarine volcanism and major movements of Earth's crust. Over the last few decades, manned and remote-controlled submarines have brought back dramatic video records of previously unseen violent eruptions, superheated water, and the exotic marine life-forms that populate these active ridges.

The active character of the midoceanic ridges and the geologic properties of the ocean floor become important to us when we study their implications in geomorphology.

The configuration of the ocean floors affects the movement of ocean water just as the topography of the land influences the movement of air. The ocean basins are filled with water that, like the air, is in constant motion. Great permanent circulation systems have developed in the oceans, and these systems affect the weather and climate on neighboring continents.

Clearly, our water-dominated planet is finally yielding secrets it has held for eons. That said, knowledge of the seafloor pales in comparison to information about the continents. The latest global terrain dataset (produced in 2014) has pixels 5 km (3 mi) in size. It represents a four-fold improvement over anything available previously, but it still provides only a fuzzy view of the world: features smaller than 10 km (6 mi) across are too small to be captured in such an image. For comparison, consider that all of Mars and 98 percent of Venus have been mapped at 100-meter resolution. Shipboard sounding devices (sonar) can map the seafloor with that precision, but no more than 10 to 15 percent of the world ocean has been mapped by sonar. Advanced sonar systems can achieve resolutions of a few meters, as would be needed to locate aircraft wreckage or small geologic features. Not even 0.05 percent of the ocean has received such coverage.

In addition to the relatively simple question about topography, there are a host of other similarly unanswered questions about the seafloor. For example, how do deep-ocean sediments vary in thickness and composition? How abundant is life on the ocean floor, and what unknown species remain to be found? How is climate change affecting physical and biological processes in the deep ocean? Despite all that has been discovered in recent years, in many ways the ocean bottom remains largely unexplored.

Human Population and Natural Processes

This book focuses on the physical systems, forces, and processes that modify natural landscapes, but at this early stage, we take note of the human factor and its impact on the natural world. As we will note later, animals, from worms to wallabies, play their roles in modifying landscapes by burrowing, digging, grazing, browsing, and even, as with beavers, felling trees and building dams. But no other species in the history of this planet has transformed natural landscapes to the degree humans have. We have converted entire regions to irrigated agriculture, terraced cultivable hill slopes, confined and controlled entire river systems, and reconstructed shorelines. We are also still deforesting vast areas, destroying soil cover, gouging huge open-pit mines, and building and paving once-natural surfaces into megacities so large that they become global-scale landscape features in their own right. In the first half of the twentieth century, geographers began to distinguish between **natural landscapes**, those areas of the planet still essentially subject to the physical processes we discuss in this book, and **cultural landscapes**, in which human intervention dominates to such an extent that physical processes have become subordinate. As the human population has grown, the cultural landscape has gained as the natural landscape has receded.

Population grew explosively during the twentieth century, from 1.5 billion in 1900 to more than 7 billion in 2015. Although the overall rate of growth has been declining in recent decades, the world still is adding about 78 million people per year. And while population in some areas of the world has begun to stabilize and even decline (in places such as Russia, Japan, and most of Europe), it continues to mushroom elsewhere, notably in South and Southwest Asia. But sheer numbers are no guide to a population's impact on the natural world. Highly developed, rich societies place demands on the resources of our planet that translate into massive intervention (in the United States, think of the Tennessee Valley Authority, the Colorado and Mississippi rivers, the farms of the Midwest, and the northeastern seaboard's megalopolis that stretches from north of Boston to south of Washington, D.C.). More populous but less developed and lower-consumption societies cannot afford to bend nature to their needs and tend to live subject to its uncertainties.

It is useful to have a sense of the spatial distribution of the world's population (Fig. 2.9). Our technological, environment-controlling capacities notwithstanding, this map still represents the historic accommodations humans made as populations entered and adjusted to habitats capable of supporting them. Two of the three great clusters—East Asia and South Asia—still represent dominantly rural, agricultural populations, but China continues to transform quite rapidly. The third and smallest of the major clusters, Europe, is the most modernized, industrialized, and urbanized of the three, and this population's impact on the region's natural environments is by many measures greater than that of the other two.

As Figure 2.9 shows, about 90 percent of the world's people live on a relatively small fraction (about 20 percent) of the land. Fertile river lowlands and deltas still contain the highest regional densities; altitudinally, more than 75 percent of all humankind resides below 500 m (1,650 ft); and nearly 70 percent of people live within 500 km (320 mi) of a seacoast. This leaves large parts of the planet with sparse human populations, including deserts such as the Sahara, high-latitude regions such as Siberia, and mountains such as the Himalayas. But even there, as we will see, humans make their impact. To the atmosphere, biosphere, lithosphere, hydrosphere, and cryosphere, should we add the *demosphere*, the sphere of demography, where humans have transformed the land?

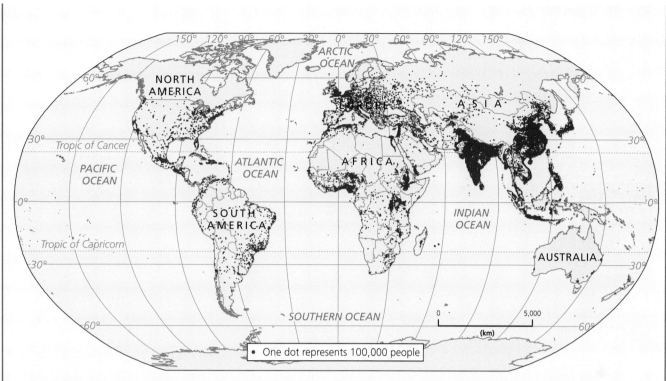

FIGURE 2.9 Spatial distribution of the world's population in 2013. The way people have arranged themselves in geographic space represents the totality of their adjustments to the environments that are capable of supporting human settlement.

Key Terms

abyssal plains *page 22*

atmosphere *page 16*

biosphere *page 17*

centrifugal force *page 15*

continental rise *page 22*

continental shelf *page 22*

continental slope *page 22*

cryosphere *page 17*

cultural landscapes *page 24*

Earth system *page 18*

ellipsoid *page 15*

hemisphere *page 19*

hydrosphere *page 16*

land hemisphere *page 20*

lithosphere *page 16*

midoceanic ridge *page 23*

natural landscapes *page 24*

topography *page 15*

water hemisphere *page 20*

Scan Here for a quick vocabulary review

ENHANCE

Review Questions

1. Why is Earth's shape closer to an ellipsoid than a perfect sphere?

2. Name and describe Earth's five spheres or interacting shells.

3. What is the difference between the land hemisphere and the water hemisphere?

4. Compare and contrast the distribution of continental landmasses in the Northern and Southern Hemispheres.

5. Describe the basic spatial patterns of the distribution of the world's population.

6. Describe the major topographic features of the world's ocean basins.

Scan Here for a quick concept review

ENHANCE

www.oup.com/us/mason

Mapping Earth's Surface

The United Kingdom was the world's (colonial) superpower when the global grid was laid out, so the British could determine where the prime meridian would lie. The Eastern and Western Hemispheres meet in this doorway of the Royal Observatory at Greenwich, near London.

OBJECTIVES

- Identify reference systems used for locations on Earth's surface
- Describe the most important characteristics of maps and the features of common classes of map projections
- Interpret isoline maps
- Discuss contemporary developments in geographic information science

When you go to an unfamiliar city, your process of learning about it begins at the hotel where you stay or in your new home. You then locate the nearest important service facilities, such as supermarkets and shopping centers. Gradually, your knowledge of the city and your activity space within it expands. Your slow pace of learning about the space in which you live repeats the experience of every human society as it has learned about Earth. Early in the learning process, directions concerning the location of a particular place are taken from or given to someone. The most common form of conveying such information is the **map**, described in Unit 1 as a model of the environment. Maps are distinguished by their graphics. Although a map might contain considerable text, unless it depicts spatial relationships visually, chances are the map fails to meet what is expected of it. We have all seen sketch maps directing us to a place for a social gathering. The earliest maps were of a similar nature, simply scratched in the dust or sand.

The origins of mapmaking are uncertain. The oldest surviving objects widely accepted as maps date from about 2500 BCE. Drawn on clay tablets in Mesopotamia (modern-day Iraq), they represent individual towns, the entire known country, and the early Mesopotamian view of the world (Fig. 3.1). The religious and astrological text above the map (top of Fig. 3.1) indicates that the Mesopotamian idea of space was linked to ideas about humankind's place in the universe. This is a common theme in **cartography**—the science, art, and technology of mapmaking and map use. Even today, maps of newly discovered space, such as star charts, raise questions in our minds of where we, as humans, fit into the overall scheme of things.

Location on Earth

Producing a map involves translating locations on Earth to an image. As discussed in Unit 2, specifying Earth locations is much simpler using a spherical model of the planet, and so we do so here. When a sphere is intersected by a plane, the lines of intersection form circles, and since the time of ancient Greece at least, circles have been used as the basis for locating points on Earth. Sometime between the development of the wheel and the measurement of the planet's circumference, mathematicians decided that the circle should be divided into 360 parts by means of 360 straight lines radiating from the center of the circle. The angle between two of these lines is called a **degree**. Each degree is divided into 60 *minutes*, and each minute is further subdivided into 60 *seconds*. Angles are expressed as a combination of degrees, minutes, and seconds (e.g., 45 30 00) or as decimal degrees (e.g., 45.5).

Having decided on the system for dividing the curved surface of Earth, the problem remained for early cartographers to choose the origin and layout of the circles. Two sets of information were used to solve this problem. First, the sense of direction gained by studying the movements of the Sun, Moon, and stars was employed. In particular, the Sun at its highest point in the sky each day is always located

A

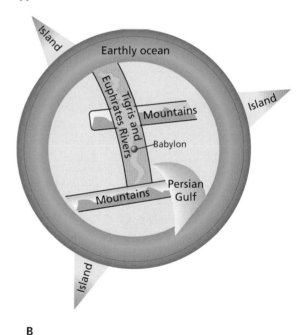

B

FIGURE 3.1 Mesopotamian world map. The original (A) was drawn in clay about 4500 years BP (before the present) with an explanatory text. An interpretive diagram (B) indicates its principal features.

in the same direction, which mapmakers designated as *south*. After establishing this, they were able to arrive at the concepts of *north*, *east*, and *west*. The division of the circle refined these directional concepts. The second set of information used was various geographical locations in the

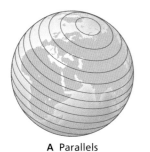
A Parallels

B Meridians

C Geographic Grid

FIGURE 3.2 Parallels (A) and meridians (B) on a globe. Parallels run east–west; meridians run north–south; together, they constitute the geographic grid (C).

Mediterranean region, which had been fixed by star measurements. These were used as reference points for the division of Earth.

Using this knowledge, cartographers were able to imagine a series of "lines" on Earth's surface (in actuality, circles around the spherical Earth), some running north–south and some running east–west, which together form a grid. The east–west lines of this grid are still called **parallels**, the name the Greeks gave them; the north–south lines are called **meridians**. As we can see in Figure 3.2, the two sets of lines differ. Parallels never intersect with one another, whereas meridians intersect at the top and bottom points (the *poles*) of the sphere. In addition, all meridians divide Earth into equal parts and in doing so span the full circumference of Earth. Meridians are therefore known as **great circles**. Except for the equator, each individual parallel splits Earth in unequal pieces, and each parallel is smaller than the Earth's circumference. Parallels are therefore known as **small circles**.

HORIZONTAL POSITION: LATITUDE AND LONGITUDE

The present-day divisions of the globe stem directly from Earth grids devised by ancient Greek geographers. The parallels lines of latitude are constructed by intersecting the sphere with planes at right angles to the axis of rotation. The largest of these, running around the middle of the globe, is the **equator**, defined as zero degrees latitude. As Figure 3.3A illustrates, **latitude** is the angular distance, measured in degrees north or south, of a point along a parallel from the equator. Lines of latitude in both the Northern and Southern Hemispheres are defined this way. For example, New York City has a latitude of 40 degrees, 40 minutes north of the equator (or 40.66 degrees), and Sydney, Australia, has a latitude of 33 degrees, 55 minutes south (33.92 degrees). The "top" of Earth, the *North Pole*, is at latitude 90 degrees N; the "bottom," the *South Pole*, is at latitude 90 degrees S.

The meridians are called lines of longitude. In 1884, the meridian that passes through the Royal Observatory at Greenwich in London, England, was established as the global starting point for measuring longitude. This north–south line is the **prime meridian** and is defined as having a longitude of zero degrees. **Longitude** is the angular distance, measured in degrees east or west, of a point along a

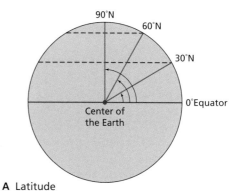
A Latitude

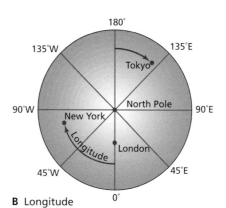
B Longitude

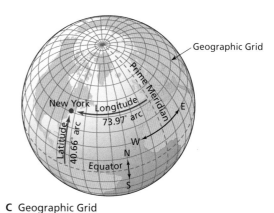
C Geographic Grid

FIGURE 3.3 Latitude and longitude. Viewed from the side (A), lines of latitude (including the equator) are horizontal parallels. Lines of longitude, the meridians, radiate from the North and South Poles. The latitude and longitude of any location, such as New York City, are the angular distances from the equator and prime meridian (C).

meridian from the prime meridian. The other meridians are ascertained as if we were looking down on Earth from above the North Pole, as Figure 3.3B illustrates. We measure both east and west from the prime meridian. Therefore, New York City has a longitude of 73 degrees, 58 minutes west of the Greenwich meridian (73.97 degrees), and Sydney has a longitude of 151 degrees, 17 minutes east (151.28 degrees).

Because meridians converge at the North and South Poles, the actual distance contained in one degree of longitude varies from 111 km (69 mi) at the equator to 0 km at the poles. In contrast, the length of a degree of latitude is about 111 km (69 mi) anywhere between 0 and 90 degrees N or S. We say "about" because, as we discussed in Unit 2, Earth is not a perfect sphere; it bulges slightly at the equator and is flattened at the poles.

Since the mid-1800s, an ellipsoid rather than a sphere has been used in most latitude/longitude systems. As we discuss in the Focus on the Science box on page 31, an ellipsoid brings a new a set of complications to the problem of creating a geographic grid. Choosing an ellipsoid and other parameters is required to establish a latitude/longitude grid, which is more properly called a *geographic coordinate system*. More than 1,000 different geographic coordinate systems have been devised over the years, and most are still in use. As discussed in this chapter's Focus on the Science: box, these differences are important only for maps of small areas where very precise specification of latitude and longitude is required. For maps of the entire globe, differences among the systems are insignificant because Earth is so close to being spherical.

VERTICAL POSITION

Latitude and longitude give horizontal position but say nothing about how high a place might be. Elevation is also a critical piece of locational information. For example, knowing that Cusco (the historic capital of the Inca Empire) has an elevation of 3,400 m (11,200 ft) seems at least as important as knowing its latitude and longitude. How is the vertical coordinate of a place, its elevation, specified?

We have all seen road signs and maps indicating elevations as some distance "above sea level" or, more correctly, as *above mean sea level (AMSL)*. This concept uses the ocean to define a reference value, with all elevations given above (or below) that reference value. The reference is known as a **vertical datum**. Historically, ocean measurements taken at a specific location were averaged over 19 years to remove tidal and other effects. The result is the average, or mean, sea level for the location. Individual nations established their own networks of tidal stations and used surveying techniques to transfer sea level values into their territories, thereby establishing a datum. In the United States, most government maps (including topographic maps described later in the unit) show elevation AMSL in one of several standard datums that have been developed over the years. *Global positioning satellite (GPS)* receivers (described later in this unit) measure elevation in a way that has nothing to

do with the ocean or sea level. Older GPS units report unadjusted elevations that can be 30 m (98 ft) different from AMSL. Newer devices convert the raw measurements into values close to AMSL.

Depicting Earth: Map Projections

If you have ever tried to cut the peel off an orange and then laid it flat on a table, you realize that this is not an easy task. At least some part of the peel must be stretched to make it completely flat. You can easily cut cylinders or cones and lay them out flat without distortion, but not a sphere. In geometric and cartographic terms, a sphere is an *undevelopable* surface, incapable of being flattened. Once the ancient Greeks had accepted the idea that Earth was a sphere, they had to determine how best to represent the round Earth on a flat surface. There is no totally satisfactory solution to this problem, but the early mapmakers soon invented many of the partial solutions that are still used commonly today.

The Greeks realized that a light placed at the center of a transparent globe would result in shadows cast by the meridians and parallels. These shadows, which form lines, could be seen in projected form on a flat piece of paper, or on some surface that could later be cut and laid out flat (Fig 3.4). The resulting series of projected lines on the new surface is a **map projection**, which is an orderly arrangement of meridians and parallels that is produced by any systematic method and used for drawing a map of Earth on a flat surface. All modern map projections are constructed mathematically rather than by a physical device and are often based on ellipsoids rather than spheres. (Just like spheres, ellipsoids are not developable and require a projection for display on a flat surface.)

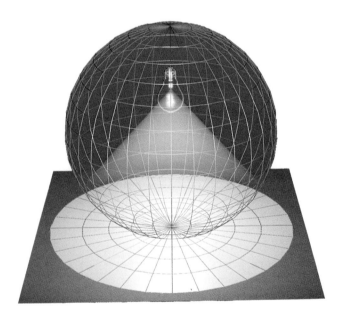

FIGURE 3.4 Shadows cast by parallels and meridians on a transparent globe are projected on a flat surface.

Where Am I, Exactly?

Modernizing societies have placed ever-more stringent demands on their mapping tools and positional data. An error of 10 m (33 ft) might have been insignificant for an eighteenth-century pioneer staking a claim in the American West, but nobody today would tolerate that kind of mistake when fencing off a suburban yard. We expect car navigation systems to show cars driving on the correct street, not through the buildings along the roadside. Problems with positional data can arise because, as mentioned in the text, a number of different geographic coordinate systems are in use. Different systems result in slightly different latitude/longitude coordinates for a single location. Equivalently, each system places the same latitude/longitude pair at a different location on Earth. The differences can be as large as several hundred meters (300 ft)! It's not hard to understand how this confusing situation arose when we consider the role of ellipsoids in mapping.

When we are producing a map of a given area, we choose an ellipsoid that comes close to the surface in the area being mapped. After all, if we are picking a model, it only makes sense to choose one that is a good approximation of the reality being modeled. As shown in Figure 3.5A, in mapmaking we can vary the size and shape of the model ellipsoid, and we also have control over the position and orientation of the ellipsoid. The ellipsoid we choose and its related parameters create a unique geographic coordinate system called a **horizontal datum**. A datum that works well for one map area is likely to be a poor choice for an area far away, which explains why there are so many datums employed worldwide (more than 1,000). Different datums lead to different horizontal coordinates for the same point (Fig. 3.5B). For example, the global positioning system uses an ellipsoid that provides a good fit for the planet as a whole but is less than optimal for individual countries and small areas. Thus, latitude and longitude provided by a GPS are not likely to agree perfectly with what is found on a map you might use for a backcountry hike. In trying to combine data from two sources on the same map, it is essential to be sure the same geographic coordinate system was used for both. If it was not, the map data must be converted to a common system. Fortunately, computer software embedded in mapping systems provides this capability, freeing users from complicated and exacting calculations.

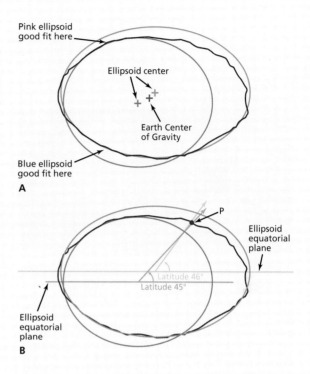

FIGURE 3.5 (A) Earth (brown line) and two approximating ellipsoids in cross-section. The best approximation to Earth depends on the location and size of the area to be mapped. (B) The latitude of point (P) depends on the ellipsoid in use.

Elevations are also not so straightforward. Consider some of the problems with the very familiar concept of sea level. Everyone knows that even in the absence of winds and tides, the sea surface would not be flat but instead would follow the shape of Earth. But exactly what shape? As discussed in Unit 2, a spinning Earth of uniform density would be an ellipsoid. A perfectly still ocean would follow the same ellipsoid shape, and gravity would cause water to flow and fill any low spots. Likewise, no high spots could be maintained against the force of gravity. However, the real Earth is not uniform. There are pockets of denser rock here and there, and continents are denser than water. In areas of greater density, the additional mass causes gravity to draw water from surrounding areas, making sea level higher (Fig. 3.6A). The "gravity surface," known as the *geoid*, undulates irregularly over the planet, as we can see in Figure 3.6B. The geoid corresponds to mean sea level, and therefore elevation is

continued

continued

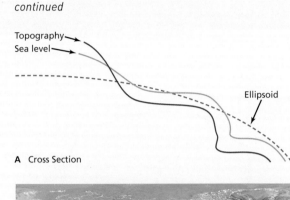

A Cross Section

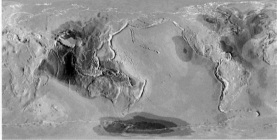

B Geoid Height Differences

FIGURE 3.6 (A) Variations in topography and in the density of underlying rock cause sea level to vary from place to place. The resulting irregular Earth "shape," defined by gravity, is known as the geoid. (B) The geoid rises and falls by more than 100 m (328 ft) relative to the smooth ellipsoid used by GPS receivers.

actually defined as height above the geoid. Global positioning system receivers measure heights relative to an ellipsoid, not elevation. However, newer devices contain detailed information about the difference between the geoid and their ellipsoid, so they can provide users with an elevation value. Notice that the geoid is not the same as sea level established by measuring the ocean and extending those measurements inland. This is why when hiking, you may have come across markers giving elevations different from what your GPS claims, and why the elevations shown on your map do not correspond to GPS elevations.

PROPERTIES OF MAP PROJECTIONS

A fundamental property of any map is its **scale**, which is the ratio of map distance to Earth distance. If a road is 1 in wide on a map and 100 ft wide in the real world, the scale of the map is 1/1,200, usually written as 1:1,200. Notice that we converted 100 ft to 1,200 in so that the units in the numerator and denominator agree. Notice also that we could just as correctly say 1 cm of map distance represents 1,200 cm of Earth distance. In other words, scale is dimensionless. Finally, notice that a map showing a large area has a small scale. In the road map example, if 1 in represented 1,000 ft on Earth, that map would show more Earth area on a page than the 1:1,200 map. The scale would be 1:12,000, smaller by a factor of 10. Maps depicting the full globe have the smallest scales of all because they show the largest area. By contrast, the maps used for backpacking are large-scale maps. Humans can't walk far in a day, so there isn't any point in carrying a map of a large area. It is more important to see fine detail in a small area, and for that, a large-scale map is needed.

Because spheres and ellipsoids are not developable, there is no way to make a map that has a uniform scale. If the map scale is correct for one location, there are other places where the scale is wrong. The scale can be correct at a single point, or along one or two lines, but not everywhere on the map. All map projections contain distortion.

Although it can't be avoided, the mapmaker does have considerable control over the distortion. She might, for example, require that the map correctly depict shapes of objects. Thus, any circle on Earth would be a circle on the map, and roads intersecting at a right angle would also be perpendicular on the map. Ensuring that shapes are preserved amounts to preserving angles, and such maps are said to be **conformal**. Another option is to require that the relative area of objects be preserved. Aptly called **equal-area** maps, these maps guarantee we can compare the size of objects on the map and not be misled about their relative size on Earth. For example, all the U.S. football fields on an equal-area map are the same size because all football fields are 120 yards long by 50.33 yards wide (including end zones). One field might appear rectangular and another trapezoidal, but their map areas are the same. Some maps are conformal, others are equal-area, but no map can be both conformal and equal-area because that would require uniform scale across the map. To minimize this problem, the Greeks followed the rule cartographers still use: select the map projection best suited to the particular geographic purpose at hand.

TYPES AND EXAMPLES OF MAP PROJECTIONS

Since the time of the Greeks, mapmakers have devised hundreds of projections to flatten the globe so that all of it is visible at once. Mathematically, it is possible to create an infinite number of map projections. In practice, however, these cartographic transformations of the three-dimensional surface of Earth have tended to fall into a small number of categories. Here we consider the four most common classes of map projections: cylindrical, conic, planar, and mathematical.

A **cylindrical projection** involves the transfer of Earth's latitude/longitude grid from the globe to a cylinder, which is then cut and laid flat. If the cylinder is aligned with the axis of rotation, the parallels and the meridians appear on the opened cylinder as straight lines intersecting at right angles. On any map projection, the least distortion occurs where the globe touches, or is *tangent* to, the geometric object it is projected onto; the greatest distortion occurs farthest from this place of contact. In the left part of Figure 3.7A, the globe and the cylinder are tangent along the parallel of the equator (the dotted red line). The parallel of tangency between a globe and the surface onto which it is projected is called the **standard parallel**. The globe in the right diagram in Figure 3.7A is larger than the cylinder, and two standard parallels result (shown in red). The projection on the right reduces distortion throughout the map and is particularly useful for representing the low-latitude zone straddling the equator between the pair of standard parallels.

The most famous cylindrical projection was devised in 1569 by Flemish cartographer Gerardus Mercator. In the **Mercator projection** (Fig. 3.8), the spacing of parallels increases toward the poles. This increase is in direct proportion to the false widening between normally convergent meridians that is necessary to draw those meridians as parallel lines. Although this produces extreme distortion in the area of the polar latitudes, it provided a tremendously important service for navigators using the newly perfected magnetic compass. Unlike any other map projection available at the time, a straight line drawn on this map is a line of true and constant compass bearing. Such lines are called *rhumb lines*. Once a navigator determines from the Mercator map the compass direction to travel, a ship can easily be locked onto this course.

Cones can be cut and laid out flat as easily as cylinders, and the **conic projection** has been in use almost as long as the cylindrical. A conic projection involves the transfer of Earth's latitude/longitude grid from a globe to a cone, which is then cut and laid flat. Figure 3.7B shows the derivation of the two most common conic projections, the one- and two-standard parallel cases, with the cone and Earth axes aligned. On the resulting maps, meridians emerge as straight lines radiating from the (North) Pole. Parallels appear as arcs of concentric circles with the same center point, which shorten as the latitude increases. Conic projections are often used for areas in the middle latitudes that cover a large expanse of longitude, such as the United States and Russia (see, e.g., Fig. 7.8).

Planar projections, in which an imaginary plane touches the globe at a single point, exhibit a wheel-like symmetry around the point of tangency between the plane and the sphere. Planar projections were the first map projections developed by the ancient Greeks. Today, they are most frequently used to represent the polar regions (Fig. 3.9). One type of planar projection, the *gnomonic*, possesses an especially useful property: a straight line on this projection is the shortest route between two points on Earth's surface. This has vital implications in intercontinental jet travel. Long international flights seek to follow the shortest routes, and these are identified on the spherical Earth by imagining the globe to be cut exactly in half by a (geometric) plane running through the origin and destination cities. Just as we saw for meridians, the resulting circle is a great circle. Long-distance air traffic usually follows great-circle routes, such as the one between New York and London shown in Figure 3.9.

Many other projections have been devised using cylinders, cones, and planes. Some projections wrap a cylinder around the North and South Poles, resulting in a geographic grid that is hardly recognizable. Some planar projections place the point of tangency at the center of the map area, which might be far removed from either pole, and in others the plane intersects the globe in a circle rather than being tangent at a point. Still other maps don't use a projection surface at all and are entirely **mathematical projections**.

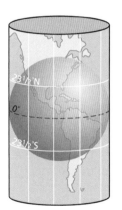

A Cylindrical Projections

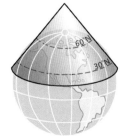

B Conical Projections

FIGURE 3.7 Construction of cylindrical (A) and conic (B) projections with one and two standard parallels.

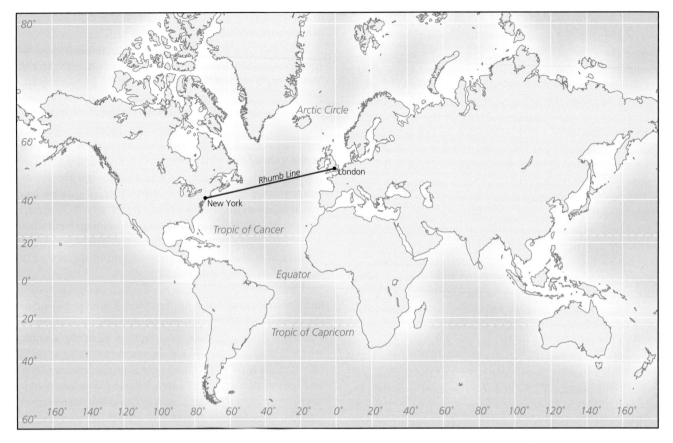

FIGURE 3.8 The Mercator projection, which produces straight parallels and meridians. Any straight line drawn on this map is a (rhumb) line of constant compass bearing, a tremendous advantage for long-distance navigation.

FIGURE 3.9 The great-circle route from New York to London is a straight line on this polar gnomonic projection. Note how much shorter this is than the rhumb line connecting the same locations.

The polar quartic map in Figure 3.10 is one example of a mathematical projection. This equal-area map preserves the relative size of the continents and does so in a way that makes them easily recognizable despite the unavoidable shape distortion. The map in Figure 3.10 possesses another feature: it is not a continuous representation but is *interrupted*. This technique helps the cartographer devote most of the map to the areas that project with the least distortion, and to de-emphasize areas of the globe that are not essential (e.g., omitting large parts of the oceans in mappings of land-based phenomena). The Robinson projection (Fig. 3.11) is also mathematical, but it is neither equal-area nor conformal. Used as the standard global base map by the National Geographic Society for most of the 1990s, the Robinson projection reflects the inventor's effort to compromise between area and shape distortion. Many other compromise projections have been developed and are in widespread use.

Map Interpretation: Isolines

One of the most important functions of maps is to communicate content effectively and efficiently. Because so much spatial information exists in the real world, cartographers must carefully choose the information to be included and deleted to avoid cluttering the map with less

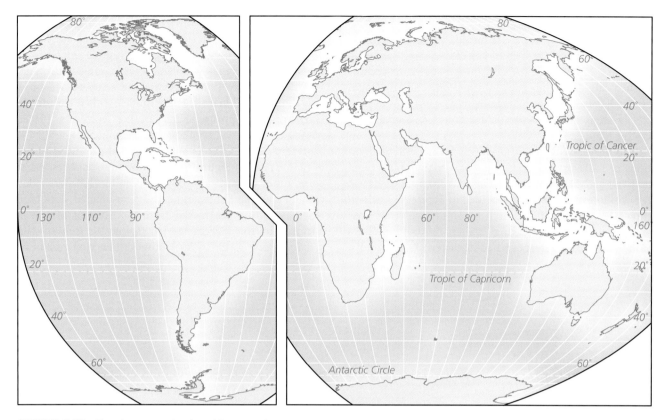

FIGURE 3.10 Equal-area projection. All mapped areas are represented in their correct relative sizes. This flat-polar quartic equal-area projection—in interrupted form—was developed in 1949 for the U.S. Coast and Geodetic Survey by F. W. McBryde and P. D. Thomas.

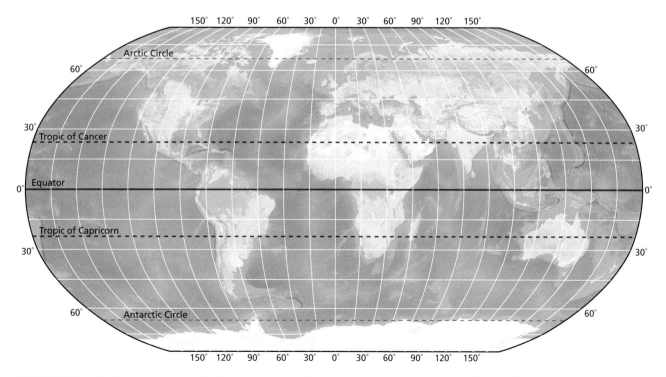

FIGURE 3.11 The Robinson projection is a compromise between equal-area and conformal projections.

relevant data. Thus, *maps are models*: their compilers simplify the complexity of the real world, filtering out all but the most essential information. Professional cartographers are experts at this, and most people are accomplished at decoding the information and messages embedded in commonly encountered maps. However, it can be difficult, especially for beginners, to understand **isoline** maps.

Isoline maps depict *surfaces*, which can be generalizations of real surfaces (e.g., the world topographic relief map in Fig. 2.5) or representations of conceptual surfaces. An example of the latter is provided in Figure 3.12, which shows Florida's yearly rainfall as a "surface" that possesses "ridges" representing higher rainfall amounts along the northwestern and southeastern coasts and a

"trough" representing lower rainfall totals extending down through the interior of the peninsula. Cartographers commonly use the mapping technique employed in this figure to solve the problem of portraying continuous surfaces like elevation, rainfall, and temperature on a two-dimensional map. The technique is called **isoline** (or **isarithmic**) **mapping**. Each isoline connects places having the same value of a given phenomenon, and so it represents a line of constant "height" above the flat base of the surface. In Figure 3.12, the boundary lines between color zones connect all points reporting that particular rainfall amount. For example, the cigar-shaped, dark blue ridge extending northward from just west of Miami is enclosed by the 152-cm (60-in) isoline (or *isohyet*—the specific name given to an isoline connecting points of

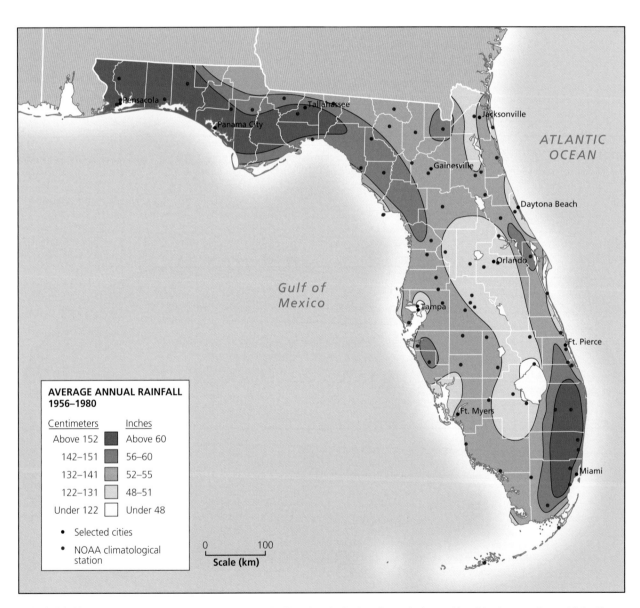

FIGURE 3.12 Average annual rainfall distribution in Florida, displaying the technique of isarithmic mapping, which allows three-dimensional volumetric data to be shown on a two-dimensional map.

identical rainfall total). Similarly, the 132-cm (52-in) isohyet encloses the lighter blue long north–south trough running south from the Orlando area through the center of the peninsula.

Perhaps the most familiar use of isolines in physical geography is the representation of surface relief by **contouring**. As Figure 3.13 demonstrates, each contour line represents a specific and constant elevation, and all the contours together provide a useful generalization of the surface being mapped. Figure 3.14 is a more practical example of contouring. The landscape portrayed in Figure 3.14A

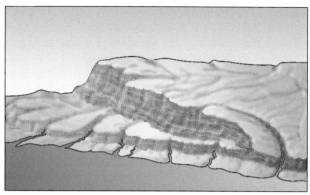

A 3-D View

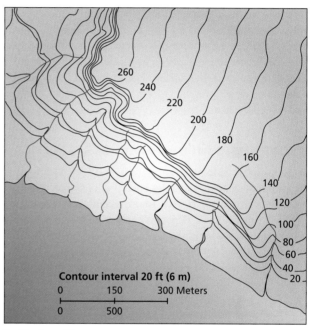

Contour interval 20 ft (6 m)

B Contour Map

FIGURE 3.14 Perspective sketch of a coastal landscape (A) and its corresponding topographic map (B), taken from U.S. Geological Survey sources. Note that this map is scaled in feet and that the contour interval—most appropriate for this map—is 20 ft (6 m).

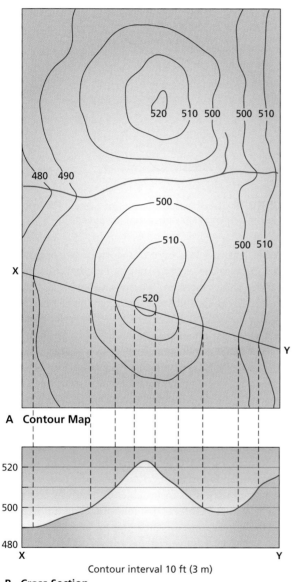

A **Contour Map**

Contour interval 10 ft (3 m)

B **Cross Section**

FIGURE 3.13 In topographic contouring, each contour line's points have identical height above sea level. The surface relief described by the contour map (A) can be linked to a cross-sectional profile of the terrain. (In part B, this is done for line X–Y on map A.) Note how contour spacing corresponds to slope patterns: the wider the spacing, the gentler the slope, and vice versa.

corresponds to the *topographic map* of that terrain depicted in Figure 3.14B. By comparing the two, we can read the contour map and understand how this cartographic technique portrays the configuration of Earth's surface relief accurately.

The land surface of the United States, except for Alaska, has been completely mapped at the fairly detailed scale of 1:24,000. The U.S. Geological Survey (USGS) has published approximately 180,000 maps for about 55,000 quadrangles at this scale. Technically, "topography" includes everything found at the surface (roads, vegetation, buildings, etc.), and USGS topographic quadrangles contain much more than elevation contours. Today, the full set of quadrangles is available online as just one part of the USGS National Map. This is a large project involving multiple agencies that contribute a wide array of spatial information. Users can customize their maps by adding or removing layers and can download the underlying data for other purposes. This is just one of the products of the technological revolution that is transforming cartography in the twenty-first century.

Evolving Cartographic Technology

The late twentieth and early twenty-first centuries saw rapid advances in computer power and speed. Cartographers quickly applied the new technology to mapmaking, and today, virtually all maps are compiled digitally. As sophisticated mapping software was pioneered, closely related breakthroughs simultaneously occurred in airborne and satellite remote sensing and GPS. This led to an explosion of new data about Earth's surface and propelled the rapid development of geographic information systems to analyze and interpret them.

GEOGRAPHIC INFORMATION SYSTEMS

A **geographic information system—GIS** for short—is a computer system that enables spatial data to be collected, recorded, stored, retrieved, manipulated, analyzed, and displayed to the user. Especially when linked to remotely sensed data from high-altitude observation platforms, this approach allows for the simultaneous collection of several layers of information pertaining to the same study area. These layers are integrated by multiple map overlays (Fig. 3.15) to assemble the components of the complex real-world pattern. One simple use of such an overlay is to locate places meeting multiple criteria. Working from the example in Figure 3.15, for example, we might seek a particular combination of vegetation, soil, and geology, such as shallow rooted plants growing on thin soils overlaying highly erodible bedrock.

For cartographers, GIS technology is particularly valuable if all digital data are referenced with respect to a known geographic coordinate system (projection and datum). This enables the data to be mapped within the framework of any map projection and to move easily from one projection to another. It also allows the collation of digital data from diverse sources, even if the original source material exists in different map projections and/or datums. Even more important than these data-management capabilities are the analytical tools provided by GIS. For example, geographic information systems allow complicated queries of spatial databases, such as "Find the average size of all water bodies having a perimeter that is at least 30 percent wetland in watersheds that are at least 70 percent cropped by fields needing application of liquid manure fertilizer." Geographic information systems also allow manipulation of data in novel ways, as illustrated in Figure 3.16. The elevation

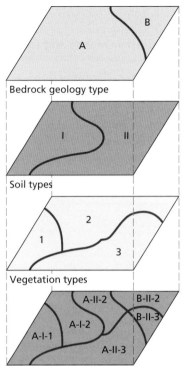

FIGURE 3.15 Geographic information system (GIS) processing of overlays to produce map composite.

values of Figure 3.16A were transformed by sophisticated processing into the wetness map of Figure 3.16B. The wetness index shows locations where water might accumulate after rainfall events and areas that might be habitually deficient in water. This is important for soil development, and in fact, this wetness map was developed as an input layer for a model that predicts the distribution of soils. Computing the wetness index involved considering potential water flow in and out of every 5-m square in the area. Such calculations, unthinkably complex to perform by hand, are routine in GIS.

Perhaps the most revolutionary aspect of GIS cartography is its break with the static map of the past. The use of GIS methodology involves a constant dialogue, via computer commands and feedback to queries, between the map and the map user. This instantaneous, two-way communication is known as **interactive mapping**, and it is expected to be the cornerstone of cartography for the foreseeable future. Online mapping systems and in-car navigation devices are only two examples of interactive mapping, and they illustrate how GIS has made cartographers of us all.

GLOBAL POSITIONING SYSTEMS

We take **global positioning systems (GPS)** for granted, but they are a relatively new technology and in many ways transformational. Developed by the U.S. military from projects started in the 1970s, the GPS did not become fully operational until 1994. Other similar systems followed

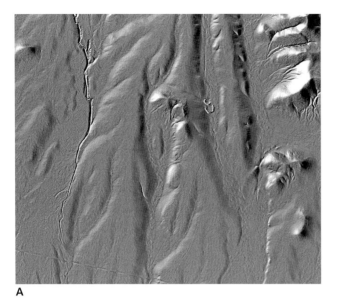
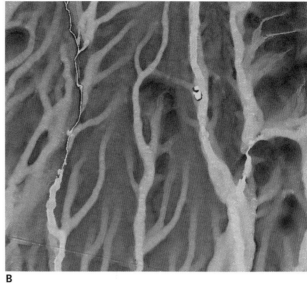

A

B

FIGURE 3.16 Elevation values (A) were used to produce measures of soil landscape wetness (B) for a small area in New Hampshire. Colors are used to show where moisture is expected to accumulate (light areas) and drain readily from the soil (dark areas).

from Russia (called GLONAS) and China (COMPASS), and the European Space Agency is building a system (Galileo) that is scheduled to be completed in 2018. As of this writing, only GPS achieves full global coverage. All systems rely on a network of satellites (24 for GPS), most of which are placed in *geosynchronous orbit*, meaning the satellites turn with Earth and remain above a fixed location. The satellites transmit several pieces of information that permit a receiver to calculate its position with amazingly high accuracy in both the horizontal and the vertical.

Each satellite transmission contains the location of the satellite and the time of transmission. The signal spreads out in all directions at the speed of light. If a receiver detects a signal and knows how long ago the signal was sent, the receiver can tell how far away the satellite is. From that and the satellite's position, the receiver knows it is somewhere

on a sphere centered on the satellite with radius equal to the satellite distance. In Figure 3.17A the receiver (indicated by the red dot) is located in Cancún, Mexico. If there is information from only one satellite, the receiver could be anywhere on the pink sphere. The receiver could be in Cancún, high in space, or even under the surface of Earth. In Figure 3.17B two spheres—representing data from two satellites—intersect in a circle; the reception of data from two satellites restricts the possible location of the receiver to the white ring passing through Cancún. In Figure 3.17C signals from three satellites insect at two points, only one of which (the red dot) is close to Earth's surface. In theory, three satellites can identify a unique locale. In practice, four or more satellites are needed, mostly because the receiver's clock is likely to be cheaply manufactured and therefore somewhat inaccurate.

A   **B**   **C**

FIGURE 3.17 A GPS receiver a known distance from a satellite could be anywhere on a sphere of that radius (A). Transmissions from two satellites intersect in a circle (B), limiting the possible locations further. Signals from three satellites are the minimum required to find the location (C).

Our explanation glosses over a number of intricate technical details. It doesn't discuss atmospheric effects, and it fails to communicate how exquisitely timed the signals must be. In fact, the computations are so sensitive that Einstein's theory of relativity is employed when setting the tick rate of satellite clocks. Relativity dictates that time for a satellite runs faster than it does on Earth because of the weaker gravity at orbital altitudes. A smaller relativistic effect, arising from the satellites' high rate of travel through space, slows time down. Overall, there is a miniscule speedup (about 1.4 seconds per century), but it must be accounted for in the time that satellites broadcast to receivers. Truly, GPS is a scientific and engineering marvel.

In principle, consumer GPS receivers are accurate to within a few meters, but in actual practice, a number of factors—such as clouds, obstructions, transmission noise, and suboptimal geometry between receiver and satellites—increase the error by a factor of five or so. Positional errors vary from moment to moment, so any statement regarding accuracy must be statistical. Most of the time, errors are given as a 50 percent circular error, meaning that half of all measurements will be within the stated distance of true position and half farther away. Expressed this way, consumer units in typical use are accurate to about 15 m (49 ft). Higher accuracy (within a few meters) has been available since 1999 in the United States from an augmented system developed mainly to meet aviation needs. A similar system has been deployed in Europe. For very exacting work, GPS units costing several thousand dollars that limit positional errors to just a few centimeters (about an inch) are available.

From its origins as a military tool, GPS has come to penetrate innumerable aspects of civilian life. This includes the practice of physical geography, where both consumer- and survey-grade GPSs are routinely used in fieldwork. Uses range from navigating to preassigned locations for soil or vegetation sampling to mapping flood deposits. In Figure 3.18, GPS is facilitating the study of Oregon's Mt. St. Helens following its major 1980 eruption. In many ways, GPS has become an essential addition to the physical geography toolbox.

REMOTE SENSING OF THE ENVIRONMENT

Maps have been used for centuries to graphically communicate information about Earth's surface. The Greek legend of Icarus, who flew too close to the Sun and melted his wax-and-feather wings, shows how badly the scientific ancestors of modern geographers wanted to view Earth from the sky. Today, this is possible, and the work of mapmakers and spatial analysts is greatly facilitated and enhanced by powerful new tools and techniques collectively known as **remote-sensing** technology, the ability to scan Earth from airborne and satellite observation platforms.

Just as it sounds, "remote" sensing involves capturing information about objects without the sensor being in direct contact with the object or scene itself. For the most part, this means acquiring data from above. Remote sensing as a science is understood to also mean data processing, storage, interpretation, and display (typically, images and maps), as well as developing methods for the same. Some remote-sensing technologies are known as *passive sensors* because they rely on radiation that is reflected or emitted by the object or area under study. *Active systems*, by contrast, emit radiation that is at least partially returned to the sensor by reflection. Photography and radar are examples of passive and active technologies, respectively. Many remote-sensing platforms contain both active and passive systems.

Aerial photography has been used since the advent of cameras early in the nineteenth century. Even though the Wright brothers did not take off until 1903, hot-air balloons and even trained birds carried cameras aloft before 1850. Over the past century, several methods were developed (many by the military to improve its advantages in ground warfare) that permitted maps to be produced directly from a series of photographs taken by survey aircraft.

At the same time, photographic technology was being perfected to expand this capability. Along with improved

FIGURE 3.18 A precision GPS unit installed to measure ground movements on Mt. St. Helens, Washington, just days before an eruption in 2004.

camera systems came ever-more sensitive black-and-white and then color films. Moreover, by World War II, ultrasensitive film breakthroughs extended the use of photography into the infrared radiation (IR) range of the *electromagnetic spectrum* (Fig. 3.19) beyond the vision capacities of the human eye. By directly "seeing" reflected and radiated solar energy, aerial infrared photography could, for the first time, penetrate clouds, haze, and smoke—and even obtain clear images of the ground at night. The U.S. Air Force further pioneered the use of IR color imagery. The "colors" obtained—known as false-color images—bore no resemblance to the natural colors of the objects photographed but could readily be decoded by analysts.

During the late twentieth century, nonphotographic remote sensing developed swiftly as new techniques and instruments opened up a much wider portion of the electromagnetic spectrum. This spectrum is diagrammed in Figure 3.19 and consists of a continuum of energy as measured by wavelength, from the high-energy shortwave radiation of cosmic rays, whose waves are calibrated in billionths of a meter, to the low-energy longwave radiation of radio and electric power, with waves that can span kilometers from crest to crest.

Figure 3.19 also shows the discrete *spectral bands* within the electromagnetic spectrum, which can be picked up by radio, radar, thermal IR sensors, and other instruments. A band is a continuous range of wavelengths and is therefore one segment of the spectrum. As remote sensing matured, scientists and engineers learned more about those parts of the spectrum and which types of equipment are best suited to studying various categories of environmental phenomena. This is important because each surface feature or object emits and reflects a unique pattern of electromagnetic energy that can be used to identify it, much like a fingerprint can identify any human individual. Remote platforms increasingly utilize multispectral systems, arrays of scanners attuned simultaneously to several different spectral bands. The combination of bands can greatly enhance the quality of observations and their interpretations.

A fairly new active technology known as **LiDAR** (light detection and ranging) has become very important in physical geography. In LiDAR, lasers emit pulses of "light." The reflections of these pulses are captured and processed to reveal the geometry of underlying surfaces. The name is a little misleading, because LiDAR uses both visible and invisible radiation at wavelengths close to the visible band to image its targets. Upward- and sideways-looking LiDAR has had a role in atmospheric research for many decades. More recently, downward-looking instruments have found favor for capturing surface topography. LiDAR stands out from other technologies for its ability to acquire information with unprecedented spatial detail: advanced LiDAR devices can image objects smaller than 10 cm (2.5 in). By measuring reflections originating at various heights, LiDAR can map both the ground surface and the structure of overlying plant canopies.

Although remote sensors can be ground based, most systems that facilitate our understanding of physical geography collect data at high altitudes. Aircraft have been and continue to be useful, but they are limited by how high they can fly (approximately 20 km [12 mi]) and the weather conditions in which they can operate. With planes unable to reach the very high altitudes required to obtain the small-scale imagery needed for seeing large areas of the surface in a single view, the advent of the Space Age in 1957 provided the needed alternative in the form of Earth-orbiting satellites.

By the 1980s, dozens of special-purpose satellites were circling Earth at appropriate altitudes and providing remotely sensed data to aid in assembling the "big picture." Among the most familiar in operation today are the National Oceanic and Atmospheric Administration (NOAA) and Geostationary Operational Environment Satellite (GOES) series. Satellites in the GOES series, first launched in 1991, are fixed in geosynchronous orbit. The GOES keeps watch on both land and ocean areas bordering the United States to track storms and support weather forecasting and research in atmospheric sciences. The NOAA

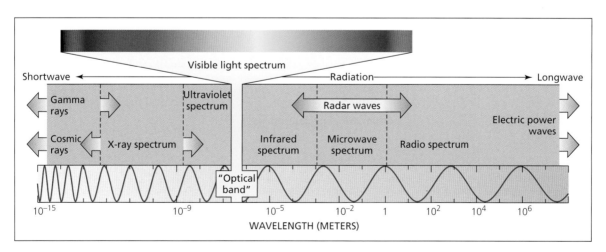

FIGURE 3.19 The complete (electromagnetic) radiation spectrum. Note the logarithmic axis for wavelength and that wavelengths are not drawn to scale.

series of polar-orbiting satellites, first launched in 1972, pass over every point on Earth at least once every six hours and provide higher-resolution images than those supplied by the GOES sensors. Images from these programs are often featured in newscasts and newspaper photos.

Also vital are satellites of the Landsat program. With launches from 1972 to 2013, Landsat is the longest-running space-based imagery program in the world. All Landsat satellites have been multispectral, capturing radiation in as many as eight bands. The newest satellite launched has eleven bands. Landsat's sensors have provided new insights into geologic structures, desert expansion, deforestation, fluctuations of glaciers and lakes, and changes in land cover and land use. At the same time, these satellites carefully monitor global agriculture, forestry, and countless other environment-impacting human activities (Fig. 3.20).

The Earth Orbiting System (EOS) of the U.S. National Aeronautics and Space Administration (NASA) is another very important program. The mission of EOS is long-term global monitoring of Earth's land surfaces, oceans, biosphere, and atmosphere. For about a decade and a half, its more than fifteen satellites have provided a diverse set of information bearing on our global habitat, ranging from gravity measurements to ocean chlorophyll, atmospheric pollutants, and even the Sun's output.

Satellite remote-sensing programs are run by other countries as well, including Brazil, Argentina, Russia, and India, as well as by the European Space Agency. There are also about a dozen commercial satellite remote-sensing operations, some based in the United States and others

FIGURE 3.20 Landsat satellite image of the Los Angeles area. The colors can be interpreted in the following manner. The built-up area is blue-gray, revealing the gridded road network. The ocean is black. Vegetated areas are red and tend to be associated with high relief, notably in the San Gabriel Mountains to the north of the city and the Santa Monica Mountains to the west. Rivers show up as thin white lines, most notably the Los Angeles River, which originates in the San Gabriel Mountains and flows southward into San Pedro Bay.

elsewhere, and innumerable companies throughout the world supply airborne data to clients on a contract basis. Clearly, we are awash in geospatial information, much of which bears directly on physical geography.

Key Terms

cartography *page 28*

conformal projection *page 32*

conic projection *page 33*

contouring *page 37*

cylindrical projection *page 33*

degree *page 28*

equal-area projection *page 32*

equator *page 29*

geographic information system (GIS) *page 38*

global positioning system (GPS) *page 38*

great circle *page 29*

horizontal datum *page 31*

interactive mapping *page 38*

isarithmic (isoline) mapping *page 36*

isoline *page 36*

latitude *page 29*

LiDAR *page 41*

longitude *page 29*

map *page 28*

map projection *page 30*

mathematical projection *page 33*

Mercator projection *page 33*

meridians *page 29*

parallels *page 29*

planar projection *page 33*

prime meridian *page 29*

remote sensing *page 40*

scale *page 32*

small circle *page 29*

standard parallel *page 33*

vertical datum *page 30*

Scan Here for a quick vocabulary review

ENHANCE

Review Questions

1. What is the difference between latitude and longitude? What are the reference lines and/or points for each measurement system?

2. Why are multiple systems of latitude and longitude in use?

3. Describe how the properties of *scale*, *area*, and *shape* relate to a map projection.

4. What are the differences between cylindrical, conical, and planar map projections?

5. What is an equal-area projection, and what mapping task(s) is it well suited to?

6. How does GPS work?

7. What are the distinguishing features of geographic information system (GIS) and remote-sensing techniques?

Scan Here for a quick concept review

ENHANCE

www.oup.com/us/mason

Earth–Sun Relationships

The winter Sun is never far above the horizon in the high latitudes. In 2013, residents of Rjukan, Norway (60°N) constructed a large mirror to reflect light into the dark village center.

OBJECTIVES

- Describe Earth's motions relative to the Sun
- Demonstrate the consequences of Earth's axis tilt for the annual march of the seasons
- Explain the time and spatial variations in solar radiation received at surface locations

Our small, fragile planet is the only member of the solar system known to contain life. In large part, this is due to Earth's nearly optimum distance from the Sun. This distance ensures that Earth experiences neither the searing heat nor the perpetual cold found on other planets. The quantity of solar energy supplied to Earth is huge; on a global average basis, it dwarfs tidal forces, volcanoes, earthquakes, and the slow movement of geothermal heat upward from Earth's interior. In addition to providing life-sustaining warmth, the Sun is the source of energy for most physical and biological processes on the planet. Solar radiation is the fuel for wind, rain, storms, ocean currents, and towering tropical clouds. It provides the energy for plant *photosynthesis* and the lift needed to raise water from the soil to the top of giant redwood trees 100 m (300 ft) above ground. Apart from a few notable exceptions, all natural phenomena ultimately result from the Sun's energy. Understanding solar radiation is crucial to understanding the workings of the physical environment. In this unit, we explore basic geometrical relationships between Earth and the Sun and the profound consequences of these relationships for the delivery of sunlight to the planet.

The discussion in this unit presents Earth–Sun relations as they are today in the twenty-first century. Unit 19 will explain that, at very long time scales, changes in Earth's orbit produce dramatic changes in climate. However, the rate of orbital change is so slow that we can ignore it when we examine seasonal and spatial variations in solar energy. Thus, for clarity's sake, this unit considers Earth–Sun relations as fixed. In a distant future edition of this book, say, 10,000 years from now, the diagrams and discussion in this unit would need to be significantly different from what you find here. Keep this point in mind, as it will be relevant in the discussion of global climate change in Part Three, Climate and Climate Change.

Earth's Planetary Motions

Like the other planets, Earth's motions through space are complex. Two of its motions are primary: revolution and rotation. A **revolution** is one complete circling of the Sun by a planet within its orbital path. Earth revolution takes exactly one year. As planets revolve, they also exhibit a second simultaneous motion—**rotation**, or spinning on an *axis*. It takes Earth almost one calendar day to complete one full rotation on its **axis**, the imaginary line that extends from the North Pole to the South Pole through the center of Earth.

As Earth revolves around the Sun and rotates on its axis, the Sun's most intense rays constantly strike different patches of Earth's surface. Thus, at any given moment, the Sun's heating (or solar energy) is unevenly distributed, always varying in geographic space and time. By time, we mean not only the hour of the day but also the time of the year as measured in the annual march of the seasons. Before examining seasonality, we need to know more about the concepts of revolution and rotation.

REVOLUTION

Earth revolves around the Sun in an orbit that is almost circular. Its annual revolution around the Sun, which determines the length of our year, takes 365.25 days. To have an even number of days every year, our calendar assigns 365 days to three consecutive years and allots one extra day to the fourth. Every fourth year, February has 29 days instead of 28, making what is called a *leap year* with 366 days. The years 2012, 2016, and 2020 are all leap years.

Like Earth itself, which is *nearly* a sphere, Earth's orbital path is *nearly* circular around the Sun, with a radius of about 150 million km (93 million mi). However, the orbital path is better approximated as an ellipse, with the Sun located away from the center at one focus of the ellipse (see Fig. 4.1). Because of the elliptical trajectory, the distance to

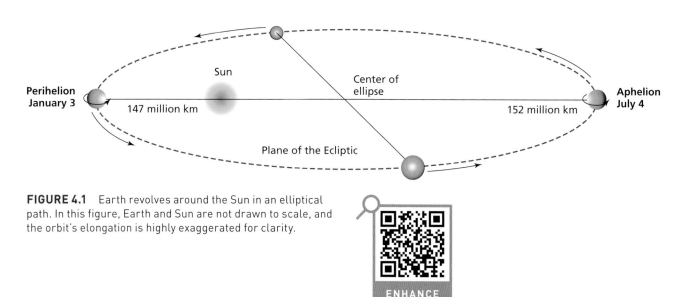

FIGURE 4.1 Earth revolves around the Sun in an elliptical path. In this figure, Earth and Sun are not drawn to scale, and the orbit's elongation is highly exaggerated for clarity.

ENHANCE

the Sun varies along the orbit, bringing Earth slightly closer to the Sun in January than in July. On January 3, when Earth is closest to the Sun, the distance between the two is about 147.3 million km (91.5 million mi). This position is the moment of **perihelion** (from the ancient Greek *peri*, meaning near; *helios*, meaning sun). From that time onward, the Earth–Sun distance increases slowly until July 4, half a year later, when it reaches about 152.1 million km (94.5 million mi). This position is **aphelion** (*ap* means away from). Thus, Earth is farthest from the Sun during the Northern Hemisphere summer and closest during the Northern Hemisphere winter (or Southern Hemisphere summer). But the total difference is only about 5 million km (3.1 million mi), or 3.3 percent, not enough to produce significant seasonal variation in the amount of solar energy received by our planet. Hence, one cannot explain seasons in terms of changing distance to the Sun.

ROTATION

Our planet ranks among the solar system's fastest-spinning bodies. This spinning produces Earth's equatorial bulging and the polar-area flattening described in Unit 2.

As Earth rotates on its axis in a west-to-east direction, the motion creates the alternations of day and night. A constantly changing one-half of the planet is always turned toward the Sun, while its other half always faces away. One complete rotation takes roughly 24 h (8,600 s). However, a variety of factors cause day length to vary slightly within the year and from one year to another. Such complications can be safely ignored in this discussion, and we therefore refer to 24 hours as the period of rotation and consequently the length of one day.

We do not notice the effect of Earth's rotation because everything on the planet—land, water, and air—rotates as well. But if we look up into the sky and watch the stars or moon rise, the reality of Earth's rotation presents itself. This is especially true concerning the stationary Sun. To us, the Sun appears to "rise" in the east (the direction of Earth's rotation) as the leading edge of the unlit half of Earth turns back toward the Sun. Similarly, the Sun appears to "set" in the west as the trailing edge of the sunlit half of Earth moves off toward the east.

Earth rotates eastward so that sunrise is always observed on the eastern horizon. From our vantage point, the Sun traverses the sky to set in the west. But of course, it is not the Sun's movement but *Earth's* movement that causes this illusion. Looking down on a model globe from directly above the North Pole (see Fig. 3.9), the rotation occurs in a counterclockwise direction.

Seasonality

Like an ellipse drawn on a chalkboard or computer screen, Earth's orbit around the Sun lies in a geometric plane

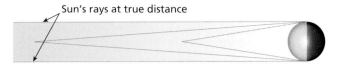

FIGURE 4.2 A diagram showing why rays of sunlight reaching Earth are nearly parallel. As a radiation source moves progressively farther from Earth the angle gets larger and the rays approach parallel.

called the **plane of the ecliptic**. The seasons occur because Earth is *tilted* with respect to the plane of the ecliptic. This is much easier to understand if we assume solar radiation arrives at Earth in the form of parallel rays. This is untrue; sunlight diverges in all directions from its source.

However, because of Earth's great distance from the Sun, the planet occupies only a small part of its orbital path (Fig. 4.2). This means rays hitting the North and South Poles diverge by only a little and are thus nearly parallel. The parallel ray assumption is only a model of reality, but it is perfectly suitable for this purpose and greatly simplifies the analysis. Although usually unstated, you will find this assumption is implicit in nearly every textbook and Internet account, and we have used it in preparing the diagrams here.

AXIS TILT

Earth's axis is always tilted at an angle of 66.5 degrees to the plane of the ecliptic. Another way of saying this is that the axis is 23.5 degrees away from perpendicular to the plane. Figure 4.3 illustrates that the amount of tilt and direction of tilt is the same no matter where Earth is in its orbit. The constant tilt of the axis is the key to seasonal changes. Sometimes, the term *parallelism* is used to describe this axial phenomenon. This term refers to the fact that the Earth's axis remains parallel to itself at every position in its orbital revolution. At one point in Earth's revolution, around June 22, the Northern Hemisphere is maximally tilted toward the Sun. At this time, the Northern Hemisphere receives a much greater amount of solar energy than the Southern Hemisphere does. When Earth has moved to the opposite point in its orbit six months later, around December 22, the Northern Hemisphere is maximally tilted away from the Sun and receives the least energy. This accounts for the seasons of heat and cold, summer and winter. Figure 4.3 summarizes these Earth–Sun relationships and shows how these seasons occur at opposite times of the year in the Northern and Southern Hemispheres.

It is worthwhile for us to examine in greater detail the particular positions of Earth in orbit illustrated in Figure 4.3. These are shown in Figure 4.4. Notice how on June 22, rays from the Sun fall vertically at noon on Earth at latitude 23.5 degrees N. This latitude, where the Sun's rays strike the surface at an angle of 90 degrees, is given the name **Tropic of Cancer**. It is the most northerly

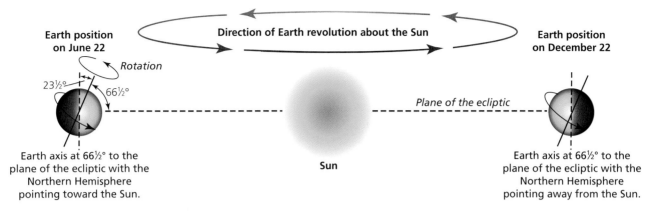

FIGURE 4.3 Extreme summer and winter positions of Earth with regard to the Sun. Earth's axis is tilted at the same angle in relation to the plane of the ecliptic throughout the year.

latitude where the Sun's noontime rays strike vertically. We can also see that all points north of latitude 66.5 degrees N, an area called the **Arctic Circle**, remain totally in sunlight during Earth's 24-hour rotation (Fig. 4.5). If a vertical pole were placed at the equator at noon on this day of the year, the Sun would appear to be northward of the pole, forming an angle of 23.5 degrees with the pole and an angle of 66.5 degrees with the ground (Fig. 4.6). Note that the summation of these two angles equals 90 degrees.

Precisely six months later, on December 22, the position of Earth relative to the Sun causes the Sun's rays to strike vertically at noon at 23.5 degrees S, the latitude called the **Tropic of Capricorn** (the southernmost latitude where the Sun's noon rays can strike the surface at 90 degrees). The other relationships between Earth and the Sun for December 22 are exactly the reverse of those described for June 22, as we can see in the right side of Figure 4.4. Accordingly, the entire area south of the **Antarctic Circle**, located at latitude 66.5 degrees S, receives 24 hours of sunlight. Simultaneously, the area north of the Arctic Circle is in complete darkness. (Note that in Fig. 4.4, the area south of the Antarctic Circle is similarly dark on June 22.)

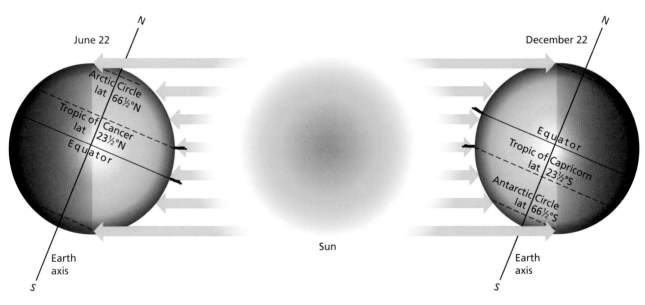

FIGURE 4.4 Relative positions of the Earth and the Sun on June 22 and December 22. Points on Earth receive the Sun's rays at different angles throughout the year.

ENHANCE

FIGURE 4.5 Looking toward the North Pole near midnight on August 1 in Alaska, the Sun is seen to approach but never cross the horizon in this time-lapse sequence.

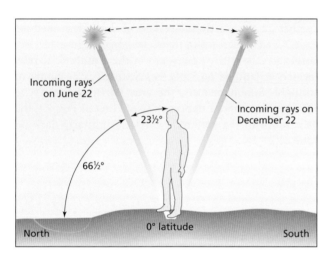

FIGURE 4.6 Incoming solar rays at noon for a person at the equator. The position of the Sun in the sky changes seasonally as the latitude where the Sun is vertical varies over the year. Compare with Figure 4.4.

SOLSTICES AND EQUINOXES

To us on Earth, it appears that the highest daily position of the Sun at noontime[1] is lower in the sky as the seasons progress from summer to fall to winter. If we were to plot the position of the noontime Sun throughout the year at a location in the middle latitudes of the Northern Hemisphere, it would seem to climb higher and higher until June 22, when its climb would appear to stop. Then it would move lower and lower, until it stopped again at December 22, before once more beginning to ascend. South of the equator, the dates are reversed, but the phenomenon is identical. The ancient Greeks plotted the apparent movement of the Sun and designated the points at which the stops occurred **solstices** (meaning "Sun stands still"). Today, we continue to use their word: the Earth–Sun position of June 22 is the **summer solstice** and that of December 22 is the **winter solstice**. In the Southern Hemisphere, these dates are reversed.

About halfway between the two solstice dates, there are two positions where the vertical rays of the Sun are found at the equator. These positions occur on or about March 21 and September 23. On these days, every parallel of latitude is half in light and half in darkness; thus, every latitude receives 12 hours of sunlight and 12 of darkness. These special positions are called **equinoxes**, which in Latin means "equal nights." In the Northern Hemisphere, the equinox of March 21 is the **spring (vernal) equinox**, and that of September 23 is the **fall (autumnal) equinox**.

THE FOUR SEASONS

We can achieve a clear idea of the causes of the seasons if we imagine we are looking down on Earth's orbit around the Sun (the plane of the ecliptic) from a point high above the solar system, a perspective diagrammed in Figure 4.7. The North Pole always points to our right. At the summer solstice, the Arctic Circle receives sunlight during the entire daily rotation of Earth, and all parts of the Northern Hemisphere have more than 12 hours of daylight. These areas receive a large amount of solar energy in the summer season. At the winter solstice, the area inside the Arctic Circle receives no sunlight at all, and every part of the Northern Hemisphere receives fewer than 12 hours of sunlight. In winter solar energy levels are at a minimum. In contrast to the winter solstice, at both the spring and fall equinoxes, the Arctic Circle, the equator, and every other parallel of latitude are equally divided into day and night. Both hemispheres receive an equal amount of sunlight and darkness, and energy from the Sun falls equally on the two hemispheres.

---

1 In saying "noontime," we are referring to the local Sun time, not human clock time. In other words, here we are ignoring the complications of daylight saving time and time zones discussed in the Perspectives on the Human Environment box on page 50.

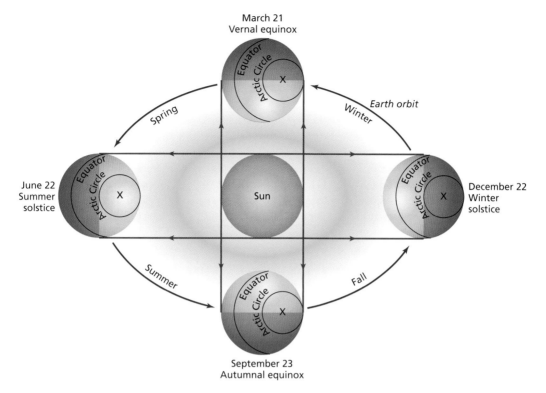

FIGURE 4.7 The march of the seasons as viewed from a position above the solar system. Seasonal terminology here applies to the Northern Hemisphere. (The Southern Hemisphere seasonal cycle is the exact opposite.)

The annual revolution of the Earth around the Sun and the constant tilt of its axis give our planet its different seasons of relative warmth and coldness. We can trace the yearly cycle of the four seasons in Figure 4.7. *Spring* begins at the vernal equinox on March 21 and ends at the summer solstice on June 22; *summer* lasts from that date through the autumnal equinox on September 23; *autumn* occurs from then until the arrival of the winter solstice on December 22; *winter* follows autumn and lasts until the vernal equinox is again reached on March 21. This cycle applies only to the Northern Hemisphere; the Southern Hemisphere's seasonal march is the mirror image, with spring commencing on the date of the northern autumnal equinox (September 23). Clearly, such seasonal labels derive from planetary positions, not temperature measurements. In fact, seasonal extremes of heat and cold typically lag behind the solstices, and many places in the tropics experience nothing recognizable as winter cold or summer warmth. We explain why this is in later units of the text.

Throughout human history, the passage of the seasons has been a basis for establishing secure reference points for measuring time (see Perspectives on the Human Environment: Measuring Time on Our Rotating Earth). The changing spatial relationships between Earth and Sun, produced by the planetary motions of revolution and rotation on a constantly tilted axis, cause important variations in the amount of solar energy entering the Earth system and ultimately received at Earth's surface. We explore those patterns at some length in Unit 5, but certain basic ideas are introduced here because they follow directly from the preceding discussion. In particular, we consider solar radiation entering the top of the atmosphere as the ultimate energy source of most physical processes.

Insolation and Its Variation

At any given moment, exactly one-half of the rotating Earth is in sunlight and the other half is in darkness. The boundary between the two halves is called the **circle of illumination**, an ever-shifting line of sunrise in the east and sunset in the west. The sunlit half of Earth is exposed to the Sun's radiant energy, which is transformed into heat at the planetary surface and, to a lesser extent, in the atmospheric envelope above it. There is considerable variation in the surface receipt of **insolation** (a contraction of the term *in*coming *sol*ar radi*ation*). Our preceding discussion hinted at one obvious factor: day length.

DAY-LENGTH VARIATIONS

As Figure 4.4 shows, day length varies by latitude according to how much of the latitude line is on the sunlit half of Earth. Similarly, because of Earth's tilt, locations experience a seasonal progression in the length of daylight. For example, the Tropic of Cancer is more on the sunlit side in June (resulting in long days) and less on the sunlit side in December (resulting in short days). The equator is unique in that it is always half in sunlight and half in darkness; therefore it has 12 hours of sunlight throughout the entire year. The complete pattern is shown in Figure 4.9. Moving up or down this

Measuring Time on Our Rotating Earth

The study of geography reminds us that the human concept of time is associated with the nature of Earth as a physical object. The measurement of time on Earth's surface is important in the study of our planet. One of the most obvious ways to start dealing with time is to use the periods of light and darkness resulting from the daily rotation of Earth. One rotation of Earth, one cycle of daylight and nighttime hours, constitutes one full day. The idea of dividing the day into 24 equal hours dates from the fourteenth century.

Historically, people in each place kept track of their own time by the Sun, and this system worked well as long as human movements were confined to local areas. But by the sixteenth century, when sailing ships began to undertake transoceanic voyages, problems arose. The Sun is always rising in one part of the world as it sets in another. On a sea voyage, such as that of Columbus in the *Santa Maria*, it was always relatively straightforward to establish the latitude of the ship. Columbus's navigator had only to find the angle of the Sun at its highest point during the day. Then, by knowing the date, he could calculate the latitude from a set of previously prepared tables giving the angle of the Sun at any latitude on a particular day.

It was impossible, however, for the navigator to calculate longitude. To do that, he would need to know precisely the difference between the time at some agreed meridian, such as the prime meridian (zero degrees longitude) and the time at the meridian where his ship was located. Until about 1750, no portable mechanical clock or chronometer was accurate enough to keep track of that time difference.

With the perfection of the chronometer in the late eighteenth century, the problem of timekeeping was addressed. But by the 1870s, as long-distance railroads began to cross the United States, a new problem surfaced. Orderly train schedules could not be devised if each town and city operated according to local Sun time, and the need for a system of time organization among different regions became essential.

The problem of having different times at different longitudes was finally resolved at an international conference held in 1884 in Washington, D.C. It was decided that all Earth time around the globe would be standardized against the time at the prime meridian (a concept introduced on p. 29), which passed through Britain's Royal Observatory at Greenwich (London). Earth was divided into the 24 time zones shown in Figure 4.8. Each zone uses the time at standard meridians located at intervals of 15 degrees of longitude with respect to the prime meridian ($24 \times 15° = 360°$). Each time zone differs by one hour from the next, and the time within each zone can be related in one-hour units to the time at Greenwich. When the Sun rises at Greenwich, it has already risen in places east of the observatory. Thus, the time zones to the east are designated as *fast*; time zones west of Greenwich are called *slow*.

This solution led to a peculiar problem. At noon at Greenwich on January 2, 2004, it was midnight on January 2 at 180 degrees E longitude (twelve time zones ahead) and midnight on January 1 at 180 degrees W (twelve time zones behind). However, 180 degrees E and 180 degrees W are the *same* line. This meridian was named the **international date line** by the Washington conference. It was agreed that travelers crossing the date line in an eastward direction, toward the Americas, would repeat a calendar day; those traveling west across it, toward Asia and Australia, would skip a day. The international date line does not pass through many land areas (it lies mainly in the middle of the Pacific Ocean), so severe date problems for people living near it are avoided. Where the 180th meridian does cross land, the date line was arbitrarily shifted so it passes only over ocean areas.

Similarly, there is some flexibility in the boundaries of other time zones to allow for international borders and even for state borders in such countries as Australia and the United States (see Fig. 4.8). Some countries, such as India, choose to have standard times differing by half or a quarter of an hour from the major time zones. Others, such as China, insist that the *entire country* adhere to a single time zone.

A further arbitrary modification of time zones is the adoption in some areas of **daylight saving time**, whereby all clocks in a time zone or subregion are set forward by one hour from standard time for at least part of the year. This was established for a number of reasons, including energy savings. Many human activities start long after sunrise and continue long after sunset. By setting the clock forward, the Sun is up later into the evening hours, and there is less need for artificial light. Studies done in the last five years suggest that actual energy savings are negligible. One study even found a slight energy cost to daylight saving time, suggesting that demand for additional heating and cooling more than offset the savings in lighting. In the United States today, most states begin daylight saving

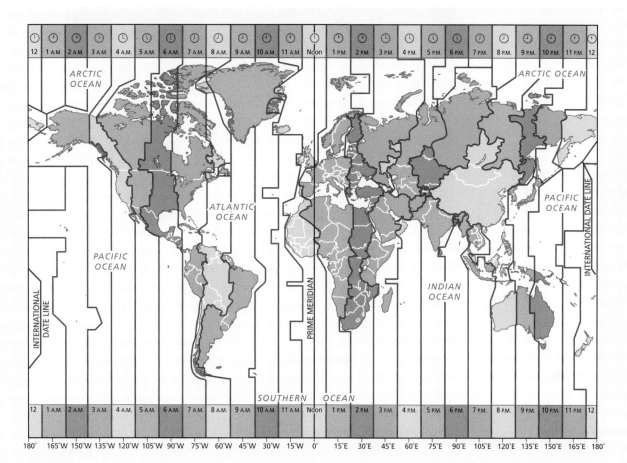

FIGURE 4.8 The time zones of Earth.

time during the second weekend in March and end it on the first weekend in November. In Russia, the government decided in 2009 to reduce the vast country's 11 time zones to nine, an arbitrary edict that made no

difference to sunrises and sunsets but inconvenienced many people in the remote east. This has led to sabotage: in China's far west, the people use "Beijing time" on wall clocks but "Xinjiang time" on wristwatches.

graph is like marching north or south across the latitudes on some day of the year. In contrast, choosing a particular latitude and moving horizontally reveals the seasonal cycle at that location. For example, moving across the diagram for 60 degrees S, the wide range in colors shows that day length varies greatly over the year. Notice first how places near the equator (the dotted yellow horizontal line) have little seasonal change in day length compared to higher latitudes. Indeed, latitudes within 15 degrees of the equator never have fewer than 11 or more than 12 hours of sunlight. Figure 4.9 also shows that there is a huge latitudinal difference in day length on days near the solstices. That is, day length decreases dramatically with latitude in the winter hemisphere and increases rapidly with latitude in the summer hemisphere. On days near the equinoxes there is little variation in day length because every place gets close to 12 hours of daylight.

Figure 4.9 shows that day length exerts a strong influence on the amount of solar radiation reaching various locations on Earth. There is, however, another very important

factor that modifies the patterns seen in Figure 4.9—namely, the position of the Sun in the sky.

SOLAR POSITION AND ITS EFFECTS

Consider first the equinoxes, when every location has 12 hours of light and dark and ask the question: Do all places receive equal solar radiation inputs? The answer is no. Insolation on top of the atmosphere is a strong function of latitude. Locations having a nearly vertical Sun at noon receive more radiation than places where the Sun is low in the sky at noon. The equator receives the greatest solar radiation because the Sun's rays strike it most directly, and every other place receives less depending on its distance north or south.

This is illustrated in Figure 4.10, which shows how the parallel rays of the Sun fall on various parts of the spherical Earth. Note that three equal columns of solar radiation strike the curved surface differently. The lower latitudes receive more insolation per unit area than the higher latitudes. At the equator, all the solar rays in column A are

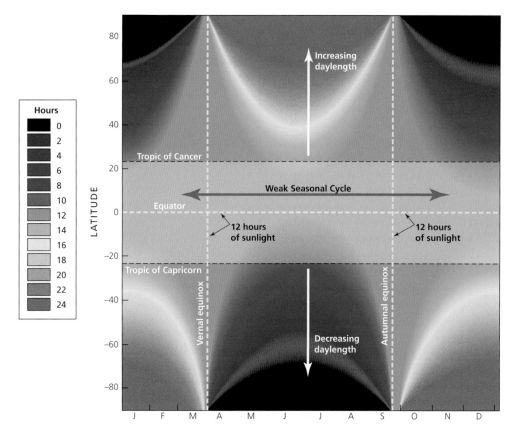

FIGURE 4.9 Hours of sunlight by latitude and season. On the equinoxes, every latitude gets 12 hours of sunlight (vertical yellow lines). The equator has 12 hours of sunlight all year long (horizontal yellow line). Notice how day length varies strongly with latitude in the middle latitudes for days near the solstices (June and December).

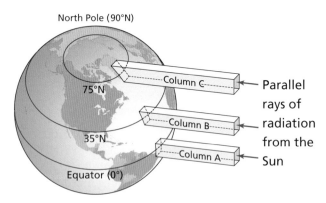

FIGURE 4.10 Reception of solar radiation with Earth in equinox position. A surface at a higher latitude receives less radiation than a surface of equal area at a lower latitude.

concentrated on a small square box; in the middle latitudes, at 35 degrees N, an equal number of rays in column B are diffused across a surface area about twice as large; and in the polar zone, at 75 degrees N latitude, that same number of rays in column C are scattered across an area more than three times the size of the box illuminated by column A.

In this example, the equator (under column A) receives the most intense insolation because the midday Sun's rays strike it vertically from the point directly overhead (known as the **zenith**), 90 degrees above the horizon. If one travels north or south from the equator, the solar radiation received decreases with progressively higher latitude as the angle of the Sun's noontime rays declines from the zenith point in the sky. The angle of the Sun above the horizon, known as the **solar altitude**, clearly has an important impact, just as day length does.

The seasonal cycle of solar elevation and daily sunlight duration are graphed for 45 degrees N latitude in Figure 4.11. Notice how the two factors work together over the course of the year. When the sun is high in the sky, days are long, and vice versa. So rather than counteract each other, day length and solar position combine and produce large seasonality in insolation. Notice also in Figure 4.11 that changes in both factors are rather slow near the solstices compared to the equinoxes. In summer, change in day length and solar position is leisurely. The spring and fall, by contrast, are times of relatively rapid change.

Curves like those in Figure 4.11 are typical of the middle latitudes in both hemispheres. In the tropics, day length is

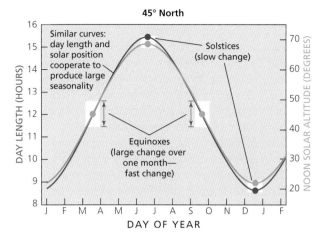

FIGURE 4.11 Day length (left axis) and solar altitude at noon (right axis) at 45 degrees N latitude. These factors work together in producing pronounced seasonality at middle and high latitudes. Notice also how the pace of change in both is much greater near the equinoxes than near the depths of summer and winter.

close to 12 hours all year long, and the noon Sun is never too far from vertical. At high latitudes it is quite different. Poleward of 66.5 degrees, day length ranges from zero hours in winter to 24 hours of light in the summer. However, the Sun is never very high in the sky. For example, at 66.5 degrees, the Sun is always within 23.5 degrees of the horizon.

INSOLATION INPUT

Figure 4.12 translates day length and solar position into radiation values, expressed as a percentage of the global average input, which is about 341 W/m². In other words, averaged over all latitudes and seasons, a square meter at the top of the atmosphere receives 341 W of solar radiation. Older incandescent-type lightbulbs typically use between 50 and 150 W. So capturing the Sun's energy might be enough for your car's stereo, but it certainly wouldn't power the vehicle.

Figure 4.12 has many features similar to the day-length plot in Figure 4.9. Just as with day length, low latitudes have little seasonal variation in radiation input. Radiation values in the low latitudes are also near or above the global average throughout the year. Once again, large seasonal differences emerge for middle and high latitudes. However, in the summer, radiation does not increase continuously with latitude as we saw day length increase in Figure 4.9. Yes, days get very long as one approaches the poles, but the progressively lower solar altitude in the polar region compensates for and reduces the intensity of radiation. The result is a broad flat "plateau" of high values extending across a large range of latitudes The take-home message is that in the summer there is little difference in daily radiation input in low latitudes compared to high latitudes. In the winter, however, there is huge variation in radiation, particularly across the middle

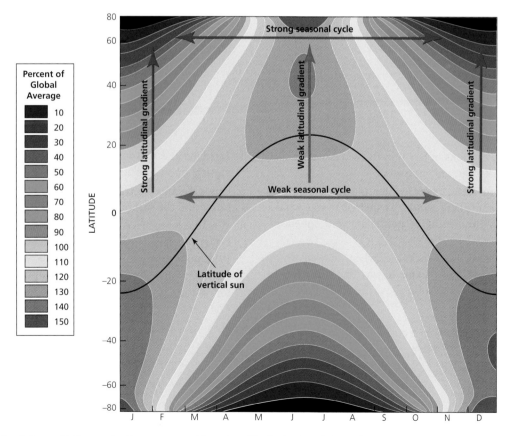

FIGURE 4.12 Solar radiation varies by season and location. The latitude axis is scaled to account for shrinking global area at higher latitudes. Compare this figure with the patterns of Figure 4.9 to understand how solar position modulates the effect of day length.

latitudes. This has large implications for temperature patterns and atmospheric motions, as we will see in Units 7 and 9.

Solar position influences the pattern in Figure 4.12, but it certainly is not the determining factor. To understand this, consider where insolation values would be largest if solar altitude were the only factor involved. In that case, on every single day of the year, radiation inputs would be greatest at the latitude where the Sun is directly overhead at noon. This latitude is indicated by the black line in Figure 4.12. We see that peak radiation is found at that latitude only on the equinoxes. This makes sense because day length is constant on the equinoxes. This constancy means that solar position is the controlling factor in the amount of radiation received on the surface. For other days, the length of day must also be considered and in some cases leads to a maximum far from the latitude of vertical Sun.

A great deal more could be said about the latitudinal distribution of incoming solar radiation shown in Figure 4.12. Please study this graph and its QR animation. Tracing patterns along various latitudes or for specific times of the year is a useful application of the concepts associated with revolution, rotation, axis tilt, seasonality, solstices, equinoxes, and insolation. This graph is a useful preface to the study of the atmosphere (which commences in Unit 6) because it marks the point of transition from astronomic controls to the terrestrial forces that shape weather and climate across the face of Earth.

Key Terms

Antarctic Circle *page 47*

aphelion *page 46*

Arctic Circle *page 47*

axis *page 45*

circle of illumination *page 49*

daylight saving time *page 50*

equinox *page 48*

fall (autumnal) equinox *page 48*

insolation *page 49*

international date line *page 50*

perihelion *page 46*

plane of the ecliptic *page 46*

revolution *page 45*

rotation *page 45*

solar altitude *page 52*

solstice *page 48*

spring (vernal) equinox *page 48*

summer solstice *page 48*

Tropic of Cancer *page 46*

Tropic of Capricorn *page 47*

winter solstice *page 48*

zenith *page 52*

Scan Here for a quick vocabulary review

Review Questions

1. Describe the motions of Earth in its revolution around the Sun and its rotation on its axis.

2. The Northern Hemisphere winter is colder than the Southern Hemisphere winter. Is this what you would expect given the timing of perihelion and aphelion? Explain.

3. Earth rotates counterclockwise when viewed from above the North Pole. Explain why the direction is clockwise when viewed from above the South Pole.

4. Describe the seasonal variation in the latitude of the vertical noontime Sun during the course of the year.

5. Compare and contrast the spring and autumnal equinoxes and the summer and winter solstices.

6. What is the international date line, and why is it a necessary part of Earth's longitude system?

7. Why is insolation at the North Pole on the day of the summer solstice greater than that received at the equator on the equinoxes?

Scan Here for a quick concept review

www.oup.com/us/mason

Radiation and the Heat Balance of Planet Earth

Douglas firs—the state tree of Oregon—are widespread in the cool moist climates of the northwestern United States and Canada. In the Grand Canyon, they are only found on north-facing slopes below the rim, where reduced solar radiation input creates a suitable microenvironment.

OBJECTIVES

- Explain heat flow processes within the Earth system
- Describe the cascade of solar energy to Earth's surface and the resulting energy exchanges between surface and atmosphere
- Link the greenhouse effect to Earth's habitability
- Summarize latitudinal differences in net radiation
- Discuss global energetics

The Sun powers essentially all of the natural world, but solar radiation is by no means directly involved in all processes. In many cases there is continual conversion of solar radiation to other forms of energy that more directly fuel our dynamic planet. The primary goal of this unit is to convey these transformations and how the various forms of heat move within the Earth system. As we will see, the atmosphere plays a large role in this system, but the oceans are also important, and neither the biosphere nor the ice-covered cryosphere can be ignored. All the spheres are important in global energy transfers, and likewise, energy transfer affects processes occurring in all aspects of the Earth system.

Energy and Heat Transfer

Energy exists in many forms. There is the *kinetic energy* possessed by a speeding bus or by hurricane winds lifting the roof off a house. (We define and discuss kinetic energy in more detail in Unit 7.) There is the *chemical energy* contained in the food you ate earlier today being consumed in life-sustaining metabolic processes. Perhaps you are reading this on a display powered by *electrical energy*. Many Earth processes depend on yet a different form of energy—**thermal energy**, or heat. The word *thermal* reflects the close connection between heat and temperature. The details of this relationship are not important in this unit. For now, it is only important to know that as the temperature

of an object increases, its internal energy (its heat content) also increases. In addition, various substances differ in the amount of heat needed to produce a given temperature change in them. A small heater can quickly warm the air in a camping tent, but that heater would have little effect on the same volume of water. Put technically, this is because water has a high *specific heat* compared to air; it takes much more energy to warm a given mass of water by one degree than it takes to warm the equivalent mass of air.

Vast quantities of heat move continuously throughout all the spheres of Earth in several different ways. In this unit, we first consider general aspects of these *modes* of heat transfer, and then we discuss their specifics as expressed in the physical environment.

RADIATION

All objects lose heat by **radiation**. For our purposes in this book, we regard radiation as a transmission of energy in the form of electromagnetic waves. In some ways, electromagnetic waves are similar to waves on a water surface: they have a given height (or wave amplitude) and a wavelength, which is the distance between two successive wave crests (Fig. 5.1). Unlike the undulations of a material water surface, electromagnetic radiation waves are electric and magnetic oscillations in free space. Radiation does not require a transfer medium (such as water), and thus it is able to move heat through the vacuum of outer space.

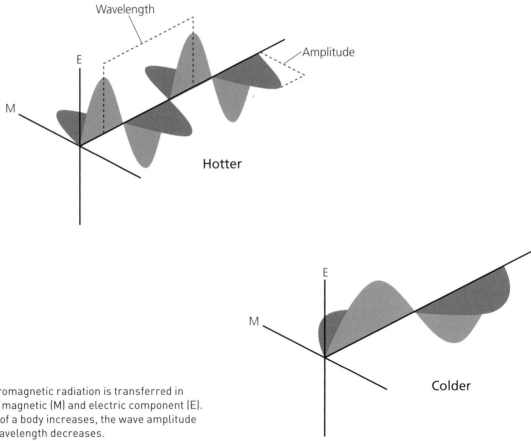

FIGURE 5.1 Electromagnetic radiation is transferred in waves having both a magnetic (M) and electric component (E). As the temperature of a body increases, the wave amplitude increases and the wavelength decreases.

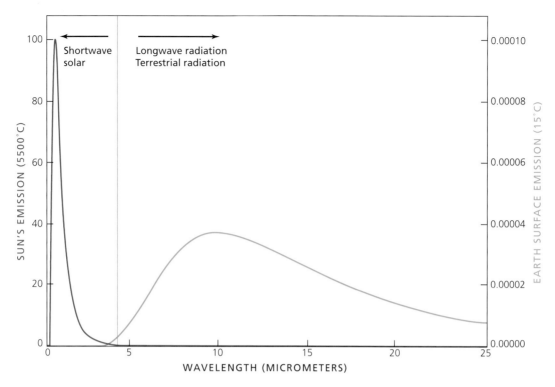

FIGURE 5.2 Emission by the Sun and the Earth on a per unit area basis. Both emit radiation over many wavelengths, but there is little overlap in their emission curves. Note that the curves are not drawn at the same scale. The terrestrial axis is exaggerated by a factor of 1 million relative to the solar curve. Even at 25 μm, a square meter of the Sun emits about 60 times as much radiation as a square meter of Earth.

All objects emit radiation, but the amount emitted and the wavelengths emitted depend strongly on temperature. First, hotter objects emit more radiation than colder objects. Thus, with a temperature of about 5500°C (9900°F), each square meter of the Sun's surface emits about 150,000 times as much as a square meter of Earth's surface, whose temperature averages only 15°C (59°F). Second, as the temperature of an object increases, a higher proportion of energy is emitted at higher-energy, shorter wavelengths. Nearly all of the Sun's emission is at wavelengths shorter than 4 μm (4 millionths of a meter). In contrast, Earth emits radiation almost entirely at wavelengths longer than 4 μm. As seen in Figure 5.2, the two emission curves hardly overlap at all. For this reason, radiation coming from the Sun is called **shortwave radiation**, and that emitted from Earth is termed **longwave radiation**. Because longwave radiation is emitted by all objects at Earth-like temperatures (classroom walls, people, the ground, the atmosphere), the term **terrestrial radiation** is often used as a counterpoint to solar radiation. The distinction is important, because solar radiation and terrestrial radiation behave very differently as they move through the Earth system.

CONDUCTION

Heat can travel within a substance by **conduction**, energy transfer accomplished by contact between individual molecules. Conduction is particularly important in soils because other modes of transfer are effectively nonexistent.

Conduction requires a transfer medium, and the rate of heat transfer depends partly on the material. Some substances (e.g., metals) are particularly good heat conductors, but others (e.g., foam beverage cups) are not. The latter allow for little conduction. In the case of soil, the amount of air and water in the ground has a large impact on conductivity. Heat conduction also depends on the prevailing *temperature gradient*, which is the change in temperature per unit distance. To understand the concept of temperature gradient, some explanation will help.

Suppose the Sun warms the ground surface to temperatures considerably above those a few centimeters below the surface. Heat flows downward by conduction toward colder temperatures. For example, imagine the surface temperature is 10°C (18°F) higher than the temperature 10 cm below the surface. In this case, the temperature gradient is 10°C per 10 cm, or 1°C per cm. If at some other time the surface temperature is only 5°C above that at 10 cm below the surface, the gradient is half as large—namely, 0.5°C per cm. Everything else being equal, the downward conduction of heat would be only half as large.

Figure 5.3 illustrates heat conduction symbolically. In the afternoon, when the surface is relatively warm, heat is conducted into the soil and the soil warms. At night, when the surface is colder, the soil below provides heat to the surface as heat is conducted upward (Fig. 5.3, left). Daily energy gains and losses are usually close in value to one another. That means that at depth only a little temperature change occurs from one day to the next. However, gains are

Ground surface

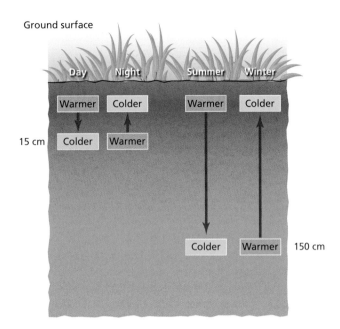

15 cm

150 cm

FIGURE 5.3 Heat is conducted from higher toward lower temperatures. On a daily basis, heat is stored in the soil during the day and lost at night. Similarly, seasonal average temperature gradients lead to heat storage and loss in the warm and cold seasons, respectively. The daily temperature cycle does not extend very far into the soil, but the day/night temperature range is large compared to the seasonal cycle.

usually somewhat larger than losses in the warm months, and therefore heat is stored for that part of the year. This heat is returned to the surface in the winter (Fig. 5.3, right). If the seasonal energy gains and losses are equal, energy conservation (recall Unit 1) ensures there is no net change in heat content or temperature from year to year.

CONVECTION

In the case of conduction, heat flows through some medium (e.g., soil or rock) with no appreciable movement of the molecules comprising the substance. If, in contrast, the material can flow, like air or water, **convection**, another mode of heat transfer, can occur. In this case the energy transfer is accomplished by the flow itself. Can you recall a time when you were buffeted by a warm blast of air or felt a cold "current" while swimming in the ocean or a lake? These examples illustrate how heat can be transported from one place to another by fluid flow (i.e., by movement of a gas or liquid). This movement of heat is convection.

Two forms of convection are important in the Earth energy balance. **Sensible heat transfer** is heat transfer that takes place when air or water at one temperature moves to a location with a different temperature. For example, if air is warmed by a hot asphalt parking lot and moves upward toward colder temperatures, heat is carried upward as well. The heat transfer is from higher to lower temperatures. The rate of transfer depends on both the temperature gradient and the vigor of the fluid motion. We all have stirred a pot of soup on the stove and in doing so ensured maximal

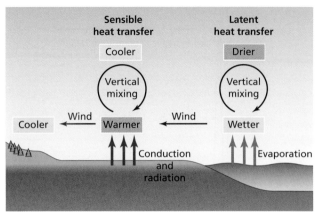

FIGURE 5.4 Sensible and latent forms of convective heat transfer. On the left, conduction and radiation from the warm surface move heat into the atmosphere. After that, vertical and horizontal motions produce a corresponding transfer of sensible heat into cooler parts of the atmosphere. On the right, water vapor is carried upward by mixing and to the left by winds. This results in both a vertical and a horizontal latent heat transfer.

sensible heat transfer from bottom to top. Notice that a net transfer of material is not needed for sensible heat transfer—in other words, soup does not accumulate at the top of the pot! What matters is that there is fluid motion in the presence of a temperature gradient. Sensible heat is depicted on the left side of Figure 5.4. Notice that sensible heat can flow vertically or horizontally.

Another form of convection is related to the evaporation of water. The conversion of water from liquid to gas requires energy (e.g., the teakettle boils only if the burner is on). Because energy is conserved, the energy consumed by evaporation does not simply disappear. Rather, it is present in latent form in the water vapor itself. If the water vapor moves, the energy of evaporation travels with it. For example, when water from a lake surface evaporates into the atmosphere and the air is stirred, heat is carried upward into the drier atmosphere above (Fig. 5.4, right). This is called **latent heat transfer**, heat transfer associated with the flow of water vapor. Latent heat transfer can be vertical, as in the example we just cited, or horizontal, if winds carry water vapor from a humid area to a drier place.

Global Energy Balance

All three modes of heat transfer—radiation, conduction, and convection—occur at scales ranging from molecular to global. For example, evaporation from tiny pores on leaf surfaces often prevents leaves from overheating in direct sun, and latent heat likewise contributes mightily to the warmth that pervades the entire region of the tropics. In this section, we examine relative magnitudes of the various transfers, first globally and then more locally. Throughout the discussion, the values given are long-term annual averages taken over all seasons. You will see that the values imply that all parts of the planet are in energy balance, which is a close approximation of reality. This ignores year-to-year variation in global temperature as well as global climate change. Global temperature change is definitely occurring, but the associated energy imbalance is minuscule compared to the mean values. We will have much more to say about climate change and energy balance in later units. In this unit, we focus on the background state within which such change is occurring. (Perspectives on the Human Environment: Solar Power Outages, which describes some of the Sun's variability, is an exception.)

SOLAR RADIATION

Measurements indicate that, on average, the intensity of sunlight is about 1,364 W/m². If we could capture a square meter's worth of energy with perfect efficiency, we could power a hair dryer. This value is known as the *solar constant*, and it pertains to a solar collector oriented perpendicular to the incoming radiation. Our nearly spherical planet hardly meets that requirement. After accounting for curvature and the fact that half the planet is turned away from the Sun, average energy input at the top of the atmosphere is reduced by a factor of 4 to about 341 W/m². This seemingly small rate of heating amounts to a huge total considering the size of Earth. The amount supplied to the whole planet in one day could supply all the world's industrial and domestic energy requirements for the next 100 years based on current rates of consumption. To make comparisons easier in this book, we express most energy transfers as percentages of incoming solar radiation. In addition, we will use the word "unit" to refer to 1 percent of incoming radiation (100 units = 100%). Thus, 100 units of radiation is 341 W/m², and 10 units is the same as just over 34 W/m².

When radiation travels through the atmosphere, several things can happen to it (Fig. 5.5). Of all the incoming solar energy, only 30 percent (30 units) travels directly to Earth's surface; this energy flow is called **direct radiation**. A substantial amount of incoming sunlight, 23 units, is reflected and scattered back into space by clouds (18 units) and gas molecules and dust particles in the atmosphere (5 units). Another 23 percent of the incoming solar rays is absorbed by clouds (4 units) and dust and other components of the atmosphere (19 units). Some of these scattered rays, 22 units, eventually find their way down to Earth's surface and are collectively known as **diffuse radiation**. Altogether, more than half (54 percent) of the solar energy arriving at the outer edge of the atmosphere reaches the surface as either direct or diffuse radiation. The rest is either absorbed by the atmosphere or scattered and/or reflected back into space.

No matter where radiation strikes Earth, one of two things happen to it. It is either *absorbed* by—and thereby heats—Earth's surface, or it is *reflected* by the surface, in

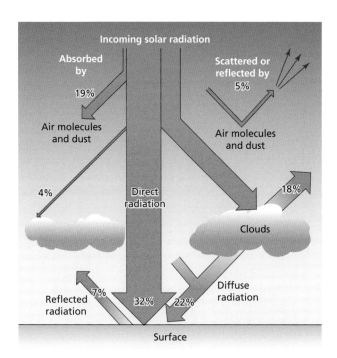

FIGURE 5.5 Solar radiation flows in the atmosphere.

which case there is no heating effect. The amount of radiation reflected by the surface depends mainly on the color and composition of the surface and the angle of the Sun relative to the slope of the surface. A ray of solar energy that strikes the ground perpendicularly is less likely to be reflected than one landing at a low, glancing angle. And if the surface is a dark color, such as black soil or asphalt, the energy is more likely to be absorbed than if the object has a light color, such as a white building.

The proportion of incoming radiation that is reflected by a surface is the surface's **albedo**, a term derived from the Latin word *albus*, meaning white. The albedo of a snowy surface, which reflects most of the incoming radiation, might be 80 percent, whereas the albedo of a dark green–colored rainforest, which reflects very little radiation, might be as low as 10 percent. Not surprisingly, albedo varies markedly from place to place (follow the QR code to see how dramatic seasonal albedo changes are in the Northern Hemisphere). The global average albedo is about 13 percent. This means that approximately 41 percent of the incoming solar radiation is absorbed by Earth's surface.

Notice that solar radiation absorbed by the surface (41 units) is about twice as large as solar radiation absorbed in the atmosphere (23 units). This is extremely important: it means the Sun mainly heats the surface. And, as we will see in the next subsections, by and large, the surface heats the atmosphere.

TERRESTRIAL RADIATION

Both Earth's surface and atmosphere emit longwave, or terrestrial, radiation. Earth's surface is warm enough that, on average, the surface emits 16 percent more radiation than Earth receives from the Sun for a total of 116 units

Solar Power Outages

Outages from powerful winds and ice storms provide routine reminders of our dependence on power-based technology to read the news, buy our books, purchase our gas, print our boarding passes, do our banking, and pay our bills. We all assume that power systems will work and that backups won't be needed. Broadcast television, for example, fell victim to cable; the time when your battery-powered, rabbit-ear TV could receive signals from a high-elevation transmitter are over. We may yet come to rue that passing.

It is generally assumed that hurricanes and other environmental crises are the most likely causes of power outages and that these interruptions will be relatively brief. However, the greatest potential danger to the entire system is not global but solar. It's not that we haven't had any warning for that; it's just that human life spans and memories are short. In truth, our Sun is anything but a steady, unwavering light. Massive flares erupt at apparently random intervals, and there is a nearly regular 11-year cycle in the giant magnetic storms we call *sunspots*. On August 28, 1859, the Sun smashed a billion-ton magnetic ball of protons through Earth's magnetic shield at several million kilometers per hour, a solar superstorm that repositioned the northern aurora over the Caribbean, created a fantastic light show, and disabled elementary telegraph systems for several days.

Were such an event to occur today, given our deep dependence on modern electronics, it would lead to total chaos. Such a magnetic superstorm would melt satellites, disrupt television, stop radio communication, immobilize computers, neutralize cell phones, degrade GPS navigation, ground airliners, and generate worldwide blackouts. Estimates of the cost of such an event vary in the trillions of dollars; a team of researchers in 2009 estimated that recovery would take from four to ten years depending on the storm's ferocity.

We know very little as yet about the causes and effects of this solar activity. In 1989 a relatively small storm caused a 24-hour blackout for the whole province of Quebec. Scientists are unsure how often events

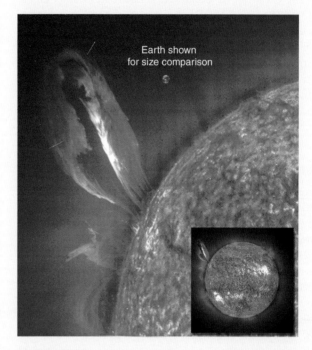

Earth shown for size comparison

FIGURE 5.6 Large solar flares disrupt satellite communications temporarily. Massive solar storms have the potential for far more serious consequences.

ENHANCE

such as the 1859 superstorm occur because we have no record of solar storms. It is not even certain just how these occurrences relate to the 11-year sunspot cycle of solar activity, whose most recent peak occurred at the end of 2013. We do know there have been some close calls: in 2012 and 2014 Earth escaped the impact of massive storms only because the Sun was pointed away from Earth when they erupted. Such events have implications for climate change as well as for a host of human activities that presume no meddling from the nearest star.

---

emitted (Fig. 5.7). This seeming impossibility is explained by realizing the surface is heated by more than just solar radiation; downwelling radiation from the atmosphere also warms the surface significantly. Of the radiation emitted by Earth's surface, the great majority is absorbed by clouds and the so-called *greenhouse gases*, the most important of

which are water vapor and carbon dioxide (see Focus on the Science: The Greenhouse Effect). On the percentage scale we have been using in this unit, 12 radiation units emitted by the surface escape to space, and 104 units are absorbed by the atmosphere. Most of the absorption occurs within 30 m (98 ft) of Earth's surface, a fact that contributes

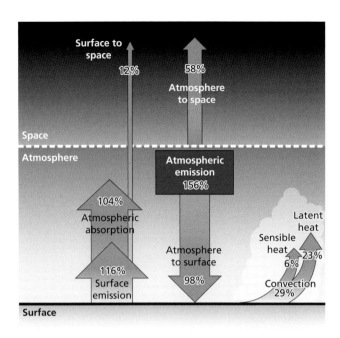

FIGURE 5.7 Global means of terrestrial radiation and convection. All values are expressed as a percentage of solar radiation entering the top of the atmosphere.

greatly to the warmth of the lower atmosphere. It is interesting to note that whereas the atmosphere is largely transparent to solar radiation (we can see great distances during the day), the atmosphere is nearly opaque to longwave radiation. If this were not the case, global average surface temperatures would be about 30°C (48°F) colder. Instead of 15°C (59°F), mean temperature would be well below the freezing point of water, at a thoroughly inhospitable temperature of only–15°C (18°F). (See Focus on the Science: The Greenhouse Effect for more about this.)

Atmospheric emission depends in large part on the temperature of the atmosphere. On a global basis, sufficient heat reaches the atmosphere to allow it to emit about 1.5 times the incoming solar amount. Most of this radiation is downward and contributes to surface heating. Known as **counterradiation**, it supplies more than twice as much heat to the surface as does the Sun. Clouds are good sources of counterradiation, which explains why morning temperatures are often higher after a cloudy night than when skies are clear. As Figure 5.7 shows, radiation from the atmosphere to space equals about 58 percent of that provided to the planet by the Sun, and many times more than the 12 units of surface emission

FOCUS ON THE SCIENCE

The Greenhouse Effect

The prospect for human-induced global warming has made **greenhouse effect** a household term, and some people conflate the two. However, the greenhouse effect existed long before humans evolved, and it is by no means exclusive to Earth. It is a physical process important on Earth and some other planets (especially Venus). On our planet, the greenhouse effect is affected by humans, but the human influence is a perturbation of a natural process we now describe.

Earth's Greenhouse Effect

Clouds, water vapor, and carbon dioxide are the main players in the greenhouse effect. Some other gases are also involved, including methane (the main constituent of natural gas) and nitrous oxide (N_2O, a gas that arises from many anthropogenic [human] and natural processes). All of these gases are relatively good absorbers of terrestrial radiation. Thus, when Earth's surface emits longwave radiation, most of that energy is absorbed by the atmosphere rather than escaping to space. Energy leaving the planet therefore originates mostly from the atmosphere rather than from the surface. Recall that because solar radiation is mostly absorbed at the surface, the surface warms

the lower atmosphere, and atmospheric temperature decreases with increasing altitude. That is, moving upward and away from the heat source, one finds progressively colder temperatures. This is true for the bottom 80 percent of the atmosphere's mass. We know that emitted radiation depends strongly on temperature. Think about what this means for radiation emitted by the atmosphere. The relatively warm base of a cloud emits more radiation than the cloud's cold top far above. Downward counterradiation from the cloud is stronger than upwelling radiation, and the same reasoning applies to the cloud-free parts of the atmosphere as well.

Two effects occur in this scenario. First, because of clouds and greenhouse gases, there is an additional source of energy available to warm Earth's surface and lower atmosphere. Second, the energy sources seen from space—cloud tops and upper atmosphere—are colder and emit less radiation than those seen from below. It follows from this second point that if a greenhouse gas is added to the atmosphere, the immediate effect of the addition of the gas is to reduce the energy emitted to space. The planet must then warm in order to emit as much radiation as

Continued

Continued

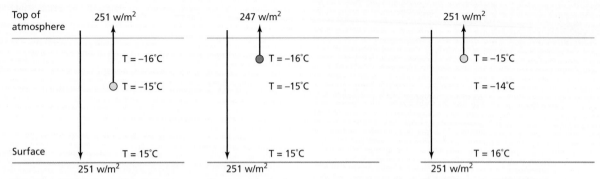

A Before greenhouse gases added   **B** Immediately after greenhouse gases added   **C** After new equilibrium

FIGURE 5.8 Schematic depiction of the greenhouse effect. Because of clouds and greenhouse gases, emission to space effectively occurs from a cold location high in the atmosphere (A). Adding greenhouse gases raises the emission altitude, resulting in lower emissions to space (B). Eventually, warming leads to a new equilibrium when emission again balances absorbed solar radiation (C).

before. In other words, because the effective emission altitude is raised to a higher, colder place, temperatures must rise to reestablish equilibrium between emitted and absorbed radiation. Figure 5.8 shows this in schematic form for an imaginary planet. In part A, before any increase, 251 W/m² are absorbed and emitted by the planet. Immediately after more greenhouse gases are added (Fig 5.8B), the effective emission altitude rises and emissions decrease. With more radiation absorbed than emitted, the surface and atmosphere warm. Eventually, the temperature at the new emission altitude is high enough for emitted radiation to once again balance incoming radiation (Fig 5.8C).

Notice that this analysis is somewhat different from many standard explanations. There is no "bouncing" of longwave radiation between the surface and the atmosphere because very little longwave radiation is actually reflected by terrestrial objects. There is no "re-radiation" of longwave radiation because longwave radiation ceases to exist once it is absorbed. Whatever is emitted depends on temperature, not incoming radiation. Although a little more complicated than a simple blanket analogy, this account will serve well when we discuss human impacts on climate in Unit 19.

Another important aspect concerns the involvement of clouds versus gases. Clouds absorb over a wide range of wavelengths in the infrared part of the spectrum. Thus, a thick cloud transmits virtually no terrestrial radiation. By contrast atmospheric gases (including greenhouse gases) are rather selective. In particular, they absorb relatively little radiation at wavelengths between 8.5 and 13 µm. This band of

wavelengths is known as the *atmospheric window* for that very reason. As humans add greenhouse gases to the atmosphere, what little absorption does occur increases. In other words, adding greenhouse gases to the atmosphere mainly affects radiation in the atmospheric window. (Radiation at other wavelengths is already nearly completely absorbed.) We can see that, according to Figure 5.2, the atmospheric window corresponds to wavelengths at which the surface is a strong emitter. This is an important coincidence—the location of the window is determined by the molecular structure of the greenhouse gases and has nothing to do with surface emission. If the atmospheric window were centered on, say, 4 µm, changes in greenhouse gas concentration would have no effect on energy balance because there is no emission at those wavelengths. This is all really quite amazing. We humans burn fuels that release gases to the atmosphere that darken the atmosphere. The darkening doesn't occur at all wavelengths—only those that happen to be terribly important to Earth's energy balance. Moreover, as we will discuss, those gases are present in such small amounts that the human additions constitute a significant change.

Warming in Greenhouses

Despite the name, Earth's greenhouse effect is quite different from the processes important in physical greenhouses. True enough, glass greenhouses and the atmosphere are similar in that shortwave radiation largely passes through both. Also, like the atmosphere, the glass in a greenhouse is nearly opaque to longwave radiation. However, greenhouses are warm mainly because heated air is confined to the house, as shown at left in Figure 5.9.

In other words, it is the suppression of convection that explains real greenhouse warmth. Louvers are used in some greenhouses to regulate excessive heat, as illustrated in the right side of Figure 5.9. Most of us have experienced how temperatures in an automobile work in much the same way. With the windows up, heat is contained and temperatures can become unbearable. At highway speed, even a small crack is often enough to maintain a comfortable temperature. Such a small opening hardly affects the radiation budget at all. If the "trapping" of sunlight explains the car's warmth, why does the interior of the car cool off when the window is open? After all, almost as much radiation is trapped after the window is opened.

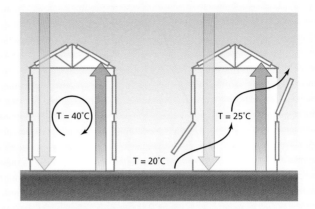

FIGURE 5.9 Greenhouses are warm because the heated air in them is confined within the building (left). A little ventilation drastically lowers temperature despite the small effect on the transmission of radiation (right).

that manage to escape to space. In other words, the planet loses energy mainly by emission from the atmosphere rather than by emission from the surface.

NET RADIATION AND CONVECTION

Both the atmosphere and the surface absorb and emit radiation, so it is natural to ask how gains compare to losses. This comparison is **net radiation**, defined simply as the difference between absorbed and emitted radiation. As Table 5.1 shows, we can calculate net radiation for Earth's surface, the atmosphere, or the entire planet, lumping together the surface and the atmosphere. For the surface, absorbed solar and atmospheric counterradiation exceed emitted longwave radiation by 29 units. Thus, the surface has a radiation surplus. The atmosphere, by contrast, emits more radiation than it absorbs and has a deficit of the same amount. Energy conservation requires that heat be moved from the surface to the atmosphere by other means, or else the surface would eventually warm beyond all measure while the atmosphere cooled.

The radiation imbalance between the surface and the atmosphere is offset by convective heat transfer, which moves 29 units of radiation from surface to atmosphere. The surface is mostly ocean, so perhaps it is not surprising that latent heat transfer is about four times larger than sensible heat flow (see Fig. 5.4, right). Considering convective heat transfer along with radiation, we see that the values given imply both the surface and the atmosphere have an energy balance.

If we look at the planet as a whole, Table 5.1 shows that absorbed solar radiation equals longwave radiation emitted. Radiation is the only significant energy transfer between Earth and outer space; thus, in this case, radiation balance also implies energy balance. The values in the table show that the surface, the atmosphere, and the entire planet

TABLE 5.1 Global Average Radiation Amounts as a Percentage of Incoming Solar Radiation

Radiation Category	Amount of Radiation (%)
Surface	
Solar Absorbed (Gain)	47
Longwave from Atmosphere (Gain)	98
Longwave Emitted (Loss)	−116
Net Radiation	29
Atmosphere	
Solar Absorbed (Gain)	23
Longwave from Surface (Gain)	104
Longwave Emitted (Loss)	−156
Net Radiation	−29
Planet	
Solar Absorbed by Surface (Gain)	47
Solar Absorbed by Atmosphere (Gain)	23
Longwave from Atmosphere (Loss)	−58
Longwave from Surface (Loss)	−12
Net Radiation	0

are all in energy balance; hence, their temperatures are unchanging from year to year. Again, for now, we are ignoring climatic change, but its inclusion would not change the values by much at all. Rather, climate change results from the cumulative effect of small changes over many years.

LATITUDINAL VARIATIONS IN NET RADIATION

Up to this point, our discussion has considered only global mean values. When net radiation is calculated separately

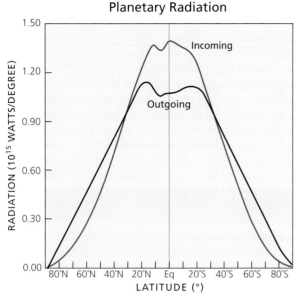

Planetary Radiation

A Radiation

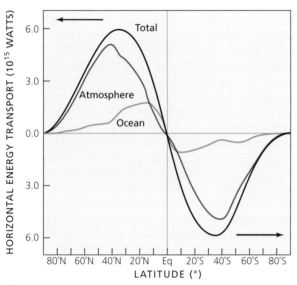

B Northward Energy Transfer

FIGURE 5.10 (A) Annual radiation gains and losses through the top of the atmosphere as a function of latitude. The vertical axis gives heat gain or loss per degree of latitude, which allows for a better comparison between latitudes. (B) Horizontal energy transport by ocean and atmosphere. Positive values are for northward heat transfer, whereas negative values denote transport to the south. The atmosphere curve includes both latent and sensible transfer.

for each latitude, some critically important differences emerge. Let us adopt the planetary view, with the surface and atmosphere taken together. Consider the energy gained and lost by a column extending from the surface to outer space. At tropical latitudes, solar heating exceeds terrestrial radiation leaving the top of the atmosphere. The reverse is true for middle and high latitudes. Thus, as Figure 5.10A shows, areas within 40 degrees latitude have a radiation

surplus, and there is a radiation deficit elsewhere. However, low latitudes do not warm as high latitudes grow colder. Rather, thanks to the motions of the oceans and atmosphere, heat moves poleward within the Earth system from lower to higher latitudes to provide an energy balance throughout. The relative importance of ocean and atmosphere varies with location (Fig. 5.10B). Generally speaking, oceanic movements are more important in low latitudes. In contrast, the atmosphere accomplishes most of the heat transport in middle and high latitudes.

Figure 5.10 correctly implies that the net effect of all atmospheric and oceanic motion is to move heat poleward. This is accomplished by prevailing winds and ocean currents and by storms, eddies, and other short-lived disturbances in both the atmosphere and the ocean. Sensible heat is important in both, as would be expected from the general decline in temperature with increasing latitude. That is, with a temperature gradient directed from equator to pole, there is heat transfer in this same direction. In fact, as Figure 5.10B shows, atmospheric heat transfer is greatest in the middle latitudes. Most of this transfer occurs in winter because the temperature gradient is especially strong in wintertime. (Recall the gradient in solar radiation of Figure 4.12.) For the atmosphere, latent heat transfer is also important in moving heat toward the poles in the middle latitudes. As we will see in Unit 9, winds in the tropics carry water vapor toward the equator, and latent heat transfer follows accordingly. Note that stating that the ocean and the atmosphere move heat toward the poles is not to say there is a net transfer of either water or air toward higher latitudes. Everywhere, any matter carried northward is balanced by returns of water and air flowing southward. As long as the heat content of northward and southward flows is different, there is a net transport of energy.

Making Sense of Global Energetics

The foregoing discussion includes ideas that are both fundamental and potentially puzzling. You might wonder, for example, why convective fluxes (flows) manage to just compensate for the radiation surplus at the surface and the radiation deficit in the atmosphere. To answer this question, you should first realize that if convection moves heat from the surface, an identical amount must be transferred to the atmosphere—there is nowhere else for it to go! Second, the convective fluxes are in large part the result of radiation imbalance. Heating at the surface leads to the evaporation and vertical mixing necessary for convection. If the Sun's output were to increase and deliver more radiation to the surface, we could expect more vigorous mixing and greater convection. This would move more heat into the atmosphere. A warmer atmosphere would emit more radiation to space, and a new equilibrium would result.

Similar reasoning applies to horizontal transports. The longitudinal gradient in net radiation is responsible for oceanic and atmospheric motions. Were the gradient to increase, we could expect more storms, stronger winds, and faster ocean currents, all of which would move additional heat poleward. Temperatures at various places would be different than they are now, but a dynamic equilibrium state would eventually prevail as the planet acquired some new arrangement of climates.

Net radiation is so important that it is tempting to think of it as an independent driver of climate. However, remember that longwave radiation depends on temperature, which is part of climate. Likewise, the distribution of clouds plays a large role in both solar and longwave radiation, and clouds are obviously also a component of climate. A more accurate view is that energy exchanges between and within the atmosphere and ocean are adjusted to seasonal and latitudinal variations in solar radiation. Those adjustments include factors such as temperature, clouds, and so forth that affect energy exchange within the Earth system. Thus, the adjustments are both a response to and a factor in the energy balance of the planet. Even nonclimatic elements of Earth are involved in this. For example, from a plant's perspective, net radiation provides energy for the evaporation of water from leaf surfaces. Without water provided by a supportive climate, the plant would wilt, and growth would be impossible. Thus, at every scale from the local to the continental, vegetation patterns are heavily influenced by net radiation and the availability of water. However, vegetation itself affects climate over a similar range of scales by influencing the amount of solar energy absorbed, providing water to the atmosphere, as well as in many other ways. To think of vegetation as purely a response to climate is wrong; it is also a shaper of climate. As we said in the introduction, Earth is a highly interconnected system.

Key Terms

albedo *page 59*

conduction *page 57*

convection *page 58*

counterradiation *page 61*

diffuse radiation *page 59*

direct radiation *page 59*

greenhouse effect *page 61*

latent heat transfer *page 58*

longwave radiation *page 57*

net radiation *page 63*

radiation *page 56*

sensible heat transfer *page 58*

shortwave radiation *page 57*

terrestrial radiation *page 57*

thermal energy *page 56*

Scan Here for a quick vocabulary review

Review Questions

1. Describe the three modes of heat transfer and how they differ.
2. Define latent heat transfer.
3. Compare and contrast solar and terrestrial radiation.
4. How much of the solar energy entering the atmosphere does the atmosphere absorb, and how much does Earth's surface absorb? How much do the atmosphere and the surface each reflect?
5. In your own words describe the meaning of the term *greenhouse effect*.
6. Define the term *albedo*. Give some examples of its application in your daily life.

Scan Here for a quick concept review

www.oup.com/us/mason

Atmospheric and Oceanic Processes

Circulation Systems in Air and Ocean

The Sun's energy, combined with forces arising from Earth's rotation, drives the ocean and the atmosphere in giant circulation systems that carry warmth from equatorial latitudes toward the poles of both hemispheres. Giant cells of roughly circular flow occupy entire ocean basins. Water travels in slow drifts and faster currents from warm tropical environs to cooler middle latitudes and beyond, returning toward the tropics with infusions of polar cold. In the atmosphere, equatorial warmth convects upward, and in the mid-latitude atmosphere, high-altitude airflows that encircle the globe bend north and south, transporting vast amounts of heat poleward. Embedded in these large-scale atmospheric systems are powerful subsystems ranging from hurricanes to blizzards. Earth's rotation influences the direction of circulation in water and air, the movement of weather systems, and even the prevalence of persistent winds. Conditions in the atmosphere and hydrosphere under certain circumstances promote storms and disturbances of all sizes. These impact the lithosphere (erosion) as well as the biosphere (with nurturing rainfall as well as the destruction of vegetation and wildlife). Furthermore, the disturbances themselves are linked to surface conditions.

UNIT 6

Composition and Structure of the Atmosphere

However thick it might seem when viewed from the surface, in fact the atmosphere is but a thin shell surrounding Earth.

ENHANCE

OBJECTIVES

- Describe the constituents of Earth's atmosphere and their relative concentrations
- Explain the origin of Earth's atmosphere and how its composition has changed through time
- Discuss the processes that add and remove important atmospheric gases and the resulting atmospheric residence times
- List the four temperature layers of the atmosphere together with their major properties

Our atmosphere, one of our most precious natural resources, constitutes a vital component in the systematic study of our planet. This thin, shell-like envelope of life-sustaining air that surrounds Earth (see the unit-opening photo on p. 68) is a place of incredible activity, as we will demonstrate in the units of Part Two. Earth's atmosphere has been called the working fluid of our planetary heat engine, and its constant motions shape the course of environmental conditions at every moment in every locality on Earth's surface. The short-term conditions of the restless atmospheric system that impinge on daily human activities are called **weather**; the long-term conditions of aggregate weather over a region, summarized by averages and measures of variability, constitute its **climate**.

The atmosphere extends from a few meters below the ground on land, or from the water's surface in oceanic areas, to a height of about 60,000 km (37,000 mi). It consists of a mixture of gases along with various liquid drops and solid particles that vary greatly from place to place in terms of type and amount. Most of the mass of the atmosphere is concentrated near the planetary surface (see Fig. 8.2 on p. 99). Physical geographers are especially interested in the lower parts of the atmosphere that lie below 50 km (31 mi) and, in particular, below 10 km (6 mi). Important flows of energy and matter occur within these lower layers, which constitute the effective atmosphere for all life-forms at the surface. In the lower layers great currents of air redistribute heat across Earth. These currents are part of the systems that produce our daily weather. Over time, weather and the flows of heat and water across Earth are eventually translated into our surface patterns of climate. People are affected by both the local climate and the larger atmosphere, and we have the power to influence each one to some degree. In the past, climatic changes occurred without human intervention, but in the future, and even today, humankind is playing a more active role in the climate.

Atmospheric Constituents

Table 6.1 lists the main ingredients of the atmosphere of today. The list includes gases, solids, and liquids. Some components are effectively constant throughout the atmosphere, but, as indicated in the second column of the table, others vary from place to place or over time (or in both space and time). Three of the gases listed are *chemical compounds*: *carbon dioxide*, *water vapor*, and *methane*. In a **chemical compound**, at least two different **elements** combine to form a single molecule. In contrast to compounds, molecules of N_2, O_2, and O_3 each consist of only one element; hence, these gases are *elemental*. Argon is also an elemental gas, but it is composed of single atoms of Ar rather than molecules. There are many other gases in the atmosphere not listed in the table, most of which are compounds.

Chemical reactions that occur within the atmosphere and near the surface transform gases into new chemicals. For example, the immense power of a lightning stroke creates two molecules of nitric oxide (NO) through the following reaction:

$$N_2 + O_2 + \text{Energy} \rightarrow 2\,NO$$

This is just one example of a multitude of reactions that occur continually throughout the atmosphere. Some reactions occur spontaneously when the right ingredients are present, whereas others require an energy source. Sunlight is an especially important source of energy for the reactions known as **photochemical reactions**. Still other reactions occur within the cells of living organisms. Considering this incessant transformation of a host of chemicals, we should think of the atmosphere as highly dynamic at invisible atomic scales, as well as at the large scales we can so easily see in high winds and violent storms.

TABLE 6.1   The Most Abundant Components of the Atmosphere

	Variation	Molecular Weight	Relative Mass of Atmosphere*	Relative Number of Molecules	Approximate Residence Time
Gases					
Nitrogen (N_2)	Constant	28.02	75.5%	78.1%	13 million years
Oxygen (O_2)	Constant	32.00	23.1%	20.8%	5,000 years
Argon (Ar)	Constant	39.95	1.3%	0.93%	Infinite
Carbon Dioxide (CO_2)	Variable	44.01	0.059%	0.04%	5–10 years
Water Vapor (H_2O)	Variable	18.02	0.03%	0.48%	10 days
Methane (CH_4)	Variable	16.04	0.0001%	0.00018%	12 years
Ozone (O_3)	Variable	48.00	0.0001%	0.00006%	~0.25 years
Other					
Water Drops and Ice	Variable	n/a	0.053%	n/a	n/a
Other Drops and Solid Particles	Variable	n/a	0.0001%	n/a	n/a

*Percentages are global averages; they do not sum to 100 because of rounding and the uncertainty of some values.

ISOTOPES

Atoms are the basic building blocks of elements, and they are made up of smaller particles called *protons, neutrons,* and *electrons.* The molecular weight of an element is determined by the number of protons and neutrons that are in each atom of that element. For example, most nitrogen is nitrogen-14, with seven protons and seven neutrons, and so most N_2 has a molecular weight of 28. Notice that in Table 6.1 the molecular weights of the gases have a fractional part. For example, nitrogen (N_2) has a molecular weight of 28.02 in the table. The fraction arises because there are various forms, called **isotopes**, of nitrogen in the atmosphere. Although most nitrogen is nitrogen-14, some slightly heavier forms also exist, the most common of which is nitrogen-15, which has one additional neutron. The molecular weight of nitrogen listed in Table 6.1 is an average based on the amounts of all nitrogen isotopes; it is not exactly 28, 29, or any other integer. Molecular weights of other gases similarly reflect the various isotopes of each element that makes up the gas. We will see the importance of isotopes later in this unit and even more in units to follow.

RELATIVE AMOUNTS

How can we describe the quantity of any given atmospheric component? One approach is to recognize that the atmosphere has a certain mass and weight. Each constituent contributes to some degree to the total weight of the atmosphere, with the most abundant and heavier constituents contributing more. The Relative Mass column in Table 6.1 shows the percentage contribution in terms of mass. Just three gases—nitrogen, oxygen, and argon—constitute about 99.9 percent of the atmosphere's mass. Notice also that the nongaseous parts—solids and liquids– contribute very little to that mass, which means that however large clouds might appear, they contain very little liquid or solid water relative to the surrounding gases. Furthermore, any precipitation falling in the form of rain, snow, or ice is an insignificant fraction of the atmosphere's mass.

As another way to think about abundance, imagine collecting a sample of atmosphere at some point. Ignoring any liquids or solids, we could at least in principle determine the number of molecules of each gas in the sample or the number of atoms for nonmolecular gases like Ar. From that, we could find the percentage of each gas in terms of number. The nitrogen percentage would be the percentage of molecules that are N_2, the oxygen amount would follow from the percentage of molecules that are O_2, and so forth. Table 6.1 lists by percentage the relative number of molecules of a particular gas in a unit of atmosphere. The information in the table indicates that out of a sample of 100 molecules, about 78 will be nitrogen, about 21 will be O_2, and perhaps just 1 atom will be argon.

The abundance of **carbon dioxide** (CO_2) is much smaller even than that of argon, and we would be unlikely to find even one CO_2 molecule in a sample of 100. On the other hand, a sample of 1 million molecules (10,000 times

larger) would contain about 400 molecules of CO_2. We would say the CO_2 concentration is 400 *parts per million* (ppm). Parts per million is a unit commonly used for gases with low concentrations. Such gases are called **trace gases**, indicating that they are present in only small amounts. Some trace gases are so rare that we use the unit *parts per billion* (ppb) to speak of them. Methane is one example; its concentration of 1.8 ppm is equivalent to 1,800 ppb. There are many other trace gases not shown in Table 6.1, all of which are even rarer than those listed.

The percent by molecule is the same as the percent by volume. That is, a cubic meter of air or any other volume is about 78 percent filled by N_2, 21 percent filled by O_2, and so forth. This is not strictly true throughout the entire atmosphere, but it holds for all but the upper reaches. It is very common to see concentrations described as percent by volume. Remember that such percentages are the same as percent by number of molecules.

The climatic importance of a gas is not related to its quantity. For example, the trace gases and water in all its forms are a miniscule part of the atmosphere, but Earth's climate would be transformed beyond recognition if they were to suddenly vanish. However inviting it might be to truncate the list after the big three (N_2, O_2, Ar), we need to consider other gases to understand our climatic environment. It would also be a mistake to think of the atmosphere as a static reservoir. Its composition is not constant over time, nor are individual molecules held in the atmosphere forever. Looking at the last column in the table, we can see that there is great variation in **residence time**. The term *residence time* refers to how long a molecule remains in the atmosphere. The next section addresses these topics, which are fundamental to understanding present-day and future climates.

The Origin, Evolution, and Maintenance of the Atmosphere

It is easy to take for granted the presence of an atmosphere, but the story of its existence is a complicated saga that begins billions of years ago. Our atmosphere was not present from the earliest of Earth's days, about 4.6 billion years ago. For tens of millions of years after the solar system took shape, collisions between the still-forming Earth and other objects were powerful enough to strip away gases that might have accumulated under the force of gravity. Earth's early atmosphere was produced by a combination of volcanic outgassing and material delivered from the outer reaches of the solar system. The early atmosphere was very different from the modern version, consisting mostly of **water vapor** (an invisible gaseous form of water) and CO_2. And most significantly, it was almost completely devoid of oxygen.

From its earliest days, our atmosphere has evolved continually in response to both physical and biological processes. As the planet cooled, water vapor was able to condense to liquid and fall out as rain. This physical process (condensation and precipitation) resulted in the oceans

Why Don't Particulates Fall?

All **particulates**—bits of solid or liquid matter suspended in the Earth's atmosphere—are denser than air, and it seems they should fall out of the atmosphere under the influence of gravity just as a lead weight falls through water. But the presence of clouds and dust that last for days suggests otherwise (Fig. 6.1). A little more thought shows that *density* (mass per unit volume) cannot be the whole story. For example, the particles in Figure 6.2 are every bit as dense as the rock from which they are cut, yet they are easily suspended in the air (Fig. 6.2).

The key is the size of the particle, which we will explain in a sequence of steps. First, realize that as an object of whatever size falls through the atmosphere, the surrounding air molecules exert a frictional drag on the object. Drag increases with speed. If an object falls fast enough, the drag force balances the force of

FIGURE 6.2 A cloud of particulates raised by a worker cutting stone.

gravity, and it falls at a constant rate. This is known as *terminal velocity*.

The amount of drag an object experiences depends in part on its surface area; everything else being equal, more surface area means more drag at any given speed. But everything else is not equal, and most important, larger objects experience a greater downward gravitational force.* That means that there are competing processes affecting an object's terminal velocity. The downward force increases with size, but so does the surface area and therefore the drag that resists the downward force.

In Table 6.2 we compare two spherical particulates made of rock with a density of 2,700 kg/m^3 (169 lb/ft^3). The spherical shape lets us compute volume and surface area using familiar formulas from geometry. We multiply volume and density to find mass. Because we are interested in atmospheric particulates, we compare small grains with diameters of 10 μm and 30 μm (0.00039 and 0.0012 in., respectively). For both objects, we list mass, surface area, and surface area per unit mass. Because the resulting numbers are small, we use scientific notation. For example, 0.000003 m = 3 × 10^{-6} m, or 3 millionths of a meter.

Notice that the ratio of area to mass is three times *larger* for the *smaller* particle in Table 6.2. This means

FIGURE 6.1 Huge amounts of desert dust are lifted to great heights from an African desert and carried westward by winds. Some will land on U.S. beaches, and even more will be deposited in the Amazon Basin.

ENHANCE

TABLE 6.2 Particle Size, Terminal Velocity, and Drag

Size	Size (m)	Volume (m³)	Surface Area (m²)	Mass (kg)	Area/Mass (m²/kg)	Terminal Velocity (m/s)
10μm	10 × 10⁻⁶	5.24 × 10⁻¹⁶	3.14 × 10⁻¹⁰	1.41 × 10⁻¹²	222.2	0.0088
30μm	30 × 10⁻⁶	1.41 × 10⁻¹⁴	2.83 × 10⁻⁹	3.82 × 10⁻¹¹	74.1	0.079

Because of rounding, some values do not sum exactly to the totals shown.

continued

continued

there should be a greater drag force per gram for the smaller object. What about gravity? On a per gram basis, it is exactly the same for both particles. So by comparing these numbers, we can see that relative to gravity, drag is much greater for the smaller object. It follows that the terminal velocity should be lower for the smaller droplet. This is exactly the case, as we can see in the last column of Table 6.2: the terminal velocities differ by a factor of about 10.

Applying this to the atmosphere, we can see that a much smaller updraft could hold the smaller 10μm particle aloft. The take-home message is that smaller objects fall more slowly. In other words, they are more easily suspended in the atmosphere. Size is the key; density differences would need to be huge to

compensate for even the modest-size differences seen in the example in Table 6.2.

Neither of our example particles falls very fast—and there are particles that fall even more slowly. Many particulates are much smaller, with terminal velocities on the order of 1 millionth of a meter per second. This is even slower than their random movement would be in perfectly still air. Thus, aerosols of this size never fall out of the atmosphere by simple gravitational settling, and they must be removed by other processes.

*If you doubt that, answer this: Could you sit on a strong cardboard box without collapsing it? Could an elephant? Why the difference? Hint: A scale that measures weight measures the force of gravity on an object. Mentally compare your weight to that of an elephant.

holding most of the world's water, with only a small fraction in the atmosphere. As they do today, falling raindrops carried carbon dioxide from the atmosphere to the ocean, where the carbon combined with other materials and ultimately formed carbon-bearing rocks. This physical process also changed the atmospheric composition, in this case reducing CO_2 values by roughly 100-fold in a billion years. Undoubtedly, the most important example of a biological process is *photosynthesis*, which is the chemical reaction by which green plants are able to grow. During photosynthesis within the cells of a plant, CO_2 and water combine in the presence of sunlight to produce new plant material and release O_2 as a byproduct. Life appeared late in Earth's history, and therefore it was billions of years before oxygen began to accumulate in any significant quantity. There are many uncertainties about Earth's oxygen history, but there is no doubt our atmosphere evolved along with life.

Photosynthesis is an example of a process that adds a gas (O_2) to the atmosphere. There are also many processes that remove gases. In the case of O_2, molecules are

consumed by forest fires, by the decay of plant material, and by students breathing as they read books like this one, to name just a few examples. Every gas has its own addition and removal processes. As Figure 6.3 shows, some of these processes are internal to the atmosphere, and others occur at the surface and outer limits of the atmosphere.

In some cases, only physical processes are involved, such as when water evaporates from a lake and increases the atmosphere's water vapor inventory. Other processes are chemical, as when lightning combines nitrogen and oxygen into a new gas. Still other processes are best thought of as biological, photosynthesis being the prime example. Regardless of the type of processes involved, for any gas, the concentration is close to constant from year to year if the production and removal processes are nearly equal. That is, individual molecules can come and go, but the number of molecules is nearly constant as long as gains equal losses. Such gases are in dynamic equilibrium, as described in Unit 1. For example, in today's atmosphere, N_2 and O_2 are in close to equilibrium conditions on human timescales. They

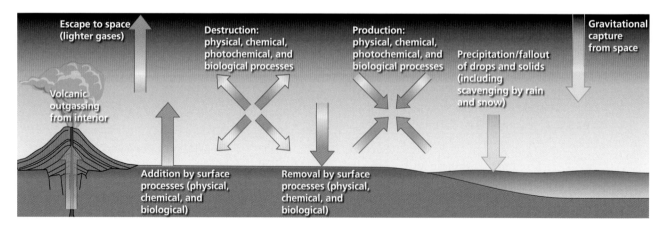

FIGURE 6.3 Schematic diagram showing how gases and other atmospheric constituents are added to and lost from the atmosphere. Not all processes are important for all constituents, and the relative importance varies greatly on geologic timescales.

FIGURE 6.4 "The apartment in Honolulu, high on the slope of Punch Bowl, afforded great daytime views over the city. But what was most remarkable, even here in the middle of the Pacific Ocean, was the daily vivid sunset, proof of the presence of volcanic and desert dust, high in the atmosphere."

stand in contrast to carbon dioxide and methane, both of which are increasing on a timescale of centuries because production exceeds removal.

With this as background, we can now discuss individual constituents: how they came to be, how they change over time, and how their concentration varies from place to place.

Contents and Arrangement of the Atmosphere

The atmosphere may be broadly divided into two vertical regions. The lower region, called the **homosphere**, extends from the surface to 80 to 100 km (50 to 63 mi) above Earth and has a more or less uniform chemical composition. Beyond this level, the chemical composition of the atmosphere changes in the upper region, known as the **heterosphere**, where turnover is so slow that gases start to settle out according to molecular weight. The homosphere contains nearly all of the atmosphere's mass and is the more

important of the two atmospheric regions for life on Earth, including humans. If we experimented by collecting numerous air samples of the homosphere, we would find that it contains three major groups of components—*constant gases*, *variable gases*, and *particulates*. This section discusses the relative nature and proportions of these components in the atmosphere (Fig. 6.4).

CONSTANT GASES

Table 6.1 labels N_2, O_2, and Ar as **constant gases**. This means that these gases show little seasonal variation and also change little from year to year. It also means that wherever we go, we will find those gases present in the percentages shown. Why does this happen? The explanation for this is twofold. First, gains and losses are close to equal, and so we observe little change over time. Second, the production and destruction rates are small relative to the amount of gas present. So even though a gas might be added or moved at Earth's surface, the stirring produced by atmospheric motions is still enough to ensure a uniform concentration. This is analogous to adding ink from a dropper to an aquarium. If the water is stirred fast enough at any moment, the color remains uniform despite continual additions from the surface.

Nitrogen (N_2) The bulk of the atmosphere we breathe consists of **nitrogen**. Atmospheric nitrogen, N_2, is almost inert and does not react easily with other chemicals. Early in the life of the solar system, most nitrogen was driven away from the inner planets by strong solar emissions, and as a result nitrogen is not one of the more abundant elements on Earth. However, because it is so nonreactive, it accumulated in the atmosphere and eventually came to dominate all other gases.

Elemental nitrogen is essential for life, and specialized bacteria have the ability to remove nitrogen from the atmosphere and make it available for themselves and other forms of life in a process known as *nitrogen fixation*. Removal and conversion to nitrogen compounds occur in other ways, including through fertilizer production and by lightning discharge, as mentioned earlier. However, fixation by microorganisms accomplishes more than 80 percent of the loss of nitrogen from the atmosphere. The total annual loss is just a few millionths of a percent of the atmosphere's nitrogen inventory, and the average N_2 molecule spends about 13 million years in the atmosphere before being removed. A nearly equal amount is returned to the atmosphere every year, mainly by other bacteria that produce N_2.

Oxygen (O_2) Oxygen gas represents about 20 percent of the atmosphere and is necessary for our survival. We absorb oxygen into our bodies through our lungs and into our blood. One of oxygen's vital functions in animals is to "burn" food so that its energy can be utilized. Digested food material is oxidized, which means it combines chemically with oxygen to form new compounds. The biological name for this process is *respiration*, and the chemical name is *oxidation*. An example of rapid oxidation is the burning of **fossil fuels** (coal, oil, and natural gas). Without oxygen, this convenient way of releasing the energy stored in these fuels would not be possible. Slow oxidation can also occur, as in the rusting of iron or the decay of dead leaves and other plant parts. Oxygen is essential not only for respiration but also for many other chemical processes. On a global basis, oxidation (which results in the loss of oxygen from the atmosphere) is roughly balanced by the production of O_2 by photosynthesis. The mass of O_2 produced and destroyed each year is small compared to the atmosphere's store, so the residence time of oxygen in the atmosphere is about 5,000 years. Recall that residence time refers to how long a molecule remains in the atmosphere; consequently this means that on average a molecule of O_2 gas remains in the atmosphere about 5,000 years.

The history of atmospheric oxygen is known only in broad terms (Fig. 6.5 shows two possible scenarios), but the importance of photosynthesis and the biosphere more

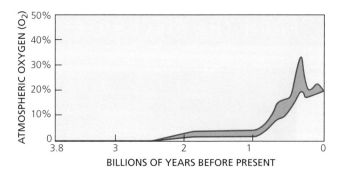

FIGURE 6.5 The history of atmospheric oxygen (O_2) concentration over geologic time. The two curves bracket estimates derived from a variety of indicators.

generally is not debated. Initially, oxygen was produced by ultraviolet radiation (uv) breaking apart water vapor into hydrogen and atomic oxygen: $H_2O + uv \rightarrow H_2 + O$. Being a light gas, hydrogen mostly escaped to space, but atoms of oxygen could combine to form O_2: $O + O \rightarrow O_2$. Formation of O_2 from the breakdown of water is slow and provided only small amounts of oxygen to the atmosphere.

Significant amounts of atmospheric oxygen appeared only after the first photosynthetic organisms evolved in the oceans billions of years ago. The increase was slow, as exposed rocks had to first be oxidized before atmospheric levels could rise. Bursts of oxygen accumulation occurred with the development of multicellular plants and colonization of the continents. The values of O_2 are believed to have reached a maximum of 35 percent about 300 million years ago. This may have been caused by an explosion of forested areas, whose woody material was slow to oxidize. The decline in O_2 that followed is unexplained.

Recall our discussion of feedback from Unit 1. Oxygen is very reactive, and its reactive nature exerts a strong negative feedback on atmospheric content that prevents excessively high levels of O_2. In particular, because materials burn more easily as oxygen increases, more forest fires occur when oxygen levels are high. Forest fires both remove oxygen through oxidation and decrease standing forests of trees, which are important terrestrial oxygen producers.

Argon (Ar) The inert gas argon appears almost entirely as the isotope argon-40. Another version—argon-36—is far more abundant in the universe generally, but like other gases, it did not survive the turbulent times early in our planet's formation. Nearly all of the argon present in the atmosphere today came later and resulted from a process called *radioactive decay* of potassium-40, in which a proton is converted to a neutron. Although the atomic mass remains 40 after this conversion, the radioactive decay results in a different element—namely, argon-40. Over billions of years, more than 90 percent of Earth's potassium-40 has decayed to argon-40; thus, little more will be produced and leak into the atmosphere. Argon is an *inert gas*, which means it does not readily interact with other chemicals. As a result, the atmospheric

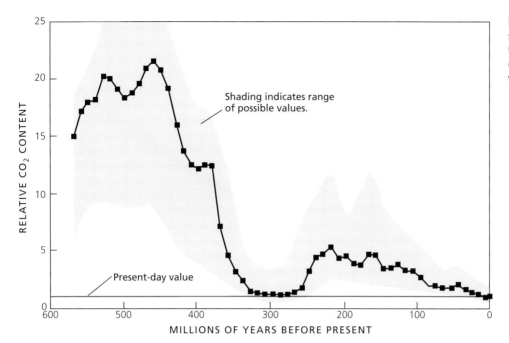

FIGURE 6.6 Mass of atmospheric CO_2 as a multiple of today's value. Shading indicates range of possible values.

content of argon is essentially constant, and the residence time is effectively infinite. Argon plays only a very minor role in the workings of environmental systems.

VARIABLE GASES

Although they collectively constitute only a tiny proportion of the air, we must also recognize the importance of certain atmospheric gases that are present in varying quantities. Three of these **variable gases** are essential to human well-being: carbon dioxide, water vapor, and ozone.

Carbon Dioxide (CO_2)   Carbon dioxide (CO_2), which on average comprises only 0.04 percent of dry air, is a highly significant constituent of the atmosphere in terms of its influence on climate. By virtue of its ability to absorb and emit longwave radiation, carbon dioxide plays a key role in the *greenhouse effect*. Recall from discussion of the greenhouse effect in Unit 5 that carbon dioxide helps keep the lower atmosphere at temperatures that permit life (which globally now average just over 15°C [59°F]).

Over eons, carbon dioxide levels in the atmosphere have varied in response to purely natural processes. As mentioned earlier, carbon dioxide declined from very high values to become a minor constituent within a billion years of Earth's formation, but even then, it was far more common than now. Figure 6.6 shows CO_2 estimates over the last 600 million years. During this time span photosynthetic plants became widespread. Earlier, we mentioned the role of vegetation in removing CO_2 from the atmosphere, but the prominent decline we see in Figure 6.6 is a little more complicated; we cannot attribute it simply to more photosynthesis and carbon storage in plants. Rather, as deeply rooted plants expanded on land hundreds of millions of years ago, they contributed to the

faster chemical breakdown of underlying bedrock. The resulting products allowed more soil carbon (originally produced by photosynthesis) to be dissolved and carried to the oceans by rivers and streams in various compounds. In the ocean, these compounds combined with other compounds to form carbon-bearing rocks that today hold most of Earth's carbon.

On much shorter timescales, human activities have significantly perturbed the atmosphere's carbon budget, with explosive growth over the past 200 years (Fig. 6.7). Land clearance associated with the expansion of agriculture was the main contributor early on, but the burning of fossil fuels has dominated for the last century or so.

The seasonal rise and fall in Figure 6.7B (indicated by the red line) reflects the Northern Hemisphere location of the observatory. Plant growth in the summer draws down atmospheric CO_2 by a few ppm, only to be restored in the winter by oxidation as leaves decay. Globally, enough carbon moves back and forth between the atmosphere and biosphere to make the residence time of carbon dioxide in the atmosphere less than a decade. This is in contrast to the constant gases we just discussed and their lengthy residence times. The rapid turnover of CO_2 compared to oxygen is explained by the meager amount of CO_2 in the atmosphere. Although roughly the same number of molecules of O_2 and CO_2 move between the surface and the atmosphere each year, the CO_2 molecules represent a much larger percentage of the atmosphere's store.

CO_2 emissions vary greatly over Earth's surface because the spatial distribution of power plants and other sources of CO_2 are far from uniform. Once in the atmosphere, CO_2 is carried by winds and mixed like ink dropped into a swirling glass of water. For example, in the winter emissions from population centers in North America, Europe, and Asia produce relatively high values

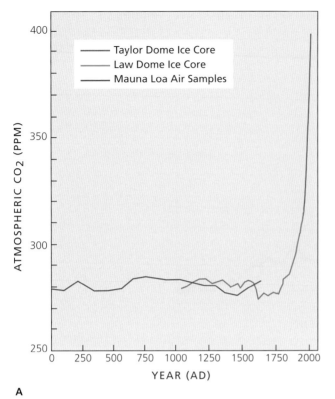

A

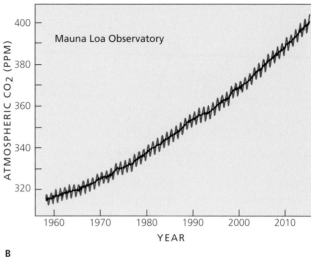

B

FIGURE 6.7 Atmospheric CO_2 over the past 2,000 years (A). Values before 1960 come from air trapped in glacial ice. The more recent record (B) is based on air samples taken at a mountain observatory in Hawai'i.

of CO_2 in downwind locations (Fig 6.8). We will have much to say about CO_2 and climate change in Units 17 and 18. For now, we simply note the unmistakable imprint of humans on atmospheric CO_2 values. That, along with natural changes, explains its position as one of the variable gases of note.

Water Vapor (H_2O) We take the existence of water on our planet for granted. But the source of Earth's original water is a scientific mystery. Until very recently the prevailing theory was that none of the water found today dates back to Earth's formation because of the very high temperatures that prevailed 4.6 billion years ago. Instead, it was believed that long after Earth's formation comets and

asteroids deposited water in the course of colliding with Earth. The most recent data from space probes sent to measure water on comets in the last few years cast doubt on their role as a major source for Earth's water, leaving asteroids as the likely delivery vehicle. In 2014 a new study concluded that Earth's water molecules are older than Earth itself and were part of rocks that formed in the cloud of dust and gas surrounding what would become the Sun. According to this theory, as these rocks collided with others during accretion of the planet, water became part of Earth and was later released by volcanic outgassing. Whatever its origin, as we discussed in Unit 5, water vapor is crucial to the energy balance of the planet. It is a powerful greenhouse gas, and it carries vast amounts of latent energy from

FIGURE 6.8 Distribution of atmospheric CO_2 as computed by the NASA GEOS-5 global computer model. Large CO_2 values over North America, Europe, and Asia reflect heavy fossil fuel use in winter.

ENHANCE

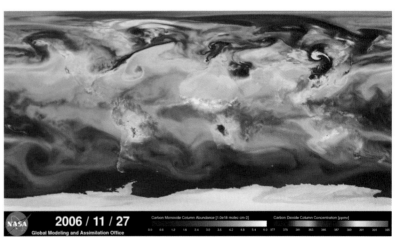

place to place as it moves along with the winds. These energy transfers are part of essential processes that keep Earth's surface temperatures moderate enough for human habitation. Without water vapor, there would be no clouds or rainfall. Most portions of the land surface would be too dry to permit agriculture. Without the great cycle in which water moves from the surface into the atmosphere and back again, which we discuss in detail in Unit 11, little life of any kind would be found on our planet.

In deserts or cold regions, water vapor makes up only a minute fraction of 1 percent of the air. But over warm oceans or moist tropical land areas, it may make up as much as 3 or 4 percent. In general, as we will see in Unit 11, the warmer the air, the more moisture or water vapor it can contain. Because the lower atmosphere has relatively high temperatures, that is where most of the water vapor occurs, and the upper atmosphere is nearly devoid of water.

Water vapor leaves the atmosphere by condensation and deposition of precipitation in falling rain, snow, and hail. It enters the atmosphere through evaporation from water surfaces and pores in the leaves of vegetation. As a plant grows, its pores must be open for CO_2 to enter, and water vapor unavoidably escapes to the atmosphere through the open pores. The atmospheric store of water vapor is fairly small (see Table 6.1) compared to the movement of water vapor in and out of the atmosphere. This results in a residence time of less than two weeks. This is an extremely fast turnover compared to other gases.

Water vapor is also unusual for its strong spatial variation, with differences by a factor of 10 possible between places just kilometers apart from one another. This shows that unlike other gases, water vapor avoids the homogenizing effects of atmospheric motion. (For example, CO_2 never shows that kind of spatial variation.) As we will see in Unit 11, precipitation is highly localized and quite efficient at removing water from the atmosphere. This permits large differences to develop over short distances despite constant atmospheric motion and a surface that is 70 percent covered with water.

Ozone (O_3) The other variable gases in the homosphere are found in much smaller quantities than water vapor. The most important of these is ozone, the rarer type of oxygen molecule composed of three oxygen atoms (O_3) instead of two (O_2). On a percentage basis, ozone amounts are highest at altitudes between 15 and 50 km (9 and 31 mi), which is an atmospheric layer known as the *stratosphere*, which we discuss later in this unit. The greatest concentrations are in the so-called ozone layer, located between about 20 and 30 km (13 and 19 mi) above the surface. Even there, ozone is very rare, ranging from just 2 to 10 parts per million. Hundreds of chemical reactions influence the abundance of ozone, and prevailing winds are also important. Most ozone forms high in the tropical atmosphere and is carried downward and poleward by circulations within the stratosphere. Ozone is created when

molecular O_2 combines with oxygen atoms in the following reaction:

$$O_2 + O + X \rightarrow O_3 + X$$

The two oxygen species collide with some other molecule (X) that carries away some energy and allows the oxygen to combine and form an O_3 molecule. This reaction is more common in the ozone layer than anywhere else, mainly because atomic oxygen (O) is produced at high altitudes.

Ozone amounts vary greatly with location and season, as seen in Figure 6.9. The maps show total ozone in the atmospheric column in millimeters. To understand how ozone can be measured in millimeters, imagine lowering a tube through the atmosphere that could somehow collect all the O_3 molecules above a point in question. Assuming all other gases are excluded, if the tube was on your desk, it would need to be just a few millimeters tall. The ozone layer amounts to perhaps 3 mm (0.12 in) of pure O_3. The maps shown use just this concept.

The maps in Figure 6.9 are for spring months, when the highest ozone amounts occur. This shows the importance of circulation, because ozone production rates are highest in summer. (In October it is spring for the Southern Hemisphere, and in April it is spring for the Northern Hemisphere). Circulation away from high production areas also explains why the largest amounts are found in the middle latitudes rather than the tropics. Notice that O_3 is very low over Antarctica in October, with values below 1.5 mm. This is the so-called **ozone hole** that appears every spring in the Antarctic stratosphere. The hole arises because humans have introduced certain chemicals containing chlorine and bromine into the atmosphere. Both of these elements react with O_3, thereby reducing ozone concentration. The Antarctic is favored because prevailing winds that circle the continent isolate that part of the atmosphere and allow extremely low temperatures to develop there. The burst of solar radiation that comes with spring frees chlorine and bromine atoms so they can go about their business of destroying ozone. The details of this process are complicated and not fully understood, but there is no doubt the ozone hole is of human origin. Humans are also responsible for longer-term changes in ozone (see Perspectives on the Human Environment: Ozone Depletion and Recovery).

Like carbon dioxide, ozone is very important despite its low concentration, and also like carbon dioxide, the importance of ozone is due to its ability to absorb radiant energy. In ozone's case, it absorbs the *ultraviolet radiation* associated with incoming solar energy. Large doses of ultraviolet radiation cause severe sunburn, blindness, and skin cancers. High levels of ultraviolet radiation are lethal for plants and animals, especially life-forms that lack skin, fur, or other protective outer layers. The ozone layer therefore provides an essential shield from excessive quantities of this high-energy radiation and is the reason Earth's

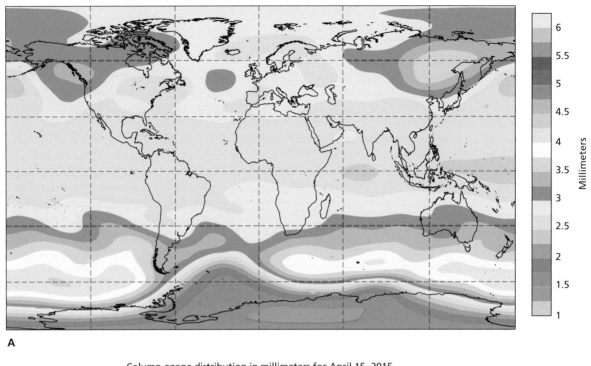

Column ozone distribution in millimeters for October 15, 2014

A

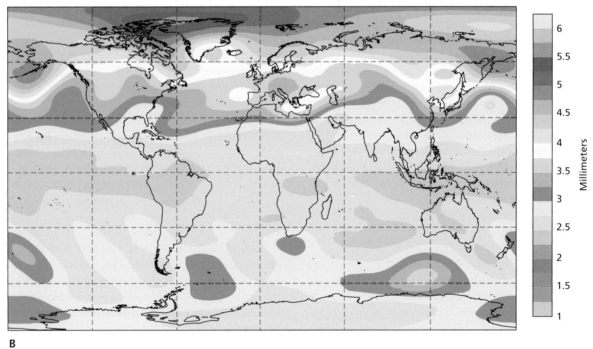

Column ozone distribution in millimeters for April 15, 2015

B

FIGURE 6.9 Column ozone distribution in millimeters for October 15, 2011 (A), and April 15, 2012 (B).

surface and the upper parts of the ocean are habitable. (Until ozone appeared, photosynthetic life was restricted to ocean depths where damaging ultraviolet radiation could not penetrate.) Ozone is also a good absorber and emitter of terrestrial radiation and belongs in the list of greenhouse gases.

Methane (CH₄) **Methane** is the final variable gas we will discuss in this unit. It is also a greenhouse gas. On a molecule-by-molecule basis, methane is even more effective at absorbing longwave radiation than CO_2. However, methane is extremely rare, with a concentration of only 1.8 ppm versus 400 ppm for CO_2; hence, its overall importance is less. Methane is produced naturally by bacteria found in wetlands and in the guts of many animals ranging from termites to cattle to, yes, even humans. Atmospheric methane has been greatly increased by humans through activities such as farming, land clearing by burning, landfill construction, and natural gas production. Human and natural releases are now about equal, and methane concentrations have grown by a factor of 6 over the past few hundred years. In the last decade, the rate of increase slowed somewhat, but methane amounts continue to rise. Methane is removed from the atmosphere primarily by reactions that ultimately produce water vapor and carbon dioxide. Methane production and decay are relatively fast, so that the residence time for methane is about 10 years or so.

Other Gases Minute quantities of many other gases are also present in the atmosphere. The most noteworthy are hydrogen, helium, sulfur dioxide, oxides of nitrogen, ammonia, and carbon monoxide. Some of these are inert (e.g., helium), and some are reactive but insignificant climatically (e.g., H_2). Some are **air pollutants** (substances that impact organisms negatively) derived from manufacturing, transportation, and other human activities. Pollutants can produce harmful effects even when their concentrations are one part per million or less. (Interestingly, when found at or near the surface, ozone loses its beneficial qualities and becomes just another pollutant.) Other pollutants are found in the form of solid particles or in liquid droplets composed of a water solution containing the pollutant.

PARTICULATES

Recall that in the atmospheric sciences, a particulate is any suspended or falling liquid or solid. Particulates include *clouds*, which are visible masses of suspended, minute water droplets and ice, as well as the rain, snow, and hail they produce. All other nongaseous components of the atmosphere (such as dust, smoke, etc.) are also particulates.

Clouds are perhaps the most conspicuous feature of the atmosphere. Like water vapor, nearly all clouds are found in the lowest 20 km (12 mi) of the atmosphere. Unit 5 discussed the importance of clouds in the planetary energy balance, and we will not repeat that discussion here. Let us instead note that particulate water (clouds and precipitation) represents only a little of the atmosphere's mass. It contributes more than water vapor but less than CO_2. It is also important to note that most clouds don't rain. Although they produce all precipitation, clouds are far more common than precipitation. We will discuss the ways that clouds generate rain and other forms of precipitation in Unit 11.

If we were to collect air samples, particularly near a city, they would likely contain a great number of particulates other than water and ice. These particulates are known as **aerosols**. Most are much smaller than cloud particles. Aerosol sizes vary by a factor of 10,000, from a billionth of a meter up to 10 μm. The larger sizes are comparable in size to cloud drops, but they are not water. Common aerosols are smoke, dust, bacteria, plant spores, and salt crystals formed by evaporation above breaking ocean waves. Aerosols are also produced in the atmosphere by the condensation of gases other than water vapor, such as sulfur dioxide emitted by volcanoes.

Typical rural air might contain about four particles per cubic millimeter. City parks often have four times that density. A business district in a metropolitan area might have 200 particles per cubic millimeter, and an industrial zone more than 4,000. We should not get the wrong impression from the high number of aerosol particles in the atmosphere. On average, only a few tenths of a gram (0.01 oz) of aerosol are suspended above each square meter (10.7 ft²) of the surface. This is 100 million times less than the mass of the gaseous atmosphere. The great majority of aerosols are very tiny, less than 0.1 μm. Their small size means that most of the mass of aerosols is found in larger-sized particles. All aerosols are too small to fall rapidly out of the atmosphere (see Focus on the Science: Why Don't Particulates Fall?), and they rely on other mechanisms for removal.

Collectively, aerosols play an active role in the atmosphere. Many of them help in the development of clouds and raindrops (see Unit 11). They are responsible for haze and can affect the color of the sky. Air and the smallest impurities scatter more blue light from the Sun than any other color. This is why the fair-weather sky looks blue. But when low-angle sunlight travels a longer distance through the atmosphere to the surface, as at sunrise or sundown, most of the blue light is scattered. We see only the remaining yellow and red light, which produces colorful sunrises and sunsets. Occasionally, when an abnormally large amount of impurities is in the atmosphere, such as after a major volcanic eruption, this process is carried to some spectacular extremes (see Fig. 6.4).

The Layered Structure of the Atmosphere

Recall that the atmosphere consists of two broad regions: a lower homosphere and an upper heterosphere. A more detailed picture of the structure of the atmosphere emerges when we subdivide it into a number of vertical layers according to temperature characteristics. Altitude has a major influence on temperature, and the overall variation of atmospheric temperature with height above the surface is shown in Figure 6.10.

The lowest 15 km (9 mi) of the atmosphere, where temperature usually decreases with an increase in altitude, is the **troposphere**. The rate of temperature change with altitude is the **lapse rate**, and in the troposphere, the average lapse rate is 6.5°C/1,000 m (3.5°F/1,000 ft). Note that a positive lapse rate denotes *decreasing* temperature with height. The upper boundary of the troposphere, where temperatures stop decreasing with height, is the **tropopause**.

Beyond this discontinuity, in the layer called the **stratosphere**, temperatures initially stay the same and then start

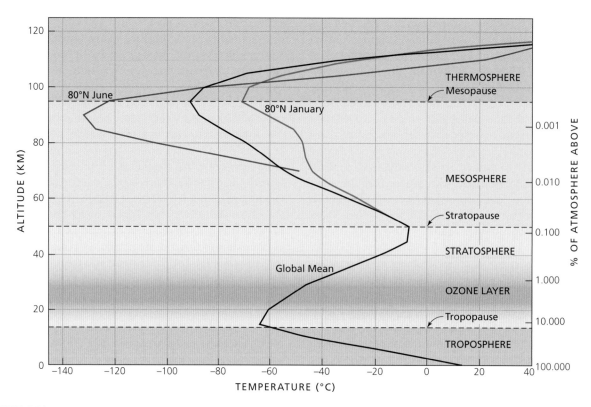

FIGURE 6.10 Variation of temperature with height. The black curve is the annual average for the entire globe. The other curves are typical of high-latitude locations in summer and winter. The right axis shows the amount of atmosphere above any altitude. For example, only 1 percent or so of the atmosphere lies above 30 km (18.5 mi).

increasing with altitude. Layers where the temperature increases with altitude exhibit negative lapse rates. These are called **temperature inversions** because they are the opposite, or inverse, of what we on the surface believe to be the normal state of temperature change with elevation—a decrease with height.

As the top of the stratosphere is approached, beyond about 50 km (32 mi) above Earth, temperatures remain constant with increasing altitude. This boundary zone is the **stratopause**. The stratopause is topped by a layer known as the **mesosphere**. In the mesosphere, temperatures again fall with height, as they do in the troposphere. Eventually, the decline in temperature stops at a boundary called the **mesopause**. This occurs at about 95 km (60 mi) above Earth's surface. Not far beyond the mesopause, temperatures once more increase with height in a layer called the **thermosphere**.

THE TROPOSPHERE

Because the troposphere is the atmospheric zone in which we live, and the layer where almost all weather occurs, we need to know quite a bit about it. We undertake a survey of its processes in Units 7 through 9, with Unit 7 focusing on temperature relationships. In anticipation of that detailed treatment, here we summarize the most significant interactions across the tropopause and the nature of the layers that lie above it.

The tropopause, as Figure 6.10 indicates, is at an average height of about 15 km (9 mi). This altitude varies with

latitude: it is lowest over the poles (about 9 km [5 mi]) and highest above the equator (about 17 km [10 mi]). The thicker tropical troposphere provides for a large expanse of declining temperature. Because of this, the coldest tropospheric temperatures are found above the sun-bathed tropics! There are usually two distinct breaks in the tropopause, which are characterized by areas of variable lapse rates. These breaks are usually at latitudes of about 25 degrees and 50 degrees N and S. The breaks, associated with fast-flowing winds in the upper atmosphere, are important because through them the troposphere and the stratosphere exchange materials and energy. Small amounts of water vapor may find their way up into the stratosphere at these breaks, and ozone-rich air may be carried downward into the troposphere through the breaks.

THE STRATOSPHERE

Above the tropopause is the thinner, clear air of the stratosphere. Jet aircraft often fly through the lower stratosphere because it provides the most favorable flying conditions. The nearly total absence of water vapor in this layer prevents the formation of clouds, providing pilots with fine visibility. And because temperature inversions limit uplift, atmospheric motion is mainly via horizontal winds, ensuring smoother flights than are possible in the troposphere. In the lower stratosphere, wind speeds are typically much less than those of the upper troposphere, but they increase with increasing altitude and can become considerably higher than the speeds we observe routinely in the

FIGURE 6.11 *Aurora borealis* ("northern lights") above North Atlantic Ocean in 2014. Such curtains of light result from the collision of ionized particles with gas molecules in Earth's atmosphere in the higher latitudes.

troposphere. The **ozone layer** (ozonosphere) lies within the stratosphere (see Fig. 6.10). Absorption of ultraviolet radiation by ozone heats the stratosphere, resulting in the negative temperature lapse rate we noted earlier and temperatures that approach those of Earth's surface in its upper reaches. The upper stratosphere is warm enough to contain significant water vapor, but little is found there or anywhere else in the stratosphere. It might seem odd that mixing processes carry CO_2 from the surface into the stratosphere and beyond but do not carry water vapor. The reason is related to the lower temperatures of the middle and upper troposphere. Water vapor condenses to liquid at low temperatures and therefore is depleted from any upward-moving currents. In effect, the cold upper troposphere acts as a lid for water vapor; hence, little moves to higher altitudes. With no significant moisture input, the entire atmosphere above the tropopause—including the stratosphere—is extremely dry.

THE MESOSPHERE

Above the stratosphere, in the altitudinal zone between about 50 and 95 km (31 and 60 mi), lies the layer of decreasing temperatures called the mesosphere. The coldest average temperatures on Earth are found in the upper mesosphere, where the annual global mean is about −90°C (−130°F). Interestingly, temperatures there are higher in winter; as Figure 6.10 shows, polar areas can be 75°C (135°F) warmer! This is doubly surprising considering the complete absence of solar radiation in winter months. The paradox is explained by persistent slow upwelling in summer and downward motion in winter. As we will see in Unit 7, temperature changes accompany such motion, and in this case, they produce air temperature contrary to what we expect when we consider only radiation. Over high latitudes in summer, the cold mesosphere at night sometimes displays high, wispy clouds, which are presumed to be sunlight reflected from meteoric dust particles coated with ice crystals. Another common phenomenon in this layer occurs when bits of

cosmic dust enter the mesosphere and burn up to produce a streak across the night sky.

In the mesosphere and beyond, sunlight strips atoms and molecules of electrons, thereby creating electrically charged particles called *ions* in a process known as **ionization**. Ionized particles concentrate in various layers that both absorb and reflect radio waves sent from Earth's surface. This can affect radio communication. For example, it can allow distant radio stations to be picked up at night. More important, because ionization occurs high in the atmosphere, dangerous ionizing radiation is effectively blocked from reaching the surface. This is another example of the atmosphere serving as a shield for the biosphere.

THE THERMOSPHERE

The thermosphere is found above 95 km (60 mi) and continues to the edge of space,[1] about 60,000 km (37,000 mi) above the surface. The temperature rises spectacularly in this layer and likely reaches 900°C (1,650°F) at 350 km (220 mi). However, air molecules are sparse at these altitudes, traveling perhaps a kilometer before colliding with one another. That means that although individual molecules are moving fast and therefore have a high temperature, there are few molecules from which to extract heat. Therefore, the high temperatures of the thermosphere do not have the same kind of environmental significance they would have in the vicinity of Earth's surface.

Ionization also takes place in the thermosphere. Intermittently, ionized particles penetrate the thermosphere, creating vivid sheet-like displays of light called the *aurora borealis* in the Northern Hemisphere and the *aurora australis* in the Southern Hemisphere (Fig. 6.11). In the upper thermosphere, further concentrations of

---

1 Where the atmosphere ends and space begins is arbitrary. For example, international law places it at 100 km for the purposes of space treaties. There is no firm boundary because the atmosphere simply becomes thinner and thinner with increasing distance.

Ozone Depletion and Recovery

It has long been known that stratospheric ozone amounts are governed by a delicate balance between processes that create O_3 and removal processes that allow O_3 to join with atomic oxygen (O) in the recombination reaction $O_3 + O \rightarrow 2O_2$. In the 1950s, scientists realized certain natural chemicals play a catalytic role in recombination; that is, they boost ozone destruction without being consumed in the process. The most important examples are hydroxyl (OH) and nitric oxide (NO). Because of their role in the destruction of O_3 molecules, our protective ozone layer is only about half as thick as it would be otherwise.

Beginning in the 1970s, several human threats to the ozone layer emerged, including anthropogenic chlorine (Cl) and bromine (Br). Chlorine and bromine were widely used for a variety of purposes, such as coolants in refrigerators and air-conditioning systems, propellants in aerosol sprays, cleaning solvents for computer components, and plastic foam. Although essentially inert in the lower atmosphere, when bombarded by ultraviolet radiation in the stratosphere, these coolants and solvents and other products break down and release Cl and Br. Both Cl and Br are catalytic ozone destroyers, and after the alarm was raised in the mid-1970s, subsequent measurements confirmed significant ozone losses.

The result was a 1987 conference in Montreal, Canada, where more than 30 countries took the first steps to limit their release of ozone-destroying gases, collectively known as **chlorofluorocarbons (CFCs)**. The Montreal Protocol established a scientific assessment panel. The panel's recommendations were followed as part of an ongoing process to counteract the depletion of atmospheric ozone. Multiple agreements in following years involving 132 countries (containing almost 85 percent of the world's population) had positive effects in combating ozone loss, as exemplified by Figure 6.12.

However, the long residence time of CFCs meant that global ozone would respond only slowly. For example, in the northern middle latitudes, ozone declined by about 7 percent per decade from 1979 to 1997. In subsequent years, ozone has leveled off or has perhaps increased slightly.

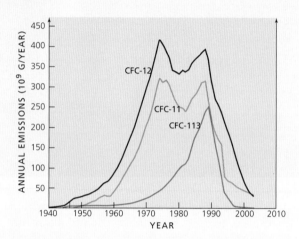

FIGURE 6.12 Global release to the atmosphere of three widely used refrigerants.

It is interesting to wonder what might have happened without the Montreal and subsequent related agreements. To answer this question, NASA scientists built a computer program to model ozone amounts for the future under various assumptions about CFC release rates. Figure 6.13 shows results for models both with and without CFC release limits in place.

The lower curve shows a two-thirds reduction of total ozone by midcentury. This curve—a global ecological calamity—amounts to a future avoided because of steps taken in the last few decades. By any measure, this part of the ozone story is a triumph for humanity and illustrates what is possible through international cooperation. However, this problem was "easy" to fix in a number of ways. The cause was clearly identified, and technical fixes were available through replacements of CFCs with other compounds that don't destroy ozone. In addition, because patents on CFCs had run out, individual producers had no incentive to argue for continued use of "their" product, and in fact, there was an active market in developing replacements. Other ecological challenges on the horizon, especially global warming, are much more difficult to address, and it remains to be seen if similar successes can be achieved.

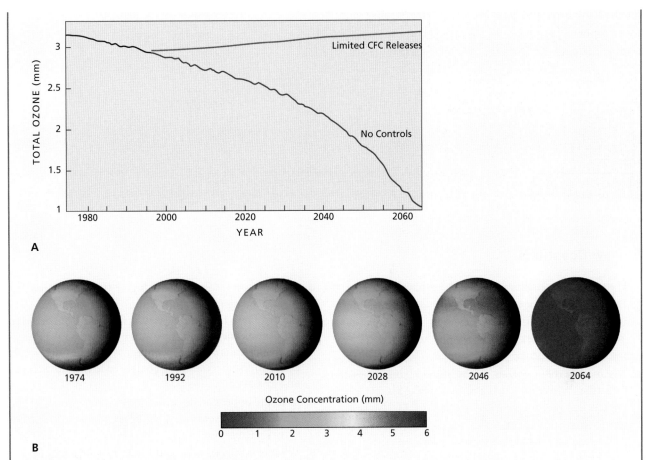

FIGURE 6.13 (A) Global ozone predicted under two CFC release scenarios—the blue line illustrates ozone rates with limited CFC releases, and the red line illustrates what would have happened if limits had not been placed on CFC emissions. (B) Ozone patterns for various years without CFC limits in place.

ions comprise the *Van Allen radiation belts*. This outermost layer of the thermosphere is sometimes referred to as the *magnetosphere* because here Earth's magnetic field is frequently more influential in the movement of particles than the gravitational field is. The thermosphere has no definable outer boundary and gradually blends into interplanetary space.

So far, our discussion of the atmosphere has been largely confined to global average conditions or has considered only the vertical dimension. We have not given much consideration to spatial variations or to atmospheric movement. The following two units address those deficiencies, starting with temperature patterns in Unit 7.

Key Terms

aerosol *page 79*

air pollutants *page 79*

carbon dioxide *page 70*

chemical compound *page 69*

chemical reaction *page 69*

chlorofluorocarbons (CFCs) *page 82*

climate *page 69*

constant gases *page 73*

element *page 69*

fossil fuels *page 74*

heterosphere *page 73*

homosphere *page 73*

ionization *page 81*

isotopes *page 70*

lapse rate *page 79*

mesopause *page 80*

mesosphere *page 80*

methane *page 78*

nitrogen *page 74*

ozone hole *page 77*

ozone layer *page 81*

particulate *page 71*

photochemical reactions *page 69*

residence time *page 70*

stratopause *page 80*

stratosphere *page 79*

Scan Here for a quick vocabulary review

Review Questions

1. What are the constituents of the atmosphere?

2. Discuss some of the ways the atmosphere's composition has changed over time.

3. What is atmospheric residence time? What factors determine the residence time of a gas?

4. Discuss the role of ozone in absorbing incoming solar radiation. Where does this absorption take place?

5. Draw your own versions of Figure 6.3 for (a) water vapor, (b) methane, and (c) hydrogen, showing only the processes that are relevant for each gas.

6. The colors used in Figure 6-8 give the impression that there are large spatial variations in atmospheric CO_2. According to the legend, what is the difference between maximum and minimum values on the map? Express this as a percentage of 400 ppm. What does this imply about atmospheric mixing compared to the rate of input and loss of CO_2?

7. Give the approximate altitudinal extent of each of the atmosphere's layers and describe the temperature structure of each.

8. What is the *ozone hole*, and what are its likely causes?

Scan Here for a quick concept review

www.oup.com/us/mason

UNIT

7

Temperatures of the Lower Atmosphere

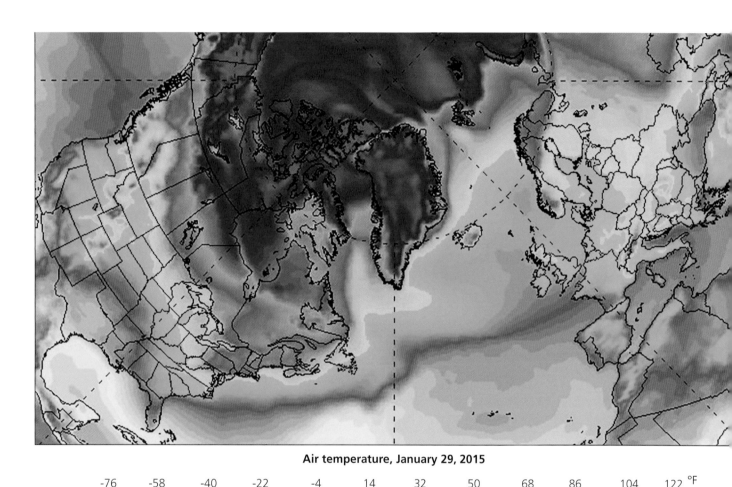

Air temperature, January 29, 2015

| -76 | -58 | -40 | -22 | -4 | 14 | 32 | 50 | 68 | 86 | 104 | 122 °F |

| -60 | -50 | -40 | -30 | -20 | -10 | 0 | 10 | 20 | 30 | 40 | 50 °C |

January of 2014 was bitterly cold for much of the Northern Hemisphere. On this day, arctic air surged south and produced freezing temperatures throughout the eastern half of the United States.

ENHANCE

OBJECTIVES

- Discuss the measurement and characteristics of temperature and heat
- Explain the adiabatic process and how vertically moving air warms and cools
- Describe the global distribution of temperature

On the rather cloudy day of June 26, 1863, British scientist James Glaisher and his assistant Coxwell climbed into the basket of a hot-air balloon in Wolverton, England. This flight, one of their 28 flights between 1862 and 1866 (Fig. 7.1), lasted an hour and a half. They ascended to 7,050 m (23,500 ft) and traveled 80 km (50 mi) before descending at Ely. En route, they encountered rain, snow, and fog. One of the main purposes of this flight was to record the temperatures along the way. In Glaisher's own words, these varied from the "extreme heat of summer" to the "cold of winter." In fact, the temperatures varied from 19°C (66°F) at the ground to –8.3°C (17°F) at 7,050 m. The pair were wary of the hazards of high altitudes: on a flight the previous year, Glaisher had fainted at 8,700 m (28,700 ft) from lack of oxygen. His assistant, arms paralyzed with cold, climbed the rigging of the balloon to release the gas control with his teeth so the craft could descend. Their extreme efforts were rewarded, however. Their flights firmly established that temperature typically decreases with an increase in altitude, at least as far as 8,700 m above the surface.

The findings about vertical temperature changes were an important addition to existing knowledge. Together with what was already known about temperature of the planetary surface, Glaisher's observations reinforced the notion that temperatures could be highly variable in any direction. Before we further examine those vertical and horizontal temperature relationships, we must briefly digress to inquire more explicitly about the concept of temperature.

What Is Temperature?

Imagine an enclosed box containing only molecules of air. The molecules move constantly in all directions in what is called *random motion*. The molecules therefore possess the energy of movement, known as **kinetic energy**. The more kinetic energy molecules possess, the faster they move. The index we use to measure their kinetic energy is **temperature**. Temperature is an abstract term that describes the energy, and therefore the speed of movement, of molecules. In a gas such as air or a liquid such as water, the molecules change their location when they move. In a solid, such as ice, the molecules only vibrate in place. Nonetheless, the speeds of all these vibrations are described by their respective temperatures.

Because they are so small, it is almost impossible to examine individual molecules, so we usually use an indirect method to measure temperature. Changes of temperature make gases, liquids, and solids expand and contract. Therefore, temperature is most commonly measured by observing the expansion and contraction of mercury in a glass tube. Such an instrument is called a **thermometer**; you are probably familiar with medical and weather thermometers.

A thermometer is calibrated according to one of three scales. The scale used throughout most of the world is the **Celsius scale** (formerly called the centigrade scale), the metric measurement of temperature we use in this book. On this scale, the *boiling point* of water is 100°C and its *freezing point* is 0°C. On the **Fahrenheit scale**, now used only in the United States, water boils at 212°F and freezes at 32°F. Scientists also employ an *absolute scale*—the **Kelvin scale**—which is based on the concept of *absolute zero*, a temperature at which there is no molecular energy and molecular motion virtually ceases completely. The kelvin is identical in size to the Celsius degree; however, on the Kelvin scale water freezes at 273 K and boils at 373 K. We do not mention kelvins again in this book. Instead we will continue to use Celsius temperatures accompanied by their Fahrenheit equivalents in parentheses.

It is important to distinguish temperature from heat. Temperature merely measures the kinetic energy of molecules. Temperature does not reflect the number of molecules in a substance or its *density* (the amount of mass per unit of volume). In contrast to temperature, the heat of an object depends on its volume, its temperature, and its capacity to hold heat. A spoonful of soup, which has a high heat capacity, might burn your tongue at the same temperature at which you could comfortably bite into a piece

FIGURE 7.1 James Glaisher and his assistant, Coxwell, during a balloon flight on September 5, 1862.

of just-baked bread. Because it contains many more molecules, a large lake with a water temperature of 10°C (50°F) contains much more heat than a cup of hot coffee at 70°C (158°F). Now that we know something about temperature, let us return to discussing the temperatures of the atmosphere.

Temperature in the Troposphere

Glaisher's observations about the lower part of the atmosphere are correct: temperature typically does decline with an increase in altitude. However, this is only true up to a point. Subsequent unmanned balloon observations showed that above about 12 km (40,000 ft), the temperature stops decreasing with height and begins to increase. Few people believed this at first, and only after several hundred balloon ascents was it finally accepted. Later ascents during the twentieth century to still-higher altitudes revealed the even more complex temperature patterns we described in Unit 6 (see Fig. 6.10). We focus here on the lowest atmospheric layer, the troposphere. Within the troposphere, we are particularly interested in temperatures very near the surface. When people say "air temperature," they typically mean a temperature measured a meter or two (3–6 ft) above the surface. Unless indicated differently, we will follow that convention.

TROPOSPHERIC TEMPERATURE AND AIR STABILITY

The troposphere is the layer of the atmosphere we live in, and it is here that the weather events and climates affecting humans occur. As illustrated in Figure 6.10, the troposphere's temperature typically decreases with increasing altitude until the tropopause is reached. Another distinctive feature of the troposphere is the possibility and frequency of vertical, as well as horizontal, movement of air. The rate of change of temperature with height—the *lapse rate*—can be a determining factor in vertical motion. As we noted in Unit 6, the average tropospheric lapse rate is 6.5°C/1,000 m (3.5°F/1,000 ft) of elevation, but large variations exist from place to place and day to day.

The lapse rate at any particular time or location is the **environmental lapse rate (ELR)**. The ELR governs the vertical temperature profile of the atmosphere. The ELR describes temperature changes in a column of stationary air. It is useful to imagine small boxes of air (called air parcels) that are able to move up and down in the air column. The ELR gives information about the environment through which such parcels move.

The ELR determines the **stability** of the air, a concept illustrated in Figure 7.2 by a wedge of wood. When the wedge is resting on its side (A), a small push at the top may move it horizontally, but its vertical position remains the same. It is therefore *stable*. When the wood rests on its curved base (B), a similar push might rock it, but it will still

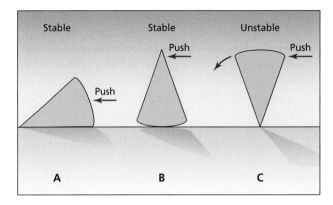

FIGURE 7.2 The concept of stability. The block of wood, like parcels of air, is considered stable as long as it returns to its original position after a small push.

return to its original position. It is still stable. However, if we balance the wedge of wood on its pointed edge (C), a small push at the top knocks it over. It does not return to its original position; it is *unstable*.

We use the same terminology to refer to the vertical movement of a small parcel of air. If it returns to its original position after receiving some upward force, we say it is stable. But if it keeps moving upward after receiving the force, then we say it is unstable. To understand air stability, it is important to first understand that air is a poor conductor of heat. Therefore, a parcel of air of one temperature that is surrounded by a mass of air at another temperature does not gain or lose heat energy in a short period of time. Air parcels are essentially isolated from the surrounding column, and this is true whether they are rising or sinking. The term **adiabatic** is used for any physical process where no heat is added or removed. Thus the rising or sinking of an air parcel is referred to as an adiabatic process.

ADIABATIC LAPSE RATES

If you have ever used a bicycle pump, you may have noticed that when the air is compressed at the bottom of the pump, the air's temperature rises and the bottom of the pump becomes hot. The opposite occurs when the volume of a given mass of air is forced to expand: the temperature of the air decreases. For example, if you let air out of a ball or bicycle tire, the air rushing out feels cool. Its temperature falls because of expansion. A similar thing happens in the atmosphere: when a parcel of air rises to a higher altitude, it expands and cools. This is an adiabatic process because that air parcel neither gains heat from nor loses heat to its surroundings. Such air-parcel lapse rates in the troposphere are called **adiabatic lapse rates**.

Adiabatic warming and cooling are a consequence of *energy conservation*, which we introduced in Unit 1. When an air parcel descends, it gains energy because work is done in compressing the parcel, just as work is done in compressing the air in the bicycle pump. If no heat is lost to the surroundings, the temperature of the parcel must increase.

Similarly, expansion during ascent represents an energy loss by the parcel, and the temperature falls accordingly.

When an air parcel is not saturated with water vapor, it cools with an increase in altitude at a constant rate of almost exactly 1°C/100 m (5.5°F/1,000 ft). This is the **dry adiabatic lapse rate (DALR)**. The DALR value depends only on the gases present in the atmosphere and the acceleration of gravity. Because these are effectively constant throughout the troposphere, the DALR is constant.

The situation is somewhat different when the air contains water vapor that is changing to water droplets as it cools. Latent heat (a concept introduced in Unit 5) is given off when the state of water changes from a gas to a liquid, a process called **condensation**. The lapse rate when condensation is occurring is less than the DALR. This lapse rate is called the *wet* or **saturated adiabatic lapse rate (SALR)**. Unlike the DALR, the value of the SALR is variable depending on the amount of water condensed and consequently the latent heat released. A typical value for the SALR at 20°C (68°F) is 0.44°C/100 m (2.4°F/1,000 ft), but it can be anything from 0.4°C to 0.9°C/100 m. The larger value corresponds to air with very slow condensation, which produces a rate almost as large as the DALR (1°C/100 m).

LAPSE RATES AND STABILITY

Let's again consider our parcel of air and see what happens to it in two different environments with two different ELRs. In Figure 7.3A, a parcel of air rises from the ground (because it is either warmed by the surface or forced upward mechanically). Because it neither gains heat from nor loses it to the surrounding air mass, it cools at the DALR, decreasing its temperature 1°C for each 100 m of ascent. In this case, the ELR is 0.5°C/100 m, so after rising 100 m, the air parcel has a temperature 0.5°C lower than its surroundings. The colder the air, the denser it is. Therefore, the air

parcel is now denser and heavier than the surrounding air and would fall back to Earth assuming the initial lifting force ceased. If mechanical lifting carried it to 300 m, it would be 1.5°C colder than its surroundings. Again, it would be denser than its surroundings and would tend to sink. We say that the air in that environment is *stable*, meaning *it resists vertical displacement*. In contrast, in Figure 7.3B the ELR is 1.5°C/100 m. Under these conditions, an air parcel rising and cooling at the DALR would be warmer than its surroundings. The warmer the air, the less dense it is, so the air parcel is lighter and less dense than the surrounding air and continues to rise. We describe the air in this environment as *unstable* because any upward movement results in continued uplift rather than a return to its original position.

What if the air parcel is saturated and cools more slowly according to the SALR? Consider first the unstable air column in Figure 7.3B. The rising air parcel was warmer than the environment without condensation, and it is even warmer with condensation. The unstable air column becomes even more unstable. That is, condensation adds to the instability, or equivalently, condensation decreases atmospheric stability. Condensation also decreases the stability of the other air column, which has an ELR of only 0.5°C/100 m. If the SALR is small enough—that is, less than the ELR—rising parcels will be warmer than their surroundings and thus unstable. Conversely, if the SALR is larger than the ELR, rising parcels will be colder than their surroundings. In this latter case, the column is stable, but less so than it was without condensation.

Putting all this together, we know that the atmosphere is unstable if the ELR is greater than the DALR. If the ELR is less than the SALR, it is stable. If the ELR is between the DALR and the SALR, it is unstable if condensation occurs and stable otherwise. In this final case, the atmosphere is said to be *conditionally unstable*.

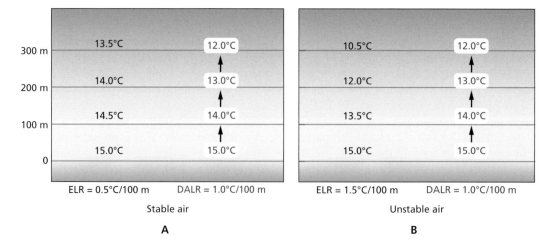

FIGURE 7.3 Environmental lapse rate conditions for a mass of stationary air (shown in blue) that surrounds an individual parcel of air (shown in white) rising through it. When the ELR is less than the DALR, an air parcel is colder and denser than surrounding air and therefore stable (A). When the ELR is greater than the DALR, an air parcel is warmer and less dense than surrounding air and therefore unstable (B).

FIGURE 7.4 Looking southward at the Florida peninsula and surrounding waters from NASA's ill-fated space shuttle *Columbia* as it passed over the Jacksonville area. In this classic summertime view, the land area of southern Florida is almost perfectly defined by its cloak of puffy clouds, the signature of atmospheric instability associated with rising hot air.

We can often tell whether or not a portion of the atmosphere is stable by looking at it. A stable atmosphere is marked by clear skies or by flat, layer-like clouds. An unstable atmosphere is typified by puffy, vertical clouds that sometimes develop to great heights. The photograph of Florida, taken from a spacecraft, in Figure 7.4 illustrates these two conditions. Over the Atlantic Ocean to the east and the Gulf of Mexico to the west, the ELR is less than the DALR, so parcels of air remain near the sea surface and the sky is clear over these bodies of water. But the higher temperatures of the land surface in daytime make the ELR higher than the DALR. Due to the resulting instability, parcels of hot air rise in the atmosphere, forming the puffy clouds as they cool.

TEMPERATURE INVERSIONS AND AIR POLLUTION

In discussing the stability of the troposphere, we are really considering the possibility—and vigor—of the vertical mixing of air within it. This has practical implications, the most important of which is how pollutants disperse when released into the atmosphere. The initial vertical (and horizontal) distribution of pollutants depends on the location of their sources. Any further spread of air pollution is associated with two main factors: (1) the stability of the air and its propensity to allow vertical mixing, and (2) how well air stability combines with the flushing effect of horizontal winds. Both are related to the temperature structure of the lower troposphere, particularly the influence of temperature inversions.

Note that, on average, conditional instability prevails—that is, the ELR is between the SALR and the DALR. Extreme stability is not the rule, and most of the time, air parcels are able to rise without difficulty (Fig. 7.5A). Hence, any pollutants in the surface layer of air disperse vertically with upward movement. At times, however, this vertical cleansing mechanism does not operate because the usual negative lapse rate is replaced by a positive one. Such an increase in temperature with height is defined in Unit 6 as a **temperature inversion** because it inverts what we, on the surface, find to be the "normal" behavior of temperature change with altitude.

Because of the nightly cooling of Earth's surface and the atmosphere near the ground, it is common for a temperature inversion—warm air lying above cold air—to develop early in the morning over both city and countryside. These inversions, which form an atmospheric "lid," can be broken down by rapid heating of the surface or by windy conditions. If these conditions do not occur, however, air pollution is trapped and intensifies beneath the inversion layer (Fig. 7.5B). This is especially true in certain urban areas, where *dust domes* frequently build up (see Perspectives on the Human Environment: Urban Dust Domes and Heating Patterns).

The subsidence (vertical downflow) of air from higher in the troposphere can also contribute to the trapping of

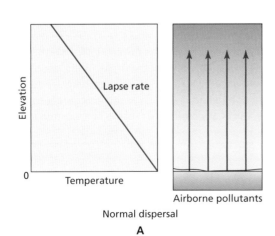

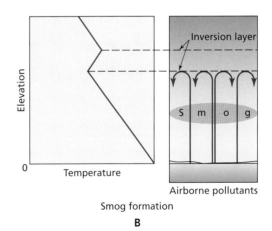

FIGURE 7.5 Effect of a temperature inversion on the vertical dispersal of atmospheric pollutants. (A) Normal dispersal; (B) smog formation.

Urban Dust Domes and Heating Patterns

A giant dome of dust and gaseous pollution buries Chicago! It's not the plot of a new horror movie. Many metropolitan areas lie beneath a **dust dome**, whose brownish haze stands out against the blue sky (as you can see in the photo on the left in Fig. 7.7). Before Chicago spawned rings of automobile suburbs, its dust dome was sharply defined. A half-century ago, the wall of dust and other pollutants began near Midway Airport, about 12 km (7.5 mi) from the city's downtown Loop. In the nearby countryside, visibility might be 25 km (15 mi), but within the dome, it was only 0.5 to 0.75 km (0.3 to 0.5 mi). A thick layer of polluted air also enshrouds Los Angeles and many other large cities. Why don't these pollutants simply blow away?

Studies of the movement of dust and gaseous pollutants over cities show that the heat generated in many urban areas forms a local circulation cell. Air currents capture the dust and mold it into a dome (Fig. 7.6). Dust and pollution particles rise in air currents around the center of the city, where the temperature is warmest. As they move upward, the air cools and diverges. The particles gradually drift toward the edges of the city and settle downward. Near the ground, they are drawn into the center of the city to complete the circular motion. Frequent temperature inversions above the city prevent escape upward, and the particles tend to remain trapped in this continuous cycle of air movement.

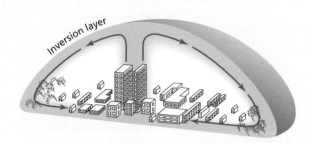

FIGURE 7.6 Air circulation within an urban dust dome.

All parts of the radiation balance discussed in Unit 5 are altered in the urban environment. The interception of incoming shortwave radiation is most apparent during the winter, when the pollution is worse and the Sun's rays are striking at a lower angle. In London, 8.5 percent of the direct solar radiation is lost when the Sun's elevation is 30 degrees above the horizon, but 12.8 percent is lost with a solar elevation of 14 degrees. On very cloudy days, 90 percent of this radiation can be lost. Consequently, the number of hours with bright sunshine is reduced. Researchers have discovered that in communities that lie outside London, it is sunny an average of 4.3 hours per day; in the center of the city, however, their findings reveal that it is sunny only 3.6 hours per day.

The amount of solar radiation absorbed by a city's surface also depends on the albedo (defined in Unit 5) of that surface. This, in turn, depends on the actual materials used in construction and varies from city to city. British cities are usually made of dark materials or materials that have been blackened by smoke. They have a lower albedo (17 percent) than agricultural land does (approximately 23 percent), and therefore they absorb more radiation. In contrast, the central areas of Los Angeles have lighter colors and therefore exhibit a higher albedo than the more vegetated residential zones.

Scientists know less about the behavior of longwave radiation in cities, but many studies show that both upward and downward longwave radiation increase in urban areas. This increased radiation can offset the decreased shortwave radiation. The result is a net radiation that is not too different from that in surrounding rural areas. Of the net radiation arriving at the city surface, some studies suggest that about 80 percent of it is lost as sensible heat warming the city air, and that the rest acts mainly as ground heat flow to warm the materials constituting the urban landscape. Very little heat appears to be involved in evaporative cooling, although that may change as rooftop gardens and other forms of city plantings become more common.

pollutants, as residents of metropolitan Los Angeles know only too well (Fig. 7.7). This air flow creates a temperature inversion that is reinforced by onshore surface winds blowing across the cold waters of the adjacent Pacific Ocean. Moreover, the cooling effect is heightened by the nighttime drainage of cold air into the Los Angeles

Basin from the mountains that form its inland perimeter. These air movements often produce and sustain temperature inversions at an altitude of approximately 1,000 m (3,300 ft), resulting in poor-quality surface-level air popularly known as *smog* (a contraction of "smoke" and "fog").

FIGURE 7.7 Central Los Angeles, with and without its shroud of smog. The Los Angeles Basin is particularly susceptible to the development of temperature inversions—and high levels of surface air pollution. Over the past decade the air quality above L.A. has improved, and contrasts like this, for which the city had developed a reputation, are now rarer.

Horizontal flushing by winds can help to relieve air pollution. We can estimate air pollution potential by calculating the vertical range of well-mixed pollutants and the average wind speed through the mixing layer. Figure 7.8 shows the average number of days per year in the United States when inversions and light wind conditions create a high air pollution potential. On average, ventilation conditions are best in the Northeast, the Midwest, and the Great Plains and poorest along the Appalachian corridor and in most of the Far West. Air pollution potential alone might be regarded as a matter of climatological luck, but the distribution of pollution sources also determines the frequency and level of air pollution. Important, too, is the role of the local topography, which does not show up at the scale of the map in Figure 7.8.

Nearby mountains can strongly counteract the flushing effect of horizontal winds. Los Angeles and Denver—two U.S. metropolitan areas that have ranked among the worst in air pollution—are classic examples. And when cities are situated in valleys near mountain ranges, wind flushing may be even further diminished. Mexico City, arguably the world's most polluted (as well as among the most populated) metropolis, not only experiences this phenomenon but also suffers because it is at a high elevation (2,240 m [7,350 ft]), where a given volume of air already contains 30 percent less oxygen than at sea level, resulting in a greater concentration of pollutants.

The Horizontal Distribution of Temperature

The spatial distribution of temperatures across the landmasses, oceans, and icecaps that constitute Earth's surface depend on a number of factors. Certainly, incoming solar radiation is one such factor. The amount of solar radiation received depends on the length of daylight and the angle of

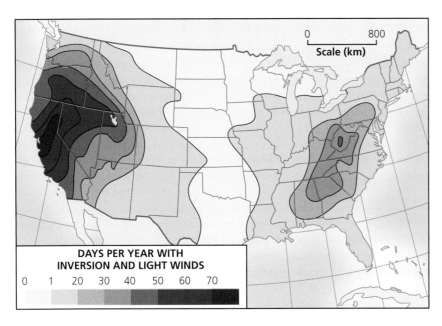

DAYS PER YEAR WITH INVERSION AND LIGHT WINDS

0 1 20 30 40 50 60 70

FIGURE 7.8 Air pollution potential in the contiguous United States, in terms of the number of days per year with light winds and inversion conditions.

the Sun's rays, which are both a function of the latitudinal position. A second major factor is the nature of the surface. Land heats and cools much more rapidly than water, and this strongly affects not only air temperatures directly above each type of surface but also adjacent areas influenced by them.

A number of less significant factors can be locally important as well. As tropospheric temperatures usually decrease with an increase in altitude, places at higher elevations tend to experience temperatures lower than those places closer to sea level. Another moderator of surface temperatures is cloudiness, and places with more extensive cloud cover generally experience lower daytime high temperatures than similar places with clearer skies. (This is also true in areas where pollution haze plays a role identical to that of clouds.) And at coastal locations and islands, air temperature is influenced to a surprising extent by the warmth or coolness of local ocean currents.

DAILY AND YEARLY CYCLES

The pattern of temperature change during a day is the **diurnal cycle**. Shortly after dawn, radiation from the Sun begins to exceed the radiant loss from Earth's surface. Earth begins to heat the air, so the air temperature rises. Temperature continues to rise as the net radiation rises. But the heating of the ground and the flow of sensible heat take some time to develop fully. Thus, maximum air temperatures usually do not occur simultaneously with the maximum net radiation peaks at solar noon, but hours later. In late afternoon, net radiation and sensible heat flow decline markedly, and temperatures begin to fall. After the Sun sets, more radiation leaves Earth than arrives at the surface, which produces a negative net radiation. The surface and the air above the surface enter a cooling period that lasts all through the night. Temperatures are lowest near dawn; at sunrise, the diurnal cycle starts again.

The pattern of temperature change during a year is the **annual cycle** of temperature. In the middle and high latitudes the annual cycle is rather similar to the diurnal cycle, with the solstices replacing sunrise and sunset. In the spring, net radiation becomes positive, and air temperatures begin to rise. The highest average temperatures do not occur at the time of the greatest insolation—the summer solstice—but usually about a month later. In autumn, decreasing solar radiation leads to progressively lower temperatures. The lowest winter temperatures occur toward the end of the period of lowest (and often negative) net radiation and when the ground has lost most of the heat it gained during summer. In spring the solar radiation once more begins its cyclical increase.

LAND/WATER HEATING DIFFERENCES

The time it takes to heat the surface at any particular location determines when the highest air temperatures occur. The difference between land and ocean is particularly noteworthy, with ocean requiring far more time to heat up than land assuming the same rate of energy input.

Several factors explain the sluggish nature of ocean temperature changes. For one thing, compared to land, radiation can penetrate the surface layer of water to a far greater depth. Rather than absorption and rapid heating of a thin layer, the energy is spread throughout a great depth. Considerable vertical mixing—driven by waves, currents, and other water movements—also constantly takes place between newly warmed (or cooled) surface water and cooler (or warmer) layers below. Moreover, as mentioned earlier, water has a high heat capacity. For example, a kilogram of water would need about three times the heat input to produce the same temperature rise as a kilogram of soil. A final reason ocean changes temperature slowly is evaporation. In the case of an ocean surface, much incoming energy is consumed in converting water from liquid to gas rather than raising the surface temperature. This is why the daily temperature range for ocean surfaces is almost always less than 1°C (1.8°F).

Not surprisingly, the ocean surface also exhibits a decidedly smaller annual temperature range compared to land (Fig. 7.9A). In the tropics, this variation averages just 1° to 4°C (2° to 7°F). Even the upper-middle latitudes record only a modest 5° to 8°C (9° to 15°F) swing between seasonal extremes. In addition, the lag between incoming solar radiation and temperature is typically larger for oceans than for continents. So, for example, the highest sea surface temperatures in the Northern Hemisphere occur in late August to mid-September and in the Southern Hemisphere in late February to early March (Fig. 7.9B).

As a consequence, the air above an ocean remains cooler in summer and warmer in winter than does the air over a land surface at the same latitude. This is reflected in Figure 7.10, which shows the annual range of temperatures for each hemisphere. Note that the Southern Hemisphere, which is only about 20 percent land, consistently exhibits smaller yearly temperature ranges than the Northern Hemisphere, where the surface is approximately 40 percent land.

In places where oceanic air is transported onto the continents, air temperatures are ameliorated accordingly; that is, they do not become extremely hot or cold. As the distance from the coast increases, however, this moderating effect diminishes (and it terminates more abruptly if high mountain ranges parallel to the shore block the inland movement of oceanic air). This is illustrated in Figure 7.11, which graphs the annual temperature regimes for San Francisco, California, and Omaha, Nebraska. These two cities are located at approximately the same latitude (40°N). Interior Omaha, in the heart of the North American continent, experiences both a hotter summer and a much colder winter, whereas San Francisco, perched on the Pacific Coast, enjoys a temperature regime that is free of extremes in both summer and winter.

The moderating influence of the ocean on air temperatures is called the **maritime effect** on climate. In the opposite case, far inland, where the ocean has a minimal ameliorating influence on air temperatures, there is a

ANNUAL TEMPERATURE RANGE

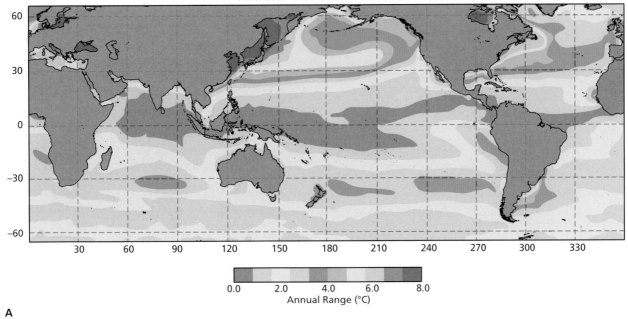

A

DATE OF MAXIMUM TEMPERATURE

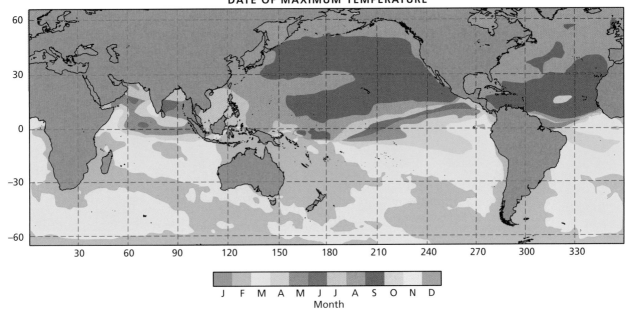

B

FIGURE 7.9 The annual range in sea surface temperature (A) and the month when the maximum temperature occurs (B).

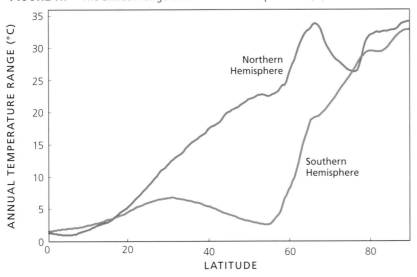

FIGURE 7.10 Average annual temperature range in degrees Celsius. The graph shows the difference between the warmest and coldest month for each latitude.

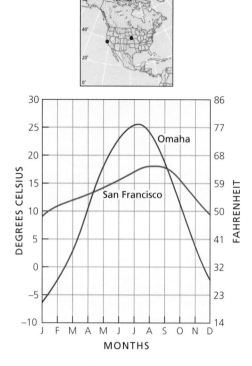

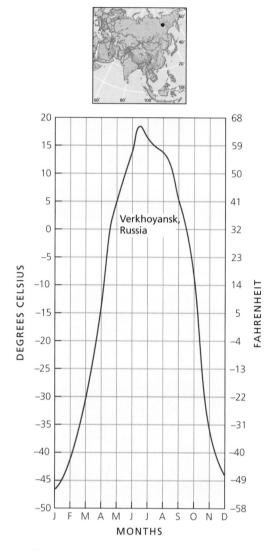

FIGURE 7.11 Annual temperature regimes in Omaha, Nebraska (41°N, 96°W), and San Francisco, California (38°N, 122°W). Note the contrast between interior continental and coastal locations.

FIGURE 7.12 Annual temperature regime in Verkhoyansk (68°N, 133°E) in Russia's far northeastern Siberia, demonstrating the extremes of continentality.

continental effect. This property of **continentality** is evident in the case of Omaha (Fig. 7.11), but the most dramatic examples are found deep inside Eurasia, in the heart of the world's largest landmass. The Russian town of Verkhoyansk, located in far northeastern Siberia, is well known to climatologists for this reason. Verkhoyansk's annual temperature regime is plotted in Figure 7.12.

Sometimes, air from outside an area has more influence on air temperatures than do local radiation and heat flows. For instance, there might be quite a large amount of net radiation at midday during a Nebraska winter, but the air temperatures may still be very low. This is because the overlying air may have come from the Arctic, thousands of kilometers to the north, where a completely different heat balance prevailed. Although the air temperature is a function of the amount of heat that makes up the local heat balance, it can also be affected by the **advection** (horizontal transport through the atmosphere via wind) of air from a region exhibiting a different heat energy balance. The results of different heat balances and the large-scale advection of air is evident in the worldwide distribution of surface air temperatures. Notice in Figure 7.9A that the annual ranges of temperature off the east coasts of Russia and Canada are surprisingly large compared to other ocean locations. This is explained by the persistent winds that blow from adjacent land areas to the west that give the ocean margin in these areas a distinctly continental flavor.

GLOBAL TEMPERATURE VARIATIONS

Figure 7.13 maps the global distribution of air temperatures. Incoming solar radiation, which is largely determined by the angle of the Sun's rays striking Earth and the length of daylight, can change significantly in most places over a period of weeks. As a result, physical geographers have always faced a problem in trying to capture the dynamic patterns of global temperatures on a map. For our purposes, the worldwide shifting of air temperatures, or their *seasonal march*, is best visualized by comparing the patterns of the two extreme months of the year—January and July—which immediately follow the solstices. The cartographic technique of isarithmic mapping (defined in Unit 3) is employed in Figure 7.13. In the figure, January and July distributions feature **isotherms**—lines connecting all points having the same temperature.

Both maps reveal a series of latitudinal temperature belts that shift toward the "high-Sun" hemisphere (Northern

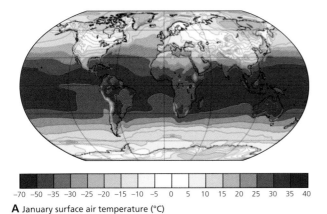

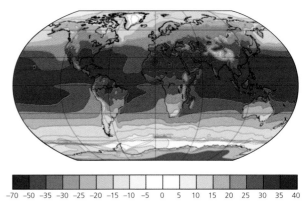

A January surface air temperature (°C)

B July surface air temperature (°C)

FIGURE 7.13 Average air temperatures in degrees Celsius for January (A) and July (B).

ENHANCE

Hemisphere in July, Southern Hemisphere in January). The tropical zone on both sides of the equator experiences the least change between January and July because solar radiation inputs remain high throughout the year (see Fig. 4.12). The middle and upper latitudes, particularly in the hemisphere experiencing winter, exhibit quite a different pattern. Here, the horizontal rate of temperature change over distance—or **temperature gradient**—is much more pronounced, as shown by the "packing" or bunching of the isotherms. Note that the isotherms are much farther apart in the tropical latitudes than they are in the middle and upper latitudes.

Another major feature of Figure 7.13 is the contrast between temperatures overlying land and sea. There is no mistaking the effects of continentality on either map: the Northern Hemisphere landmasses vividly display substantial interior annual temperature ranges. Clearly visible, too, are the moderating influence of the oceans and the maritime effect on the air temperatures over the land surfaces near them. Where warm ocean currents flow, as in the North Atlantic just west and northwest of Europe, the onshore movement of air can decidedly ameliorate winter temperatures; note that northern Britain, close to 60 degrees N, lies on the

5°C (41°F) January isotherm, the same isotherm that passes through North Carolina, which is at about 35 degrees N on the northern edge of the U.S. Sun Belt! In general, we observe a poleward bending of the isotherms over all the oceans, an indicator of their relative warmth with respect to land at the same latitude. The only notable exceptions occur in conjunction with cold ocean currents, such as off the western coasts of Africa and South America south of the equator or off the northwestern coast of Africa north of the equator.

The flow patterns associated with ocean currents and the movement of air above them remind us that the atmosphere and hydrosphere are highly dynamic entities. Both the atmosphere and the hydrosphere contain global-scale circulation systems that are vital to understanding weather and climate. Now that we are familiar with the temperature structure of the atmosphere, we are ready to examine the forces that shape these regularly recurring currents of air and water.

Key Terms

adiabatic *page 87*
adiabatic lapse rate *page 87*
advection *page 94*
annual cycle *page 92*
Celsius scale *page 86*
condensation *page 88*
continental effect *page 94*
continentality *page 94*

diurnal cycle *page 92*
dry adiabatic lapse rate (DALR)
  page 88
dust dome *page 90*
environmental lapse rate (ELR)
  page 87
Fahrenheit scale *page 86*
isotherms *page 94*
Kelvin scale *page 86*

kinetic energy *page 86*
maritime effect *page 92*
saturated adiabatic lapse rate
  (SALR) *page 88*
stability *page 87*
temperature *page 86*
temperature gradient *page 95*
temperature inversion *page 89*
thermometer *page 86*

Scan Here for a quick vocabulary review

ENHANCE

Review Questions

1. Discuss the relationship between temperature and heat.
2. What are the major differences among the Celsius, Fahrenheit, and Kelvin temperature scales?
3. What is the difference between stable and unstable air?
4. Describe the adiabatic process. Define the DALR and the SALR.
5. Why is the SALR less than the DALR? Why is the SALR variable?
6. It is tempting to think the upper troposphere is cold because rising parcels of air cool adiabatically with increasing altitude. But Unit 5 says that temperature lapses upward because the surface is the heat source for the troposphere. What is wrong with the adiabatic explanation? Hint: How would the temperature lapse change if there were no uplift and no vertical mixing? Do vertical motion and mixing lead to more uniform or less uniform temperature?

7. What is a *temperature inversion*, and what are the negative consequences for an urban area affected by this atmospheric condition? How does a temperature inversion influence the ventilation of air pollution?
8. What factors shape the spatial distribution of temperature across Earth's surface?
9. What are the main differences in annual temperature regimes between maritime and interior continental locations?

Scan Here for a quick concept review

ENHANCE

www.oup.com/us/mason

UNIT

8

Air Pressure and Winds

Differences in air pressure are the ultimate cause of all winds, but other factors also affect their speed and direction.

OBJECTIVES

- Explain atmospheric pressure and its altitudinal variation
- Relate atmospheric pressure to wind flow at the surface and aloft
- Explain the forces affecting the movement of air on large scales
- Describe how the operation of small-scale local wind systems is affected by topography and surface type as well as by variations in atmospheric pressure

In our previous discussions of the atmosphere, we have likened it to a blanket and a protective shield. In this unit we add yet another analogy: an ocean of air surrounding Earth. Such imagery is often used in physical geography because the atmosphere resembles the world ocean in its circulation patterns. This unit is about atmospheric pressure and other dynamic forces that shape the regularly recurring global movements of air (and water). Units 9 and 10 focus on the patterns that result from these forces—the atmospheric and oceanic currents that constitute the general circulation systems affecting our planetary surface and form the framework for weather and climate.

The primary role of the general circulation of the atmosphere is to redistribute heat and moisture across Earth's surface. Were it not for the transport of heat from the equator to the poles, most of Earth's surface would be uninhabitable because it would be too hot, too cold, or too dry. Atmospheric circulation accounts for approximately 75 percent of this heat redistribution, and oceanic circulation accounts for the remainder. The ultimate cause of these circulations is latitudinal variation in the amount of net radiation received at the surface of the planet. The resulting temperature differences created by the net radiation imbalance produce the global wind system, which sustains the surface ocean circulation. We will begin to examine this system by detailing the relationships among air pressure, heat imbalances, winds, the rotational effects of Earth, and the effects of surface friction.

Atmospheric Pressure

Wind, the movement of air relative to Earth's surface, is a response to an imbalance of forces acting on air molecules. This is true whether the air is moving horizontally or vertically, and indeed, these two movement dimensions are related through the concept of atmospheric pressure. In this unit, we first consider atmospheric pressure, its fundamental cause, and the vertical distribution it produces. We then link atmospheric pressure to wind flow by considering the additional forces that come into play once motion begins and the patterns of air circulation that result.

THE CONCEPT OF PRESSURE

The primary force exerting an influence on air molecules is the **gravitational force**. The atmosphere is "held" against Earth by this force, and our planet would have no atmosphere at all were it not for gravity. Gravity causes the air to press down against the surface of Earth. The combined weight of all the air molecules in a column of atmosphere exerts a force on the surface of the planet. Over a given area of the surface, say, 1 cm² (0.39 in²), this force produces a **pressure**. Although several different units are used to measure atmospheric pressure, the standard unit of pressure in atmospheric studies is the *millibar (mb)*. The average pressure of the atmospheric column pressing down on Earth's surface (referred to as *standard sea-level air pressure*) is 1,013.25 mb, equivalent to a weight of 14.7 lb/in².

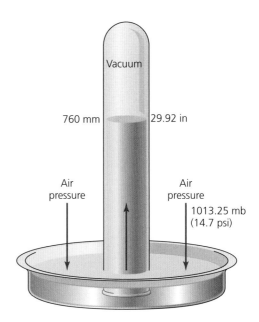

FIGURE 8.1 The mercury barometer invented by Torricelli. The greater the atmospheric (air) pressure, the higher the column of mercury. The values in the figure give standard sea-level pressure in several commonly used pressure units.

Atmospheric pressure is commonly measured as the height of a column of liquid it will support. In 1643, Italian scientist Evangelista Torricelli (1608–1647) performed an experiment in which he filled a glass tube with mercury and then placed the tube upside down in a dish of mercury. Figure 8.1 depicts his experiment. Instead of the mercury in the tube rushing out into the dish, the atmospheric (or air) pressure pushing down on the mercury in the surrounding dish supported the liquid still in the tube. The height of the column in the tube was directly proportional to the atmospheric pressure—the greater the air pressure, the higher the column in the tube. A hydrostatic balance exists between the gravitational force pulling the mercury column downward and the pressure-gradient force pushing the column upward. (Note that standard sea-level air pressure produces a reading of 760 mm [29.92 in] for the height of the mercury column, as illustrated in Fig. 8.1.) Torricelli had invented the world's first pressure-measuring instrument, known as a **barometer**.

ATMOSPHERIC PRESSURE AND ALTITUDE

Once scientists found they could measure atmospheric pressure, they began investigating its properties. They soon discovered that atmospheric pressure does not vary all that much horizontally but does decrease rapidly with increasing altitude. Measurements of atmospheric pressure from both higher land elevations and balloons showed dramatic results. The standard pressure at sea level (1,013.25 mb) decreases to about 840 mb at Denver, Colorado, the "Mile-High City," whose elevation is 1,584 m (5,280 ft). On top of Mt. Whitney in California's Sierra Nevadas, at 4,418 m (14,495 ft), the air pressure is approximately 600 mb. On top of the world's

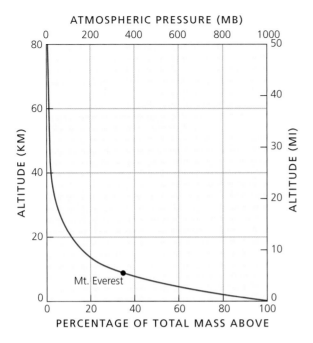

ATMOSPHERIC PRESSURE (MB)

FIGURE 8.2 Mass of the atmosphere as a function of altitude. A greater proportion of the atmospheric mass is concentrated near Earth's surface. Atmospheric pressure depends on the overlying mass of air, so pressure also decreases with altitude.

tallest peak—Mt. Everest in South Asia's Himalayas, at an elevation of 8,850 m (29,035 ft), the pressure is only 320 mb, less than one-third of what is found at sea level.

Because air is compressible, overlying air compresses the atmosphere below, leading to higher density near the surface. We may deduce that most of the molecules are concentrated near Earth's surface. This is confirmed in Figure 8.2, which graphs the percentage of the total mass of the atmosphere below certain altitudes. For example, 50 percent of the air of the atmosphere is found below 5 km (3.1 mi), and 85 percent lies within 16 km (10 mi) of the surface. Atmospheric pressure varies in a similar manner with increasing altitude.

Air Movement in the Atmosphere

Since the days when it became common for sailing ships to make transoceanic voyages, people have known that the large-scale winds of the planet flow in certain generalized patterns. This information was vital in planning the routes of voyages that might take two or three years. However, it was often of little assistance in guiding ships through the more localized, smaller-scale winds that fluctuate from day to day and place to place. It is useful to separate large-scale air movement from smaller-scale movement, even though the two are related to the same phenomenon—atmospheric pressure. In this unit, we begin by considering the causes, and resultant patterns, of the large-scale movement of air that is in contact with the surface of Earth. We then go on to discuss large-scale wind systems and small-scale wind systems in turn.

CAUSES OF ATMOSPHERIC CIRCULATION

Three basic factors explain the circulation of air in the atmosphere: (1) Earth receives an unequal amount of solar radiation at different latitudes, (2) Earth rotates on its axis, and (3) flowing air experiences frictional drag from the surface of Earth. If we examine the amount of incoming and outgoing radiation by latitude, as shown in Figure 8.3A for the Northern Hemisphere (with the Southern Hemisphere's general pattern being nearly identical), we find that there is a marked surplus of net radiation between the equator and the 35th parallel. At latitudes poleward of 35 degrees N, outgoing radiation exceeds incoming radiation. As explained in Unit 4, the main reason for this is that rays of energy from the Sun strike Earth's surface at higher angles, and therefore at greater intensity, in the lower latitudes than in the higher latitudes (see Fig. 4.10). As a result, the equator receives about three times as much annual solar radiation as the poles.

If this latitudinal imbalance of energy were not somehow balanced, the low-latitude regions would be continually heating up and the polar regions continually cooling down. Energy, in the form of heat, is transferred toward the poles. The total amount of heat transported is indicated in Figure 8.3B. We can see from the total heat-transport curve that the maximum transport occurs in the middle latitudes. The poleward transfer of heat is dominated by atmospheric transfer in the middle latitudes, where vigorous mixing of poleward-moving tropical air and equatorward-moving polar air occurs. The poleward transfer of heat in the equatorial and tropical zones is dominated by ocean transport.

FORCES ON AN AIR MOLECULE

Wind is the movement of air from regions of higher to lower pressure. We experience horizontal and vertical winds as well as surface- and upper-level winds. Three forces determine the speed and direction of surface- and upper-level winds: (1) the *pressure-gradient force*, (2) the *Coriolis force*, and (3) the *frictional force*.

Observations of station pressure (mb) are first converted to sea-level pressure (mb) to remove the strong influence of elevation on surface pressure and are then mapped as lines of equal sea-level pressure called **isobars**. These maps of sea-level pressure are the basis for examining surface winds. In analyzing wind at high altitude, it is easier to examine maps that show the altitude at which a given pressure occurs. Such maps show the spatial distribution of heights for a single pressure level and are called constant pressure surface maps. For example, one such map, the 500-mb map, shows heights for the 500-mb surface and is used to study winds about 5 km (16,000 ft) above the surface.

Recall that gravity causes the air to press down against the surface of Earth; this is atmospheric pressure. The pressure may be different at two different surface locations. The difference in surface pressure over a given distance between two locations is the *pressure gradient*. When there is a

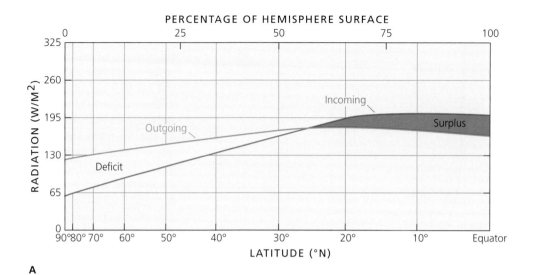

PERCENTAGE OF HEMISPHERE SURFACE

A

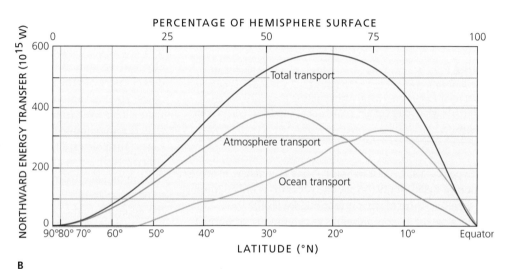

PERCENTAGE OF HEMISPHERE SURFACE

FIGURE 8.3 (A) Latitudinal profiles of solar radiation absorbed and terrestrial radiation emitted per unit area of Earth's surface, averaged over all Northern Hemisphere longitudes. (B) The consequent poleward transfer of total heat and the separate atmospheric and ocean transfer components.

B

pressure gradient, it produces a force that causes air to move from the place of higher pressure to that of lower pressure. This force, called the **pressure-gradient force**, increases as the difference in air pressure across a specified distance increases. The direction of the pressure-gradient force is always oriented from high to low pressure at a 90-degree angle to isobars, and the magnitude of the pressure-gradient force is inversely related to the spacing of isobars. The pressure-gradient force is the fundamental cause of all air movement. A horizontal difference in air pressure often arises from a difference in air temperature, which in turn causes a difference in air density (Fig. 8.4). As discussed

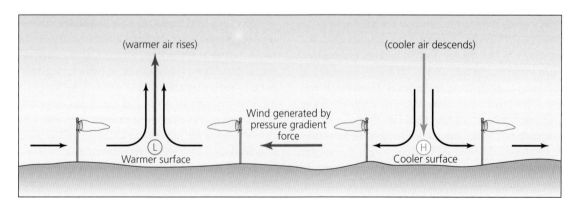

FIGURE 8.4 Air movement is always from areas of higher pressure (H) toward areas of lower pressure (L). The larger the pressure difference between H and L, the greater the pressure-gradient force and the faster the wind.

more fully later in the section on sea and land breezes, warm air is less dense and tends to rise, lowering surface pressure; cold air is denser and tends to sink, raising surface pressure. These thermally induced pressure gradients can be generated by local-scale convection, shown in Figure 8.4, or by global-scale convection resulting from the equator-to-pole net radiation imbalance (see Fig. 8.3). Horizontal pressure gradients can also be generated by forces associated with the movement of air itself. These are known as *dynamically produced gradients*, as opposed to the *thermal (temperature-induced) gradients*. The pressure-gradient force between high- and low-pressure regions can thus be initiated by thermal causes as well as by dynamic causes.

The rotation of Earth adds an interesting complication to the movement of wind (and ocean currents) across the globe that a simple example can illustrate. Suppose you have two children playing on a merry-go-round that is rotating counterclockwise. One child is at the center of the rotating surface directly over the axis of rotation, and a second child is standing along the edge. A third child is standing on the ground adjacent to the merry-go-round. If the child in the center tries to toss a ball to a friend riding along the edge of the merry-go-round, the ball will appear to take the curved path shown in Figure 8.5B. This happens because between the time the ball is released by the first child and the time it reaches the edge of the merry-go-round, the position of the second child has rotated counterclockwise. To both children, it would appear that the path of the ball was deflected to the right, resulting in an apparent curved path for their frame of reference. The third child standing on the ground adjacent to the rotating merry-go-round, however, would have observed that the ball traveled in a straight path. The two different perceived paths of movement are created by the two different frames of reference, one on and the other off the rotating surface.

By convention, winds are described as motion relative to the surface, which means we have chosen to use a rotating frame of reference. The merry-go-round example helps us see that air moving in a straight line relative to a fixed position is seen as a curving wind relative to the surface. We account for this apparent deflection, which occurs for any freely moving object like the ball on the merry-go-round or air particles, by introducing an apparent force called the **Coriolis force**. Although not a true force in the physical sense, the effect of rotation is customarily described using this term, and we will do so as well. The direction of the Coriolis force deflection is to the right of moving objects in the Northern Hemisphere and to their left in the Southern Hemisphere (Fig. 8.6). This is because, when viewed from above, Earth's surface rotates in a counterclockwise direction in the Northern Hemisphere and in a clockwise direction in the Southern Hemisphere. Thus, if a wind blows from 45 degrees N toward the North Pole, it is deflected to the right and becomes a westerly wind (see Fig. 8.6). (Note that this westerly wind blows toward the east: *winds are always named according to the direction from which they come*.) The magnitude of the Coriolis force depends on the

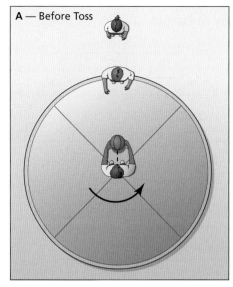

FIGURE 8.5 For the two children on the rotating merry-go-round, the ball appears to curve to the right after it is tossed. The child on the adjacent ground observes the ball to move in a straight path. The apparent deflection of an object moving over a rotating surface is created by the different frames of reference and is accounted for by the Coriolis force.

rate of Earth's rotation (which is effectively constant) and varies directly with the wind speed. The Coriolis force is only significant over fairly long time periods (≥ 3 hours) because Earth turns only once on its axis every 24 hours. The Coriolis force also varies with latitude and is at a maximum at the poles and at a minimum (zero) at the equator.

Finally, some of the air motion in the atmosphere takes place near Earth's surface, where it encounters frictional drag. The closer individual air molecules are to the surface, the more they are slowed by surface drag, which creates a **frictional force**. The direction of the frictional force is

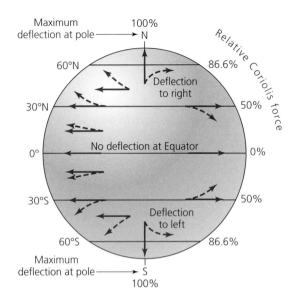

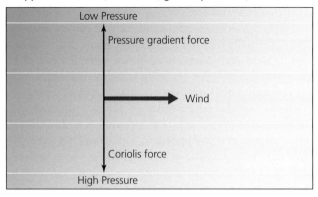

A Upper-level wind (no friction, geostrophic flow)

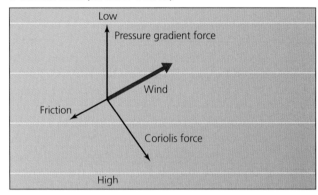

B Surface wind (effect of friction)

FIGURE 8.6 Latitudinal variation in the Coriolis force for winds of uniform speed. Deflection is zero at the equator, increases with latitude, and is most pronounced at each pole. Deflection is to the right in the Northern Hemisphere and to the left in the Southern Hemisphere. (Note the percentages along the right of the globe.)

FIGURE 8.7 Effect of the pressure-gradient, Coriolis, and frictional forces in producing geostrophic wind in the upper atmosphere (without frictional drag) (A) and surface wind in the surface layer (with frictional drag) (B). The Northern Hemisphere pattern is shown.

always opposite to the direction of air movement, and the magnitude of frictional force depends primarily on the "roughness" of the surface. There is less frictional drag with movement across a smooth snow or water surface than across mountainous terrain, a forested landscape, or the rough skyline of a metropolitan area. Air slowed by friction with the surface creates frictional drag on faster-moving air above. As a result, friction is important for some depth of atmosphere rather than just for air actually in contact with the surface. The **surface layer** in which frictional force must be considered extends 500 to 2,000 m (1,640 to 6,560 ft) into the atmosphere. In the **upper atmosphere** above this level, the effects of the frictional force can be ignored.

We are now ready to examine how the pressure-gradient, Coriolis, and frictional forces act together to determine horizontal air motion in the upper atmosphere and surface-air layers. Collectively, they determine the pattern of wind flow within any area.

Large-Scale Wind Systems

Except for small-scale winds (which persist for only a few hours) and those near the equator, the wind never blows in a straight path from an area of higher pressure to an area of lower pressure. Once the pressure-gradient force initiates air movement, the Coriolis and frictional forces come into play and heavily influence the direction and speed of wind flow.

GEOSTROPHIC WINDS

Once a molecule of air starts moving under the influence of the pressure-gradient force, the Coriolis force deflects it to the right if it is in the Northern Hemisphere. Figure 8.7A diagrams the path that results if we ignore friction for the moment: eventually, the pressure-gradient force and the Coriolis force acting on the wind balance each other out. The resultant wind, called a **geostrophic wind**, has a constant speed and follows a relatively straight path that minimizes deflection. Geostrophic winds occur in the "free" or upper atmosphere above the surface layer, where friction with the surface is important. Because of the balancing of forces, a geostrophic wind always flows parallel to the isobars, with high pressure to the right of the direction of air flow in the Northern Hemisphere. In the Southern Hemisphere, high pressure is to the left of the direction of geostrophic winds. Geostrophic wind speed is directly related to the pressure gradient, so that if we have a map showing the distribution of atmospheric pressure well above the surface, we can quickly determine the direction of geostrophic wind movement and the spatial distribution of relative wind speeds.

SURFACE WINDS

Near the surface of Earth, especially below an altitude of about 1,000 m (3,300 ft), frictional force comes into play and disrupts the balance represented by the geostrophic wind in Figure 8.7A. Frictional force acts opposite to the wind direction and therefore reduces wind speed, as illustrated in Figure 8.7B. This weakens the Coriolis force, which varies directly with wind speed, and results in less deflection. Thus **surface winds** blow *across* the isobars instead of parallel to the isobars. This produces a flow of surface air out of high-pressure areas and into low-pressure areas at an oblique angle to the isobars. Surface wind speed is determined by the pressure gradient and the roughness of the surface.

Since surface-pressure systems are often somewhat circular when viewed from above, we can deduce the general circulation around cells of low and high pressure (Fig. 8.8). Surface winds converge toward a **cyclone** (a low-pressure cell—L in Fig. 8.8A). This converging air has to go somewhere, so it rises vertically in the center of the low-pressure cell. The reverse is true in the center of an **anticyclone** (a high-pressure cell—H in Fig. 8.8B): diverging air moves outward and draws air down in the center of the high-pressure cell (see also Fig. 8.4). Thus, cyclones are associated with *rising air* at their centers, and anticyclones are associated with *subsiding air* at their centers. This simple vertical motion produces different weather associated with each type of pressure system, as we will see in later units. Note that the words "cyclone" and "anticyclone" refer to pressure patterns, not weather patterns. Thus cyclones are not necessarily stormy, as might be inferred from everyday use in news media.

Figure 8.8 is potentially confusing, because although we have said air is deflected to the right by the Coriolis force, the wind seems to be turning to the left as it flows around the cyclone. Notice, however, that the pressure-gradient force is everywhere directed inward toward the low-pressure area. The arrows don't follow the pressure-gradient force but in every case show deflection to the right of the perpendicular. This results in an overall counterclockwise circulation, which could be called left-turning. However, left-turning is definitely *not* the same thing as leftward deflection from the pressure-gradient force.

Looking at surface winds on any day (such as those found through the QR link), we find air streams that curve out of high-pressure cells and converge on low-pressure ones. In some places, pressure gradients and winds are weak, whereas in other places they are strong. It goes almost without saying that wind flow impinges on a wide range of human activities (see Perspectives on the Human Environment: Air Pressure and Wind in Our Daily Lives).

Small-Scale Wind Systems

Small-scale wind systems are often a significant part of the local climate because they respond to much subtler variations in atmospheric pressure than the variations depicted in Figure 8.8. For small-scale wind systems having a time

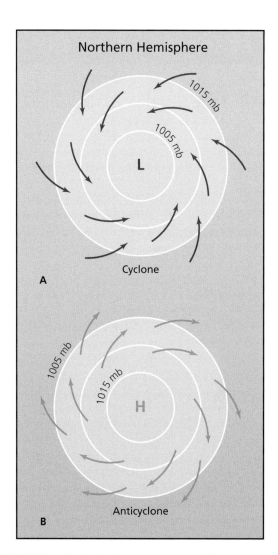

FIGURE 8.8 Air circulation patterns associated with a cyclonic low-pressure cell (A) and an anticyclonic high-pressure cell (B) in the Northern Hemisphere. In the Southern Hemisphere, air circulation around low- and high-pressure cells is in the opposite direction (clockwise toward cyclones and counterclockwise away from anticyclones).

ENHANCE

period of three hours or less, the deflection associated with the Coriolis force is minimal, although it is an important force for small-scale wind systems of longer duration. A number of common small-scale wind systems serve to illustrate how topography and surface type can influence the pressure gradient and its resultant wind flow.

SEA/LAND BREEZE SYSTEMS

In coastal zones and on islands, two different surface types are in close proximity—land and water. As we have noted, land surfaces and water bodies display sharply contrasting thermal responses to energy input. Land surfaces heat and cool rapidly, whereas water bodies exhibit a more moderate temperature regime.

Air Pressure and Wind in Our Daily Lives

Even a 98-pound weakling can support 10 tons. This is not an ad for a bodybuilding program: we all bear the weight of the atmosphere pressing down on us! At sea level, this weight ranges from 9 to 18 metric tons depending on our size. Like the deep-sea creatures that live their entire lives with the weight of much of the ocean above them, we have adapted to functioning and moving efficiently in our particular environmental pressure zone. Only a sharp change—such as the one experienced by a Florida sea-level flatlander who takes a vacation trip high in Colorado's Rocky Mountains—reminds us of our adjustment to, and dependence on, a specific atmospheric environment. Two factors influence this sensitivity to altitude.

The first is the density of air molecules, particularly oxygen, at any altitude. At higher elevations, the air is "thinner"; that is, greater distance exists between the oxygen molecules, and consequently there are fewer of them in any given air space. We have to do more breathing to get the oxygen necessary to maintain our activity levels. When the 1968 Olympics were held in Mexico City, the lower density of oxygen at that elevation (2,240 m [7,350 ft]) was the decisive factor in the unimpressive competition times recorded by most of the participating athletes. In contrast, Bob Beamon shattered the world record in the long jump in the less dense atmosphere at the same Olympics.

The second factor is the response of our internal organs to changes in atmospheric pressure. Our ears may react first as we climb to higher elevations. The "pop" we hear is actually the clearing of a tiny tube that allows pressure between the inner and middle ears to equalize, thereby preventing our eardrums from rupturing. At very high altitudes, of course, we must travel in pressurized aircraft. And even on the surface, the world's highest railroad—China's recently built line to Lhasa across the Tibetan Plateau (with an average elevation of greater than 4,500 m [15,000 ft])—is required to use pressurized passenger cars.

Winds play a constant role in our lives as well, because they are a major element of local weather and climate. In coastal areas or on islands, atmospheric conditions can vary considerably over short distances. Oceanfront zones facing the direction of oncoming wind experience more air movement, cloudiness, and humidity than nearby locations protected from this air flow by hills or mountains. Places exposed to wind are called **windward** locations; areas in the "shadow" of protecting topographic barriers are known as **leeward** locations.

There are countless examples of associations between wind and human activities (Fig. 8.9). The horizontal flushing effects of wind are vital to maintaining acceptable air quality in urban areas, where large quantities of pollutants are dumped into the local atmosphere. Many outdoor sporting events are affected by wind conditions during games. For example, certain baseball stadiums are infamous for their unpredictable wind currents. San Francisco's Candlestick

A

B

FIGURE 8.9 Winds flowing over the complex terrain of Rio de Janeiro, Brazil, can be a challenge for paragliders (A). By contrast, funneling effects in mountain passes near Tehachapi, California, help create a valuable energy resource harvested by more than 5,000 turbines in a large wind farm (B).

During the day, a land surface heats up quickly, and the air layer in contact with it expands vertically in response to the increased air temperature. Thus, above the surface, isobars slope downward toward the cooler water, as illustrated in Figure 8.10A. This means a pressure gradient is directed toward the ocean, and air moves in response to that gradient. As molecules leave the column over land, the surface pressure on land falls and correspondingly rises over the ocean. At low levels, a small pressure-gradient force is directed from ocean to land, and the air near the surface moves in response to that force. Consequently, coastal zones generally experience air moving from water to land during the day—a **sea breeze** (Fig. 8.10A). At night, when the temperature above the land surface has dropped sufficiently, the circulation reverses because the warmer air (and lower pressure) is now over the water. This results in air moving from land to water—a **land breeze** (Fig. 8.10B). The shallow, thermally induced pressure variations associated with the sea breeze/land breeze circulation are most strongly developed on clear summer days with calm winds in the surrounding area.

Note that when generated, sea and land breezes produce a circulation cell composed of the surface breeze; rising and subsiding air associated with the lower- and higher-pressure areas, respectively; and an air flow in the direction opposite to that of the surface (Fig. 8.10). Although it modifies the wind and temperature conditions at the coast, the effect of this circulation diminishes rapidly as one moves inland. Note also that we use the word *breeze*. This term accurately depicts a rather gentle circulation in response to a fairly weak pressure gradient. The sea breeze/land breeze phenomenon is easily overpowered if stronger pressure systems are nearby.

MOUNTAIN/VALLEY BREEZE SYSTEMS

Mountain slopes, too, are subject to the reversal of day and night local circulation systems. This wind circulation is also thermal, meaning that it is driven by temperature differences between adjacent topographic features. During the day, mountain terrain facing the Sun tends to heat up more rapidly than do shadowed, surrounding slopes. This causes lower pressure to develop over the mountain ridges, spawning a **valley breeze** that flows upslope during the day. At night, greater radiative heat loss from the mountain slopes cools them more sharply, producing a small increase in pressure and a **mountain breeze** that flows downslope. Figure 8.11 illustrates the operation of this type of oscillating, diurnal (day/night) wind system in a highland valley that gently rises away from the front of the diagram. Up-valley and down-valley winds are also generated along the axis of the valley during the day and night, respectively.

OTHER LOCAL WIND SYSTEMS

Another category of small-scale wind systems involves **cold-air drainage**, the steady downward oozing of heavy, dense, cold air along steep slopes under the influence of gravity. The winds that result are known as **katabatic winds** and are especially prominent under calm, clear conditions where the edges of highlands plunge sharply toward lower-lying terrain. These winds are fed by large pools of

A Sea breeze

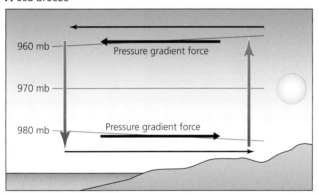

B Land breeze

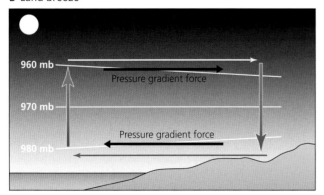

FIGURE 8.10 Sea breeze/land breeze air-circulation systems. These reversing, cell-like air flows develop in response to minor pressure differentials associated with day/night temperature variations along the land/sea coastal zone.

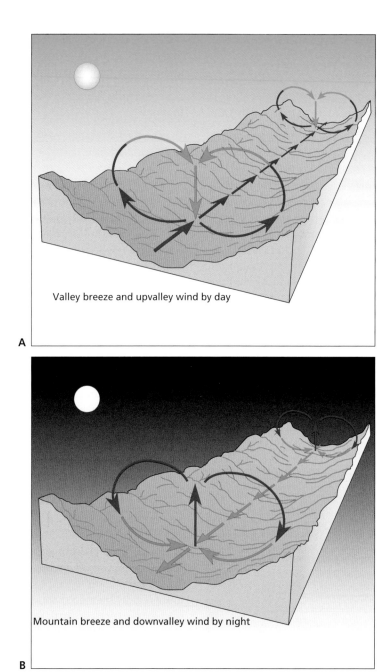

FIGURE 8.11 Formation of valley (daytime) and mountain (nighttime) breezes. Red arrows represent upslope and up-valley winds (warmer); blue arrows indicate downslope and down-valley winds (cooler).

extremely cold air that collect above highland zones. They are also common around major ice sheets, such as the huge, continental-scale glaciers that cover most of Antarctica and Greenland.

Katabatic winds can attain destructive intensities when the regional air flow steers the cold air over the steep edges of uplands, producing a cascade of air much like water in a waterfall. The most damaging winds of this type occur where local topography channels the downward surge of cold air into narrow, steep-sided valleys. The Rhône Valley of southeastern France is a notable example. Each winter, it

experiences icy, high-velocity winds (known locally as the *mistral* winds) that drain the massive pool of cold air that develops atop the snow-covered French and Swiss Alps to the valley's northeast.

Yet another type of small-scale wind system is associated with the forced passage of air across mountainous terrain (which we discuss in more detail in Unit 12). Briefly, this cross-mountain movement wrings out much of the moisture in the original mass of air, and it also warms the air adiabatically (i.e., with no heat added) as it descends after its passage across the upland. Thus, the lee side of the mountain experiences dry and relatively warm winds, which taper off with increasing distance from the highland. Occasionally, such winds can exceed hurricane-force intensity (greater than 120 km/h [75 mph]). This happens when the winds are reinforced by an anticyclone positioned upwind from the mountains that feeds air into a cyclone located on the downwind side of the upland.

The best-known U.S. example of this type of system is the **chinook wind**, which occurs along the eastern (downwind) side of the Rocky Mountains on the western edge of the central and northern Great Plains. Another well-known example is the **Santa Ana wind** of coastal Southern California, an occasional hot, dry air flow whose unpleasantness is heightened by the downward funneling of this wind from the high inland desert (where it is generated by an anticyclone) through narrow passes in the mountains that line the coast. Such winds can exacerbate wildfires during the dry season.

In many locations, as we have just seen, small-scale winds can at times become more prominent than large-scale air flows. But the regional expressions of the global system of winds still play a far more important role overall. That is the subject of Unit 9, which investigates the general circulation of the atmosphere.

Key Terms

anticyclone *page 103*
barometer *page 98*
chinook wind *page 107*
cold-air drainage *page 105*
Coriolis force *page 101*
cyclone *page 103*
frictional force *page 101*
geostrophic wind *page 102*

gravitational force *page 98*
isobar *page 99*
katabatic wind *page 105*
land breeze *page 105*
leeward *page 104*
mountain breeze *page 105*
pressure *page 98*
pressure-gradient force *page 100*

Santa Ana wind *page 107*
sea breeze *page 105*
surface layer *page 102*
surface wind *page 103*
upper atmosphere *page 102*
valley breeze *page 105*
wind *page 98*
windward *page 104*

Scan Here for a quick vocabulary review

Review Questions

1. Define atmospheric pressure and describe its vertical structuring within the atmospheric column.
2. What are the forces that determine wind speed and direction?
3. Define the Coriolis force and describe its operations in both the Northern and Southern Hemispheres.
4. What is a geostrophic wind? Where does it occur and why?
5. Describe the air-circulation patterns associated with cyclones and anticyclones.
6. Describe the operation of the sea breeze/land breeze small-scale wind system.

Scan Here for a quick concept review

www.oup.com/us/mason

UNIT
9

Circulation Patterns of the Atmosphere

The systematic arrangement of pressure and wind on Earth ensures that North Africa cannot experience towering rain systems that dot Equatorial latitudes.

OBJECTIVES

- Describe the simple model of the global atmospheric circulation presented in the text
- Discuss the pressure systems and wind belts of the circulation model and the

complications that arise when the model is compared to the actual atmospheric circulation

- Explain the basic workings of the upper atmosphere's circulation

U nit 8 introduced the basic causes of air movements in the atmosphere. In this unit, we focus on the global pressure and wind systems that constitute the general atmospheric circulation. In the short run, these prevailing wind patterns carry along, and to a certain extent cause, the weather systems that affect us daily. The longer-term operation of this general circulation, in conjunction with atmospheric energy flows, produces the climates of Earth.

The workings of the general atmospheric circulation are very complex, and a clear physical explanation of some of its most basic elements is still lacking. Nonetheless, we can deduce many of these features using our knowledge of the basic forces affecting the movement of air. In the process, we can develop a model to describe and explain the major processes. We begin by considering the atmosphere's surface circulation and making some observations regarding wind flow in the upper atmosphere.

A Model of the Surface Circulation

As a starting point, imagine the probable arrangement of the global atmospheric circulation on a uniform, nonrotating Earth. This hypothetical situation would result in a single thermally induced convection cell in each hemisphere. There would be a single narrow belt of low pressure around the equator, where air would rise, and a cell of high pressure at each pole, where air would subside. The surface winds on such a planet would be northerly in the Northern Hemisphere and southerly in the Southern Hemisphere, moving directly from high pressure to low pressure across the pressure gradient, as shown in Fig. 9.1. Such a simple model is, in fact, a good approximation of the surface circulation on Titan, the largest of Saturn's 62 moons, and the only moon in our solar system known to have a dense nitrogen atmosphere. The slow rotational rate of Titan—it takes 16 Earth days to complete one rotation about its axis—results in a very weak Coriolis force. As a result, a single thermal cell extends from the subsolar point in one hemisphere to the pole of the opposite hemisphere on Titan.

We might further speculate that adding the rotation of Earth to this simple model would produce surface northeasterly winds in the Northern Hemisphere (as the air moving toward the equator is deflected to the right) and southeasterly winds in the Southern Hemisphere. In fact, such a model could not occur for several reasons. To cite just one, everywhere on the planet the atmosphere would be moving against the direction of Earth's rotation. This situation would result in a small slowing of Earth's rate of spin and a cessation of wind for the atmosphere, which is much less massive than Earth itself. As it turns out, we do have areas of low pressure at the equator and high pressure at the poles, but the situation in the

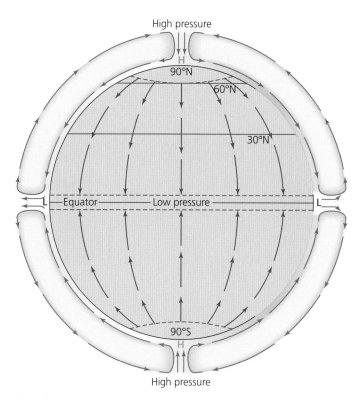

FIGURE 9.1 Hypothetical atmospheric circulation on a featureless, nonrotating Earth. Polar high pressure and equatorial low pressure would result in northerly surface winds in the Northern Hemisphere and southerly surface winds in the Southern Hemisphere. The rotation of Earth, the annual variation in the latitude of the vertical noon Sun, the distribution of oceans and continents, and land/water heating contrasts at the surface prevent this simple general circulation from existing.

midlatitudes is more complex than the situation in the simple model shown in Figure 9.1.

We now can introduce the actual effects of rotation on our hypothetical planet, but, for the time being, let's ignore seasonal heating differences and the land/water contrast at the surface. The model that results is an idealized but reasonable generalization of the surface circulation pattern, which is illustrated in Figure 9.2. As we can see in the figure, in some places, the wind flows mainly parallel to lines of latitude in what is called **zonal flow**. In purely zonal flow, a parcel of air remains in the same latitude "zone." This contrasts with **meridional flow**, which refers to north–south flow along a meridian of longitude. Of course, actual winds are seldom purely east–west or north–south, and thus they have both meridional and zonal components.

THE EQUATORIAL LOW AND SUBTROPICAL HIGH

Year-round heating in the equatorial region produces a thermal low-pressure belt in this latitudinal zone. That belt of rising air is called the **equatorial low** or **intertropical convergence zone (ITCZ)**. (The reason for this latter

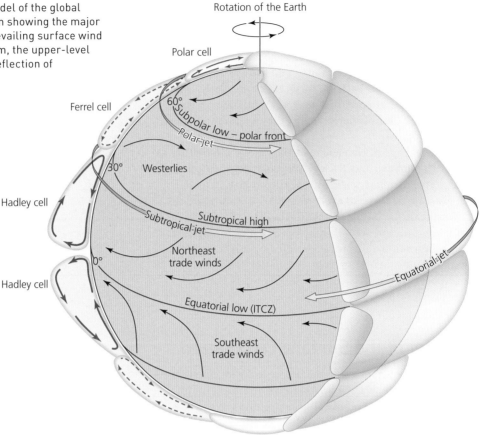

FIGURE 9.2 A conceptual model of the global atmospheric circulation pattern showing the major surface pressure belts, the prevailing surface wind systems that develop from them, the upper-level jet streams, and the Coriolis deflection of surface winds.

terminology will become evident shortly.) The air rises from the surface to the top of the troposphere (see Unit 6), where the tropopause resists further vertical lifting and forces the air to diverge poleward in both the Northern and Southern Hemispheres.

Much of this now-cooled, poleward-moving air descends in a broad region centered approximately at latitudes 30 degrees N and 30 degrees S. The descending air produces a belt of high pressure at the surface near both these latitudes. These two high-pressure belts, understandably, are termed the **subtropical highs**.

THE TRADE WINDS AND THE WESTERLIES

Remembering that surface winds diverge out of areas of high pressure, we can deduce the nature of the wind movement, both equatorward and poleward, from this high-pressure belt. Air returning toward the equator in the Northern Hemisphere is deflected to the right and forms a belt of northeasterly winds called the **Northeast Trades** (see Perspectives on the Human Environment: The Sailor's Legacy: Naming the Winds). Air returning toward the equator from the subtropical high in the Southern Hemisphere is deflected to its left to form the **Southeast Trades**. As we can see in Figure 9.2, these two

wind belts converge along the equator, and this is the rationale for calling this low-latitude area the intertropical convergence zone.

Air moving poleward from the two subtropical highs acquires increased Coriolis deflection (see Fig. 8.6) and forms two belts of generally west-to-east-flowing winds (one in the Northern Hemisphere, one in the Southern Hemisphere) known as the **westerlies**. These prevailing winds form broad midlatitude belts from about 30 degrees N to 60 degrees N and about 30 degrees S to 60 degrees S (see Fig. 9.2).

THE POLAR HIGH, POLAR EASTERLIES, SUBPOLAR LOW, AND POLAR FRONT

Let us leave the westerlies for a moment and consider the winds emanating from the **polar highs**, large cells of high pressure centered over each pole. Here, air moving toward the equator is sharply deflected to become the **polar easterlies**. You can see in Figure 9.2 that the polar easterlies flowing out of the polar highs meet the westerlies flowing out of the subtropical highs. The atmospheric boundary along which these wind systems converge is called the **polar front**. Along the polar front in each hemisphere (located equatorward of 60°N and 60°S, respectively), less-dense warm and moist air from the subtropical high is

FIGURE 9.3 "Flying eastward over the Sierra Nevada Mountains, beautiful clear blue skies allowed us a spectacular view of Yosemite Valley. As our little commuter plane crested the spine of the mountain range and began its descent toward the Mammoth Lake airport, it was caught in violent motions that at times seemed strong enough to rip the plane's wings off. Heads bouncing off walls, we watched through the thrown-open cockpit door as the pilot and co-pilot fought to get the plane safely on the ground. Three bounces later they did, and a white-faced flight attendant wished us the best for the rest of the day.

"We had been caught in a *rotor*, a rolling cylinder of wind that forms on the lee side of mountains (such as the Sierras and the Andes) that run perpendicular to prevailing westerly wind. Air flowing up over the mountains curves downward on the lee side, and if the wind is strong enough, the resulting wave contains a rotor. Sometimes mountain waves and rotors are revealed by stationary clouds, but other times they can't be detected until encountered by airplanes like ours."

forced to rise over denser cold and dry polar air. This rising air produces a belt of low pressure at the surface called the **subpolar low**.

Note that Figure 9.2 contains three types of **jet streams**, which are like rivers or tubes of more rapidly moving air flowing within surrounding air moving in the same general direction. These three types are the **polar jet**, the **subtropical jet**, and the **equatorial jet**. All three jets in Figure 9.2 (which we will discuss in more detail later in the unit) are upper-level jets, meaning they are found in the upper troposphere.

Overall, our now-completed conceptual model of the surface circulation for our uniform, rotating planet has seven pressure features (an equatorial low, two subtropical highs, two subpolar lows, and two polar highs) and six intervening wind belts (the Northeast and Southeast Trades plus the westerlies and polar easterlies of each hemisphere). Note that the circulation patterns of the Northern and Southern Hemispheres are identical except for the opposite Coriolis deflection (see Fig. 9.2).

The Actual Surface Circulation Pattern

In contrast to the idealized pattern of the surface-pressure and circulation model shown in Figure 9.2, the actual pattern (Fig. 9.4) is considerably more complex because it incorporates two influences ignored in the model. Recall from Unit 4 that the location of maximum solar heating shifts throughout the year as the latitude of the (noonday) vertical Sun changes from 23.5 degrees N at the Northern Hemisphere's summer solstice to 23.5 degrees S at its winter solstice. This causes a seasonal north and south movement of pressure and wind belts. It is significant that the continents respond more dramatically to this seasonal variation in heating than do the oceans (as we noted in Unit 7). The result is individual pressure cells (which we can call *semipermanent highs and lows*) rather than uniform, continuous belts of low and high pressure.

Moreover, recall that the distribution of land and water differs significantly between the two hemispheres (Fig. 9.5).

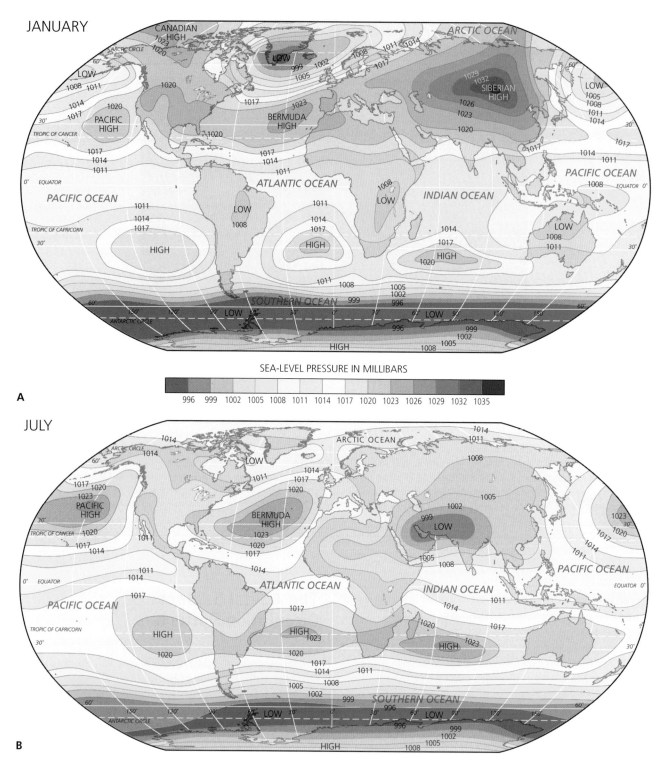

FIGURE 9.4 Global mean sea-level pressure (mb) patterns in January (A) and July (B).

The Northern Hemisphere contains two large landmasses and has a greater percentage of land area than the Southern Hemisphere. The North Pole is in an ocean surrounded by landmasses, and in the Northern Hemisphere land area is concentrated in the middle latitudes. The Southern Hemisphere is mostly water, has a continental pole surrounded by ocean, and has minimal land area within the middle latitudes. Because of their different geography, the two hemispheres exhibit somewhat different atmospheric circulations. Nonetheless, their circulations are similar enough to consider together.

In examining Figure 9.4, bear in mind it is a map of sea-level pressure, not surface pressure. The variation in pressure with height is so great that a surface map would not be

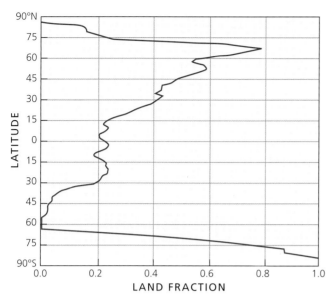

FIGURE 9.5 Fraction of surface area covered by land as a function of latitude.

pressure gradients and resulting winds. It is, however, necessarily an artificial depiction of the distribution of pressure, especially for high-elevation locations.

Before discussing the actual surface circulation, secondary surface circulations, or upper-level circulations in more detail, it is important to consider several critical constraints under which the atmospheric (and oceanic) circulation must operate. The equator-to-pole net radiation imbalance (see Fig. 8.3), which drives the atmospheric circulation, must be balanced by a net poleward transfer of heat. The global water balance must be maintained (see Units 10 and 11) so that the poleward-moving water mass is balanced by an equal equatorward-moving water mass. A similar balance of poleward- and equatorward-moving atmospheric mass must also be achieved. Finally, Earth's angular momentum must be balanced. Because of frictional coupling between surface winds and Earth's surface, the momentum imparted to the surface by westward-moving and eastward-moving surface winds must be in approximate balance for the rotation rate of the planet to remain constant.

very helpful, as it would essentially be a map of elevation. To create the maps such as the ones in Figure 9.4, the illustrator transforms surface pressures to sea-level values using a presumed relationship between height and pressure. The resulting maps allow better assessment of horizontal

THE EQUATORIAL LOW (ITCZ)

The **Hadley cell** is without question the best developed of the three cells shown in Figure 9.2. This thermally induced cell, shown in greater detail in Figure 9.6, features meridional

PERSPECTIVES ON THE HUMAN ENVIRONMENT

The Sailor's Legacy: Naming the Winds

During the colonial era Spanish sea captains headed to the Caribbean and the Philippines in search of gold, spices, and new territory for the crown. They depended on a band of steady winds to fill the sails of their galleons as they journeyed westward in the tropics. Those winds were named the trade winds, or the *trades*. These were the winds that first blew Christopher Columbus and his flotilla to the New World in 1492.

In the vicinity of the equator, the Northern and Southern Hemisphere trades often converge in a zone of unpredictable breezes and calm seas. Sailors dreaded being caught in these so-called *doldrums*. A vessel stranded here might drift aimlessly for days. That was the fate of the ship described in these famous lines from Samuel Taylor Coleridge's *The Rime of the Ancient Mariner*:

> Day after day, day after day,
> We stuck, nor breath nor motion;
> As idle as a painted ship
> Upon a painted ocean.

Ships also were becalmed by the light and variable winds in the subtropics at latitudes around 30 degrees N and 30 degrees S. Spanish explorers who ran afoul of the uncertain breezes in these often-steamy regions forced their horses overboard to lighten their loads and save water for the crew. According to one theory, the trail of floating corpses caused navigators of the seventeenth century to label this zone the *horse latitudes*.

In the middle latitudes of the Southern Hemisphere, ships heading eastward followed the strong westerly winds between 40 degrees S and 60 degrees S. These winds were powerful but much stormier than the trades to the north, so, depending on their approximate latitude, they became known as the *Roaring Forties*, the *Furious Fifties*, and the *Screaming Sixties*. Thus, some important terminology still applied to wind belts today dates from the early days of transoceanic sailing.

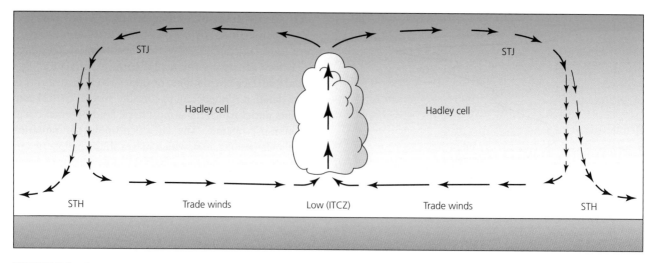

FIGURE 9.6 A cross-section of the Hadley cell circulation. Surface convergence of the trade winds feeds convective lifting in updraft zones that are embedded along the equatorial low (the ITCZ). Westerly acceleration of upper-level winds leads to the formation of the subtropical jet stream (STJ), which initiates convergence of air and general subsidence at about 30 degrees latitude. The resulting subtropical high (STH) is a broad and continuous belt of high pressure that encircles Earth.

flow in all seasons. Strong surface heating by the high sun produces a surface trough of low pressure along the equator featuring intense localized convective lifting in places, interspersed with areas of less intense lifting and localized descending air. The rising branch of the Hadley cell diverges in the upper troposphere and spreads horizontally, creating a horizontal poleward flow of air in each hemisphere. As this air flow moves poleward, it accelerates (in order to preserve angular momentum) and is turned eastward by the Coriolis force. This acceleration of upper-level westerly wind produces the subtropical jet stream that is positioned near the poleward limb of each Hadley cell (see Fig. 9.6). Convergence of air near the subtropical jet stream initiates a broad descending limb of air at about 30 degrees latitude. This general subsidence of air produces a surface high-pressure system, called the subtropical high. The subtropical high is a nearly continuous belt of high pressure centered at about 30 degrees; it supports both the trade wind and westerly surface wind systems. The return flow of the opposing trade winds creates a broad region of surface convergence that feeds the updrafts along the equatorial low.

Careful inspection of Figure 9.4 reveals several modifications to the simplified picture seen in Figure 9.2. The equatorial low, or ITCZ, migrates into the "summer" hemisphere (the Northern Hemisphere during July, the Southern Hemisphere during January), a shift most prominent over landmasses (Fig. 9.7). Note that in July, the equatorial low is located nearly 25 degrees north of the equator in the vicinity of southern Asia. But ITCZ migration is subdued over oceanic areas because bodies of water are much slower to respond to seasonal changes in solar energy input. Note, too, that the migration of the equatorial low into the Southern Hemisphere in January (see Fig. 9.7) is far less pronounced. This makes sense, because there are fewer large landmasses in the tropics south

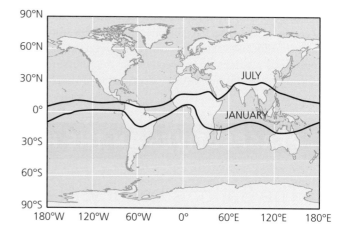

FIGURE 9.7 Mean positions of the intertropical convergence zone (ITCZ) in January and July.

of the equator; the equatorial low is, however, considerably displaced poleward over Africa and Australia and, to a lesser extent, above South America.

THE BERMUDA AND PACIFIC HIGHS

The **polar cell** in Figure 9.2 does exist as a shallow and weak thermally direct cell, but its rate of meridional (north–south) flow is much smaller than that of the Hadley cell. The **Ferrel cell** in Figure 9.2 is associated with negligible meridional flow and, as we will see shortly, is dominated by strong zonal (west–east) flow. It is best viewed as an indirect circulation cell produced by the two stronger adjacent thermally direct cells.

In contrast to our model's subtropical high-pressure belts (straddling 30°N and 30°S), the actual semipermanent highs at these latitudes are broader belts of high pressure, with cells of higher pressure centered over the subtropical oceans (Fig. 9.4). There are five such cells on the map, one above each subtropical ocean (North Pacific, South Pacific, North Atlantic, South Atlantic, South Indian). These cells are present all year but contrast most strongly with surrounding continents in the "summer" hemisphere. In the Northern Hemisphere, the North Atlantic's high-pressure cell is called the *Bermuda* (or *Azores*) *High*, and the North Pacific cell is referred to as the *Pacific* (or *Hawai'ian*) *High*. These cells also shift north and south with the Sun, but to a much smaller extent latitudinally than the equatorial low. This again is due to the more subdued response of water to seasonal changes in solar energy receipt. Meteorologists have yet to provide a clear explanation of why the subtropical highs persist year-round over the subtropical oceans, or why they are most strongly developed in summer, while the Hadley cell circulation, from which they are derived, is most intense in winter.

THE CANADIAN AND SIBERIAN HIGHS

In the polar regions, our simplified picture of a polar high centered over each pole needs considerable revision, particularly in the Northern Hemisphere, where the large landmasses of Eurasia and North America markedly protrude into the higher latitudes. It is here, rather than over the more northerly Arctic Ocean, with its floating polar ice cap, that the seasonal cooling is most extreme. As a result, the winter polar high in the Northern Hemisphere is actually two separate cells, a weaker cell centered above northwestern Canada (the *Canadian High*) and a much more powerful cell covering all of northern Asia (the *Siberian High*). In summer, the polar high in the Northern Hemisphere is centered over 90 degrees, where surface air temperatures are lowest. Note, too, that these features are much stronger in the winter and at their weakest during the summer (see Fig. 9.4). In the Southern Hemisphere, the simplified model's single cell of high pressure over the pole (see Fig. 9.2) is reasonably accurate; the South Pole lies in an interior zone of Antarctica, the coldest region on Earth because it is in the center of an ice-covered high-elevation landmass.

THE ALEUTIAN, ICELANDIC, AND SOUTHERN HEMISPHERE SUBPOLAR LOWS

Finally, we need to example the actual configuration of the subpolar low between the polar and subtropical highs. The equatorial low, subtropical highs, and polar highs shown in Figure 9.2 are semipermanent and stationary features. Their existence is nearly continuous over long periods of time, and their positions remain relatively fixed over short timescales. The subpolar lows, in contrast, are the result of transient and migratory low-pressure systems. As we will see in Unit 13, these systems form continuously along the polar front with an average life cycle of approximately one week. Furthermore, they migrate in the general westerly flow pattern within the Ferrel cell. Consequently, the subpolar low regions should be viewed as statistical features representing seasonally averaged pressure patterns rather than as map patterns likely to be present on any particular day.

In the Northern Hemisphere in January (see Fig. 9.4A), when both the polar and subtropical highs are apparent, the subpolar low is well defined as two cells of low pressure, one over each subarctic ocean. These are the *Aleutian Low* in the northeastern Pacific off Alaska and the *Icelandic Low* in the North Atlantic, centered just west of Iceland. They represent regions with a high frequency of converging and ascending air around low-pressure systems that form and move along the polar front. We cover these features in greater depth in the discussion in Unit 13 of air masses and storm systems of the midlatitudes. Such transient and migratory low-pressure systems are much weaker during the summer, when the equator-to-pole temperature gradient drives less heat poleward. Consequently, the seasonally averaged pressure patterns shown in Figure 9.4B indicate the near disappearance of the subpolar lows in summer.

As for the Southern Hemisphere, note on the map that the subpolar low is evident in both winter and summer, indicating year-round presence along the polar front. The persistence of the polar high over Antarctica makes this possible. Moreover, the absence of landmasses in this subpolar latitudinal zone (see Fig. 9.5) causes the subpolar low to remain beltlike rather than forming distinct cells over each ocean.

Secondary Surface Circulation: Monsoonal Wind Flows

The global scheme of wind belts and semipermanent pressure cells we have just described constitutes the **general circulation** (or *primary circulation*) system of the atmosphere. At a more localized scale, there are countless instances of "shifting" surface wind belts that create pronounced winter/summer contrasts in weather patterns. Here, to illustrate such regional (or *secondary*) circulation systems, we confine our attention to one of the most spectacular examples: the Asian monsoon.

A **monsoon** (derived from *mawsim*, the Arabic word for season) is a seasonal reversal of onshore and offshore winds from regional landmasses as a result of seasonal pressure changes, which are themselves created by differential heating of land and water. Monsoonal circulation occurs most prominently across southern and eastern Asia, where seasonal wind reversals produced by the shifting systems cause alternating wet and dry seasons. Specifically, the moist onshore winds of summer bring the *wet monsoon*, whereas the offshore winds of winter are associated with the *dry monsoon*.

The South Asian monsoon circulation over the Indian subcontinent results from a particularly complex interaction of seasonal shifts in the general circulation, the different heating characteristics of land and water, topographic influences of the Himalaya Mountains upon regional air flow, and the linkages between surface and upper-level winds. It provides an excellent example of how air flow patterns in the upper atmospheric circulation exert a decisive influence on what happens at the surface.

The equatorward shift in the general circulation during the fall allows the subtropical jet stream to shift south, where it splits and flows both north and south of the Himalayas (Fig. 9.8). This southward shift in the subtropical jet

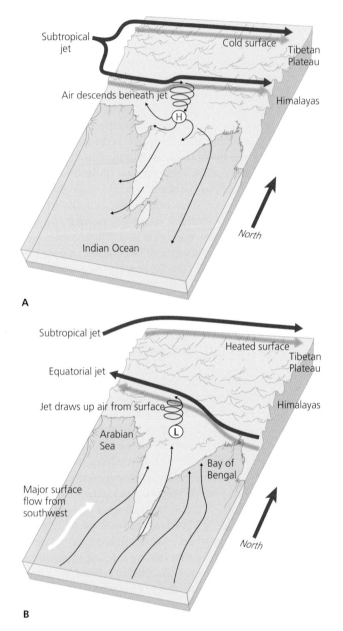

A

B

FIGURE 9.8 General air flow patterns over the Indian subcontinent during the winter (A) and summer monsoon season (B). The regional circulation system involves strong seasonal changes in the heating patterns over the Indian subcontinent and Tibetan Plateau, seasonal reversal of offshore and onshore winds over India, shifts in the position of the subtropical and equatorial jet streams, and distinct winter and summer precipitation patterns.

ENHANCE

stream initiates a series of steps that mark the end of the wet season. In winter, declining solar insolation and cooling of the landmasses produce a thermal high pressure over the interior of southern Asia (see Fig. 9.4) as well as locally over the Indian subcontinent (see Fig. 9.8). Although sea-level pressure is particularly high over Tibet in winter, the high is shallow, and air does not flow out and over the Himalayas. Instead, the offshore flow is supplied by air carried in by the jet stream. The northeasterly surface winds that result over much of the region bring cool continental air that contains little moisture, so precipitation during winter is at a minimum.

With the transition from spring to summer, solar insolation increases, causing the Tibetan Plateau to warm and develop a low-pressure center and weakening the high pressure over the Indian subcontinent as well. The ITCZ (equatorial low) shifts far northward to about 25 degrees over northern India (see Fig. 9.7), and the Southern Hemisphere subtropical high also shifts equatorward. The heated Indian subcontinent develops a thermal low-pressure system, which encourages an onshore low-level flow of air. As a result, the air flow from the Southeast Trades crosses the equator and is recurved—by the opposite Coriolis deflection of the Northern Hemisphere—into a southwesterly flow onto the Indian subcontinent. This air passes over most of the warm tropical Indian Ocean and is therefore very moist. Surface heating by insolation and low-level convergence around the low-pressure system bring uplift and precipitation. The release of latent heat further intensifies the low-level convergence and onshore flow onto India and strengthens the Hadley cell circulation, which intensifies the subtropical high. A positive feedback results, in which stronger trade winds lead to further advection of warm and moist air, which intensifies the convergence, precipitation, and release of latent heat. The equatorial jet stream also moves to a position at about 15 degrees over central India and carries away air imported by low-level flow. The shift in the subtropical jet stream to a position entirely north of the Himalayas, and its replacement by the equatorial jet stream, are sudden, and so there is a rather dramatic shift from the dry to the wet season.

The arrival of this saturated air over the Indian subcontinent by late June marks the onset of the wet summer monsoon, and precipitation is frequent and heavy. The world record for highest total precipitation in a single month is held by the town of Cherrapunji in the hills of northeastern India (see p. 214, Fig. 16.15), where in July 1861, 930 cm (30 ft) of rain fell. During the winter months of the dry monsoon, however, average precipitation values at Cherrapunji are normally on the order of 1 to 2 cm (0.5 to 1.0 in). The Himalayas provide a barrier to the northward penetration of the monsoon circulation, and monsoon rains are confined south of them, although an eastern branch of the monsoon circulation brings rain to East Asia.

The southwestern wet monsoon consists of two main branches, as Figure 9.8 shows. One branch penetrates the Bay of Bengal to Bangladesh and northeastern India, where it is pushed westward into the densely populated Ganges Plain of northern India by the Himalayan Mountain wall (Fig. 9.9). A second branch to the west of the subcontinent arrives from the Arabian Sea arm of the Indian Ocean. The rains from both branches gradually spread across much of the subcontinent, soak the farm fields, and replenish the wells.

These wet-monsoon rains, however, are not continuous, even during the wettest of years. Once the onshore wind movement is established, rains depend on the recurrence of smaller-scale low-pressure cells within the prevailing southwesterly air flow. These depressions reinforce the lifting of the moist air and enhance the formation and continuation of rain throughout the summer months.

Variations in the timing of the onset, duration, and reversal of the Indian monsoon, as well as of its precipitation intensity and amount, can have dramatic effects upon this densely populated region through the production of damaging droughts and floods. Higher-than-normal monsoon precipitation during July and August 2010 led to extensive flooding in the Indus River Valley of Pakistan, killing thousands and leaving an estimated 20 million persons homeless. Figure 9.10 shows maps of accumulated rainfall (mm) over an identical 54-day period in both 2009, which was a more typical year, and 2010. Graphs of daily rainfall depth (mm) over the same period are also shown. The extensive flooding in 2010 resulted from the greater area that received precipitation as well as more intense precipitation, since the number of rain days was similar between the two years.

Circulation of the Upper Atmosphere

Recall from Unit 8 that wind flow in the upper atmosphere is largely geostrophic (perpendicular to the pressure gradient, as shown in Fig. 8.7), with the higher pressure on the right looking downwind in the Northern Hemisphere and on the left looking downwind in the Southern Hemisphere. Furthermore, the general circulation aloft is much simpler without the effects of the land/water contrast that make the surface pattern decidedly cellular. The specific nature of the upper atmospheric circulation is complex, however, so discussion here focuses on key generalizations.

All winds are truly three-dimensional in nature and possess a zonal (west–east), meridional (north–south), and vertical (upward–downward) component. In discussing the conceptual model of the general circulation of the atmosphere shown in Figure 9.2 we noted that meridional flow was strong in the Hadley cell, weak in the polar cell, and negligible in the Ferrel cell. In contrast, the upper atmospheric circulation is dominated by strong zonal flow.

Figure 9.11 shows the zonal mean wind speeds in the Northern Hemisphere averaged over longitude bands for multiyear periods during summer and winter (left and right panels). The shaded area provides a cross-section of wind speed from the surface (1000 mb) to the top of the atmosphere (0 mb). An analogous contour diagram for the Southern Hemisphere would be very similar. The winds of the upper atmosphere generally blow from west to east throughout a broad latitudinal band. Essentially, wind flow is westerly poleward of 15 degrees N. A deep region of easterly winds extends from the surface to the top of the atmosphere equatorward of about 15 degrees N, corresponding to the surface trade winds and equatorial jet stream of the Hadley cell (see Fig. 9.2). A shallow layer of surface easterly winds is also found for the polar cell, but this cell is capped by westerly winds at the mid-troposphere and above. The pressure gradient in the upper atmosphere is not uniform, and two zones of concentrated westerly flow occur in each hemisphere: one in the subtropics and one along the polar front (see Fig. 9.2). These concentrated, high-altitude jets are discontinuous accelerations of air flow within the core of the upper-level westerlies. Jet stream winds are anything

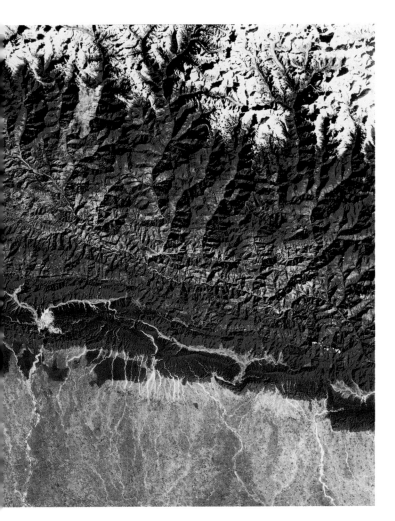

FIGURE 9.9 One of Earth's steepest topographic gradients links the ice-capped Himalayas (top) to the plain of the Ganges River (bottom). This gigantic mountain barrier plays an important role in the Indian monsoon circulation, where in summer low-level winds are steered onto the Gangetic lowland, bringing torrential rains that nourish one of humanity's largest and most heavily populated agricultural regions.

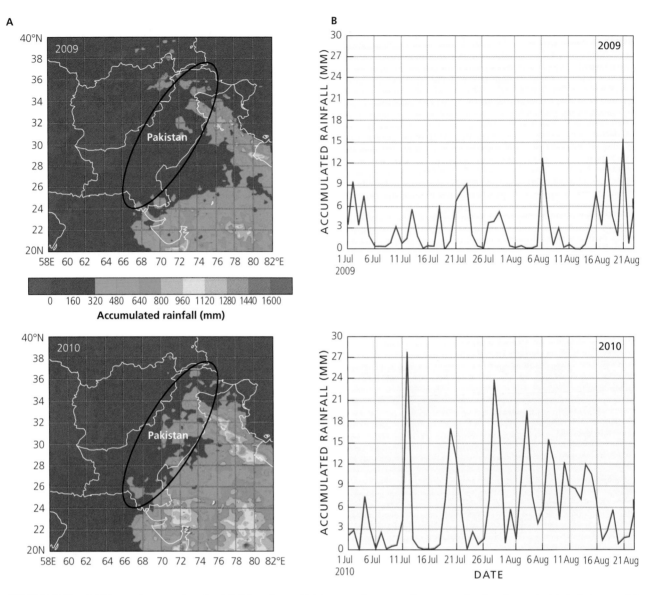

FIGURE 9.10 (A) Maps of accumulated rainfall (mm) over Pakistan and surrounding countries for the period July 1–August 23 for 2009 (top) and 2010 (bottom). (B) Graphs of daily accumulated rainfall (mm) for the same period for a representative area for 2009 (top) and 2010 (bottom). All data are from the Tropical Rainfall Measuring Mission satellite (TRMM).

but constant; they speed up and slow down in response to thermal and dynamic processes in the atmosphere.

The cross-sectional profile of the atmosphere between the equator and the North Pole shown in Figure 9.2 reveals that these two jet streams—appropriately called the subtropical jet stream (discussed earlier in the unit) and the polar jet stream—are located near the tropopause (12 to 17 km [7.5 to 10.5 mi]). The diagram also shows the existence of a third jet stream, the equatorial jet stream, which is a major feature of the opposite, east-to-west flow in the upper atmosphere of the equatorial zone south of 15 degrees N. Interestingly, the equatorial jet stream occurs in the Northern Hemisphere only, whereas the subtropical and polar jet streams exist in both hemispheres. The large north–south meanderings of the two latter jet streams are instrumental in moving large quantities of heated air from the equatorial to higher latitudes. They usually flow at extremely high rates of speed

(occasionally reaching 350 km/h [220 mph]) and can thus achieve the heat and volume transfers that could not be accomplished at the far more moderate velocities associated with the circulations of the cells depicted in Figure 9.2.

The westerly winds addressed in Figure 9.11 extend throughout most of the atmosphere in the middle and polar latitudes and reach maximum speeds at about the 200-mb level. They are strongest and nearest the equator during winter, when the Hadley cell circulation is most strongly developed and the equator-to-pole temperature gradient is greatest. Recall that this region of westerly zonal flow includes two jet streams—the subtropical jet stream and the polar jet stream—yet the patterns in Figure 9.11 show that the zonal winds are dominated by the subtropical jet stream. As will be discussed later, the polar jet stream is typically stronger than what is implied by the average speeds shown in Figure 9.11. The subtropical jet stream, which is at the

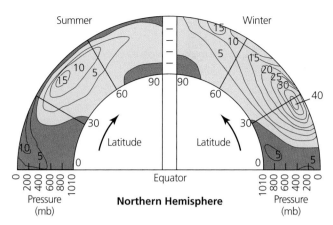

FIGURE 9.11 Zonal mean wind speeds (m/s) by latitude and altitude averaged across all longitudes for the Northern Hemisphere in summer and winter. Easterly zonal winds are dark blue; westerly zonal winds are lightly shaded.

poleward limit of the Hadley cell (see Fig. 9.6), is caused by the conservation of angular momentum. Its strength and latitudinal position are more constant in space and time, and as a result, this stream is particularly apparent in Figure 9.11. The polar jet stream is much more variable, and thus it appears as a more subdued feature of mean circulation. The two jet streams also arise for different reasons. The polar jet stream is caused by the strong temperature gradient present in the middle latitudes.

Figure 9.12 shows the height at which the 500-mb surface is reached in the Northern Hemisphere atmosphere in January; high 500-mb heights indicate high pressure 5 to 6 km (3 to 4 mi) above the surface, and lower heights correspond to lower pressure aloft. (Height maps are particularly useful because the pressure-gradient force can be deduced

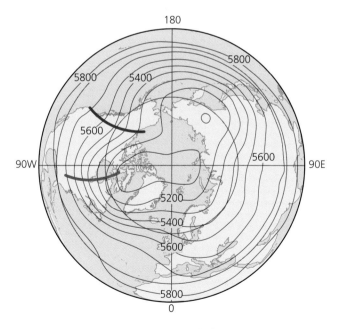

FIGURE 9.12 Mean 500-mb height (m) in the Northern Hemisphere for January. The thick lines show locations of an upper-air ridge (red) and trough (blue).

directly from the contour spacing. If isobars are used instead, it is necessary to also consider density in deriving the pressure-gradient force.) The close spacing of the height contours in the middle latitudes implies a strong pressure-gradient force and strong winds. This is closely related to the sharp horizontal temperature gradients that occur along the polar front. These thermal contrasts cause accelerations in the air flow that produce the polar jet stream.

Another important generalization concerns the frequency of deviations from zonal wind flow in the midlatitudes, particularly above the heart of North America. Here, it is common to find great arcs in the trajectory of upper-air motions. These airstreams curve north and south as they move air generally eastward. These are called **Rossby waves** (or longwaves), which develop as a dynamic response to the equator-to-pole temperature gradient and are also affected by Earth topography and land/ocean contrasts. These waves consist of alternating sequences of *troughs* (areas of low pressure) and *ridges* (areas of high pressure) that cause the geostrophic wind to flow northwesterly and southwesterly around them (Fig. 9.12). These deviations from westerly air flow are important because they amount to substantial meridional (north–south) air exchange. Poleward-moving air carries heat from lower latitudes, and similarly, equatorward-moving air brings cold air from higher latitudes. Thus, Rossby waves help correct the radiation imbalance between the polar and tropical regions.

Rossby waves, and their polar jet stream core, are transient features. Their position and strength fluctuate constantly over short-term as well as seasonal periods of time. They move equatorward and strengthen in winter, when the latitudinal heat gradient is greater, and shift poleward and weaken in summer with the reduced need for poleward energy transfer. Because of this variability in strength and position, they are not easily captured by average values such as those depicted in Figure 9.12. Short periods of high speed are overwhelmed in the average by lengthier periods of slower motion. Only the most persistent waves appear in Figure 9.12, and on any given day, an observer sees wave patterns considerably different from those shown. It is also worth noting that episodes of pronounced meridional flow produce unusual weather for the surface areas they affect. The next time you notice an unseasonable weather event, it is likely to be a result of waves developing in the upper atmospheric pressure pattern.

In a sense, it is misleading to treat the surface and upper atmospheric pressure patterns and their resultant wind flows separately because they are always interrelated. Indeed, strong surface pressure gradients are invariably reinforced by strong upper atmospheric pressure gradients. This is not to say that lower- and upper-level patterns are similar, and in fact, the pattern aloft is usually very different from that at the surface. Still, particular upper-air patterns often lead to particular surface features, and surface features can in turn maintain upper-level patterns. It is also important to keep in mind that there is another component of this cause–effect relationship: the circulation of the world ocean. The enormous influence of the circulating atmosphere on the ocean surface has yet to be examined. That is our next task.

Key Terms

equatorial jet stream *page 111*

equatorial low *page 109*

Ferrel cell *page 114*

general circulation *page 115*

Hadley cell *page 113*

intertropical convergence zone (ITCZ) *page 109*

jet streams *page 111*

meridional flow *page 109*

monsoon *page 115*

Northeast Trades *page 110*

polar cell *page 114*

polar easterlies *page 110*

polar front *page 111*

polar high *page 110*

polar jet stream *page 111*

Rossby waves *page 119*

Southeast Trades *page 110*

subpolar low *page 111*

subtropical high *page 110*

subtropical jet stream *page 111*

westerlies *page 110*

zonal flow *page 109*

Scan Here for a quick vocabulary review

ENHANCE

Review Questions

1. List the seven semipermanent pressure features of the surface atmospheric circulation and give their approximate locations.

2. List the six intervening wind belts that connect these semipermanent highs and lows.

3. Discuss the shifting of these wind and pressure systems in relation to the seasons of the year.

4. What are some of the key differences between the ideal model and the actual pattern of surface atmospheric circulation?

5. Describe the mechanisms of the monsoonal circulation of South Asia.

6. Describe the zonal circulation pattern of the upper atmosphere, its relation to the jet streams, and why meridional flow occurs.

Scan Here for a quick concept review

ENHANCE

www.oup.com/us/mason

UNIT 10

Circulation of the World Ocean

Hydrosphere meets lithosphere on the north shore of Oahu, Hawai'i.

OBJECTIVES

- Describe the relationship of surface ocean circulation to the general circulation of the atmosphere

- Explain the role of the ocean circulation in the transport of heat at Earth's surface

- Describe the major currents that constitute the surface ocean circulation

- Identify the significant water currents that occur below the surface layer of the sea

- Discuss El Niño, the periodic, large-scale abnormal warming of the sea surface at low latitudes of the Pacific Ocean

In this unit, we focus on the large-scale movements of water, known as **ocean currents**, that form the oceanic counterpart to the atmospheric system of wind belts and semipermanent pressure cells we introduced in Unit 9. The two systems are closely integrated, and we examine that relationship in some detail. Ocean currents affect not only the 71 percent of the planet covered by the world ocean but the continental landmasses as well. We must become familiar with the global oceanic circulation because it is vital to our understanding of the weather and climate of Earth's surface.

The Coupled Ocean–Atmosphere System

The solar radiation that is absorbed at the surface of Earth provides the energy to heat both the atmosphere and the oceans. The troposphere (see Fig. 6.10) is heated mostly by the absorption of longwave radiation emitted from Earth's surface and by convection from the surface. Air temperature decreases with vertical distance from these heat sources, which results in a naturally unstable air layer. Consequently, the troposphere experiences active vertical mixing between the warm surface air and cooler air at higher altitudes. Absorption of solar radiation also heats the oceans, but it produces a pattern in which warm water lies on top of progressively colder water (Fig. 10.1). This stable temperature profile suppresses vertical mixing in most of the world's ocean waters.

OCEAN TEMPERATURE ZONES

The world's oceans are divided into three zones based on temperature (see Fig. 10.1). The thickness and even existence of the zones vary considerably from place to place, but it is useful to consider global average values. The **mixed layer** (or surface layer) extends to about 75 m (245 ft) in depth and contains about 3 percent of the total ocean water

volume. Active mixing by wind-generated waves and surface ocean currents results in nearly constant temperature and salinity conditions within this thin layer. The temperature of the mixed layer approximates the average temperature of the near-surface air layer. The **thermocline** is the layer that extends beneath the mixed layer to a depth of about 1,000 m (3,300 ft) and is marked by relatively sharp changes in temperature with depth. The **deep-ocean layer** occurs at depths greater than 1,000 m and includes about 97 percent of the total ocean water volume. Water temperatures are nearly uniform with depth, varying between 1° and 3°C (33.8° and 37.5°F). All three layers can be observed in the tropical, midlatitude, and polar oceans of the world, although the depths and steepness of the thermocline within each layer vary with latitude (see Fig. 10.1).

The atmosphere and world ocean are closely coupled through multiple interacting processes. Wind-flow patterns in the atmosphere create surface ocean currents and, through cold temperatures at higher latitudes, have a strong influence on the deep-sea circulation. The ocean also affects the atmosphere through the transport of heat from low to higher latitudes and the capacity of the world ocean to store huge amounts of heat.

Like the global atmospheric circulation, the world ocean is a significant transporter of heat from equatorial to polar regions. On a global basis, oceanic circulation accounts for approximately 25 percent of the total transport of heat from low to higher latitudes; atmospheric circulation is responsible for the other 75 percent. However, in the broad zone between about 30 degrees N and S of the equator, the oceans account for the majority of the poleward heat transfer (see Fig. 8.3).

Perhaps the single most important influence of the world ocean on the atmosphere is due to the water's heat-storage capacity. (You can review the concept of heat capacity in Unit 7.) The oceans act as a vast heat reservoir, and the mixed layer actively exchanges heat, mass, and momentum with the atmosphere. The top 10 m (33 ft) of the ocean contain the equivalent mass of the entire atmosphere. Because of the greater density of water compared to air and the much higher specific heat of water, the heat capacity of the mixed layer is nearly 30 times greater than that of the atmosphere. The total heat capacity of the oceans is about 1,000 times the heat capacity of the atmosphere. Heat exchanges between the ocean and the atmosphere take place over short timescales of days to weeks, as well as longer timescales of seasons to centuries, and are an important internal driver of global weather and climate.

SURFACE CURRENTS

Through the circulation of water masses in large-scale currents, the world ocean plays a vital role in constantly adjusting Earth's surface heat imbalance. Although the sea contains numerous horizontal, vertical, and even diagonal currents at various depths, almost all of the oceanic heat-transfer activity takes place via the operation of horizontal currents in the uppermost 75 m (245 ft) of water. Thus, most of our attention in

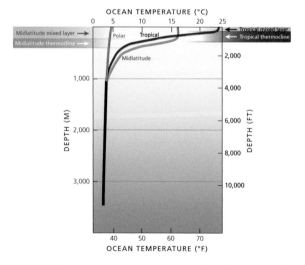

FIGURE 10.1 Typical temperature profiles illustrating the mixed, thermocline, and deep-ocean layers in tropical, midlatitude, and polar oceans.

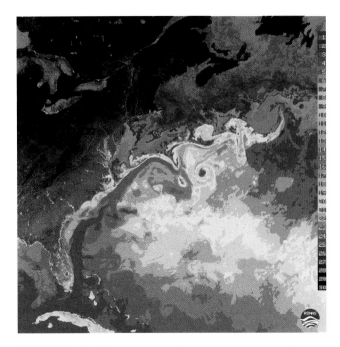

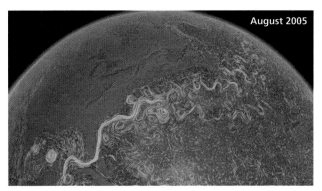

FIGURE 10.3 Ocean surface currents in the western North Atlantic Ocean in August 2005. The Gulf Stream follows a narrow continuous path until North Carolina, where it begins to meander more actively, ultimately turning into the more diffuse northeasterly-trending North Atlantic Drift. Multiple warm-core and cold-core eddies are evident in the animation.

ENHANCE

FIGURE 10.2 This sea surface temperature image of the western North Atlantic Ocean shows a warm current, the Gulf Stream, as it skirts the North American coast from Florida to North Carolina. The Gulf Stream flows northeastward into the open ocean and becomes the less clearly defined North Atlantic Drift. This drift cools as it moves into the North Atlantic, yet the warmth that is carried north- and northeastward into the middle latitudes has a major influence on atmospheric environments in Western Europe. A cold current, the Labrador Current, flows south and is wedged between the Gulf Stream and the northeastern United States. The color chart at right indicates approximate temperatures in degrees Celsius.

this unit is directed toward the 3 percent of the total volume of the world ocean that constitutes this mixed layer.

Although they transport massive volumes of water, most global-scale ocean currents differ only slightly from the surface waters through which they flow. So-called warm currents, which travel from the tropics toward the poles, and cold currents, which move toward the equator, usually exhibit temperatures that deviate by only a few degrees from those of the surrounding sea, although larger ranges in temperature can be found at the regional scale (Fig. 10.2). Although small, these temperature differences are often sufficient to markedly affect atmospheric conditions over a wide area because of complex ocean–atmosphere interactions. Compared to a cold water surface, warm ocean waters provide more longwave radiation as well as sensible and latent heat to the overlying atmosphere. These changes make the lower atmosphere more unstable, which ultimately strengthens surface winds and storm systems. The lower atmospheric heating associated with cold ocean waters stabilizes the atmosphere and weakens storm systems.

In their rates of movement, too, most currents are barely distinguishable from their marine surroundings. Currents tend to move slowly and steadily and are often embedded within larger pools of more slow-moving water. The Gulf Stream (see Fig. 10.2), for example, is only several hundred

meters deep and travels at an average speed of only 6.4 km/h (4 mph). Ocean currents are often called **drifts** because they lag far behind the average speeds of surface winds blowing in the same general direction. Faster-moving currents are usually found only where narrow straits squeeze the flow of water, such as between Florida and Cuba (Fig. 10.3) or in the Bering Sea between Alaska and northeasternmost Russia.

Generation of Ocean Currents

Ocean currents are generated in several ways. The most important factor is the kinetic energy of the prevailing surface winds. (Recall from Unit 7 that kinetic energy is the energy of movement.) The kinetic energy of surface winds imparts momentum to the ocean surface, which in turn generates the kinetic energy of surface water movement. Once set in motion, as is the case with moving air, the water is subjected to the deflective Coriolis force (see Unit 8). That is, the water experiences deflection relative to the wind flowing over the water. As a rule, the Coriolis force steers the surface current to flow at an angle of about 45 degrees to the right of the wind in the Northern Hemisphere and at approximately 45 degrees to the left in the Southern Hemisphere. This surface motion also influences waters to a depth of around 100 m (330 ft). Within this column, the motion in each underlying water layer is increasingly to the right in the Northern Hemisphere (and to the left in the Southern Hemisphere) as depth increases and exhibits a decreasing speed of flow.

Another source of oceanic circulation, which largely affects the deep-ocean zone below the mixed layer, is variation in the density of seawater. (Recall from Unit 6 that density is mass per unit volume.) Density differences can arise from temperature differences or from variations in salinity. Freshwater has a maximum density at 4°C (39°F), and its density declines above and below this value. If a lake or a low-salinity sea cools enough, the coldest water will be at the surface. For most of the ocean, however, density increases with progressively decreasing temperature. The salinity relationship is direct: higher salinity content always increases the density of water. When evaporation exceeds precipitation, which occurs in the subtropical oceans, a local increase in the density of surface ocean water occurs. When precipitation outpaces evaporation, which occurs in the middle and polar latitudes, there is a local decrease in surface water density. When the ocean freezes salt is not incorporated into the ice but is instead left behind in the surrounding seawater. Thus freezing increases the salinity and reinforces the increase in density caused by cooling. Low temperatures and the seasonal exclusion of salt from sea-ice formation in polar oceans during winter results in the formation of cold, saline, dense surface water in the high latitudes that slowly sinks. We discuss this phenomenon later in this unit in more detail. The slow downward diffusion of dense water is primarily driven by salinity differences, not temperature differences. This is because the water temperature is nearly constant in the polar ocean during winter (see Fig. 10.1).

Wind-driven Flow

In our discussion of the ocean circulation so far, we have for the most part been dealing with models that describe ideal situations. But it would be erroneous to presume that a large-scale ocean current is an unswerving river of water that follows the same exact path and exhibits constant movement characteristics. Deviations from the "norm" occur all the time.

Most ocean currents develop riverlike *meanders*, or curving bends, not unlike the polar jet stream we discussed in Unit 9. These meanders can become so pronounced (especially after the passage of storms) that many detach and form localized *eddies*, or loops, that move along with the general flow of water. These phenomena are most common along the boundaries of currents, where opposing water movements heighten the opportunities for developing whorl-like local circulation cells. Figure 10.3 illustrates such a situation along the western edge of the warm Gulf Stream current off the Middle Atlantic coast of the United States. The narrowing of the current as it accelerates between Florida and Cuba, and the formation of clockwise-rotating warm-core eddies and counterclockwise-rotating cold-core eddies to the north and south, respectively, of the Gulf Stream are clearly visible.

GYRE CIRCULATIONS

Prevailing winds, the Coriolis force, and sometimes the configuration of bordering landmasses combine to channel ocean currents into cell-like circulations that resemble large cyclones and anticyclones. In the ocean basins, these continuously moving loops are called **gyres**, a term used for both clockwise and counterclockwise circulations. Gyres, in fact, are so large that they can encompass an entire ocean. Because ocean basins are usually more extensive in width than in length, most gyres assume the shape of elliptical cells elongated in an east–west direction.

The ideal model of gyre circulation in the world ocean, shown in Figure 10.4, displays a general uniformity in both

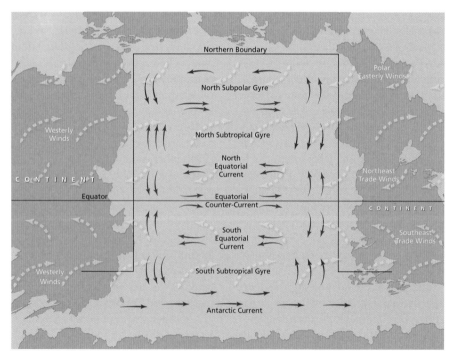

FIGURE 10.4 Generalized pattern of currents in a typical ocean basin, showing the major circulation cells (gyres) and their influencing wind systems. Warm currents are shown in red; cold currents are in blue. Note that the Northern Hemisphere differs from the Southern Hemisphere because of the variation in continental landmass configurations.

the Northern and Southern Hemispheres. As with the model of the general circulation of the atmosphere (Fig. 9.2), these hemispheric flow patterns essentially mirror one another. Each hemisphere contains *tropical* and *subtropical gyres*. Differences in high-latitude land/water configurations, which are shown in Figure 10.4, give rise to a fifth (subpolar) gyre in the Northern Hemisphere that is not matched in the Southern.

Subtropical Gyres   The **subtropical gyre** dominates the oceanic circulation of both hemispheres. In each case, the subtropical gyre circulates around the subtropical high-pressure cell that is stationed above the center of the ocean basin (see Fig. 9.2). The two clockwise-circulating gyres of the Northern Hemisphere are found beneath two such high-pressure cells: the Pacific (Hawai'ian), centered high over the North Pacific Ocean, and the Bermuda (Azores), centered high above the North Atlantic Ocean. In the Southern Hemisphere, three subtropical gyres each exhibit a counterclockwise flow trajectory; these are located beneath the three semipermanent zones of subtropical high pressure, respectively centered over the South Pacific Ocean, the South Atlantic Ocean, and the southern portion of the Indian Ocean (see Fig. 9.2).

The broad centers of each of these five gyres are associated with subsiding air and generally calm wind conditions and are therefore devoid of large-scale ocean currents. The currents are concentrated along the peripheries of the major ocean basins, where they constitute the various segments, or *limbs*, of the subtropical gyre. Along their equatorward margins, the subtropical gyres in each hemisphere carry water toward the west. The strong tropical sun gradually heats this water as it slowly moves westward. These currents diverge as they approach land. Some of the water is reversed and transported eastward along the equator as the Equatorial Countercurrent, but most of the flow splits and is propelled poleward as **warm currents** along the western edges of each ocean basin.

As Figure 10.4 illustrates, when polar waters are encountered in the upper midlatitudes, the currents swing eastward across the ocean. In the Northern Hemisphere, these eastward drifts remain relatively warm currents because landmasses block the colder waters of the subpolar gyre to their north from mixing with this flow. When they reach the eastern edge of the ocean, the somewhat-cooled waters of the subtropical gyre turn toward the equator and move southward along the continental coasts. These **cold currents** parallel the eastern margins of the ocean basins and finally converge with the equatorial currents to complete the circuit and once again form the westward-moving (and now warming) equatorial stream.

Gyres and Wind flow   The wind arrows in Figure 10.4 remind us that a subtropical gyre circulation is continuously maintained by the operation of the wind belts above it. The trade winds in the tropical segments of the eastern margin of an ocean work in conjunction with the westerlies in the midlatitude segments of the western periphery to propel the circular flow. If you blow lightly across the edge of a cup of coffee, you will notice that the liquid begins to rotate. Similarly, with two such reinforcing wind sources to maintain circular flow, it is no wonder that these gyres never cease.

The circulatory interaction between the sea and the overlying wind flows is more complicated in the case of the Northern Hemisphere **subpolar gyres**, where landmasses and sea ice interrupt surface ocean flows. Along the southern limbs of these gyres, westerly winds drive a warm current across the entire ocean basin. This flow remains relatively warm because, as indicated earlier, the southward penetration of cold Arctic waters is largely blocked by continents. The cold waters squeeze through the narrow Bering Strait in the northernmost Pacific and the slender channel separating Canada and Greenland in the northwestern North Atlantic. In the more open northeastern Atlantic between Greenland and Europe, a branch of the warm eastward-moving drift enters the subpolar gyre. This warm drift is frequently driven across the Arctic Circle by a reinforcing southwesterly tailwind associated with storm activity along the polar front. The subpolar gyres are nonexistent in the Southern Hemisphere. Instead, an eastward-moving cold current is propelled by the westerlies, which, in the absence of continental landmasses, create a circumpolar vortex encircling Antarctica.

Upwelling   Another feature of the subtropical gyre circulation is represented in Figure 10.4 by the wider spacing of arrows that show currents along the eastern sides of the ocean basins. This wider spacing indicates that surface waters are not squeezed against the eastern edge of the basins, as they are against the west. It also suggests the presence of an additional influence that reinforces the actions of prevailing winds—*upwelling*. The subtropical highs produce equatorward surface winds along the eastern edges of the subtropical oceans (the western edges of the subtropical continents), as shown in Figure 9.2. As mentioned earlier, the surface winds produce a surface ocean current that is deflected at a 45 degree angle to the right of the wind direction in the Northern Hemisphere. This surface motion also influences the waters below to a depth of around 100 m (330 ft). Within this column, the motion in each underlying water layer is deflected increasingly to the right (or, in the Southern Hemisphere, to the left) as depth increases and exhibits a decreasing speed of flow. The net effect of this process over the 100-m column of water is a net transport of water to the right in the Northern Hemisphere (and to the left in the Southern Hemisphere) (Fig. 10.5). This *Ekman transport*, as it is called, produces **upwelling**, or the rising of cold water from the ocean depths to the surface, and westward divergence of coastal waters from continental coastlines. As warmer surface waters are transported out to sea, they are replaced by this cold water, which lowers the surface air temperatures and the local rate of evaporation.

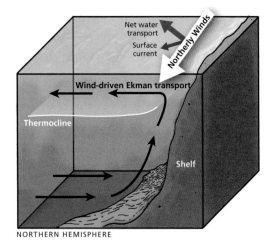

FIGURE 10.5 Schematic view showing how currents along a north–south-trending coast generate upwelling through Ekman transport. Winds are equatorward in both cases. The Coriolis force deflects ocean movements to the right (in the Northern Hemisphere) and left (in the Southern Hemisphere) of the wind at progressively larger angles with increasing depth. For the top 100 m (330 ft), the net transport is at a 90-degree angle, which means the currents move water westward in both hemispheres.

Not surprisingly, some of the driest coastal areas on Earth are associated with upwelling, particularly in latitudes under the influence of the semipermanent subtropical high-pressure cells.

Subtropical west coasts of continents adjacent to upwelling zones experience aridity. Much of coastal northwestern Mexico, as well as the parts of northern Chile and Peru that border the Pacific, exhibit desert conditions. Moreover, the areas of northwestern and southwestern Africa bordering the Atlantic upwelling zones contain two of the driest deserts in the world—the Sahara and the Namib, respectively.

Upwelling does produce one important benefit for humans: it carries nutrients to the surface that support some of the most productive fishing grounds in the world

ocean. Because the productivity of ocean phytoplankton is primarily limited by nutrients, upwelling from the deeper ocean along coastal upwelling zones produces localized areas of extremely high productivity (Fig. 10.6). Areas with consistently strong surface winds, such as the winds that occur within zones of low surface air pressure, are also areas of relatively high productivity. Areas that experience weak or calm winds and *downwelling* associated with semipermanent high-pressure cells are largely devoid of nutrients and are biological deserts.

The Geography of Ocean Currents

The basic principles of oceanic circulation are now familiar to us, and we can apply them at this point to the actual distribution of global-scale surface currents mapped in Figure 10.7. These currents do respond to seasonal shifts in the wind belts and semipermanent highs and lows. However, those responses are minimal because seawater motion changes quite slowly and usually lags weeks or even months behind shifts in the atmospheric circulation. The geographic pattern of ocean currents in Figure 10.7 reflects the average annual position of these flows. But most currents deviate only slightly from these positions. On the world map, the only noteworthy departures involve the reversal of smaller-scale currents under the influence of monsoonal air circulations near the coasts of southern and southeastern Asia.

Look closely at Figure 10.7 and take note of one surprising feature: the existence of eastward-moving water near the equator. Called *Equatorial Countercurrents*, this water runs counter to the prevailing trade winds and the surrounding westward *Equatorial Currents*. Equatorial Countercurrents also run counter to our intuition because, although they are wind driven, they carry water against the prevailing winds. The explanation of this behavior is too complex for an introductory text, but we can say that the Northern Hemisphere versions result from interplay between a slight slope in the water surface and variations in the Coriolis force with latitude. The southern Equatorial Countercurrents are not so well understood.

Pacific Ocean Currents The Pacific Ocean's currents closely match the model of gyre circulations displayed in Figure 10.4. In both the Northern and Southern Hemisphere components of this immense ocean basin, surface flows are dominated by the subtropical gyres. In the North Pacific, the limbs of this gyre are the clockwise flow of the North Equatorial, Japan (Kuroshio), North Pacific, and California Currents. Because the Bering Strait to the north admits only a tiny flow of Arctic seawater to the circulation, all of this gyre's currents are warm except for the California Current. That current is relatively cold as a result of upwelling and its distance from the tropical source of warm water. The lesser circulations of the tropical and

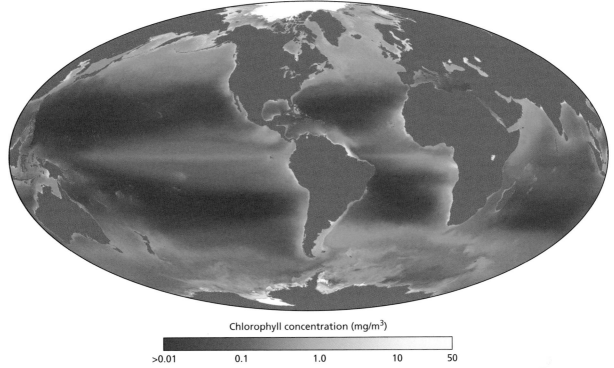

ENHANCE

FIGURE 10.6 Global chlorophyll concentrations (a primary indicator of the presence of phytoplankton) averaged over a 9-year period from 1997–2006 as measured by the SeaWiFS satellite sensor. Highest chlorophyll concentrations (in yellow) are found in upwelling zones; areas of moderate chlorophyll concentration (light blue) correspond to regions of strong surface winds or shear zones between opposing surface ocean currents; dark blue areas show regions of very limited ocean productivity and correspond to subtropical high-pressure cells.

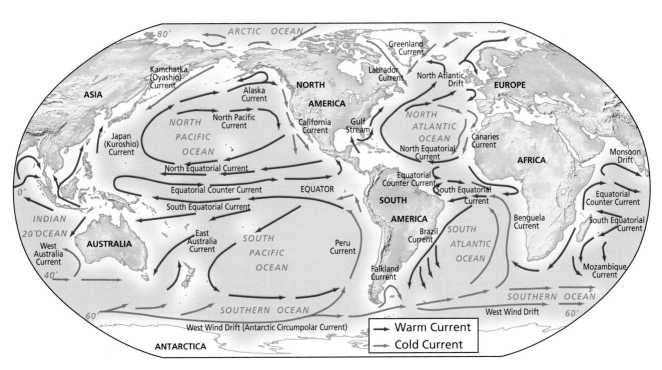

FIGURE 10.7 World distribution of surface ocean currents, showing average positions and relative temperatures in each of the ocean basins.

subpolar gyres are also evident in the North Pacific. The **tropical gyre** encompasses the low-latitude Equatorial Countercurrent and the North Equatorial Current. The subpolar gyre consists of the upper midlatitude loop of the North Pacific and the Alaska and Kamchatka (Oyashio) Currents.

The South Pacific Ocean's flow pattern is a mirror image of the Northern Hemisphere flow pattern except for the fully expected replacement of the subpolar gyre by the globe-encircling movement of the West Wind Drift (the Antarctic Circumpolar Current), where the Pacific gives way to the Southern Ocean at approximately 45 degrees S. The strong subtropical gyre that dominates the circulation of the South Pacific includes the South Equatorial, East Australian, West Wind Drift, and Peru Currents. The South Pacific's tropical gyre comprises the Equatorial Countercurrent and the South Equatorial Current.

Atlantic Ocean Currents The Atlantic really consists of two ocean basins because of the hourglass-like narrowing between South America and Africa in the vicinity of the equator. Nevertheless, the overall pattern of Atlantic currents is quite similar to that of the Pacific. The subtropical gyre—composed of the North Equatorial Current, Gulf Stream, North Atlantic Drift, and Canaries Current—dominates circulation in the North Atlantic Ocean. The subpolar gyre, as we noted earlier in this unit, is modified by sea ice and high-latitude land bodies, but the rudiments of circular flow are apparent on the map. In the equatorial latitudes, despite the east–west proximity of continents, both tropical gyres have enough space to develop their expected circulations.

In the South Atlantic Ocean, the subtropical gyre is again dominant, with the warm waters of the Brazil Current bathing the eastern shore of South America and the cold waters of the Benguela Current, reinforced by upwelling, paralleling the dry southwestern African coast. South of 35 degrees S, the transition to the Southern Ocean and its West Wind Drift is identical to the pattern of the South Pacific.

Indian Ocean Currents The Indian Ocean's circulatory system is complicated by the configuration of the surrounding continents. There is an eastward opening to the Pacific in the southerly low latitudes, and the northern part of the Indian Ocean is closed off by the South Asian coast. Still, we can observe some definite signs of subtropical gyre circulation in the Indian Ocean's Southern Hemisphere component. Moreover, a fully developed pair of tropical gyres in the latitudinal zone straddling the equator operates for a good part of the year. During the remaining months, as we explained in the discussion of the wet monsoon in Unit 9, the intertropical convergence zone is pulled far northward onto the Asian mainland. This temporarily disrupts the air and even the oceanic circulations that prevail north of the equator between October and June.

DEEP-SEA CURRENTS

Recall that significant water movements occur below the mixed layer of the sea. Another complete global system of currents exists below that level, where the bulk (approximately 97 percent) of the world ocean's water lies. This deep-ocean circulation operates in sharp contrast to the surface circulation, which interfaces with the atmosphere and is largely driven by momentum transfer from prevailing surface winds.

The deep-sea system of oceanic movement is categorized as a **thermohaline circulation** because it is controlled by differences in the temperature and/or salinity of water masses. Thermohaline circulation involves the flow of currents driven by differences in water density. The temperature and salinity differences that trigger thermohaline circulation are most strongly developed in the North Atlantic Ocean. The poleward-moving Gulf Stream waters are cooled by emission of longwave radiation as well as by sensible and latent heat transfer to the overlying polar air (see Unit 5). As the surface water cools and becomes denser, some of the water sinks and enters the deep circulation; the remainder moves equatorward to feed the cold surface currents on the eastern side of the oceans. The freezing of surface seawater in winter, which increases the salinity—and density—of the water just under the ice (which consists mainly of nonsaline freshwater), is another important driver of the sinking motion. The cold sunken water moves equatorward at depth as a slow, persistent meandering current that is channeled by the seafloor topography and adjacent cold deep-ocean layers.

Thermohaline circulation forms a global system often called the **global conveyor belt** that moves masses of water throughout the world ocean. As Figure 10.8 illustrates, each ocean basin has its own deep-water circulation, with the only interoceanic exchanges occurring in the depths of the Southern Ocean. Note, too, that cold bottom water from each polar oceanic zone flows into different oceans: the deep-sea currents of the North and South Atlantic emanate from the Arctic Ocean, whereas those of the Pacific and Indian Oceans are generated in the waters surrounding Antarctica. Figure 10.8 also shows that the deep-sea limb of the global conveyor belt is connected to a surface-layer limb of warm water that transports vast amounts of heat energy among the same ocean basins. The rate at which cold surface waters sink in the North Atlantic and flow within the deep-ocean circulation varies over time and is believed to contribute to both short- and longer-term climate change. The conveyor belt system illustrated in Figure 10.8 is a highly simplistic model. Although useful for organizing our thinking about ocean circulation, motions at depth are far more complex and not known in detail even today.

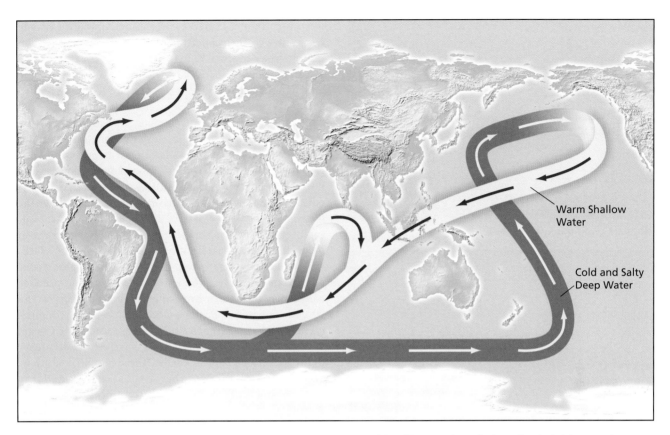

FIGURE 10.8 A global conveyor belt is often used as a conceptual model to illustrate how the largely wind-driven surface ocean currents are linked to the largely density-driven deep-ocean currents. The mixed layer features shallow, narrow, and relatively fast-moving horizontal currents, while the deep-ocean currents move as slow, diffuse surges with both horizontal and vertical movement.

El Niño–Southern Oscillation

The global climate responds to external forces, such as variations in the energy output of the Sun, as well as internal variations from interaction between numerous components, such as the land, oceans, ice, and biosphere. Linkages between the coupled ocean–atmosphere system are currently the subject of intensive climatological research. Much of this research has focused on the oscillating ocean–atmosphere system known as the **El Niño–Southern Oscillation (ENSO)** that occurs within the largest stretch of equatorial water—the vast central Pacific that sprawls across 160 degrees of longitude between Indonesia and South America.

Intermittent anomalies in seawater temperature off the coasts of Peru and Ecuador in northwestern South America have long been known, but their larger significance became apparent only during the 1980s. The warming of coastal Pacific waters in this region by about 2°C (4°F) is, in fact, a yearly occurrence. The fish catch temporarily is reduced when a local southward-drifting warm current suppresses the usually present upwelling (see Fig. 10.6) that transports nutrients crucial to the surface-layer food chain. Because this three-month phenomenon usually arrives around Christmastime, it is called *El Niño* ("the child"), a

reference to the Christ child. The seasonal El Niño pattern can sometimes linger up to three years and be accompanied by an expanded zone of warm coastal waters. During an El Niño event, the upwelling of cold water associated with the Peru Current ceases, and the absence of nutrients results in massive fish kills and the decimation of the bird population (which feeds on the fish). In addition, Ecuador and Peru experience increases in rainfall that can result in crop losses as well as severe flooding in the heavily populated valleys of the nearby Andes Mountains. These localized El Niño effects are but one symptom of a geographically much wider anomaly in ocean–atmosphere interactions in the equatorial Pacific Ocean.

Climatologists first examined this phenomenon using the Southern Oscillation Index, a measure based on surface-pressure differences between Darwin, Australia, in the western equatorial Pacific and Tahiti in the central equatorial Pacific. More recently, they have used a variable called the Oceanic Niño Index (ONI) derived from sea surface temperature to study the feature. The nature of ocean–atmosphere coupling in the equatorial Pacific Ocean during three ranges of values for the ONI is illustrated in Figure 10.9.

Under so-called *neutral conditions*, the eastern equatorial Pacific Ocean is an area of higher atmospheric

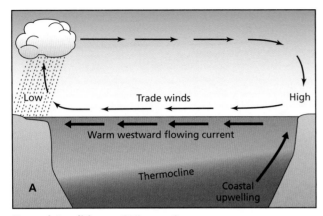

Normal Conditions – ONI near 0

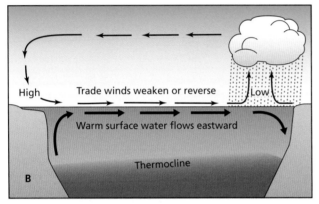

El Niño Conditions ONI Positive (> 0.5)

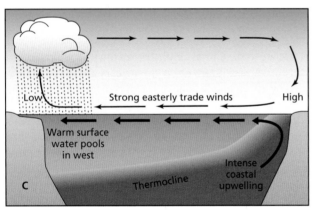

La Niña Conditions ONI Negative (< -0.5)

FIGURE 10.9 *Ocean–atmosphere coupling and atmospheric interaction in the equatorial Pacific Ocean during the normal or standard pattern (A), El Niño (cold) phase (B), and La Niña (warm) phase of the ENSO Index (C).*

ENHANCE

pressure than the western equatorial Pacific Ocean. Intense sinking of air along the eastern edges of the subtropical high-pressure cells (see Fig. 10.7) produces converging trade winds along the ITCZ (see Unit 9) that propel the westward-flowing Equatorial Current. The cold Peru and California Currents also converge along this zone and flow

westward. The cold surface currents and upwelling of cold deep water create a condition of atmospheric stability that inhibits the air from rising to form the equatorial low-pressure trough. The result is clear skies, low precipitation, and high surface pressure. A thermal circulation cell called the *Walker circulation* develops, not unlike the Hadley cell circulation. The westward-flowing surface winds of the Walker circulation drive a westerly equatorial surface ocean current toward the semipermanent low in the western equatorial Pacific that is positioned over Indonesia and northern Australia. As Figure 10.9A shows, a cell of air circulation forms above the equator, with air rising above the western low, flowing eastward at high altitude, and subsiding over the eastern Pacific. Sea surface temperatures in the western equatorial Pacific may be as much as 8°C (14°F) warmer than surface waters in the east. The high sea surface temperatures heat the lower atmosphere and produce convective instability and intense precipitation in the western equatorial Pacific. Ocean–atmosphere coupling causes the thermocline to rise in the eastern Pacific Ocean, and water levels as much as 0.5 to 1.0 m (1.5 to 3.0 ft) higher in the western equatorial Pacific. These normal conditions occur when ONI values are between –1.0 and 1.0, and they are an example of a positive feedback process (see Unit 1). A strengthening Walker circulation leads to stronger convection in the western equatorial Pacific. This in turn drives a more vigorous upper-level return flow and more intense sinking of air in the eastern equatorial Pacific. This process further strengthens the Walker circulation because the building high-pressure system leads to stronger easterly trade winds and surface ocean currents and enhanced coastal upwelling.

For reasons that are still not fully understood, there is a natural collapse of both this pressure difference between the eastern Pacific high and the western Pacific low and the resultant westward surface wind flow. The weakening trade winds set into motion a series of steps that produce a reversal in the flow of equatorial water and wind as the Walker circulation cell operates in the opposite direction (Fig. 10.9B). These steps also involve positive feedback processes. The piled-up warm water in the western Pacific surges back to the east as the greatly enhanced Equatorial Countercurrent, and the eastern Pacific equatorial zone becomes overwhelmed by water with temperatures as much as 8°C (14°F) higher than normal. Upwelling of cold water in the west and the surge of warm water in the east first weaken and then reverse the western Pacific low and the eastern Pacific high, bringing dry conditions to the normally wet western Pacific and heavy rain to the eastern Pacific. Moreover, as Figure 10.9B illustrates, these events are accompanied by a subsurface infusion of warm water, which causes the Peruvian upwelling to cease and reinforces the anomalous heating of the ocean surface. These conditions, referred to as an *El Niño event*, correspond to ONI values greater than 0.5.

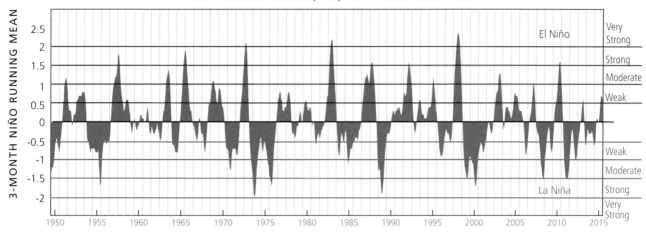

OCEANIC NIÑO INDEX (ONI) 1950 - FEBRUARY 2015

FIGURE 10.10 Variation in the Oceanic Niño Index from 1950 to present. Values above 0.5 indicate El Niño conditions. Values below –0.5 are considered La Niña events.

The effects of ENSO spread far beyond the equatorial Pacific because they are associated with tremendous variations in the amount of latent heat and moisture added to the atmosphere. Heat and moisture transfer to subtropical and extratropical regions through the atmospheric circulation and affect weather and climate in various parts of the globe. Climatologists today rank the phenomenon as a leading cause of disturbance in global weather patterns and the major source of interannual climatic variability within about 30 degrees of the equator. Figure 10.10 shows the occurrence of Niños over the last 65 years and the irregular intervals at which they come and go. As of this writing, the El Niño of 1997–1998 was the strongest ever observed. Sea surface warming between Indonesia and Peru more than doubled the expected ENSO temperature anomaly. Heavy rains and disastrous flooding occurred in Ecuador and Peru, eastern China, and parts of the U.S. Pacific Coast. The other abnormality caused by El Niño was drought, which affected interior Australia, Indonesia, northern India, southern Africa, and northeastern South America.

ENSO-related studies now recognize a third phase of the ENSO oscillation called **La Niña** (Fig. 10.9C). This phase corresponds to an amplification of normal ENSO conditions and occurs at ONI values less than –0.5. These *cold events* represent an opposite extreme to a fully developed El Niño and are associated with their own impacts upon distant weather patterns. The map of cumulative rainfall from a five-month period of a strong La Niña phase in Figure 10.11 illustrates this pattern.

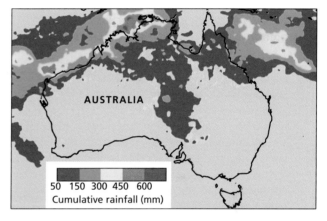

FIGURE 10.11 Cumulative rainfall totals from eight tropical cyclones during January–April 2011 over Australia. Northwestern Australia received over 600 mm (23.6 in) of rainfall during this extreme La Niña phase of the Southern Oscillation.

Studies of ENSO prompted scientists to search for other large-scale anomalies that may affect global weather patterns. They are particularly interested in such influences on the climate of heavily populated regions in the middle latitudes (see Focus on the Science: The North Atlantic Oscillation). This work continues today on a number of fronts and has forged a relatively new subfield of climatology known as *teleconnections*—the study of long-distance linkages between weather patterns.

The North Atlantic Oscillation

When ENSO came to be better understood, questions soon arose about possible "oscillations" elsewhere above the global ocean, perhaps with similar influence over regional climate. The **North Atlantic Oscillation (NAO)** is now recognized as the dominant driver of variations in winter weather for eastern North America and Europe. The NAO is an alternating pressure gradient between the subpolar low-pressure system identified in Unit 9 as the Icelandic Low and the subtropical high-pressure system known as the Bermuda High, especially its eastern segment centered above the Azores Islands and referred to as the Azores High.

The NAO Index is based on the pressure difference between average sea-level pressure in Iceland, near the Icelandic Low, and in Lisbon, Portugal, near the Azores High. A positive NAO Index value is associated with higher than average pressure differences between the pair of stations, producing the circulation patterns and climate anomalies shown in Figure 10.12. The enhanced pressure gradient intensifies the polar jet stream, creating strong upper-level winds, a more zonal flow pattern, and a northward shift in the position of the polar jet stream. Strong surface westerlies across the North Atlantic cause storms to track poleward, resulting in cool and wet summers and warm and wet winters across northern Europe and dry conditions in the Mediterranean Basin. Fewer cold-air outbreaks occur across the interior of eastern North America, bringing mild and wet winters to the eastern United States. A negative NAO Index value weakens the pressure difference between the Icelandic Low and the Azores High, causing the polar front jet to shift southward, weaken, and assume a more meridional flow pattern (see Unit 9). Storm systems track across southern Europe and the Mediterranean Sea, bringing cool and wet conditions, while cold and dry weather is present across northern Europe. The stronger meridional flow across North America results in more frequent cold-air outbreaks and cold and snowy weather in the eastern United States.

The NAO follows multiyear and slowly changing oscillations, but no regular temporal pattern has been identified as yet. Seasonal seesawing does occur, but there are also times when the Icelandic Low develops late, or not at all, so that the westerlies are all but stopped, and cold air has free reign over Europe, even in summer. Conversely, the Icelandic Low may persist for several years, right through the winters, giving Europe warm, moist winters and stormy summers. The NAO and other recently discovered oscillations are

known to be important sources of interannual to decadal climate variability, yet a clear physical explanation of their cause has yet to be discovered.

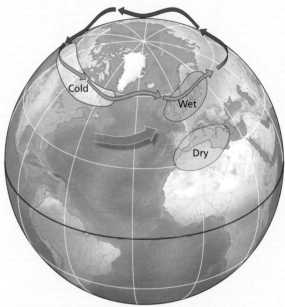

Positive NAO phase

Negative NAO phase

FIGURE 10.12 General circulation patterns and climate anomalies associated with the positive (top) and negative (bottom) phases of the North Atlantic Oscillation. The NAO is the most important internal driver of climatic variability affecting Europe and northeastern North America.

Key Terms

cold current *page 125*

deep-ocean layer *page 122*

drift *page 123*

El Niño–Southern Oscillation (ENSO) *page 129*

global conveyor belt *page 128*

gyre *page 124*

La Niña *page 131*

mixed layer *page 122*

North Atlantic Oscillation (NAO) *page 132*

ocean current *page 122*

subpolar gyre *page 125*

subtropical gyre *page 125*

thermocline *page 122*

thermohaline circulation *page 128*

tropical gyre *page 128*

upwelling *page 125*

warm current *page 125*

ENHANCE

Scan Here for a quick vocabulary review

Review Questions

1. Describe the ways ocean currents develop.
2. What is the relationship between the subtropical gyres and the overlying atmospheric circulation?
3. Name the major currents of the Pacific Ocean and their general flow patterns.
4. Name the major currents of the North and South Atlantic Oceans and their general flow patterns.
5. Describe the general pattern of deep-sea currents. What factors influence those currents?
6. Briefly describe what is meant by El Niño–Southern Oscillation and discuss this phenomenon's major mechanisms.

ENHANCE

Scan Here for a quick concept review

www.oup.com/us/mason

UNIT 11

Atmospheric Moisture and the Water Balance

The hydrologic cycle in action—moist air meets dry land near the highest summit of the Hawai'ian island of Maui.

OBJECTIVES

- List the various physical states of water and describe the heat transfers that accompany changes between states
- Explain the various measures of atmospheric humidity, how they are related, and the processes responsible for condensation
- Outline the hydrologic cycle and the relative amounts of water that flow in each stage within this cycle
- Summarize the process of evaporation
- Relate the conditions necessary for the formation of clouds
- Explain the processes responsible for precipitation
- Describe Earth's surface water balance and its variations

Our physical world is characterized by energy flows and mass transfers. Energy and matter are never destroyed; they continually pass from one place to another. One of the best examples of such cyclical motion is the flow of water on Earth. The *hydrosphere* encompasses the global water system and the movement of water that occurs in the world ocean, on and within the land surface, and in the atmosphere. Unit 10 addressed the oceans, and in Part Five, we examine water movements on and beneath the surface.

In this unit we focus on water in the atmosphere and its relationships with Earth's surface. Our survey considers the continual movement of water among various Earth spheres that leads to a balance of water at the planetary surface. This circulation of water is powered by radiant energy from the Sun, a form of heat that both evaporates liquid water and drives the atmospheric and oceanic motions that carry water in its various forms from place to place.

The Physical Properties of Water

Human bodies are 70 percent water. Each of us requires 1.4 liters (l) (1.5 qt) of water a day to survive, and our food could not grow without it. Water is everywhere on Earth. It covers 71 percent of our planet's surface. We drink it, bathe in it, travel on it, and enjoy the beauty of it. We use it as a raw material, a source of electric power, a coolant in industrial processes, and a medium for waste disposal.

The single greatest factor underlying water's widespread importance is its ability to exist in three physical states within the temperature ranges encountered near Earth's surface (Fig. 11.1). The solid form of water—ice—is composed of molecules bound to one another by strong intermolecular attractions. This results in a highly ordered lattice pattern that gives ice its solid structure. The bonds that link molecules of ice in this rigid pattern can be weakened by heat energy. When enough heat is applied, ice loses its highly ordered pattern, and the molecules take on the liquid form, which we will simply call "water" in accordance with everyday language. Because bonds are constantly breaking and reforming in water molecules, they lack a rigid structure and are free flowing. The introduction of additional heat completely frees individual liquid molecules from one another, and they begin to move like other gases, with no intermolecular attraction. These airborne molecules constitute a gas known as **water vapor**, which was discussed in more detail in Unit 6.

Figure 11.1 is a schematic view of the amount of energy that is consumed (labeled "heat addition" in the figure) or liberated ("heat release") when water moves between various states. Water is so common and consistent in its physical behavior that it was once used as the basis for an energy unit, with 1 *calorie* (cal) defined as the amount of heat energy required to raise the temperature of 1 gram (g) (0.04 oz) of water by 1°C (1.8°F). Today, the *joule* (J) is the unit used for energy, with 1 cal representing about 4.18 J. It takes about 334 J (80 cal) to change 1 g of water from the solid state to the liquid state, a process we call **melting**. The heat involved in melting is called the **latent heat of fusion**. Similarly, it takes 2,260 J (597 cal) to change the state of 1 g of water at 0°C (32°F) from a liquid to a gas. This change is called **evaporation** or *vaporization*, and the heat associated with it is known as the **latent heat of vaporization**. Sometimes ice changes directly into water vapor. In this process, called **sublimation**, the heat required (2,594 J [677 cal]) is the sum of the latent heats of fusion and vaporization.

These physical processes are completely reversible. Water vapor changes back into water in the **condensation** process, and water changes into ice through **freezing**. Water vapor can change directly into ice as well, a process that is called **deposition**. Each reverse process releases the identical quantity of latent heat as is consumed by its corresponding forward process.

Changes in the physical states of water occur continuously in the global environment and result in the exchange of large amounts of heat energy. The latent heat required for melting, vaporization, and sublimation is often obtained from the local environment. This results in local cooling. Similarly, the release of latent heat through freezing, condensation, and deposition supplies additional heat to the environment, which results in local warming. If you doubt that condensation releases heat, consider that people comfortably sit in saunas with temperatures near 100°C (212°F), the sea-level boiling point of water. However, placing one's finger in the plume of water vapor from a boiling teakettle quickly results in a serious burn as water vapor condenses to liquid. The large amount of latent heat released and transferred to the skin easily damages exposed tissues (illustrated by the condensation arrow in Figure 11.1).

Measuring Water Vapor

Lord Kelvin (1824–1907), the inventor of the absolute temperature scale that bears his name, once said that we do not know anything about anything until we can measure and

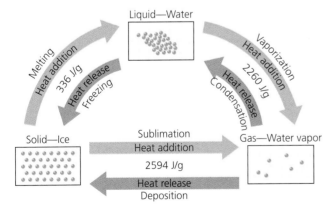

FIGURE 11.1 Schematic view of the molecular structure of water in its three physical states and heat-energy exchange among those states. The heat-exchange numbers between the arrows are explained in the text (values are for 0°C).

specify its quantity. *Humidity*, which is the amount of water vapor present in a column of air, can be expressed by several different measures. One measure simply reports the pressure of water vapor in a volume of air. The term **vapor pressure** refers to the pressure exerted by the molecules of water vapor in the atmosphere. Other atmospheric measurements of water vapor are also of value, particularly *dew-point temperature*, *specific humidity*, and *relative humidity*. We discuss these measurements in more detail in the material that follows.

In thinking about these measures it is helpful to realize that the water vapor in the air can, in principle, always be converted to a corresponding amount of liquid water. **Saturated air** has the maximum number of water vapor molecules possible at its given temperature. Condensation takes place when a parcel of saturated air is cooled or when more water vapor is transferred from the surface to the atmosphere. For example, condensation often happens around the surface of a cold soft-drink bottle or can. The air in contact with the container is cooled, which is equivalent to moving horizontally to the left in the graph in Figure 11.2. Eventually, the *saturation vapor pressure* is reached, and the vapor in the cooled air condenses into liquid water droplets on the container.

This is the process that occurs when Earth cools at night, subsequently cooling and saturating the air next to its surface. The water vapor beyond the saturation level condenses into fine water droplets, on surfaces at and near the ground, that we call **dew**. Accordingly, the temperature at which air becomes saturated, and at which condensation occurs, is called the **dew-point temperature**, or simply dew-point. Dew-point temperature is a temperature value (specified in degrees Celsius or in another temperature unit), but it provides information about humidity. If the dew-point temperature is close to the air temperature, we know the air is moist because it hardly needs to be cooled at all to reach saturation. Conversely, if the dew-point temperature is far below air temperature, the air is dry.

Another measure used to assess the amount of water vapor in the air needs less explanation. *Specific humidity* is the ratio of the weight (mass) of water vapor in the air to the combined weight (mass) of the water vapor plus the air itself. If water vapor is added to the air, the mass of water vapor must increase, and specific humidity rises accordingly. Vapor pressure, dew-point temperature, and specific humidity are all absolute measures of humidity; the higher their values, the greater the moisture content of the air.

Relative humidity tells us how close a given parcel of air is to its dew-point temperature, or saturation level. (Recall from Unit 7 that an air parcel is a small volume of air.) Consequently, we define **relative humidity** as the proportion of water vapor present in a parcel of air relative to the maximum amount of water vapor that could be present in the air at the same temperature. Relative humidity is expressed as a percentage; air of 100 percent relative humidity is saturated, and air of 0 percent relative humidity is completely dry.

Relative humidity is considered a relative measure of humidity because it only tells you the degree of air saturation.

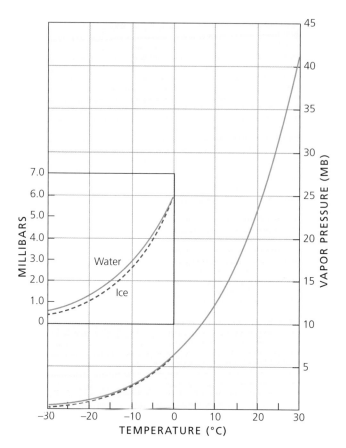

FIGURE 11.2 Variation of saturation vapor pressure (mb) with temperature (°C). Notice that the curve is not a straight line, and that vapor pressure increases at an ever-increasing rate with temperature. The inset shows that at temperatures below 0°C, saturation values over supercooled water are greater than those over ice.

The absolute moisture content of the air can't be determined from relative humidity alone; we must also know the air temperature. Relative humidity typically varies throughout the day in opposition to temperature. It is usually lower in early and midafternoon when temperatures are high and thus the saturation point is also high. At night, when the temperature falls, the now-colder air has a lower saturation vapor pressure, and so the relative humidity increases as the air approaches its dew-point temperature.

Now that we have some basic terminology and an understanding of the ways water changes from one physical state to another and how we measure water vapor in the air, we can proceed to find out how it circulates through the Earth system. The natural cycle describing this circulation is called the hydrologic cycle.

The Hydrologic Cycle

Early scientists believed that the wind blew water from the sea through underground channels and caverns and into the atmosphere, removing the salt from the water in the process. We now talk in terms of the **hydrologic cycle**, whereby water continuously circulates between the

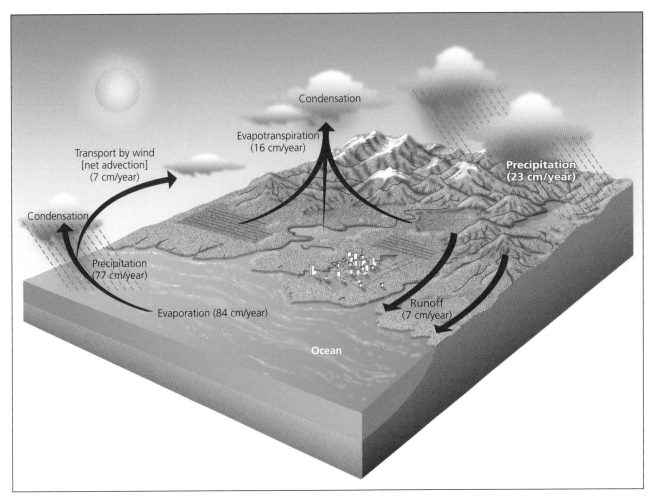

FIGURE 11.3 Hydrologic cycle. The amount of water moving between reservoirs is expressed as the volume of water transferred in cubic centimeters per year divided by Earth surface area in square centimeters. This means the values are the equivalent depth of water transferred in centimeters per year. All of these numbers can be directly compared to the global average precipitation rate, which is about 100 cm/y (1 cm = 0.4 in).

atmosphere, land surfaces, plants, the oceans, and freshwater bodies. Figure 11.3 illustrates the complete cycle and shows the relative depths of water involved in each stage of the cycle. The following observations are important to note when studying Figure 11.3:

1. The largest amounts of water transferred in the global cycle are those involved in the direct evaporation from the sea to the atmosphere and in precipitation back to the sea. As noted previously, evaporation is the process by which water changes from the liquid to the gaseous (water vapor) state. **Precipitation** includes any liquid water or ice that falls to the surface through the atmosphere.

2. The passage of water to the atmosphere through leaf pores is called **transpiration**, and the term **evapotranspiration** encompasses the combined total of water that transpires from plants and water that evaporates from the land surfaces, including open water surfaces such as lakes and rivers. Evapotranspiration and the precipitation of water

onto the land surface play a quantitatively small, but important, part in the hydrologic cycle.

3. If precipitation at the land surface does not evaporate or transpire, it is removed via the surface network of streams and rivers, a phenomenon called **runoff**. In many places, water infiltrates (soaks in) and travels through the ground before emerging and contributing to stream flow and the transfer from land to ocean.

The hydrologic cycle operates as a closed system in which water is continuously moved among the component spheres of the Earth system. The system can be split into two subsystems, one consisting of the precipitation and evaporation over the oceans and the other involving evapotranspiration and precipitation over land areas. The values in Figure 11.3 indicate that evaporation exceeds precipitation over the oceans, and precipitation is greater than evapotranspiration over the land surfaces.

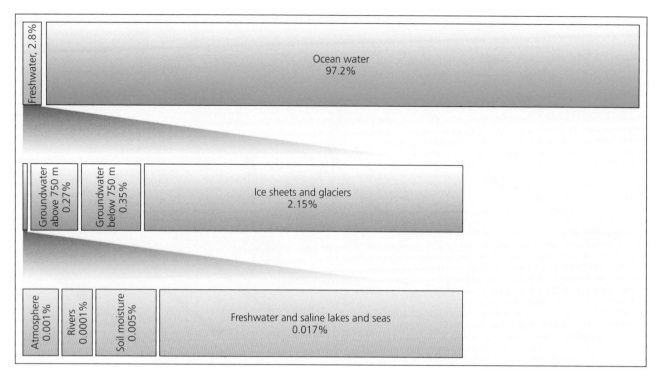

Ocean water
97.2%

Freshwater, 2.8%

Groundwater above 750 m 0.27%

Groundwater below 750 m 0.35%

Ice sheets and glaciers
2.15%

Atmosphere 0.001%

Rivers 0.0001%

Soil moisture 0.005%

Freshwater and saline lakes and seas
0.017%

FIGURE 11.4 Distribution of water in the hydrosphere. The middle and lower bars show the percentage distribution of the 2.8 percent of total hydrospheric water that is fresh. Of that freshwater component, very little is easily available to humans.

The two subsystems are linked by the horizontal movement of water from the oceans to the land surfaces in the atmosphere (known as *advection*) and by the return of water from the land surfaces to the oceans by surface runoff. Because latent heat is absorbed in evaporation and released in condensation, the imbalance between evaporation and precipitation over the ocean and land surfaces means that the hydrologic cycle produces a net transfer of energy from oceanic to continental areas, as mentioned in Unit 8.

Water circulates between the lower atmosphere, the upper lithosphere, the plants of the biosphere, and the oceans and freshwater bodies of the hydrosphere, or it is sequestered (held) in the ice, snow, and permafrost of the cryosphere for varying lengths of time. The time required for water to traverse the full hydrologic cycle can be quite brief. A molecule of water can pass from the ocean to the atmosphere and back again within a matter of days. Over land, the cycle is less rapid. Precipitation goes into the soil or subsurface and can remain there as soil moisture or groundwater for weeks, months, or years. The circulation is even slower where water in the form of ice in the cryosphere is concerned. Significant water has been locked up in the major ice sheets and glaciers of the world for many thousands (and in some cases, millions) of years.

Water is quite unequally distributed within the hydrosphere, as Figure 11.4 reminds us. The world ocean contains 97.2 percent of all terrestrial water, but the high salt content makes it of little direct use to people. Of the remaining 2.8 percent, which constitutes the world's freshwater supply, three-quarters is locked up in ice sheets and glaciers. The next largest proportion of freshwater, about

one-eighth, is accessible only with difficulty because it is deep groundwater located below 750 m (2,500 ft).

Therefore, the freshwater needed most urgently for our domestic, agricultural, and industrial uses must be taken from the relatively small quantities found at or near the surface—in the rivers, lakes, soil layer, and shallow groundwater. These storage areas contain less than 1/200th of the 2.8 percent of freshwater in the hydrosphere, and that finite supply is increasingly taxed by growing human consumption and abuse (see Perspectives on the Human Environment: Water Usage in the United States).

Evaporation

We cannot see evaporation occurring, but its results are sometimes visible. When you see mist rising from a lake that is warm in comparison to the cold air above it, you are seeing small liquid water droplets that have already condensed. With this image in mind, it is not too hard to imagine molecules of invisible water vapor rising upward in the same way.

CONDITIONS OF EVAPORATION

Evaporation occurs when two conditions are met. First, heat energy must be available at the water surface to supply the heat needed to change the liquid water to a vapor (see Fig. 11.1). Most of this heat is provided by the radiant energy from the Sun. Longwave radiation emitted by the atmosphere is another source of energy that contributes to evaporation. Horizontal advection of warm air can also serve as an important source of energy for evaporation, particularly in arid and semiarid environments. If the advected air is warmer than the underlying surface, a downward transfer of heat to the surface will occur that provides additional heat energy for the vaporization of water.

The second condition is that the air must not be saturated. This is because in order for evaporation to occur, the air must be able to absorb the evaporated water molecules. For example, air near a water surface normally contains a large number of vapor molecules. The vapor pressure, caused by the density and movement of the vapor molecules, is high. In contrast, air at some distance above the water surface has fewer molecules of vapor and, consequently, a lower vapor pressure. In this case, we say there is a *vapor-pressure gradient* between the surface and air. A vapor-pressure gradient exists above a water surface as long as the air has not reached its saturation level (dewpoint temperature). A large vapor-pressure gradient enhances the rate of surface evaporation. Such large vapor-pressure gradients are associated with conditions of low relative humidity and strong winds. The strong winds help remove the evaporated water upward and away from the open water surface, thus maintaining a large vapor-pressure gradient.

Knowing the requirements for evaporation—a heat source and a vapor-pressure gradient—we can infer the kind of atmospheric conditions that would lead to maximum rates of surface evaporation, assuming open water is available for vaporization. The combined effect of these atmospheric conditions is known as *atmospheric demand*, which determines the maximum rate of evaporation from open water surfaces. Because solar radiation from the Sun is the leading heat source, large amounts of evaporation can be expected where there is a great deal of sunlight. This is particularly true of the tropical oceans. Water molecules also move most easily into dry air because of the sizable vapor-pressure gradient. So the drier the air, the more evaporation will occur, as will be seen later in the unit when we discuss evaporation from subtropical oceans.

PERSPECTIVES ON THE HUMAN ENVIRONMENT

Water Usage in the United States

In the United States today, freshwater—drawn from lakes, rivers, and groundwater—is consumed in prodigious quantities at a daily rate of nearly 1.3 billion cubic meters (336 billion gal), or about 4.2 cubic meters (1,100 gal) per person. About 80 percent of the total is from surface water, with the remaining 20 percent from groundwater. That is the highest rate of usage in the world—more than twice as high as in the countries of the European Union and hundreds of times higher than in most developing nations. Given this massive rate of consumption, we—and the rest of humankind—are facing a critical shortage of usable freshwater in the foreseeable future.

The spatial distribution of total water withdrawal for the United States is shown in Figure 11.5. Surprisingly, the heavily populated (and urbanized) Northeast and Midwest are comparatively moderate water consumers. Much greater regional water use—and overuse—occurs in the drier western United States, most notably in the croplands of central and southeastern California, western Texas, eastern Colorado, Nebraska, and southern Idaho, where the reliance on irrigation systems is extreme. All too often water use becomes a divisive political issue. For example, in 2011 a severe and prolonged drought began in California. By 2015 urban populations were forced to ration water while the vastly more consumptive agricultural uses were permitted to continue unchecked. Some crops were demonized for their heavy water needs (such as almonds, which require about 1 gallon per nut and alone consume 10 percent of the state's water), and city dwellers were castigated for prioritizing green lawns and swimming pools over beleaguered farmers.

Among the major sectors of the economy, two account for about 80 percent of all national water withdrawals. The largest by far is the energy sector's thermoelectric power industry, which in any given year withdraws about 49 percent of the water supply to produce electricity, mainly by converting heated water into steam for use in power generation. The second leading user is agriculture, which claims about 31 percent to water the nation's ever-expanding irrigated crop acreage. The remaining 20 percent is withdrawn for public water-supply systems (~11 percent), domestic use (~1 percent), livestock consumption (~1 percent), aquaculture (~2 percent), industrial use (~4 percent), and mining operations (~1 percent). Interestingly, public-supply withdrawals decreased by about 5 percent from 2005 to 2010, which was the first decrease since records began in 1950. That drop in water use is doubly impressive considering that population *increased* by 4 percent over the same 5-year period. Agriculture practices also became more water-efficient, with irrigation withdrawals falling to the lowest levels since 1965 despite an increase in total irrigated area.

We must emphasize that despite many water conservation successes, current water use rates far exceed inputs to groundwater supplies and are therefore not

continued

continued

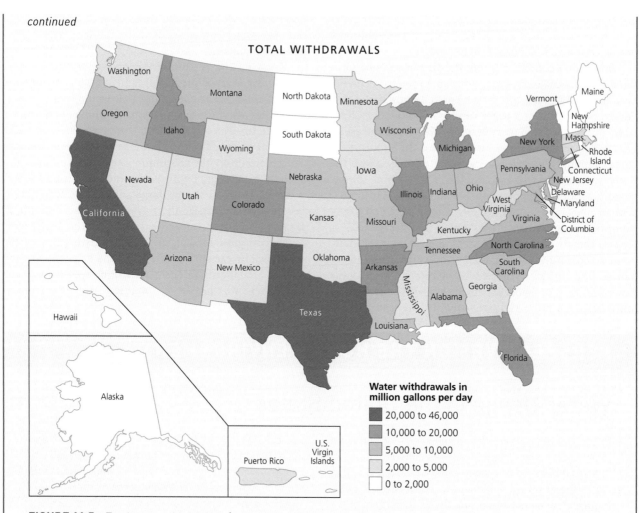

TOTAL WITHDRAWALS

Water withdrawals in million gallons per day

- 20,000 to 46,000
- 10,000 to 20,000
- 5,000 to 10,000
- 2,000 to 5,000
- 0 to 2,000

FIGURE 11.5 Total water withdrawals (millions of gallons per day) for the United States in 2010.

sustainable. As a whole, the country is rapidly drawing down reserves built up over thousands of years. In the High Plains, for example, groundwater withdrawals at present are about 6 times larger than the recharge rate. Approximately 30 percent of the groundwater reserve has already been taken, and 40 percent more will be withdrawn by 2060, leaving 30 percent for subsequent years. With declining availability, withdrawals from the High Plains Aquifer are projected to decrease into the next century, eventually falling below the recharge rate.

If fully depleted, an average of 500 to 1300 years would be needed to refill the underlying storehouse according to recent calculations. Concerns like these are not merely hypothetical speculations about some imagined, far-off future. For example, today in California, where drought has added to normal irrigation needs, residents of some cities have literally seen their source of drinking water disappear because of groundwater depletion. There are other impacts of excessive groundwater use, some of which are described in Unit 38.

EVAPOTRANSPIRATION

Most of the evaporation into the atmosphere occurs over the ocean. Over land, water evaporates from lakes, rivers, wetlands, swamps, marshes, damp soil, and any other surface in which water is directly exposed to the atmosphere. The water that plants lose to the air during transpiration is another major source of atmospheric water.

Transpiration from plants is not only controlled by the atmospheric demand described previously in this unit but also by a complex set of controls that are part of a plant's physiology. Liquid water vaporizes from the soft interior

plant tissues and diffuses to the atmosphere as water vapor through openings in leaf surfaces called *stomates*. Plants control the size of these stomatal openings or even completely close them to regulate water loss. Plants face a dilemma of sorts. Their stomatal openings must be open so carbon dioxide can diffuse into the leaf interior to be consumed during photosynthesis, but this very process allows water vapor to diffuse into the atmosphere!

Leafy plant canopies also have a different surface area than do open water surfaces. The ratio of the total leaf area to the corresponding ground area, called the *leaf area index*

(m^2/m^2), can be as large as 4 for a dense canopy. With several times as much surface area, vegetated surfaces are capable of losing much more water than are open water surfaces under the same conditions of atmospheric demand.

Physical geographers draw a distinction between potential evapotranspiration and actual evapotranspiration. **Potential evapotranspiration (PE)** is the maximum amount of water that can be lost to the atmosphere from a land surface when the availability of surface water is not a limiting factor. **Actual evapotranspiration (AE)** is the amount of water that is lost to the atmosphere from the surface. Actual evapotranspiration can be less than the potential rate because of either soil-moisture conditions or plant physiological controls. Actual evapotranspiration can also equal potential evapotranspiration when transfer processes within the soil and plant environments are able to keep pace with atmospheric demand. Because soil and plant processes are involved, determining evaporative transfers from surface to the atmosphere is more difficult for continental areas than for oceans.

Once water has entered the atmosphere via evaporation, we can see how it makes its way through that component of the hydrologic cycle and moves back to the surface. We begin with the formation of clouds and then trace the development of precipitation.

Condensation and Clouds

CLOUD FORMATION

Clouds are visible masses of suspended, minute water droplets and/or ice crystals. Two conditions are necessary for the formation of clouds:

1. The air must be saturated, which normally happens by cooling below the dew-point temperature (which causes water vapor to condense). Parcels of air most often cool to produce condensation when they are vertically lifted in the atmosphere and experience cooling by expansion as a result of being exposed to the lower atmospheric pressure. Air also cools when it contacts and mixes with colder air or exchanges sensible heat with a colder surface.

2. There must exist a substantial quantity of small airborne particles called **condensation nuclei**, around which liquid droplets can form when water vapor condenses. The saturation vapor pressure relationship shown in Figure 11.2 applies only to a flat surface of pure water. Condensation on curved surfaces requires much higher pressures and would not easily occur on rounded cloud drops if they were pure water. Indeed, relative humidities of perhaps 200 percent would be needed for condensation in the absence of condensation nuclei. Condensation nuclei both provide a surface for condensation and, when dissolved in water, create an impure solution that allows easier condensation. Fortunately, condensation nuclei are almost always present in the atmosphere in the form of dust, salt, or other forms of particulate matter.

The greater the moisture content of cooling air, the greater the condensation rate and the development of the cloud mass. Although most clouds form and remain at some altitude above Earth's surface, clouds can come into direct contact with the surface. When masses of fine water droplets suspended in air concentrate near the ground, they produce what we commonly call **fog** (Fig. 11.6). Fog produces very little precipitation and is far less widespread and so less important globally than other clouds, but given its location near the surface, it is the one cloud most of us are likely to encounter physically.

Fog can form in various ways. At night, if skies are clear, the surface and lower atmosphere cool efficiently by radiation, and a **radiation fog** is the result. For radiation fog to develop winds must be light enough to prevent mixing with warmer air from above, and so radiation fogs are common with still or nearly still air (Fig. 11.7). By contrast, an **advection fog** occurs when fairly moist air moves over a colder surface and loses enough heat to reach the condensation point. This can happen when air moves from ocean to land, like in the famous fogs of the San Francisco Bay area. It can also happen over the ocean if air moves from an area of warm water to cold, as in the fogs that form above the cold Labrador Current (discussed in Unit 10).

Besides being the source of all precipitation, clouds play a key role in the atmosphere's heat balance. Clouds are the single largest contributor to the albedo of the planet (see Unit 5). They reflect some incoming shortwave radiation back to space at their tops and scatter another part of this incoming solar radiation upward and downward before it can strike the surface directly. The downward-scattered solar radiation reaches the surface as diffuse radiation (see Fig. 5.4). At the same time, clouds absorb a large part of the longwave radiation emitted by Earth's surface. This absorbed radiant energy heats the atmosphere through the greenhouse effect, and the atmosphere then emits most of that energy either upward toward space or downward back toward the land and the sea.

CLOUD CLASSIFICATION

Cloud-type classification, which is helpful for understanding the processes responsible for any given particular cloud, is based on the criteria of general structure, appearance, and altitude (Fig. 11.8). One major cloud-type grouping encompasses **stratus clouds**. As this term implies, stratus clouds have a layered appearance. They also are fairly thin and normally cover a wide geographic area. Stratus clouds are classified according to their altitude. Below 3 km (1.9 mi or 10,000 ft), they are simply called *stratus* clouds—or *nimbostratus* if precipitation is occurring. Between 3 and 6 km (10,000–20,000 ft), they are designated *altostratus* clouds, and above 6 km, *cirrostratus* clouds.

A second major cloud-type category consists of **cumulus clouds**, which are thick, puffy, billowing masses that often develop to great heights. Like stratus clouds, cumulus clouds are also subclassified according to their position in the lower,

FIGURE 11.6 "When you are in the Maasai Mara to study wildlife, you get up early—before the sunrise. Here, not far from the Kenya–Tanzania border, the elevation is enough to make for some chilly nights, and you start the day in a sweater (later the sun will make you wish for a patch of shade). So cool does the ground surface become that, even in this arid environment, the slight amount of moisture in the air will condense and a ground fog develops. It is an unusual sight here in tropical Africa, and it does not last long after sunrise. But it moistens the grass just enough to enhance its vigor and to make it more palatable to the grazers of the morning."

FIGURE 11.7 Spectacular radiation fog in Dubai, United Arab Emirates. This fog is common in the spring and fall when air in this desert location cools to the dew-point.

middle, and upper altitudinal levels, proceeding in ascending order through *cumulus* or *stratocumulus* to *altocumulus* and finally *cirrocumulus* (see Fig. 11.8). Very tall cumulus clouds, extending from 500 m (1,600 ft) at their base to about 12 km (7.6 mi or 40,000 ft) at their anvil-shaped heads, are called *cumulonimbus*. These are often associated with violent weather, including heavy rain, high winds, lightning, and thunder.

Cirrus clouds, a third common cloud type, are thin, wispy, streak-like clouds that consist of ice particles rather than water droplets. They invariably occur at altitudes higher than 6 km (3.8 mi or 20,000 ft) (see Fig. 11.8).

Precipitation

Condensation and cloud formation are necessary prerequisites for precipitation, but they do not always lead to precipitation. This is confirmed by the everyday observation that clouds are far more common than rain or snow. When water droplets within clouds first form, the droplets are so small that the slightest upward air current keeps them airborne

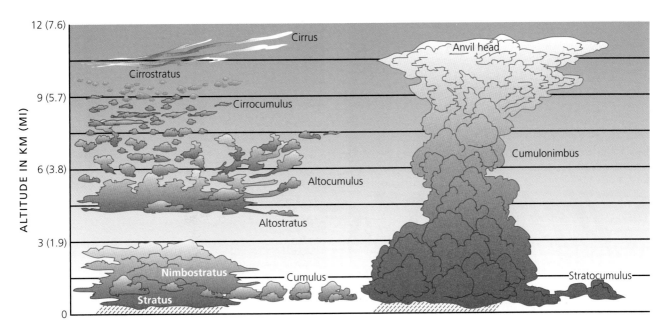

FIGURE 11.8 Schematic diagram of the different cloud types: stratus, cumulus, and cirrus, arranged by their typical altitudes.

for long periods. Even if the air were perfectly still (which never happens), the small cloud droplets would evaporate before they reached the ground when they fell through unsaturated air below the cloud. Cloud droplets and tiny ice crystals first must grow by a factor of about 1 million if they are to fall fast enough to survive descent to the surface. We now examine the processes by which that happens.

COLD CLOUDS AND THE ICE-CRYSTAL PROCESS

Probably the most common precipitation-forming process, the *ice-crystal process* was first identified in the 1930s by meteorologists Tor Bergeron (1891–1977) and Walter Findeisen (1909–1945). This process requires the presence of both liquid droplets and ice particles in a cloud. Ice particles are normally present if the temperature is below 0°C (32°F) and if there are small particles called *freezing nuclei*. Freezing nuclei perform an analogous function for ice particles that condensation nuclei perform for water droplets. That is, freezing nuclei allow freezing to occur at temperatures near 0°C rather than only at much colder temperatures.

Recall that the saturation vapor pressure over ice is less than the saturation vapor pressure over water at the same temperature (see Fig. 11.2). When a cloud contains both ice particles and water droplets, the water droplets tend to experience greater rates of evaporation than condensation, causing them to lose mass and add water vapor content to the surrounding air. The resultant water vapor causes the ice crystals to have greater rates of deposition as water changes state from vapor to solid, and therefore the crystals gain mass. Because the vapor pressure over the ice crystal is lower than that over the water droplet, the ice crystals grow at the expense of the cloud droplets (see Figure 11.9A).

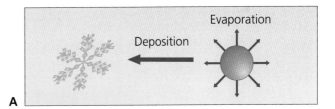

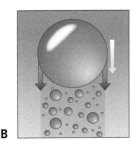

FIGURE 11.9 Processes of precipitation formation. (A) The ice-crystal process in cold clouds. (B) The collision-coalescence process.

Ice particles are also an excellent source of *ice nuclei*. Once the ice crystals begin to fall through a cloud, they collide with water droplets and grow as the water freezes onto the ice particles. This process, called *riming*, produces larger ice crystals that fall more rapidly, which promotes further growth by riming. Another process, called *aggregation*, also leads to larger ice crystals. Ice crystals are often covered by a thin film of water that acts as an adhesive when two ice crystals collide. Through a combination of these processes, ice crystals become larger and join together to form snowflakes that eventually become heavy enough to drop out of the clouds. Most rainfall and snowfall in the middle and high latitudes are initiated by the ice-crystal process, but in the tropics, the temperature of many clouds

A

B

C

D

FIGURE 11.10 Four major forms of precipitation. (A) A rainstorm douses the Ponderosa pine forest near Flagstaff, Arizona. (B) Falling snow accumulates in south-central Alaska. (C) Freezing rain forms an icy coating on pine needles on a golf course in Wawona, near Yosemite National Park, California. (D) Golf-ball-sized hailstones litter the countryside following a storm in northern Texas.

does not necessarily drop below the freezing point. In locations like these, a second process—coalescence—operates to make raindrops large enough to fall from clouds.

THE COALESCENCE PROCESS

The *coalescence process* (sometimes called the collision-coalescence process) requires some liquid droplets to be larger than others, which happens naturally since condensation nuclei are a wide range of diameters. As they fall, the larger droplets collide and coalesce with the smaller ones, producing even larger water droplets (Fig. 11.9B). At the same time narrowly missed smaller droplets continue to be caught up in the wake of the larger ones and drawn to them. The larger droplets ultimately grow at the expense of

the smaller ones and soon become heavy enough to fall to Earth. This process dominates in warm clouds, such as those in the equatorial regions, where the entire cloud depth may have temperatures above freezing, but it also works in combination with the ice-crystal process in cool and cold clouds if part of the cloud is above freezing.

FORMS OF PRECIPITATION

In the tropical latitudes where most of the troposphere is above freezing, precipitation occurs primarily by the coalescence process and can only reach the surface as **rain**. In the middle and high latitudes, the ice-crystal process dominates, yet precipitation reaches Earth's surface in several forms, as the photos in Figure 11.10 illustrate. The form of the precipitation reaching the surface in these latitudes depends on the temperature profile between the cloud and the surface. If the ice crystals do not encounter an air layer above freezing while falling to Earth's surface, they will reach the ground as **snow**. If the ice crystals encounter a deep layer of above-freezing air while falling, they will melt and reach the surface as rain. Falling rain may encounter a subsequent deep layer of air below freezing, which causes the rain to freeze and arrive as pellets of ice called **sleet**. If the layer of below-freezing air is not deep enough to freeze the falling rain, the rain will reach the surface as liquid but then freeze after reaching the ground to form a layer of ice where it lands. This form of precipitation, called **freezing rain** (or *glaze*), is a tremendous hazard affecting transportation, utility lines, trees, and structures. **Hailstones** result when ice crystals are blown about within a tall and very moist cloud (see Fig. 11.10D). As the crystal travels through regions containing numerous water droplets, a layer of ice forms from collision and the subsequent freezing of liquid water. In other parts of the cloud ice is added to the hailstone by the ice-crystal process.

The Surface Water Balance

As far as we humans are concerned, the most crucial segment of the hydrologic cycle occurs at the planetary surface. At the interface between Earth and atmosphere, evaporation and transpiration cool the surface, add moisture to the atmosphere, and enable plants to photosynthesize and grow. At the surface, precipitation also provides the water needed for that evapotranspiration. Because the balance between the atmospheric demand and the supply of surface water is critical to so many physical processes, it has become common to use the **water balance** as an integrating approach in geography. An accountant keeps a record of financial income and expenditures and ends up with a bottom-line balance. The balance is positive when profits have been earned and negative when excess debts have been incurred. We can describe the balance of water at Earth's surface in similar terms, using methods devised by climatologist C. Warren Thornthwaite (1899–1963) and his associates.

At the local scale, water can be gained at the surface by precipitation or snowmelt and can be lost by evapotranspiration or through surface runoff. As mentioned earlier, potential evapotranspiration represents the maximum uptake of water by the atmosphere. Everywhere, even in moist tropical environments, periods of time occur during which the actual evapotranspiration falls below PE because of limited water availability. In contrast, AE will equal PE when precipitation exceeds PE; stored soil moisture can provide an additional supplement to maintain AE at PE rates for a limited period of time. Extended periods of time during which PE exceeds precipitation, however, will cause AE to eventually fall below PE. Similarly, should precipitation exceed PE, soil moisture will increase until the soil is fully recharged with water. Continued excess precipitation above PE will then lead to a surplus of water and surface

runoff. In the following material, we present a few examples to illustrate the dynamic interaction of these processes in real-world physical environments.

THE RANGE OF WATER BALANCE CONDITIONS

Baghdad, Iraq, exemplifies places that have a water balance deficit (Fig. 11.11A). Potential evapotranspiration exceeds the water gained in precipitation for all but a few months in winter. Limited soil-moisture recharge occurs during the brief winter rainy season, but this water is quickly withdrawn and consumed with the arrival of warm spring temperatures. Because a severe soil-moisture deficit is present for most of the year, the vegetation in this region is sparse, except where irrigation is possible.

In Tokyo, Japan, the situation is reversed (Fig. 11.11B). During every month of the year, rainfall exceeds the amount of water that can be lost through evapotranspiration. The surplus water provides all that is needed for luxuriant vegetation and still leaves a large amount to run off the land surface.

An intermediate situation exists in Faro, Portugal (Fig. 11.11C). From November to March, precipitation exceeds PE, but from April through October, there is a water deficit. Starting in April, when PE surpasses precipitation, soil moisture is withdrawn from storage and used to supplement AE. By midsummer, most of the stored soil moisture has been consumed, and a large soil-moisture deficit exists. This continues until the end of October, when rainfall once more exceeds PE and the soil water is recharged. By mid-January, the soil moisture has been recharged by winter storms, and a brief period of surplus occurs.

The amount of runoff in any location cannot exceed the amount of precipitation, and usually there is much less runoff than precipitation because significant quantities of water are always lost through evaporation. In the United States, runoff approaches the amount of precipitation only in parts of western Oregon and Washington State in the Pacific Northwest. In the much drier southwestern states, annual runoff is usually only 1–2 percent of annual precipitation.

WATER BALANCE VARIATIONS AND LATITUDE

Across the globe as a whole, the values of precipitation, evaporation, and runoff vary greatly with latitude. As Figure 11.12 shows, annual precipitation and runoff are highest near the equator. Evaporation is also high in this low-latitude zone, but it is greatest in the subtropical latitudes (20 to 35 degrees). The upper midlatitudes (45 to 60 degrees) exhibit water surpluses. Precipitation, runoff, and evaporation are lowest in the highest latitudes. The forces underlying this distribution of the three variables, which is graphed in Figure 11.12, have been discussed in Units 7–10.

The graph in Figure 11.12 mainly shows the results of differences in the radiant energy that reaches different latitudinal zones and global-scale atmospheric and oceanic

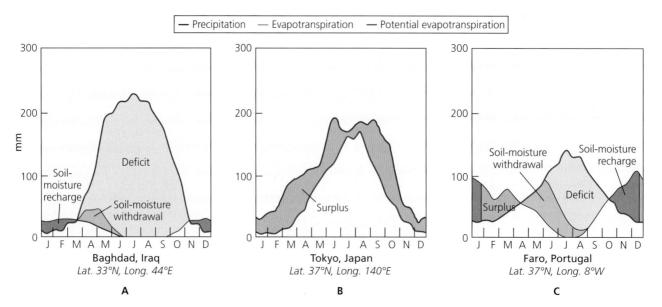

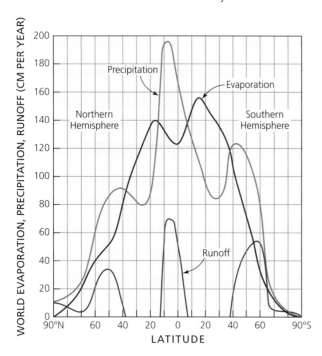

FIGURE 11.11 A range of water balance conditions found at the surface of Earth. (A) Baghdad, Iraq, experiences a constant deficit because potential evapotranspiration normally exceeds precipitation. (B) In Tokyo, Japan, the situation is reversed, and a constant water surplus is recorded. (C) In Faro, Portugal, the intermediate situation occurs, with a combination of surplus and deficit that occur at different times of the year.

FIGURE 11.12 Average annual latitudinal distribution of precipitation, evapotranspiration, and runoff in centimeters per year.

circulation. The dominant role of the Hadley cell in the global climate system discussed in Unit 9 is clearly visible. Convergence of the trade winds along the ITCZ produces copious amounts of precipitation. Near the equator, strong radiant energy from the Sun leads to high evaporation rates and causes air to rise, cool, and thereby yield large quantities of precipitation. Although evapotranspiration is high in an absolute sense, it is smaller than in the subtropical regions as a result of their extensive cloud coverage. In

subtropical areas, the descending limb of the Hadley cell produces atmospheric warming, the semipermanent high-pressure cells, and clear weather. Here, there are high rates of evaporation over the oceans but little evaporation over the land surfaces because of the scarcity of moisture to be evaporated.

In the higher midlatitudes, eastward-moving storms (driven by the westerlies and the polar front jet stream) provide moderate amounts of precipitation in most areas, but smaller quantities of radiant energy evaporate less of that water than would be the case at lower latitudes. In the high latitudes, the cold air contains little water vapor even when saturated. Consequently, there is little precipitation, and given the low amounts of radiant energy received here, rates of evapotranspiration are minimal. These relationships should be kept in mind when we consider and compare the world distributions of evapotranspiration (Fig. 11.13) and precipitation (Fig. 11.14).

WATER BALANCE AND ATMOSPHERIC RIVERS

We should not think the water balance of a location is maintained by steady inputs and withdrawals of water. The graphs and maps shown previously depict average climatic conditions and completely ignore the hydrosphere's variability. As a result, the figures rarely represent what a place experiences at a single moment in time, or how rapidly the water balance can change. **Atmospheric rivers** are especially important in this regard. These are narrow, winding corridors about 400 km (250 mi) wide and up to 2,000 km (1,240 mi) long carrying vast amounts of water vapor within the atmosphere. The word "river" is appropriate, because the flow rate is comparable to that of the Amazon

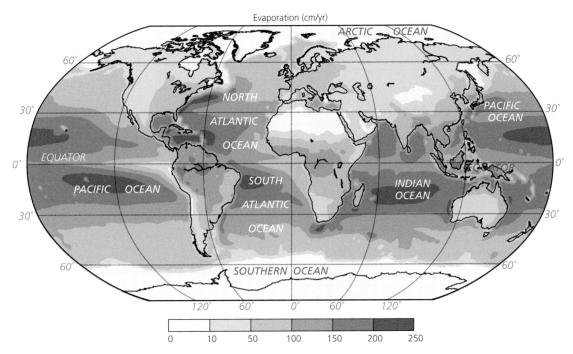

FIGURE 11.13 Global distribution of annual evaporation and evapotranspiration in centimeters.

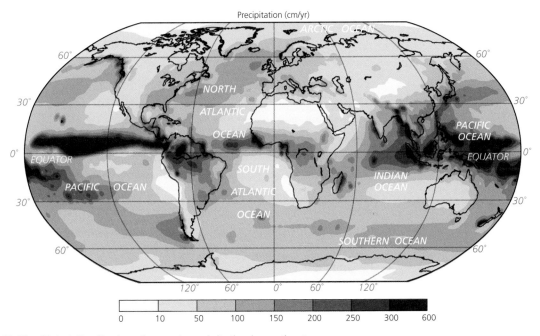

FIGURE 11.14 Global distribution of annual precipitation in centimeters.

and other rivers on land. However, atmospheric rivers last only a week or two, and they move water vapor rather than liquid water (see Fig. 11.15). About 130 of these plumes come and go throughout the year, and yet they are collectively responsible for about 90 percent of north–south transport in the middle latitudes.

With their high flow rates, atmospheric rivers can sustain heavy, prolonged rainfall in storms, and so they are often implicated in flood events. The infrared satellite image in Figure 11.15 might look particularly dramatic, but in fact such occurrences are routine. This means that

the hydrologic cycle and local water balances are highly variable in both time and space.

POPULATION AND THE WATER BALANCE

As a final exercise in this unit, let us reconsider the distribution of Earth's population (see Fig. 2.9) in the context of the global patterns of evapotranspiration (see Fig. 11.13) and precipitation (Fig. 11.14). This comparison reveals that most people live in areas where there are neither great surpluses nor great deficits in the water balance.

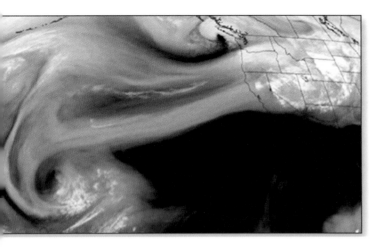

FIGURE 11.15 An atmospheric river of water vapor on September 30, 2013. The subtropical Pacific is supplying large amounts of moisture to the atmosphere, which in turn is carried eastward by westerly motion.

ENHANCE

For example, the great fertile zones of North America, Europe, and much of China are concentrated in mid-latitude areas, which, year in and year out, do not usually experience excessive conditions of precipitation or evapotranspiration. Thus, 75 cm (30 in) of annual rainfall in the U.S. Corn Belt or the North China Plain is far more effective for raising crops than the 200 cm (80 in) received at a tropical location, where much of the moisture is removed by evapotranspiration.

The variability of precipitation is a phenomenon that concerns farmers around the world. In general, variability increases as the yearly precipitation total decreases. We pursue this matter in our survey of the dry climates in Unit 15, which features a map of the global distribution of annual precipitation variability (see Fig. 15.13).

Before we study the geography of climate, however, we need to become familiar with the organized weather systems that continually parade above Earth's surface. These systems play a major role in shaping the temperature and moisture regimes of many climate types. We begin in Unit 12 by relating the atmospheric moisture flows we have just learned about to the formation of weather systems.

Key Terms

actual evapotranspiration (AE) *page 141*

advection fog *page 141*

atmospheric rivers *page 146*

cirrus clouds *page 142*

cloud *page 141*

condensation *page 135*

condensation nuclei *page 141*

cumulus clouds *page 141*

deposition *page 135*

dew *page 136*

dew-point temperature *page 136*

evaporation *page 135*

evapotranspiration *page 137*

fog *page 141*

freezing *page 135*

freezing rain *page 145*

hailstones *page 145*

hydrologic cycle *page 136*

latent heat of fusion *page 135*

latent heat of vaporization *page 135*

melting *page 135*

potential evapotranspiration (PE) *page 141*

precipitation *page 137*

radiation fog *page 141*

rain *page 144*

relative humidity *page 136*

runoff *page 137*

saturated air *page 136*

sleet *page 145*

snow *page 145*

stratus clouds *page 141*

sublimation *page 135*

transpiration *page 137*

vapor pressure *page 136*

water balance *page 145*

water vapor *page 135*

Scan Here for a quick vocabulary review

ENHANCE

Review Questions

1. Describe the energy requirements for the melting of ice and the vaporization of water.

2. Describe the various measures of atmospheric humidity and their relation to one another.

3. How does evaporation differ from evapotranspiration?

4. How does potential evapotranspiration differ from actual evapotranspiration?

5. Describe the two processes of raindrop formation.

6. Describe the necessary conditions for surface runoff within the context of the water balance.

Scan Here for a quick concept review

www.oup.com/us/mason

Precipitation, Air Masses, and Fronts

A sharp line of clouds mark the boundary between two distinctly different masses of air.

OBJECTIVES

- Discuss air masses—their character, origin, movement patterns, and influence on precipitation

- Describe the four basic lifting mechanisms for producing precipitation: convergent-lifting, frontal, convectional, and orographic

On a day-to-day basis, the atmosphere is organized into numerous weather systems that blanket Earth. In this unit, we make the connection between the atmospheric moisture flows covered in Unit 11 and the formation of those weather systems. We first introduce the concept of air masses—large uniform bodies of air that move across the surface as an organized whole—and the weather contrasts that occur along the advancing edges of these air masses. We next turn our attention to the forces that cause the atmospheric lifting of moist air, thereby producing precipitation.

Air Masses in the Atmosphere

In large, relatively uniform expanses of the world, such as the snow-covered arctic wastes or the warm tropical oceans, masses of air that are in contact with the surface may remain stationary for several days, often beneath a high-pressure system having calm or light winds. As air hovers over these areas, it takes on the properties of the underlying surface, such as the icy coldness of the polar zones or the warmth and high humidity of the maritime tropics. Extensive geographic areas with relatively uniform characteristics of temperature and moisture form the **source regions** where air masses are produced.

An **air mass** is a very large body of air with relatively uniform qualities of temperature, density, and humidity. By "large" we mean about 1,500 km (932 mi) across, or closer to the size of a continent rather than the size of a state or province. The word "uniform" in our definition means air masses don't have significant horizontal variation in their properties. Moving vertically within an air mass there is usually substantial variation in temperature and moisture, and of course density always decreases vertically. Air masses travel as distinct entities, covering many hundreds of kilometers as they are steered away from their source regions by the air flows of large-scale atmospheric circulation.

CLASSIFYING AIR MASSES

Air masses have letter codes that reflect the type of surface and the general location of their source regions. These codes for the leading types of air masses are as follows: maritime (**m**) or continental (**c**) and tropical (**T**) or polar (**P**). The air above a source region reaches an equilibrium with the temperature and moisture conditions of the surface. In a warm source region, such as the Caribbean Sea, the equilibrium of temperature and moisture in an overlying air mass is established in two or three days. Unstable air in either a maritime tropical (**mT**) air mass or a continental tropical (**cT**) air mass is warmed to a height of about 3,000 m (10,000 ft).

In contrast, within cold source regions, air masses can take a week or more to achieve equilibrium with their underlying surfaces. Only a relatively shallow layer of air, up to about 900 m (3,000 ft), cools in continental polar (**cP**) or

maritime polar (**mP**) air masses. The air above that level is cold as well, but for reasons other than contact with the surface. It takes a relatively long time for a cold air mass to become established because cooling at its base stabilizes lapse rates (see Unit 6), thereby preventing vertical mixing and prohibiting efficient heat exchange.

In addition to the four leading types of air masses (**mT**, **cT**, **mP**, and **cP**), two more are recognized. The first is found in the highest latitudes and is called continental Arctic (**cA**); paradoxically, **cA** air masses form at higher latitudes than the polar air masses do. The final type of air mass is found in the lowest latitudes astride the equator and is labeled maritime equatorial (**mE**).

MOVEMENTS OF AIR MASSES

People who live in the Midwestern United States can testify to the bitterness of the polar air that streams southward from Canada in the winter and the oppressiveness of the warm humid air that flows northward from the Gulf of Mexico in the summer. Air masses affect not only their source areas but also the regions across which they migrate. As an air mass moves, it becomes somewhat modified by contact with the surface or by changes within the air, but many of its original characteristics remain identifiable far from its source region. This is a boon to weather forecasters because they can predict weather conditions more accurately if they know the persisting characteristics of a moving air mass as well as its rate and direction of movement.

Figure 12.1 shows the principal air masses affecting North America. The arrows in the figure represent the most common paths of air mass movement away from the source regions indicated on the map. The size, physical properties, and movement of the air masses change on a seasonal basis with the migration of the vertical Sun and seasonal shifts in the position of the jet streams and prevailing surface-pressure systems. The position of the polar front and the strength and location of the polar and subtropical jet streams are especially important in the movement of the air masses from their primary source regions into the middle latitudes on an annual as well as a weekly basis. The main current of a river does not let eddies pass from one bank to another, and so it is with the polar and subtropical jet streams, which provide unseen barriers to advancing air masses. This is a key to understanding weather in most of the United States, and our discussion of the polar jet stream, begun in Unit 9, is developed further in Unit 13.

Lifting Mechanisms That Produce Precipitation

Precipitation originates from parcels of moist air that have been adiabatically cooled below their condensation level (dew-point temperature; see Unit 11). This cooling is accomplished through the lifting of air from the vicinity of

FIGURE 12.1 Source regions and common paths of the principal air masses that affect the continental United States.

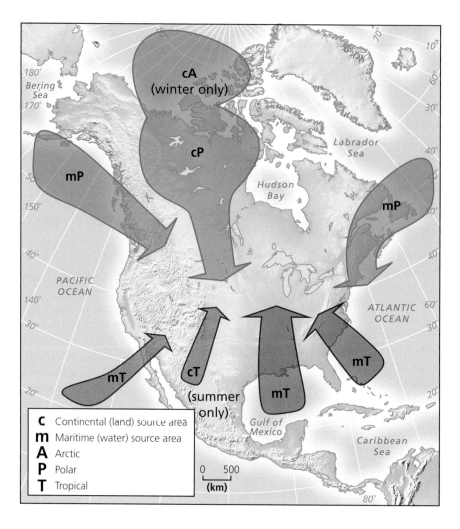

the surface to higher levels in the atmosphere. The occurrence of at least one of four processes is necessary to induce the rising of moist air that results in significant precipitation. On many occasions, more than one process occurs, which may enhance the production of precipitation. These four precipitation-producing mechanisms involve: (1) the forced lifting of air where low-level wind flows converge (which results in convergent-lifting precipitation), (2) the forced uplift of air at the edges of colliding air masses (which results in frontal precipitation), (3) convectional uplift caused by atmospheric instability (which results in convectional precipitation), and (4) the forced uplift of moving air that encounters mountains (which results in orographic precipitation). We discuss each of these processes in detail in the material that follows.

CONVERGENT-LIFTING PRECIPITATION

Where airstreams come together around a low-pressure system at or near the surface, the horizontal convergence of air produces vertical uplift. This follows from one of the conservation principles discussed in Unit 1—namely, mass conservation. As air converges (flows in) at low levels, it cannot disappear into nothingness; it must either accumulate or leave the surface by flowing upward. If the air were to

accumulate, it would fill the low levels, and convergent flow would cease. It follows that if low pressure and convergent flow near the surface are to persist, there must be rising motion to carry mass away. Convergent lifting is behind many large storms such as tropical hurricanes and blizzards in the middle latitudes. Such weather systems are so important they are covered in their own unit immediately following this one. Wind flow into the zone of low pressure along the equator is another important source of convergent lifting. Indeed, the most prominent and durable tropical weather systems marked by this process of **convergent-lifting precipitation** lie where the trade winds from the Northern and Southern Hemispheres come together at the intertropical convergence zone.

The Intertropical Convergence Zone In our discussion of global wind belts and semipermanent pressure zones in Unit 9, we noted that the Northeast and Southeast Trade Winds converge along the equatorial trough of low pressure. The rising air that results is responsible for the cloudiness and precipitation that mark the ITCZ. The ITCZ occurs at low latitudes all around Earth, most notably as a narrow band above the oceans, particularly the equatorial Pacific (Fig. 12.2). More intense surface heating above the continents and islands produces a broader region of

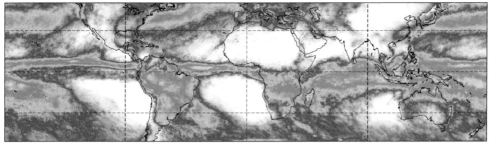

January

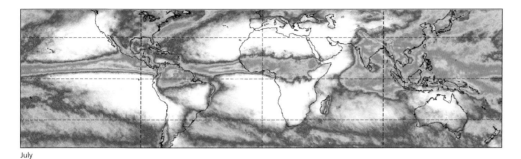

July

FIGURE 12.2 Average daily rainfall rates (mm/day) for January and July based on measurements from the Tropical Rainfall Measuring Mission (TRMM) satellite for the years 1998–2007.

Average Rainfall (mm/day)

0   5   10   15   20

precipitation that is vital to millions of inhabitants of tropical lowlands and hillsides, especially in the Indonesian archipelago of Southeast Asia and the Congo Basin in western equatorial Africa. The precipitation-producing clouds in the ITCZ have a seemingly random distribution and occupy only a small part of the area at any given moment. Thus, rather than continuous uplift everywhere, the ITCZ has highly localized towers of rising air separated by much larger areas of no uplift and slowly descending air. However, because uplift is possible almost everywhere for some part of the year, the nearly continuous time-averaged precipitation patterns shown in Figure 12.2 result.

The ITCZ changes its location throughout the year, generally following the latitudinal corridor of maximum solar heating. Accordingly, as Figure 9.7 shows, the average July position of the ITCZ lies at about 10 degrees N, but in January, at the opposite seasonal extreme, it is hundreds of kilometers to the south. Although the ITCZ owes its origin to the position of the overhead Sun, this is not the only factor that determines its location. The distribution of land and sea as well as the flows of the tropical atmosphere are also important. As a result, at any given moment, the ITCZ may not be where generalized theory tells us it ought to be.

Closer inspection of Figure 12.2 shows that in July the ITCZ is entirely absent in the Southern Hemisphere, when it is drawn northward to a latitudinal position beyond 20 degrees N over southern Asia. (Recall that this was pointed out in our discussion of the wet monsoon in Unit 9.) Yet in January, the ITCZ ranges from a southernmost extreme of 20 degrees S, above northern Australia, to a northernmost position within the Northern Hemisphere tropics in the oceans adjoining South America. Not unexpectedly, the moisture regimes of areas lying inside the broad latitudinal precipitation bands shown in Figure 12.2 are boosted as the ITCZ shifts across this region during the annual cycle.

FRONTAL PRECIPITATION

A second process that generates precipitation is associated with the collision of air masses of significantly different temperatures. Such activity is a common occurrence in the middle and upper-middle latitudes, where the general atmospheric circulation causes poleward-flowing tropical air to collide with polar air moving toward the equator. Because converging warm and cold air masses possess different densities, they do not readily mix. Rather, advancing cold and denser air injects itself beneath lighter warm air and pushes it upward. Similarly, if the warmer air approaches a mass of cooler air, it rides up over the cooler air. In both situations, warm, often moist air rises. And when this air cools sufficiently by adiabatic expansion, it produces condensation, clouds, and precipitation.

Air masses are, by definition, relatively uniform in terms of temperature, density, and humidity. In contrast, boundaries between differing air masses are zones of sharp gradients in density, humidity, and especially temperature. Transition zones between masses are known as **fronts**. Although fronts can be stationary, they usually move as the contrasting air masses migrate. A moving front is the leading edge of the air mass built up behind it. Whether moving or stationary, fronts and the uplift associated with them lead to cooling and very often precipitation termed **frontal precipitation**.

Recall that when a warm air mass infringes upon a cooler one, the lighter warmer air overrides the cooler air.

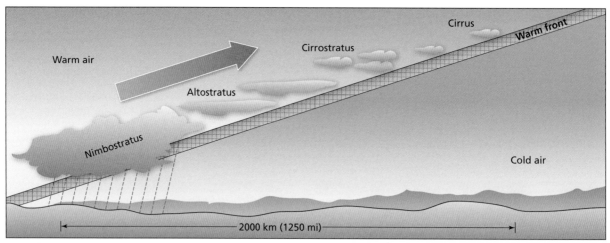

Warm front

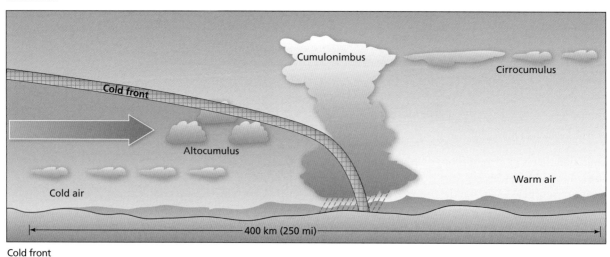

Cold front

FIGURE 12.3 Vertical cross-section through a warm front (top) and cold front (bottom). The vertical scale is greatly exaggerated. Warm fronts typically have a 1:200 vertical-to-horizontal ratio; the ratio for cold fronts is approximately 1:70.

This produces a boundary called a **warm front** (Fig. 12.3, top). Warm fronts, because of their gentle upward slope, are associated with wide areas of light to moderate precipitation that extend out well ahead of the front. As indicated in the figure, warm fronts involve the entire sequence of stratus clouds, ranging from upper level cirrostratus down through altostratus and nimbostratus layers (see Fig. 11.8).

The behavior of a **cold front** is far more dramatic. A cold front occurs when the cold air, often acting something like a bulldozer, hugs the surface and pushes other air upward as it wedges itself beneath the preexisting warmer air (Fig. 12.3, bottom) and causes that warm air to rise. A cold front assumes a steeper slope than a warm front, causing more abrupt cooling and condensation in the warm air that is uplifted just ahead of it. This produces a smaller but much more intense zone of precipitation. The zone usually exhibits the kind of stormy weather associated with the convectional precipitation process we will discuss shortly.

Cold fronts exhibit a variety of cumulus-type clouds, particularly the clusters of cumulonimbus clouds that are the signature of severe thunderstorms. But once the swiftly passing violent weather ends, the cool air mass behind the front produces generally fair and dry weather.

In the middle and high latitudes, frontal precipitation often occurs in combination with convergent precipitation associated with midlatitude cyclones. However, the two lifting processes do not always occur together. In our treatment of weather systems in Unit 13, we examine the life cycle of these storm cells and their fronts, extending the discussion begun here. For now, we close by mentioning that our discussion of fronts is highly idealized. In the same way that atoms with their protons, neutrons, and electrons are a model of matter, a bulldozing cold front producing uplift is a model of atmospheric behavior. This frontal model dates to World War I (hence the militaristic terminology) and is a useful tool. However, modern theory shows that uplift occurs along fronts for a variety of reasons, none of which are linked to wedging or bulldozing. Presentation of that theory is well beyond an introductory text, and so we retain the much simpler model described here.

CONVECTIONAL PRECIPITATION

Convection is spontaneous vertical air movement in the atmosphere as a result of buoyancy. During summer, heating by solar radiation can lead to a layer of warm, less dense air near the surface. Overlain by denser air aloft, the surface air experiences a buoyant force and rises like a cork through water. With sufficient moisture, uplift, and cooling, the result is **convectional precipitation**. The process of convection is generally localized, usually covering only a few square kilometers of the surface. It occurs preferentially over dark surfaces with low albedo where there is a greater absorption of solar radiation. Because land surfaces warm easily, convection is also more common over land than over adjacent water bodies. Even with uniform heating, convection typically is localized. When uplift begins in some spot for essentially random reasons, air flows in from the surrounding area at low levels to replace the vacating mass, and the convection at that spot is reinforced. (You can see something analogous if you study a pot of water as it reaches the boiling point. Bubbles form in random locations on the bottom of the pot. Once started, those locations are the sites of continuing plumes of rising bubbles. If you allow the water to cool and start over, you'll see the same behavior, but with plumes forming in different locations.) A rising column of unsaturated air, known as a *convection cell*,

develops quickly. The temperature of this upward-flowing air cools at the dry adiabatic lapse rate of 1°C/100 m (5.5°F/1,000 ft). (Recall from Unit 7 that the DALR is the cooling rate of rising unsaturated air.) This cooling rate continues as long as the relative humidity of the rising air remains below 100 percent. When the dew-point temperature is reached, condensation commences and a small cumulus cloud forms. That cloud soon mushrooms as the added energy of the latent heat released by the condensing water vapor makes the air even more buoyant, resulting in an even more powerful updraft.

Air Mass Thunderstorms Very large individual cumulus clouds of the towering cumulonimbus type develop if three conditions are met. First, there must be sufficient water vapor in the updraft to sustain the formation of the cloud. Second, the immediate atmosphere must be unstable so that the upward airflow originally triggered in the surface-layer convection cell is able to persist to a very high altitude. And third, there must be relatively weak winds aloft. When these conditions are satisfied, strong surface heating is likely to produce *air mass thunderstorms*, which are also called *single-cell thunderstorms*.

Air mass thunderstorms are common in both the low and middle latitudes and are found within a single air mass, normally the warm, moist, and unstable maritime equatorial (**mE**) or maritime tropical (**mT**) air masses. Thunderstorms are most common during late afternoon hours in the warmest months of the year. Like the larger-scale storm systems discussed in Unit 13, thunderstorms have a distinct life cycle, which is illustrated in Figure 12.4.

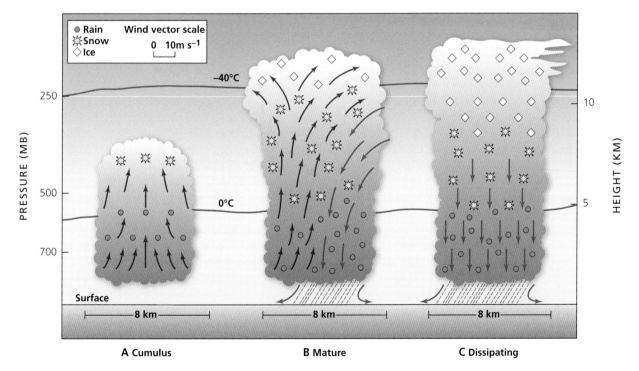

FIGURE 12.4 The three stages of the life cycle of an air mass thunderstorm: cumulus, mature, and dissipating. Updrafts dominate the developing, or cumulus, stage, both updrafts and downdrafts are found during the mature stage, and only downdrafts are found during the dissipating stage. The approximate width at each stage of the thunderstorm is shown at the bottom.

FIGURE 12.5 Massive thunderstorm over the Indian Ocean east of Madagascar. With a diameter of nearly 100 km (60 mi) and an anvil cloud reaching 16 km (10 mi) in height, a storm of this type generates powerful wind flow at the surface and strong vertical updrafts resulting in heavy rainfall.

The **life cycle of an air mass thunderstorm** begins with the onset of the convection process in an unstable atmosphere as moist heated air rises, undergoes condensation, and releases large quantities of latent heat. This heat energy makes the air much warmer than its surroundings, and the resulting updrafts of warm air attain speeds of about 5–20 m/s (10–45 mph). During this *developing* (or *cumulus*) *stage*, raindrops and ice crystals may form, but they do not reach the ground because of the updrafts (Fig. 12.4A).

When the middle *mature stage* is reached, the updrafts continue, producing towering cumulonimbus clouds and intense precipitation. Heavy rain begins to fall from the bottom of the cloud (Fig. 12.4B). The falling raindrops produce downdrafts of cold air as they drag surrounding air molecules toward the ground. The falling air must be replaced, so drier air from outside the cloud is entrained and carried downward. Cooling by evaporation of cloud and raindrops further contributes to downdrafts, which are often felt at the surface as a refreshing gust of cool air just before the rain begins. The sinking cold air spreads out at the ground, while the top of the cloud is drawn out by upper air winds to form an *anvil top*, so named because it resembles the shape of a blacksmith's anvil. Downdrafts dominate the air mass thunderstorm during the *dissipating*

stage, when the downdrafts first weaken and then suppress the updraft of warm, moist air that supplies the energy for the thunderstorm (Fig. 12.4C). The final two stages of the air mass storm may take place within only 30 minutes, and the entire life cycle within 45 minutes to an hour.

Severe Thunderstorms Other lifting mechanisms can also produce thunderstorms. All thunderstorms are triggered by the uplifting of moist, unstable air and the release of enough latent heat to fuel continuing uplift. Some of the most severe thunderstorms are generated by particularly abrupt uplift along boundaries of different air masses. For example, a rapidly moving cold front separating continental polar and maritime tropical air can result in a dramatic storm, as can a dry line separating continental tropical and maritime tropical air masses.

These small-scale weather systems can attain sizable dimensions. The massive thunderclouds shown in Figure 12.5 were easily discerned by astronauts as they orbited above the tropical Indian Ocean to the east of the African continent.

A thunderstorm is classified as a *severe thunderstorm* if it has wind gusts of 93 km/h or greater (58 mph), generates hail 2.5 cm (1 in) or more in diameter, or produces a tornado. Air mass thunderstorms are seldom severe, and so severe thunderstorms usually differ from the storm we described in the previous section in several notable ways. First, severe thunderstorms usually do not occur as the single-cell thunderstorms shown in Figure 12.4, but as multiple cells possessing unique spatial arrangements. Second, severe thunderstorms can form and persist in the presence of strong vertical wind shear, meaning there is strong vertical variation in wind with altitude. Third, they generate a wider variety of weather conditions. Finally, severe thunderstorms can produce heavy and persistent precipitation for hours on end.

The longer duration of precipitation from severe thunderstorms is due to their unique structure, which we show in Figure 12.6. The vertical wind shear tilts the severe thunderstorm cell so that rainfall within the downdraft zone does not interfere with the updraft zone that provides the inflow of warm and moist air fueling the system (represented in the figure with red dashed lines). In fact, the downdraft of cold air (the blue lines in the figure) forms a gust front that wedges beneath the updraft and helps lift air into the cell. The downdrafts and updrafts do not compete with one another but, in fact, complement and reinforce each other and allow the thunderstorm to persist for much longer.

Radar studies have shown that individual severe thunderstorms, ranging from 2 to 8 km (1.2 to 5 mi) in diameter, are discontinuous lines of cells perhaps 80 km (50 mi) in advance of and paralleling a cold front. Such *squall-line thunderstorms* may extend for several hundred kilometers. As the squall line advances, new cells develop to replace the old ones that dissipate. Multiple thunderstorm cells can occur over a large area, forming a *mesoscale convective complex* that can be as large as 100,000 km² (38,600 mi²).

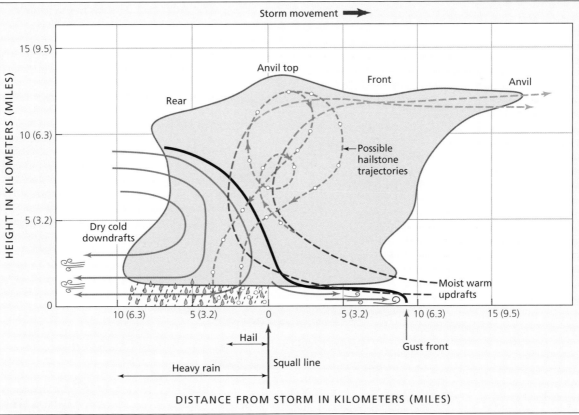

FIGURE 12.6 Photo and conceptual model of a severe thunderstorm producing sizable hailstones. The red dashed lines are the warm updrafts, the blue lines show the cold downdrafts, and the green lines represent the movement of hailstones. Hailstones do not always follow circular paths. The thunderstorm in this diagram is moving from left to right.

Occasionally, smaller thunderstorms expand over a much wider area and assume their own rotation, forming *supercell thunderstorms* that can be 32 km (20 mi) in diameter.

Thunderstorm-Related Weather Phenomena *Lightning* and *thunder* are related phenomena produced by both air mass and severe thunderstorms. A separation of positive and negative charges is produced (by a mechanism that is not fully understood) within thunderstorm clouds by the vertical air currents that produce collisions between ice particles and ice-covered snowflakes. The cold tops of thunderstorm cells acquire a positive charge, while the cooler cloud bottoms have a negative charge. Air is a poor conductor of electricity, but when voltages exceed 9,000 V/m (3,000 V/ft), an electrical discharge—lightning—occurs. Nearly 80 percent of electrical discharges occur either within a single cloud or between adjacent clouds, but some lighting strokes reach the ground or underlying water surface, producing spectacular electrical storms on warm summer nights. Lightning strokes rapidly heat the surrounding air, which expands explosively, producing thunder.

Satellites now allow global measurement of the frequency and spatial distribution of lightning (Fig. 12.7). The highest average lightning counts are found along the ITCZ, where convergent and convective lifting form thunderstorms year-round. In the eastern United States, both summer air mass thunderstorms and warm-season severe thunderstorms are found. Thunderstorm occurrence drops off sharply as one moves from land to ocean surfaces. Note the sharp line of lightning occurrence in northern India, where the Himalaya Mountains both provide uplift and restrict the movement of maritime tropical air masses onto the Tibetan Plateau.

A precipitation phenomenon associated with thunderstorms in the midlatitudes is *hail*. As Figure 12.6 illustrates, in the mature stage of a storm, an ice crystal might be caught in a circulation that continually moves it above and below the freezing level. This circulation pattern can create a small hail embryo. Subsequent motions within the turbulent cloud produce further growth. This occurs both by dry deposition in the ice-crystal process and by wet deposition as the hailstone collides with water droplets (Unit 11). Many a farmer's crop has been destroyed because of these conditions, when hailstones as large as golf balls can bombard the ground (see Fig. 11.10D). Severe thunderstorms can generate even more severe weather in the form of **tornadoes** (see Perspective on the Human Environment: The Historic 2011 Tornado Season).

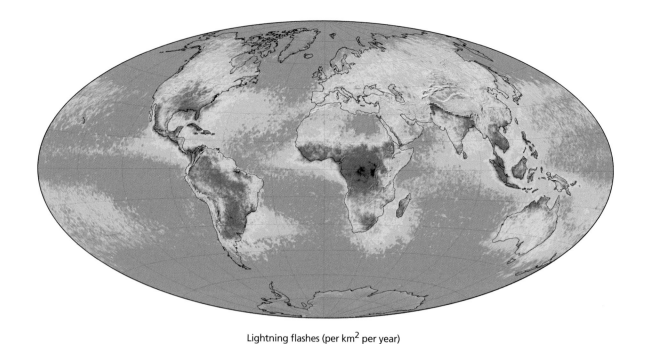

Lightning flashes (per km² per year)

0.1   0.4   1.4   5   20   70

FIGURE 12.7 Average yearly number of lightning flashes observed by satellites between 1995 and 2002. Widespread convergence and convection in the tropics and along the equator produce air mass thunderstorms that are present year-round. Air mass thunderstorms and severe thunderstorms along frontal boundaries occur during the warm season in the middle latitudes. Thunderstorms are much less common in the high latitudes because of their weak surface heating and more limited air mass contrasts.

ENHANCE

Though they certainly do not occur everywhere, thunderstorms of all types are very common in a global sense. Estimates suggest that about 1,800 thunderstorms are active at any given time, which means that they occur on the order of 20 million per year. On average, there are more than 100,000 thunderstorms in the United States each year. Thunderstorms are important to the planet's energy balance, particularly in tropical latitudes where immense quantities of latent heat are released by condensation of warm, moist air.

OROGRAPHIC PRECIPITATION

Mountains, and highlands in general, strongly influence air and moisture flows in the atmosphere. *Orographic uplift* (*oros* is the Greek word for mountain) occurs when a moving mass of air encounters a mountain range or other upland zone and is forced to rise over the obstacle by propelling surface winds. When a mountain range is close to and parallels the coast, and when prevailing onshore winds carry maritime air laden with moisture, even a small amount of cooling triggers condensation during the forced ascent of this air flow. The rain (and often snow) that results from this lifting process is called **orographic precipitation**.

The precipitation on the range's windward slopes, however, is not matched on the inland-facing leeward slopes. Quite to the contrary, the lee side of the range is marked by dryness. Once the mass of air crests the range in a stable atmosphere that does not promote continued uplift, the air immediately begins to descend to lower altitudes, warming rapidly and reducing its relative humidity to below 100 percent.

A classic example of this orographic effect exists along most of the West Coast of the conterminous United States, where the Sierra Nevada and the Cascade Mountains form

---

PERSPECTIVES ON THE HUMAN ENVIRONMENT

The Historic 2011 Tornado Season

The United States experiences more severe thunderstorms than any other land area on Earth, and each year these more than 10,000 storms spawn an average of 1,274 tornadoes. The tornadoes most frequently develop from supercell thunderstorms over the Great Plains, the Midwest, and the interior South, particularly in the north–south corridor known as Tornado Alley, which extends through central Texas, Oklahoma, Kansas, and eastern Nebraska. Here, dry air from the high western plateaus moves eastward into the Plains and forms an elevated inversion that inhibits vertical lifting of air. Cool and dry air moving south from Canada meets warm, moist, low-level air swept northward from the Gulf of Mexico. When the maritime tropical Gulf air collides with the cooler and drier continental polar air, the sharply different temperature and humidity characteristics of these air masses create a localized region of extreme instability. The convective lifting is often so strong that it can break through the upper-level inversion, leading to rapid growth of cumulonimbus clouds that can unleash some of the most violent weather in the world.

The 1,884 tornadoes reported during calendar year 2011 have gone down in the annals of U.S. tornado history as the largest annual total ever recorded, surpassing the previous all-time high of 1,717 twisters in 2004. The map shown in Figure 12.8C provides the salient details. What people will most remember about 2011 is the number of major tornadoes that struck populated areas. As a result, 552 deaths were directly attributed to those tornadoes, making this the second-deadliest tornado-related year in history (exceeded only by the more than 800 fatalities in 1925, when the infamous Tri-State Tornado killed 695 Midwesterners in a single day). In the disastrous 2011 tornado season, two cities were forever changed by the massive tornadoes that devastated them only 25 days apart: Tuscaloosa, Alabama, and Joplin, Missouri.

Tuscaloosa, home of the University of Alabama, was nailed on April 27 amid the biggest single outbreak of tornadoes in American history, which lasted nearly three full days from April 25 to 28. A huge swath of the city was struck by an EF-4 tornado (The EF, or Enhanced Fujita, Scale is a widely used system for tornado damage, with EF-5 being the highest storm rating), with winds topping out at 305 km/h (190 mph). The death toll was 43. More than 1,000 were injured, and the costs of the destruction in the city's immediate vicinity exceeded $2 billion. Figure 12.8A is a photo of the type of destruction that was typical in the wake of the storms. Scientists have recently determined that the April outbreak was likely intensified by smoke particles that blew in from fires in Central America intentionally set to clear land for agriculture. According to computer simulations, the smoke lowered cloud height and created greater wind shear at low levels, both of which make such storms more likely. Here, then, is another example of interconnections within the physical Earth system.

Joplin, the leading regional center of southwestern Missouri, took a direct hit from a rare EF-5 tornado less

continued

---

continued

A

B

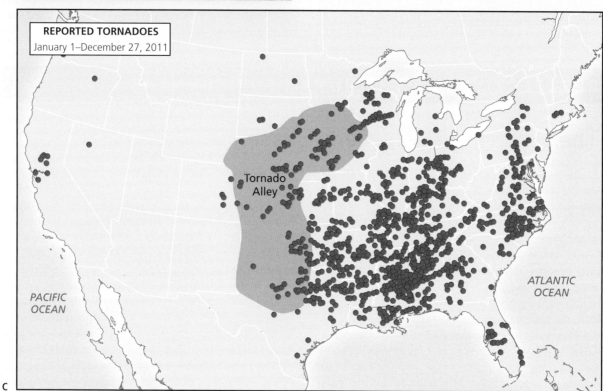

REPORTED TORNADOES
January 1–December 27, 2011

Tornado Alley

PACIFIC OCEAN

ATLANTIC OCEAN

C

FIGURE 12.8 The powerful tornado that struck Tuscaloosa, Alabama, on Wednesday, April 27, was one of many in a three-day period (A). Less than a month later an even stronger tornado, which is believed to be the largest single U.S. tornado in the last 60 years, laid waste to Joplin, Missouri (B). The year 2011 as a whole was remarkable for tornado count and severity, and was also unusual for the large number of tornadoes found to the east of Tornado Alley (C).

than a month later on May 22. One hundred and fifty-nine people died in this disaster, and the damages totaled almost $3 billion. The photo of the utter devastation in Figure 12.8B was taken in the heart of the storm's destruction corridor, estimated to be 9 km long and half a kilometer wide (6 mi by 0.5 mi). The peak wind speed of an EF-5 tornado surpasses 320 km/h (200 mph). The winds that shattered this urban landscape were among the most powerful our atmosphere can produce.

Even for strong tornadoes the destruction path is usually only 10 km (6.2 mi) in length and 150 m (492 ft) in width. The strongest are on the ground for only

10–15 minutes, and most tornadoes last just 3–4 minutes. The average count of 1,274 tornadoes per year in the United States sounds like a lot, but considering the size of the potentially affected areas, this is a low total frequency. Even in the heart of Tornado Alley, a significant tornado occurs only once every 4,000 years on average. In other words, if a spot has been hit by a tornado, thousands of years will likely elapse before a tornado strikes that same location again. If this seems confusing, think about how someone always wins the lottery, but the chances of any particular person winning are extremely low.

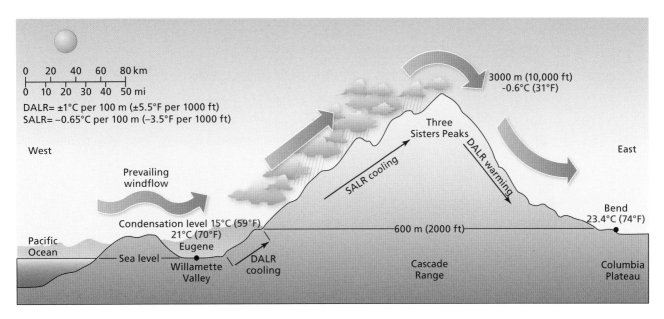

FIGURE 12.9 Orographic precipitation on the upper windward slope of the Cascade Mountains in west-central Oregon. Note the significant temperature and moisture differences between the windward and leeward sides of this major mountain barrier. (The vertical scale is greatly exaggerated.)

a north–south wall extending from Southern California well into Canada. Figure 12.9 illustrates the eastward-moving passage of moist Pacific air across the Cascades of central Oregon, specifically between Eugene in the near-coastal Willamette Valley and Bend on the interior Columbia Plateau. Between these two cities lie the Three Sisters Peaks, a complex of high summit ridges atop the Cascades, reaching an elevation of approximately 3,000 m (10,000 ft).

Imagine that a parcel of air with a temperature of 21°C (70°F) arrives at the near sea-level western base of the Cascades in Eugene. This parcel of air begins its forced ascent of the mountain range and starts to cool at the DALR (1°C/100 m [5.5°F/1,000 ft]). At this rate, the air parcel cools to 15°C (59°F) by the time it reaches 600 m (2,000 ft). Now assume that the air reaches its dew-point temperature at this altitude, and condensation begins. Clouds quickly form, and continued cooling leads to heavy rainfall as the air parcel continues to make its way up the windward slope.

Because saturation has occurred and condensation has begun, however, the air above 600 m cools more slowly at the saturated adiabatic lapse rate, which we will assume averages 0.65°C/100 m (3.5°F/1,000 ft) in this example. By the time the uplifted air reaches the mountaintop at an altitude of 3,000 m (10,000 ft), its temperature is −0.6°C (31°F), which is some 15.6°C (28°F) lower than it was at 600 m. Since this temperature is below the freezing point, precipitation near the summit falls as snow. (In winter freezing and snowfall will commence at a lower elevation on the mountainside.)

Having crested the range in an atmospheric environment presumed to be stable, the air parcel returns to its original level and swiftly descends to a lower elevation. The air is no longer saturated, so the air parcel warms at the DALR. When it reaches the bottom of the 2,400-m (8,000-ft) leeward slope, its temperature will have risen by 24.0°C (44°F) to 23.4°C (75°F). Thus, our air parcel is now noticeably warmer on the lee side of the range, even though the elevation around Bend is 600 m (2,000 ft) higher than it is at Eugene. It is also much drier, as explained earlier. The generally dry conditions on the leeward side of the range caused by this orographic process are termed the **rain shadow effect**.

In the absence of new moisture sources, rain shadows can extend for hundreds of kilometers. This is the case throughout most of the far western United States, because once Pacific moisture is largely removed from the atmosphere, it cannot be locally replenished. Figure 12.10 vividly illustrates the rain shadow covering the eastern half of Oregon. This rain shadow is interrupted only by secondary orographic precipitation effects associated with smaller mountain ranges in the state's northeastern corner. Today, we associate Oregon with the green trees of the state license plate, but in the 1800s its eastern moisture desert was a considerable obstacle to pioneer wagon trains heading westward. The travails of one party were dramatized by the 2010 movie *Meek's Cutoff*, but with considerable understatement—rather than a happy ending, in actuality about 50 people died along the way or shortly thereafter.

Areas on the lee side of mountains frequently experience the rapid downward movement of warm, dry air known as a *foehn* (pronounced *fern*) wind. Above the western plateaus of the United States and Canada, these air flows are called *chinook* winds and can sometimes reach sustained speeds

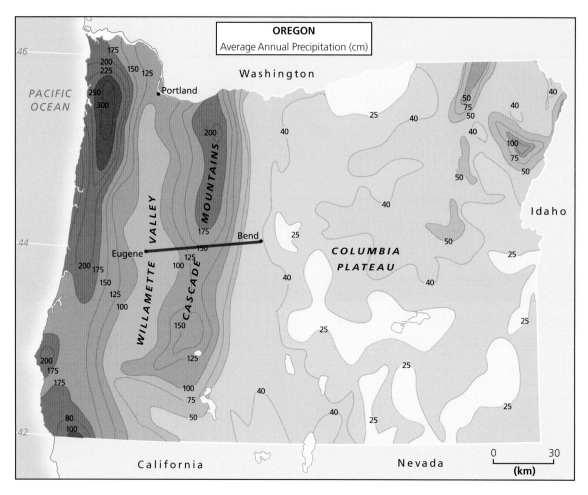

FIGURE 12.10 Oregon's precipitation pattern is dictated by westerly winds off the Pacific that are forced across the north–south-trending Cascade Mountains. The predominantly north–south orientation of isohyets illustrates the pronounced orographic effect. The transect across the Cascades between Eugene and Bend, diagrammed in Figure 12.9, is marked by a blue line.

approaching 50 m/s (110 mph). Their sudden onset, strong winds, and high temperatures produce a variety of effects, ranging from train derailments to snowpacks that vanish overnight.

Key Terms

air mass *page 151*

cold front *page 154*

convection *page 155*

convectional precipitation *page 155*

convergent-lifting precipitation *page 152*

front *page 153*

frontal precipitation *page 153*

life cycle of an air mass thunderstorm *page 156*

orographic precipitation *page 159*

rain shadow effect *page 161*

source region *page 151*

tornado *page 158*

warm front *page 154*

Scan Here for a quick vocabulary review

ENHANCE

Review Questions

1. What conditions result in convectional precipitation? What are some of the more severe types of weather associated with this type of precipitation?

2. Describe the geographic pattern of precipitation on the windward and leeward flanks of a mountain range.

3. Describe the general precipitation characteristics associated with a warm front.

4. Describe the general precipitation characteristics associated with a cold front.

5. Describe the characteristics of the following four major air mass types: **mT**, **cT**, **mP**, and **cP**.

Scan Here for a quick concept review

www.oup.com/us/mason

UNIT
13

Weather Systems

Swirling clouds mark the approach of a midlatitude cyclone as Western Europe still enjoys the clear skies of an anticyclone. In a few hours, the front will strike Ireland and then move across Great Britain and the mainland.

OBJECTIVES

- Demonstrate the importance of migrating weather systems in the global weather picture
- Discuss the significant weather systems of lower latitudes, including tropical storms and hurricanes
- Explain the weather systems of middle and high latitudes, including the formation of cyclones and the weather patterns associated with them
- Examine the transformation of energy and moisture within weather systems

Weather systems are organized phenomena of the atmosphere with inputs and outputs and changes of energy and moisture. Unlike the semipermanent pressure cells and wind flows of the general circulation, transient weather systems are secondary features of the atmosphere and are far more limited in their magnitude and duration. Recurring weather systems, together with the constant flows of moisture, radiation, and heat energy, make up our daily weather. This unit focuses on the migrating atmospheric disturbances we call **storms**. We begin by looking at the storm systems of the low latitudes, which are typically associated with heavy precipitation. We then shift our focus to the storm systems that punctuate the weather patterns of the middle and higher latitudes.

Low-Latitude Weather Systems

The equatorial and tropical latitudes are marked by a surplus of heat, which provides the energy to propel winds and to evaporate the large quantities of water that are often carried by the winds. The abundance of water vapor, convergence of the trade winds, and frequent convection produce heavy rains in the tropics. Although the tropical troposphere is marked by weak horizontal temperature gradients, significant vertical cloud development and strong weather patterns often occur. Tall cumulonimbus towers only 2–3°C (3.6–5.4°F) warmer than the surrounding air can penetrate the troposphere and produce near–gale force winds and precipitation amounts of up to 100 mm/day (4 in/day). Several of these tropical weather systems are shown in Figure 13.1. Recall that in previous units we covered two major low-latitude weather systems. In Unit 12, we discussed the largest system, the intertropical convergence zone, and in Unit 9, we examined the related monsoonal circulation of Asia. In this unit, we investigate the smaller-scale moving weather systems that recur in tropical regions. This includes weak systems, known as easterly waves and tropical depressions, and "gale force" tropical systems with sustained winds above 17.5 m/s (39 mph), including tropical storms and tropical cyclones.

EASTERLY WAVES AND TROPICAL DEPRESSIONS

For centuries, people have known about the trade winds, the nearly constant easterly surface flow of tropical air between 30 degrees N and 30 degrees S latitude that generates convergence along the ITCZ. In the zone outside the ITCZ, cumulus clouds and precipitation occur as cloud clusters that move westward within the surface trade wind regime. These cloud clusters move in association with weak troughs in sea-level pressure that are called **easterly waves** (or *tropical waves*) because of their wavelike pattern on surface weather maps (Fig. 13.2A).

The easterly wave axis (shown in blue on the map) marks a zone of westward-moving low pressure around the wave. These zones often reveal strongly asymmetrical weather patterns. In this Northern Hemisphere example air

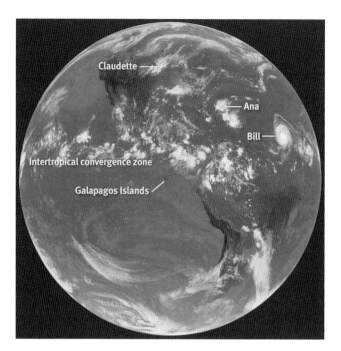

FIGURE 13.1 Full-disk image from the GOES-14 weather satellite at 1:31 p.m. EST on August 17, 2009. This thermal infrared image measures radiation emission from Earth at a wavelength of 10.7 μm; the coldest temperatures are bright white, and the hottest temperatures are black. The Galapagos Islands mark the approximate location of the equator. The ITCZ appears as a discontinuous band of thunderstorms. Tropical Storm Claudette is rapidly losing energy along the eastern Gulf Coast, Tropical Depression Ana is approaching Puerto Rico, and Hurricane Bill is building in the central North Atlantic Ocean.

flowing through the wave turns counterclockwise (cyclonically) as it bends around the axis of low pressure. The westward-moving air converges on the upwind side of the wave and is forced to rise. This convergence produces condensation, towering cumulus and cumulonimbus clouds, and often heavy rainfall. Divergence occurs on the western side of the wave trough, producing descending air, adiabatic warming, and clear weather. Easterly waves are a primary source of precipitation for oceans, islands, and coastal areas in the tropical latitudes.

Weather systems associated with easterly waves are most frequently observed in the Caribbean Basin. Similar phenomena are also common in the west-central Pacific and in the seas off China's central east coast. These systems travel toward the west at about 18 to 36 km/h (11 to 22 mph) and are relatively predictable in their movement. But for reasons that are still not well understood, each year certain easterly waves increase their intensity. As these weak low-pressure troughs deepen, they begin to assume a more organized, rotating, cyclonic pattern. Such a disturbance is called a **tropical depression**. About 100 tropical disturbances form each year in the North Atlantic Ocean, of which approximately 70 reach the Caribbean Sea. About 25 of these intensify and develop greater structure and cyclonic rotation to become tropical depressions.

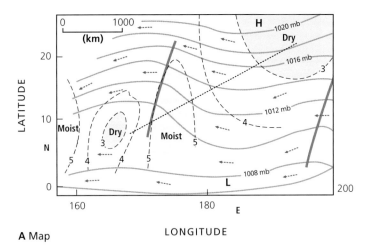

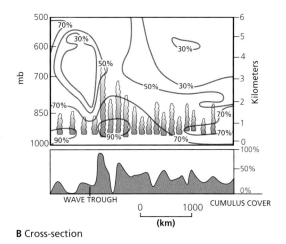

A Map

B Cross-section

FIGURE 13.2 (A) A weather map of an easterly wave in the central Pacific Ocean approaching the Marshall Islands. The trough axis is represented by the leftmost bold blue vertical line. A second trough axis lies about 3,000 km to the east at the far right of the map. Surface winds flow through the wave and show cyclonic (counterclockwise) rotation across the trough axis. Shading shows where column totals of water vapor content as a proportion of atmospheric mass are above 0.5 percent (blue) and below 0.3 percent (tan). (B) Cross-section along the dotted line showing the average vertical heights of cumulus clouds. Isolines of relative humidity (%) are shown, as is the percentage of cumulus cloud cover.

TROPICAL STORMS AND TROPICAL CYCLONES

If a tropical depression intensifies and sustained wind speeds surpass 17.6 m/s (39 mph), it becomes a **tropical storm**. Further development is possible, and if a fully formed tropical storm exhibits sustained winds in excess of 33 m/s (74 mph), a **tropical cyclone** is born. An average of eight tropical cyclones occur each year in the North Atlantic.

An intensely developed tropical cyclone is one of the most fascinating and potentially destructive features of the atmosphere. These severe storms go by different names in different regions of the world. In the western Atlantic and eastern Pacific Oceans, they are called **hurricanes**. Elsewhere, they are known as *typhoons* (western North Pacific) or *cyclones* (Indian Ocean). We employ the terms *hurricane* and *tropical cyclone* interchangeably in this text. A hurricane is a tightly organized, moving low-pressure system, normally originating at sea in the warm, moist air of the low-latitude atmosphere. Like all cyclonic storms, hurricanes have roughly circular wind and pressure fields. In Figure 13.3, a satellite image of Hurricane Mitch, which ravaged parts of Central America in 1998, the circular wind flow extends vertically, and the pinwheel-like cloud pattern reminds us of the storm's cyclonic origin.

Hurricanes have wind speeds greater than 33 m/s (74 mph) and central surface pressures that commonly reach 950 mb or lower. In the most severe tropical cyclones, central pressures are below 900 mb, and wind speeds approach 90 m/s (200 mph), invariably leading to massive destruction when they reach land. The hurricane's structure is diagrammed in Figure 13.4. The open **eye** dominates the middle of the cyclonic system. This "hole in the doughnut" is surrounded by tall clouds reaching an altitude of 16 km (10 mi). The strongest winds and heaviest rainfall (up to 50 cm

FIGURE 13.3 When this photo was taken, the eye of Hurricane Mitch was still over water, but this 1998 storm was poised to strike Central America and become the costliest natural disaster in the modern history of the Western Hemisphere. Honduras was hit hardest, with nearly 10,000 deaths and the loss of more than 150,000 homes, 34,000 km (921,000 mi) of roads, and 335 bridges. In the wake of Mitch, nearly one-quarter of the country's 6.4 million people were left homeless, and most of the agricultural economy was ruined. Honduras and adjacent areas of neighboring Nicaragua will carry Mitch's scars for generations to come.

ENHANCE

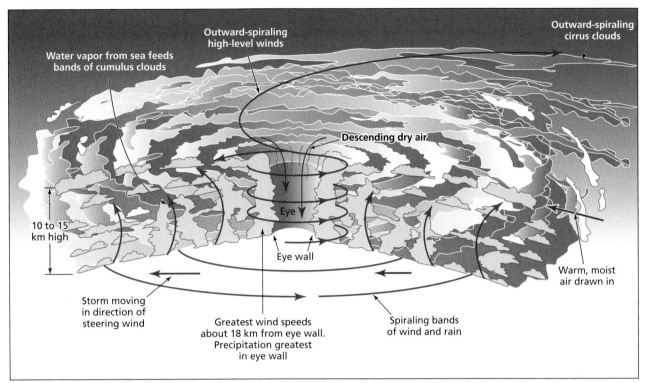

Water vapor from sea feeds
bands of cumulus clouds

Outward-spiraling
high-level winds

Outward-spiraling
cirrus clouds

Descending dry air

Eye ▼

10 to 15
km high

Eye wall

Warm, moist
air drawn in

Storm moving
in direction of
steering wind

Greatest wind speeds
about 18 km from eye wall.
Precipitation greatest
in eye wall

Spiraling bands
of wind and rain

FIGURE 13.4   Cross-sectional view of a hurricane.

ENHANCE

[20 in] per day) are found in the **eye wall**, the towering ring of cumulonimbus clouds that surrounds the eye. Within the eye itself, which extends vertically through the full height of the storm, winds are light and there is little rain as air descends down the entire length of this central column. At the top of the storm, winds and clouds spiral outward as shown in Figure 13.4. Near the surface temperatures vary little across the tropical cyclone, but aloft the center of the storm can be 10°C (18°F) warmer than surrounding areas outside the storm. This warmth is explained by the release of latent heat and gives rise to the term *warm-core low*.

The diameter of a hurricane can vary from 80 km (50 mi) to about 2,500 km (1,550 mi). In general, the intensity and destructive potential of a hurricane are more closely associated with the central surface pressure than with the diameter of the system. Surface air pressure may decrease by about 35 mb from the hurricane's outer edge to the inner core of the system, but then it drops rapidly, perhaps by another 20 mb, toward the hurricane eye. This sudden and large surface-pressure decrease, which may approach 1 mb/km in the hurricane core, is unmatched by that of any other storm system (except tornadoes) and is most responsible for their destructive power.

Hurricane Development   Hurricanes are efficient machines for drawing large quantities of excess heat away from the ocean surface and transporting them into the upper atmosphere. Accordingly, hurricanes are most likely to form in late summer and autumn when tropical sea surfaces reach their peak annual temperatures. Many aspects of hurricane formation are still only partially understood, particularly the exact triggering mechanism that transforms fewer than one of every 10 easterly waves into a tropical cyclone. Several mechanisms have been identified as being capable of causing this intensification. Easterly waves are the most common precursor. Other possible mechanisms are summer low-pressure systems that form over West Africa and are carried westward by the trade winds, convergence along the poleward limit of the ITCZ, and the trailing cold fronts along midlatitude cyclones that extend unusually far equatorward.

Once a tropical cyclone has developed, it becomes self-sustaining. Vast quantities of heat energy are siphoned from the warm ocean below and transported aloft as latent heat. This energy is released as sensible heat when condensation occurs. The sensible heat provides the system with potential energy, which is partially converted into **kinetic energy**, causing the hurricane's violent winds. The warm core leads to upper-level divergence at the top of the hurricane, which feeds vertical lifting within the eye wall, and promotes surface convergence. As long as the hurricane remains over warm water, the system is self-sustaining, which explains the long life of some hurricanes.

Hurricanes originate between 5 and 25 degrees latitude in all tropical oceans except the South Atlantic and the southeastern Pacific, where the ITCZ seldom occurs. The general conditions necessary for hurricane formation include a location 5 degrees or more poleward of the equator, little change in wind speed or direction with height, warm sea surface temperatures of 26.5°C (80°F) or greater, an unstable troposphere, and high atmospheric moisture content. Equatorward of 5 degrees, the Coriolis force is so weak that surface lows easily fill with converging air, and thus hurricanes seldom form there. Poleward of 25 degrees, sea surface temperatures are too cool. Once formed, the movement of hurricanes is controlled by the steering effects of larger-scale air currents in the surrounding atmosphere. These can be erratic, and a hurricane is often likened to a block of wood floating in a river with complicated currents.

In the Northern Hemisphere, hurricanes usually travel first westward and then to the northwest before curving around to the north and east, where they come under the influence of the westerly winds of the middle latitudes. The paths of typical Atlantic hurricanes are mapped in Figure 13.5. The figure illustrates the vulnerability of the U.S. southeastern coast, from the Chesapeake Bay south to the Florida Keys, as well as the entire rim of the Gulf of Mexico. Hurricanes usually advance with forward speeds of 16 to 24 km/h (10 to 15 mph). If they pass over a large body of land or cooler waters, their energy source—the warm ocean—is cut off, and they gradually weaken and die.

Hurricane Destruction Before these tropical cyclones die, normally within a week of reaching hurricane intensity, they can cause considerable damage on land. The Saffir–Simpson Hurricane Wind scale, which ranks hurricane intensity from 1 to 5 (Table 13.1), is used to estimate the potential property damage and flooding expected from a hurricane landfall. Much of this damage is caused by high winds and torrential rains. Near coastlines waves and tides

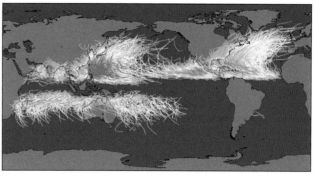

Saffir-Simpson hurricane intensity scale

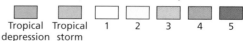

Tropical   Tropical   1   2   3   4   5
depression   storm

FIGURE 13.5 Storm tracks of weak and severe tropical cyclones for the past 150 years. Weaker systems, shown in blue and yellow, occur along the equator during their early stages of development and over land and in the middle latitudes as they lose energy and weaken. These systems are steered by the easterly trade winds within the tropics and the westerlies as they move into the middle latitudes. The most powerful hurricanes (Categories 4 and 5) are most common in the western North Pacific Ocean.

also rise to destructive levels, particularly when wind-driven water known as a **storm surge** (which can surpass normal high-tide levels by more than 10 m [32 ft]) is hurled ashore. The amount of energy released as kinetic energy in the winds of a typical hurricane is equivalent to about 50 percent of the energy production in all the world's power plants combined. This makes it easy to imagine the destructive potential of a hurricane. With the exception of tornadoes—which sometimes are spawned in the inner-spiral rain bands of the strongest hurricanes—tropical cyclones are the most dangerous of all atmospheric weather systems.

The most powerful hurricanes to strike the United States rank among landmark weather events of the past century. Before World War II (and the development of sophisticated weather tracking), hurricanes in 1900 and 1935 caused enormous destruction on the coasts of Texas and the Florida

TABLE 13.1 Saffir–Simpson Hurricane Wind Scale

Category	Sustained Wind Speed	Description of Effects
1	119–153 km/h (74–95 mph)	Damage primarily to unanchored mobile homes, shrubbery, tree branches; power lines blown down
2	154–177 km/h (96–110 mph)	Damage to roofing material, doors, and windows; small trees blown down; unprotected marine craft break their moorings
3	178–209 km/h (111–130 mph)	Damage to small buildings; large trees blown down; mobile homes destroyed; flooding near coast destroys many structures
4	210–249 km/h (131–155 mph)	Extensive roof failures on houses and smaller commercial buildings; major damage to doors, windows, and lower floors of near-shore structures; flooding up to 9 km (6 mi) inland from coast
5	> 249 km/h (155 mph)	Widespread roof failures and destruction of many smaller-sized buildings; major damage to all structures less than 4.5 m (15 ft) above sea level; flooding up to 16 km (10 mi) inland from coast

Source: NOAA, National Hurricane Center.

FIGURE 13.6 Some of the worst devastation caused by Hurricane Andrew in the suburbs south of Miami on August 24, 1992. This is part of Cutler Ridge, located just inland from where the eye wall passed over the Florida coast.

Keys, respectively; in 1938, much of southern New England was devastated. After 1945, hurricanes were named in alphabetical order according to their occurrence each year. Among the most destructive of these have been Hazel (1954), Camille (1969), Agnes (1972), Hugo (1989), Andrew (1992), and Katrina (2005). Hurricane Sandy, in 2012, developed into a devastating extratropical storm, described in the Perspectives on the Human Environment: October Disaster box.

One of the costliest hurricanes to strike the United States was Hurricane Andrew in 1992 (Fig. 13.6). In its path across southernmost Florida, this relatively small but ferocious hurricane was the first to directly hit one of the nation's largest metropolitan areas—Miami—with its full force (average wind speed 75 m/s [165 mph]). The final accounting was truly staggering: 95,400 private homes completely destroyed; a total of $16.04 billion paid to settle 795,912 property insurance claims; more than 125,000 people left homeless and 86,000 jobless; and a cleanup effort involving the removal of 35 million tons of debris at a cost of $600 million.

Hurricane Andrew's claim to the title of costliest U.S. hurricane did not stand for long. In 2005, Hurricane Katrina made landfall near New Orleans, Louisiana, with a central pressure of 920 mb. This storm claimed approximately 1,500 lives and exceeded $100 billion in damage, mostly in Louisiana and Mississippi. Continued catastrophic losses from hurricanes are certain to continue along the U.S. Gulf Coast as a result of the growth of coastal urban centers, increased capital investment, sea-level rise, and weak and ineffective coastal management.

In developing countries, it is not unusual for the death toll for a single storm of that strength to be in the thousands. In fact, the second-greatest natural disaster of the twentieth century was the tropical cyclone that smashed into Bangladesh in 1970, killing upward of 300,000 people. Perhaps as many as half that number perished when another cyclone struck the country in 1991. Beyond the toll in human lives, hurricanes can also cause catastrophic disruptions in the social and economic life of a nation. In recent years, no country has suffered more than Honduras, which since 1998 has been struggling to recover from taking the brunt of Mitch, the most powerful hurricane to make landfall in the recorded history of Central America (see Fig. 13.3).

Although the tropics are generally marked by monotonous daily weather regimes, violence and drama accompany the occasional storms that are spawned in the energy-laden air of the low latitudes. Weather systems in the middle and higher latitudes are typically less dramatic, but they are usually more frequent, involve wider areas, and affect much larger populations.

October Disaster

The month of October is a time of declining hurricane threat in the western Atlantic: fewer storms and lower intensities indicate that, at least statistically, the season's worst is over. When a tropical storm named Sandy made its appearance in the central Caribbean in mid-October 2012 it struggled to achieve hurricane status as uncooperative high-altitude winds threatened to tear it apart. A somewhat-cool ocean surface only reluctantly provided the necessary heat and moisture. When it did develop into a Category 1 hurricane that killed at least 69 on the islands of Hispaniola and Cuba, it was an unusually severe October impact. However initial forecasts suggested that the storm would bypass the North American mainland and head off into the central Atlantic.

And then the computer-generated forecast tracks began to reveal that a disaster was in the making. After skirting Florida and South Carolina and doing damage in coastal North Carolina and Virginia, Sandy made a westward turn and made landfall between Delaware and New York's Long Island. As the storm moved northward it merged with and received a boost from a midlatitude disturbance to become an extratropical cyclone of extraordinary proportions. Superstorm Sandy bore the mark of both its parents. From the tropical cyclone it inherited a deep central low pressure, with strong convergence at low levels. From its midlatitude parent it gained a new energy source (via a strong cold front) and size. As a result its eye wall resembled that of a Category 3 storm, and it covered much of the eastern United States (Fig. 13.7A). Wind strength would not exceed Category 1, but Sandy's arrival at high tide greatly amplified the effects of its 9-foot storm surge and created havoc along deeply indented, low-elevation shorelines.

Sixty million people were potentially vulnerable to Sandy's impact. Evacuations began and preparations accelerated, but

A

B

FIGURE 13.7 Seen in its post-tropical state at landfall, Superstorm Sandy was much larger than a hurricane (A). Seawater pushed ashore by the storm cascades into the construction site for the Ground Zero Memorial, Manhattan, New York (B).

nothing could have mitigated the terrible assault that began on Sunday, October 28, and continued into Monday. Howling winds and rising waters created total devastation near shores and far inland; in New York City, where the death toll was heaviest, subways and tunnels filled with salt water, and people drowned, died under falling trees, or were swept away in their cars (Fig. 13.7B). Half the fatalities and the worst devastation in New York's five boroughs occurred on Staten Island. Next hardest hit was the borough of Queens, where electrical sparks from downed wires started a fire that destroyed more than 100 homes. In New Jersey, which bore the brunt of the impact although fewer lives were lost, entire communities were swept away. Rising waters in Newark Bay flooded electrical stations, leaving nearly 3 million without power.

Within a few days after Sandy's impact, the first estimates of the damage (in addition to over 100 deaths) left no doubt that Sandy ranked second in terms of cost only to Hurricane Katrina's 2005 devastation of New Orleans.

Weather Systems of the Middle and Higher Latitudes

The general atmospheric circulation of the middle and higher latitudes exhibits some basic differences from that of the tropical latitudes. The principal differences result from the strong latitudinal temperature gradients that produce steeply sloping pressure surfaces and the presence of large wavelike patterns (Rossby waves; see Unit 9) within the polar jet stream (Fig. 13.8). Both factors lead to the frequent interaction of dissimilar air masses. In the higher latitudes, high-pressure and low-pressure systems moving eastward are carried along by the westerly winds of the upper air. In the middle latitudes, air of different origins—cold, warm, moist,

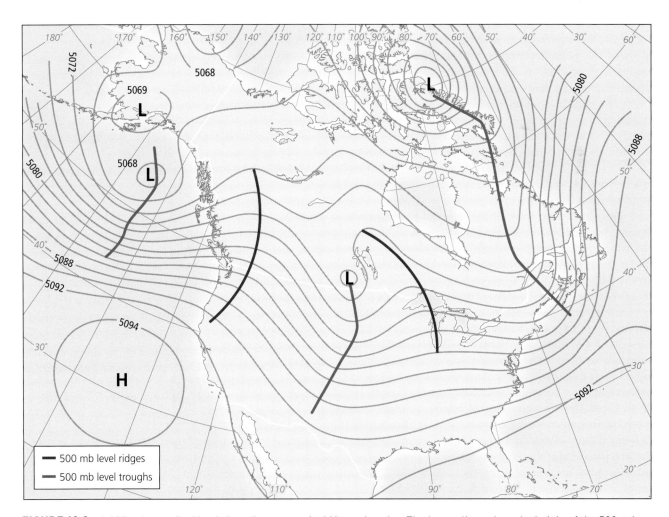

FIGURE 13.8 A 500-mb map for North America on a typical November day. The brown lines show the height of the 500-mb level in meters above sea level. Ridge axes are shown in red; trough axes are in blue.

dry—constantly comes together, and fast-flowing jet streams are associated with sharp differences in temperature.

In addition to the general pattern of horizontal temperature gradients produced by the decreasing solar radiation that accompanies increasing latitude, contrasting surface conditions generate strong horizontal temperature gradients on a local scale. These include warm ocean currents along continental barriers, such as the Kuroshio Current along eastern Asia, the Gulf Stream off the eastern United States, and the Alaska Current along the Alaskan coast; the Mediterranean Sea along southern Europe (winter only); and the North Atlantic Drift next to floating sea ice in the North Atlantic (winter only). Both the strength and position of the jet stream and the surface temperature gradients change seasonally with the migrating latitude of the overhead Sun. Therefore, the weather in the middle latitudes is much more variable over distance and changeable over time than the weather in the tropics.

THE POLAR JET STREAM

Unit 9 explained how the fast-flowing upper air currents of the jet streams are related to the process of moving warm air from the tropics and cold air from the high latitudes. Unit 12 pointed out that jet streams are found along the boundaries between large areas of cold and warm air. One of them, the **polar jet stream**, although not in itself a weather system, controls the formation and movement of the most important midlatitude weather systems.

In the middle latitudes, isotherms and pressure surfaces slope away from the tropics toward the polar regions (see Fig. 13.8). In addition, surface temperatures drop abruptly, and warm and cold air face each other horizontally. As we know, such a narrow zone of contact constitutes a front. The *polar front* separates relatively cold polar air from relatively warm tropical air. The jet stream associated with this temperature divide is called the polar jet stream. As noted in Unit 9, at the core of the wavelike westerly winds of the upper troposphere, the polar jet stream snakes its way around the globe in large wave-like meanders, producing alternating ridges of high pressure and troughs of low pressure (see Fig. 13.8).

These meanders develop in response to fairly complex physical laws that govern the movement of high-velocity air currents in the upper atmosphere. As upper-level winds approach an upper-level trough axis from the west (see Fig. 13.8), the winds slow down and move closer together horizontally in a process called **convergence**. To the east of an upper-level trough axis, the winds speed up and spread farther apart horizontally, resulting in **divergence**. This upper atmospheric convergence and divergence cause the air to subside (under the area of convergence) or ascend (under the area of divergence). As a result, these zones of convergence and divergence have important implications for the formation and movement of surface-pressure systems and attendant weather patterns. Convergence and divergence also occur when the jet

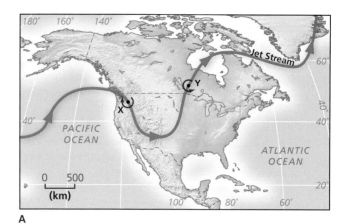

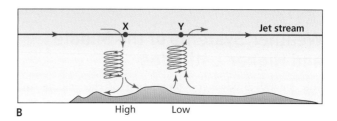

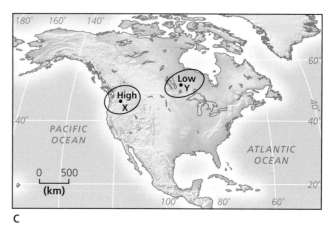

FIGURE 13.9 Relationship between the polar jet stream and surface-pressure patterns. (A) Position of the jet stream 9,000 m (30,000 ft) above North America. (B) Associated air movement around points *X* and *Y* in vertical cross-section. (C) Resulting surface-pressure conditions.

stream is alternately squished and stretched as it crosses over and beyond a major mountain barrier. For all these reasons the polar jet stream is key to the process of **cyclogenesis**—the formation, evolution, and movement of midlatitude cyclones.

MIDLATITUDE CYCLONES

The weather at Earth's surface in the midlatitudes is largely determined by the position of the polar jet stream. In Figure 13.9A, the jet stream passes across North America 9,000 m (30,000 ft) above the ground. The air in the jet stream converges at point *X*, as shown in cross-section (Fig. 13.9B), and it descends in a clockwise downspiral. The descending air causes high pressure at the surface, as shown in Figure 13.9C, and spreads out or diverges from the center of the surface anticyclone. At point *X*, therefore, we expect fair weather because of the descending air.

Farther downwind in the eastward-flowing jet stream, at point *Y*, the air in the jet stream diverges. The diverging air aloft must be replaced, and air is drawn up from below. As air is drawn up the column, a converging cyclonic circulation develops at the surface, with its characteristic low pressure. The rising air cools and forms clouds that soon produce precipitation. In this manner, the jet stream becomes the primary cause of fair weather when it generates upper atmospheric convergence in the middle latitudes and stormy weather when it produces divergence.

Weather Sequence in a Mature Midlatitude Cyclone

The midlatitude cyclone, like its low-latitude counterpart, goes by several names. It is often called a *depression* or a *wave cyclone*, or just a *surface low*. The unit-opening satellite image (p. 164) shows a huge, spiraling midlatitude cyclone over the North Atlantic Ocean northwest of the British Isles. These are the most common large-scale weather systems found outside the tropics. Midlatitude cyclones are characterized by circular wind flow and a low-pressure field as well as by sharply contrasting air mass properties.

The sequence of weather along a cross-section of a mature midlatitude cyclone appears in Figure 13.10, which shows a wave cyclone above the southeastern United States. The most prominent features of Figure 13.10A are the roughly circular configuration of the isobars and the positions of the fronts. Note that the isobars are generally circular within the cold-air zone but are closer to straight within the *warm sector*, the wedge of warm air enclosed by the cold and warm fronts. In the mature stage, the center of the surface low is generally centered on the apex where the two fronts meet. Also noteworthy is the *kinking*, or sharp angle, the isobars form at the fronts themselves.

The wind arrows tend to cross the isobars at a slight angle—as we would expect when friction at Earth's surface partially upsets the geostrophic balance (see Fig. 8.7). The wind arrows indicate a large whirl of air gradually accumulating at the center of the low-pressure cell. This converging air rises and cools, especially when it lifts at the fronts, and its water vapor condenses to produce a significant amount of precipitation. Maritime tropical (**mT**) air, warmer than

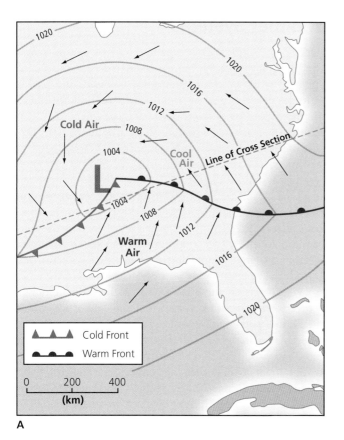

A

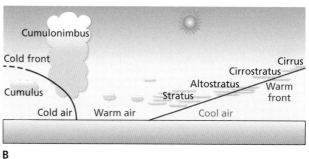

B

Post Cold Front	Warm Sector	Pre Warm Front	Sector Weather Condition
Heavy rain, then clearing	Showers	Rain and fog	Precipitation
↘ ↘	↗ ↗	↖ ↖	Surface Wind Direction
Rising	Low steady	Falling	Pressure
Lower	Higher	Lower	Temperature

C

FIGURE 13.10 Mature stage of a midlatitude cyclone over the southeastern United States. (A) Pressure fields, wind flows, and fronts. (B) Cross-sectional view along the dashed line mapped in (A). The vertical scale is greatly exaggerated. Cold fronts typically rise only 1 m for every 70 m horizontal extent; for warm fronts, the ratio is about 1:200. (C) Summary of surface weather conditions along the line of cross-section shown in Part A.

ENHANCE   ENHANCE

the surface below it, forms the warm sector, and cold continental polar (**cP**) air advances from the northwest.

In the cross-sectional view in Figure 13.10B, the warm front has the expected gentle slope. As the warm air moves up this slope, it cools and condenses to produce the characteristic sequence of stratus clouds. The cold front, in contrast, is much steeper than the warm front. This forces the warm air to rise much more rapidly and creates towering cumulus clouds, which can yield thunderstorms and heavy rains.

Although no two wave cyclones are identical, there are enough similarities between all cyclones that an observer on the ground can predict the pattern and sequence of their atmospheric events. The events, which are summarized in Figure 13.10C, are as follows:

1. As the midlatitude cyclone approaches along the line of cross-section shown in Figure 13.10A, clouds thicken, steady rain falls, and the pressure drops.
2. As the warm front passes, the temperature rises, the wind direction shifts, pressure remains steady, and the rain lets up or turns to occasional showers with the arrival of the warm sector.
3. Then comes the frequently turbulent cold front with its higher intensity but shorter duration, rain, another wind shift, and an abrupt drop in temperature.

The entire system usually moves eastward, roughly parallel to the steering upper-air jet stream. A weather forecaster must predict how fast the system will move and whether the wave cyclone will intensify or lose energy. Those two predictions are often difficult to make and account for most of the forecasts that go awry.

Conveyor Belt Model Depictions using surface and 500-mb maps do not adequately represent the three-dimensional nature of midlatitude cyclones, which involve both vertical and horizontal air flows in so-called **conveyor belts**. Figure 13.11 is a satellite image of a massive winter wave cyclone and an oblique view of the warm and cold fronts of a mature cyclone with the conveyor belt model added. Three major flows make up the model in Figure 13.11B. The **warm conveyor belt** consists of warm and humid air originating in the lower troposphere of the warm sector that rises up and over the warm front, eventually sweeping eastward in the upper troposphere. This belt supplies most of the moisture source for the system and produces the most clouds and precipitation in the cyclone. In strong cyclones the warm belt is fed by an atmospheric river, as described in Unit 11. The **cold conveyor belt** is in the lower troposphere ahead of the warm front. It flows toward the center of the low, parallel to the warm front. The cold conveyor belt then makes a broad anticyclonic sweep as it ascends to the middle troposphere, where it too joins the broad westerly flow aloft. It produces much of the middle troposphere cloud cover associated with the comma-shaped cloud that dominates the cyclone. The **dry conveyor belt** brings cold and dry air from the middle and upper troposphere into the cyclone. Some of the air sinks to the surface, creating the dry slot often seen wedged between the cold front and the center of the low.

The bulk of the air rises and merges with the general westerly upper-level flow.

Because they are so influential and common, midlatitude cyclones are a dominant climatic control outside the tropics. Cyclone formation and movement are closely tied to surface temperature gradients, the position of the polar jet stream, the presence of mountains, and the location of surface boundaries. Therefore midlatitude cyclones frequently develop in specific zones and travel along common paths. The location of these cyclogenetic zones and storm tracks moves toward the poles in the summer and toward the equator in the winter. Cyclones move generally eastward with speeds from 0 to 60 km/h (0 to 35 mph) in winter; in summer, the velocities range from 0 to 40 km/h (0 to 25 mph). These wave-shaped cyclones may be anywhere from 300 to 3,000 km (200 to 2,000 mi) in diameter; they can range from 8 to 11 km (5 to 7 mi) in height and can have a central surface pressure as low as 960 mb. They provide the greatest source of rain in the midlatitudes, and they are a primary mover of energy across the middle latitudes from tropical climates to the higher latitudes.

Energy and Moisture Within Weather Systems

Every day, a multitude of weather systems parade across the planetary surface. These systems include the atmospheric disturbances of the tropical, middle, and higher latitudes discussed in this unit, which are fundamental to an understanding of the climates of those zones. They also involve the small-scale systems that make up the daily texture of the atmosphere, such as monsoon showers (covered in Unit 8) and thunderstorms (covered in Unit 12). Each weather system, though distinctive in detail, is energized in a broadly similar way by transformations of energy, and in every case water moves between its gaseous and nongaseous phases. These transformations are shaped by the conservation of energy and mass, and we can gain insight into weather systems by appealing to those principles, which we introduced in Unit 1.

We know that the Sun's energy is the ultimate power source for any weather system, but the connection is not direct, like water spilling over a dam and turning a waterwheel. After all, the Sun can shine brightly on the calmest of days, and violent storms can occur at night with no solar radiation input at all. Rather, the connection is indirect. Solar energy is stored, concentrated spatially, and later released in bursts of kinetic energy, some of which are strong enough to tear roofs from buildings and topple railroad cars.

Fundamentally, all weather systems represent transformation of **potential energy** (PE) into kinetic energy (KE). Part of an air column's PE is the internal energy of its molecules, which is directly proportional to temperature. In a warmer column, there is more potential energy that could be transformed to energy of motion than there is in a cooler column. An air column's potential energy is also partly gravitational. Imagine sitting on a beach and tossing pebbles

A

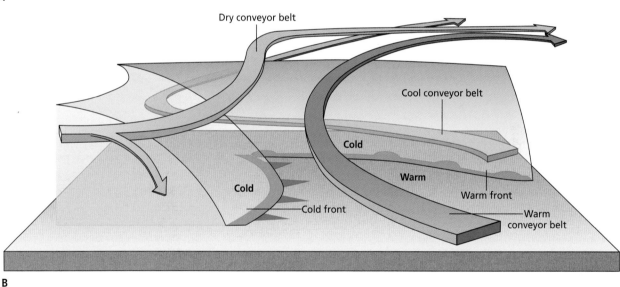

Dry conveyor belt

Cool conveyor belt

Cold

Warm

Warm front

Cold

Cold front

Warm
conveyor belt

B

FIGURE 13.11 (A) February 2, 2011, GOES-13 satellite image of a massive winter midlatitude cyclone moving across the eastern United States. The cyclone is about 2,000 km (1,240 mi) in diameter; a large comma cloud extends from the Midwest to New England. A dry slot formed by the dry conveyor belt is faintly evident to the west of the cold front. (B) This view of a midlatitude cyclone shows the relative positions of the warm conveyor belt, the cold conveyor belt, and the dry conveyor belt.

into a pile. You are lifting the rocks against the force of gravity, and doing so requires energy. That energy can't disappear but rather is stored in the pile as **gravitational potential energy**. If you keep tossing, eventually the pile will at least partially collapse, and pebbles that were stationary will begin moving downward. A pebble's kinetic energy does not come out of nothingness; rather, the kinetic energy arises through transformation of potential energy, and in this example, only gravitational potential energy is transformed.

How does potential energy accumulate in the atmosphere? Consider what happens when the Sun rises, say, after a calm, clear night. It warms the atmosphere both directly through absorption by gas molecules and indirectly through the surface energy transfers discussed in Unit 5 (upwelling longwave radiation plus sensible and latent heat flows). The air column expands vertically in response to the warming, and its center of gravity becomes higher than before (Fig 13.12). Both the internal and gravitational

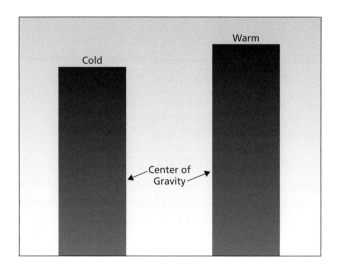

FIGURE 13.12 Warming an air column causes it to expand vertically, raising the center of mass (which for these columns occurs at the 500-mb level). Inflation by warming is greatly exaggerated in the figure.

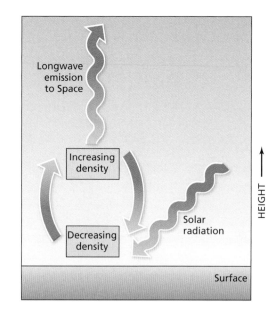

A

FIGURE 13.13 Schematic diagram depicting accumulation of potential energy in a convective thunderstorm (A) and over a large expanse in the middle latitudes (B). The sloping motions in (B) occur at different longitudes and are not stacked above each other as this cross-section unavoidably implies.

components of PE increase, and the total PE is larger. Similar considerations apply to layers of air within the column: heating and inflation raise the layer's center of gravity, whereas cooling and deflation do the opposite.

The atmosphere's store of PE is vast, but only a small fraction is available for conversion to KE. Under most circumstances, an air parcel can't simply start sinking, because it is supported by underlying parcels. What is needed is some way for unusually dense air to accumulate above less-dense air and thereby generate the vertical motion and overturning that lies at the heart of all disturbances. Let's consider how this happens for storms at two extremes of spatial scale: a convective air mass thunderstorm and a midlatitude cyclone. In the former (Fig. 13.13A), cooling at high levels by radiation to space occurs above solar-induced heating at lower levels. The effect is to increase the density of the air aloft while lower levels become less dense, and the stage is set for updrafts and downdrafts. Technically, we are not comparing actual densities aloft and below. Instead, we imagine bringing air parcels to the same level and compare those "potential" densities. If the upper air has greater potential density, overturning occurs. By contrast, midlatitude cyclones depend on air warmed at lower latitudes moving poleward and upward while higher-latitude air does the reverse (Fig. 13.13B). This motion takes place on gradually sloping trajectories over thousands of kilometers, and it is accomplished by the great north–south meanderings of westerly waves across the middle latitudes.

Upward and downward motions, whether nearly vertical as in a convective air mass thunderstorm (see Fig. 13.13A) or sloping as in a midlatitude cyclone (see Fig 13.13B), move lighter air up and heavier air down and thus lower the atmosphere's center of mass. The fall amounts to a decrease in gravitational PE, and so PE is converted to KE. Within updrafts, cooling air eventually condenses and releases the latent heat of vaporization. This warms the air, encourages

stronger uplift (Unit 12), and partially offsets the storm's decrease of gravitational potential energy.

Eventually, every storm dies out when available potential energy is exhausted. Formation of new storms requires recharging the PE reservoir, whether by heating and inflation for thunderstorms or by warm and cold fingers of air pushing poleward and equatorward in westerly waves. Calculations show that recharge times are measured in hours for thunderstorms, whereas four or five days are needed for midlatitude cyclones. This agrees with everyday experience regarding these storms. Notice that recharge/discharge cycles require the existence of thresholds within the system. That is, rather than continuous mild leakage, potential energy accumulates until it quite suddenly bursts forth in what we call a "disturbance." We saw similar threshold behavior in our pebble example, with the sudden onset of conversion to kinetic energy as the slope failed and stones began tumbling down.

In closing, let us make two final points regarding atmospheric energetics. First, however dramatic and damaging the resulting winds might be, the conversion of potential energy to kinetic energy is highly inefficient. For example, if an air parcel sank from 5 km (3.1 mi) to the surface and all of its PE was converted to KE, it would hit the ground at over 300 m/s (671 mph)! Descents of 5 km are common, but such downdraft speeds are obviously absurd. This tells us that little of the PE goes to KE; in fact, more than 99 percent of the PE is shed through collisions with other parcels and conversion to heat energy. Second, transformation from PE to KE is not something that happens only in weather systems.

Prevailing winds and systems such as the Hadley circulation are sustained by that same equally inefficient conversion, although in a much more continuous fashion.

Over the whole planet, only a small part of absorbed solar radiation, on the order of 1.4 percent, is converted to available PE. Remarkably, this seemingly trivial amount—just 3 W/m² —is able to power wind and ocean belts that span the globe and storms whose violence is almost beyond imagination. This happens because the Earth system is like a child in a swing who continually receives nudges from a parent or sibling. Small energy inputs offset frictional losses in the hinges, and the swinging motion is maintained with little effort. In the atmosphere's case, the motion appears as both unrelenting broad-scale general circulation and, when available PE is concentrated in small areas, weather disturbances whose KE production greatly exceeds the global average.

Key Terms

cold conveyor belt *page 174*

convergence *page 172*

conveyor belt *page 174*

cyclogenesis *page 172*

divergence *page 172*

dry conveyor belt *page 174*

easterly wave *page 165*

eye *page 166*

eye wall *page 167*

gravitational potential energy
  page 175

hurricane *page 166*

kinetic energy *page 167*

polar jet stream *page 172*

potential energy *page 174*

storm *page 165*

storm surge *page 168*

tropical cyclone *page 166*

tropical depression *page 165*

tropical storm *page 166*

warm conveyor belt *page 174*

Scan Here for a quick vocabulary review

ENHANCE

Review Questions

1. Describe the general conditions necessary for the formation of a hurricane.

2. Describe the internal structure of a fully developed hurricane.

3. After Sandy hit his neighborhood and his building lost heat, a New Jersey man kept his apartment warm by using his oven and boiling water on the stove. Evaluate this strategy using what you know about energy conservation from Unit 1. Does boiling water help?

4. Explain why tropical storms get a boost from evaporation, but yet the New Jersey man mentioned in Question 3 was no better off for it.

5. How does the polar jet stream contribute to the formation of a midlatitude cyclone?

6. Describe the internal structure of a midlatitude cyclone.

7. Describe the general weather changes at a location experiencing the complete passage of a midlatitude cyclone.

8. Explain how potential energy relates to weather systems.

Scan Here for a quick concept review

ENHANCE

www.oup.com/us/mason

PART THREE

Climate and Climate Change

Sep 17 Average, 1981–2010

UNITS

Climate and Climate Change

The processes that deliver energy and moisture to Earth's surface, which we discussed in Parts One and Two, are quite dependable, so much so that the concept of climate is part of our everyday considerations. For example, we instinctively know that air conditioning is a must for coping with summer in Phoenix, Arizona. Some summers are cooler than others in Phoenix, but nobody expects June temperatures there to be like those in San Francisco, California.

This part begins with a discussion of climate classification, whereby we turn vague notions of "climate" into more precise concepts that capture the atmospheric characteristics of a place. The vocabulary and approach developed in Unit 14 provide a way to describe the climates found today on Earth, which we do in the units that follow. The final two units take up the grand topic of climatic change and related questions: How has our climatic environment changed in the past (Unit 18), and how might human activity affect climates in the coming decades (Unit 19)? As we will see, feedback processes in the Earth system have endowed the planet with a remarkable constancy in climate, but humankind is testing the limits of that resiliency by our increasing reliance on fossil fuels and the associated release of greenhouse gases.

Climate Classification and Regionalization

Rainfall in many tropical locations is highly seasonal, with heavy amounts falling when the intertopical convergence zone is near, and corresponding dry conditions when the ITCZ has moved away. A characteristic mix of trees and grasses is found in such climates, well illustrated by this February (wet-season) scene in the Serengeti Plains of Tanzania.

OBJECTIVES

- Define climate and discuss the general problems of climate classification based on dynamic phenomena

- Outline the useful climate classification system devised by Köppen based on temperatures and precipitation amounts and timing

- List and describe climate types and their regional distribution

- Apply the modified Köppen classification system to Earth and briefly describe major climate regions as they appear on a hypothetical continent and the world map

Climatology is the study of Earth's regional climates, whereas meteorology is the study of short-term atmospheric phenomena that constitute weather. To understand this distinction, we may cite the rule that weather happens now, but climate goes on all the time. Thus, **climate** involves long-term, aggregate weather conditions we expect to experience in a particular place. If we live in Florida or Hawai'i, we expect to wear light clothes for most of the year; if we live in Alaska, we expect to wear heavy jackets. Climate, therefore, is a synthesis of the succession of weather events we have learned to expect in any particular location.

In more concrete terms, we can define the climate of a place as the statistical properties of the atmosphere at that location. The average temperature of a place is an element of climate, as is the average snowfall amount, the average wind speed, and so forth. In using an average value, we are implicitly acknowledging that there is variability from one year to the next that must be discounted to calculate "typical" conditions. The convention is to use a recent 30-year average, known as the **climatic normal**. The choice of 30 years rather than some other number represents a trade-off between enough years to smooth interannual variability (the differences between individual years) but not so many as to include data from potentially different climates of the past. The climatic normal is meant to capture present-day climate, and therefore it is updated every decade to reflect changing conditions. Currently, the normal is computed from the years 1981–2010, which replaced the previous 1971–2000 period.

Seasonal changes are a critical part of climate for most locations, so annual averages are seldom sufficient. Most commonly, climate is captured by looking at monthly normals; 30 Januaries are averaged to give the January normal, 30 Februaries for the February normal, and so on. Statistical properties other than averages are also part of climate. For example, it is important to know how consistent rainfall is from one year to the next. Two places might have similar July normal rainfall, but perhaps one place receives amounts close to normal every year, whereas the other place varies wildly. That difference in variability could have huge impacts for farmers who depend on rain-fed agriculture for their livelihood. The frequency of extreme events is also part of a place's climate. For example, how often are temperatures above 40°C (104°F)? How many freezing events are expected each year?

Because a degree of regularity exists in the heat and water exchanges at Earth's surface and in the general circulation of the atmosphere, a broad and predictable pattern of climates occurs across the globe. However, this pattern is altered by additional *climatic controls* such as the location of land and water bodies, ocean currents, and mountain ranges and other highlands. Units 15 to 17 examine the ways broad patterns and fine details create distinctly different climates across Earth. This unit provides a framework for that investigation by discussing the classification of climate types and their global spatial distribution.

Classifying Climates

Because of the great variety and complexity of recurring weather patterns across Earth's surface, atmospheric scientists reduce countless local climates to a relative few that possess important unifying characteristics. Such classification is the organizational foundation of all the modern sciences. Where would chemistry be without its periodic table of the elements, geology without its timescale of past eras and epochs, and biology without its Linnaean system of naming plant and animal species?

The ideal climate classification system would achieve five objectives:

1. It should clearly differentiate among all the major types of climates that occur on Earth.
2. It should show the relationships among these climate types.
3. It should apply to the entire planet.
4. It should provide a framework for further subdivision to cover specific locales.
5. It should demonstrate the controls that cause any particular climate.

Unfortunately, in the same way that no map projection can simultaneously satisfy all of our requirements (see Unit 3), no climate classification system can simultaneously achieve all five of these objectives. There are two reasons for this. First, so many factors contribute to climate that we must compromise between simplicity and complexity. We have values for radiation, temperature, precipitation, evapotranspiration, wind direction, and so forth. We could use just one variable—as the Greeks used latitude to differentiate torrid, temperate, and frigid zones—but that would produce a classification system that is too simple to be useful. At the other extreme, we could use all the values to gain infinite detail—but at the expense of overwhelming complexity.

The second reason that all of these requirements cannot be satisfied is that Earth's climates form a spatial continuum. Sharp areal breaks between the major types of climate occur rarely because both daily and generalized weather patterns change from place to place gradually, not abruptly. Yet any classificatory map forces climatologists to draw lines separating a given climate type from its neighbors, thereby giving the impression that such a boundary is a sharp dividing line rather than the middle of a broad transition zone.

When confronted with these problems, climatologists focus their compromises on a single rule: the development or choice of a climatic classification must be determined by the particular use for which the system is intended. Because there are many potential uses, it is not surprising that many different classification systems exist. Whereas qualitative and/or subjective criteria could be used to distinguish climates (see Perspectives on the Human Environment: Climate in Daily Human Terms), the practicalities of contemporary physical geography demand an objective, quantitative approach. Moreover, for our purposes, the classification system must be reasonably simple and yet reflect the full diversity of global climatic variation. The **Köppen climate classification system** provides us with that balanced approach, offering a descriptive classification of world climates that brilliantly negotiates the tightrope between simplicity and complexity. To Wladimir P. Köppen (1846–1940), the key to classifying climate was plant life.

The Köppen Climate Classification System

In 1874, Swiss botanist Alphonse de Candolle (1806–1893) produced the first comprehensive classification and regionalization of world vegetation based on the internal functions of plant organs. To his everlasting credit, Wladimir Köppen recognized that a plant, or assemblage of plants, at a particular place represents a synthesis of the many variations of the climate prevailing there. He therefore looked to de Candolle's classification of vegetation to solve the puzzle of the global spatial organization of climates.

Köppen (sometimes spelled Koeppen; pronounced *Ker*-pin with the *r* silent) compared the global distribution of vegetation mapped by de Candolle with his own maps of the world distribution of temperature and precipitation. He soon identified several correlations between the atmosphere and biosphere and used them to distinguish one climate from another. For example, Köppen observed that in the high latitudes, the boundary marking the presence or absence of extensive forests closely coincides with the presence or absence of at least one month in the year with an average temperature of 10°C (50°F). He noted many other correlations like this, and in 1900, he published the first version of his classification system. This regionalization scheme was subsequently modified by Köppen and other researchers (notably Rudolf Geiger [1894–1981]) and became the most widely used climatic classification system.

Figure 14.1 presents a simplified version of the Köppen system that we use in this book. Just as code letters are used

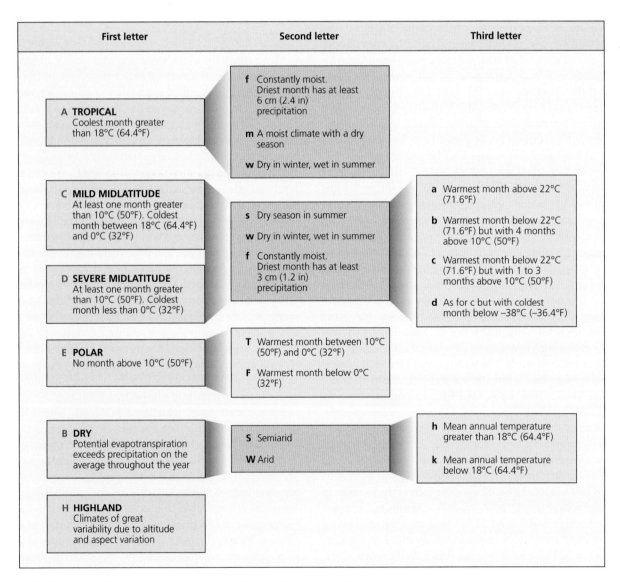

FIGURE 14.1 Simplified version of the modern Köppen climate classification system.

to describe air masses (see Unit 12), the Köppen classification uses a shorthand notation of letter symbols to distinguish different characteristics of the major climates. The six major climate groups (column 1 in Figure 14.1) are **A**, **B**, **C**, **D**, **E**, and **H**. The major tropical (**A**), mild midlatitude (**C**), severe midlatitude (**D**), and polar (**E**) climates are differentiated according to temperature. The variable highland climates (**H**) are grouped into a separate major category. The **C** and **D** climates are also called *mesothermal* and *microthermal*, respectively. *Mesothermal* implies a moderate amount of heat, and *microthermal* implies a small amount; both refer to winter temperature, as the criteria in their respective boxes in Figure 14.1 indicate. The major dry or arid (**B**) climates are distinguished by potential evapotranspiration exceeding precipitation. Recall from Part Two that potential evapotranspiration represents atmospheric demand

for water (evaporation and transpiration from a well-watered surface). The term had not been invented when Köppen did his pioneering work, so his system does not include the direct calculation of potential evapotranspiration. However, he understood that rain falling in hot areas with abundant solar radiation would quickly pass from liquid water to the gaseous state on all surfaces (including plants) and that plants transpire water from leaf surfaces into the atmosphere. Köppen therefore knew that such rain would be less effective in maintaining surface moisture than the same rainfall amount in a cool climate. His formulas for **B** climates capture this effect by considering precipitation amounts relative to temperature.

As Figure 14.1 shows, we further divide the major climate groups in terms of heat or moisture by the use of a second letter (column 2) and in some cases a third letter

---

PERSPECTIVES ON THE HUMAN ENVIRONMENT

Climate in Daily Human Terms

1. "The butter when stabbed with a knife flew like very brittle toffee. The lower skirts of the inner tent are solid with ice. All our [sleeping] bags were so saturated with water that they froze too stiff to bend with safety, so we packed them one on the other full length, like coffins, on the sledge."

2. "The rain poured steadily down, turning the little patch of reclaimed ground on which his house stood back into swamp again. The window of this room blew to and fro: at some time during the night, the catch had been broken by a squall of wind. Now the rain had blown in, his dressing table was soaking wet, and there was a pool of water on the floor."

3. "It was a breathless wind, with the furnace taste sometimes known in Egypt when a *khamsin* came, and, as the day went on and the sun rose in the sky it grew stronger, more filled with the dust of Nefudh, the great sand desert of Northern Arabia, close by us over there, but invisible through the haze."

4. "The spring came richly, and the hills lay asleep in grass—emerald green, the rank thick grass; the slopes were sleek and fat with it. The stock, sensing a great quantity of food shooting up on the sidehills, increased the bearing of the young. When April came, and warm grass-scented days, the flowers burdened the hills with color, the poppies gold and the lupines blue, in spreads and comforters."

5. "There was a desert wind blowing that night. It was one of those hot dry Santa Anas that come down through the mountain passes and curl your hair and

make your nerves jump and your skin itch. On nights like that every booze party ends in a fight. Meek little wives feel the edge of the carving knife and study their husbands' necks. Anything can happen. You can even get a full glass of beer at a cocktail lounge."

What do these rather poetic excerpted selections have in common? They attempt to convey a sense of climate, the impact a particular climate—especially a demanding and harsh climate—has on people who attempt to explore and live in it. Description of climate is one of the cornerstones of the literature of exploration and travel. Perhaps nothing conveys a sense of an environment as efficiently and dramatically as notation of its winds, precipitation, and temperatures. And, as these skillful writers show, climate tells us a great deal about the "feel" of a place.

Now we'll relieve your curiosity. Selection 1 is from Robert Falcon Scott's memoirs of his explorations in the Arctic, published posthumously in 1914. Selection 2 is by Graham Greene, who, after being stationed in West Africa during World War II, wrote *The Heart of the Matter.* Selection 3 is by Thomas Edward Lawrence (Lawrence of Arabia), the British soldier who organized the Arab nations against the Turks during World War I. Selection 4 is by John Steinbeck, writing in *To a God Unknown* during the 1930s about the Mediterranean climate of central California. Number 5 is fiction, a famous passage from Raymond Chandler's detective story *Red Wind* that surely resonates with any Los Angeles native.

(column 3). In the second column, the lowercase letters **f**, **m**, **w**, and **s** tell us when precipitation occurs during the year and are applicable to **A**, **C**, and **D** climates. Capital letters **S** and **W** indicate the degree of aridity in dry (**B**) climates, with **S** designating semiarid conditions and **W** full aridity. Letters referring to moisture conditions are shown in the green-colored boxes in Figure 14.1; letters referring to heat are shown in the peach-colored boxes. The heat modifiers of the third column—lowercase letters **a**, **b**, **c**, and **d**—provide details on the temperatures of **C** and **D** climates. The third letters **h** and **k** do the same for the **B** climates, and the second letters **T** and **F** subdivide the temperatures in polar (**E**) climates.

We can define many different climates with this shorthand code, including all that can be represented and mapped at the world regional scale. As the next three units proceed through each major climate, you will be able to decode any combination of Köppen letter symbols by referring back to Figure 14.1. For instance, an **Af** climate is a tropical climate in which the average temperature of every month exceeds 18°C (64.4°F) and total monthly precipitation always exceeds 6 cm (2.4 in).

Before considering each of the six major climate groups, we need to establish their spatial dimensions both in terms of a hypothetical model and in reality. As follows from our emphasis on the human environment, the discussion here will focus on terrestrial climates. Of course, Köppen's rules apply wherever the necessary monthly temperature and precipitation values are available, including the oceans of our planet. Ignoring ocean climates would be a huge mistake, as they cover more than 70 percent of Earth's surface.

The Size and Regional Distribution of Climate Types

The relative size of each climate group is more easily gauged by percentages than by area values such as square kilometers. These percentages are shown in Table 14.1 as a fraction of total Earth area and as fractions of Earth land and ocean area. The last two columns in the table show the percentage of each climate that occupies land and ocean, respectively. For example, tropical (**A**) climates cover almost 30 percent of Earth (29.4 percent), with about four times as much ocean area as land (23.9 percent vs. 5.6 percent). The dry (**B**) climates are almost as extensive and likewise are mainly ocean climates, as are the mild (**C**) climates. They stand in contrast to the severe midlatitude (**D**) group, which covers less than 7 percent of Earth and is almost entirely a continental group for reasons we will explain in the following material. Polar (**E**) climates are another large group (covering about 18 percent of Earth), and they are more than 80 percent oceanic. The Highland (**H**) fraction represents only a little more than 1 percent of the planet and is of course entirely continental.

TABLE 14.1 Areal Coverage of the Six Major Climates in the Modified Köppen Classification

Climate Type	Percent of Earth Area			Percent of Climate Area	
	Total	Land	Ocean	Land	Ocean
A—Tropical	29.4	5.6	23.9	19.0	81.0
B—Dry	25.0	8.3	16.7	33.3	66.7
C—Mild Midlatitude	20.1	3.9	16.2	19.4	80.6
D—Severe Midlatitude	6.5	6.0	0.6	91.4	8.6
E—Polar	17.8	2.9	14.9	16.4	83.6
H—Highland	1.2	1.2	0.0	100.0	0.0
Total	100	27.9	72.1	n/a	n/a

Because of rounding, some values do not sum exactly to the totals shown.

In addition to the amount of area they cover, the pattern of climates in space is a critical element of the global environment. The easiest way to picture the distribution of world climates is first to imagine how they would be spatially arranged on a **hypothetical continent** of low, uniform elevation. Such a model is shown in Figure 14.2. This figure, which we refer to as the "world island" in the material that follows, is a generalized representation of the world's landmasses. A comparison with the actual global map (Fig. 14.3) reveals that this hypothetical continent tapers from north to south in close correspondence with the overall narrowing of land areas between 50 degrees N and 40 degrees S. Note that on the world map, the bulk of both Eurasia and North America is greatest in the latitudinal zone between 50 degrees and 70 degrees N. Conversely, at 40 degrees S, the only land interrupting the world ocean is the slender cone of southern South America and small parts of Australia and New Zealand.

Five of the six major climate groups—**A**, **B**, **C**, **D**, and **E**—are mapped on the hypothetical continent in Figure 14.2. **H** climates are not shown beyond a token presence because this model de-emphasizes upland areas. The hypothetical continent model represents an abstraction of the more complex actual distribution of climate regions (shown in Fig. 14.3), which, like all regional schemes, is itself a simplification of the still-greater complexity that exists across Earth's surface.

The following discussion of world climate regions is based upon a side-by-side consideration of Figures 14.2 and 14.3. The hypothetical continent in Figure 14.2 is useful for gaining a general understanding of the latitudinal extent and continental position of each climate type. The observed Köppen climate map in Figure 14.3 illustrates the connections to the specific regions that blanket the continental landmasses and surrounding oceans. The emphasis in the remainder of this unit is on broad patterns. Units 15, 16, and 17 consider each major climate type in greater detail.

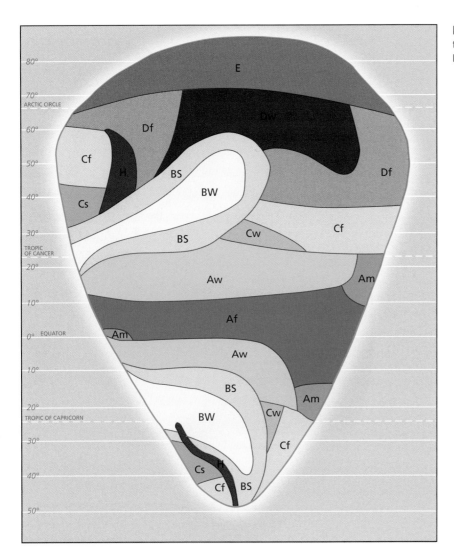

FIGURE 14.2 Distribution of climate types across a hypothetical continent of low, uniform elevation.

THE DISTRIBUTION OF TROPICAL (A) CLIMATES

The tropical **A** climates straddle the equator, extending to approximately 25 degrees latitude in both the Northern and Southern Hemispheres (see Fig. 14.2). These climates are warm all year; they experience nothing one would call winter based on temperature. Thus, when we refer to winter in **A** climates, we mean the time when the overhead Sun is far away (December, January, and February for Northern Hemisphere locations and June, July, and August for Southern Hemisphere locations). The heart of the **A** climate region is constituted by its wet subtype, the *tropical rainforest* climate (**Af**), named for the vegetation it nurtures. On the poleward margins of the tropical rainforest, forming transitional belts between the **A** and **B** climates, lies the **Aw** climate, which exhibits a distinct winter dry season. This is called the *savanna* climate because of the tall grasses that grow there and dominate the spaces between clumps of trees and/or thorny bushes. A third **A** climate, the **Am**, or *monsoon*, variety of the rainforest climate is found at coastal corners of the **Af** region

and is subject to quite pronounced seasonal fluctuations in rainfall while maintaining the high annual total characteristic of **Af** areas.

Closer inspection of the tropical climate zone on the hypothetical continent shown in Figure 14.2 reveals that areas of **A** climate are widest on the eastern side. This occurs because the trade winds blow onshore from the northeast and southeast, making the latitudinal extent of the **A** region about 20 degrees greater on the east coast of a continent than on the west coast.

The real-world Köppen map (see Fig. 14.3) displays a more intricate pattern but is consistent with the generalizations we have just made with respect to the distribution of **A** climates. In the Americas, the **A** climates are centered around the equator, reaching to about 25 degrees N and S on the eastern coasts and islands and to about 20 degrees N and S in western Central and South America, respectively. The African pattern is similar south of the equator, but to the north, the **A** climates are compressed because the moist Northeast Trades are blocked by the landmass of southern Asia. For reasons identified in Unit 9, the Asian tropical

WORLD CLIMATES
After Köppen-Geiger

A Tropical Climate

Af	No dry season
Am	Short dry season
Aw	Dry winter

B Dry Climate

BS	Semiarid	h = hot
BW	Arid	k = cold

C Mild Midlatitude Climate

Cf	No dry season
Cw	Dry winter
Cs	Dry summer

a = hot summer
b = cool summer
c = short, cool summer
d = very cold winter

D Severe Midlatitude Climate

Df	No dry season
Ds	Dry summer
Dw	Dry winter

E Polar Climate

E	Tundra and ice

F Highland Climate

H	Unclassified highlands

0 — 2000 (km)

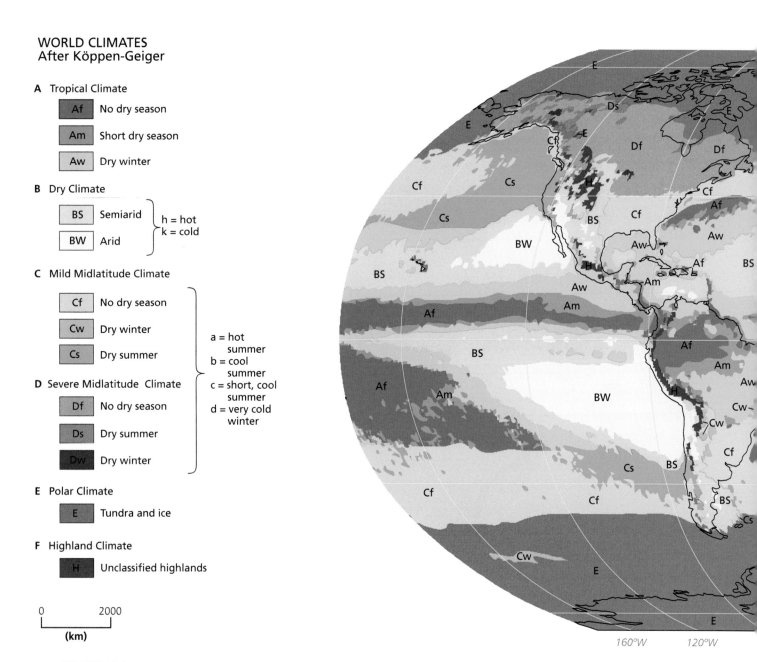

FIGURE 14.3 Global distribution of climates according to the modified Köppen classification system.

climates are dominated by the monsoon effect. The prevalence of oceans and seas between Southeast Asia and northern Australia has sharply reduced the extent of tropical climates over land, but many of the islands of the lengthy Indonesian archipelago still exhibit rainforest climates over most of their surfaces. The large expanse of ocean found at tropical latitudes ensures that this group is largely oceanic.

THE DISTRIBUTION OF ARID (B) CLIMATES

The dry **B** climates are located poleward of the **A** climates on the western sides of continents. These climates are associated with the subsiding air of the subtropical high-pressure zones, whose influence is greatest on the eastern side of the

oceans. This leads to a major intrusion of arid climates onto neighboring continents in both the Northern and Southern Hemispheres. As the hypothetical continent model shows (see Fig. 14.2), this occurs at the average latitudes of the subtropical high between 20 degrees and 30 degrees N and S. Once inland from the west coast of the continent, however, the **B** climate region curves poleward, reaching to about 55 degrees N and 45 degrees S, respectively. As we will discuss in Unit 15, continentality (see Unit 7) also plays a significant role in the occurrence of **B** climates, particularly on the Eurasian landmass. Despite that, dry climates are about two-thirds oceanic, as follows from the large area of subtropical ocean found in both hemispheres.

The heart of the **B** climate region contains the driest climatic variety, true *desert* (**BW**, with **W** standing for

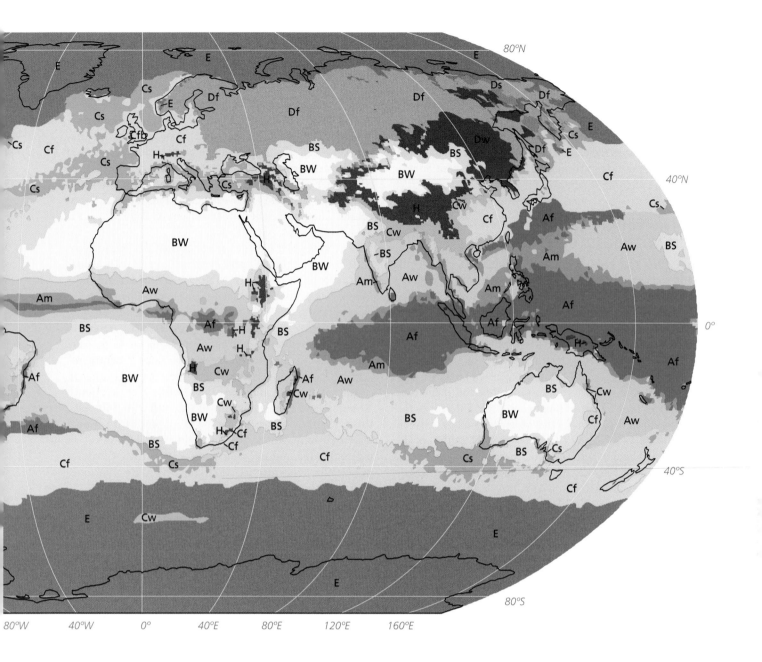

wüste, the German word for desert). The **BW** region is surrounded by the semiarid short-grass prairie, or *steppe* (**BS**), which is a moister transitional climate that lies between the **BW** core and the more humid **A**, **C**, and **D** climates bordering the dry-climate zone.

On the world map (see Fig. 14.3), we can again observe the consistencies between model and reality. The most striking similarity occurs with respect to the vast bulk of subtropical-latitude dryness found in all oceans and on land that stretches east–west across northern Africa and then curves northeastward through southwestern and central Asia. The desert climate (**BW**) dominates here and forms a more or less continuous belt between northwestern Africa and Mongolia, interrupted only by narrow seas, mountain corridors, and a handful of well-watered river

valleys. Included among these vast arid basins are North Africa's Sahara, the deserts of the Arabian Peninsula and nearby Iran, and China's Takla Makan and Gobi deserts. On land these **BW** climates extend north of 35 degrees N and become colder (as indicated by the third Köppen letter **k** replacing **h**). Note that all the deserts are framed by narrow semiarid or steppe (**BS**) zones that also change from **h** to **k** in the vicinity of the 35th parallel.

In North America, the arid climates occur on a smaller scale but in about the same proportion and relative extent vis-à-vis this smaller continental landmass. The rain shadow effect (see Unit 12) plays an important role in ensuring the dryness of the western United States and Canada, where high mountains parallel the Pacific Coast and block much of the moisture brought onshore by prevailing westerly winds. In the Southern Hemisphere, only in Australia is distribution of the **B** climate reminiscent of distributions north of the equator. This is because this island continent constitutes the only large body of land within the subtropical latitudes. On the tapering landmasses of South America and southern Africa, the arid climate zones are compressed but still correspond roughly to the regional patterns of the hypothetical continent in Figure 14.2. As in the Northern Hemisphere, dryness in the subtropics extends great distances westward from Southern Hemisphere continental areas.

THE DISTRIBUTION OF MILD MIDLATITUDE (C) CLIMATES

As Figure 14.2 reveals, the mild midlatitude (moderate winter temperature) **C** climates are situated where we expect given their name. The land/water differential between the Northern and Southern Hemispheres, however, becomes more pronounced poleward of the subtropics, and the hypothetical continent no longer exhibits north–south patterns that mirror each other. North of the equator, where the world island's greatest bulk is approached, the absence of oceanic-derived moisture in the continental interior produces an arid climate zone that interrupts the east–west belt of **C** climates that lie approximately between the latitudes of 25 degrees and 45 degrees N. In the Southern Hemisphere, where the mesothermal climatic belt is severely pinched by the tapering model continent, **C** climates reach nearly across the remaining landmass south of 40 degrees S.

The subtypes of the **C** climate group reflect the hemispheric differences just discussed, but some noteworthy similarities exist as well. On the western coast, adjacent to the **B** climate region, small but important zones of *Mediterranean* (**Cs**) climate occur where the subtropical high-pressure belt has a drying influence during the summer. Poleward of these zones, *humid climates with moist winters* (**Cf**) owe their existence, especially on the western side of the continent, to the storms carried by the midlatitude westerly winds. **Cf** climates are also found on the eastern side of the continent, where they obtain moisture from humid air on the western side of the adjoining ocean's subtropical high-pressure zone. Another mild midlatitude subtype, sometimes called the *subtropical monsoon* (**Cw**) climate, found in the interior of the **C** climate region represents locations where winter storms do not bring sufficient precipitation to create the even seasonal distribution found in **Cf** climates.

The world Köppen map (see Fig. 14.3) again demonstrates the applicability of the distributional generalizations of the model continent. Both Eurasia and North America exhibit the central **B** climate interruption of the mild midlatitude climatic belt, with the **C** climates less prominent on the Eurasian landmass, where distances from maritime moisture sources are much greater. Southern Hemisphere patterns are also generally consistent with the model,

although the **C** climates are often confined to narrow coastal strips that result from rain shadow effects produced by South America's Andes Mountains, South Africa's Great Escarpment, and Australia's east coast Great Dividing Range. Note how the midlatitude ocean of both hemispheres is covered by a continuous belt of **C** climates. This is a result of the water's moderating effect on temperature.

The individual **C** climates further underscore the linkages between the hypothetical continent and the real world. Mediterranean (**Cs**) climates are located on every continent's west coast in the expected latitudinal position (around 35 degrees). They even penetrate to Asia's "west coast," where the eastern Mediterranean Sea reaches the Middle East.

The west coast marine variety of the **Cf** climate is generally observed on the poleward flank of the **Cs** region, but mountains paralleling the ocean often restrict its inland penetration. The clear exception to this pattern is Europe, where blocking highlands are absent and onshore westerlies flow across relatively warm ocean waters to bring mild winter conditions hundreds of kilometers inland. The **Cf** climates on the eastern sides of continents are particularly apparent in the United States (as a result of the moderating effects of the Gulf of Mexico), South America, and the Pacific rim of Australia. In East Asia, this **Cf** subtype is limited by the more powerful regional effects of monsoonal reversals. Where it does occur in eastern China and Japan, it is associated with a massive population concentration that contains more than one-eighth of humankind.

The **Cw** climates of the Northern Hemisphere are confined mainly to the subtropical portions of southern and eastern Asia and are closely related to the monsoonal weather regime. The Southern Hemisphere **Cw** climate regions are transitional in both Africa and South America, between the equatorial savannas to the north and the subtropical arid and humid climates to the south.

THE DISTRIBUTION OF SEVERE MIDLATITUDE (D) CLIMATES

The microthermal **D** climates, with their severely cold winters, occur exclusively in the Northern Hemisphere. In the Southern Hemisphere, no large landmasses exist between 50 degrees and 70 degrees S, so latitude and the effect of continentality do not support the cold winters of the **D** climates. On the world island (see Fig. 14.2), the severe midlatitude climates are associated with the large landmasses in the upper-middle and subpolar latitudes. They occupy an east–west belt between approximately 45 degrees and 65 degrees N. This climatic band of severe winters is at its widest in the interior of the hypothetical continent in Figure 14.2, moderating only near the coasts (particularly the west coasts), where the influences of maritime air penetrate inland.

We can observe two subtypes of the **D** (humid cold) climate group. The **Df** variety experiences precipitation throughout the year. It can exhibit a fairly warm summer near the **C** climate boundary and certain coastal zones, but in the interior and toward the higher latitudes, the summers are shorter and much colder. The other subtype, **Dw**, occurs where the effects of continentality are most pronounced. Annual moisture totals there are lower, and the harsh winters are characterized by dryness. (So little dry-summer **Ds** climate occurs on Earth that we do not show any on the hypothetical continent.) The extreme cooling of the ground in midwinter in **D** climates is associated with a large anticyclone that persists throughout the cold season and blocks surface winds that might bring in moisture from other areas.

The actual distribution of **D** climates on the world map (see Fig. 14.3) accords quite well with the model continent (see Fig. 14.2). The greatest east–west extents of both Eurasia and North America are the heart of the **D** climate region, which dominates the latitudes between 50 degrees and 65 degrees N. The only exceptions occur along the western coasts, where the warm North Atlantic Drift and Alaska Current bathe the subpolar coastal zone with moderating temperatures. In the Eurasian interior, extreme continentality pushes the microthermal climates to a higher average latitude. This pattern is not matched in eastern Canada, where the colder polar climates to the north penetrate farther south. The small area of **D** climates that are found over oceans are mostly in the Pacific Ocean to the east of Siberia, where northwesterly flow from the continent overwhelms the marine influence. A similar, but less extensive, pattern appears in the Canadian North Atlantic near Nova Scotia.

As for the **D** subtypes, they correspond closely to the model. However, note that North America, wide as it is, does not contain sufficient bulk to support the **Dw** climate or even the coldest variant of the **Df** climate (**Dfd**). Only Eurasia exhibits these extreme microthermal climates, which are concentrated on the eastern side of that enormous landmass. Two reasons underlie the eastward deflection: (1) the overall elevation of the land surface is much higher in the mountainous Russian Far East and northeastern China, and (2) northeastern Asia is farthest from the warm waters off western Eurasia that pump heat and moisture into the prevailing westerly winds that sweep around the globe within this latitudinal zone. Looking closely at Figure 14.3, we can see that only a few very small areas of dry-summer (**Ds**) climates exist.

THE DISTRIBUTION OF POLAR (E) CLIMATES

In both hemispheres, land areas poleward of the Arctic and Antarctic Circles (66.5°N and S, respectively) with deficits of net solar radiation exhibit *polar* (**E**) climates. The hypothetical continent (see Fig. 14.2) and world Köppen maps (see Fig. 14.3) display similar patterns, with **E** climates dominant in Antarctica, Greenland, and the poleward fringes of northernmost Eurasia and North America. As for subtypes, the *tundra* (**ET**) climate, where the warmest

month records an average temperature between 0°C (32°F) and 10°C (50°F), borders the warmer climates on the equatorward margins of each polar region. The highest latitudes in the vicinity of the poles themselves—as well as in most of interior Greenland and high-lying Antarctica—experience the *icecap* or *frost* (**EF**) climate, in which the average temperature of the warmest month fails to reach 0°C (32°F). Note that like the **A**, **B**, and **C** climates, polar climates are mostly oceanic (80.6 percent, according to Table 14.1). This is explained by the large amount of ocean at high latitudes, where solar energy inputs are too low to produce even one month with an average temperature above 10°C (50°F).

The **E** climates are also often associated with the high-altitude areas of lower latitudes, particularly in the upper reaches of mountainous zones that are too cold to support vegetation. In these situations, they are labeled **H**, or *highland*, climates. We survey these climates together with the **E** climates in Unit 17.

Extent and Boundaries of Climate Regions

The classification system developed by Köppen and his associates has been criticized because it does not consider the causes of climate and because some of its climate–vegetation links are weak. Yet it remains the most often-used classification system, and it has proven itself both useful and appropriate for the purpose of introducing students to the complexity of global climatic patterns. Earlier in this unit, we discussed the shortcomings of any geographical classification of climate. We conclude by again stating that the lines that look like sharp regional boundaries on the hypothetical continent and world Köppen maps are really zones of transition from one climate type to another. This applies equally to land surfaces and oceans.

In what way can the Köppen map of world climates have practical use? Here's an instance. In the late 1980s, one of the authors of this book was invited by a group of bankers to make a presentation on geography's relevance for business and industry. Using a series of maps, including the Köppen climate map, he highlighted the geographic context of China's explosive economic growth. After the lecture, an obviously excited executive said that he "had to have that map" and wondered how he could get a copy as soon as possible.

The author asked the executive why this map had generated such urgent interest. The executive replied that his bank was sending employees to work for its expanding network all over the world, and these employees were always asking, "What's the climate like over there?" Now, he said, this map would go into the bank's annual report so that at a glance employees could tell that the weather in Shanghai was going to be like Charleston, South Carolina; that those moving to Cape Town (South Africa) would feel like they'd moved to San Francisco; and that moving to Moscow would be like living in Ottawa. "I never thought of geography as relevant to banking," said the executive. Well, think again!

Key Terms

climate *page 181*
climatic normal *page 181*

hypothetical continent
page 184

Köppen climate classification
system *page 182*

*Scan Here for a quick
vocabulary review*

Review Questions

1. How is climate distinct from weather?

2. What is a climatic normal?

3. What are the thermal criteria for tropical, mild midlatitude, severe midlatitude, and polar climates in the Köppen system?

4. Outline the interpretation of the second and third letters of the codes used in the Köppen system. For example, how does an **Af** climate differ from an **Aw** climate?

5. Many forested areas receive less precipitation than desert locations in other parts of the world. Why does this happen?

6. Describe two advantages of the Köppen climate classification system.

7. Describe two problems or shortcomings of the Köppen climate classification system.

8. Why should Köppen regional-scale boundaries not be used to ascertain climate zone boundaries at the local scale?

9. All the Köppen climates except for the **D** climates are mostly oceanic. If climates were randomly located on land and ocean, how much of each type would be oceanic? (Hint: How much would be oceanic on a planet completely covered by ocean?) Viewed in this way, which climates are more oceanic than expected from chance alone?

*Scan Here for a quick
concept review*

www.oup.com/us/mason

Tropical (A) and Arid (B) Climates

*A **BW** climatic zone expands: desertification in the Somali borderland of Kenya, East Africa.*

OBJECTIVES

- List and describe the major types of tropical climates
- List and describe the major types of arid climates
- Explain tropical (A) and arid (B) climates using climographs developed from data recorded by weather stations

- Summarize climate-related environmental problems within tropical and arid climate zones
- Discuss the causes and consequences of tropical deforestation and desertification

Unit 14 set the stage for a global survey of the principal climate types based on the Köppen classification and regionalization system. The more detailed treatment begins in this unit with the **A** (tropical) and **B** (dry or arid) climates whose spatial distribution has been extracted from the global map in Unit 14 and drawn in Figure 15.1. In this unit, we provide greater detail about each of the principal subdivisions of the tropical and arid climates. This unit also focuses on the leading environmental problem of each major climate group. In the **A** climates, **tropical deforestation**, the clearing and destruction of tropical rainforests, is a worsening crisis with far-reaching environmental consequences. In the **B** climates, our principal concern is with **desertification**, a form of land degradation in rather dry areas where increasing aridity eventually leads to desert conditions. We cover the rest of the major climate types in Units 16 and 17.

The Major Tropical (A) Climates

Contained within an almost continuous east–west belt astride the equator is the warmth and moisture of the **tropical climate** (**A**). Warmth is derived from proximity to the equator (To be an A climate temperatures must average higher than 18°C [64.4°F], even in the coolest month of the year); moisture comes from the rains of the intertropical convergence zone, wet-monsoon systems, easterly waves, and hurricanes. As Figure 15.2 shows, this climate occupies more than half of Earth's area within 15° of the equator, and more than 90 percent in places close to the equator. In places where it ranges beyond the subtropics, it does so because warm ocean currents maintain high temperatures year-round.

The tropical climates are subdivided into three major types: the *tropical rainforest*, the *monsoon rainforest*, and the *savanna*. The percent of Earth area occupied by these three subdivisions is shown in Table 15.1. Recall from Unit 14 that these subtypes differ with respect to precipitation regime. The first two are moist climates with high annual totals, whereas the savanna has a dry season long enough to produce seasonal drought.

THE TROPICAL RAINFOREST (Af) CLIMATE

The tropical rainforest zone (**Af**) exhibits the greatest effects of abundant heat and moisture of the three tropical climates. In areas undisturbed by human intervention, tall evergreen

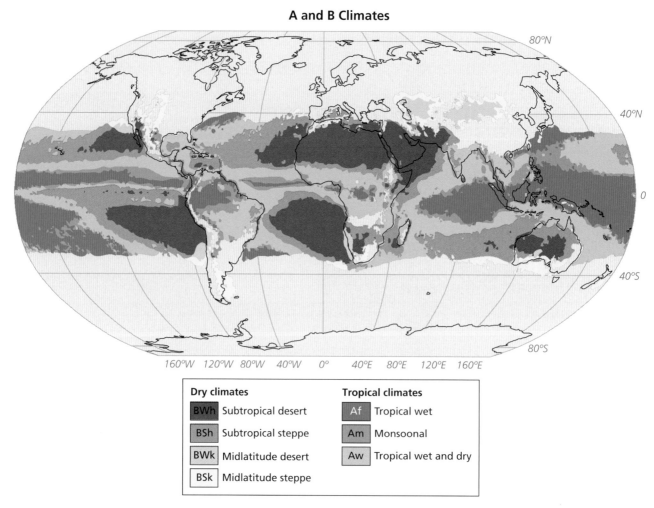

A and B Climates

Dry climates

BWh	Subtropical desert
BSh	Subtropical steppe
BWk	Midlatitude desert
BSk	Midlatitude steppe

Tropical climates

Af	Tropical wet
Am	Monsoonal
Aw	Tropical wet and dry

FIGURE 15.1 The distribution of tropical (**A**) and dry (**B**) climates.

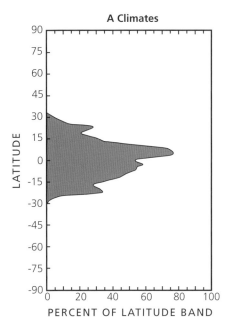

A Climates

LATITUDE

PERCENT OF LATITUDE BAND

FIGURE 15.2 Percent of Earth covered by **A** climates as a function of latitude.

trees completely cover the land surface (Fig. 15.3). This dense forest has little undergrowth because the dense canopy of the tall trees allows just a small amount of sunlight (if any) to reach the ground. Any serious discussion of this natural environment today must take into consideration the appalling destruction of rainforests throughout the tropical latitudes (see Perspectives on the Human Environment: Deforestation of the Tropics).

Figure 15.4 illustrates the climatic characteristics of the tropical rainforest with three graphic displays for a typical weather station in the **Af** zone. São Gabriel da Cachoeira is a Brazilian village situated almost exactly on

TABLE 15.1   Areal Coverage of Tropical Climates

Climate Subtype	Percent of Earth Area
Tropical rainforest (**Af**)	10.5
Monsoon rainforest (**Am**)	4.5
Savanna (**Aw**)	14.6

FIGURE 15.3   "Traversing the Panama Canal turned out to have several geographic benefits, in addition to learning about the history of the project and understanding better the changing relationship between Panama and the United States. I was unaware that the forests flanking the Canal had been used to train U.S. military forces in advance of their assignment to Vietnam in the Indochina War. Nor did I expect to see pristine rainforest of the kind shown here, on Barro Colorado Island."

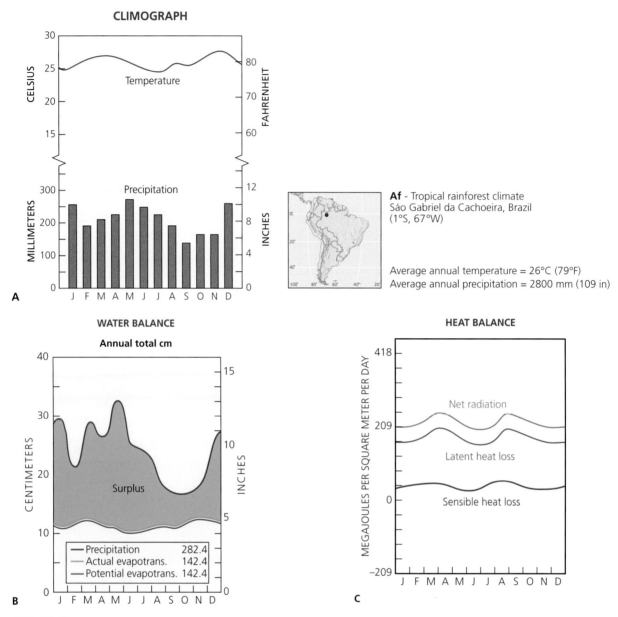

FIGURE 15.4 Climograph and related graphic displays for a representative weather station in the tropical rainforest (**Af**) climate zone.

the equator deep in the heart of the Amazon Basin (see the inset map in the figure). Of the three graphs, the most important for our purposes is the first—the climograph, a device we refer to repeatedly in this unit, as well as Units 16 and 17. A **climograph** for a given location simultaneously displays its key climatic variables of average temperature and precipitation, showing how they change month by month throughout the year. The heat balance graph shows how net radiation supplied to the surface is partitioned into dry (sensible) and evaporative (latent) heat transfer.

The climograph in Figure 15.4A reveals that average monthly temperatures (shown with the red line) in São Gabriel are remarkably constant, hovering between 20°C (50°F) and 26°C (79°F) throughout the year. In fact, the diurnal (daily) temperature may vary as much as 6°C (11°F); in guessing the temperature, your wristwatch would be a

better guide than the calendar! Rainfall (shown by the bars) also tends to occur in a diurnal rather than seasonal rhythm. This normal daily pattern consists of relatively clear skies in the morning followed by a steady buildup of convectional clouds from the vertical movement of air with the increasing heat of the day. By early afternoon, thunderstorms burst forth with torrential rains. After several hours, these stationary storms play themselves out as temperatures decline a few degrees with the approach of night. This pattern repeats itself day after day.

In São Gabriel, no month receives less than 130 mm (5.1 in) of rain, and in May, about 10 mm (0.4 in) of rainfall on an average day. The average total rainfall for the year is 2,800 mm (109 in). Such large amounts of rain keep the water balance in perpetual surplus, as the water balance graph in Figure 15.4B indicates. Other **Af** climates with a greater range in annual rainfall do not have enough moisture to meet atmospheric demand, even after drawing on stored soil water. As a result, they experience a short season of moisture deficit, with actual evapotranspiration less than potential evapotranspiration. In its length and magnitude, the deficit pales in comparison to the wetter season's surplus, and the climate is decidedly moist despite that somewhat drier period each year. In all Af climates abundant moisture at the surface means evaporative (latent) heat losses exceeds sensible heat losses throughout the year (Fig. 15C).

THE MONSOON RAINFOREST (Am) CLIMATE

The monsoon rainforest (**Am**) climate is restricted to tropical coasts that are often backed by highlands. It has a distinct dry season in the part of the year when Sun angles are lower, but the dry season is almost always a short one; a compensating longer season of heavy rainfall generally prevents any soil-moisture deficits. These characteristics are seen in the representative climograph for this climate type (Fig. 15.5) from the city of Trivandrum, located near India's southern tip on the narrow Malabar Coast at the base of the Cardamom Uplands.

In some places, such as the hills leading up to the Himalayas in northeastern India, the highlands add an orographic effect (see Unit 12) to the wet-monsoon rains. The station of Cherrapunji (see Fig. 16.15) in this hilly region

CLIMOGRAPH

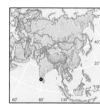

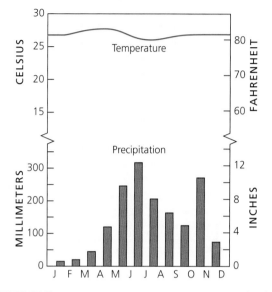

Am - Tropical monsoon climate
Trivandrum, India
(8°N, 77°E)

Annual average
temperature = 27°C (81°F)

Annual average
precipitation = 1835 mm (72 in)

FIGURE 15.5 Climograph of a monsoon rainforest (**Am**) weather station.

Deforestation of the Tropics

Only a little more than a generation ago, lush rainforests clothed most of the land surfaces of the equatorial tropics and accounted for 10 percent of Earth's vegetation cover. Since 1980, however, industries and millions of residents of low-latitude countries have been destroying billions of those trees through rapid tropical deforestation, largely for agricultural expansion and lumber. In Brazil's Amazon Basin, an area about the size of New Jersey is burned annually, half of it virgin forest and the other half jungly regrowth, to briefly replenish the infertile soil. If this prodigious removal rate endures, the planet's tropical rainforests—already effectively reduced to two large patches in western equatorial Africa and the central Amazon Basin—will disappear by the middle of this century.

As rainforests are eradicated, they do not just leave behind the ugly wastelands shown in Figure 15.6. The smoke from their burning releases particles and gases that rise thousands of meters into the atmosphere and are transported thousands of kilometers

FIGURE 15.6 While the burning of the Amazonian rainforest is the subject of many news stories and much scientific research, other remaining areas of rainforest and their inhabitants also suffer. The loss of these regions in Africa is accelerating, the situation in Borneo is deteriorating, and even in remote Papua New Guinea (shown here) the rainforest is receding.

continued

continued

around the world. A number of atmospheric scientists believe that the effects of such fires, together with the removal of massive stands of carbon dioxide–absorbing trees, could have an increasingly harmful effect on global climate.

The causes of deforestation in tropical countries lie in the swift reversal of perceptions about tropical woodlands in recent years. Long regarded as useless resources and obstacles to settlement, rainforested areas suddenly assumed new importance as rapidly growing populations ran out of living space in the late twentieth century. In Brazil, in particular, the government began to build roads and subsidize huge colonization schemes, instantaneously attracting millions of peasants who were able to buy land cheaply for the first time in their lives. Even though the deforested soil soon proved infertile, enough affordable land was available locally to allow these farmers to relocate to newly cleared plots every few years and begin the cycle over again.

In addition to this shifting-cultivation subsistence economy, many commercial opportunities arose as well. Cattle ranchers were only too happy to buy the abandoned farmlands and replace them with pasture grasses, leading to a mushrooming of the ranching industry in Amazonia, Central America's Costa Rica, and elsewhere. At the same time, lumbering became a leading activity in rainforested areas as developed countries, especially Japan and South Korea, turned from their exhausted midlatitude forests to the less expensive and much more desirable trees of the tropics as sources for paper, building materials, and furniture. By the mid-1990s, hundreds of millions of workers in developing countries had come to depend on the rainforests for their livelihoods. That total steadily mounts each year.

Deforestation poses many serious threats to the global environment as well, not the least of which are mass extinctions of plant and animal life (estimated to be 100 species per day), because the tropical rainforest is home to as many as 80 percent of the planet's species. As mentioned earlier, rainforest destruction may also be linked to Earth's climatic patterns. Researchers continue to explore possible connections and longer-term consequences. Two major climatic problems associated with the wholesale burning of tropical woodlands have been identified, and both are likely to reach global proportions as deforestation continues to intensify.

The first of these problems involves the loss of the rainforest over large areas. This loss modifies regional climates, particularly in the direction of dryness. This shift occurs because approximately 80 percent of the rainfall in areas covered by tropical rainforests comes from transpiration and evaporation. In a healthy rainforest climate, transpired water vapor rises in the atmosphere and is precipitated back as rainfall. In deforested regions, this cycle is broken, and most of the rainfall is lost to runoff.

The second climatic problem is that the trees of the tropical rainforest have always absorbed large quantities of atmospheric carbon dioxide, thereby helping to maintain a stable global balance of this greenhouse gas. Climatologists are actively studying the long-term implications of the destruction of this important reservoir for atmospheric CO_2. Some predict a rapid rise in atmospheric concentrations of CO_2. Others see a worldwide intensification of pollution because rainforests have been known to function as large-scale cleansing mechanisms for effluents that enter its ecosystems through the atmosphere and hydrosphere.

averages an annual total of 11,450 mm (451 in) of rain—most of it occurring during the three summer months! (Until the 1950s, Cherrapunji was believed to be the rainiest place on Earth, but that distinction now belongs to Mt. Waialeale on the Hawai'ian island of Kauai, where 11,990 mm [472 in] of rain falls in an average year.) The vegetation of the monsoon rainforest zone consists mainly of evergreen trees, with occasional grasslands interspersed. The trees, however, are not as dense as they are in the remaining areas of true tropical rainforest.

THE SAVANNA (Aw) CLIMATE

The **savanna** (**Aw**) climates are in the transitional, still-tropical latitudes between the subtropical high-pressure and equatorial low-pressure belts. Although these areas often receive between 750 and 1,750 mm (30 and 70 in) of rain per year, there is an extended dry season in the months

when the angle of the Sun is at its lowest. Moisture deficits occur during the dry season, whose length and severity increases with the area's distance from the equator. The climograph for Nkhata Bay, Malawi, in southeastern Africa (Fig. 15.7) shows the typical march of temperatures and precipitation totals in a normal year. Do not be misled by the decline of both variables in the middle months of the year, because, on this occasion, we deliberately selected a Southern Hemisphere station to remind you of the opposite seasonal patterns that exist south of the equator.

In the region of East Africa to the north of Malawi, the pastoral Maasai people (Fig. 15.8) traditionally move their herds of cattle north and south in search of new grass sprouted in the wake of the shifting rains that correspond to the high Sun. The vegetation of this region, in addition to tall, coarse seasonal grasses, includes clumps of trees or individual trees and thorn bushes (see Figs. 15.8 and 25.5). Some savannas have tree densities as large as those of moist

CLIMOGRAPH

Aw - Tropical savanna climate
Nkhata Bay, Malawi
(15°S, 35°E)

Average annual
temperature = 23°C (74°F)

Annual average
precipitation = 1676 mm (66 in)

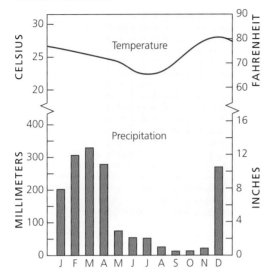

FIGURE 15.7 Climograph of a tropical savanna (**Aw**) weather station.

FIGURE 15.8 Tree-dotted savanna on Kenya's Maasai Mara game reserve.

tropical forests, but canopies in the savanna are so sparse that abundant sunlight still reaches the ground below, allowing grasses to provide continuous cover. There are extensive areas of savanna climate in the world, most prominently covering parts of tropical South America, South Asia, and Africa (see Fig. 14.3). Along the poleward margins of this climate zone, rainfall is so unreliable that these peripheral areas are considered too risky for agricultural development in the absence of irrigation systems.

The Major Arid (B) Climates

The **arid climates** (**B**) cover more of Earth than any other climate, occupying about 25 percent of the planet and 30 percent of the continents (Table 14.1). Figure 15.9 illustrates how they dominate the subtropics and extend into the middle latitudes of both hemispheres.

B climates provide a striking visual contrast to the **A** climates (Fig. 15.10) because their occurrence requires that potential evapotranspiration exceed the moisture supplied by precipitation. Dry climates are found in two general groups (see Fig. 15.1). The first group consists of two interrupted bands near 30 degrees N and 30 degrees S, where the semipermanent subtropical high-pressure cells are dominant. The largest areas of this type are the African–Eurasian desert belt (stretching from the vast Sahara to western India) and the arid interior of Australia. The driest of these climates are often located near the western coasts of continents. Weather stations in northern Chile's Atacama Desert, for instance, can go 10 years or more without recording precipitation.

The second group of dry climates is attributed to two different factors. First, a continental interior remote from any moisture source, experiencing cold high-pressure air masses in winter, can have an arid climate (the central Asian countries east of the Caspian Sea are a good case in point). And second, rain shadow zones in the lee of coast-paralleling mountain ranges often give rise to cold deserts. The interior far west of North America and southern Argentina's Patagonia are examples of this phenomenon.

Average annual temperatures in arid climates are usually typical of those that might be expected at any given latitude—high in the low latitudes and low in the higher

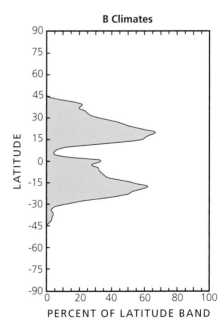

FIGURE 15.9 Percent of Earth covered by **B** climates as a function of latitude.

FIGURE 15.10 One of the world's driest landscapes—a rocky desert in the western Sahara near the border between Morocco and Algeria.

latitudes. However, the annual range of temperature is often greater than might be expected. In North Africa's central Sahara, for example, a range of 17°C (30°F) at 25 degrees N is partly the result of a continental effect. The diurnal range of temperature can also be quite large, averaging between 14° and 25°C (25° and 45°F). On one unforgettable day in the Libyan capital city of Tripoli, the temperature went from below freezing (20.5°C [31°F]) at dawn to 37°C (99°F) in midafternoon! Rocks expanding through such extreme changes of heat sometimes break with a sharp crack; North African soldiers in World War II occasionally mistook these breaking rocks for rifle shots.

The **B** climates are not necessarily dry all year round. Their precipitation can range from near 0 to about 625 mm (25 in) per year, with averages as high as 750 mm (30 in) in the more tropical latitudes. The wetter locations are those with greater evapotranspiration; thus, they remain in the **B** category. In summer, these equatorward margins of the **B** climates are sometimes affected by the rains of the ITCZ as they reach their poleward extremes. In winter, the margins of the arid areas nearest the poles may be subject to midlatitude cyclone rains. There are also rare thunderstorms, which occasionally lead to flash floods. At Hulwan, just south of Cairo, Egypt, seven of these storms in one 20-year period yielded a total of 780 mm (31 in) of rain. Moreover, along coasts bathed by cold ocean currents, rather frequent fogs provide some moisture; Swakopmund, in southwestern Africa's Namibia, adjacent to the Benguela Current (see Fig. 10.7), experiences about 150 days of fog each year.

THE DESERT (BW) CLIMATE

For the most part considerable dryness is the rule in **B** areas (with the areas we just discussed being exceptions). This aridity is exemplified by Yuma, Arizona, where the average annual temperature is 23.5°C (74.3°F), with an annual range of some 23°C (41°F). The climate here is **BWh**—hot, dry desert. The small amount of rainfall shown in Figure 15.11 (89 mm [3.4 in]) comes either from occasional winter storms or from the convectional clouds of summer. The water balance diagram below the climograph in Figure 15.11 shows a marked deficit throughout the year: potential evapotranspiration greatly exceeds precipitation. Most water that falls evaporates quickly back into the dry atmosphere. But as the heat balance diagram for Yuma indicates, there is little water for evaporation, so latent heat loss is negligible. Consequently, most of the heat input from net radiation, which is markedly seasonal, is expended as sensible heat—adding even more warmth to the already-hot air. (Daily highs between May and October almost always exceed 38°C [100°F].)

THE STEPPE (BS) CLIMATE

As the world climate map (see Fig. 15.1) shows, the most arid areas—deserts (**BW**)—are always framed by semiarid zones. This is the domain of the **steppe** (known in western North America as the *short-grass prairie*) or **BS** climate, which is a transitional type between fully developed desert conditions and the subhumid margins of the **A**, **C**, and **D** climates that border **B** climate zones. The maximum and

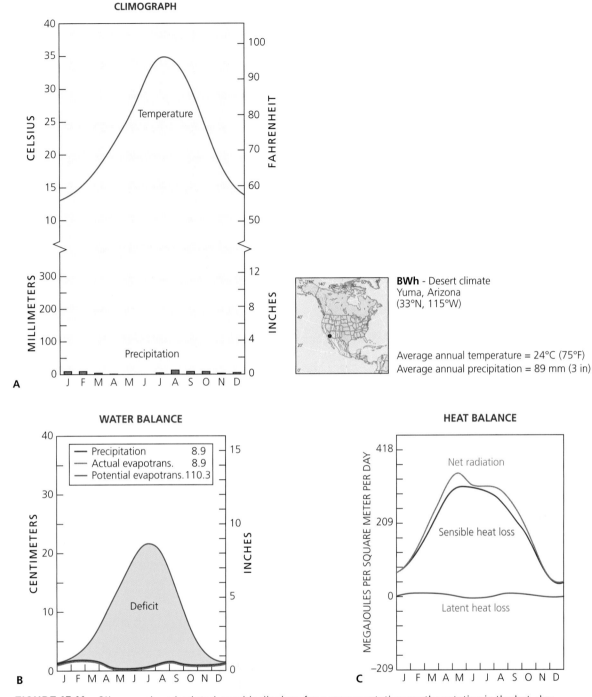

FIGURE 15.11 Climograph and related graphic displays for a representative weather station in the hot, dry desert (**BWh**) climate zone.

minimum amounts of annual rainfall that characterize semiarid climates tend to vary somewhat by latitude. In general, rainfall in lower-latitude **BS** climates ranges from 375 to 750 mm (15 to 30 in) yearly; in the middle latitudes, a lower range (from 250 to 635 mm [10 to 25 in]) prevails. Astana, the capital of the central Asian republic of Kazakhstan, which lies at the heart of the Eurasian landmass, is a classic example of the cold subtype of the steppe climate (**BSk**). The climograph for this city (Fig. 15.12) attests to its proximity to the Turkestan Desert, which lies just

to the south: rainfall totals only 279 mm (11 in) per year. As for the monthly temperature curve, an enormous range of 39°C (70°F) is recorded, reflecting one of the most pronounced areas of continentality on Earth's surface.

The problem of drought is an ever-present environmental hazard in all semiarid areas. For example, the Great Plains region of the United States has suffered many severe dry spells since it was settled by farmers in the late nineteenth and early twentieth centuries. The global distribution of precipitation variability mapped in Figure 15.13

CLIMOGRAPH

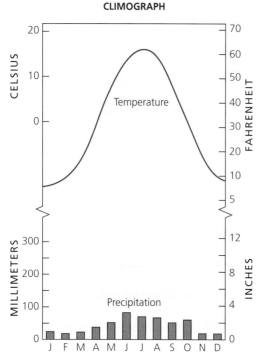

BSk - Steppe climate
Astana, Kazakhstan
(51°N, 71°E)

Average annual temperature = 2°C (35°F)
Average annual precipitation = 279 mm (11 in)

FIGURE 15.12 Climograph of a midlatitude steppe (**BSk**) weather station.

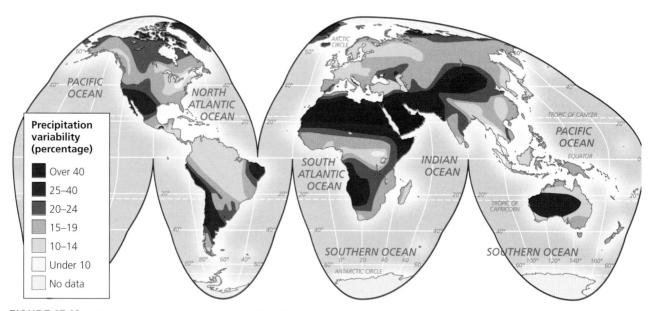

FIGURE 15.13 Global distribution of annual precipitation variability shown as percentage departures from normal.

underscores this dilemma. Perversely, for human purposes, rain falls most reliably where it is abundant, whereas the arid areas experience the greatest variations from year to year. (Compare Fig. 15.13 with Fig. 11.14, showing global precipitation amounts.) Throughout the human presence on this planet, dry climates have proven inhospitable, restricting settlement to the rare water sources of oases or to the valleys of rivers that originate outside a given zone of aridity. In certain cases, such as Egypt's Nile Valley, irrigation can maximize these scarce water resources, but the practice often tends to wash important minerals from the soil. Over the past quarter-century, environmental abuse has been a growing problem at the margins of arid zones, and *desertification* constitutes a serious crisis in many such places (see Perspectives on the Human Environment: Human Activity and Desertification).

Table 15.2 breaks down dry climates into the four subtypes discussed previously. We see that for Earth as a whole, the arid and semiarid regions are of roughly the same size (11.8 and 13.1 percent, respectively). Second, we

TABLE 15.2 Areal Coverage of Dry Climates

Climate Subtype	Percent of Earth Area
Subtropical Desert (**BWh**)	10.5
Subtropical Steppe (**BSh**)	10.8
Midlatitude Desert (**BWk**)	1.3
Midlatitude Steppe (**BSk**)	2.4

see that the hot deserts and steppes are about eight times as large as the midlatitude dry climates (21.3 percent vs. 3.7 percent of Earth). This makes sense when we look back at Figure 15.9, and it follows logically from the large surface area found in subtropical latitudes compared to midlatitudes and from the much greater evaporative demand in lower latitudes.

PERSPECTIVES ON THE HUMAN ENVIRONMENT

Human Activity and Desertification

Desertification may be an awkward word, but it is on target in what it defines—the process of desert expansion into steppelands, largely as a result of climatic change and human degradation of fragile semiarid environments (Fig. 15.14). The term was added to the physical geographer's lexicon in the 1970s following a disastrous famine, caused by rapid and unforeseen desiccation, that struck the north-central African region known as the **Sahel**. *Sahel* means shore in Arabic and refers specifically to the southern shore of the Sahara, an east–west semiarid (**BSh**) belt straddling latitude 15 degrees N and ranging in width from 320 to 1,120 km (200 to 700 mi) that stretches across the entire continent of Africa (see Fig. 15.1). During the height of the tragic starvation episode, which lasted from 1968 to 1973, paralyzing drought destroyed both croplands and pasturelands, and at least 250,000 human lives were lost, as were 3.5 million head of cattle.

This catastrophe prompted an international research effort, and one of its products was a world map of spreading deserts (Fig. 15.15). Note that the most hazardous zones of this map coincide closely

FIGURE 15.14 The desert encroaches on human settlement near Nouakchott, the capital of Mauritania in West Africa. A dune-stabilization effort in the foreground gives temporary protection to the poorest of the city's inhabitants, who live in the squatter settlement just beyond.

continued

continued

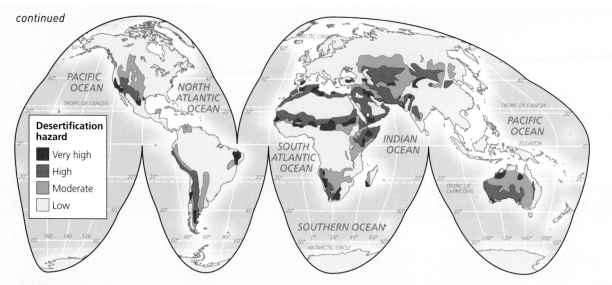

FIGURE 15.15 Degree of desertification hazard according to the scientists of the United Nations Environmental Program. The red brick–colored areas indicate a critical situation; the orange-colored zones experience serious effects; the tan-colored areas represent a significant problem.

with the semiarid regions mapped in Figure 15.1. Another accomplishment was the creation of a global desertification monitoring network, which has provided some alarming data. At least 15 million km² (5.8 million mi²) of the world's arid and semiarid drylands—an area almost the size of South America—have become severely desertified, thereby losing over 50 percent of their potential productivity. Another 34.8 million km² (13.4 million mi²)—an area greater than the size of Africa—have become at least moderately desertified, losing in excess of 25 percent of their potential productivity. Moreover, it is estimated that an area half the size of New York State (64,800 km² [25,000 mi²]) is being added to the total of severely desertified land each year. More than a billion people (about one-seventh of humankind) are already affected by desertification, and the world's arid-land population is growing at a rate 50 percent faster than the population in adequately watered areas.

When desertification was first studied, it was believed that climate change played a dominant role. Today, however, we know the situation is more complicated than that. British geographer Andrew Goudie has noted the particular susceptibility of semiarid areas to desertification resulting from agricultural overuse. Rainfall in semiarid areas is sufficient to swiftly erode carelessly used soils, and farmers often misinterpret short-term success during wetter-than-normal years as a sign of long-term crop-raising stability. Desperate for more food-producing, income-earning space, farmers in developing countries readily expand their cash cropping and cattle herding into marginal lands during wet years, ploughing furiously and introducing more grazing animals than the limited grass/shrub vegetation can support in dry years. Once these bad land-use practices are established,

a depletion of vegetation occurs which sets in [motion] such insidious processes as [soil erosion and the blowing away of loosened topsoil by the wind]. The vegetation is removed by clearance for cultivation, by the cutting and uprooting of woody species for fuel, by overgrazing [cattle], and by the burning of vegetation for pasture and charcoal. (Andrew Goudie, *The Human Impact on the Natural Environment*, 5th ed. [Cambridge: MIT Press, 2000], p. 75)

As shown in Figure 15.14, the final stage of the desertification process is nothing less than the total denudation of once-productive land. In the more extreme cases, the cultural landscape may even be obliterated by the encroaching desert sands. In the western Sahel, encroachment is now so far advanced that some predict that much of the upper West African country of Mali will soon become uninhabitable. Most of its permanently settled adult residents already shovel sand every day to keep daily life from grinding to a halt.

With desertification heavily tied to artificial causes, planners and policymakers are focusing on human activities in their search for solutions to the crisis. One highly promising avenue is the new technology of *fertigation*, currently being perfected by the Israelis in their large-scale development of the Negev Desert. Here, plants have been genetically engineered to be irrigated by the brackish underground water that is available throughout the region. (Most semiarid areas are underlain by similar water resources.) To avoid evaporation, the water is pumped through plastic pipes directly to the roots of plants, where it combines with specially mixed fertilizers to provide exactly the right blend of moisture and nutrients. Today, more than half a million people are thriving in the Negev's agricultural settlements and producing enormous quantities of high-quality fruits and vegetables for domestic and foreign markets.

Key Terms

arid climate *page 197*
climograph *page 194*
desertification *page 192*

Sahel *page 201*
savanna *page 196*
steppe *page 198*

tropical climate *page 192*
tropical deforestation *page 192*

Scan Here for a quick vocabulary review

Review Questions

1. Describe the general latitudinal extent of **A** climates.
2. Describe the general latitudinal extent of **B** climates.
3. Contrast the rainfall regimes of **Af**, **Aw**, and **Am** climates and provide an explanation for their differences.

4. What is the primary difference between **BS** and **BW** climates?
5. Why is ongoing tropical deforestation regarded as a catastrophe by so many natural scientists?
6. Describe the circumstances leading to desertification.

Scan Here for a quick concept review

www.oup.com/us/mason

Mild Midlatitude (C) Climates

Natural vegetation found on the island of Crete is adapted to the characteristically dry hot summer of the Mediterranean region.

OBJECTIVES

- Describe the relatively mild C climates, which affect the lives of billions of people in the midlatitude zones
- Relate the effects of pressure zones and air masses to the spatial distribution and on-the-ground climographs of the perpetually moist mesothermal climates

- Explain the reasons for the unique dry-summer climates
- Describe the dry-winter, "subtropical," monsoonlike C climates and their surprising climographs

On a global scale, mesothermal climates are transitional between those of the tropics and those of the upper midlatitude zone, where polar influences begin to produce climates marked by harsh winters. The **mild midlatitude** (or **mesothermal**) **climates** (**C**) dominate the equatorward sides of the middle latitudes and are generally aligned as interrupted east–west belts (Fig. 16.1). As we can see in Figure 16.1, **C** climates are nearly continuous in the southern midlatitudes but occupy no more than 65 percent of any latitude in the Northern Hemisphere, where large landmasses interrupt their extent.

The specific limiting criteria for a **C** climate are (1) an average temperature below 18°C (64.4°F) but above 0°C (32°F) in the coolest month, and (2) an average temperature of not less than 10°C (50°F) for at least one month of the year. We can tell from these limits that the annual range in temperature in these regions is large enough to legitimately think of their winters as cold, but not so cold and long that these are considered snow climates. As a result, **C** climates stand in contrast to the year-round warmth of **A** climates and the severe winters of the **D** and **E** groups.

Mild midlatitude climates are associated with that great lower atmosphere battleground, the latitudinal belt where polar and tropical air masses collide and mix (see Unit 9, p. 119. As a result, changeable weather patterns are the rule in **C** climates, as meandering jet streams periodically steer a parade of anticyclones and cyclonic storms across the midlatitudes from west to east. Because so many of the world's large population concentrations are in the **C** climate zone, the episodes of extreme weather are of particular concern in daily human terms. These departures from average conditions can also occur over longer time periods, as we see in

TABLE 16.1 Areal Coverage of Mild Midlatitude Climates

Climate Subtype	Percent of Earth Area
Perpetually Moist (**Cf**)	13.9
Dry Summer (**Cs**)	5.0
Dry Winter (**Cw**)	1.2

our focus throughout this unit on **drought** as a major environmental problem of the mesothermal climates.

We distinguish three major subtypes of mild midlatitude climate according to their pattern of seasonal precipitation. The **Cf** climates lack a dry season and are accordingly called *perpetually moist*, the **Cs** climate has a dry season in the summer, and the **Cw** climate is dry in the winter. All together the **C** climates cover about 20 percent of Earth (Table 16.1). Over half is **Cf**, with most of the rest in the dry-summer (**Cs**) category. This leaves only about 1 percent for the dry-winter variety.

The Perpetually Moist (Cf) Climates

The perpetually moist mesothermal (**Cf**) climates occur in two major regional groupings, both of which are near a major source of water (Fig 16.2). In the material that follows, we describe each in detail.

THE HUMID SUBTROPICAL (Cfa) CLIMATE

The warmer, perpetually moist climate, usually called the *humid subtropical* (**Cfa**) climate, is situated in the southeastern portions of the five major continents. It owes its existence to the effects of the warm, moist air traveling northwestward (in the Northern Hemisphere) around the western margins of the oceanic subtropical high-pressure zones. Additional moisture comes from the movement of midlatitude cyclones toward the equator in winter and from tropical cyclones in summer and autumn.

Miyazaki, located on the east coast of the southern Japanese island of Kyushu, exhibits the fairly strong seasonal temperature effects of the humid subtropical climate (Fig. 16.3). Average monthly temperatures there range from 6.8°C (44°F) in mildly cold January to 26.7°C (80°F) in hot, humid August; the mean annual temperature is 15.7°C (63°F). Rainfall for the year almost equals that of the tropical rainforest—2,560 mm (100 in). Rainfall peaks in early summer. The water balance at Miyazaki is never at a deficit, not even during the relatively dry winter. Throughout the year, actual evapotranspiration always equals the potential evapotranspiration. The heat balance diagram in Figure 16.3 illustrates the humid nature of this climate. Most of the net radiation is used to evaporate water, and a relatively small proportion passes into the air as sensible heat.

The similarities of the humid subtropical climates in diverse parts of the world are underscored in the climographs

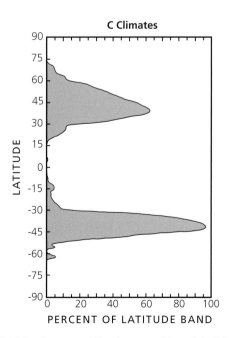

FIGURE 16.1 Percent of Earth covered by mild midlatitude (**C**) climates as a function of latitude.

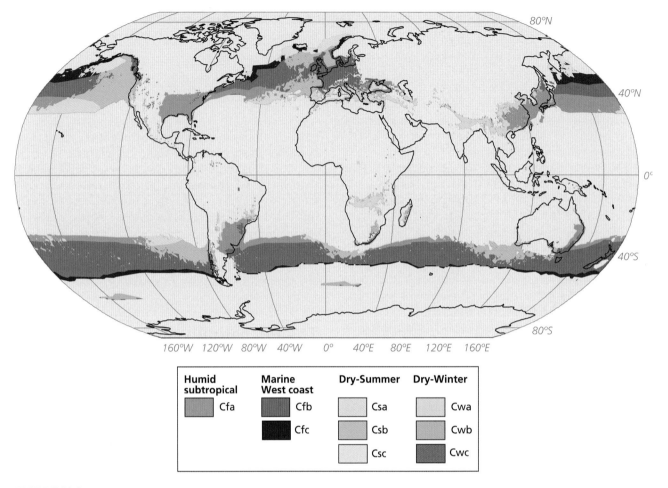

FIGURE 16.2 Global distribution of mild midlatitude climates.

for Charleston, South Carolina (Fig. 16.4), and Shanghai, China (Fig. 16.5). Both cities are on the southeastern seaboard of a large landmass (North America and Asia, respectively) near latitude 32 degrees N. Compared to Miyazaki (see Fig. 16.3), Charleston and Shanghai exhibit significantly lower annual precipitation totals. This is undoubtedly a function of stronger wind flows off the continents lying to their west. In contrast, offshore Kyushu is more constantly bathed by the moist southeasterly winds generated by the North Pacific's subtropical high.

We can also detect slight differences in a comparison of all three **Cfa** climate example stations (see Figs. 16.3, 16.4, and 16.5). For example, whereas summer temperatures are strikingly alike, Charleston has a noticeably milder winter than Shanghai because of the smaller bulk of North America (which does not develop as cold and durable a winter high-pressure cell as interior Asia) and the proximity of the warm Gulf of Mexico to the southwest. Note that the overall similarity of climatic characteristics between Charleston and Shanghai is not replicated in the cultural landscapes of the southeastern United States and eastern China. In fact, they differ as much as any two

cultures on Earth (Fig. 16.6), and they demonstrate the completely different uses that human societies can make of nearly identical natural environments.

THE MARINE WEST COAST (Cfb, Cfc) CLIMATE

The second group of perpetually moist mesothermal climates, where coasts are caressed by the prevailing westerlies year-round, is usually called the **marine west coast (Cfb, Cfc) climate**. Figure 16.7 shows one such area in Western Europe; others occur in the Pacific Northwest of the United States, the adjacent west coast of Canada, southern Chile, southeastern Australia, and New Zealand. As the map suggests, this relatively mild climate covers much of the midlatitude oceans as well, and the warmer **Cfb** subtype is by far the more common of the two variants. In continental locations, midlatitude cyclones and maritime polar air masses bring a steady flow of moist, temperate, marine air from the ocean onto nearby land surfaces. The degree to which the **Cfb** and **CFc** climates extend inward depends on topography. Where few orographic barriers exist, such as in

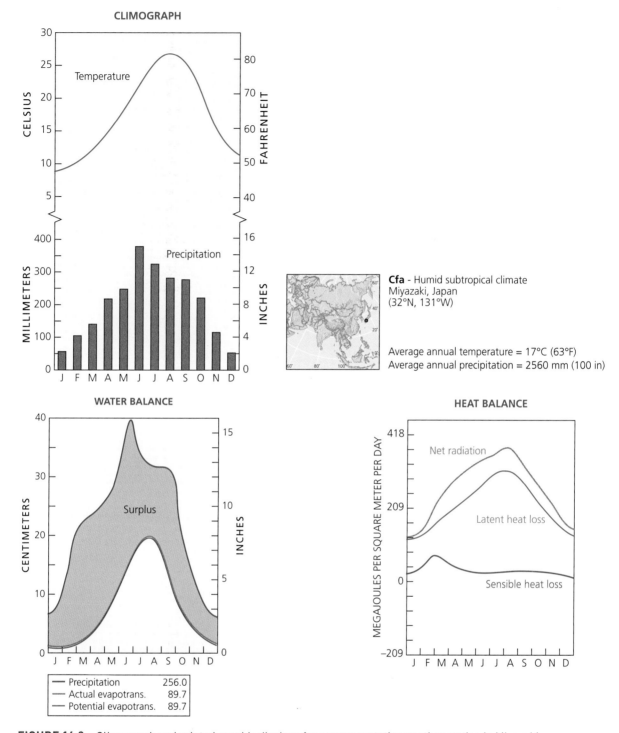

FIGURE 16.3 Climograph and related graphic displays for a representative weather station in Miyazaki, Japan, in the humid subtropical (**Cfa**) climate zone.

Europe north of the Alps, the **Cfb** climate reaches several hundred kilometers eastward from the North Atlantic. However, where mountains block the moist onshore winds, such as in the far western United States and Chile in southernmost South America, **Cfb** and **CFc** climates are confined to coastal and near-coastal areas.

Extremes of temperature in marine west coast climates are rare. Average monthly temperatures never exceed 22°C (71.6°F), and at least four months record mean temperatures above 10°C (50°F) in the **Cfb** zones (the cooler **Cfc** subtype experiences fewer than four months of mean temperatures above 10°C). Winters are quite warm considering the regions' latitude, usually about 15°C higher than average for their latitude band. The **Cfb** patterns are clear in the climographs for Seattle (Fig. 16.8) and London (Fig. 16.9). Note that despite its popular image as a rainy city,

CLIMOGRAPH

Cfa - Humid subtropical climate
Charleston, South Carolina
(33°N, 80°W)

Average annual
temperature = 188° (64°F)

Average annual
precipitation = 1247 mm (49 in)

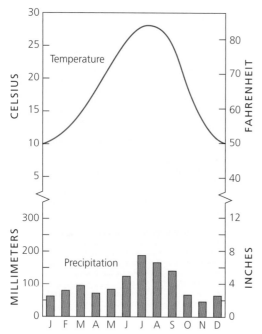

FIGURE 16.4 Climograph of Charleston, South Carolina, a humid subtropical (**Cfa**) location with a particularly mild winter.

CLIMOGRAPH

Cfa - Humid subtropical climate
Shanghai, China
(32°N, 122°E)

Average annual
temperature = 16°C (61°F)

Average annual
precipitation = 1135 mm (45 in)

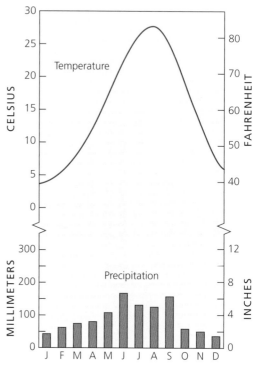

FIGURE 16.5 Climograph of Shanghai, China, a humid subtropical (**Cfa**) location with a pronounced winter, enhanced in this case by the dry monsoon of interior Asia.

FIGURE 16.6 These farming areas are located in similar **Cfa** climatic zones, but the agricultural landscapes could not be more different. (A) In the southeastern United States, large fields, independent farmsteads, and highway transport prevail. (B) In east-central China, small rectangular plots, clustered villages, and water transport dominate.

FIGURE 16.7 "The name Ireland is practically synonymous with **Cfb** climate. I was reminded of this while sailing the Irish coast in day after day of dreary, rainy weather. When the sun broke through as we approached the little harbor of Dunmore East near Waterford in the southeast, you could see why this is known as the Emerald Isle. All was bright green as far as the eye could see; grass grew even atop the cliffs."

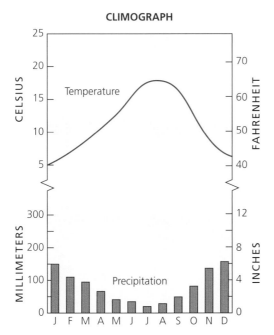

Cfb - Marine west coast climate
Seattle, Washington
(48°N, 122°W)

Average annual
temperature = 12°C (54°F)

Average annual
precipitation = 1013 mm (40 in)

FIGURE 16.8 Climograph of a marine west coast (**Cfb**) weather station in Seattle, Washington, in the U.S. Pacific Northwest.

London receives on average only 583 mm (22 in) of precipitation annually. This amount is equivalent to the yearly total in a moist semiarid climate. But London does indeed experience a great number of cloudy and/or rainy days each year (the hallmark of many a marine west coast climate location), with much of its precipitation occurring as light drizzle. The cooler **Cfc** pattern is shown in the climograph for Reykjavik, Iceland (Fig. 16.10), which records only three months with mean temperatures at or above 10°C (50°F).

The **Cfb** climates are situated precisely in the middle of the midlatitude prevailing westerly wind belts. Cooler **Cfc** climates develop at higher latitudes, but almost exclusively in the Northern Hemisphere, where strong warm ocean currents sweep poleward along western continental shores (e.g., off Iceland, Norway, and Alaska). Western civilization moved to the **Cfb** climates of northwestern Europe after leaving its Mediterranean hearth. The adequate year-round precipitation naturally produces forests of evergreen conifers and broadleaf trees that shed their leaves in winter. Although dry spells are not unknown in these areas, the soil rarely has a moisture deficit for a long period. When it does, the deficit can lead to serious drought and to a

CLIMOGRAPH

Cfb - Marine west coast climate
London, United Kingdom
(52°N, 0°)

Average annual
temperature = 10°C (50°F)

Average annual
precipitation = 583 mm (22 in)

FIGURE 16.9 Climograph of a marine west coast (**Cfb**) location in maritime Europe. Despite its image as a rainy city, London annually receives only about half the precipitation of Seattle (compare with Fig. 16.8).

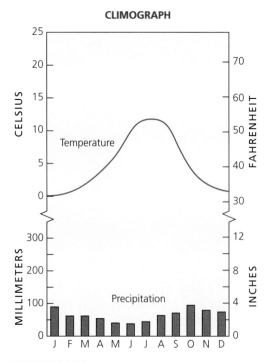

Cfc - Marine west coast climate
Reykjavik, Iceland
(64°N, 22°W)

Average annual
temperature = 5°C (41°F)

Average annual
precipitation = 805 mm (32 in)

FIGURE 16.10 Climograph of a weather station in Reykjavik, Iceland, that experiences the cooler variety of the marine west coast (**Cfc**) climate.

resulting cascade of disruption in human activities (see Perspectives on the Human Environment: Australia's Dreadful Decade of Drought).

The Dry-Summer (Cs) Climates

The second major mild midlatitude subtype is the dry subtropical or **Mediterranean climate** (**Cs**). It is frequently described, particularly by people of European descent, as the most desirable climate on Earth. This is the climate that attracted filmmakers to Southern California, so famous for its clear light and dependable sunshine. The warmer variety of Mediterranean climate, **Csa**, prevails in the Mediterranean Basin itself as well as in other interior locations. The

cooler variety, **Csb**, is found on coasts near cool offshore ocean currents, most notably in coastal areas of California, central Chile, southern and southwestern Australia, South Africa, and Europe's Iberian Peninsula (northern Portugal and northwestern Spain).

In the **Cs** regime, rainfall, largely produced by midlatitude cyclones, arrives in the cool season. Annual precipitation totals are moderate, ranging from approximately 400 to 650 mm (16 to 25 in). The long, dry summers are associated with the temporary poleward shift of the oceanic subtropical high-pressure cells. In particular, when a subtropical high moves to a higher latitude, subsiding air on the eastern side dominates the Cs climate zone. Thus, extended rainless periods are quite common in Mediterranean climates. San Francisco, California, illustrates several aspects of the "perfect" climate (Fig. 16.11). The cool offshore California Current, with its frequent fogs (see Fig. 10.7), keeps average monthly temperatures almost constant throughout the year, with a range of only 6.6°C (12°F) around an annual mean of 13.6°C (56.5°F). The average annual rainfall total is 551 mm (22 in), with the rains mostly emanating from winter storms steered southward by the polar front jet stream. In winter, actual evapotranspiration reaches the potential amount, but in summer, it falls well short.

Csb - Mediterranean climate
San Francisco, California
(38°N, 122°W)

Average annual temperature = 14°C (57°F)
Average annual precipitation = 551 mm (22 in)

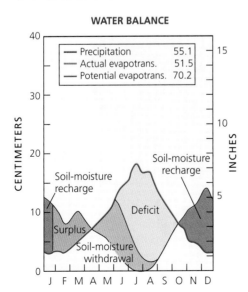

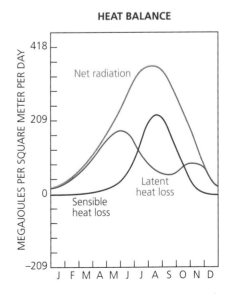

FIGURE 16.11 Climograph and related graphic displays for a representative weather station in the Mediterranean (**Csb**) climate zone. The energy balance graph is based on data from Sacramento, located 150 km (95 mi) inland from San Francisco.

Australia's Dreadful Decade of Drought

Köppen's map of Australia's climates paints a daunting picture (see Fig 14.3). The heart of the landmass, about half of its total area, is dominated by **BWh** (hot desert) conditions. Encircling this torrid center is a wide zone of **BSh** (hot steppe). Only a mere fringe of humid temperate (**C**) and tropical (**A**) climates exist in the east and north. In the east, the Murray-Darling river system, augmented by thousands of wells and numerous dams, sustains farmers and city dwellers alike.

As the map suggests, long-term average precipitation in this eastern zone justifies its designation as a **Cf** (no dry season) climate. But the close proximity of preponderant high-variability dry climates creates an ever-present risk. For decades, as the population grew and farming expanded, the region, nicknamed the "Lucky Country," was just that. When the luck ran out at the turn of this century, a drought began that lasted year after year, dried up tributaries and wells,

desiccated farmlands, devastated Australian farming, and caused urban water shortages that generated ugly disputes over policy (Fig. 16.12). Forest fires swept by blistering winds scorched tens of thousands of hectares and killed dozens of people. It was as if the desert had taken control of the green.

In verdant southeastern Australia, water levels in the Murray-Darling river catchment region fell to their lowest level ever recorded during the drought. By 2009, Lake Hume, at the Murray River's largest dam, had fallen to one-third of capacity. Farms were abandoned, their irrigation systems ruined. Scientists pointed to Australia's vulnerability to El Niño events and to global warming as possible causes, but the most obvious warning can be found in the Köppen map. The **BW** and **BS** climates that border the **Cf** eastern zone are marked by low precipitation and high variability, and nature draws no sharp lines of demarcation between zones. Australia's decade of drought

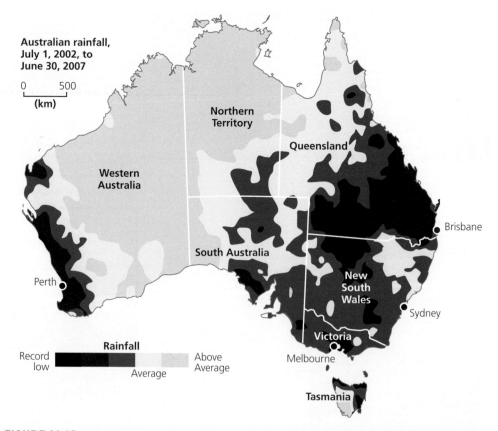

FIGURE 16.12   Australian rainfall for a five-year period of prolonged drought. The values mapped express departures from the long-term average rather than absolute amounts. For example, even during this extreme drought, Melbourne received more rain than central Australia.

will not be the last time its moister margins fall victim to the power of desert regimes.

How did this environmental story end? Ten years of nearly constant aridity broke in 2010 with devastating floods that struck the east from Queensland to New South Wales. Thousands of people were driven from their inundated homes, crocodiles and snakes swam in streets that had turned into streams, and erosion swept millions of tons of drought-loosened soil into the ocean. It was the ultimate manifestation of climatic variability.

FIGURE 16.13 California's agriculturally rich San Joaquin Valley (the southern component of the Central Valley) depends on irrigation for its productivity. Cotton is particularly water consumptive, as illustrated by these lush fields bordering the dry foothills that lead up to the Sierra Nevada on the valley's eastern margin.

The effect of the dry summer of the **Cs** climates is reflected in the energy balance diagram for nearby Sacramento, located about 150 km (95 mi) northeast of San Francisco. During winter, much of the net radiation is expended as latent heat loss. But in the summer, dry soil conditions necessitate that a large proportion of the net radiation be used in the form of sensible heat to warm the air. With intensive fruit and vegetable agriculture dominating California's Central Valley south of Sacramento, widespread irrigation is an absolute necessity (Fig. 16.13). For purposes of comparison, Figure 16.14 shows the climograph for Athens, Greece, a classic example of the warmer **Csa** Mediterranean climate variety. Note the similarities to the San Francisco temperature and precipitation pattern (see Fig. 16.11); both locations display the summer heat and winter rain sequence. But Athens, though coastal, is affected by a Mediterranean Sea that is much warmer than California's offshore waters, and Athens also gets Saharan summer heat—hence the temperature spike evident in Figure 16.14.

The Dry-Winter (Cw) Climates

Most areas of the **Cw** climate, the third major subtype of mesothermal climate, differ little from the tropical savanna (**Aw**) climate discussed in Unit 15. The only departures are the amounts of rainfall and a distinct cool season, with the average temperature of at least one month falling below 18°C (64.4°F). The **Cw** climate therefore is the only subdivision of the **C** climates located extensively in the tropics. For this reason, the **Cw** climate often shares with the **Aw** climate the designation *tropical wet-and-dry* climate.

In many cases, the winter dry season of **Cw** climates is a result of offshore-flowing winds that accompany a winter monsoon. Indeed, the **Cw** climate is sometimes called the **subtropical monsoon climate**. In South and Southeast Asia as well as in northeastern Australia, rainfall is provided by the wet summer monsoon. In other locales, such as south-central Africa, the winter dry season occurs because the rains of the ITCZ migrate into the Northern Hemisphere with the Sun.

The **Cw** climates are also associated with higher elevations in the tropical latitudes, which produce winter temperatures too cool to classify these areas as **A**. Cherrapunji, in the Khasi Hills of northeastern India, is a good example (Fig. 16.15). Because of its location near the subtropics at latitude 25 degrees N, combined with its altitude of 1,313 m (4,300 ft), this highland town exhibits the characteristics of the cooler variety of this climatic subtype (**Cwb**). Cherrapunji's enormous annual rainfall (which averages 11,437 mm [450 in]), as explained in the discussion of the **Am** climate in Unit 15, is the result of wet-monsoon winds that encounter particularly abrupt orographic lifting on their way inland from the Bay of Bengal to the blocking Himalaya Mountain wall. Finally, there are small pockets of **Cw** climate on the poleward side of **Cf** climates. Colder in winter than the subtropical version of **Cw**, these higher-latitude **Cw** climates result when winter storms do not bring sufficient precipitation to ensure a perpetually moist climate.

Csa - Mediterranean climate
Athens, Greece
(38°N, 24°E)

Average annual
temperature = 18°C (64°F)

Average annual
precipitation = 402 mm (16 in)

Cwb - Subtropical monsoon climate
Cherrapunji, India
(25°N, 92°E)

Average annual
temperature = 17°C (63°F)

Average annual
precipitation = 11,437 mm (450 in)

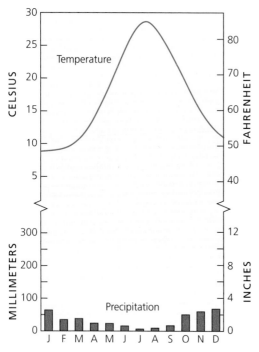

FIGURE 16.14 Climograph of a weather station in Athens, Greece, that experiences the warmer variety of the Mediterranean (**Csa**) climate.

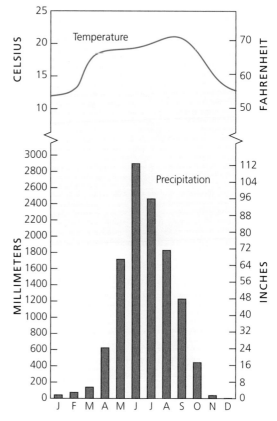

FIGURE 16.15 Climograph of a weather station in Cherrapunji, India, in the **Cwb** zone, sometimes called the subtropical monsoon climate.

Key Terms

drought *page 205*
marine west coast climate *page 206*

Mediterranean climate *page 211*
mild midlatitude (mesothermal)
   climate *page 205*

subtropical monsoon climate
   page 213

Scan Here for a quick vocabulary review

Review Questions

1. What climatic variable distinguishes **Cfa** from **Cfb** and **Cfc** climates?

2. How do **Cw** climates differ from **Aw** climates?

3. What seasonal precipitation cycles are associated with Mediterranean (**Cs**) climates? What is the explanation for this?

4. What are the two types of perpetually moist **C** climates? What factors explain their climatic characteristics and locations?

Scan Here for a quick concept review

www.oup.com/us/mason

Higher-Latitude (D, E) and High-Altitude (H) Climates

*Climatic boundary on the ground. The moist **H** (highland) region of the Kenya Highlands, foreground, meets the **BS** (steppe) climate on the Rift Valley floor west of Nairobi.*

OBJECTIVES

- Describe the severe, frigid, bitter climates described by geographers as subarctic (D), polar (E), and high altitude (H)

- Discuss the harsh *microthermal* (low-temperature) D climates with their cold winters and short summers

- Examine the climographs of the Point Barrow (E) tundra climate and describe the underlying climatic factors

- Explain how the effect of altitude creates the frigid H climate—even in the tropics

In this unit, we survey the remaining climates in the Köppen global classification and regionalization system. These climates range from the upper-middle to the polar latitudes and into the highest elevations capable of supporting permanent human settlement. We begin with the severe midlatitude or *microthermal* (**D**) climates that dominate the Northern Hemisphere continents poleward of the mild midlatitude (**C**) climates. Some large population clusters are located within the milder portions of the microthermal climate zone. The major environmental problem we consider here—*acid precipitation*—is associated with the burning of fossil fuels and associated emissions of sulfur and nitrogen compounds. Our attention next turns to the polar (**E**) climates that blanket the highest latitudes of both hemispheres. Finally, we survey the highland (**H**) climates, some of which are reminiscent of **E**-type temperature and precipitation regimes because they lie at altitudes high enough to produce polar-like conditions regardless of latitude.

The Major Severe Midlatitude (D) Climates

Northern Hemisphere continental landmasses extend in an east–west direction for thousands of kilometers, as you can see by looking at the world map in Figure 17.2. In northern Eurasia, for instance, the distance along 60 degrees N between Bergen on Norway's Atlantic coast and Okhotsk on Russia's Pacific shore is almost 9,000 km (5,600 mi). Russia alone is so longitudinally wide that the summer Sun rises above the eastern Pacific shoreline before it has set over the Baltic Sea in the west—nine time zones away! Consequently, vast areas of land in the middle and upper latitudes are far away from the moderating influence of the oceans. The resulting continentality shapes a climate in which the seasonal range of temperature is carried to extremes, with warm summers balanced by harsh, frigid winters.

The **severe midlatitude** (or **microthermal**) **climates** (**D**) are distinguished by a warm month—or months—when the mean temperature is above 10°C (50°F) and at least one month when the mean temperature is below freezing (0°C [32°]). Because they depend on large landmasses at high middle latitudes, **D** climates are found only in the Northern Hemisphere (Fig. 17.1), and because there is so much land at these locations the climate covers more than 50 percent of the Earth in some places (and all of the land area to be found in those places). The people who live in **D** climates have acquired lifestyles as varied as their wardrobes to cope with such conditions.

As we just noted, the cold-winter **D** climates occur mainly in the middle and higher-latitude continental expanses of Eurasia and North America (Fig. 17.2). These areas are sometimes punctuated by regions of the colder **E** climates, especially in northeastern Asia and the Canadian Arctic. The **D** climates also cover certain Northern Hemisphere midlatitude uplands and the eastern sides of

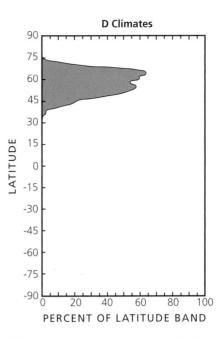

FIGURE 17.1   Percent of Earth covered by **D** climates as a function of latitude.

continental landmasses. These microthermal climates represent the epitome of the continentality effect. Although **D** climates cover only 6.5 percent of Earth's surface overall, they cover no less than 21 percent of its land area (Table 14.1).

The severe midlatitude climates are usually subdivided into two groups. In the upper-middle latitudes, where more heat is available, we identify the **humid continental** (**Dfa/Dwa/Dsa** and **Dfb/Dwb/Dsb**) **climates**. Farther north toward the Arctic, where the summer net radiation cannot raise average monthly temperatures above 22°C (71.6°F), lie the **subarctic** (**Dfc/Dwc/Dsc** and **Dfd/Dwd/Dsd**) **climates**, also known as **taiga**. (*Taiga* means "snowforest" in Russian.) Both groups of climates usually receive enough precipitation to be classified as moist all year round (**Df**). But in eastern Asia, the cold, low-humidity air of winter produces a dry-winter season (**Dw**). There are also some very small pockets of dry-summer (**Ds**) climates, but they occupy less than 0.5 percent of the Earth. As indicated in Table 17.1, the colder of the two subtypes—the subarctic—represents about two-thirds of the total area in this group.

All **D** climates show singular extremes of temperature throughout the year. For example, the heat balance for the northern Siberian town of Turukhansk is such that in July the temperature averages a pleasant 14.7°C (58.5°F), but the January mean plunges to –31°C (–24°F). However brutal the latter reading might seem, Turukhansk is merely an example of the **Dfc** climatic subtype, which is not the most extreme category. The harshest subtype, **Dwd**, is exemplified by Verkhoyansk, a remote far northeastern Russian village 2,100 km (1,300 mi) northeast of Turukhansk, whose particularly extreme temperatures were discussed in Unit 7 (see Fig. 7.12). As the climograph in Figure 17.3

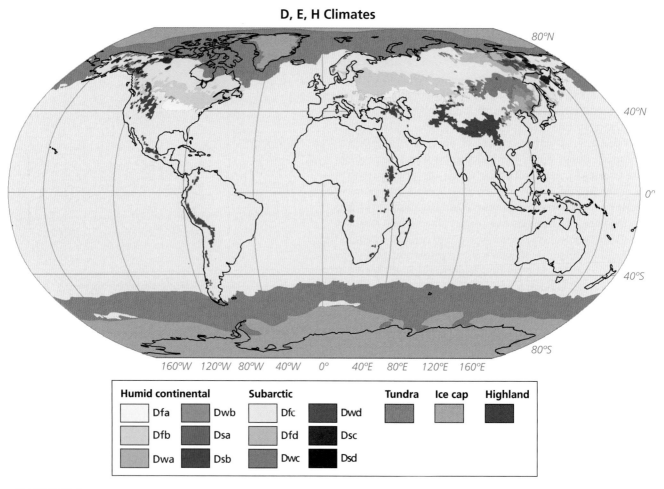

FIGURE 17.2 The distribution of midlatitude severe, polar, and highland climates.

TABLE 17.1 Areal Coverage of Severe Midlatitude Climates

Climate Subtype	Percent of Earth Area
Humid Continental (**Dfa, Dwa, Dsa, Dfb, Dwb, Dsb**)	2.3
Subarctic (**Dfc, Dwc, Dsc, Dfd, Dwd, Dsd**)	4.2

shows, the average January temperature in Verkhoyansk is an astonishing −46.8°C (−52°F). Also fascinating is the range of temperature, the difference between the highest and lowest monthly averages: 45.7°C (82°F) for Turukhansk and 62.5°C (112.5°F) for Verkhoyansk. These are among the greatest annual temperature ranges occurring anywhere on our planet's surface.

Such severe temperature ranges have a far-reaching impact. There is a short, intense growing season because the long summer days at these high latitudes entail substantial solar radiation. Special quick-growing strains of vegetables and wheat have been developed to make full use of this brief period of warmth and moisture (Fig. 17.4). On the Chukotskiy Peninsula, the northeasternmost extension of Asia where Russia faces Alaska's western extremity across the narrow Bering Strait, one variety of cucumbers grows to full size within 40 days. But agriculture is severely hampered by the thin soil (most of which was removed by the passing of extensive ice sheets thousands of years ago) and permafrost. **Permafrost**, which we will discuss at length in Unit 46, is a permanently frozen layer of the subsoil that sometimes exceeds 300 m (1,000 ft) in depth and is found in many parts of the subarctic.

Although the top layer of soil thaws in summer, the ice beneath the soil presents a barrier that water cannot permeate. As a result, the surface is often poorly drained. Moreover, the yearly freeze–thaw cycle expands and contracts the soil, making construction of any kind difficult, as Figure 17.5 illustrates. Because of the soil's instability, the trans-Alaska oil pipeline built during the 1970s is elevated above ground on pedestals for much of its 1,300-km (800-mi) length. This prevents the heated oil in the pipeline from melting the permafrost and causing land instability that might damage the pipe.

Long-lasting snow cover has other effects. It reflects most of the small amount of radiation that reaches it so that little is absorbed. It radiates longwave radiation effectively, which helps cool the lower atmosphere and

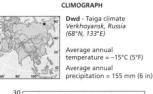

CLIMOGRAPH

Dwd - Taiga climate
Verkhoyansk, Russia
(68°N, 133°E)

Average annual
temperature = –15°C (5°F)

Average annual
precipitation = 155 mm (6 in)

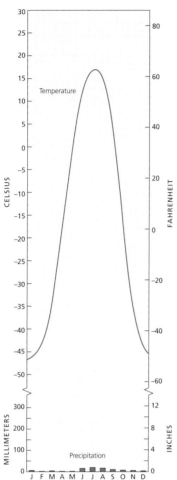

FIGURE 17.3 Climograph of a weather station in Verkhoyansk, Russia, in the **Dwd** zone, the harshest extreme of the severe midlatitude climate regions.

contributes to the production of areas of high atmospheric pressure. This presents difficulties for human transport and other activities, but paradoxically, it does keep the soil and dormant plants warm because snow is a poor conductor of heat. Measurements in St. Petersburg, Russia's second-largest city, have shown that ground temperatures below a layer of snow can be as high as –2.8°C (27°F) when the air overlying the cover is a bitter –40°C (–40°F).

In the microthermal climates, most of the precipitation comes in the warmer months. This is not only because the warmer air has greater moisture capacity. In winter, large anticyclones (see Unit 8) develop in the lower layers of the atmosphere. Within the anticyclones, the air is stable, and these high-pressure cells tend to block midlatitude cyclones. In summer, surface heating by solar radiation promotes convection, so midlatitude cyclones can pass through **D** climate areas more frequently during that time

FIGURE 17.4 Short growing season, luxuriant growth. These giant cabbages are among many varieties of vegetables grown during the brief summer in Alaska.

of year. In some places, there is also a summer monsoon season. China's capital city of Beijing (**Dwa**) experiences a pronounced wet season sometimes referred to as a monsoon (Fig. 17.6), which is typical of upper midlatitude coastal areas on the East Asian mainland.

The more detailed features of the severe midlatitude climate type are seen in the temperature, precipitation, water balance, and heat balance data in Figure 17.7. These graphs are for the southern Siberian city of Barnaul, located in a **Dfb** region within the heart of Eurasia at 53 degrees N, near the intersection of Russia and China with the western tip of Mongolia. The extreme seasonal change of Barnaul's climate is readily apparent. Average monthly temperatures range from 20°C (68°F) in July to –16.7°C (0°F) in January; the annual mean temperature is 1.4°C (34.4°F). Most of the precipitation falls during summer, but it is not enough to satisfy the potential evapotranspiration rate. Although the climate is labeled humid (i.e., this is not a **B** climate), excessive moisture is not the rule during the warm season, and this climate bears little resemblance to the muggy humid subtropical regime.

The heat balance diagram illustrates the large amount of net radiation that supports Barnaul's intense summer growing season. The lack of naturally available surface water means that only a little of this energy is used in evaporation in summer; most of it warms the air as sensible heat. As in

FIGURE 17.5 A once-smooth road after several annual cycles of freezing and thawing in the underlying surface layer of soil. This valley is located near Mt. Drum in south-central Alaska.

Turukhansk and Verkhoyansk, the low winter temperatures are the result of net radiation deficits that are not balanced by a significant flow of sensible heat to the surface.

Barnaul, a manufacturing center, is located near the Kuznetsk Basin, one of the largest heavy-industrial complexes in Russia. Such regions are important features of the economic geography of the middle latitudes. They involve physical geography as well because industrial areas are the sources of pollutants that produce **acid precipitation**—abnormally acidic rain, snow, or fog resulting from high levels of the oxides of sulfur and nitrogen in the air. This is a serious environmental problem that plagues many **D** climate zones (see Perspectives on the Human Environment: Acid Rain).

The Polar (E) Climates

Beyond the Arctic and Antarctic Circles (66.5° N and S, respectively), summer and winter become synonymous with day and night. Near the poles, there are six months of daylight in summer, when the monthly average temperature might "soar" to –22°C (–9°F). In winter, six months of darkness and continual outgoing radiation lead to the lowest temperatures and most extensive ice fields on the planetary surface. The lowest temperature ever recorded on Earth, –89°C (–129°F), was on a memorably chilly day in 1983 at one of the highest altitude stations in central Antarctica, the Russian research facility of Vostok.

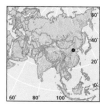

Dwa - Humid continental climate
*Beijing, China
(40°N, 117°E)*

Average annual
temperature = 12°C (54°F)

Average annual
precipitation = 623 mm (25 in)

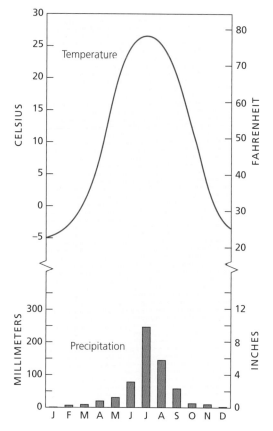

FIGURE 17.6 Climograph of a weather station in Beijing, China, in the **Dwa** climate zone, which also experiences a marked summer monsoon.

Polar climates (E) have a mean temperature of less than 10°C (50°F) in the warmest month. This group is found in the high latitudes of both hemispheres, and, as Figure 17.9 shows, these regions cover all of Earth poleward of about 70 degrees. As a result they cover a surprisingly large area of the planet, almost 18 percent (see Table 14.1). The group's two subtypes, known as **tundra** (EF) and **icecap** (ET), are differentiated on the basis of the warmest month. Where the average temperature exceeds 0°C (32°F), there is a seasonal thaw, and the climate is tundra. About two-thirds of polar climates meet that standard (Table 17.2).

TABLE 17.2 Areal Coverage of Polar Climates

Climate Subtype	Percent of Earth Area
Tundra (**ET**)	11.4
Icecap (**EF**)	6.4

WATER BALANCE

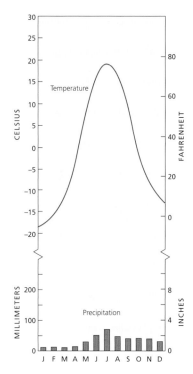

Precipitation	33.7
Actual evapotrans	33.7
Potential evapotrans	52.0

Soil-Moisture Withdrawal

Deficit

Soil-Moisture Recharge

Surplus

Surplus

HEAT BALANCE

Net Radiation

Sensible Heat Loss

Latent Heat Loss

CLIMOGRAPH

Dfb - Humid continental climate
Barnaul, Russia
(53°N, 84°E)

Average annual temperature = 1° (34°F)
Average annual precipitation = 337 mm (13 in)

Temperature

Precipitation

FIGURE 17.7 Climograph and related graphic displays for a representative weather station in Barnaul, Siberia, in the **Dfb** climate zone.

PERSPECTIVES ON THE HUMAN ENVIRONMENT

Acid Rain

In the mid-twentieth century, the term *acid rain* became common shorthand for a growing problem: the precipitation of sulfuric and nitric acid from polluted water vapor in the atmosphere onto plants (including crops), animals (including livestock), lakes, and aquatic life. Such destructive effects came not only from rain but also from snow, fog, and even dust particles in the atmosphere. Thus, although rain was the main culprit, the term *acid precipitation* is more accurate.

When fossil fuels (oil, coal, and natural gas) are burned in power plants and factories, they release substantial amounts of sulfur dioxide (SO_2) and nitrogen oxides (NO_x) into the air. These gases react chemically with water vapor in the atmosphere and are transformed into acid solutions in clouds as well as in fog, rain, and snow. At the surface, acidity and alkalinity are measured using the **pH scale**, which ranges from 14 (the alkaline maximum) to 0 (the acid extreme). The neutral level is 7.0, and in humid environments, lakes may be slightly acidic (6.5), while salty ocean water has a pH of about 7.8. When acid precipitation increases the acidity of freshwater lakes, causing a drop in pH from, say, 6.5 to 5.9, changes in the phytoplankton can eliminate entire fish species, and when it reaches 5.6, large patches of slimy algae start to choke off the oxygen supply, block sunlight, and destroy the food chain. If pH reaches 5.0, no aquatic species can survive.

continued

continued

Large-scale damage can also be observed in once-healthy woodlands and on farms, where acid-sensitive crops are threatened, and even in cities and towns where acid rain attacks bricks and mortar. As soon as the effects of acid precipitation were recognized, it also became clear that winds can carry the fallout far and wide so that power plants and factories in one country damage natural environments in another. This led to some angry arguments between neighbors in North America (Fig. 17.8) and Europe, with persistent wind patterns being part of the evidence.

Although much has been done to mitigate the negative impacts of acid precipitation, industrial and transport-related sources, now augmented by fast-developing economies from Brazil to China, continue to emit huge quantities of sulfur dioxide and nitrogen oxides into the atmosphere, and the fallout persists.

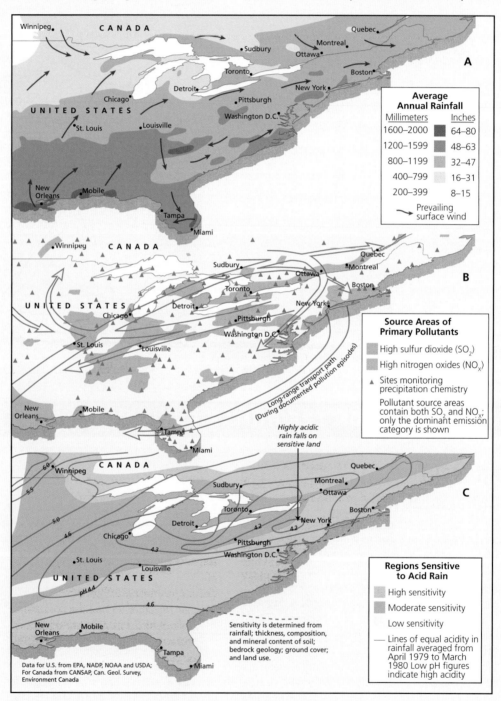

FIGURE 17.8 Eastern North America weather patterns (A), pollution sources and associated wind flows (B), and land sensitivity related to acid rain occurrence (C).

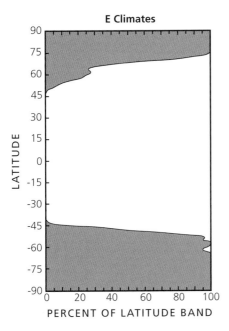

E Climates

LATITUDE (y-axis): 90, 75, 60, 45, 30, 15, 0, -15, -30, -45, -60, -75, -90

PERCENT OF LATITUDE BAND (x-axis): 0, 20, 40, 60, 80, 100

FIGURE 17.9 Percent of Earth covered by **D** climates as a function of latitude.

THE TUNDRA (ET) CLIMATE

The tundra climate is named after its associated vegetation of mosses, lichens, and stunted trees (Fig. 17.10). Practically all of the **ET** climates are located in the Northern Hemisphere, where the continentality effect yields particularly long and bitter winters yet allows enough summer warmth for a seasonal thaw. Furthermore, many of the **ET** areas border the Arctic Ocean, which provides a moisture source for frequent fogs when the water is not frozen. Permafrost occurs throughout these regions, and in summer poor drainage leads to stagnant water, which becomes breeding grounds for enormous swarms of flies and mosquitoes. Precipitation, mainly from frontal midlatitude depressions in the warmer months, seldom exceeds 300 mm (12 in) for the year. Combined with relative warmth and long days, this is enough to sustain low-growing plant forms, most of which are nonwoody. The climograph for Point Barrow at Alaska's northern tip (Fig. 17.11) displays the typical **ET** temperature and moisture regimes.

FIGURE 17.10 Snow-capped Mt. McKinley, the highest mountain in North America (altitude 6,194 m [20,320 ft]), is located near the center of the Alaska Range. Known to Native Americans as Denali (the High One), the mountain is flanked by tundra (foreground). The boundary between **ET** and **EF** climate lies along the foothills between the snowy mountain massif and the lower plain.

ET - Tundra climate
Point Barrow, Alaska
(71°N, 157°W)

Average annual
temperature = −12°C (10°F)
Average annual
precipitation = 153 mm (6 in)

EF - Ice cap climate
Mirnyy Station, Antarctica
(67°S, 93°E)

Average annual
temperature = −11°C (12°F)
Average annual
precipitation = No Data

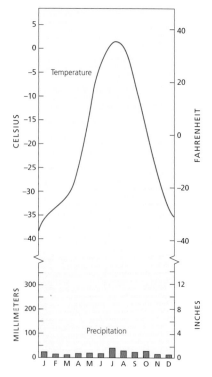

FIGURE 17.11   Climograph of a high-latitude tundra (**ET**) weather station at Point Barrow, Alaska.

THE ICECAP (EF) CLIMATE

In icecap (**EF**) climates, the second major polar subtype, we find the lowest annual temperatures on Earth. In a given year, about 90 mm (3.5 in) of precipitation, usually in the form of snow, fall onto the barren icy surface (Fig. 17.12). The inhospitable climate has made it difficult to collect data from these areas, which are mainly in Antarctica and interior Greenland as well as atop the Arctic Ocean's floating icecap.

The Russian station of Mirnyy is located in Antarctica just inside the Antarctic Circle. It is at a relatively low latitude and so exhibits temperatures that are rather moderate for an **EF** climate, but the mean annual temperature of −11°C (12°F) is not high, and the warmest monthly average does not rise above the freezing point—the hallmark of an **EF** climate (Fig. 17.13). Year-round snow makes it almost impossible to obtain accurate water balance data, but the Russians have measured the heat balance components (see Fig. 17.12), which vividly characterize the frigid **EF** climate. There is significant positive net radiation at Mirnyy for only about four months of the year; during other months, more radiation leaves Earth than arrives. Sensible heat flow throughout the year is directed from the air toward the surface, the final result of the general circulation of the atmosphere moving

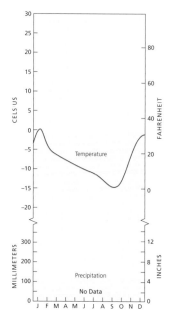

HEAT BALANCE

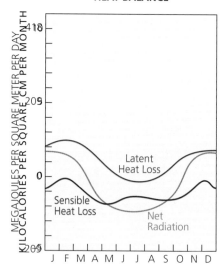

FIGURE 17.12   Climograph and heat balance diagram for a representative weather station in the icecap (**EF**) climate zone. There are no precipitation data available for this remote Antarctic location.

heat toward the poles. Sensible heat and net radiation can provide energy for some evaporation in the warmer months, but in the winter, condensation of moisture onto the surface provides only a minor source of heat.

The extreme conditions notwithstanding, a number of scientists now live and work in icecap climates, studying the environment or searching for oil and other secrets of Earth. But they depend for survival on artificial heating

FIGURE 17.13 Very low evaporative demand allows vast ice sheets to accumulate in the **EF** climate despite meager additions from precipitation.

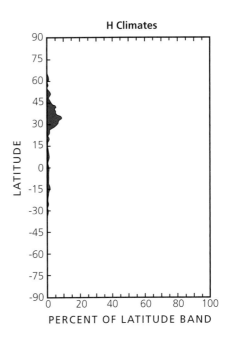

FIGURE 17.14 Percent of Earth covered by **E** climates as a function of latitude.

and food supplies flown in from the outside world, without which human life could not exist in the coldest climate on the planet.

High-Altitude (H) Climates

Our overview of the mosaic of climates covering Earth's land surface would not be complete without some mention of the climates of the highland regions. The mountains of the highest uplands reach into the lower temperatures and pressures of the troposphere. The drop in temperature and the corresponding changes in climate that occur with increasing elevation replicate those observed in a horizontal passage from equatorial to progressively higher latitudes. There is no universally agreed-upon definition of the highland (**H**) group. We use a convention that the location must be above 1,500 m (4,921 ft) in elevation and the Köppen temperature group must be at least one group "colder" than would be found at sea level in the same location.[1] Such places cover little of the planet, just slightly more than 1 percent. Most of them lie in the Northern Hemisphere, where the largest areas of high plateaus and mountains are found, but even there they don't exceed 10 percent of any latitude (Fig 17.14).

One of the outstanding features of **highland climates** (**H**) is their distinct **vertical zonation** according to altitude. Nowhere is this better demonstrated than in the tropical Andes Mountain ranges of northwestern South America. Figure 17.15 combines the characteristics

of many of these highlands into a single model. The foothills of the Andes lie in a tropical rainforest climate in the Amazon Basin to the east and in arid climates tempered by a cool Pacific Ocean current to the west. Above 1,200 m (4,000 ft), tropical climates give way to the subtropical zone. At 2,400 m (8,000 ft), the mesothermal climates appear, with vegetation reminiscent of that growing in Mediterranean climatic regions. These in turn give way to microthermal climates at 3,600 m (12,000 ft), and above 4,800 m (16,000 ft), permanent ice and snow create a climate like that of polar areas (Fig. 17.15). It is sometimes said that if one misses a bend at the top of an Andean highway, the car and driver will plunge through four different climates before hitting bottom!

These altitudinal zones mainly reflect the way temperature drops as elevation rises. But wind speeds also tend to increase with height, as can rainfall (and snowfall), fog, and cloud cover. Moreover, the radiation balance is markedly altered by altitude. Because less shortwave radiation is absorbed by the atmosphere at higher elevations, greater radiation values are recorded at the surface. This is especially true of the ultraviolet radiation responsible for snow blindness as well as the suntans of mountaineers and skiers. Where there are snow-covered surfaces, much of the incoming radiation is reflected and not absorbed, which further acts to keep the temperatures low.

Mountainous areas are often characterized by steep slopes, and these slopes have different orientations, or *aspects* (they may face in any compass direction). In the Northern Hemisphere, southerly aspects receive far more solar radiation than do north-facing slopes, which may receive no direct solar radiation for much of the winter

---

1   To determine this, temperatures are projected to sea level by adding 6.5°C (11.7°F) for every 1 km (3,280 ft) of elevation. If the Köppen category changes to a warmer group (e.g., from **E** to **D**), the climate is considered highland.

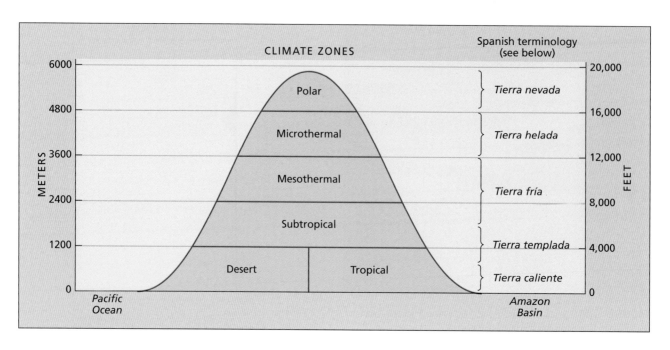

Vertical zone	Elevation range	Average annual temperature range
Tierra nevada	4800+ m (16,000+ ft)	<-7°C (<20°F)
Tierra helada	3600–4800 m (12,000–16,000 ft)	-7°–13°C (20°–55°F)
Tierra fría	1800–3600 m (6000–12,000 ft)	13°–18°C (55°–65°F)
Tierra templada	750–1800 m (2500–6000 ft)	18°–24°C (65°–75°F)
Tierra caliente	0–750 m (0–2500 ft)	24°–27°C (75°–80°F)

FIGURE 17.15 Highly generalized west-to-east cross-section of the Andes in equatorial South America. The vertical climatic zonation shown on the mountain corresponds to the Spanish terminology listed.

season. These differences in radiation are often manifested as vegetation contrasts on different slope aspects. South-facing slopes are commonly drier and exhibit sparser vegetation cover. North-facing slopes are typically lusher, because temperatures, and hence evaporation rates, are lower. Rugged terrain also influences local wind-flow patterns that may have an effect on climatic conditions. Finally, the climate of a particular upland area depends on its location with respect to the global-scale factors of climate, such as the general circulation of the atmosphere.

A broadly simple pattern of climates occurs because of the heat and water exchanges at Earth's surface and because of the spatial organization of the general circulation of the atmosphere. This pattern is altered in detail by the specific location of land and water bodies, ocean currents, and upland regions. The resulting mosaic of climates may be classified in different ways. We have primarily followed the system first devised by Köppen, who divided the climates of Earth into six major types and a number of additional subtypes.

Units 18 and 19 explore two additional aspects of the geography of climate: (1) natural climatic variations over the long span of Earth history, and (2) human impacts on the climatic environment.

FIGURE 17.16 "Watching the sun rise over Kilimanjaro and its ice-capped peak, Kibo, is one of the most memorable experiences a traveler in East Africa can have, and I am here as often as I can [be]. In the far distance to the left (east, since we are in Kenya and viewing from the north) you can see Mawensi, the eroded peak that once looked like Kibo does today. Up there beyond 5,800 m (19,000 ft), the climate is frigid and polar, although we are within sight of the equator. Two German geographers were the first Europeans to reach the summits: Hans Meyer climbed Kibo (5,895 m [19,340 ft]) in 1889 and Fritz Klute scaled Mawensi (5,355 m [17,564 ft]) in 1912. Global warming is in the process of shrinking the permanent ice and snow on Kibo."

Key Terms

acid precipitation *page 220*
highland climate *page 225*
humid continental climate *page 217*
icecap *page 220*
permafrost *page 218*

pH scale *page 221*
polar climate *page 220*
severe midlatitude (microthermal) climate *page 217*
subarctic climate *page 217*

taiga *page 217*
tundra *page 220*
vertical zonation *page 225*

Scan Here for a quick vocabulary review

ENHANCE

Review Questions

1. How do humid continental climates differ from taiga climates?

2. What is permafrost, and how does it form?

3. What is acid precipitation, and why is it often found so far from its source areas?

4. Why do polar climates invariably exhibit low precipitation values?

5. Describe the effects of increasing altitude on temperature, precipitation (types and amounts), and wind speeds.

Scan Here for a quick concept review

ENHANCE

 www.oup.com/us/mason

UNIT
18

Natural Climate Change

Water-carved canyons in Egypt's Sinai Peninsula are evidence of climate change occurring over thousands of years. The rock itself was formed at the bottom of a sea that covered the area millions of years ago, and reveals an ancient climate even more different from the parched desert found in the area today.

OBJECTIVES

- Discuss the origin and nature of climate change
- Discuss important indicators and evidence of climate change
- Explain the natural mechanisms that cause climate change
- Recount Earth's climatic history prior to the industrial era

Massive polar ice sheets and reliable wind patterns covering huge areas of the planet give an impression of climate permanence that is wholly misleading. Our planet has an incredibly dynamic history, and in fact, today's climate bears little resemblance to what has been the common pattern throughout Earth's existence. Some changes can be traced to conditions that ended billions of years ago (e.g., a molten surface, frequent collisions with large objects), but many major changes occurred long after the turbulent times of Earth's formation had calmed. Our main purpose in this unit is to explore *natural* climate change—namely, that which occurs without human influence. Thus, our focus is on changes throughout geologic time up to the Industrial Revolution that began in the mid-1700s. More recent changes are covered in Unit 19 as part of the discussion of humanity's influence on modern and future climates. Industrialization provides a convenient break point in time, but of course natural processes did not suddenly cease and cede control of climate to humanity; natural processes continue to operate and influence the planet even in the presence of human drivers. In addition, we should mention that humans are affecting climate in ways that are not related to "industry" as commonly defined. And, as will be discussed in Unit 19, the human imprint may well go back thousands of years, long before mechanization took hold.

The Nature of Climatic Change

Recall that climate refers to statistical properties of the atmosphere that attempt to capture typical atmospheric conditions or behavior (Unit 14). It follows that climatic change is a change in a statistical value. That is, a single extreme year does not amount to a climatic change. One year might be consistent with what happens in a new climatic regime, but by definition, *climate* refers to conditions that persist for some length of time, as captured by averages taken over a number of years. Remember also from Unit 14 that although air temperature and rainfall are very important components of climate, all variables both near the surface and throughout the atmosphere (wind, clouds, net radiation, etc.) are part of climate as well. Thus, persistent changes in any atmospheric variable constitute climatic change.

Perhaps the most important point to realize about climate variation is that change occurs on multiple timescales simultaneously. This is a mouthful, but, in many respects, it is a familiar concept. For example, in developed countries, people's wealth typically increases over their lifetime. But within that long-term trend, there are times of decreasing assets as a result of unemployment or extensive spending. Some periods are characterized by extreme variability, such as the college years when summer jobs often recharge depleted bank accounts. There can be instantaneous pulses of wealth when a long-lost relative bequeaths a prized painting. The resulting increase might be permanent, or there could be a delayed return to previous values if the painting is sold and the money is squandered on wild living. The essential point is that if someone asks if you are getting richer or poorer at any moment in time, the answer depends on the timescale. Is the questioner interested in how things are going for the decade or how things are going in that year or that month?

Climate change is much the same, and statements about change must be tied to a particular timescale. For example, right now Earth is getting warmer on a scale of centuries but colder over tens of thousands of years. There have been cooling trends within times of warming, and vice versa. There have been shocks that move climate to a new state, sometimes followed by relaxation to older patterns. Some climatic periods have been characterized by much more variability than others. It must also be noted that unlike the prosperity example, with its overall progression of increasing wealth, Earth's climate is not trending toward some particular long-term state. It continues to be subject to many processes, giving rise to change at a multitude of scales.

A second very important point is that not all places experience the same magnitude or even direction of climate change. There is a tendency for the tropics to have much smaller changes than the middle and high latitudes, but even that statement needs qualification. For example, over the last million years, the greatest climatic changes have been allocated differently across the equator, with the Northern Hemisphere showing much larger changes.

Climate Indicators

Climate change is the norm, not the exception, on our planet. But how do we know this? How is it possible to reconstruct climates of millions, even hundreds of millions, of years ago? The evidence exists in various forms. When glaciers expand, they scour the surface, breaking off, grinding up, and carrying away part of the bedrock below. Later, when they melt away, they leave behind their telltale work in the form of deposits that leave no doubt as to their former presence (see Units 43 to 45). We know that particular types of soils form under certain environmental conditions (e.g., thick tropical soils and thin desert soils). When such soils are buried by subsequent geologic events, they form a durable record of climatic conditions at the time of their formation. Occasionally, entire diagnostic landforms are buried and preserved—for example, fossilized sand dunes, which may tell us more than simply revealing that it was dry in the time and place of formation. From their shape, the fossilized dunes may reveal the prevailing winds of the past, even millions of years ago. Many times, the character of rocks can provide clues to past global environments. For instance, distinctive rock formations rich in iron oxide appeared when atmospheric oxygen began to accumulate about 2 billion years ago (Fig. 18.1). Likewise, many of the coal beds mined today are derived from dead plant material growing under warm, moist, and heavily vegetated conditions 300 million years ago.

Evidence from sediments deposited over long periods on ocean floors and lake bottoms is also crucial in this scientific detective work. Samples ("cores") of these deposits are tested

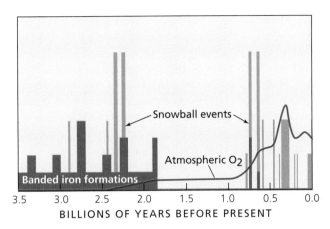

ENHANCE

FIGURE 18.1   Banded iron formations (purple bars) reveal an early atmosphere with little oxygen. Multiple lines of evidence indicate Earth's climate has been punctuated by ice-house conditions numerous times, with huge icecaps like those we find at high latitudes today (blue bars).

for accumulations of the fossils of tiny sea creatures, for mineral composition, and even for volcanic ash and plant pollen (Fig. 18.2). Different types of pollen can be readily identified under a microscope and used to infer the mix of vegetation and climatic environment for the region. Ice from existing polar and mountain glaciers in the tropics can also be cored and used to uncover past climates. The composition and thickness of each year's layer of ice can be analyzed, as can tiny air bubbles within the surrounding ice.

For long-term climate reconstructions, variations in isotopes of oxygen and carbon are the most valuable of all indirect measures. Recall from Unit 6 how elements are found in various forms, which are called isotopes, that differ in their

number of neutrons yet are chemically almost the same. For example, 99.76 percent of oxygen atoms have eight neutrons, but about 0.2 percent have 10 neutrons. Both types have eight protons, so the atomic weights of the two isotopes are 16 and 18, and they are known as O-16 and O-18, respectively. In the case of carbon, the common form is C-12, but there is another stable isotope, C-13, as well as a radioactive isotope, C-14. The relative amount of a rare isotope is expressed by the ratio of that isotope to the most common form. Thus, rather than work with O-18 amounts themselves, the ratio O-18 divided by O-16 is normally used for analysis and is referred to as the *O-18 ratio*. Variations in oxygen and carbon isotope ratios are usually expressed in relative terms, as changes, or "departures," from established standards.

Any two isotopes of the same element are chemically quite similar but not identical. Different isotopes enter reactions at different rates, and this is what makes them so helpful in climate reconstructions. In the case of oxygen, the water molecules containing the heavier form evaporate just a little less readily than those composed of the lighter O-16. An expanding glacier is fed by ocean water evaporating and precipitating on land. This results in the ocean becoming somewhat richer in O-18 as lighter H_2O is preferentially taken up by the atmosphere. Variations in the ocean's O-18 ratio are recorded in the remains of organisms that incorporate dissolved oxygen into their shells. By coring the seafloor and measuring the abundance of the two oxygen isotopes in sequentially older layers of rock, scientists produce environmental records going back many millions of years.

Carbon isotopes are used in a variety of ways. For example, by comparing the C-14 content of surface waters with those at depth, scientists can determine how long it takes for water to reach the depths of the sea and thus obtain information about the rate of overturning for the world ocean. Carbon-13 is useful in a different way. Because C-12

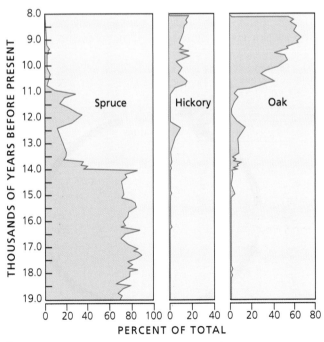

FIGURE 18.2   Under a microscope, the distinctive shapes of plant pollen grains emerge (A) and can be used to identify the type and relative abundance of vegetation in layers within a lake core (B).

is preferentially taken up by plants during photosynthesis, deposits of organic (plant) carbon have more C-12 and less C-13 than those derived from inorganic sources. The details are complicated, but changes in the abundance of C-13 give a picture of how carbon has moved back and forth between rock formations and the more active ocean and atmosphere reservoirs.

Many other climatic indicators exist, and all are similar to those mentioned in several ways. First, they are indirect, or *proxy*, measures of climate and therefore do not give precise values for variables of interest like temperature, precipitation, and wind speed. Second, they represent samples taken at a particular location, and it can be challenging to know how large an area they represent. Finally, the spatial coverage is highly nonuniform, which greatly complicates attempts to map past climates and make statements for the globe as a whole.

Climate Change Mechanisms and Feedbacks

Because we know that climate changes on many timescales, it shouldn't be surprising that there is not a single cause of climatic change. Many factors affect climate, some over relatively short spans of time, and others more slowly. We identify some specific agents of such change in the next section, but in general, we can separate external factors from processes that are internal to the Earth–atmosphere system. Any change in solar radiation, for example, obviously arises outside Earth and changes global temperatures directly by perturbing the planetary energy balance. Asteroid impacts are another example of external factors. A huge cataclysm 65 million years ago darkened the skies and quite possibly drove dinosaurs into extinction. Perhaps not so obviously, this category also includes the slow movement of continents that occurs over hundreds of millions of years and thereby results in constantly changing configurations of land and ocean. Not only do climates of drifting continents change as

they move into higher or lower latitudes, but because oceanic and atmospheric circulation are altered, global climate changes as well. Although this tectonic movement is obviously an Earth process, the atmosphere and ocean don't participate in causing these movements, and thus, from a climate change standpoint, tectonic movement is considered an external factor. Similar considerations apply to volcanic eruptions, periods of mountain building, and other such "nonatmosphere" planetary processes.

These external factors stand in contrast to a variety of internal feedbacks that affect climate. Probably the most important of these internal factors is **ice-albedo feedback**, which exerts a strong amplifying effect on climate change. With its high albedo, ice strongly reflects solar radiation compared to other surface veneers (see Unit 5). To see why this matters, suppose an increase in solar output leads to warming and less land area covered by ice. With lower albedo, more solar radiation is absorbed. This leads to further warming and amplifies the initial temperature rise (Fig. 18.3A).

Suppose, on the other hand, volcanic activity increases and surface temperatures fall because of shadowing by the overlying veil of particulates. Lower temperatures might lead to an expansion of ice and a consequent decrease in absorbed solar radiation, which in turn produces lower temperatures. Here again, the feedback amplifies the initial change and therefore is a positive feedback (Fig. 18.3B).

Another positive feedback (discovered in 2015) concerns the breakup of Arctic Ocean ice by waves. Thick ice covering the water absorbs the energy of waves generated in the open ocean. But when the climate warms and ice is thinner, waves more easily break the floating ice. As ice area shrinks, the corresponding larger expanse of water allows winds to whip up large waves. Wave energy increases and penetrates farther into the ice pack, creating still more open water. This positive feedback operates in the opposite direction during cold periods: thicker ice resists breaking and can expand, which reduces wave size. Notice that positive means amplifying, not warming, and is not necessarily beneficial.

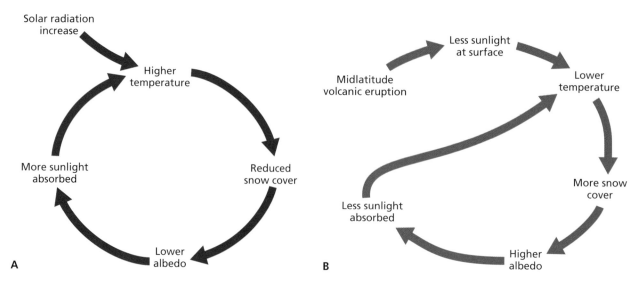

FIGURE 18.3 Ice-albedo feedback amplifies both warming (A) and cooling (B) regardless of the cause.

Earth's Thermostat: CO_2–Weathering Feedback

Most hotel rooms have a wall thermostat designed to keep the room at a more or less constant temperature. If the air gets too cold, the thermostat signals a heater that warms the room. The thermostat doesn't heat the room; rather, it triggers processes that eventually lead to warming and recovery of the original temperature. The thermostat amounts to a negative feedback because it counteracts the initial cooling. The Earth system has a somewhat similar negative feedback that regulates global temperature. Unlike the room thermostat just described, this so-called **carbon dioxide–weathering feedback** actively resists warming as well as cooling.

Weathering is the breakdown of rock, and in this case, we are talking about chemical decomposition of rocks containing calcium, silicon, and oxygen. Such rocks are very abundant on Earth, as we discuss in Unit 27. When CO_2 is dissolved in rainwater, it forms a weak acid. This acid dissolves rock, creating new compounds and ions that are free to enter chemical reactions. Most important, calcium can combine with carbon and oxygen in the ocean to form calcium carbonate (limestone), a completely different rock. The overall process is known as the *Urey reaction*, which is shorthand for a number of reactions and biological processes that in the end bury carbon in carbonate rock formations. The source of that carbon is atmospheric CO_2, so the Urey reaction in effect removes carbon from the atmosphere and hides it away in rocks.

Carbon is supplied to the atmosphere by many processes. For example, recall from Unit 6 that we discussed the importance of respiration and decay of plants on a seasonal basis. On longer timescales, volcanic eruptions and other deep-Earth processes release carbon from the interior to the atmosphere.

This *outgassing* acts as a source of CO_2, counter to the Urey reaction.

Carbon removal through weathering is far more variable than outgassing, and because of that, CO_2-induced weathering can serve as a thermostat. The speed of the Urey reaction depends on the precipitation rate, as rainfall delivers the acid needed to weather rock. Recall that when saturated, a warmer atmosphere contains more water vapor than a cold atmosphere, and the tropics are generally rainy for just this reason. Similarly, a warm climate typically has greater precipitation rates than a cold climate. In other words, if the climate warms, more water in the atmosphere means rainfall will increase, assuming all else is equal. With more rainfall, more carbon is removed by the Urey reaction, and atmospheric CO_2 falls, counteracting the initial warming. As illustrated in Figure 18.4, the same logic applies to cooling.

The CO_2-weathering feedback we just described is a negative feedback on temperature; therefore, it amounts to a planetary thermostat. However, it acts only on long timescales, because carbonate rock formation is slow. If atmospheric CO_2 were to suddenly increase, after about 100,000 years only half of that carbon would be sequestered. It would take several hundred thousand years for the climate system to fully respond. This is like a room thermostat that can keep up with nighttime cooling but can't instantly counteract an icy blast if a window is suddenly opened. Over eons, the planetary thermostat has been doing its work, and this has greatly subdued climate change on the long timescales the thermostat can affect. Changes on shorter timescales slip through like dust through a screen.

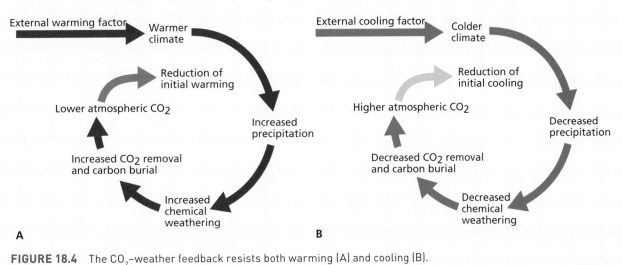

FIGURE 18.4 The CO_2–weather feedback resists both warming (A) and cooling (B).

continued

continued

Just like the room thermostat, the CO_2–weathering feedback "setting" can be changed to higher or lower temperatures. Suppose, for example, that mountain building leads to more exposed rock and weathering while everything else is unchanged (including volcanic outgassing). The long-term burial rate would increase, leading to lower atmospheric CO_2 and a colder effective temperature setting. By contrast, a higher setting would result if volcanic activity were to increase, because higher temperatures would be needed for the burial rate to balance the new inflow. There are other ways the setting can change. For example, if large continental areas drifted out of the tropics to drier subtropical latitudes (Figure 18.5), there would be less weathering and less carbon burial, moving the setting to warm. Conversely, if more land moved to moist climates, there would be more weathering, and CO_2 would fall. A long-term fall in sea level would expose more land for weathering and similarly move the setting to a colder position.

The huge shifts between ice-house and hot-house climates discussed in the text occurred on timescales much longer than the thermostat response time, and therefore they can't be explained by the thermostat swinging Earth's climate back and forth. The timing of those changes was set by the time needed for mountain building, continental movements, and sea-level change, and if anything, those changes moved the thermostat to new settings.

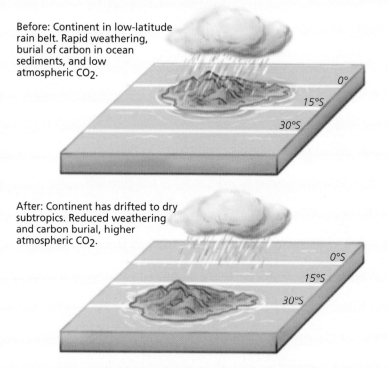

Before: Continent in low-latitude rain belt. Rapid weathering, burial of carbon in ocean sediments, and low atmospheric CO_2.

After: Continent has drifted to dry subtropics. Reduced weathering and carbon burial, higher atmospheric CO_2.

FIGURE 18.5 Example of how the CO_2–weathering thermostat can be set to lower or higher temperatures.

Negative feedbacks are also common, but they have a stabilizing effect on climate. That is, they resist, rather than amplify, climate change. As a very simple example, consider that as the planet warms it emits more longwave radiation to space. Thus, an increase in solar radiation might warm the planet, but eventually enough radiation will be emitted that Earth comes into equilibrium with the new solar input. Indeed, if Earth couldn't shed increasing amounts of radiation, it would warm indefinitely in response to the Sun's additional input. This is an example of a feedback process that operates instantly: as soon as Earth warms, more radiation is emitted. Another negative feedback, the **carbon dioxide–weathering feedback**, acts on roughly a 100,000-year scale. As described in Focus on the Science: Earth's Thermostat: CO_2–Weathering Feedback, warmer conditions lead to lower atmospheric CO_2, which in turn causes cooling (and vice versa). Between these two extremes, there are many other feedbacks that both amplify and resist climate change. Because of the differing lag times, the response to a change in a given external factor varies through time, and there is no guarantee climate will ever be in equilibrium with a changing external factor.

Considering this, it is not surprising that the one constant in Earth's climate is ceaseless variation.

Earth's Climate History

We turn now to the broad strokes of climate change that have occurred over the course of Earth's history, and we cover some important events and periods. Keep in mind that we view all these events through the lens of proxy data, and because of this, only qualitative descriptions are possible in many cases. Furthermore, causes are impossible to identify with total certainty; the best climate science can do is to unearth mechanisms that are consistent with apparent changes. Of course, with better records and more accurate dating for the recent past, the picture is clearer for times closer to the present in terms of both what happened and why.

ICE-HOUSE AND HOT-HOUSE CLIMATES

Most of Earth's history has been largely free of year-round, or "permanent," ice like the kind that is present today in Greenland and Antarctica. On timescales of billions of years, the best description of Earth's climate is a comparatively warm or **hot-house climate**, punctuated by a fairly small number of glacial events (see Fig. 18.1). During the warm episodes, tropical conditions were far more extensive than they are now, and global temperatures were anywhere from 5° to 15°C (9° to 27°F) warmer than they currently are. For example, about 100 million years ago, corals were growing about 15 degrees poleward of their current locations, and dinosaurs roamed north of the Arctic Circle and in Antarctica as well. Much of this warmth can be attributed to elevated CO_2, which measured 1,120 ppm at the time. In addition, the oxygen concentration is thought to have been only 10 percent, which is about half of today's value. At this level significantly less radiation would be reflected to space, further contributing to the period's warmth. Most hot-house periods lasted about 100 million years, but, as seen in Figure 18.1, there was one very long warm period without glaciers that lasted for more than 1 billion years during Earth's middle age. Glacial or **ice-house climates** have been comparatively brief, lasting on the order of only tens of millions of years or less. We find ourselves today in just such an ice-house climate, and in fact, ice-house conditions have prevailed for the last 15 million years.

The most extreme ice-house climates are known as **snowball Earth** events, aptly named because nearly the entire planet is covered with ice during these times. One or two of these events might have occurred about 700 million years ago. Glacial deposits are found at tropical latitudes, and carbon isotopes suggest an almost complete shutdown of the biosphere. Nearly global ice coverage is consistent with both of these pieces of proxy evidence. The cause of such events is unknown, but ice-albedo feedback would have amplified whatever cooling agent was operating. Some calculations suggest that once ice came within 30 degrees of the equator, full ice conditions were inescapable as the planet plunged into runaway cooling.

During snowball conditions, the seas would have become depleted of oxygen, setting up conditions for banded iron rocks to form as photosynthesis increased, just as they did billions of years earlier. Such rocks did appear (see Fig. 18.1), which provides further evidence for the snowball hypothesis. How did this snowball event end? Excessive ice cover during this or any other snowball event slows the removal of carbon by weathering from the atmosphere, and so atmospheric CO_2 eventually rises high enough for Earth to recover by greenhouse warming.

THE MOST RECENT HOT-HOUSE AND THE PETM

Earth's climate was in a broad warm period from about 60 to 40 million years ago (Fig. 18.6), during geologic time divisions known as the Paleocene and Eocene (see the geologic timescale in Fig. 28.11 on p. 362). Notice that during the run-up to the maximum, there was a short spike in warmth about 55 million years ago. Known as the **Paleocene–Eocene Thermal Maximum (PETM)**, it was too brief to be clearly drawn in at the scale of Figure 18.6.

Looking at a smaller time slice of just 2 million years (Fig. 18.7), you can see the PETM was driven by an abrupt infusion of carbon to the atmosphere. With more CO_2 on average, the planet warmed by about 5°C (9°F), as did the deep ocean represented in the figure. Following the carbon pulse, the weathering feedback began its work, and both CO_2 values and temperatures recovered within about 200,000 years. The PETM is significant because it is the largest natural warming event known. In just 20,000 years, about 2,000 billion tons of carbon were added to the atmosphere. In geologic time frames, this was a planetary gusher equivalent to about half of the remaining fossil fuel reserves on Earth, and the climatic consequences were huge. However, it is less than half the input and hundreds of times slower than what we humans will do, assuming we burn the easily recoverable coal reserve over the next century or so. A century is but a blink of an eye compared to PETM timescales. (Note that even a 1,000-year period is almost invisible in Figure 18.7.)

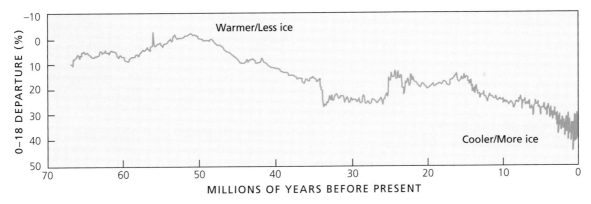

FIGURE 18.6 Climate of the last 70 million years, as indicated by oxygen isotope records. The curve reflects changes in both ice amount and temperature.

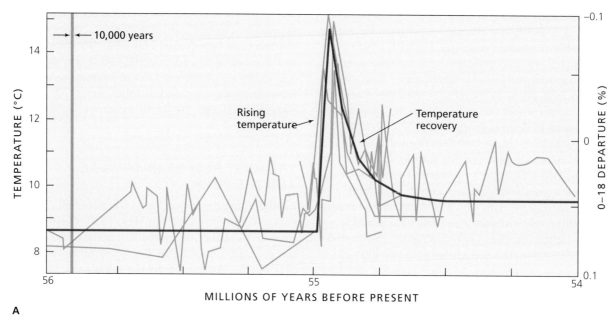

A

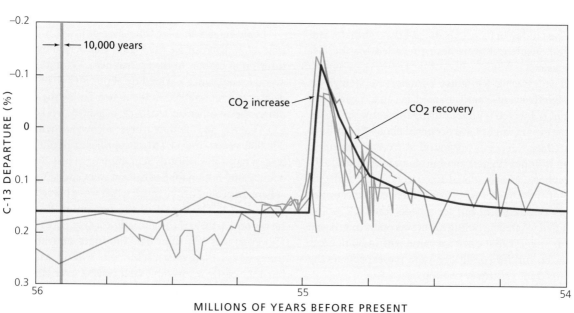

B

FIGURE 18.7 The Paleocene–Eocene Thermal Maximum. (A) Deep-sea temperatures based on oxygen isotope records from Atlantic and Pacific Ocean seabed cores. (B) Corresponding C-13 records suggest that this maximum was driven by the release of about 2,000 billion metric tons of carbon to the atmosphere.

TODAY'S ICE-HOUSE CLIMATE: GLACIAL/INTERGLACIAL CYCLES

The last hot-house period crested about 50 million years ago, and a slow, tottering shift to ice-house conditions followed. As Figure 18.6 illustrates, there is no definite starting point to this period, and in fact, different parts of Earth became covered with ice at different times. That said, by any

reckoning the planet has been in the depths of an ice-house climate for the past million years. Looking at the figure, notice how climatic variability has increased in the latter part of the record. These changes reflect repeated advances and retreats of land ice called **glacial/interglacial cycles**. The last six of these cycles are depicted in Figure 18.8 along with atmospheric methane and CO_2 changes.

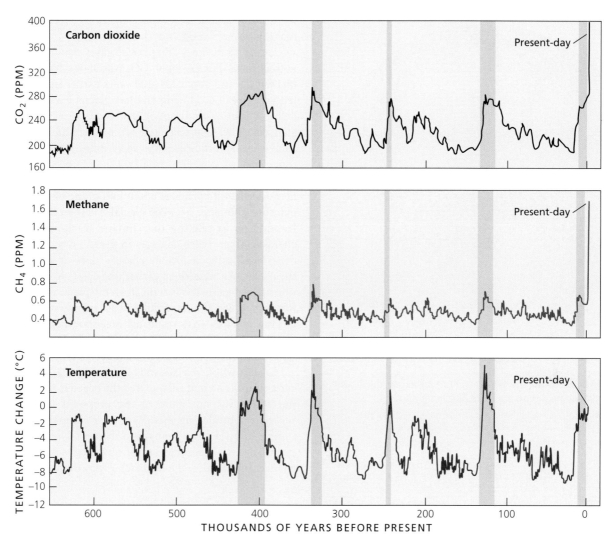

FIGURE 18.8 Ice-core records for the last 600,000-plus years. Shading shows periods of elevated temperature, carbon dioxide, and methane.

During glacial times, ice advances toward the equator from higher latitudes, with ice volume increasing by a factor of three or so in the form of ice sheets 3,500 to 4,000 m (2 to 2.5 mi) thick. Sea level falls by about 120 m (394 ft) as about 2.5 percent of the ocean evaporates and is stored on continents. For example, Figure 18.9 shows Northern Hemisphere ice about 20,000 years ago, during the so-called **last glacial maximum (LGM)**. Notice that ice sheets were much larger in the Western Hemisphere than they are now, and yet western Siberia was largely ice-free. "Glacial" means more ice, but it does not mean there are glaciers everywhere; not even all high-latitude locations are ice covered. Ice retreats toward the poles during the warm interglacial periods, with a spatial distribution similar to what we find today. Modern society developed and has existed entirely within an interglacial period, and our climate is only warm compared to the icehouse climate within which it is embedded.

Revisiting Figure 18.8, we can see clearly that glacial/interglacial cycles are not perfectly rhythmic waves. In fact, when curves such as the one shown in Figure 18.6 are analyzed statistically, several different cycles are found. Many records have cycles of about 100,000 years, as well as of about 40,000 years and 20,000 years. The 100,000-year cycle is visible in Figure 18.8 as interglacial peaks appearing at roughly that interval. The other cycles are much harder to discern visually, but they are present as well.

The underlying cause of glacial/interglacial cycles is variation in Earth's orbit. Our annual path around the Sun is prone to cyclical changes in shape, and there are also cyclical changes in the degree and direction of tilt of Earth's axis. The orbital cycles have periods of 100,000 years, 40,000 years, and 20,000 years, just like the climate changes. These so-called **Milankovitch cycles** lead to variations in the seasonal and spatial distribution of sunlight reaching the planet and thereby have the potential to influence climate. But what evidence suggests such variations do in fact affect climate? It turns out there is a strong match between the climate cycles and radiation cycles. The best agreement is found using the radiation curve for the northern midlatitudes. With so much land in that part of the

Generalized map of
glacial extent

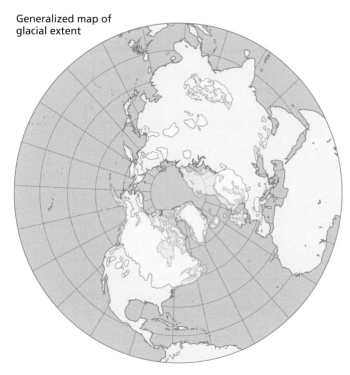

FIGURE 18.9 Northern Hemisphere ice extent during the last glacial maximum about 20,000 years ago.

world, the ice-albedo feedback operates effectively there, making the correlation not that surprising.

Figure 18.8 shows greenhouse gases rising and falling with climate. For example, CO_2 was about 280 ppm both in 1750 CE and about 120,000 years earlier in the last interglacial period. During the last glacial cycle, CO_2 was below 200 ppm for about 50,000 years. However, careful analysis shows that CO_2 lags behind the climate changes somewhat. This indicates that variations in CO_2 are not the root cause of, but can amplify, the climate change. The 80-ppm change (between 280 ppm and 200 ppm) just mentioned could produce about half of the observed temperature difference between glacial and interglacial times. In addition to making the climatic response to orbital changes larger, these changes in greenhouse gases make the orbital-induced climate change more uniform across the planet. As mentioned earlier, the cycles in solar radiation affect mostly the Northern Hemisphere midlatitudes where the ice-albedo feedback is strongest. However, mixing ensures that resulting greenhouse gas changes appear almost equally everywhere, and changes in the longwave radiation budget are felt everywhere.

Exactly why CO_2 rises and falls with glacial/interglacial cycles is unknown. It is clear that the ocean is the source of the variability, but none of the proposed mechanisms succeeds in producing changes of the right magnitude and timing, and the cause remains a mystery. Notice that the carbon cycle is once again exerting a feedback on climate change. In this case, the feedback is positive, which is just

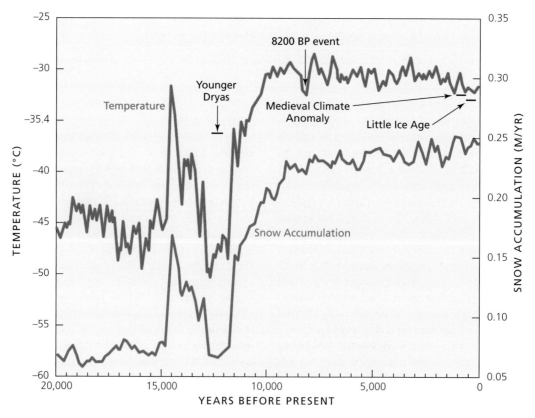

FIGURE 18.10 Temperature and snow accumulation for central Greenland.

the opposite of how the much slower CO_2–weathering thermostat works.

The LGM ended with warming that began about 15,000 years ago. After only 2,000 years, the warming was interrupted by a return to nearly full glacial conditions that lasted for 1,200 years (Fig. 18.10). Called the **Younger Dryas**, it is generally considered to have been caused by changes in North Atlantic Ocean circulation (an internal process). The Younger Dryas represented the last gasp of the glacial period and was followed by sharp warming into the current interglacial.

The Greenland record graphed in Figure 18.10 indicates that more snow accumulates on glaciers in warmer periods than in colder times. This is one example of how warmer air masses contain more water vapor and can support heavier precipitation. More important, note how rapidly conditions can change. For the Younger Dryas, the graph shows 5 to 10°C (9 to 18°F) changes in surface temperature over just a few decades leading up to and following the period.

Notice that we avoided the term *ice age* in our discussion despite its widespread use in the popular media. We've done so because there is no consistency in how the term is used by the media. Sometimes *ice age* refers to an ice-house climate; sometimes it refers to a glacial period; and sometimes it is used, usually capitalized, to mean the most recent glacial episode.

THE CURRENT INTERGLACIAL

The most recent interglacial is known in geologic terms as the **Holocene epoch**, and it extends from the present back about 11,700 years. Although warmer than the preceding glacial period, Holocene climate has hardly been constant. The Greenland record shows a cooling about 8,200 years before present (BP). The 8,200 BP event is reflected in other North Atlantic climate records and was quite possibly caused by a pulse of glacial meltwater from North America. Such additions can interfere with deepwater formation as described in Unit 11, and thereby affect ocean currents that bring warmth from the tropical Atlantic to Europe.

Figure 18.11 shows global climate for the last 2,000 years based on both proxy measures and direct measurements for the last 150 years or so. There is clearly no consensus among the curves, but the period 900–1200 CE does appear to have been somewhat warmer than the time immediately before and after it. This is called the **medieval climatic anomaly,** and it corresponds to somewhat warmer conditions in the North Atlantic area. Vikings used the warmth to their advantage and established settlements in Greenland that would be abandoned in the 1300s as colder climates returned and unbearable hardships ensued (Figure 18.12). Had the settlers been willing to adopt new cultural practices in the face of climate change (e.g., seal

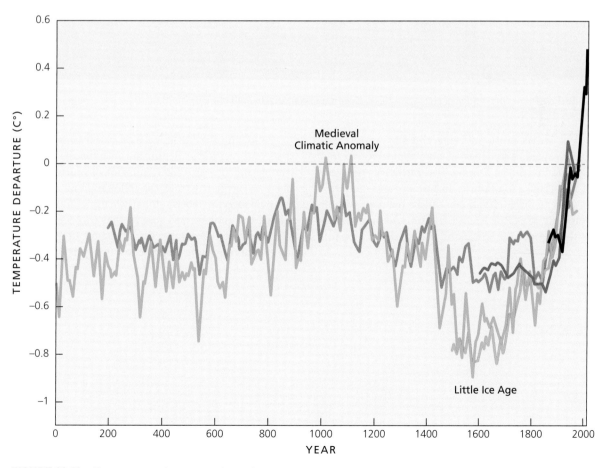

FIGURE 18.11 Temperature departures from the long-term mean for the last 2,000 years as estimated by various proxy measures (colored curves) and by thermometers (black) since 1854.

FIGURE 18.12 Ruins of a medieval Norse village in Greenland. Relatively mild climatic conditions meant traditional farming practices could succeed, at least initially.

hunting), it seems much suffering and starvation might have been avoided. A long period from about 1400 to 1850 was also cold in western Europe. This **Little Ice Age** was a time of numerous crop failures in Europe, and mountain glaciers grew to lower elevations. Unlike the medieval climatic anomaly, the Little Ice Age is expressed in mountain records throughout the world and is considered global in scope.

Following the Little Ice Age, a warming trend began that continues today. Beginning with the age of industrialization and the very likely imprint of human activity on Earth at the global scale, the following unit explores both the evidence for recent global warming and the theories related to humankind's role in that warming. As will be seen, the speed of changes likely to come in the near future far exceeds anything discussed in this unit.

Key Terms

Scan Here for a quick vocabulary review

Review Questions

1. Cite three examples from everyday life of natural or social phenomena that vary on different timescales.

2. Describe in general terms the climate changes that have occurred over Earth's history.

3. What drives glacial/interglacial cycles?

4. The level of atmospheric oxygen was a contributor to the warm period 100 million years ago. Should atmospheric oxygen be considered a greenhouse gas? Explain.

5. Name three things that could shift the CO_2–weathering thermostat to a colder planetary temperature.

6. Explain how ice-albedo feedback works.

7. Is a negative feedback one that leads to cooling? Explain.

8. Compare the Little Ice Age to the last glacial maximum.

9. What is the evidence for snowball Earth?

10. Is the CO_2–weathering feedback a solution for global warming resulting from fossil fuel use? Can we count on it to prevent negative impacts from human-induced warming? Explain.

Scan Here for a quick concept review

www.oup.com/us/mason

UNIT 19

Global Warming and Human Impacts on Climate

An oil slick from the 2010 Gulf of Mexico spill illustrates one reason coal, oil, and natural gas have been called "humanity's terrible treasure." The prospect of sweeping climate changes resulting from continued reliance on fossil fuels provides another reason for ambivalence.

OBJECTIVES

- Explain the mechanisms by which humans affect climate, especially greenhouse gas warming
- Describe observed global climate changes in the industrial period
- Discuss the evidence for humanity's role in global warming

- Describe general circulation models, their use in climate change science, and their limitations, and outline possible future twenty-first-century climates projected by GCMs

Modern societies affect climate on a continuum of spatial scales. At the local scale, on a hot summer day, a small urban pocket park provides cooling relief from the adjacent concrete jungle. Yet at the same time, the city as a whole is likely warmer than surrounding rural cropland. Connected urban areas such as those seen in Southern California or the Washington–New York–Boston corridor have larger regional climatic effects. At the largest scale, there is the prospect of humankind affecting *global* climate, which is the focus of this unit.

In particular, in this unit we examine what is usually called **global warming**, a general term that refers to an observed temperature increase over the past 150 years; the extent to which humans might be responsible for that warming; and scenarios for climate change in the next century or two. The term **anthropogenic climate change** is also used as a way of separating human-caused changes from those arising naturally (see Unit 18). As we will see, over the past decade or so, the evidence for human impacts on global climate has reached the point that scientists now speak of our planet having entered a new unit of geologic time. This new era is called the **anthropocene** (from *ánthrōpos*, meaning human in Greek), marking the time when humanity's impact on global ecosystems has become significant.

Human Impact on Climate: Processes

In the most general terms, there are three ways humans can potentially affect climate. Perhaps the most obvious is the direct addition of heat to the atmosphere. Residential and commercial heating in winter clearly releases energy, as does the burning of fuel to generate electricity. In addition, every powered device (hair dryers, air conditioners, etc.) and transportation vehicle (cars, ships, etc.) releases energy as heat to the surroundings. At present, worldwide energy production is a seemingly impressive 15 trillion watts. However, this translates to only 0.1 watt per square meter of Earth's surface area, or about 0.02 percent of the solar energy input. Thus, although energy inputs can be important locally, on a global scale we can rule this out as a mechanism for anthropogenic climate change.

By contrast, two other mechanisms have the potential to effect global change. The first is modification of land surface characteristics. For example, when forests are replaced by agricultural lands, the surface energy and moisture balances change. There are changes in albedo (and thus the amount of solar energy absorbed), leaf area (and thus evapotranspiration), and plant canopy architecture (affecting both longwave and shortwave radiation). Urbanization means impervious surfaces replace porous ground, thereby impacting surface energy and mass exchange. Urbanization, deforestation, and land-use changes in general also affect wind patterns in the planetary boundary layer, with corresponding impacts on vertical and horizontal heat and moisture transport.

The other mechanism for affecting climate is the most important of all: changing atmospheric composition. Because trace greenhouse gases such as carbon dioxide and methane are such a small percentage of total atmospheric gases, we can significantly affect their concentrations and emissions. In addition, as the ozone hole illustrates (Unit 6), humans can add gases that in turn affect other important atmospheric constituents. And we can also add atmospheric particulates that play important radiative roles and have indirect effects on cloud formation as well.

Notice that these last two categories of anthropogenic change (land use and atmospheric composition) do not involve significant energy input by human action. There may be energy released in the process, but the resulting climate change is not driven by that release of energy. In addition, notice that these two categories are not necessarily independent. For example, burning tropical forests changes surface properties and also supplies CO_2 and particulates to the atmosphere. Overgrazing in semiarid regions alters the surface energy balance and also increases the atmosphere's dust content as winds scour surfaces that were formerly more heavily vegetated. In thinking about how human activities affect climate, we need to be aware of joint and possibly interacting impacts from multiple actions.

An example of this is provided by the so-called **early anthropocene hypothesis**, which asserts that humans began modifying global climate 8,000 years ago. Ice-core data suggest that, compared to previous glacial cycles, carbon dioxide and methane began an abnormal increase thousands of years ago (Fig. 19.1), about the time agriculture began to expand. According to this hypothesis, deforestation released CO_2 through the combustion of woody material. Somewhat later, the growth of rice paddy agriculture is believed to have increased methane concentration through the action of single-celled organisms living in submerged fields. The implication is that if these greenhouse gas increases had not occurred, global climate would have cooled soon after the most recent warming, following the same timetable as the previous interglacials. That is, the planet would have begun a slide toward glacial conditions about 2,000 years ago had humans not intervened.

Evidence for the early advent of the anthropocene period has strengthened in the last few years as estimates for prehistoric land clearing have been revised upward and computer-modeling studies have provided increasing support. That said, the hypothesis remains controversial. Regardless of how the question is resolved, it serves to demonstrate the feasibility of climate modification through anthropogenic inputs of greenhouse gases. More recent anthropogenic inputs during the industrial period are unquestioned, are far larger, and, as we shall see, have correspondingly greater potential for global change.

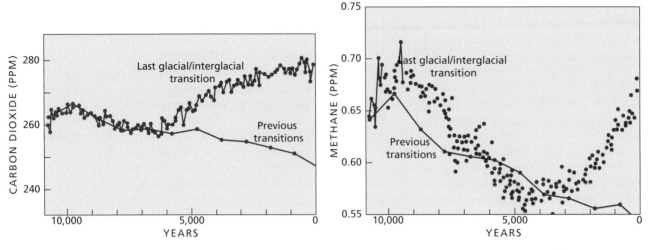

FIGURE 19.1 The presence of carbon dioxide and methane in the atmosphere has increased in the last few thousand years (dots). In contrast, a composite constructed by averaging data over the last six inter-glacials shows falling values (solid line).

Global Warming Observed: Temperature Change in the Industrial Period

Routine observation of air temperature and precipitation goes back to the early 1800s, with the establishment of fledgling weather networks. However, until the twentieth century, there were far too few weather stations for reliable computation of a global average of these important climatic elements. Even now, despite many thousands of stations on land, countless shipboard observations, and sophisticated satellite techniques, global average precipitation isn't known to within 10 percent of the true value. Past values of precipitation are even more uncertain and suggest no consistent trend over the industrial period.

We therefore focus on air temperature, whose variation in time and space is much better constrained by direct observation and for which there is unambiguous evidence of an overall change. As Figure 19.2 shows, temperatures have increased by about 1°C (1.8°F) since the early twentieth century. The increase has not been uniform, and indeed, global warming was interrupted from about 1945 until the mid-1970s. Compared to the first half of the twentieth century, warming has been particularly rapid in the last few decades.

Figure 19.2 contains data collected through the end of 2014, which was the warmest year of record. On average the last 15 years have been warmer than any other comparable period, and in fact, except for 1998, the warmest years on record have occurred since 2000. But even over

that time span, temperature has not increased uniformly, and for a few years in succession there appeared to be very little warming. Indeed, by the end of the decade in 2010, there was talk of a **global warming pause** in the popular press and a **warming hiatus** in scientific circles. The term *hiatus* seemed apt, because strong warming in the decades prior to 1998 (the fifth-warmest year on record) had apparently been interrupted. With renewed warming in the last few years, and improved temperature reconstructions, we now know the hiatus was not real. In actuality, the rate of warming in the twenty-first century has been at least as great as that of the last half of the twentieth century and possibly larger.

Can the temperature data behind Figure 19.2 be trusted? In short, yes. First, there is little uncertainty in global averages for the last 60 years; the error for individual years is only about 0.05°C (0.09°F). This uncertainty is more than a factor of 10 *smaller* than the overall change. Second, certain satellite radiation measurements can be used to estimate tropospheric temperatures, and these indicate that warming has been occurring since the record was begun in 1975. These measurements provide independent support for the latter part of the thermometer record in Figure 19.2.

In addition, many other climate indicators have been observed that are consistent with a rise in global mean temperature. For example, from 1901 to 2002, the extent of seasonally frozen ground decreased by 7 percent in the Northern Hemisphere. The freeze-up date for ice in rivers and lakes has occurred later in the year by an average of 5.8 days per century, and the subsequent ice melt and

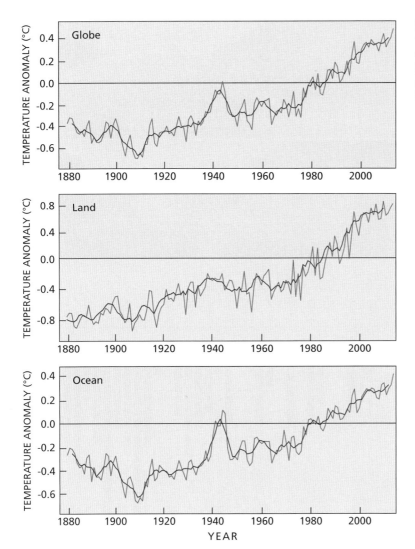

FIGURE 19.2 Temperature trends from 1880–2015, as reconstructed by NOAA. The values plotted are departures from average (temperature anomalies). In this case the averaging period is 1971–2000, which is also the period used for the current climatic normal.

breakup has moved earlier by an average of 6.5 days per century. Similarly, over much of the midlatitudes, the number of days with frost has decreased. In recent decades the area and thickness of Arctic Sea ice has decreased significantly, as one would expect with a warming climate. For example, between 1979 and 2014, September sea ice decreased from about 8 million km² to about 5 million km², and some predict an effectively ice-free summer within a decade.

Worldwide, there has been a decrease in the number of extreme cold events and an increase in the number of extreme warm events. Perhaps most dramatically, many mountain glaciers have shrunk by strikingly large amounts (Fig. 19.3). Globally, glaciers have had a negative mass balance every year since 1993, meaning they have lost more water to evaporation and melting than they have gained from precipitation. Over the last three decades, the average glacier has lost the equivalent of 15 m (49 ft) of ice depth.

As mentioned earlier, global warming doesn't mean that every year is warmer than the year before. In addition, global warming does not mean that every place experiences the same amount or even direction of change in a given year. For example, if we consider the map for 1998 (Fig. 19.4A), we see that many oceanic locations in that year were a little cooler than average, and many were very near average. Most of the warming was in the Northern Hemisphere, with North America being particularly warm. Note also that the pattern in 2010 (Fig. 19.4B) is different than the 1998 pattern. This difference indicates that warm years are by no means identical. (If you use the animation accompanying Figure 19.4, you can see that the same holds for cold years—no two are alike.)

Finally, global warming does not mean that every place has warmed during the industrial period. Consider Figure 19.5, which maps the trend from 1979 to 2005 as the rate of temperature change per decade. A place with a mapped value of 1°C per decade warmed by an average of 0.1°C per year over the period. We see large areas of the Southern Ocean that warmed hardly at all. The same is true for the eastern Pacific in both hemispheres. On the other hand, very little land area escaped warming, except perhaps for Antarctica, where data are insufficient to determine a pattern one way or the other.

A

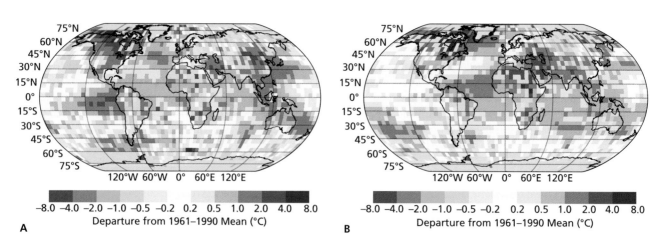

FIGURE 19.3 Mountain glaciers have long been used as a sensitive indicator of global warming. The Vadret da Morteratsch in the Swiss Alps retreated by about 400 m (1,310 ft) between 1985 and 2007 (A). Worldwide, the cumulative effect of consistent water loss amounts to about 15 m (49 ft) of ice thickness (B).

B

A

Departure from 1961–1990 Mean (°C)

B

Departure from 1961–1990 Mean (°C)

FIGURE 19.4 Temperature anomalies for 1998 (A) and 2010 (B). Areas with insufficient data appear in gray.

ENHANCE

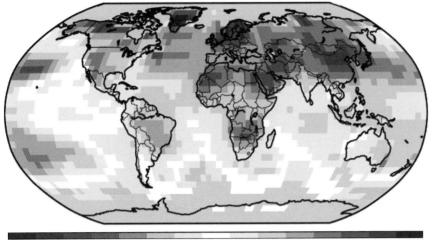

FIGURE 19.5 Rate of temperature change from 1979–2005 in degrees Celsius per decade. Areas with insufficient data appear in gray.

-0.75 -0.65 -0.55 -0.45 -0.35 -0.25 -0.15 -0.05 0.05 0.15 0.25 0.35 0.45 0.55 0.65 0.75
°C per Decade

FOCUS ON THE SCIENCE

Weather Extremes and Climate Change

The first six months of 2012 were unusual for U.S. climate. The worst drought in 60 years gripped many of the lower 48 states, and at times more than 60 percent of the area was in a state of moderate to extremely severe drought. In many locales, corn crops were declared a complete loss, and foliage was mowed down to be sold as a substitute for hay, which refused to grow under the parched conditions. Extreme heat accompanied the drought in late spring, setting thousands of records for high temperature and disrupting numerous human activities. The first half of the year was the warmest on record for 28 states, and for another 10, it was the second warmest.

Unusual atmospheric occurrences such as these nearly always lead to questions about climatic change, particularly anthropogenic change. This is true for individual events like a single strong hurricane or major flood and also for summary measures like average temperature and annual tornado counts. Some people are quick to blame anthropogenic change for such occurrences. They might claim that a map pattern similar to that depicted in Figure 19.6 could never arise

FIGURE 19.6 Numerous heat waves hit the continental United States in the first half of 2012, setting records throughout the eastern half of the country. The map ranks statewide average temperatures based on 118 years of data.

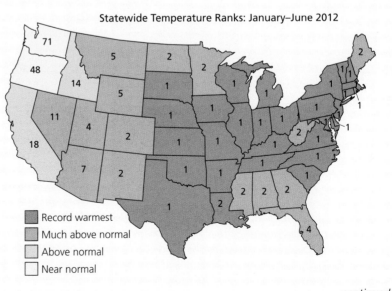

Statewide Temperature Ranks: January–June 2012

- Record warmest
- Much above normal
- Above normal
- Near normal

continued

continued

naturally, and therefore humans must be responsible. Other people shrug such things off with the adage "records are meant to be broken," or point to a similarly unusual event in the past as proof that climate change can't be responsible for whatever the latest incident might be. For example, knowing the Dust Bowl years of 1930s were even drier than the present, they wonder how anybody could think climate change is connected to the 2012 drought. With just a little thought, it's easy to see that neither viewpoint can be supported.

By definition, climate is a statistical concept that implicitly acknowledges variability from year to year. For any climate—in the present, future, or distant past—we would never expect the same number of hurricanes from one year to the next, the same intensity for the strongest hurricane, or the same annual precipitation total. Every climate variable exhibits interannual variation (i.e., variation from one year to the next), and thus, for every variable, a range of values is possible, with some values more likely than others.

Figure 19.7 illustrates this for temperature. The vertical axis in the graphs represents the frequency, or probability, of occurrence in any given year. When the distribution is symmetric around the most likely value, the middle value is also the average. Usually, we define **extreme events** as temperatures above or below some threshold, such as 90°F for extreme heat or –20°F for cold snaps. These thresholds provide a basis of comparison for different climates. For instance, we can ask how two places compare in the frequency of extreme heat or compare climates at different times.

Viewed in this way, climate change can manifest itself in several ways. For example, temperature can possibly become more variable with no change in average conditions (Fig. 19.7A). In this case, extreme cold and extreme heat are both more common than before, but near-average conditions decrease in occurrence. Alternatively, there can be a change in the mean with no change in variability. For example, if the average temperature increases, as in Figure 19.7B, the frequency of hot events increases while cold events become less common. Or both types of change could occur (Fig. 19.7C), which in this example results in many more instances of extreme heat and a small increase in cold events.

These diagrams reveal both the logic and the defect in the skeptic's viewpoint. The skeptic is correct in saying that nothing "new" has occurred since the climate changed, and thus there is no way to associate a single event with climate change. However, the probability of occurrence has changed, and the increase or decrease in the frequency of extreme events is certainly a result of climate change. It is a

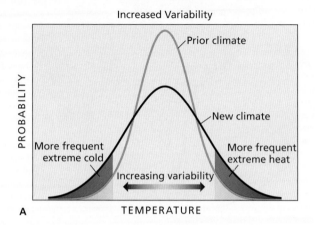

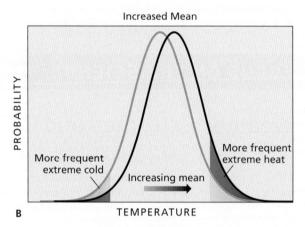

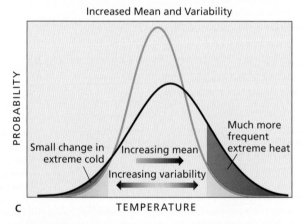

FIGURE 19.7 Schematic plots for hypothetical changes in temperature. Climate change alters the occurrence of extreme events in various ways depending on how the probability distribution changes.

little like trading a biased coin for a fair coin. The only possible outcomes are still heads and tails, but the proportion is no longer 50-50; therefore, a long run of heads or tails becomes likelier.

This same reasoning applies to climate variables such as heavy rains, the number of thunderstorms with large hail, hurricane counts, or any other measure. The general principle is that large changes in the frequency of extreme events can occur with small changes in the mean. Note that, in many cases, extreme values have much greater ecological significance than the average values. For example, the power delivered by wind is proportional to the cube of wind speed, and so the damage caused by extremes is more than would be expected from a change in the mean. Many biological and physical processes are nonlinear in temperature, and there can be thresholds across which certain processes are activated or deactivated. In a similar vein, the health consequences of heat waves can be disastrous. For example, the 2010 Russian heat wave is believed to have killed 10,935 people in Moscow alone. The summer of 2003 was the hottest in Europe since records began in 1540 and is estimated to have resulted in 70,000 deaths. That said, we should remember that this principle about the importance of extremes certainly does not apply to every Earth system. Most of the work done by streams, for example, is accomplished by common flows (annual or semiannual), not the unusual flood that appears every century or so.

Detecting changes in extremes is challenging from an observational standpoint for the simple reason that they are unusual. Because they don't appear often, longer records are needed for counts to be high enough to yield a good estimate of their frequency. Another complicating issue is that extreme events often have multiple causes. The 2003 European heat wave, for instance, was partly related to drought. Drier than normal soil caused more solar energy to be expended in sensible heat rather than in evapotranspiration. A different heat wave might result almost entirely from changes in atmospheric motion, with no significant contribution from surface conditions. Treating these as the same event could be misleading, especially in a climate change context, where it would be unreasonable to expect the same mix of causal factors.

Because they can be run indefinitely (i.e., for many years) with constant forcing values (see p. 251), general circulation models don't suffer from the same observational problems, and not surprisingly, they have been widely used to assess the role humans have played in extreme events. The frequency distribution for any computed variable can be "observed" in the model output and examined for changes as forcing variables are added and removed. In addition, model output can be analyzed to determine what internal processes are responsible for changes in extremes. For the 2003 heat wave, general circulation model experiments suggest that anthropogenic factors significantly increased the probability of occurrence. According to the models, were it not for human impacts, such an event would occur very rarely, perhaps only once every 10,000 years.

Implicating Humans in Observed Global Warming

In 2013, an authoritative report from the Intergovernmental Panel on Climate Change (IPCC) stated that most of the observed global warming since the middle of the twentieth century is virtually certain to be the result of elevated greenhouse gases resulting from human activities. The IPCC is a standing panel commissioned by the World Meteorological Organization charged with the task of periodically summarizing scientific knowledge related to climate change and its impacts. It conducts no original research but instead evaluates and collates information from previously published studies. The 2013 publication was the fifth in a series of reports that began in 1990. It was authored by hundreds of experts in climate science from around the world and reviewed by many hundreds more.

Like the previous reports, the intent of the 2013 IPCC report was to capture the scientific consensus of what is known and unknown about climate change. So, as was true for preceding publications, the 2013 report used carefully constructed terms to express uncertainty. For example, the term *virtually certain* means at least 99 percent probable. Thus, according to the IPCC, there is no more than a 1 percent chance that the claim about the role of humans in global warming is wrong. Notice the IPCC did not say warming was virtually certain to have occurred. There is no doubt about that; the 1 percent uncertainty refers to the cause of warming, and the IPCC asserts that human activities are almost certainly the agent. Other organizations and individuals have made similar claims about anthropogenic warming over the years, but IPCC statements such as this are particularly important and influential for policymakers.

On what basis can the IPCC say anthropogenic greenhouse gases are almost certainly behind the bulk of global warming? To be convincing, one would need to know greenhouse gases have increased, human activities are responsible for those increases, and the observed warming could plausibly result from those increases.

THE HUMAN ROLE IN ELEVATED GREENHOUSE GASES

The eighteenth and nineteenth centuries were times of extensive agricultural expansion and land clearing in Europe and the Americas. As carbon formerly held in

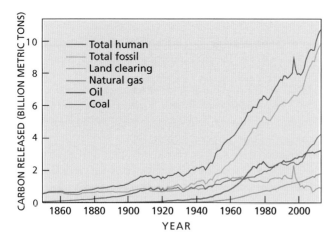

FIGURE 19.8 Anthropogenic carbon emissions.

vegetation was oxidized, it constituted a significant anthropogenic release of carbon to the atmosphere. But by the 1920s, the biospheric source was exceeded by fossil carbon releases from burning coal, oil, natural gas, and (to a lesser extent) cement manufacturing (Fig. 19.8). Land clearing remains important, but fossil carbon emissions have increased almost exponentially over the past century, growing a few percentage points per year on average. (Periods of world war and the oil embargo of the 1970s are notable exceptions.)

To put this growth in perspective, consider that the current fossil fuel release rate is about 100 times larger than the rate of natural outgassing by volcanoes and deep-sea vents. As we discussed in Unit 6, atmospheric CO_2 grew from about 280 ppm in pre-industrial times to 400 ppm at present and showed a similar roughly exponential increase (see Fig. 6.7). The carbon isotopes C-13 and C-14 confirm that this agreement is no mere coincidence. Fossil carbon is depleted in C-13 and has no C-14. As a result, fossil fuel–generated inputs to the atmosphere leave a characteristic isotopic signature that can be traced to the source. For this reason, we know most of the historic atmospheric CO_2 increase came from fossil fuels.

This is not to say that all of the anthropogenic carbon released has remained in the atmosphere, and in fact, the growth in atmospheric carbon is only half as large as would be expected considering the emissions alone. So the question is not why has atmospheric CO_2 increased (an outcome for which humans are responsible), but rather, why hasn't it increased more? It must be that about half of the fossil fuel carbon released to the atmosphere is taken up by the oceans or the continental areas or both. Current thinking estimates that about one-third is taken up by oceans and two-thirds by continents, but, as we discuss later in the book, this question of the "missing carbon" remains a matter of intense research. Other greenhouse gases have also increased with industrialization, including methane, nitrous oxide, and ozone-destroying chlorofluorocarbons (CFCs) (Fig. 19.9). Notice that CFC concentrations have stabilized

and started to decrease following worldwide agreements to limit their use.

TEMPERATURE RESPONSE TO GREENHOUSE GAS EMISSIONS

Human activity has led to elevated levels of greenhouse gases, but why should we believe increasing greenhouse gases are behind the observed warming? After all, climate varies continually for purely natural reasons, and perhaps the observed warming is nothing more than that. The strongest evidence for greenhouse warming comes from atmosphere–ocean **general circulation models** (GCMs).

General circulation models are programs running on very powerful computers that simulate the Earth system in all three dimensions. They use the conservation principles discussed in Unit 1 and other physical laws to calculate the motions and state (temperature, pressure, etc.) of the atmosphere and oceans for the entire globe. In a typical run, the model is supplied with external factors known as **forcings**, such as Earth's topography, solar radiation, and atmospheric composition. Some forcings are constant for a model run, whereas others are variable (e.g., CO_2 concentration, volcanic aerosol load). The computer model calculates the state of the atmosphere, ocean, and land surfaces forward in time for the duration of the model run. The response of climate to individual agents of climate change can be examined by making multiple model runs with various forcings included and excluded.

Because of the huge computational resources needed, many compromises are made in GCMs. They have limited spatial detail and greatly simplify physical and biological processes. For example, although GCMs calculate soil water gains and losses, vegetation is only crudely represented, and the vertical movement of water within the soil column is often completely ignored. More than 30 GCMs are in use at various research institutes and universities throughout the world, and each contains different approximations and therefore differs from the others in many important ways. Because of this, they generate different climates when supplied with the same external factors. However, they all share one very important trait in that they attempt to represent physical processes with as much realism as possible. That is, they do not rely on extrapolation from the past, nor do they use purely statistical relations to estimate the climatic response to some forcing.

The IPCC used output from many GCM runs to assess the importance of greenhouse gas emissions in the observed warming. Figure 19.10 shows model results obtained from dozens of models with and without anthropogenic forcings. In the first case (Fig. 19.10A), the models included natural forcings (e.g., variations in solar radiation and volcanic aerosols) and anthropogenic influences. The latter included greenhouse gas emissions as well as other human agents such as smokestack aerosols and land-use change. We see that the model average tracks observed temperature changes quite well. By contrast, if anthropogenic factors

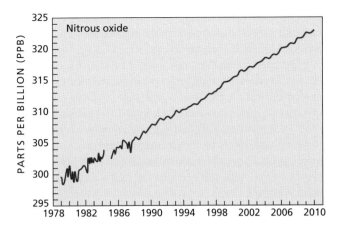

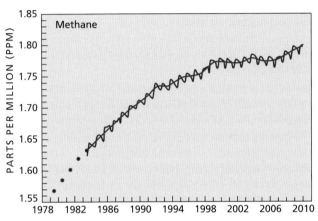

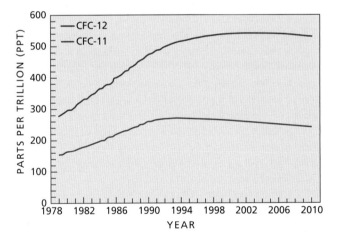

FIGURE 19.9 Recent trends in methane, nitrous oxide, and chlorofluorocarbons (CFCs). Note the different units for each gas.

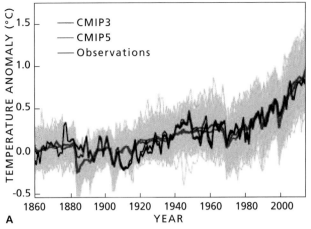

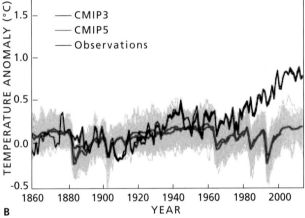

FIGURE 19.10 Global mean temperature as observed (black curves) and simulated by a suite of GCMs used in the 2013 IPCC report. In (A), both natural and anthropogenic forcings are included, whereas in (B), only natural factors are present.

greenhouse gases in the computed warming and have found similarly consistent evidence for a fossil carbon fingerprint in other aspects of twentieth-century climate change. As we discussed in Unit 1, nothing is certain in science, and that is certainly true with regard to human influence on past climates. However, the strong consensus among experts leaves little room for doubt that our species has already caused detectable global change. The prospect for much larger changes in the future is also likely, as we detail in the next section.

Future Human Impacts on Climate

Public debate over future global warming is seemingly everywhere and highly politicized, with voices that range from denial to assertions of an impending (but preventable) global catastrophe. Our goal in this section is not to examine that dialogue or the motives of participants. Instead, we aim to help readers achieve a science-based perspective on this question of enormous ecological and social importance. The challenge is to determine how much atmospheric greenhouse

are excluded (Fig. 19.10B), the rapid warming of recent decades disappears, and the model climates cool. These models cannot reproduce the observed temperature change without including anthropogenic greenhouse gases, and in fact, they suggest that Earth would have cooled somewhat were it not for the human influence on global climate.

This is but one of the ways that GCMs are used to evaluate the role of greenhouse emissions. Many other far more sophisticated analyses have confirmed the dominance of gases might increase as a result of continued fossil fuel use and the possible climatic response to those increases.

FUTURE GREENHOUSE GAS EMISSIONS

With respect to fossil carbon releases, there is no doubt about the potential for elevated greenhouse gas concentrations. The amount of fossil fuels remaining to be harvested is not known exactly, as it depends not only on the resources in the ground but also on what technologies are available for future extraction. A common estimate is that about 5,000 billion metric tons (gigatons, or GT) of carbon remain, mostly in the form of coal. If this were released instantaneously and remained in the atmosphere, atmospheric CO_2 would grow by a factor of six. That will not happen, of course. Society will withdraw carbon at a variable rate over time in response to energy demand, costs relative to nonfossil sources, wars, and other unknowable factors.

Releases are only part of the story; we also need to make assumptions about the fate of carbon added to the atmosphere. As mentioned previously, at present the carbon cycle acts as a buffer for anthropogenic carbon and in effect reduces the impact of human perturbations by a factor of two. In the face of such uncertainties, the best one can do is construct scenarios like those depicted in Figure 19.11. These so-called **representative concentration pathways (RCPs)** were prepared for use in GCMs using assumptions about population growth, industrialization rates for developing countries, per capita energy demand, land-use changes, and so forth. An RCP reflects multiple greenhouse gases as well as non-greenhouse factors. For comparison purposes, an RCP can be expressed as a curve of equivalent CO_2 concentration, as in the left axis of Figure 19.11, or as the equivalent increase in net radiation at the top of the troposphere (right axis). The RCPs are standardized and used by modelers throughout the world. Each is named according to the approximate peak increase in downward radiation. In Figure 19.11, there are

peaks of approximately 8.5 W/m², 4.5 W/m², and 2.6 W/m², and the corresponding pathways are known as RCP8.5, RCP4.5, and RCP2.6, respectively.

The three scenarios represented by the three colored lines in Figure 19.11 represent very different futures. At one extreme there is RCP2.6, a low-use scenario in which greenhouse gas emissions peak later this decade and are reduced thereafter. This assumes an intentional transition to clean and efficient energy sources and unprecedented international cooperation working toward environmental stability. The high-use scenario is RCP8.5, which represents continued rapid growth in fossil fuel use and no attempts to limit greenhouse gas emissions. In other words, RCP8.5 is essentially business as usual. The resulting CO_2 values would be huge and probably unmatched on Earth in the past 50 million years.

GCM RESPONSE TO ANTHROPOGENIC EMISSIONS

General circulation models have been widely used to assess the climate effects of elevated greenhouse gases and are undoubtedly the best tool for this task. You might think that they aren't necessary, and that one could simply compare past CO_2 excursions and the corresponding climates to discover the relationship between the two. However, as we noted earlier, ancient climates are unfortunately known only in general terms, and the present-day anthropogenic pulse of carbon is vastly faster than anything nature could produce. As a result, we can't use past conditions to reliably predict the effect of human-induced changes in greenhouses. General circulation models have another advantage in that their workings can be examined to identify the physical mechanisms behind changes in computed climates. Using proxy data alone, we might determine that some place was drier or wetter in CO_2-elevated times, but it would be very hard say how that change arose.

The 2013 IPCC report compared a suite of models running under a variety of anthropogenic forcings. Table 19.1 gives the average global temperature change at various times for the low-, medium-, and high-use RCPs shown in Figure 19.11, and Figure 19.12 shows yearly changes through the end of this century.

These results suggest we could expect global mean temperatures to rise by more than 1°C (1.8°F) by mid-century

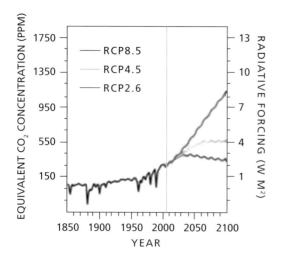

FIGURE 19.11 Three IPCC representative concentration pathways (RCPs) expressed as an equivalent concentration of carbon dioxide (left axis) and in terms of the corresponding change in downward shortwave and longwave radiation at the top of the troposphere.

TABLE 19.1 Projected Average Global Temperature Change

	2046–2065		2081–2100	
Scenario	Mean	Likely Range	Mean	Likely Range
RPC2.6	1.0°C	0.4°C to 1.6°C	1.0°C (5.63°F)	0.3°C to 1.7°C
RPC4.5	1.4°C	0.9°C to 2.0°C	1.8°C (4.77°F)	1.1°C to 2.6°C
RPC8.5	2.0°C	1.4°C to 2.6°C	3.7°C (3.22°F)	2.6°C to 4.8°C

Average global temperature change computed by a group of GCMs used in the IPCC report, expressed as a departure from the 1986–2005 average

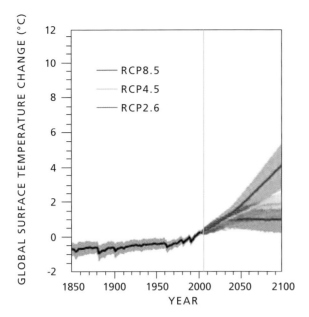

FIGURE 19.12 Global mean temperature change computed by GCMs for three emission scenarios. Shaded areas depict the range over the suite of models used. The 1980–1999 average is used as the reference temperature.

regardless of which scenario is followed. By the year 2100, the scenarios diverge significantly, and the high-use RCP scenario (RCP8.5) produces a warming of close to 4°C (7.2°F). The implication is that actions taken to control greenhouse emissions can have little effect in the next few decades, but climates in the late twenty-first century (and beyond) would be strongly impacted. Differences among the models (seen in the width of shaded areas) grow larger over time, as would be expected, given that they all have the same starting point.

As would be expected from observed changes in temperature, the geographic expression of *computed* warming is not uniform; warming is not consistent across the globe. As we can see in Figure 19.13, in all scenarios higher latitudes of the Northern Hemisphere, especially continental locations, warm the most. The models also suggest strong warming in the interiors of North and South America, northwestern Africa, and western Australia. The least amount of warming is seen for oceans outside the tropics. In viewing these maps, pay particular attention to the shading patterns, which indicate the degree of certainty. The dot pattern indicates places of high certainty, where the models agree and the change stands out sharply from year-to-year variability. Slanted lines indicate places where the change cannot be confidently separated from yearly variation.

As the GCM atmospheres warm, the moisture content increases because more energy is available for evaporation and the saturation point of air is higher when it is warmer. On a global basis, precipitation rises, as would be expected with increasing atmospheric moisture, but the precipitation change is anything but globally uniform (see Fig. 19.13, right column). In most models, precipitation increases in the tropics and middle to high latitudes, but the already dry

subtropics become even drier. Comparing these maps to the temperature warm change maps in Figure 19.7, we see there is considerably more uncertainty about precipitation change, especially for RCP2.6.

As you know from Unit 11, precipitation only partly determines the moisture availability at a location. We must also consider evapotranspiration (evaporation plus transpiration, which we abbreviate as ET). A locale might receive more rainfall in the future, but if ET increases more, that location becomes drier. So, in addition to precipitation (P) and ET alone, we are interested in the difference ET minus P. Figure 19.14 shows projected changes in this value, along with the resulting change in soil moisture. In the maps, we see that evapotranspiration increases for most of the planet, and that the increase in evapotranspiration more than offsets any precipitation increase throughout much of the subtropics and lower midlatitudes. In other words, the drying suggested by the precipitation pattern is amplified when evapotranspiration is considered. Not surprisingly, soil moisture declines in most areas (see Fig. 19.14, bottom).

The comparisons we just presented are simple differences between projected climates and the present. More meaningfully, one might want to consider how a given change compares to interannual variability. Recall that interannual variations are the changes that occur from one year to the next. If the mean change is large relative to differences that occur naturally from year to year, it is arguably of greater ecological significance because it is outside the experience of organisms found there. This approach measures the dissimilarity between current and projected climates after standardizing for natural variability, and it provides an alternative to differences in average values. This approach also allows for multiple variables to be considered in assessing how much a climate changes.

Figure 19.15A shows a map based on eight GCMs used in the fourth IPCC report, which was published in 2010. All models were run using a high-use CO_2 scenario. The maps illustrate how different future climates might be from today based on projected changes in seasonal precipitation and temperature. The maps show a standardized measure of difference, and high values, especially those above three or four, correspond to places where projected late twenty-first-century climates are projected to be very different from today's.

In striking contrast to what we saw for temperature, in 2080–2099 large areas of the tropics are projected to have the greatest dissimilarity from 1980–1999 climates. Although absolute changes for the tropics tend to be smaller than those for middle and high latitudes, interannual variation at low latitudes is much less, which means that the standardized changes are larger. As we can see in Figure 19.15B, at some locations the present-day climate is so dissimilar from climates found anywhere in the late twenty-first century that their climate is said to "disappear." Collectively, these are known as **disappearing climates**. For this scenario, the land area with disappearing

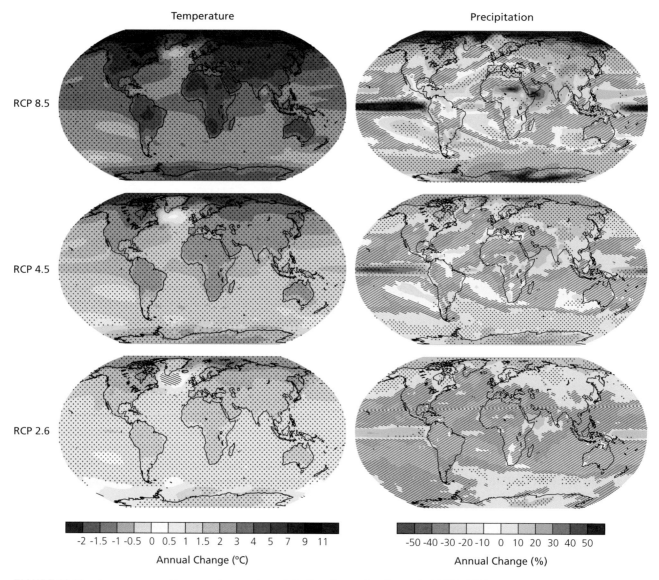

Temperature   Precipitation

RCP 8.5

RCP 4.5

RCP 2.6

-2 -1.5 -1 -0.5 0 0.5 1 1.5 2 3 4 5 7 9 11

Annual Change (°C)

-50 -40 -30 -20 -10 0 10 20 30 40 50

Annual Change (%)

FIGURE 19.13 Average annual temperature and precipitation change for the period 2081–2100, computed by GCMs for three RCPs. Changes are expressed as departures from the 1986–2005 mean.

climates ranges from 10 to 48 percent of the total land area, depending on the model. For a low-use scenario (not shown in the figure), the study found a range of 4 to 20 percent. The concern is that natural vegetation in disappearing climates will be particularly ill suited to new conditions, because new climates will be so unlike the climates to which they are now adapted.

It is important to realize that evaluations of potential climate change (where, how much, etc.) depend very heavily on what measures we choose to employ. Simply tabulating the expected difference in one or two variables may not be sufficient. In addition, the importance of a climate change is not limited to changes in its average value. Changes in the frequency of extreme events must also be considered. Such events are unusual or infrequent, and they are not necessarily linked to changes in the mean (see Focus on the Science: Weather Extremes and Climate Change). In addition, changes in persistence, such as longer

runs of hot days or abnormally wet or dry periods, are also important. Table 19.2 lists some of the climate changes discussed in the IPCC report and possible impacts on human and natural systems.

EVALUATING GCM PROJECTIONS

What should we make of these projections? Can the GCMs be trusted to give accurate pictures of what might come with continued fossil fuel use? These are justifiable questions. So much of the expected climatic response to elevated greenhouse gases is indirect, and GCMs involve so many assumptions and approximations. But we do know that the immediate (direct) effect of a CO_2 doubling is to add about 4 W/m^2 to the planetary energy budget, or about 1.6 percent of absorbed solar radiation. The resulting drift toward global warming is amplified by feedbacks, the most important of which are related to water vapor and surface albedo.

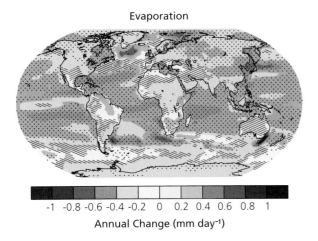

Evaporation

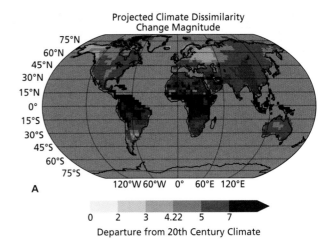

Projected Climate Dissimilarity
Change Magnitude

A

0   2   3 4.22 5   7

Departure from 20th Century Climate

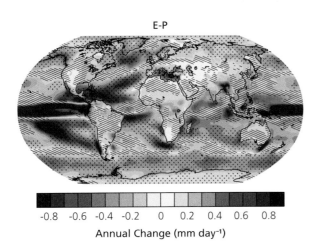

E-P

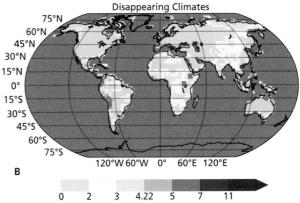

Disappearing Climates

B

0   2   3 4.22 5   7   11

Departure from Most Similar 20th Century Climate (Any Location)

FIGURE 19.15 Average dissimilarity between 1980–1999 and 2080–2099 climates as projected by eight GCMs under a high-use CO_2 scenario. (A) depicts dissimilarity for individual locations and thus represents the change projected to occur for each grid cell. In (B), the future world is examined for the climate most similar to the present. Yellow through red colors depict disappearing climates projected to no longer occur anywhere in the future.

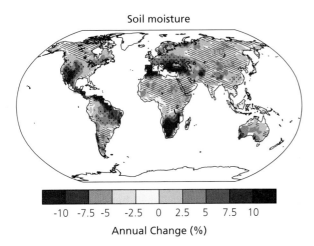

Soil moisture

-10 -7.5 -5 -2.5 0 2.5 5 7.5 10

Annual Change (%)

FIGURE 19.14 Projected annual changes in evapotranspiration (top), the difference between evapotranspiration and precipitation (middle), and soil moisture (bottom) for 2081–2100. All maps correspond to the high-use concentration pathway RCP8.5.

In the case of water vapor, as mentioned earlier, the warmer atmosphere has a higher saturation point, and therefore the specific humidity increases. This is similar to what we observe now, where the tropical atmosphere is much moister than that of the higher latitudes. Signs of this

are also seen in an observed moistening over the last few decades that has accompanied the rise in global temperature. Water vapor is a powerful greenhouse gas, and therefore the increase in moisture contributes to anthropogenic warming in a positive feedback. There is another, more complicated **water vapor feedback** that acts in the opposite direction. As the atmosphere warms, the saturated adiabatic lapse rate decreases (see Unit 7). Thus, everything else being equal, because of this, a saturated parcel at some elevation high in the troposphere is warmer and emits more radiation to space. This emission acts contrary to warming, and hence, it is a negative feedback. But the positive water vapor feedback is larger, so the net effect is that increasing water vapor warms the planet. The best estimate is that fully half of the global temperature response to a CO_2 doubling comes from water vapor feedback.

We discussed ice-albedo feedback in Units 5 and 7 and earlier in this unit. Recall that it is also a positive feedback and that it amplifies warming. Obviously, calculating its

TABLE 19.2 Climate Change and Its Consequences

Very Likely Change (More than 90% Probability)	Impacts
Higher maximum temperatures and more heat waves in continental areas	• Increased heat-related deaths and illness • Increased heat stress in farm animals and wildlife • Increased risk of lower yields for some crops • Increased cooling demand and reduced electrical energy supply
Higher minimum temperatures, fewer cold waves in continental areas	• Decreased cold-related deaths and injury • Decreased risk of damage to some crops • Extended range and activity for some crop pests • Reduced energy demand for heating
More high-intensity precipitation events	• Increased damage from flooding and landslides • Increased soil erosion

Likely Climate Change (More than 66% Probability)	
Increase in tropical cyclone peak winds, total precipitation, and maximum precipitation intensity	• Increased cyclone-related deaths and infectious diseases • Increased damage to coastal buildings, roads, and other infrastructure • Increased coastal erosion and damage to coastal ecosystems
Decrease in precipitation in most subtropical areas	• Decreased crop yields • Decreased water resource quantity and quality • Increased frequency of forest fires

role in the response to elevated greenhouse gases depends on correctly simulating the seasonal and geographic distribution of snow cover and sea ice. These are challenging problems for the current generation of GCMs.

Remember that although the latest GCMs are very comprehensive, they are still extreme simplifications of the natural world. Their limited spatial detail won't allow thunderstorms and hurricanes to appear, let alone individual clouds. The treatment of clouds is especially problematic considering the importance of clouds in Earth's radiation balance. Clouds are very effective at absorbing longwave radiation emitted from the surface, and therefore increasing cloud amounts might lead to warming. But thick clouds strongly reflect incoming solar radiation, which drives temperatures down. Which process is more important depends on the structure of the cloud, including particle sizes

and the vertical distribution of ice and water. It follows that to say convincingly whether clouds will amplify or reduce anthropogenic warming, a GCM needs to correctly calculate cloud changes in detail, which is something they cannot do at present.

Anthropogenic aerosols are another source of model uncertainty. Those forming from sulfur dioxide emissions have a net cooling effect; they reflect and scatter sunlight. On the other hand, black carbon aerosols found in soot absorb sunlight, so those particulates are a warming agent. Since the advent of the industrial period, reflecting aerosols have dominated, and anthropogenic particulates as a category have slowed global warming. Whether or not that trend will continue is highly uncertain, partly because emissions depend critically on assumptions about efforts to control pollution, and partly because these agents of change have such a short atmospheric lifetime. Thus, unlike CO_2, the concentration of anthropogenic particulates changes rapidly after any change in emission. In addition to their direct radiative effect, aerosols have an indirect effect on climate through action as condensation nuclei. This affects both the albedo and lifetime of individual clouds, and these in turn involve micro-scale processes within the cloud and large-scale patterns of moisture and motion. Not surprisingly, the indirect effect of aerosols is poorly understood. Similar uncertainties exist in many other aspects of GCMs, including behavior of the oceans and the terrestrial biosphere.

So, if GCMs are so crude, can't we ignore their prescriptions for the future? Might they grossly overestimate the importance of greenhouse gases? Are there important omitted processes and simplifications that, if corrected, would lead to different conclusions? A number of different suggestions have been offered, none of which has stood up to careful analysis. For skeptics of anthropogenic warming, the best candidate is a cloud process conceivably operating in the moist tropics. The idea is that with global warming, convective cloud towers will be more efficient at precipitating liquid water, leaving less water vapor to form high cirrus clouds. For thin ice clouds like these, the longwave warming effect is more important; they help warm the surface. The argument is that the quantity of those clouds will decrease and partially offset warming. After about a decade of debate, this particular process has been discredited for lack of evidence. However, as we have noted, clouds remain a large source of uncertainty in GCMs, and many scientists are working to understand their influence using both observations and models.

It also bears mentioning that uncertainty cuts both ways: perhaps future climate changes have been underestimated. After all, greenhouse gas emissions in the last few years have been greater than even the most pessimistic scenario imagined. In addition, simulations assume that the carbon cycle will continue to operate as it has recently—as a brake on climate change by taking up some of the carbon released by humans. But the high northern latitudes are warming faster than projected, and they contain the largest amounts of soil carbon in the form of frozen peat deposits. With summer melting penetrating deeper into the soil, there is a prospect for additional release of carbon not presently considered in the scenarios. The oceans have taken up about 30 percent of the carbon released by human activity since the Industrial Revolution, but their ability to do so diminishes as they warm. Thus, they constitute a feedback of uncertain size. Finally, GCM defects could mean the computed response to increasing greenhouse gas concentrations is too small. For example, perhaps the cloud feedback is strongly positive, and climate will warm significantly more than suggested by the current generation of GCMs.

In the press and the other popular media, the question of anthropogenic climate change is usually discussed as a prediction problem. It's like asking: What will happen? Or, even more simply: Should we believe in climate change? Anthropogenic change is better seen as a two-part question of risk. First, has science identified a credible threat? And if so, are the associated risks to social disruption acceptable?

In response to the first question—the prospect for significant climate change—the science-based answer is clearly yes. Again, this is not to say climate science will ever predict exactly what will happen with certainty or that climatologists are certain about the future. As individuals and as a society, we often take precautions for events we don't believe will occur with certainty. Most people wear seatbelts when driving, and emergency plans are in place for all sorts of natural and human disasters. We don't believe a bomb-carrying terrorist will attempt to board any particular airplane, but the cost of disaster is so great that we employ expensive procedures to check passengers and luggage on every flight. Given our desire for insurance against damage in other areas, it seems odd to be preoccupied with counting the ranks of believers and nonbelievers in anthropogenic change. The far more appropriate question is whether society is willing to accept the risks that come with the likely climate changes that will follow from continued reliance on fossil fuels.

Key Terms

anthropocene *page 243*

anthropogenic climate change
    page 243

disappearing climates *page 253*

early anthropocene hypothesis
    page 243

extreme events *page 248*

forcings *page 250*

general circulation models *page 250*

global warming *page 243*

global warming pause *page 244*

representative concentration
    pathways (RCPs) *page 252*

warming hiatus *page 244*

water vapor feedback *page 255*

Scan Here for a quick vocabulary review

Review Questions

1. What variables other than temperature indicate global warming during the past century?

2. Has global warming during the industrial period been roughly uniform across the planet? If not, what places have warmed more or less than the global average?

3. Carbon dioxide and methane have increased over the last 200 years. How do we know humans are responsible?

4. How (by what activities) have societies added carbon to the atmosphere?

5. What is the most convincing evidence suggesting greenhouse gases are responsible for most of the observed warming over the past century?

6. Discuss some of the uncertainties in general circulation models.

7. Other than warming, what do general circulation models suggest about future climates?

Scan Here for a quick concept review

www.oup.com/us/mason

The Biosphere

Exotic-looking giant baobab trees characterize this ecosystem in Madagascar, but it operates through the same basic processes we can observe across all of Earth's biosphere. The baobab trees and other plants capture energy from sunlight in the form of carbon compounds, while drawing nutrients and water from soils; that energy is passed on through food webs linking a host of animals and microorganisms.

Biosphere

The biosphere is the youngest of Earth's great environmental realms, where the lithosphere, hydrosphere, and atmosphere converge and interact in the formation and sustenance of all forms of life. Numerous processes and cycles link the lithosphere and the biosphere. Some of the most important of those connections involve chemical elements that are crucial to the growth of living organisms. These elements move between the lithosphere, hydrosphere, atmosphere, and biosphere along a complex web of pathways that we describe as *biogeochemical cycles*.

A key process that influences biogeochemical cycles in many ways is the development and deepening of soil. Precipitation, heat, and the activities of plants and animals combine to transform rock into soil, so it is not surprising that elements of the global climatic map are reflected in the distribution of major types of soil, referred to *as soil orders*.

The many combinations of climate and soil support a great variety of plant and animal communities, which together make up the great *biomes* such as the tropical rainforest or temperate grasslands. Within each biome, both soils and living organisms store carbon, nitrogen, and other biologically important elements for various periods of time, as these elements travel along their pathways through the environment. Tropical forests contain vast quantities of carbon in living plants, while northern coniferous forests and tundra store large amounts of carbon in their soils, emphasizing again the importance of climate. But climate changes over time, and the biosphere also changes in response. The future holds the prospect of rapid climate change resulting from increased levels of carbon dioxide and other greenhouse gases in the atmosphere. As the biosphere adjusts to the changing climate, the amount of carbon held in plants and soil may change, which in turn can alter the carbon dioxide content of the atmosphere as well as the rate of climate change. Human impacts and natural processes also greatly affect the global cycles of nitrogen and other elements. Like all environmental maps and diagrams, those of soils, biomes, and biogeochemical cycles are but still pictures of an ever-changing planet. Understanding the causes and rates of change is as important as mapping the present natural geography of the world.

Biogeochemical Cycles

Sand dunes on the Tibetan Plateau. On the right, bare sand moves freely in the wind at the interface between atmosphere and lithosphere. On the left, vegetation has stabilized this critical interface and allowed a soil to form. The vegetation and soil exchange carbon, nitrogen, and other elements with each other, and with the atmosphere above and the lithosphere below.

OBJECTIVES

- Define the major fields of study that focus on the geography of soils and living organisms
- Explain the importance of physical geography to the cycles of biologically important chemical elements
- Summarize the global cycles of carbon and nitrogen, the role of the biosphere in those cycles, and the response of these cycles to global environmental change

In Part Three, we focused on one major aspect of physical geography: climate. In this part, we turn to the geography of soils and living organisms. The geography of soils is a part of the science of **pedology**, a term that derives from the ancient Greek word *pedon*, meaning ground. The geography of plants and animals defines the field of **biogeography**, a combination of biology and geography. Biogeography is further divided into two subfields: **phytogeography**, the geography of plants (**flora**); and **zoogeography**, the geography of animals (**fauna**). In the units that follow, we will examine in detail the internal structures of soils and their spatial patterns across the globe, as well as the diverse communities of plants and animals found in various regions of the world. Above all, though, we will emphasize connections: between soils and plants, between plants and climate, and even between natural environments and human societies. This emphasis on interrelationships is a hallmark of the field of geography as a whole.

There are many good reasons to emphasize connections between various environmental systems, rather than simply mapping or cataloguing the climates of the world, soils, and plant and animal communities as we observe them today. One of the best reasons is that geographic patterns of climate, soils, and living organisms are dynamic and can change dramatically over time. Just a few thousand years ago, green pastures, lakes, and wetlands existed where desert conditions prevail today. Animals that are now extinct roamed countrysides that are today empty and barren. A mere 10,000 years ago, glaciers were melting and receding after covering much of the U.S. Midwest and the heart of Europe. Today we face the likelihood that Earth's climate will warm substantially over the next century in response to increased greenhouse gases in the atmosphere (as discussed in Unit 19). This climatic change portends serious and sometimes unforeseeable effects on plants and animals, including those we depend on for our food supply. In short, maps of climates, as well as of other elements of the environment, are still pictures of an ever-changing world.

To understand, predict, and respond to those changes it is essential to study how processes of the atmosphere, biosphere, and lithosphere interact. For example, the course of climate change over the next century depends in part on how the biosphere responds to both rising greenhouse gas levels and temperatures, and how that response itself affects greenhouse gases and climate through positive and negative feedbacks. Just as important, we need to understand how human activities of all kinds have altered processes of the biosphere and atmosphere.

With that in mind, this first unit of Part Four addresses a particular kind of connection between the biosphere, atmosphere, and lithosphere involving the movement of certain chemical elements—most notably carbon (C) and nitrogen (N)—along a complex set of pathways through the global environment. These elements are essential building blocks of the organic compounds that make up living organisms, but they also occur in nonliving matter,

in soils, in the atmosphere, and even in rocks deep beneath Earth's surface. In this unit we introduce the **biogeochemical cycles** of these key elements. Changes in biogeochemical cycles are implicated in many kinds of environmental change, from the loss of diversity in aquatic ecosystems to the global climate change underway at present.

Building Blocks of Life

Living organisms are made up mainly of **organic** compounds, that is, chemical compounds with basic structures built from carbon (C) atoms (note that this is a very different definition of "organic" than is used in terms like "organic farming"). Some of these carbon atoms are almost always bonded to hydrogen (H) atoms. Oxygen (O) is also an important building block of organic compounds, which are synthesized within plants, animals, and other organisms. Other elements are essential parts of specific organic compounds that play crucial roles in life. Nitrogen (N) is a key component of amino acids, which in turn combine to form proteins that facilitate and control chemical reactions within organisms. In addition, nitrogen is an essential component of the DNA that encodes an organism's genetic information and allows it to be passed on to offspring. Phosphorus (P) is needed for organic chemicals that transfer energy within organisms. Many other elements are considered essential to plants, animals, or both, although they are often only required in tiny quantities.

Each of these essential elements moves through the global environment along a web of pathways. These pathways connect *pools*, which we can think of as storage reservoirs. For example, the atmosphere forms an important pool of both carbon and nitrogen. To call these locations of storage "pools" is to use a fairly crude analogy. When we refer to the pools of carbon and nitrogen in terrestrial (land-based) vegetation, we are actually referring to carbon stored in many millions of individual organisms of a great variety of species growing together in diverse plant communities around the world (Fig. 20.1). However, imagining global vegetation as a single pool, or container, of carbon is exactly the kind of simplification we need to understand the big picture of carbon moving through the global environment.

We refer to the pathways and pools of a single chemical element as its biogeochemical cycle (*bio* indicates the connection with life, while *geo* emphasizes the fact that all of these elements reside at times in the rocks, sediments, and soils of solid Earth). In this unit we will focus on the cycles of carbon and nitrogen, because they are so important to understanding past and present changes in the environment.

Not only is carbon the most basic building block of organic compounds, but one of its *pools*, or storage reservoirs, is in the atmosphere, where it is mainly in the form of carbon dioxide (CO_2) gas. Recall from Unit 5

FIGURE 20.1 The diverse plants of this forest on the coast of Washington State are constructed almost entirely from compounds of carbon, hydrogen, and oxygen, combined with smaller quantities of nitrogen and other elements. The trees, ferns, and other plants visible in this photo contain a small portion of the global pools of carbon and nitrogen in terrestrial vegetation.

that carbon dioxide is a *greenhouse gas* that effectively absorbs longwave radiation. The abundance of carbon dioxide in the atmosphere strongly influences global temperatures (methane, another carbon-containing greenhouse gas, is much less abundant in the atmosphere but also has a significant effect on climate). Furthermore, the total pool of carbon in the atmosphere is fairly small compared to the amounts that move into or out of the pool each year. This means that even fairly small changes in the amount of carbon moving along those pathways can have major effects on atmospheric carbon dioxide and on Earth's climate as well. Not surprisingly, scientists have studied the carbon cycle intensively in recent years.

In contrast to carbon, the oxygen cycle is not a major topic of research. Oxygen is part of many organic compounds, and all animals and plants need it for the key process of respiration (which we will discuss in Unit 24). While the atmosphere's oxygen content has varied considerably over Earth's history (see Unit 6), those changes were driven by biological developments of great magnitude, like the first appearance of organisms carrying out photosynthesis. Today, the pool of oxygen in the atmosphere is very large, and any likely changes in flows of oxygen to and from the atmosphere will not have substantial effects on climate or living organisms.

Nitrogen, like carbon, is a biologically important element that has a cycle closely linked to both natural environmental change and human impacts. For nitrogen, those links are more subtle and complex, however. Nitrogen gas (N_2) is even more abundant in the atmosphere than oxygen gas, making up a little more than 78 percent of dry air. It may seem strange, then, to learn that a limited supply of nitrogen is often the most important factor holding back growth of plants. This is because plants cannot use the nitrogen gas that is so abundant in air. Instead, plants must take up other chemical forms of nitrogen from the soil or water in which they are growing. Major human impacts on the global cycle of nitrogen in recent centuries have had little impact on the large pool of nitrogen gas in the atmosphere but have significantly affected the abundance of the nitrogen forms that plants and algae can access in many soils and water bodies. Because of those major impacts on important ecosystems, the nitrogen cycle, like the carbon cycle, has been the subject of a great deal of recent research.

Furthermore, human activities have affected the abundance of *nitrous oxide* (N_2O), another nitrogen-containing gas in the atmosphere. While nitrous oxide is a very minor component of air, it is a greenhouse gas that is far more effective than carbon dioxide on a molecule-by-molecule basis. Over the past two centuries, the expansion of agriculture and the increasing use of fossil fuels have caused an increase of about 20 percent in the abundance of nitrous oxide in the atmosphere, which contributes to current warming of the global climate.

Although we focus on nitrogen and carbon in this unit, it's important to note that a variety of other elements such as phosphorus (P), sulfur (S), and iron (Fe) all have biogeochemical cycles that each play a role in certain kinds of environmental change. Phosphorus fertilizer carried into lakes by runoff can change the aquatic ecosystem drastically. Sulfur compounds released by volcanic eruptions cause short-term changes in the global climate. Some researchers have suggested that iron-rich dust deposited on the oceans during past glacial periods enhanced the growth of algae and indirectly caused a reduction in the carbon dioxide content of the atmosphere.

Soil and Vegetation: A Critical Interface

As we discuss the cycles of carbon and nitrogen, you will notice that important parts of both cycles take place in the atmosphere, which we have investigated in depth in the preceding units of this book. However, many of the more complex and important processes involved in these cycles occur where soils and terrestrial vegetation form the *interface* between the lithosphere and the atmosphere. This interface controls the interaction between rocks and their

FIGURE 20.2 A freshly dug pit exposes the soil beneath a young forest in northern Minnesota. The soil stores nutrients and water essential for the forest's growth. At the same time, the leaves that fall each year supply the organic matter that gives the surface soil layer its dark color and makes it a particularly rich store of nutrients. Color variations deeper in the soil are related to weathering and the downward movement of minerals carried by percolating water.

minerals, on the one hand, and the air with its moisture and heat, on the other (see opening photo of this unit).

At this interface, soils and vegetation develop over time and continually interact with each other (Fig. 20.2). Precipitation falling from the atmosphere to Earth's surface enters through the open pores of the soil, the spaces between the solid mineral masses. Water in the soil pores promotes the decay of minerals within and below the soil layers. Heat from the atmosphere also facilitates the process of decay. As energy and matter enter the soil from the overlying atmosphere, the soil evolves over long periods of time. We will discuss the processes of soil formation and the properties that result from them in more detail in Units 21 through 23. One of the most important consequences of soil development, at least in its early stages, is the increasing availability of essential nutrients for plants. Some of these nutrients are released through

the breakdown of minerals, while others are added as dissolved material in rainfall or as wind-blown dust particles. The soil stores all of these nutrients, along with water, which can then be used as needed by plants.

In this way, the soil plays a vital role for land-based vegetation. Similarly, vegetation plays a crucial part in soil development. Above all, plants carry out *photosynthesis*, and the products of photosynthesis ultimately become part of the soil, greatly influencing its properties. As we will discuss in Unit 24, photosynthesis involves the capture of energy from solar radiation and the use of that energy to make organic compounds from carbon dioxide and water. Photosynthesis is also important for the carbon cycle, because it moves carbon from one pool (the atmosphere) to another (vegetation). Plants use energy captured in photosynthesis to convert its initial products into a range of organic compounds, including some containing nitrogen or other elements brought up from the soil.

Eventually, all parts of the plant die; dead leaves fall to the surface of the soil, and dead roots are buried within it. A host of different animals and microorganisms living in the moist soil environment feed on those plant parts. Part of the organic matter they contain is converted to carbon dioxide, which returns to the atmosphere, another key link in the carbon cycle. The other part builds up in the soil, giving its upper layer a distinctive dark color (see Fig. 20.2). Nitrogen from decaying plant material is stored in that soil organic matter or released in forms the plants can use again. Certain soil microorganisms can capture nitrogen in its gaseous form (N_2) and convert it into a plant-accessible form. Overall, then, the zone of soil and vegetation just above and below Earth's surface is critical for the operation of most biogeochemical cycles.

The Global Carbon Cycle

The **global carbon cycle** is often represented by a kind of map depicting the most important pools and the pathways between them (Fig. 20.3). Carbon is present in numerous varied forms and locations in the environment, but to simplify our analysis, we concentrate on several especially important global pools. Figure 20.3 includes current estimates of the total amount of carbon stored in each pool in petagrams (Pg). One petagram is 1 billion metric tons of carbon (a little more than 1 billion U.S. tons); therefore, some of the pools contain vast amounts of carbon.

By far the largest pool is made up of carbon in the rocks that make up *Earth's crust*. Carbon occurs in sedimentary rocks such as limestone, dolomite, and shale, which we will discuss in Unit 28. The pool of carbon in the crust also encompasses fossil fuels found in rocks, including coal, oil, and natural gas, all of which originate from organic compounds produced by photosynthesis in past periods of geologic time. The next-largest pool includes all forms of carbon in the *ocean*; much of this pool is carbon dioxide

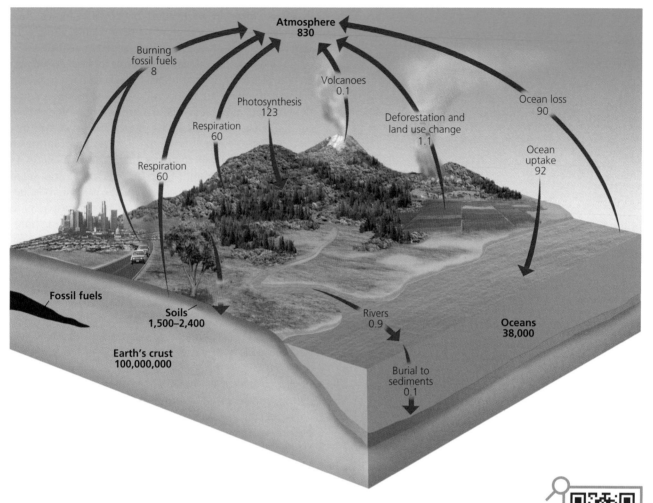

FIGURE 20.3 The modern global cycle of carbon, including the major pools in which carbon is stored, with the estimated size of each pool indicated in petagrams (1 Pg = 1 billion metric tons) of carbon. The arrows represent processes moving carbon between pools, and the number beside each arrow represents the yearly flux, the petagrams of carbon moved along that pathway each year. Carbon buried in deep permafrost is also probably a large pool but is not shown because of uncertainty about its size and flows in and out of it.

ENHANCE

dissolved in ocean water. A very small fraction of the ocean pool is made up of living marine organisms, which are not shown separately in Figure 20.3. Other pools include the carbon found in the *atmosphere*, *soils*, and *terrestrial biomass* (organisms living on land). As mentioned earlier, most carbon in the atmosphere is in the form of carbon dioxide, although much smaller quantities are present in other forms, especially methane gas (CH_4). In vegetation, almost all carbon is in organic compounds, while in soils, carbon is found in both organic compounds and carbonate minerals (see Unit 27, page 350).

The pathways shown in Figure 20.3 are specific processes that move carbon between pools. Some of these processes are fairly easy to understand; for example, burning fossil fuels produces carbon dioxide that enters the atmosphere. Volcanic eruptions also release carbon dioxide, although at a much lower average yearly rate. This is because

large eruptions quickly release large amounts of carbon dioxide, but they are relatively rare and short-lived, whereas fossil fuel use continually produces carbon dioxide at high rates. Carbon dioxide molecules in the air overlying the ocean are continually dissolved in ocean water, while other molecules of the same gas are released into the air.

The most important pathways linking the atmosphere, biosphere, and soils are more complex, and we will examine them in detail in Units 21 through 24. These pathways include photosynthesis and the addition of dead organic matter from plants to soils. Another important process is *respiration*, in which organic compounds are broken down through a reaction with oxygen, which releases carbon dioxide back into the air. Respiration is carried out by plants, animals, and microorganisms, including those that feed on dead organic matter in the soil. A final set of processes that have become especially important today involve land-use changes. For example, when humans cut down forests and replace them with pastures or agricultural land, most of the carbon that was stored in the trees is released to the atmosphere through burning or decomposition of the dead trees.

Figure 20.3 also shows the estimated rate of carbon movement, or *flux*, between pools, in petagrams per year. Note that movement of carbon in or out of the pool in Earth's crust is relatively slow, even if we count the burning of fossil fuels, and it is scarcely changing the size of that very large pool. Most fossil fuel use only began with the Industrial Revolution of the late 1700s and 1800s, and before that time fluxes of carbon in or out of the rocks of the crust were very slow. Because the atmospheric pool of carbon is far smaller than the pool in Earth's crust, the amount of carbon now being added to the atmosphere by fossil fuel combustion is enough to cause a fairly rapid increase in the atmospheric pool.

The movement of carbon between plants, soils, and the atmosphere each year is also quite substantial compared to the size of the pools involved. Plants take up a bit less than one-sixth of the carbon in the atmosphere each year through photosynthesis. The best estimates we have today suggest that the amount of carbon returned by plants and soils to the atmosphere through respiration almost balances that removed from the air by photosynthesis. This makes sense, as most of the organic matter produced each year is ultimately broken down in respiration. Here, a scientific puzzle arises, however. There *is* a small but important imbalance. About 2–3 Pg more carbon per year is removed from the atmosphere than is returned. It cannot be accounted for by land-use changes such as the regrowth of forest on old farm fields; that regrowth is more than balanced out by deforestation in other areas. Scientists are increasingly certain that the missing carbon is taken up by the biosphere through increased growth of vegetation, especially in forests (see Focus on the Science: Biogeographic Processes and the Carbon Cycle on p. 312 in Unit 24).

The fluxes each year between the atmosphere and the ocean are also quite large but mostly cancel each other out, with almost as much carbon moving from the ocean to the atmosphere as in the other direction (most moves in the form of carbon dioxide, as noted earlier). There is a small but significant imbalance between these large flows of carbon as well, though. It represents a net loss of a little more than 2 Pg of carbon per year from the atmosphere, a bit more than one-fourth of the carbon added to the atmosphere by fossil fuel combustion. Clearly, without this ocean "sink" of carbon, the rate of global warming could be faster. The ocean cannot indefinitely take up dissolved carbon dioxide, however, since there is an upper limit to the *solubility* of this gas in ocean water. This is because the solubility of a gas in water represents the maximum content of the dissolved gas at a given temperature; as Earth's climate warms, the solubility of carbon dioxide in the ocean decreases somewhat. Some carbon continually flows into and out of the small part of the ocean pool made up of living organisms as they take up dissolved carbon dioxide for use in photosynthesis and release it in respiration, but those flows are not shown in Figure 20.3 because they take place within the ocean pool as a whole.

Figure 20.3 portrays most aspects of the carbon cycle that are important for understanding the recent rise of atmospheric carbon dioxide. There is also a far smaller quantity of carbon—less than 2 ppm of the atmosphere—present in the atmosphere as methane. Methane is a highly effective greenhouse gas that has much more influence on climate than its very minor abundance would suggest, and it has its own distinctive pathways into and out of the atmospheric pool. Methane is produced naturally in wetland soils and in the digestive systems of grazing animals such as cattle. A significant amount is even produced in the guts of termites! From all these sources, methane is released to the atmosphere. Humans have enhanced the production of methane in several ways, including the development of rice paddies (artificial wetlands) and the expansion of cattle raising. The decomposition of organic material in landfills also releases methane, and large quantities leak from natural gas pipelines. These activities have raised atmospheric methane well above the level it was at 300 years ago.

The Global Nitrogen Cycle

The pathways of nitrogen through the environment form a more complex web than those of carbon and are difficult to portray in a single diagram. We first turn to the most important processes of the cycle, as they occur on land in a setting with natural vegetation, before any significant human impacts (Fig. 20.4). The same key processes depicted in this figure occur in aquatic systems (the ocean and lakes). Like carbon, nitrogen exists in a number of very different forms. The vast majority of all nitrogen atoms in the atmosphere are in the form of nitrogen gas (N_2). The fluxes of nitrogen into and out of this huge atmospheric pool are quite small, but they are critically important for life on Earth.

Before the twentieth century, nitrogen left the pool of N_2 gas in the atmosphere primarily through biological **nitrogen fixation**, a natural process in which certain bacteria and cyanobacteria capture nitrogen from the air and produce *ammonium* (NH_4^+) ions (*ions* are atoms, or groups of atoms, with positive or negative charges and usually dissolved in water). Nitrogen-fixing bacteria on land often live in a symbiotic (mutually beneficial) relationship with plants. The best examples involve bacteria that live in small round growths, or nodules, on the roots of wild or cultivated legumes (e.g., common beans, soybeans, and alfalfa). Legumes absorb the ammonium produced through nitrogen fixation. While legumes and certain other plants can use ammonium directly as a source of nitrogen, in most cases, one more step is needed. Another group of bacteria carry out the process of *nitrification*, which converts

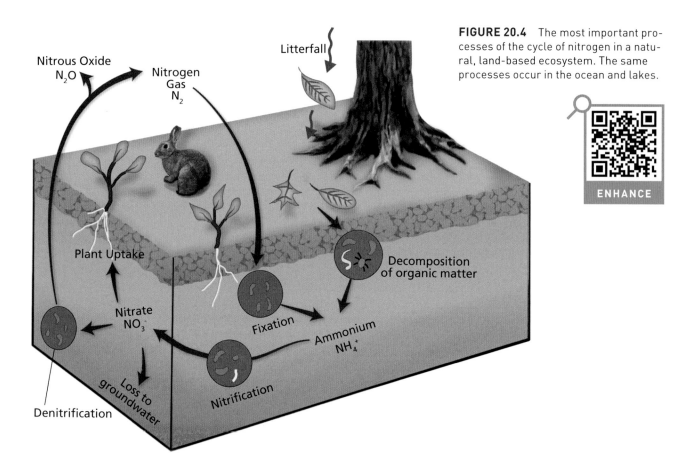

FIGURE 20.4 The most important processes of the cycle of nitrogen in a natural, land-based ecosystem. The same processes occur in the ocean and lakes.

ENHANCE

ammonium to nitrate (NO_3^-) ions, an accessible form of nitrogen used by most plants and algae. In the oceans, colonies of cyanobacteria fix nitrogen that can eventually be used by algae. Overall, then, microorganisms that we can't see and rarely think about are essential in supplying nitrogen from the atmosphere to Earth's natural ecosystems.

Once nitrogen is taken up by plants, or by algae in the ocean and lakes, it is built into a great variety of nitrogen-containing organic compounds such as proteins and DNA. Animals satisfy their need for nitrogen by feeding on plants, or on other animals that feed on plants; in this way nitrogen moves up through the food chain. As shown in Figure 20.4, however, the nitrogen in living organisms is eventually returned to the surrounding environment. All plants and animals die, and plants regularly shed dead leaves (a process called *litterfall*) and older roots. As a result, nitrogen-bearing organic matter is added to the soil or to water bodies. Then yet another crucial group of microorganisms feed on the organic matter, carrying out *organic matter decomposition* and releasing nitrogen, mainly in the form of ammonium. The nitrifying bacteria turn much of the ammonium into nitrate, and growing plants or algae can take up these forms of nitrogen once again. This part of the nitrogen cycle comes close to forming a closed loop, in which nitrogen is recycled continually.

At least small quantities of nitrogen continually leak out of this loop, however. One way this happens is through **denitrification**, the conversion of nitrate ions to nitrogen

gas. Again, this is a process carried out by microorganisms that live in certain soil and aquatic environments. Denitrification releases nitrogen gas back to the large pool in the atmosphere and more or less balances the nitrogen that flows out of that pool through nitrogen fixation. Nitrous oxide gas (N_2O) is sometimes produced and released to the atmosphere during denitrification as well (see Fig. 20.4). This is one of the most important natural processes producing nitrous oxide gas, which is a rare but important greenhouse gas.

Humans began to alter the nitrogen cycle thousands of years ago when legumes were first cultivated as crops, replacing other vegetation that did not have nitrogen-fixing microbes attached to its roots. Today, humans have far greater impacts on several parts of the cycle, as illustrated in Figure 20.5. The two parts of that figure show modern flows of nitrogen in various forms, not only those that occur naturally, but also those that have emerged as humans have expanded their agricultural fields, built power plants, and manufactured chemical fertilizers. Figure 20.5A focuses on the processes we have already discussed, especially fixation and denitrification, which move large quantities of nitrogen into and out of the large atmospheric pool of nitrogen gas.

Large areas of the world are now covered with fields of soybeans and alfalfa. These legumes are roughly doubling the rate of biological nitrogen fixation from its natural level. Even more dramatic human impacts have followed

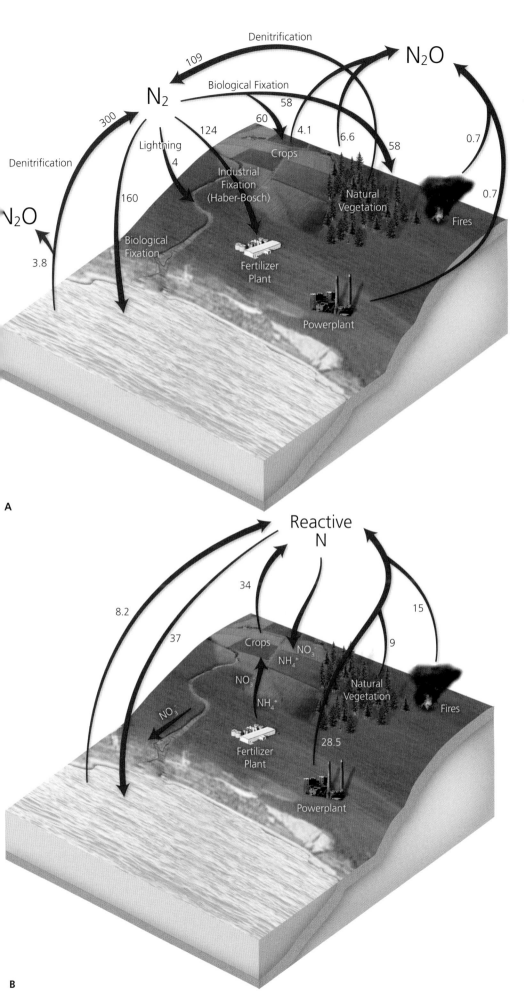

FIGURE 20.5 The modern cycle of nitrogen, including both natural processes and human impacts. Arrows indicate flows between nitrogen pools, with estimated rates in teragrams (1 Tg = 1 million metric tons) of nitrogen per year. (A) The major processes of fixation and denitrification. (B) Some of the more significant natural and human sources of *reactive nitrogen* in the atmosphere, as well as the deposition of that reactive nitrogen into soils and water bodies.

ENHANCE

the discovery of methods for fixing nitrogen in chemical plants by the chemists Fritz Haber and Carl Bosch in 1909–1910. Before that time, the only way to supply nitrogen to crops and increase their growth was to apply animal manure or fertilizer made from bird droppings, which were shipped around the world for that purpose. The Haber-Bosch process made large-scale industrial production of nitrogen fertilizer possible, allowing enormous increases in crop yields. The Haber-Bosch process also helped make enormous destruction possible in World War I, soon after its discovery, because fixed nitrogen is a key ingredient of explosives used in ammunition and bombs. Today, industrial nitrogen fixation is at least twice the total rate of biological fixation. The rates of both biological and industrial fixation far exceed the rate of another natural fixation process, the process that occurs during lightning strikes (discussed in Unit 6).

Researchers generally believe that cultivation of soils for agriculture has also increased the global rate of denitrification. This means that as the growth of agriculture and fertilizer manufacturing increases the removal of nitrogen from the air through fixation, more is also being returned to the atmosphere. Changes in the global rate of denitrification are of interest partly because they affect the level of nitrous oxide in the atmosphere, as mentioned previously. Figure 20.5A highlights the fact that combustion of fossil fuels also releases nitrous oxide into the air, which is yet another human impact on the nitrogen cycle.

Finally, in Figure 20.5B, we can examine a variety of other effects of industry and agriculture on so-called *reactive nitrogen*. This is a catch-all term for any form of nitrogen other than nitrogen gas (N_2). Besides nitrous oxide, reactive nitrogen includes nitrate and ammonium ions and an array of other compounds that go through complex chemical reactions in the atmosphere. For example, recall from Unit 6 that the heat of lightning strikes can cause nitrogen to combine with oxygen to form NO (nitrogen oxide, a form of reactive nitrogen). Similarly, when fossil fuels are burned in power plants or car and truck engines, reactive nitrogen is produced and emitted into the atmosphere. Agricultural fields, sewage treatment plants, and fires give off reactive nitrogen as well.

Reactive nitrogen from all of these sources can eventually fall to Earth, where it becomes nitrate or ammonium in soils or water bodies and essentially acts as a nitrogen fertilizer. Reactive nitrogen can be deposited from the air onto wild landscapes far from cultivated fields, or onto the open ocean, so in effect we are unintentionally adding nitrogen fertilizer to those ecosystems just as we intentionally add nitrogen fertilizer to croplands. Finally, nitrate ions can be easily washed off the soil surface in agricultural fields and carried downward to aquifers by percolating rainwater, further spreading human impacts on the global nitrogen cycle. Perspectives on the Human Environment: Fertilizer and the Cycle of Nitrogen explores these impacts in more detail.

From this brief overview of the nitrogen cycle, it is clear that human activity has affected it in ways that are as profound as the better-known human impacts on the carbon cycle. As humans strive to understand and cope with the effects of the changes we have wrought in global biogeochemical cycles, the justification for geographers' long-standing interest in human–environment interactions becomes clear.

PERSPECTIVES ON THE HUMAN ENVIRONMENT

Fertilizer and the Cycle of Nitrogen

Global population growth raises a painful question: Can we feed more than 7 billion people? To do so, Earth's limited amount of agricultural land must produce more and more food. One way to meet the challenge is to use fertilizer. Fertilizer has long been used by humans to increase agricultural productivity. We know that some eastern Native Americans increased soil fertility by planting a fish with each seed hundreds of years ago. In many other cultures, farmers have applied organic fertilizer—primarily manure—for centuries. In the twentieth century, it became possible to cheaply produce chemical fertilizers that supply all of the primary nutrients: nitrogen, phosphorus, and potassium. Along with new hybrid or genetically modified crops and pesticides, the use of chemical fertilizers has allowed dramatically higher yields of major crops such as corn, rice, and wheat (the yield of a crop is the amount of grain or other plant material produced per hectare or acre). The quantities of fertilizer used are enormous, however. At the start of the twenty-first century, 140 million metric tons of nitrogen were used worldwide, which equals a 23-kg (50-lb) bag of chemical fertilizer for each person on Earth. That much nitrogen cannot be added to soils without altering the global cycle of nitrogen.

Until recently, many forests and other natural ecosystems were almost closed systems with regard to nitrogen, the most important plant nutrient. Vegetation containing nitrogen died and decayed on the forest floor, and the nitrogen released by decomposition was quickly taken up again by plants. The relatively small amount of nitrogen lost from the system each year was made up mainly by biological nitrogen fixation. Today, nitrogen compounds produced by burning fossil fuels are deposited on forests even in remote parts of the world, potentially overloading them with nitrogen. However, most of the nitrogen in these ecosystems still cycles between soils and vegetation.

Agricultural systems rarely retain and recycle most of the nitrogen added to them, unlike natural forest ecosystems. Where does all of the nitrogen added as fertilizer go? Some of this nitrogen quickly leaves the system in the form of nitrate ions, which are readily dissolved in percolating water and carried downward to groundwater aquifers. The rapidly growing crop takes up a great deal of the nitrogen, and part of that nitrogen is removed with the harvest each year. Only a fraction of the total added nitrogen remains in the plant material that is not harvested and becomes organic matter in the soil. That nitrogen in soil organic matter is slowly released for plant use, in a process similar to the process that supplies nutrients to forest plants. Agricultural fields are prone to soil erosion, however, so even some of this stored nitrogen can be carried away to streams and lakes with the eroded soil. Overall, the amount of nitrogen lost each year is quite large and must be replaced by more fertilizer to maintain high crop yields.

One consequence of excessive soil fertilization is the *eutrophication* of water bodies that receive runoff carrying nitrogen and phosphorus from farm fields or other sources, such as sewage or food-processing waste. Algae respond to the added nutrients by growing rapidly. The algae in turn deplete the oxygen needed by fish and other aquatic life. Eutrophication probably began long ago in areas of high population and intensive agriculture, but it occurs on a much more dramatic scale today. For example, the "dead zone" in the Gulf of Mexico is a large area where eutrophication has been caused by nutrients carried down the Mississippi, mainly from farm fields. The dead zone is located along the coasts of Louisiana and Texas, where it threatens a rich aquatic ecosystem. This clearly illustrates how humans have drastically modified the global cycles of nutrients.

FIGURE 20.6 Tanks holding anhydrous ammonia at an agricultural co-op on the Great Plains provide a vivid illustration of human impacts on the nitrogen cycle. The ammonia is produced by the Haber-Bosch process and is injected into soils to replenish the supply of nitrogen required by fast-growing crops. Only 150 years ago, the same soils supported natural grassland ecosystems in which most nitrogen was recycled between plants and soil.

Key Terms

biogeochemical cycles *page 261*
biogeography *page 261*
denitrification *page 266*
fauna *page 261*

flora *page 261*
global carbon cycle *page 263*
nitrogen fixation *page 265*
organic *page 261*

pedology *page 261*
phytogeography *page 261*
zoogeography *page 261*

Scan Here for a quick vocabulary review

Review Questions

1. Which subjects are encompassed by *pedology*? By *biogeography*?

2. What do we mean when we refer to *pools* of an element such as carbon, and what is their role in biogeochemical cycles?

3. Why is understanding vegetation and soil important for understanding connections between the atmosphere and the lithosphere?

4. Describe the various forms of nitrogen that occur in the soil, and how plants make use of nitrogen.

5. Explain why the pools of carbon in terrestrial biomass and soils can have important effects on the amount of carbon in the atmosphere, even if these pools are much smaller than carbon pools in rocks or the oceans.

6. Describe three ways in which human activities have altered the cycles of carbon and nitrogen.

Scan Here for a quick concept review

www.oup.com/us/mason

Formation of Soils

Soil on a forest-covered hillside in Minnesota. The contrast between the soil and underlying bedrock debris illustrates the profound effects of soil-forming processes. The shovels and depth scale are essential tools of field soil scientists

OBJECTIVES

- List the factors that influence soil formation and summarize the major soil-forming processes
- Describe and explain a typical soil profile and the processes responsible for the formation of soil horizons
- Discuss the concept of soil regimes based on climate and other factors

In an untouched forest or a cultivated field, the connection between soil and life is clear. Soil supports plants and supplies them with water and nutrients. Out of sight below the ground, the soil itself is home to a great variety of life-forms, including the organisms responsible for the decay of plant material and the release of nutrients used by growing vegetation. The U.S. Natural Resource Conservation Service defines **soil** as "a mixture of fragmented and weathered grains of minerals and rocks with variable proportions of air and water; the mixture has a fairly distinct layering; and its development is influenced by climate and living organisms." We will examine *weathering*, the chemical alteration and physical disintegration of earth materials by the action of air, water, and organisms, more closely in Unit 36. In this unit, we concentrate on the major controls governing soil formation.

The term *soil* means different things to different people. Agricultural scientists regard soil as the top few layers of weathered material that hold most plant roots. Geologists, however, use the term to refer to all materials produced by soil-forming processes at a particular site. Applying this definition, geologists consider as soil those soils that were produced thousands of years ago and are now covered by layers of other material, even though it is impossible to grow plants in them (Fig. 21.1). Alternatively, civil engineers look upon soil as something to build on and, in general, as anything that does not have to be blasted away. Although soil particles are often eroded and transported by wind or water, our concern here is with soil that is still in place, where it formed from weathered rock or sediment. The general location of soil is at the interface of the atmosphere, hydrosphere, biosphere, and lithosphere.

We may not normally think of it as such, but soil is one of our most precious natural resources. Geographers classify Earth's resources as either **renewable resources**, those that can regenerate as they are exploited, or **nonrenewable**

FROM THE FIELDNOTES

FIGURE 21.1 "As we crossed the pasture toward an old dirt road in western Nebraska, the dark-colored band across the roadcut face immediately caught our attention: It was a buried soil, formed at the land surface and then covered with more than 6 m (19 ft) of *loess*, a deposit of wind-blown dust. The farmers who once hauled their crops to town along this road probably paid little attention to the buried soil; after all, it can no longer support a crop of wheat. To us, however, it was a window into a landscape of the past, a time 13,000 to 10,000 years ago when an unusually wet climate allowed lush grasses to cover this often dry and dusty place. We even found the burrows of gophers, prairie dogs, and insects that lived in this soil before it was buried." (For more on this site, see E. Marín-Spiotta, et al., "Long-Term Stabilization of Deep Soil Carbon by Fire and Burial During Early Holocene Climate Change," *Nature Geoscience* 7 [2014]: 428–432.)

resources, such as metallic ores and petroleum, which when consumed at a certain rate will ultimately be used up. Renewable resources, however, are not inexhaustible. Schools of fish, for example, form a renewable resource: harvested at a calculated rate, they restore themselves continuously. But as all fishing people know, this balance can easily be disturbed by overfishing. And once a population has been overexploited, the fish may not return for many years, if ever.

So it is with soil. Soil is a renewable resource; we use it and deplete it, but it continues to regenerate. But like fish, soil is not inexhaustible. Damaged beyond a certain level, a soil may be entirely lost to erosion or made completely unproductive for crops or natural vegetation. For example, farmers know the risks involved in cultivating steep slopes by the wrong methods. Terraces must be created, but, more important, they must be *maintained* (Fig. 21.2). Once neglected, a terraced slope is ripe for soil erosion, and once gullies appear, the process may be irreversible.

The Formation of Soil

Soils of some kind must have formed as soon as Earth's atmosphere contained enough moisture to begin the breakdown of rocks by weathering. As the atmosphere became enriched with oxygen and plants and animals colonized Earth's landmasses, soil formation progressed through processes like those still at work today.

SOIL COMPONENTS

Soil contains four components (Fig. 21.3): minerals, organic matter, water, and air. A soil's minerals make up its tiny rock particles, the organic matter is of various types, water usually clings to the surfaces of the rock particles, and air fills the intervening gaps. We now look at each component more closely.

Minerals are naturally occurring chemical elements or compounds that possess a crystalline structure and are the

FIGURE 21.2 "Driving upcountry from Colombo to Kandy in Sri Lanka, you see areas where the whole countryside is transformed by human hands into intricate, meticulously maintained terraces. It is the product of centuries of manipulation and maintenance, with long-term success depending on careful adjustment to nature's cycles and variations. Scenes like this (similar ones can be observed in Indonesia, the Philippines, and other rice-growing societies) reflect a harmonious human–environment relationship evolved over countless generations."

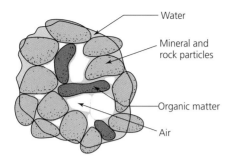

FIGURE 21.3 Four major components and their relative positions within a tiny clump of soil.

constituents of rocks (as discussed in more detail in Unit 27). The most common elements in Earth's crust—silicon, aluminum, iron, calcium, sodium, potassium, and magnesium—are the building blocks for many minerals. The most prevalent element of all is oxygen, and it is found in countless combinations with other elements. Rocks contain assemblages of minerals formed from thousands of such element combinations. When rocks break down into soils, some of these mineral components become available to plants as nutrients.

Another soil component is *organic matter*, the material that forms from living matter. In the upper layers of the soil, the decaying remains of plant leaves, stems, and roots accumulate. There is also waste matter from worms, insects, and other animals. The processes of decay are carried out mainly by an astronomical number of microscopic bacteria and fungi. All of these contribute to the organic content of the soil (recall from Unit 20 that these processes together comprise an important pathway within the global carbon cycle).

Soil contains life-sustaining *water*, which usually originates as rainfall or melted snow. There is an electrical attraction between the mineral particles and the water molecules surrounding them. In a moist soil, water fills much of the space between the mineral particles. But even in dry soils, the attraction is so persistent that there may be a thin film of water, just a few molecules thick, around the mineral particles. The water in soils is not pure but exists as a weak solution of the various chemicals naturally found in the soil. Without water, the many chemical changes that must occur in the soil could not take place.

Air fills the spaces among the mineral particles, organic matter, and water. It is not exactly the kind of air we know in the atmosphere. Soil air contains more carbon dioxide and less oxygen and nitrogen than atmospheric air.

It is straightforward enough to deduce that these four soil components came from the other spheres of the Earth system. The more interesting task is to discover how. Five important factors will direct us to that understanding.

FACTORS IN THE FORMATION OF SOIL

Soil scientist Hans Jenny (1899–1992) described the factors influencing soil formation as *cl*imate, *o*rganisms, *r*elief, *p*arent material, and *t*ime—often abbreviated as *clorpt*. We now consider each of these factors, though in a slightly different order, starting with the **parent material**, the rocks and sediments of Earth that are the parents of soil.

Parent Material   The simplest kind of soil formation involves what is known as **residual parent material**. In this case, the soil forms directly from the residue left by weathering of the local bedrock. The properties of soils formed from residual parent material are often directly related to characteristics of the original rock, even though soil and rock may look quite different. For example, in the southeastern United States, many thick red soils have formed from rocks that are not red themselves but do contain abundant iron. As the rocks are broken down by weathering, the iron is released to form reddish-brown iron oxide minerals that give the soils their distinctive color. Such cases are quite common throughout the world, though reddish-brown soils also form from a variety of other parent materials.

In many other cases, the soil is formed from **transported parent material** deposited by wind, water, or glacier ice. A soil formed from transported material may be totally independent of the underlying solid rock. During the most recent Ice Age, large quantities of pulverized rock were transported by glaciers thousands of kilometers and deposited in new areas, such as in large parts of the northeastern and Midwestern United States. These glacial deposits became the parent material for new soil formation.

The sediments deposited in stream valleys are an important type of parent material transported by water. This type of parent material often creates fertile soils, as in the case of the lands bordering the Mississippi River. In other instances, the wind carries and eventually deposits thick blankets of fine matter that form the parent material of new soil. Such deposits of windblown dust are called *loess* (see Unit 47). Much of the central Great Plains region of the United States is underlain by loess deposits up to 60 m (200 ft) thick.

Climate   In the Piedmont areas of central Georgia and Maryland, the soil developed from a rock called granite—yet the resulting soils are not the same, because the two states have different climates. The warmer, moister climate of the southeast has engendered more advanced chemical change in the soils of Georgia. As far as the development of soils is concerned, the most important elements of climate are moisture and temperature. The amount of soil moisture is determined by the amount of precipitation and evaporation at a particular location. Both moisture and high temperatures accelerate chemical reactions in the soil. For this reason, we find thick, well-developed soils in stable landscapes of the humid tropics. The soils of desert regions have their own distinctive characteristics, such as accumulation of calcium carbonate.

Organisms   In 1881, near the end of his life, Charles Darwin published a book with the curious title *On the Formation of Vegetable Mould Through the Action of Worms*. The book presented Darwin's conclusion, based on decades of careful observation and experiments, that earthworms and other burrowing animals are responsible for the formation of soil (which he referred to as "vegetable mould"). Darwin highlighted the role of worms in bringing fine material to the ground surface and forming a loose, porous soil mantle enriched with organic matter. Pedologists (scientists who study the formation and geography of soils) now recognize that soils form through a host of processes, but among those processes, the action of worms—and many other animals and plants—clearly plays a vital role.

Vegetation is the original source of most organic matter in the soil. Leaves and dead roots shed by plants are added to the soil, where they become food sources for the countless decomposer organisms that live there. Many invertebrate animals such as earthworms are important in the initial stages of decomposition, the breakdown of dead organic matter. For example, some earthworms pull leaves from the forest floor into their burrows, where they become accessible to other decomposers. We would need a microscope to see the bacteria and fungi that play the most important roles in this process. During decomposition, some of the organic matter is continually converted to carbon dioxide and lost to the atmosphere, while the remainder becomes **humus**, forming a dark layer at the top of the soil (Fig. 21.4). Soils that are rich in humus form an important part of the global pool of carbon held in soils. Humus is also a key location where nitrogen is retained in ecosystems (see Unit 20).

In undisturbed natural forests and grasslands, soil and vegetation form a nearly closed system. Substances that are essential for plant growth—such as nitrogen compounds, phosphates, and potassium—circulate through this system and are collectively known as *plant nutrients*. These nutrients are present in dead leaves and roots added to the soil. Decomposing microorganisms release them from organic matter in simple forms that can enter plants through their roots. Plants use these nutrients to grow, and when they die, decomposing bacteria return the nutrients to the soil to continue the cycle. The system is not completely closed, because some nutrients are continually lost, and some are added. However, the amounts lost are quite small compared to the quantity of each nutrient that is held in the vegetation and soil.

Besides the role they play in nutrient cycles, animals ranging from ants to badgers have more direct physical effects on soils simply because they move vast quantities of soil as they burrow through it. An important effect of burrowing is to mix together varied layers of parent material. Animals can also sort material by particle size, bringing fine soil to the surface and burying coarse rock fragments. Most burrowing loosens soil and creates additional open space (pores) for water and air, a process appreciated by gardeners.

The effect of animal burrowing may be most noticeable when the population of burrowers suddenly changes. Recently, it has been discovered that the system of nutrient cycling in northern hardwood forests of the United States has been disturbed not by farming but by earthworm species accidentally introduced from Europe. These exotic earthworms (actually, the night crawlers often used for fish bait) have accelerated the decomposition process by mixing leaf litter into the soil. The earthworm invasion "front" is sometimes clearly visible on the forest floor, where it is marked by loss of the leaf-litter layer. Unfortunately, this loss makes the forest floor a less hospitable environment for many native plants, including spring wildflowers.

Topography Another factor that affects the formation of a soil is its location with respect to Earth's terrain. Although Jenny used the word *relief* for this factor, this term usually refers only to the local range of elevation, and many other components of topography can influence soil development. The aspect of a slope—the direction it faces—partially determines its receipt of radiation and therefore the amount of moisture evaporating from it. Soils on a shaded slope often contain more moisture than those on the opposite slopes facing the sun, and this moisture may allow more rapid soil formation. Windward and leeward slopes receive varying amounts of precipitation. Their steepness also affects runoff and thus the amount of moisture that penetrates to the lower layers of the soil. In these ways, topography can control the amounts of moisture and heat in the developing soil's environment to a large degree. A hillside might have a relatively thin layer of soil because

FIGURE 21.4 Humus, which gives upper soil layers their dark color, is formed by partial decomposition of leaves and other organic matter.

its slope allows more runoff of water and more rapid erosion of the soil. In contrast, a less well-drained valley bottom receiving a large amount of moisture is an optimal location for the chemical processes of soil formation. Material eroded from steep slopes is often added to soils at the foot of hills, creating a deeper soil mantle.

Time Whether a soil is deep or shallow, it may need a long time to form. Some processes of soil formation are slow, and thus time is an important factor. An example of relatively rapid soil formation comes from the Indonesian volcanic island of Krakatau. In this tropical climate, 35 cm (13.5 in) of soil developed on newly deposited rock materials within 45 years. The same process often takes much longer in colder climates: some of the organic matter in arctic soils at Point Barrow, Alaska, is still not thoroughly decomposed, even though it is nearly 3,000 years old. Often, the first clear evidence of soil formation appears within a century, but a more mature soil requires thousands of years to evolve. This tells us how serious any damage to a soil can be. The destruction caused in a few decades of careless farming (e.g., on overly steep slopes) can take centuries to repair.

PROCESSES OF SOIL FORMATION

As we just discussed, soil formation results from a large set of processes, all occurring simultaneously within the soil's developing layers. We have considered some specific examples, such as burrowing by animals, in our discussion of the factors that influence soil-forming processes. In 1959, soil scientist Roy W. Simonson (1908–2008) published a general theory of soil formation that provides a useful framework for classifying and understanding all of these varied processes. Simonson noted that many of these processes occur in all soils during their formation but are very active in some soils and nearly dormant in others. He also observed that at a given point in the landscape, soil properties change considerably with depth below the surface. These changes are distinct enough that we can observe a sequence of layers referred to as **soil horizons**. Simonson concluded that the development of such layers in soils can be ascribed to four sets of processes (Fig. 21.5):

1. **Addition** refers to the gains made by the soil when organic matter is added from decaying plant material or when windblown dust or material moved downslope by water comes to rest on the soil. Soils often have a dark-colored upper horizon that results from the addition of organic material.
2. **Transformation** denotes the weathering of rocks and minerals and the decomposition of organic material in the soil. Weathering of minerals from the parent material produces new clay minerals, iron and aluminum oxides, and dissolved material. Organic matter decomposition results in humus formation.
3. **Depletion** results from the loss of dissolved soil components as they are carried downward by water, plus

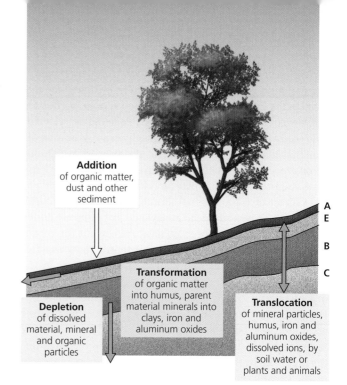

FIGURE 21.5 Four types of soil-forming processes. Together, these processes differentiate the parent material into soil horizons, labeled **A**, **E**, **B**, and **C** on the right side of the figure.

the loss of mineral or organic matter particles in suspension as the water percolates through the soil or runs off over the ground surface.
4. **Translocation** refers to the movement of dissolved and suspended particles from one depth to another within the soil. A common example is the movement of clay particles downward through soil pores, which creates clay-rich subsoils. Animal burrowing can move soil material in all directions.

These four fundamental groups of processes take place within all soils, but not at the same rate or with the same degree of effectiveness. In fact, abrupt boundaries between soil horizons suggest that conditions influencing the effectiveness of the four processes change rapidly over only a few tens of centimeters below the ground surface. This makes sense when we consider that upper horizons of the soil at a particular location are much more frequently wetted by rainfall than are deeper parts of the same soil. The upper horizons also undergo more frequent temperature fluctuations and experience much greater animal burrowing and root growth. On the other hand, when we move horizontally across the landscape, soils tend to change more gradually, in response to broader geographic changes in the factors of soil formation we have just discussed—parent material, climate, organisms, topography, and time.

Soil Profiles

We now have some information about the factors involved in soil formation and the major processes that go on within soils. The time has come to dig a hole in the soil in a certain location and see what lies below the surface. After a few hours of vigorous digging, we have made a hole or pit about 1.5 m (5 ft) deep. Even before we finish, we can easily recognize a series of horizons, revealed by changes in color and by the "feel" or texture of the materials we encounter. Often, we can even recognize certain horizons by how hard it is to dig through them. It is important to recognize that soil horizons do not represent layers of material deposited on the soil, one after another. Instead, soil horizons develop in place as soil-forming processes differentiate the original parent material into zones with different properties at various depths. Some soils have quite sharply defined horizons; in others, horizons are difficult to identify. All the horizons, from top to bottom, are known as the **soil profile**. A soil profile is to a soil geographer what a fingerprint is to a detective. Soil scientists, of course, cannot take the time to dig holes wherever they study soils, so they often use a device called a soil auger, a hollow, sharp-edged pipe that is used to retrieve a sample of the upper horizons (or, in a shallow soil, the entire profile).

When we look at a 1.5-m (5-ft) profile of a well-developed, mature soil, it is usually not difficult to identify the major horizons. Soil scientists for many years used a logical scheme to designate each horizon in a model profile: they divided the profile into three segments named alphabetically. The **A** horizon lies at the top (often darkened by organic material); the **B** horizon is in the middle, often receiving dissolved and suspended particles from above; and the **C** horizon lies at the bottom, where the weathering has had only minor effects on the parent material.

This **A**–**B**–**C** designation was eventually found to need some modification. For instance, many soil profiles include an uppermost horizon—*above* the **A** horizon—consisting mostly of organic material in various stages of decomposition. Where this occurs, it is identified as an **O** horizon. Many forest soils have a thin, dark-colored **A** horizon, with a much lighter-colored, less humus-rich layer just below. The lower, lighter zone was initially considered a type of **A** horizon, but eventually, its distinctive qualities were recognized, and it was designated as the **E** horizon (see Fig. 21.5). At the base of the soil, we apply the term **R** horizon to refer to relatively unweathered rock.

A typical soil profile—for example, representing a soil found in the moist, midlatitude Piedmont region of the southeastern United States—looks like the illustration in Figure 21.6. Note that lowercase letters are often added to indicate specific types of the major horizons; for example, **Oa** and **Bt** designate special types of the **O** and **B** horizons. In this section, we discuss a few of the many possible kinds of soil horizons using the profile depicted in Figure 21.6.

The profile in Figure 21.6 contains horizons marked by six capital letters—**O**, **A**, **E**, **B**, **C**, and **R**. These are the *master horizons*. The **O** horizon actually consists of two layers: a bed of not-yet-decomposed leaves and twigs on top (**Oi**), and beneath this a layer of highly decomposed organic matter (**Oa**). This is a good example of the use of lowercase letters to subdivide master horizons into more specific categories. Immediately below the **Oa** layer is the

A horizon. Although the **A** horizon is derived mainly from the same mineral parent material as the lighter-colored horizons below, it is colored dark brown by the organic matter added by fallen leaves and roots that is concentrated near the surface. Soon, however, the soil becomes lighter, and about 25 cm (10 in) down, it shows evidence of depletion—the removal by percolating water of material in solution or suspension. This lighter material forms the **E** horizon (**E** standing for the process known as **eluviation**, the general term for such removal, meaning "washing out").

Below the **E** horizon is the **B** horizon. When a soil is young or its development is limited for some other reason, it may not yet possess a clearly expressed **B** horizon. Figure 21.7 illustrates the development of soil horizons on a transported parent material. Note that a **B** horizon develops after the **A** horizon and only appears after the soil has evolved for some time (see the third panel in Fig. 21.7). When the **B** horizon first develops, it is differentiated from the parent material mainly by its structure and color, and is not yet enriched by particles carried downward from the **O** and **A** horizons. Such weakly developed **B** horizons are labeled **Bw** (**w** for weak). After enough time has elapsed, translocation is in full force (see the final panel in Fig. 21.7). At this point, an **E** horizon has often developed through eluviation, and the **B** horizon matures. The term **illuviation** signifies the deposition of dissolved material or particles carried downward by percolating water. Thus, eluviation (removal) from the **A** and **E** horizons is matched by illuviation (deposition) into the **B** horizon.

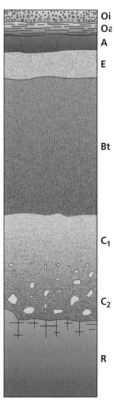

FIGURE 21.6 Soil profile typical of the humid midlatitudes showing soil horizons.

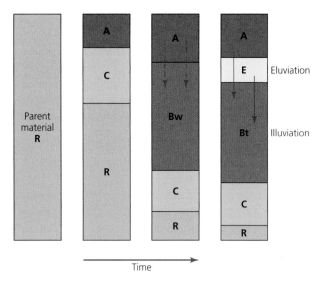

FIGURE 21.7 Stages of soil horizon evolution on a sedimentary parent material.

The symbol **Bt** identifies a **B** horizon formed by illuviation of clay particles, an especially important process of translocation. This movement of clay is enhanced by chemical changes in the soil. For instance, the sedimentary materials from which many soils of the U.S. Midwest are formed, consisting of rock pulverized by glaciers, contain calcium carbonate ($CaCO_3$), or lime. The presence of abundant dissolved calcium in these materials causes clay particles to *flocculate*, or cling to each other and to sand or silt grains. When rainwater percolates downward through such a soil, the $CaCO_3$ is slowly dissolved in the water and carried away (a process of depletion). This eventually makes the topsoil acidic, which in turn frees the clay particles to move with water flowing through soil pores and form a **Bt** horizon.

What happens when the parent material contains little or no clay? There is little clay illuviation into the **B** horizon, and no **Bt** horizon develops. However, the decaying organic materials in the **O** horizon do acidify rainwater, causing chemical reactions with aluminum (Al) and iron (Fe) in the upper soil. Oxides of Al and Fe are then illuviated into the **B** horizon along with some organic matter, creating the reddish-brown **Bhs** horizon of many sandy forest soils. This is another process of translocation. In the United States, soils with **Bhs** horizons occur in forested areas from Maine to Florida. Yet another process of **B** horizon development involves the accumulation of $CaCO_3$, which precipitates from soil water. Horizons produced in this manner are designated **Bk**.

In the **C** horizon, the parent material is in the earliest stages of being transformed by weathering into soil particles and lacks many of the distinctive properties of soil. The parent materials of soil range widely, of course, from already-pulverized glacial sediments and river-deposited alluvium to hard bedrock. Figure 21.6 assumes a hard-bedrock parent material. Accordingly, the **C** horizon in Figure 21.6 is divided into a lower zone (**C2**), where pieces

of bedrock still lie interspersed with weathered soil material, and an upper zone (**C1**), where the weathering process is more advanced. The **R** horizon in Figure 21.6 denotes solid rock, which is barely affected by soil formation. In this horizon, the transformation process occurs only where cracks and other zones of weakness in the bedrock are loosened and opened.

Soil Regimes

The soil profile in Figure 21.6 is only one of thousands of such profiles charted by soil scientists around the world. Imagine the range of possibilities. Take the complex pattern of global geology, superimpose a map of climates, overlay this with the diversity of biogeography, add local variations of relief and slope, and insert the factor of time. The resulting soil map is infinitely complicated, and understanding it is quite a challenge.

Soil geographers often use the notion of **soil regimes** to comprehend spatial patterns of soil formation. The term *soil regime* refers to a particular environment of soil formation that produces a specific set of soil characteristics. Even if their parent material remained constant throughout the world, soils would differ because they form under different temperature, moisture, biogeographic, and other conditions or regimes. Recall from Part Two that soils may form under continuously moist, seasonally moist, or dry conditions and under continuously warm, seasonally warm, continuously cool, seasonally cold, or continuously frigid conditions—or at any point in between. Early in the study of soils, it became clear that certain soils with particular characteristics prevailed under certain regimes: soils with abundant calcium carbonate and **Bk** horizons under arid climates; soils with **Bt** or **Bhs** horizons under forests in moist climates of the middle and high latitudes; and soils rich with iron and aluminum oxides under moist, warm climates in tropical areas.

By studying soils developed under such contrasting regimes, soil scientists learned how soil-forming processes vary in intensity from region to region. Under arid climates, eluviation is constrained by the limited amount of water available for downward percolation. The soil is rarely wetted deeply, and as it dries out after wetting, $CaCO_3$ forms in shallow horizons. This often results in light-colored, cemented **Bk** horizons (Fig. 21.8). Soils that develop under such regimes are appropriately called *Aridisols*.

Soils that form under coniferous forests in the cool, moist climate of higher latitudes reveal the effects of quite different processes. Under forest cover, the soil acquires an **O** horizon made up of a mat of conifer needles. When rainwater or meltwater seeps through the **O** horizon, it extracts organic acids. This acidic solution enters the **A** and **E** horizons, enhancing weathering processes. Iron and aluminum dissolve under these conditions and eluviate downward, often attached to organic compounds. In the **E** horizon, mineral grains stripped of iron and other coloring agents give the soil a light gray, ashlike color (Fig. 21.9).

FIGURE 21.9 The gray, ashlike color of the **E** horizon is a common feature of the soil type classified as *Spodosol*.

FIGURE 21.8 An Aridisol in the Mojave Desert of southern Nevada. The light color starting at a depth of 60 cm (24 in) is a **Bk** horizon formed by the accumulation of calcium carbonate.

Deeper in the soil, iron, aluminum, and organic matter all accumulate to form a dark reddish-brown **Bhs** horizon. The resulting soils are called *Spodosols*, based on the Greek word for "wood ash" and referring to the appearance of the **E** horizon.

Another kind of soil regime prevails in moist equatorial and tropical areas, where temperatures are high year-round and moisture is abundant. Bacteria and fungi are especially active under these conditions. Organic matter is decomposed quickly, and relatively little humus accumulates despite the large quantities of organic matter that are added each year from tropical forests. Weathering of minerals is intense, leaving behind compounds of iron and aluminum that increasingly come to dominate older soils of the tropics. The ultimate result is often a very thick soil, poor in nutrient materials but rich in iron and aluminum oxides, which color it with a characteristic reddish tinge (see Fig. 23.6). Such soils are referred to as *Oxisols*.

These are just three combinations of regimes producing characteristic soils—three among thousands. Before we try to make sense of the global and regional distribution of soils (Unit 23), we must examine the more important physical and chemical properties that mark them (Unit 22).

Key Terms

addition *page 276*

depletion *page 276*

eluviation *page 277*

humus *page 275*

illuviation *page 277*

nonrenewable resources *page 272*

parent material *page 274*

renewable resources *page 272*

residual parent material *page 274*

soil *page 272*

soil horizon *page 276*

soil profile *page 277*

soil regime *page 278*

transformation *page 276*

translocation *page 276*

transported parent material *page 274*

Scan Here for a quick vocabulary review

Review Questions

1. Taking into account the five major factors in soil formation, explain why two soils forming in the same climate under similar vegetation might still cnd up with quite different horizons and other characteristics.

2. If you were studying a soil profile in the field, which characteristics might allow you to identify the **O**, **A**, **E**, **B**, and **C** soil horizons?

3. Describe at least one process that produces each of the following horizons: **O**, **A**, **E**, and **B**. For each process, state whether it represents addition, transformation, depletion, or translocation.

4. What properties and horizons of soils might be different if there were no land plants or animals (which was the case early in Earth's history)?

Scan Here for a quick concept review

www.oup.com/us/mason

Properties of Soil

Large cracks formed as this soil dried, an indication of high clay content. Soil structure is formed in part by such shrinkage and the swelling that occurs when the soil is wetted again by rainfall.

OBJECTIVES

- Define terminology used to describe soil characteristics
- List key physical properties that affect plant growth and practical uses of soil
- Discuss chemical properties of soil and how they influence the supply of nutrients to plants
- Illustrate the likely arrangement of soil characteristics in a hypothetical landscape
- Describe soil development as a system with positive and negative feedbacks

The next time you take a daytime highway trip, you can make it much more interesting by stopping at some road cuts and examining the exposed rock and soil. In Part Five, we discuss some of the rock types and structures you may be able to see; if you are lucky, you might even find an unusual mineral or fossil. But the soil alone makes a stop worthwhile. Especially in relatively fresh road cuts, you may be able to see several soil horizons, humus, plant roots, and even the imprints made by worms and other inhabitants of the soil.

Do not expect to recognize all the soil properties the road cut reveals. It is one thing to understand a soil profile from a textbook, but in the field those horizons might not be so well defined and evident. It does, however, help to know some basic terminology, and you can measure some of the soil properties yourself. If the road cuts through a hill, you may note that the soil's depth changes from the top of the hill to the bottom, often becoming thinnest on the steepest part of the slope where the rate of erosion is greatest. Individual soil horizons may also vary considerably down the slope.

Sol and Ped

The terms *sol* and *ped* appear frequently in soil studies either alone or in some combined form. *Sol* is a shortened version of a Latin word for soil or ground (*solum*), and *ped* is derived from the Greek word *pedon*, which has a similar meaning as sol. We already have encountered both terms, in Oxi*sol* in Unit 21 and in *ped*ology, the science of soils, which we introduced in Unit 20. In examining the classification of soils in Unit 23, we identify soil orders by means of a dominant characteristic followed by *sol*. For example,

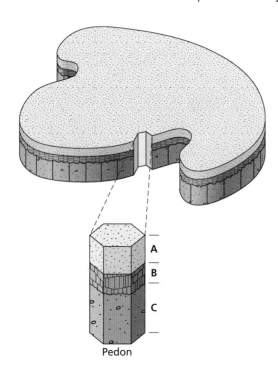

FIGURE 22.1 Complete soil column, or pedon.

soils forming in dry areas are known as *Aridisols: arid* plus *sol*. We use the term *ped* by itself to identify a naturally occurring aggregate, or "clump," of soil and its properties. The properties of naturally formed peds are important, and we discuss them later in this unit.

The original Latin and Greek words *solum* and *pedon* are also used in pedology, but with special meanings. The **solum** of a soil consists of the **A** and **B** horizons and constitutes that part of the soil in which the processes of soil development have been most active. Below the solum, in the **C** horizon, parent material still has most of its original properties. The **pedon** is a column of soil drawn from a specific location, extending from the ground surface all the way down into the **C** horizon (Fig. 22.1). That is, a pedon is a soil column representing the entire soil profile.

Physical Properties of Soil

Some of the most basic physical properties of a soil can be directly observed or estimated in the field, including the soil's *texture*, *structure*, and *color*. By studying those basic properties, we can often learn a great deal about other physical characteristics of the soil, such as whether it can take up and store enough water to sustain rapid plant growth, or whether it can support the weight of a building without settling too much.

SOIL TEXTURE

Descriptions of soils often feature terms such as *clay*, *loam*, or *silt*. These terms have quite specific meanings and refer to the sizes of the individual particles that make up a soil (or one of its horizons). If you were to rub a tiny clump of soil between your thumb and forefinger, you would be left with the individual grains that make up the clump. These grains may be quite coarse and look or feel like sand, or they may be very fine, like dust on your fingertip. The size of the particles in soil, or its *texture*, is important because particle size is related to the closeness with which soil particles can be packed together, the amount of water the soil can retain, how easily roots can penetrate it, and other aspects of soil behavior and performance.

Soils often contain particles of several very different sizes. For instance, that clump on your forefinger may contain some sand and some much finer particles. In your fieldnotes, therefore, you might call it a *sandy clay*—a mostly fine-grained, sticky soil with a gritty feel indicating the presence of some coarser particles. The coarsest grains in a small clump of soil are *sand* (not counting larger gravel and boulders sometimes found in soil as well). Sand particles, by the official definition of the Natural Resources Conservation Service of the U.S. Department of Agriculture (USDA), range in size from 2 down to 0.05 mm (0.08 to 0.002 in). The next-smaller particles are called *silt* (0.05 to 0.002 mm [0.002 to 0.00008 in]). There are still-smaller grains than silt—namely, *clay* particles. The smallest clay particles are less than a one-hundred-thousandth of a millimeter in diameter.

A single soil usually contains grains of all three size categories: sand, silt, and clay. Such a soil is called a **loam** if all three are present in roughly similar proportions. Unlike sand, silt, and clay, therefore, loam refers not to a size category but to a certain combination of variously sized particles. The USDA has established a standard system to ensure that such terms as *loam, sandy clay,* and *silt loam* have objective meanings; they are defined by specific proportions of sand, silt, and clay (Fig. 22.2).

When you rub some soil between thumb and forefinger, you can tell that particles of different sizes are in the clump, but you can only estimate in what proportions. A more accurate determination of these proportions requires laboratory analysis, and then the numerical results must be checked against the official USDA graph (Fig. 22.2). Note that this triangular chart shows the percentages of all three components. A soil that is about one-third sand, one-third silt, and one-third clay falls in the clay loam area. A soil with 45 percent silt, 45 percent clay, and 10 percent sand is a silty clay. You can choose other combinations yourself and see what they are called by using the chart.

Soil texture is related to the parent material from which the soil was derived. Some types of bedrock or sedimentary parent materials yield sand-rich soils, whereas others give rise to clayey soils. As noted earlier, texture is a critical factor determining the pore spaces in soil. The *porosity* of a soil is the proportion of the total soil volume made up of pore space, which can be occupied by either air or water. Texture has a particularly strong influence on the typical size of soil pores: large pores occur between sand grains, while the pores between clay particles are minuscule. Pore size in turn controls the soil's *permeability*, the rate at which water or air can flow through the soil. Large pores allow much more rapid flow than small pores.

Small pores also increase the soil's **field capacity**, the amount of water it can hold against the force of gravity. To visualize this property of soils imagine a sponge that is wetted and then held up so that the water can drain out freely. A portion of the water is always retained in the sponge. Similarly, when soils are thoroughly wetted by rainfall, some of the added moisture is retained even after the soil has freely drained for a day or two. The amount of moisture that is retained is the field capacity, and it is important in determining how much water will be available for plant use. In fact, fine-grained soils can retain water in their pores much more effectively than a sponge, and even plant roots cannot extract some of the water retained in the smallest soil pores.

Thus, water does not readily flow into or out of the tiny pores of clay-rich soils, but what does enter the soil is tightly held. Rain that falls on a clay soil often runs off over the surface rather than entering the soil. In addition, the small and often water-filled pores of a clay soil reduce the circulation of nutrient- and oxygen-carrying solutions, with negative effects on plant growth. Clay soils shrink as they dry, opening up large cracks, and then swell again when wetted. This behavior can damage buildings and roads.

In contrast, a sandy soil with much larger pores allows water to percolate downward under the influence of gravity, readily taking up rainwater but also draining (and drying up) rapidly. Plants with roots in sandy soils have little opportunity to absorb water because such soils are quickly drained; therefore, vegetation on sandy soils often suffers stress from lack of moisture. On the other hand, sandy soils do allow an adequate supply of oxygen to plant roots, they are easily worked with farm machinery, and they provide a stable base for buildings.

From what you have just learned, you would probably guess that loams—mixtures of sand, clay, and silt—present the best combination of textural properties for most practical purposes. Indeed, that is the case. Loams do not stay waterlogged, nor do they yield their water content too rapidly. Pore spaces are large enough to keep the soil well oxygenated around plant roots. Good drainage, which is directly related to soil texture, is a key to successful crop cultivation. Indeed, many farmers say texture is more important than nutrient content in selecting good farmland. Nutrients can be supplemented artificially, but soil texture cannot easily be changed.

Soil consistence refers to a group of properties that are closely related to texture. These properties help determine how easy or difficult it will be to work the soil with agricultural implements or construction equipment and whether structures built on the soil will be stable. Consistence is described using the results of simple tests done in the field—for example, by rolling some moist soil in the hand and observing its behavior. After subjecting a bit of soil to this test, some of it will have stuck to the skin, indicating its *stickiness*. The rolled-up soil may form a small rope and then break up, or it may attain a thin, twine-like shape. This reveals its *plasticity*, which is the degree to which the soil can be worked into a new, permanent shape. The greater the clay content, the more tightly the soil particles will bind

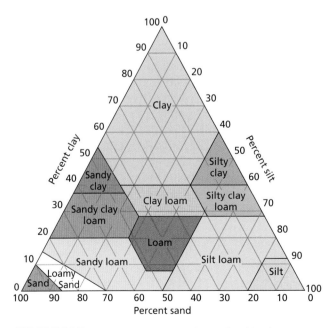

FIGURE 22.2 Soil texture categories defined by the percentages of sand, silt, and clay found in a soil sample.

together and the longer the rope- or worm-like roll will be. When dry, a clay-rich soil will be very hard. Conversely, the greater the sand content, the more likely the soil is to be highly *friable*, that is, to easily crumble into small pieces.

SOIL STRUCTURE

Recall that peds are naturally occurring clumps of soil that tend to form and stay together unless they are purposely broken up. Again, you will discern this tendency in soil when you examine it: dislodge a bit of soil and it will not disintegrate, like loose sand does, but will instead form small clumps. Only when you rub these peds in your hand do they break up into individual grains. Peds develop because soil particles are packed together as roots grow through the soil or when the soil shrinks as it dries out. The closely packed soil particles may also be bound together by films of clay or humus. Peds give soil its *structure*, and they are quite important because they affect the circulation of water and air throughout the soil. Soil structure—that is, the nature of its peds—affects the soil's vulnerability to erosion, its behavior under cultivation, and its durability during dry periods.

Soils exhibit four basic structures: *platy*, *prismatic*, *blocky*, and *granular* (see Figs. 22.3 through 22.6).

1. **Platy structure** (Fig. 22.3), as the term suggests, involves peds that look like thin plates stacked horizontally. Platy structure is most often observed in **E** horizons but can occur in other parts of the profile.

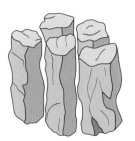

A

B

FIGURE 22.4 Prismatic soil structure.

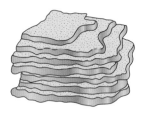

A

B

FIGURE 22.3 Platy soil structure.

2. **Prismatic structure** (Fig. 22.4) involves peds arranged in vertical columns and often occurs in **B** horizons. Individual prismatic peds range from 0.5 cm (0.2 in) to as large as 10 cm (4 in) and often break up into smaller blocky peds when some force is applied to them.

3. **Blocky structure** (Fig. 22.5) consists of irregularly shaped peds that fit together roughly like building blocks. Angular blocky structure is made up of sharp-edged blocks that fit closely together, while the peds in subangular blocky structure are a little more rounded. Blocky structure is typical of **B** horizons but can occur anywhere in the soil profile.

4. **Granular structure** (Fig. 22.6) displays peds that are usually small and often nearly round in shape so that the soil looks like a layer of breadcrumbs. Soils with such structure are highly porous and have high permeability if the peds remain intact. Small granular peds the size of sand grains are susceptible to erosion by wind or water. Granular structure is much more common in **A** horizons than deeper in the soil profile.

Not only the shape of the peds but also their size and strength are important, and these characteristics are included in standard soil profile descriptions. (Note the objects included to show the scale of each of the photographs in Figs. 22.3 through 22.6.)

A

B

FIGURE 22.5 Blocky soil structure.

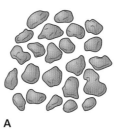

A

B

FIGURE 22.6 Granular soil structure.

SOIL COLOR

Soils generally exhibit a range of colors. Only rarely do soils not display some color variation along their profile. Soil color is a most useful indicator of the processes that prevailed in soil development. A dark brown to black color often reflects the presence of humus; brown and red colors usually result from iron oxides. A soil's color may range from nearly black in the **A** horizon at the top of the profile to reddish-brown in the **B** horizons to light brown in the **C** horizon. These changes indicate the decrease in humus from the **A** to the **B** horizon and the production of iron oxides by weathering of the parent material, adding color to the **B** horizon. Soils that are commonly waterlogged often have distinctly grayer colors than well-drained soils.

Soil scientists use the Munsell Soil Color Chart to describe soil colors objectively. This chart contains hundreds of colors, each with an alphabetical and numerical code. To use the chart, we find the color in the chart that best matches the color of a sample of soil and record its code. If the job is done carefully, we can be confident that a coded soil from the **A** horizon of a U.S. Great Plains pedon is approximately the same color as one from an **A** horizon in Ukraine without having to put the samples side by side. Wetting causes soil colors to become much more vivid, so it is important to ensure that soil samples have about the same moisture content before comparing their colors.

Chemical Properties of Soil

To determine the chemical properties of soils, it is usually necessary to bring samples into the laboratory. While this is time-consuming, it is essential for judging how effectively a soil can supply nutrients to plants, or whether it will limit plant growth because of its acidity or alkalinity.

IONS, ACIDITY, AND ALKALINITY

Ions, atoms or groups of atoms that have electrical charges, are always present in soils, both dissolved in water in the soil pores and bound to the surface of soil particles. Positively charged ions, such as those of calcium (Ca^{2+}) or hydrogen (H^+), are known as **cations**. Ions often form from minerals that dissolve in soil water, part of the process of weathering, which we discuss in more detail in Unit 36. Calcite ($CaCO_3$, or calcium carbonate), for example, is a common mineral that breaks up into calcium and bicarbonate ions when it dissolves. Percolating rainwater can add other ions to the soil as well.

Hydrogen (H^+) ions are especially important because their abundance in the soil solution defines the **acidity** of the soil. The acidity of soil is measured by its pH value, which decreases as H^+ ions become more abundant. A pH value of 7.0 is considered neutral. Lower values (normally between 3.5 and 7.0) indicate acidic soils, whereas higher values (7.0 to 11.0) indicate *alkaline* soils. Different plants and microorganisms are adapted to varying pH levels, and the patterns of vegetation often provide soil surveyors with clues about whether the underlying soils are acidic or alkaline.

The processes that control soil acidity are complex. Carbon dioxide from the atmosphere dissolves in soil water, leading to chemical reactions that release H^+ ions. Organic acids are released into the soil from decaying leaves and needles. In the short term, these processes can be counteracted by mineral weathering, which consumes H^+ ions and keeps the soil from becoming acidic. In the long term, however, most soils of humid regions tend to become considerably more acidic over time. As part of this long-term acidification, important plant nutrients are lost from the soil or converted to forms that plants cannot use, so old, highly acidic soils have low fertility. Many centuries ago, it was discovered that adding lime (calcium carbonate) to such soils makes them much more productive. We now know this is because rapid weathering of the lime consumes H^+ ions and raises soil pH. This should remind us that fertilizers and other manipulation of the soil's chemistry have long been an integral part of the human use of Earth.

Alkaline soils are common in arid regions, where little if any water percolates through the soil profile. Here, the natural acidification that occurs in humid soils does not take place, and calcium carbonate (lime) often forms naturally in soil horizons, maintaining the pH at high levels. Extreme soil alkalinity is usually related to the presence of sodium salts and is toxic to many plants.

COLLOIDS AND CATION EXCHANGE

Growing plants must take up essential nutrients from the soil. How can a nutrient be held in the soil and then released to enter plant roots? One of the important processes by which nutrients are stored and released involves nutrients that occur as cations. Three essential nutrients that form cations are calcium, potassium, and magnesium. Although we did not discuss them in Unit 20, each of these elements, like carbon and nitrogen, has a *biogeochemical cycle* that involves release from rocks by weathering, along with cycling back and forth between soil and plants.

Since ions are dissolved, you might expect that they would be quickly carried away as water percolates down through the soil, but in fact, they are often retained on tiny particles called **colloids**. Colloids are generally smaller than 1 μm (0.000001 m [0.00004 in]) in diameter, and they include both minerals and humus.

Many colloid particles possess a distinctive property: they have negative electrical charges on their surfaces, which attract positively charged cations, including calcium, magnesium, and potassium. The colloids hold these cations but can release them slowly as plants use up the supply of nutrient cations that are freely available in the soil water. The cations released from colloids are replaced by other cations from the soil water, so the overall process

is referred to as **cation exchange**. Cation exchange provides a slow but steady supply of nutrients to plants. Without the colloids and their charged surfaces, these nutrients would quickly be carried away by water percolating through the soil.

Soil scientists often use a lab procedure to measure the total quantity of a cation, such as calcium, that can be released from soil colloids. A soil sample is mixed with water containing abundant cations of some other element, such as sodium. The sodium cations displace almost all of the calcium cations into the water, where they can be measured. A soil sample with abundant *exchangeable calcium* can supply even the most rapidly growing plants with all the calcium they need.

Soils of Hills and Valleys

Topography strongly influences soil formation, mainly through its effects on the flow of water and soil erosion. In a flat or gently undulating (rolling) countryside, most rainfall tends to infiltrate into the soil, and as it percolates downward it drives the major soil-forming processes, such as weathering and illuviation. In this type of landscape, soil profiles tend to develop more fully and are often fairly similar across large areas. In areas where the landscape is hilly, local variation of the soils tends to increase. More of the rainfall runs off over the surface from steeper slopes and narrow ridge tops, and soil erosion is also more rapid in those locations. As a result, the soil profile is often thinner and less well developed on steep slopes and narrow hilltops. Especially thick soils are found at the foot of the slopes where surface runoff infiltrates and eroded soil material accumulates. As one observes the soils in a local area, it is soon possible to predict how the soils will vary on a particular slope based on patterns that are repeated across the landscape.

Soil scientists noted this kind of pattern while carrying out soil surveys in East Africa in the 1930s and suggested the term **soil catena** (pronounced kuh-TEE-na) to describe it. The word *catena* is from Latin, meaning chain or series. A soil catena is a sequence of soil profiles appearing in regular, predictable succession down the slopes of a local region with uniform geology. Figure 22.7 illustrates a catena similar to one of those originally studied in East Africa. Thin soils with many rock outcrops are found near the top of the hills. Over time soil particles moved some distance downhill by overland flow during rainstorms or by the burrowing of animals to form a material known as *colluvium*, in which distinctive red soils developed. Near the foot of the hills, flowing water carried material washed off the higher ground and sorted the grains by size. Larger sand grains were deposited first, forming sandy soils, and thick dark-colored clay soils are found at the bottom of the catena. Pedologists have found the concept of the catena useful in describing the geography of soils in many regions, but it is important to remember that each catena is unique, depending on the local topography, bedrock, vegetation, and climate.

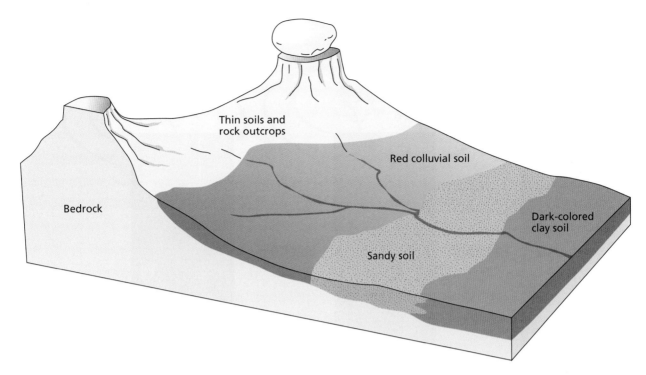

FIGURE 22.7 Soil catena in East Africa.

Development of the Soil System

We are now at the point where we can sketch out major processes in the development of soil and its properties, regarding the soil and vegetation as a *system* (Fig. 22.8). One important characteristic of many natural systems is the presence of *feedbacks*. A feedback occurs when a process in the system produces results that, in turn, alter the rate of the original process. A positive feedback makes the process more rapid and/or more effective, while a negative feedback slows down the process.

The lithosphere furnishes the soil parent material (see A in Fig. 22.8). Moisture is added from the atmosphere in the form of precipitation, plants become rooted in the soil, animals burrow through it, and microorganisms colonize it. Plants add organic matter, and the growth of their roots together with animal burrowing and wetting and drying cycles produce soil structure (see B in Fig. 22.8). Beneath the surface, water and heat from the atmosphere begin to break down rock particles through *weathering*, releasing dissolved ions and also producing some newly formed clay and iron or aluminum oxides (weathering is discussed in more detail in Unit 36). Weathering changes the soil texture, though very slowly.

In the early stages of soil development (B in Fig. 22.8), chemical elements that are essential for plant growth build up within the plant–soil system. The vital nutrient nitrogen is captured from the air through fixation and accumulates within both the living vegetation and the soil organic matter (see the discussion of the nitrogen cycle in Unit 20). Other nutrients such as phosphorus,

potassium, and calcium are released from rocks through weathering and are retained in organic matter or attached to clay particles, which are also products of weathering.

Now an important positive feedback emerges. As the store of nutrients builds up, plant productivity increases. More organic matter may be added to the soil, root growth may be more extensive, and more diverse animal life supported by richer plant growth may burrow and mix the soil more effectively. Furthermore, as soil structure develops, more water can percolate down through the soil to greater depths, increasing the rates of weathering and other soil-forming processes. In this way, initial plant growth and soil development lead to enhanced plant growth and more effective soil development.

Much later in the system's development, negative feedbacks can set in (see C in Fig. 22.8). As weathering continues, more and more of the soil is made up of resistant mineral grains that are not easily decomposed and do not yield many dissolved nutrients. As the soil becomes more acidic over time, certain nutrients are depleted or immobilized. Overall, the availability of nutrients to plants either remains stable or decreases, limiting plant productivity. Buildup of clay in subsurface horizons can slow the deep percolation of water. Through all of these effects, the major soil-forming processes work at slower and slower rates, and eventually the soil changes little over time. In other words, as soil formation progresses, it can eventually become self-limiting.

Over geologic time, the development of the plant–soil system can shift back to an initial or intermediate stage

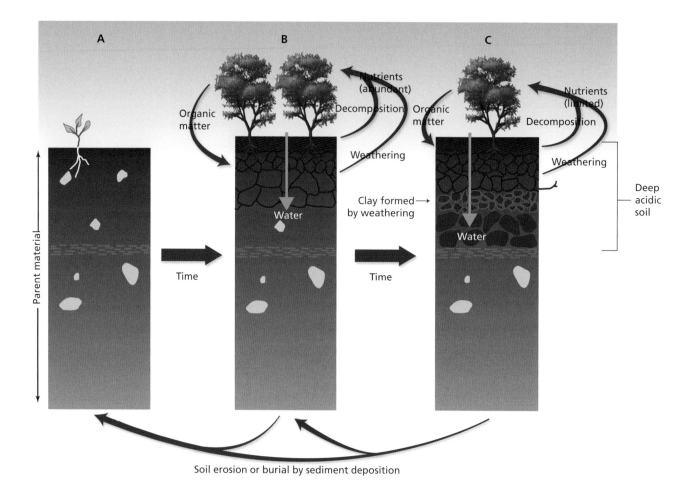

FIGURE 22.8 Development of the soil over time, viewed as a system with positive and negative feedbacks. Initial development, from A to B, often favors increased availability of plant nutrients, more productive plant growth, and more effective soil development. Later stages of soil development (B to C) can limit nutrient availability and plant growth, and slow the rate of soil-forming processes.

(A or B, Fig. 22.8). A landslide may remove not only vegetation but the entire soil profile from a hillside, or slower erosion may remove only the upper, most weathered part of the soil. Climate change may reinvigorate some soil-forming processes or allow expansion of glaciers and ice sheets that erode existing landscapes and deposit new soil parent material. Then the development process, with its positive and negative feedbacks, begins again.

Key Terms

acidity *page 285*

blocky structure *page 284*

cation *page 285*

cation exchange *page 286*

colloids *page 286*

field capacity *page 283*

granular structure *page 284*

ion *page 285*

loam *page 283*

pedon *page 282*

platy structure *page 284*

prismatic structure *page 284*

soil catena *page 286*

solum *page 282*

Scan Here for a quick vocabulary review

Review Questions

1. How does the *solum* of a soil differ from a *pedon*?

2. What is meant by the *permeability* of a soil, and why might a sandy soil and a soil that has well-developed granular structure both have high permeability?

3. Which soil would probably have a greater field capacity, one that has a silty clay loam texture or one that is a loamy sand? Explain your answer using the textural triangle shown in Figure 22.2.

4. How is soil acidity defined, and how does it generally change over long periods of time in a humid-region soil?

5. Why does the soil vary so much along a catena, such as the one shown in Figure 22.7?

6. Describe a positive and a negative feedback involved in the development of the soil system over time.

Scan Here for a quick concept review

www.oup.com/us/mason

UNIT 23
Classification and Mapping of Soils

A Northern Wisconsin landscape with soils developed under deciduous forest, still present at right. On the left, the forest has been cleared, and the plowed soil has the light color typical of cultivated Alfisols.

OBJECTIVES

- Relate a brief history of pedology, highlighting problems in achieving a universal soil classification scheme
- Outline the current system of soil classification, the Soil Taxonomy
- Describe the 12 soil orders in the Soil Taxonomy and discuss their regional patterns on both the U.S. and the world map

In this unit, we focus on the geographical perspective in order to discover how soils are distributed across the United States and the rest of the world. As noted at the beginning of Part Four, the map of world climates continually reemerges as we discuss soils, vegetation, and animal life on Earth's surface. In addition to reflecting climates, soil maps also record the effects of the other factors that influence soil formation, which we discussed in Unit 21. Because some of those factors vary greatly even on a single hillside, a small area can contain a bewildering variety of soils. For more than a century, pedologists have been working to devise acceptable systems of classifying soils in ways that can be represented on maps.

Classifying Soils

Until about 1850, soils were believed to be simply weathered versions of the underlying bedrock. The Russian scholar Vasily V. Dokuchaev (1846–1903) was the first pedologist to demonstrate that soils with the same parent material develop differently under different environmental conditions. Dokuchaev and his students traveled extensively through the steppes and forests of Russia, noting how the soils changed with vegetation and climate and laying the foundations for our modern understanding of soil geography.

Soil scientists in the United States had been grappling with many of the same problems and eventually learned of Dokuchaev's school of thought through a book published in German by one of his students. At that time, the most influential American soil scientist was undoubtedly Curtis F. Marbut (1863–1935), director of soil surveys for the U.S. government. He directed the development of the first U.S. system of soil classification, published in 1938, which often employed Russian ideas and Russian terminology, but with a particularly strong emphasis on the role of climate. The Marbut system classified soils based on assumptions about the type of mature soil that should normally be found in a given climatic zone. Unfortunately, judging whether a soil fit that idealized type was somewhat subjective, and soils that clearly did not fit the expected type were often neglected. Soil surveyors working for the U.S. Department of Agriculture (USDA) soon recognized that a variety of different soil types can occur in a single climatic zone, and that certain distinctive kinds of soil occur in a wide range of climates.

To more accurately represent the complex patterns of soil geography they observed in the field, the USDA soil survey staff sought to move away from assumptions about what kinds of soils should be typical of particular environments. Instead, they developed a system in which soils are classified using well-defined properties that can be directly observed and measured. The presence or absence of certain types of *diagnostic horizons* is especially important. Diagnostic horizons are soil horizons that are strictly defined in terms of their range of thickness, color, clay content compared with overlying horizons, and other key characteristics. After numerous revisions, a fairly complete version of the system was published in a manual called ***Soil Taxonomy*** in 1975, and that is the term commonly applied to the system of soil classification used in the United States today. Although other countries have developed their own soil classifications, most soil scientists worldwide are familiar with the Soil Taxonomy system.

Soil Taxonomy

The highest level of classification in the Soil Taxonomy is the **soil order**. A soil order is a general grouping of broadly similar soils based on the presence or absence of specific diagnostic horizons or other important properties. The 12 soil orders used in the Soil Taxonomy today are listed in Table 23.1. Below the soil order, the Soil Taxonomy has five

---

TABLE 23.1 Organization of the Soil Taxonomy

Level	Description
Order	The most general class. Soils of a given order have a similar degree of horizon development, degree of weathering or leaching, gross composition, and presence or absence of specific diagnostic horizons, such as an oxic horizon in an Oxisol. The 12 orders are: 1. Aridisols 7. Gelisols 2. Mollisols 8. Entisols 3. Alfisols 9. Inceptisols 4. Ultisols 10. Histosols 5. Oxisols 11. Vertisols 6. Spodosols 12. Andisols
Suborder	Distinguished by the climate where a soil occurs, how wet the soil typically is, distinctive parent materials, or other important soil properties (see examples in Table 23.2).
Great Group	Distinguished by the kind, array, or absence of diagnostic horizons.
Subgroup	Determined by the extent of development or deviation from the major characteristics of the great groups.
Family	Distinguishing features include soil texture, mineral composition, temperature, and chemistry.
Series	A collection of individual soils that might vary only in such aspects as slope, stoniness, or depth to bedrock.

Source: U.S. Department of Agriculture.

levels of organization, which divide soils into progressively more specific categories.

For our purposes, we need consider only the 12 soil orders, three of which (Aridisols, Spodosols, and Oxisols) were already encountered in Unit 21. Our objective here is to study the general distribution of these soil orders.

SOIL DISTRIBUTION IN RELATION TO CLIMATE AND VEGETATION

Since the time of Dokuchaev, soil geographers have emphasized the strong relationships among climate, vegetation, and distinctive soil types. Although the Soil Taxonomy was designed partly to give factors other than climate, such as parent material, the weight they deserve, relationships with climate and vegetation are still quite clear for seven of the 12 soil orders. Figure 23.1 illustrates where each of those seven orders—Aridisols, Mollisols, Alfisols, Ultisols, Oxisols, Spodosols, and Gelisols—typically occurs within broad zones of latitude in the Northern Hemisphere, with the warm tropics at the bottom and the cold Arctic at the top.

Within the tropical and temperate zones, there is a wide range of moisture conditions, from humid at the left side of the diagram to arid at the right. The type of vegetation varies with both latitude and moisture. This pattern of variation is the major topic of Unit 25.

The location on this diagram of the three soil orders we discussed in Unit 21 should make sense. Spodosols are found under coniferous forest in the cold subarctic zone, whereas Aridisols occur in deserts of the subtropical and temperate zones. Oxisols occur in the tropical rainforest, where both temperature and moisture are high and soils are affected by intense weathering. Of course, Figure 23.1 is a simplification of the real world, like any such diagram that fails to take into account local effects of topography, parent material, and soil age. Keeping that caution in mind, the relationships shown in Figure 23.1 are a good starting point for understanding the global geography of soils. Note that five soil orders—Entisols, Inceptisols, Histosols, Vertisols, and Andisols—are not placed on this diagram at all. This is because their spatial patterns result mainly from factors other than climate.

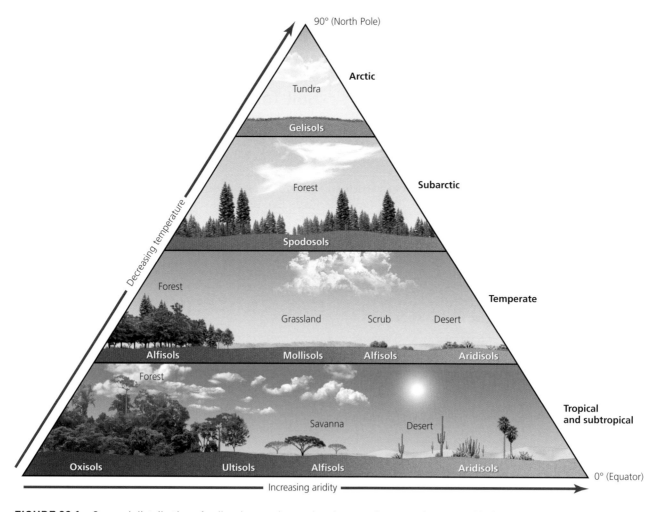

FIGURE 23.1 General distribution of soil orders and associated vegetation types in zones of latitude from the tropics northward to the Arctic. Soils and vegetation vary with both temperature (from bottom to top of the diagram) and moisture (left to right).

TABLE 23.2 Suborders in Soil Taxonomy (Examples for Two Soil Orders)

Order	Suborder	Characteristics
Entisol	Aquent	Periodically saturated with water
	Arent	Lacks horizons because of plowing or other human activity
	Fluvent	Formed on floodplains and other areas of active sediment deposition
	Orthent	Occurs in many settings where rapid erosion or other factors limit soil development
	Psamment	Formed in sandy parent material, such as dune sand
	Wassent	Continuously saturated with water
Alfisol	Aqualf	Periodically saturated with water
	Cryalf	Occurs in cold areas, including high altitudes
	Udalf	Occurs in humid climates
	Ustalf	Dry for significant part of the year, occurs in monsoon and other climates with wet summers
	Xeralf	Experiences a pronounced summer dry period (mainly in Mediterranean climates)

Source: U.S. Department of Agriculture, *Keys to Soil Taxonomy*, 11th ed. (2010).

THE SOIL ORDERS

To better understand the location of the soil orders shown in Figure 23.1 and the five that cannot be shown on such a diagram, we need to acquaint ourselves in some detail with the properties of all 12 soil orders (Table 23.1). As you read the text that follows, you may want to consult the Focus on the Science: Soil Taxonomy: What's in a Name? box (p. 295), which explains how the names of the soil orders were derived. Although our main focus is on the soil orders, it is worth noting that the names of suborders are formed by adding a prefix to a root that is part of the soil order name. For example, the suborder of *Aquents* is made up of wet (*aqu-*) or poorly drained Entisols (*-ent*). Other examples are listed in Table 23.2. The most important characteristics used to place soils in suborders are usually related to climate and periodic saturation with water.

We first discuss the seven soil orders that are clearly associated with particular climates and vegetation types, which are the ones shown in Figure 23.1. As you read the descriptions of these soil orders, the reasons for their location on that figure should become clear.

1. Aridisols

Soils classified as **Aridisols** must occur in arid climates, but they also must contain some type of **B** horizon. The **B** horizons in many Aridisols are fairly well developed and formed by accumulation of clay, calcium carbonate, or gypsum. (Recall that accumulation of calcium carbonate was discussed in Unit 21; see Fig. 21.8.) In an arid climate where little if any water moves through the soil profile, this kind of **B** horizon development probably requires a long period of soil formation on stable, slowly eroding landforms. Therefore, many desert landscapes contain a mixture of Aridisols on stable landforms and more weakly developed soils (Entisols) in areas subject to greater erosion or deposition (Fig. 23.2).

If you have spent much time in deserts, you have probably seen windblown dust. Not surprisingly, dust deposition is an important process in Aridisol formation: dust contains fine silt and clay particles and calcium carbonate that contribute to the development of the **B** horizon. Aridisols are usually farmed only in areas where they can be irrigated. Livestock grazing is a common land use on Aridisols, but the sparse desert shrubs and grasses are easily damaged by overgrazing, resulting in water or wind erosion.

2. Mollisols

Mollisols are the characteristic grassland soils of the temperate latitudes, particularly where the grasses' productivity is high. They are the soils of the steppes, the grass-covered plains of south-central Russia and the U.S. Great Plains that lend their name to the **BS** climatic zone (Unit 15). Mollisols feature a characteristic thick, dark, humus-rich surface layer, often with strong granular structure (Fig. 23.3). Several processes combine to produce this horizon. First, organic matter is added to the soil deep below the ground surface from the abundant grassroots as older roots die and are replaced by new growth. Burrowing animals that thrive in grasslands also mix organic matter deep into the soil. Last but not least, decomposition of the added organic matter is slow because the soils experience long dry periods and cool winters in which microorganisms are inactive. Because of the dry climate, weathering is limited and base cations are retained in the surface horizon. Calcium carbonate accumulation in the **B** horizon of Mollisols is common. Mollisols occasionally occur in other environments, including some tropical grasslands and even deciduous forests, but thick, dark **A** horizon formation is not generally favored in those environments. The heart of the Corn Belt in the Midwestern states of Iowa, Illinois, and southern Minnesota is dominated by Mollisols formed under tall-grass prairie. Mollisols in drier climates support most of the world's wheat crop in the United States, Canada, Russia, Argentina, and northern China.

FIGURE 23.2 Great Basin Desert, Nevada. Entisols are found on steep slopes and on recent deposits of streams; the rest of the landscape is covered by Aridisols.

FIGURE 23.3 Profile of a Mollisol, the dominant soil beneath the grasslands of the Great Plains.

FOCUS ON THE SCIENCE

Soil Taxonomy: What's in a Name?

Some of the names in the Soil Taxonomy can seem quite odd, and you may wonder why they were chosen. When this system was first proposed, the major goals were (1) to replace terms used in older soil name systems, (2) to produce words that could be remembered easily and that would suggest some of the properties of each kind of soil, and (3) to provide names that would have some meaning in all languages derived from either Latin or Greek. (The majority of modern European and American languages are so derived.) The use of Latin or Greek roots is a long tradition in science, going back to the time when almost all European scholars and scientists wrote their major works in Latin, and Greek was also widely known. The Latin scientific names used for plant and animal species originate from this tradition. In fact, the USDA sometimes called upon classical scholars to provide the Latin and Greek roots that would match the soil characteristics. The etymology of the names finally chosen is as follows:

Order Name	Derivation	Root/Key
1. Aridisol	*aridus*, Latin for "dry"	**arid**
2. Mollisol	*mollis*, Latin for "soft"	**molli**fy
3. Alfisol	*al*, from aluminum; *ferrum*, Latin for "iron"	ped**alf**er (former soil name)
4. Ultisol	*ultimus*, Latin for "last"	**ulti**mate
5. Oxisol	*oxide*, French for "containing oxygen"	**oxi**de
6. Spodosol	*spodos*, Greek for "wood ash"	**pod**sol (Russian term)
7. Gelisol	*gelare*, Latin for "to freeze"	**gel**ation (solidify by freezing)
8. Entisol	None; coined for this purpose	rec**ent**
9. Inceptisol	*inceptum*, Latin for "beginning"	inception
10. Histosol	*histos*, Greek for "tissue"	histology
11. Vertisol	*verto*, Latin for "turn"	in**vert**
12. Andisol	*ando*, volcanic ash	**and**esite (a volcanic rock)

3. Alfisols All **Alfisols** exhibit noteworthy clay accumulation in the **B** horizon, one of the diagnostic features of this soil order. The largest areas of Alfisols developed under deciduous forests of the temperate regions, in moister, less continental climates than those in which the Mollisols developed. In forests, most addition of organic matter occurs not below ground, as in grasslands, but as leaves fall to the soil surface. As a result, Alfisols forming under forest develop only a thin humus-enriched **A** horizon, underlain by a light-colored **E** horizon (Fig. 23.4). When Alfisols are plowed, the **A** and **E** horizons mix to form a light-colored surface layer, easy to distinguish from the dark-colored surface of plowed Mollisols. Forested Alfisols can become quite acidic but are less weathered and retain more base cations than Ultisols. They can support productive farming, especially where lime is applied and nutrients are added with fertilizers. In Europe, forest-covered Alfisols were cleared and put into use as farmland centuries ago, and the same process was repeated more recently in the Midwestern United States. Another significant group of Alfisols occurs in much drier environments. These include grasslands in Mediterranean climates (the Xeralfs suborder; see Table 23.2), the subtropical savannas bordering the Sahara and Kalahari Deserts (Ustalfs), and dry grasslands such as the short-grass plains of western Texas (also Ustalfs). Although we might expect Mollisols to form in these dry environments, addition of organic matter is apparently too meager to produce a thick humus-enriched horizon.

FIGURE 23.4   Profile of an Alfisol, a soil type associated with productive forests and farmland in the Midwestern United States.

FIGURE 23.5   Profile of an Ultisol, a deep, highly weathered, and acidic soil with a **B** horizon rich in clay.

Cattle or sheep grazing and irrigated fields of grain or cotton are common uses of these dry-region Alfisols.

4. Ultisols   Warmer, wetter climatic zones host the Ultisols. Like Alfisols, Ultisols are distinguished by a **B** horizon of strong clay accumulation, but they are more weathered than Alfisols (Fig. 23.5). The native vegetation of Ultisols is forest or, more rarely, moist savanna grassland. Ultisols are highly acidic, and their base cations have been greatly depleted by long-term weathering in warm, humid climates. In the eastern United States, Ultisols lie to the south of the southern border of Pleistocene glaciation, and Alfisols lie mostly to the north. This indicates that Ultisols are older by hundreds of thousands or millions of years; along with the climatic factor, this greater age helps explain their weathered nature. Ultisols are also widespread

in the humid subtropical climate of southern China and parts of Africa and South America. Ultisols have often been used for farming, but their fertility is rapidly depleted under cultivation. In the southeastern United States, many Ultisols that were once used to grow cotton or tobacco have now been planted with trees because they became so unproductive as farmland. In southern China, Ultisols have been farmed for much longer, but only through intensive replenishment of nutrients using manure or other fertilizers.

5. Oxisols   As noted in Unit 21, Oxisols are restricted to tropical areas with high rainfall. High temperatures all year and abundant water flow through the soil have resulted in extreme weathering of the parent material. Many of the common minerals have been depleted, leaving iron

FIGURE 23.6 "What should have been one day in Suva, the capital of Fiji, turned into a week due to schedule complications, and we had an opportunity to study Viti Levu's interesting human as well as physical geography. Driving through the interior was challenging because maps and reality seemed to differ; but homesteads and villages were interesting and welcoming. The island has some fertile soils, but much of the center was dominated by Oxisols which, when fully developed, are deep (often dozens of meters) and characteristically red-colored due to the preponderance of iron and aluminum in the oxic horizon. The upper part of this profile of a local Oxisol shows the virtual absence of humus in these soils. Note also the virtual absence of color change from very near the top of the soil all the way down."

and aluminum oxides that do not dissolve readily and therefore accumulate in the soil profile. Oxisols tend to have **B** horizons that are colored bright red or orange by iron oxides (Fig. 23.6). The natural vegetation on most Oxisols is tropical rainforest or savanna. Within the soil, important nutrients such as phosphorus are present in low levels or in forms that are not usable by plants. Therefore, most nutrients are in living vegetation. When leaves fall to the forest floor, they are quickly decomposed, and plants take up the released nutrients. Because of the low availability of nutrients in Oxisols, subsistence farmers working them traditionally have employed a pattern of *shifting* *cultivation* in which the forest is cut down and burned, releasing nutrients that can be used for a few years of farming. Then the forest is allowed to regenerate, slowly rebuilding the store of nutrients within the ecosystem. Population pressures in some parts of the world have altered this agricultural system, and the rapid deterioration of this soil's already-limited fertility is usually the result.

6. Spodosols Spodosols are characterized by a **B** horizon with an illuvial accumulation of iron and aluminum oxides and humus, usually reddish-brown to black in color. Spodosols display little if any clay accumulation in the **B** horizon. A characteristic ash-gray **E** horizon usually overlies the diagnostic **B** horizon (see Fig. 21.9). The light color of the **E** horizon is that of the bare mineral grains, because materials such as humus or iron oxides that often coat the grains and give them color have been moved downward into the **B** horizon. Spodosols are found only in humid regions under forest, most often in sandy parent materials. The largest areas of Spodosols are in the northern coniferous forests of Eurasia and North America. Spodosols also occur in patches within subtropical and tropical forests—for example, in the coastal plain of the southern United States and in the Amazon Basin. These Spodosols of warm climates formed in parent materials that are nearly pure sand and are highly resistant to weathering. Because of the association of forest cover with Spodosols, lumbering is one of the most important human activities that occurs on these soils. Agricultural use is limited by the acidity and infertility of most Spodosols, but some areas are farmed in northern Europe, Canada, and New England.

7. Gelisols Gelisols are the newest soil order, added to the Soil Taxonomy in 1998 to distinguish soils that contain permafrost (permanently frozen ground, discussed in Unit 46). Permafrost is found within 1 or 2 m (40 to 80 inches) below the surface of Gelisols, and these soils often display evidence of cryoturbation (frost-churning). In the United States, Gelisols are found primarily in areas of Alaska poleward of 60 degrees N (Fig. 23.7). Gelisols often contain large amounts of organic matter that have accumulated because the microorganisms that normally decompose organic debris are inactive at such low temperatures. In fact, Gelisols and Histosols together hold a large portion of the entire pool of carbon in soils (see Fig. 20.3) because of their high organic matter content. A warming climate at high latitudes will speed up decomposition and convert organic carbon to carbon dioxide, which is then added to the atmosphere. This serves as a large positive feedback to the global warming process. Clearly, the geographic distribution of Gelisols is important information for understanding change in the global carbon cycle (Unit 20).

The five remaining soil orders (Entisols, Inceptisols, Histosols, Vertisols, and Andisols) are not closely associated with certain climatic zones but instead are strongly influenced by parent material, topographic setting, or soil age.

FIGURE 23.7 A Gelisol near Fairbanks, Alaska, with permafrost visible below the knife near the bottom of the profile. The dark-colored horizons in the upper 40 cm (16 in) contain abundant organic matter preserved because of the cold, wet condition of this soil.

8. Entisols The Entisol soil order contains all the soils that do not fit into any of the other orders. Entisols are excluded from most other soil orders because they lack any kind of **B** horizon; their profiles usually contain an **A** horizon directly overlying a **C** or **R** horizon. Pedologists regard such soils as very weakly developed, even less well developed than the Inceptisols, which have a weakly expressed **B** horizon. Many Entisols are of recent origin; they are young soils in which there has not been time for **B** horizon development. On steep slopes, floodplains, and windblown sand dunes, frequent erosion or deposition of sediment allows only short periods of soil formation (Fig. 23.8). Other Entisols form in parent materials that are unusually resistant to weathering and other soil-forming processes. Because there are numerous reasons for the absence of well-developed horizons, Entisols are found in many different climates.

9. Inceptisols Inceptisols take their name from the Latin word for "beginning," because they have the beginnings of a **B** horizon. This horizon is noticeable mainly because it is somewhat browner or redder than the parent material and has developed some soil structure.

FIGURE 23.8 The Entisols on these recently deposited sand dunes are capable of supporting some adapted vegetation. But the vegetation is vulnerable to disturbance, and when it is lost the sands will move in the wind, and the young soils will be destroyed. This is why many dune areas are protected and people are asked to keep off.

More advanced soil development, such as accumulation of clay, has not occurred. Inceptisols are less weakly developed than Entisols, but their occurrence can be explained by similar factors. That is, Inceptisols may be relatively young soils, or they may occur on weathering-resistant parent material or steep slopes. Inceptisols are distributed from the Arctic to the tropics and are often found in highland areas as well as under many types of vegetation (Fig. 23.9).

10. Histosols Histosols are made up of a mixture of organic matter and mineral material, but they contain enough organic matter to completely dominate the soil properties (other soils contain organic matter in much lower proportions). Histosols usually have such high organic matter content because they are water-saturated for much of the year, which limits the decomposition of organic material by microorganisms. Common terms for these soils are *peat* or *muck*, and they occur mainly in wetlands (Fig. 23.10). Histosols are unique among the soil orders because they can be totally destroyed over time by drainage, natural or artificial. Drainage speeds up the decomposition process and destroys much of the organic matter that distinguishes the Histosols. As with warming of Gelisols, this decomposition releases soil carbon to the atmosphere in the form of carbon dioxide. It is difficult to generalize about the geographic distribution of Histosols because wetlands occur in so many environments. Vast areas of Histosols occur in the boreal forests of North America and Eurasia and in tropical lowlands. Drained Histosols are used for cultivation of specialized vegetable crops such as carrots and sod for landscaping.

FIGURE 23.9 Profile of an Inceptisol. The **B** horizon, near the middle of the rock hammer, is thinner than in many other soil profiles and marked only by a reddish-brown color and a weak structure; therefore, it is considered weakly developed.

11. Vertisols You may have seen areas where clay soils develop large cracks as they shrink in the dry season and then swell with moisture when rain returns (Fig. 23.11). The Vertisols represent an extreme case of such shrinking and swelling because of their high content of clay, which is usually well above 35 percent. The clay is normally derived from the parent material, so Vertisols are found where clay-producing materials are present. They are most extensive in Australia, India, Sudan, and other regions with mesothermal or tropical climates and periodic dry and wet seasons. However, there are also Vertisols in colder climates—for example, in the lake clays of the Red River Valley in North Dakota and Manitoba. Vertisols are hard to use for most human purposes, particularly construction. When they shrink and swell, fences and telephone poles may be thrown out of line. Pavements, building foundations, and pipelines can all be damaged by the movement of these "turning" soils.

12. Andisols The soil order of Andisols was established in 1990 to include soils developed on volcanic ash, along

FIGURE 23.10 Profile of a Histosol near East Lansing, home of Michigan State University and its Department of Geography, in central Michigan. In many parts of the world, such organic soil is dug up, cut into shoebox-sized pieces, dried, and used as fuel.

FIGURE 23.11 The sequence of wetting and drying of the soil creates the cracks characteristic of Vertisols.

with a few other soils that have similar properties. The Andisols were mostly classified as Inceptisols in the past, and they have similar weakly developed **B** horizons. The volcanic ash parent material results in distinctive soil properties, such as high organic matter content, low density, and a great ability to retain water. Because of these properties, Andisols are quite fertile and are used for intensive farming in many regions. These soils lie principally in the Pacific Ring of Fire (see Unit 30), Hawai'i, and other volcanic zones (Fig. 23.12). They occupy less than 1 percent of the world's land surface.

SOILS OF THE UNITED STATES

The descriptions of soil orders and Figure 23.1 provide some indication of the environments in which we expect to find soils of each order. However, the spatial patterns of soils in the real world can be quite complicated and difficult to interpret unless careful consideration is given to all the factors affecting soil formation. Figure 23.13 shows the spatial pattern of soil distribution for the conterminous United States. The climatic relationships are quite clear with soils such as Spodosols and Aridisols, but complications arise because other soils, such as Inceptisols, form where other factors are more important than climate. For example, Appalachian soils are often Inceptisols because they form on steep slopes, and the large patch of Entisols in Nebraska marks a field of sand dunes bearing young soils. In the Rocky Mountains, traveling the short distance from a dry

FIGURE 23.12 Road cut on the island of Hawai'i revealing the profile of an Andisol atop layered volcanic ash.

basin floor to the summit of a mountain range takes us through several zones of quite different climate and vegetation and through a wide range of bedrock types. Therefore, it is not surprising that this region is a complex patchwork of Alfisols, Mollisols, Aridisols, Inceptisols, and Entisols.

THE WORLD SOIL MAP

At the global scale, there are many distinct relationships between soils and climate, but there are also some differences in detail, as Figure 23.14 illustrates. The relationship between soils and climates is quite clear in central Eurasia.

Soil types vary from south to north, changing from Aridisols to Mollisols, and then to Alfisols, Spodosols, and, finally, Gelisols. These variations parallel changes in climate from desert through steppe and microthermal climates to polar climates (see Fig. 14.3). It is noteworthy that the importance of climate as a factor in soil formation was first recognized in this part of the world by Dokuchaev and other Russian pedologists. A similar progression may be observed in corresponding parts of North America. Perhaps the most obvious relationship at the global scale is that between the desert climates and the Aridisols and Entisols.

Despite the clear relationships between climate and soil types at this macro scale, similar climates do not always correspond to similar soil types. The humid subtropical climate (**Cfa**) is a case in point. This climate type prevails in the southeastern parts of the five major continents (see Fig. 14.3), but another look at Figure 23.14 shows that the soil types in these areas vary. In the southeastern United States and southeastern China, areas with this climate are completely dominated by Ultisols. Alfisols predominate in southeastern Africa and in east-central Australia. Southeastern parts of South America with **Cfa** climates around the Rio de la Plata are covered with a mixture of Mollisols, Alfisols, Ultisols, and Oxisols. These contrasts are partly related to differences in soil age and parent material, because each region has a somewhat different geologic history. Recent vegetation history may also play a role. All the factors involved in soil formation must be considered to explain the complex geographic patterns of soils at local and global scales.

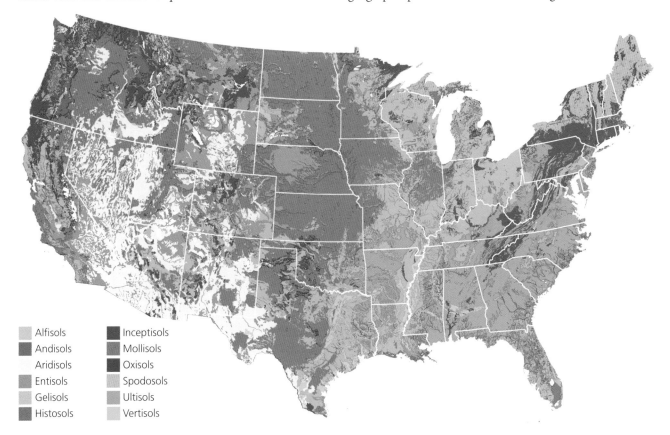

▨ Alfisols	▨ Inceptisols	
▨ Andisols	▨ Mollisols	
▨ Aridisols	▨ Oxisols	
▨ Entisols	▨ Spodosols	
▨ Gelisols	▨ Ultisols	
▨ Histosols	▨ Vertisols	

FIGURE 23.13 Spatial distribution of the soil orders within the conterminous United States.

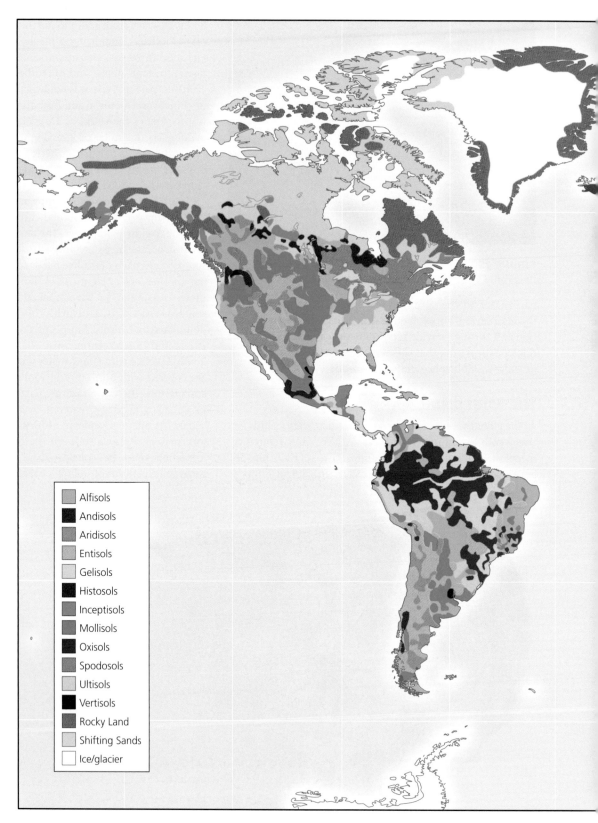

FIGURE 23.14 Global distribution of the soil orders.

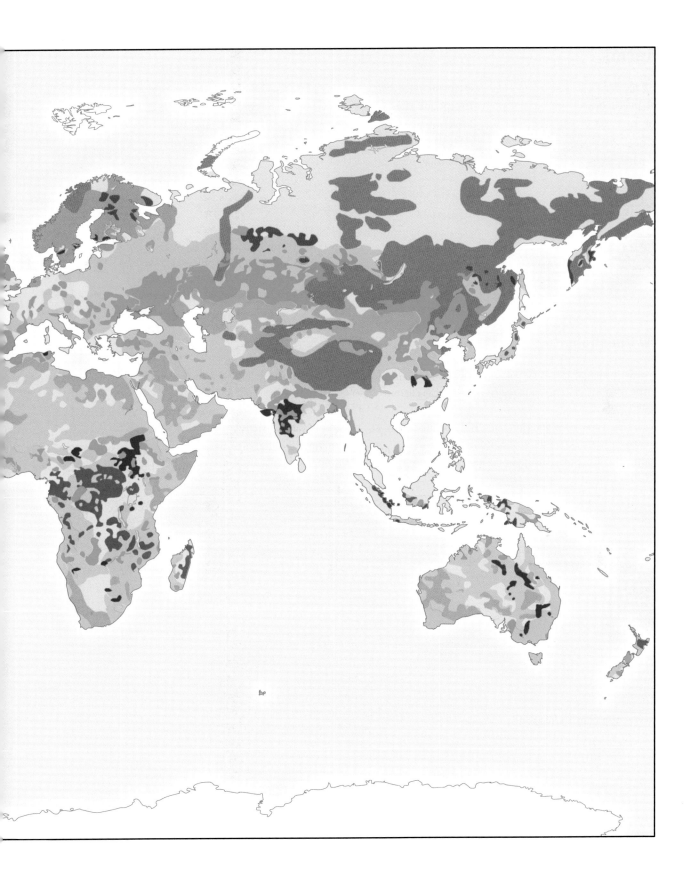

Key Terms

Alfisol *page 295*

Andisol *page 299*

Aridisol *page 293*

Entisol *page 298*

Gelisol *page 298*

Histosol *page 299*

Inceptisol *page 298*

Mollisol *page 293*

Oxisol *page 296*

soil order *page 291*

Soil Taxonomy *page 291*

Spodosol *page 298*

Ultisol *page 296*

Vertisol *page 299*

Scan Here for a quick vocabulary review

Review Questions

1. Why was it logical for early soil classification systems (e.g., the Marbut system) to place so much emphasis on climate?

2. Describe the types of vegetation usually associated with Mollisols, Ultisols, Oxisols, and Spodosols.

3. Explain why Entisols, Inceptisols, and Histosols can occur in a range of different climates, whereas Oxisols are found almost entirely in the tropics and Gelisols occur mainly in the polar regions.

4. Which soil orders are more strongly influenced by parent material than by any other single factor?

5. Explain the difference between the type of **B** horizon found in Alfisols and the type of **B** horizon that characterizes Spodosols.

Scan Here for a quick concept review

www.oup.com/us/mason

UNIT
24
Biogeographic Processes

Death and life in the African savanna—vultures on an elephant carcass in Tsavo West National Park, Kenya.

OBJECTIVES

- Describe the processes of photosynthesis and respiration, and the limitations on photosynthesis imposed by climate and other factors

- Define the concept of ecosystems and highlight the important energy flows within them

- Examine the factors influencing the geographic dispersal of plant and animal species within the biosphere

Soil is located at the base of the biosphere. In fact, life exists in many forms within the soil layer itself. However, in this unit and the next, we turn to more obvious components of Earth's "life layer"—vegetation, with its stems and leaves above the soil and its roots extending down into it, together with animals and other organisms that make up the living components of ecosystems.

Dynamics of the Biosphere

The story of the development of life on Earth parallels that of the formation of the atmosphere. The earliest atmosphere, about 4 billion years ago, was rich in gases such as methane, ammonia, carbon dioxide, and water vapor. More than a half-century ago, a scientist named Stanley Miller filled a flask with a mixture of gases similar to this early atmosphere. He then subjected the contents to electrical discharges (to simulate lightning) and to boiling (to simulate the conditions on the early crust, much of which was volcanic and red hot). This experiment resulted in the formation of molecules called amino acids, the building blocks of protein, which in turn are constituents of all living things on Earth. Miller's experiments were designed to replicate the events that actually happened on Earth between 3.5 and 4 billion years ago that led to the formation of the first complex molecules in the primitive ocean. While Miller's simple experimental setup did not reproduce the exact set of conditions that allowed the first organic molecules to form, it showed that simple physical processes could have led to emergence of the earliest forms of life—single-celled microscopic organisms classified as bacteria and archaea.

PHOTOSYNTHESIS

When the first bacteria floated on the surface of Earth's evolving ocean, a process began that would enable life to advance—**photosynthesis**. Recall that this process was introduced in Unit 20 as a pathway of the global carbon cycle; here, we will consider it in more detail. Photosynthesis captures energy from solar radiation and uses it, together with carbon dioxide and water, to produce organic compounds. Those compounds in turn fuel the activity of living organisms. Photosynthesis is the most basic process of energy flow through the biosphere.

In ancient oceanic bacteria, as in plants today, solar energy was used to drive the reaction of water (H_2O) and carbon dioxide (CO_2), yielding *carbohydrates* (organic compounds and food substances) and oxygen (O_2). Starting about 2 billion years ago, the production of oxygen gas by photosynthetic organisms increased the oxygen content of the atmosphere from 1 percent to 20 percent. The atmosphere became enriched not only in oxygen gas but also in high-altitude ozone (O_3). As a result, more complex life-forms were able to evolve as the developing ozone layer increasingly afforded protection from solar ultraviolet radiation. Organisms eventually evolved from an aquatic environment and began to colonize the landmasses. The first land plants colonized Earth more than 470 million years ago (Fig. 24.1).

In order for photosynthesis to occur, a pigment, *chlorophyll*, must be present in the part of the plant where the process is taking place. Chlorophyll absorbs red and blue light but reflects green light, giving plant leaves their characteristic green color. The energy of the absorbed radiation drives the chemical reactions that produce carbohydrates. Therefore, plants often have adaptations related to the competition

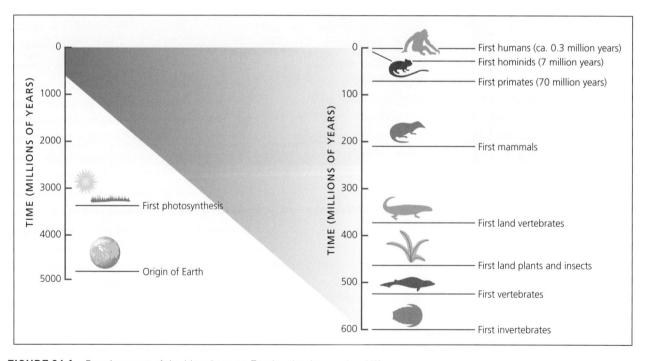

FIGURE 24.1 Development of the biosphere on Earth—the timescale of life.

FIGURE 24.2 An expanse of rainforest in eastern Bolivia. A nearly continuous canopy of leafy vegetation characterizes the tropical rainforest, as seen from the air (or the ground). The tall trees spread their crowns until they interlock, competing for every ray of sunlight.

for sunlight. For instance, some plants can survive in low-light environments. This allows them to take advantage of other resources not available to plants that need more light to survive and reproduce. Many other plants have growth forms or life cycles that allow them to maximize the light they can capture. Where solar energy arrives on Earth in the greatest quantities, the tall trees of the equatorial and tropical forests soar skyward, spreading their leafy crowns in dense canopies as they vie for every ray of the Sun (Fig. 24.2).

The capture of solar energy through photosynthesis is critical to virtually all life on Earth. At the most basic level, plant material produced by photosynthesis directly or indirectly provides a food supply for all animals, including humans. We not only use plants for food but also depend on them for fiber, building materials, and fuel. We have increasingly used fossil fuels—and materials made from them, such as synthetic fibers—to meet some of the same needs, but fuels such as oil, coal, and natural gas actually originate as organic compounds produced through photosynthesis. The importance of organic matter accumulation in forming the soils that sustain land-based ecosystems was discussed in Unit 21; this process would not occur at all if the organic matter was not originally produced by photosynthetic plants. Photosynthesis removes carbon dioxide from the atmosphere and adds oxygen to the air in return. Humans could not live as they do in an oxygen-poor atmosphere, and had photosynthesis not altered Earth's primitive envelope of gases, the evolution of life would have taken a very different course.

RESPIRATION

Oxygen is particularly important for the process of cellular **respiration**, carried out by plants, animals, and microorganisms. In respiration, living organisms use oxygen to break down carbohydrates and release the biochemical energy stored in them. This energy is then available for vital activities ranging from producing proteins to hunting prey or reading this textbook. Respiration also releases water

and carbon dioxide. As plants respire, they return some but not all of the carbon dioxide they have taken up from the atmosphere through photosynthesis. The rest is eventually returned, however. Animals that feed on plants return more carbon dioxide through respiration, and the bacteria and fungi that decompose dead plant material also convert much of it to carbon dioxide through respiration.

Overall, then, photosynthesis removes carbon dioxide from the atmosphere and releases oxygen to it; respiration takes up oxygen and returns carbon dioxide to the atmosphere. Together, photosynthesis and respiration represent by far the most important processes by which carbon moves between atmosphere and biosphere in its global biogeochemical cycle (see Fig. 20.3).

LIMITATIONS OF PHOTOSYNTHESIS

Combining what we have learned about climatic patterns with what we know about photosynthesis, we can deduce where the process occurs most productively. Solar energy is most abundant in the tropical latitudes, and in general terms, photosynthesis is most active there. However, some significant limitations must be recognized, especially when we focus on plants that grow on land. In equatorial and tropical latitudes, the high solar energy also generates heat, and heat increases the plants' rates of respiration. When leafy plant surfaces become hot, therefore, heightened respiration offsets the production of carbohydrates by photosynthesis.

Another limiting factor for land plants is the availability of water. Carbon dioxide from the atmosphere must be dissolved in water within leaves before the plant can use it in photosynthesis. For this reason, continued photosynthesis requires moisture. Plants also require water to maintain the strength of their stems and leaves (wilting plants indicate a lack of available water). Small holes in the leaf surface, called *stomata*, are openings through which the plant takes up carbon dioxide for photosynthesis. The larger the opening, the greater the amount of carbon dioxide that can be taken up. At the same time, however, water in the leaf can evaporate at high rates when the stomata are open, sometimes faster than it can be replenished by uptake of water from the soil. Plants respond to this stress by partially or completely closing the stomata, but that also slows down photosynthesis. This trade-off is most important in hot, dry conditions, when evaporation rates are high.

As we learned in Unit 11, water loss through the leaf pores of plants is called transpiration, part of the overall process referred to as *evapotranspiration*. The amount of moisture lost in transpiration is closely related to the production of living organic plant matter because water loss and uptake of carbon dioxide for photosynthesis both take place through the stomata. As a consequence our knowledge of the global map of moisture (see Fig. 11.11) is an important aid in understanding where photosynthesis occurs at higher or lower rates.

BIOMASS AND NET PRIMARY PRODUCTIVITY

The total living plant material in a given area can be collected, dried, and weighed, giving us the total **biomass** of the local vegetation in grams per square meter. Technically, the plant matter alone should be called the *phytomass*, while total biomass includes animals and microorganisms, but plant matter often completely dominates the total biomass. Within a local area, plant biomass is continually depleted as leaves fall and are decomposed or as animals feed on plants and then release the stored energy through respiration. At the same time, however, new biomass is produced by photosynthesis. Forests contain the largest biomass of all vegetation types by far. Replace a tropical forest with a banana plantation, and biomass will be much less.

Plants that carry out photosynthesis are *primary producers*, the original source of biomass for almost all life-forms. **Net primary productivity** is the amount of biomass produced annually by the primary producers minus the amount of biomass that is lost when the primary producers themselves carry out respiration. More simply, it is the new growth of plants and other primary producers over the course of a year, including the new growth eaten by animals. Figure 24.3 maps, in very general terms, the global distribution of net primary productivity. Note the overall similarities between certain climatic patterns and this configuration. Productivity is greatest in well-watered equatorial and tropical lowlands; it is at a minimum in desert, highland, and high-latitude zones. Note that this is true only for the *natural* vegetation, not for cultivated crops. Net primary productivity of crops varies widely, from very high values in well-watered and well-fertilized fields to values as low as those of the deserts shown in Figure 24.3.

FOOD CHAINS AND ENERGY FLOW THROUGH ECOSYSTEMS

An **ecosystem** includes all of the living organisms in a particular area; their physical environment, such as soil or lake water; and all of the pathways by which matter and energy flow between the living organisms and their environment. Ecosystems are open systems as far as energy is concerned: solar energy is absorbed through photosynthesis and stored in the form of organic compounds, which then pass from organism to organism in the ecosystem, generating a **food chain**. Eventually, all of the stored energy is put to use through respiration and ultimately becomes heat that dissipates in the environment; carbon dioxide is the byproduct of this process.

Figure 24.4 illustrates this concept of energy flow using a highly simplified version of a grassland ecosystem. In this ecosystem, the most important primary producers are the perennial grasses. The grasses are *autotrophs*, which means that they provide their own food source. The net primary productivity of the ecosystem, largely the work of these grasses, represents the initial input to the food chain. The grasses and other plants are eaten by **herbivores** such as bison and many other animals from gophers to grasshoppers. In turn, the **carnivores** feed on the herbivores, and dead organic matter shed by plants, herbivores, and carnivores is consumed by decomposer organisms in the soil. Each major group along this food chain—the primary producers, herbivores, carnivores, and decomposers—is called a **trophic level**.

You can observe quite a different ecosystem in action in a sunlit freshwater pond (Fig. 24.5) and find that energy flows through it in much the same way. Energy from sunlight is taken up in photosynthesis mainly by algae called *phytoplankton* (these are sometimes visible in the aggregate as a greenish sheen on the water). The phytoplankton are fed on by small larvae and other tiny life-forms in the pond,

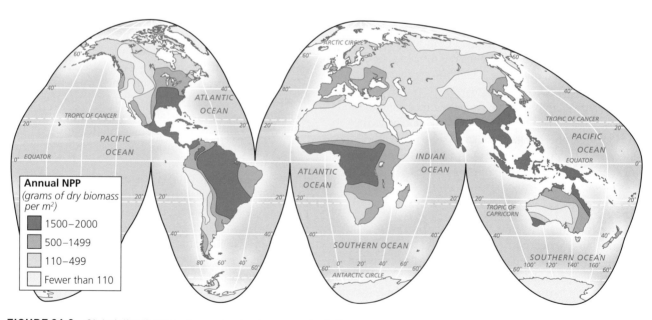

FIGURE 24.3 Global distribution of annual net primary productivity.

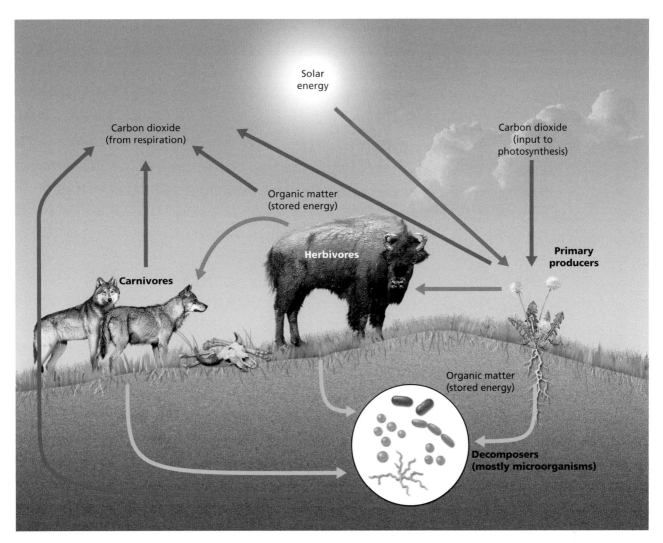

FIGURE 24.4 Energy flow through an ecosystem (a highly simplified version of the Great Plains grasslands of North America).

ENHANCE

collectively called *zooplankton*. Zooplankton are eaten by small fish, and these fish are later consumed by larger fish. Meanwhile, plants and animals die in the pond and decompose, sustaining a large number of microorganisms.

As energy flows through a food chain, much of that input is lost at each trophic level. For example, in the freshwater pond, the biomass of each higher level is far smaller than that of the level below it (Fig. 24.5). A mere 15 percent of the biomass produced by the phytoplankton—the food energy the phytoplankton capture from sunlight—is passed on to the herbivores (the zooplankton). Only 11 percent of the zooplankton biomass is taken up by the carnivores (the smallest fish), and a small fraction of that goes on to the larger fish (the top carnivores).

Several important consequences arise from the loss of energy between levels of the food chain. First, there must

always be a large number of autotrophs to support smaller quantities of herbivores and even fewer carnivores. The total biomass at each higher trophic level is much lower than in levels below, as illustrated in Figure 24.6. Because only a small part of the energy produced in the form of food is passed from one stage to another, to obtain enough food the animals at the higher trophic levels must have large territorial areas that provide an adequate supply of the species on which they feed. This explains why large carnivores such as lions require a wide territorial range.

Another consequence is that because food and energy move along a chain in only one direction, the whole system can be altered dramatically if earlier links of the chain are broken. For example, in the grassland ecosystem shown in Figure 24.4, if the grasses and other natural vegetation are depleted by drought or replaced by wheat fields, the

FIGURE 24.5 "A pond like this is a crucial part of the local ecosystem. The Sun's energy reaches the surface of this lake in Britain's Lake Country, where phytoplankton (microscopic green plants) convert it through photosynthesis into carbohydrate, a food substance. Tiny life-forms, zooplankton, feed on this carbohydrate, and small fish eat the zooplankton. Larger fish feed on the smaller fish, and the food chain continues."

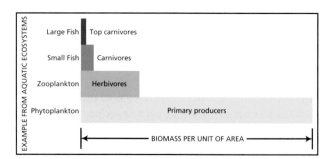

FIGURE 24.6 Mass of living materials per unit of area in different trophic levels of an ecosystem.

population of herbivores and carnivores cannot be sustained. A related consequence of the chainlike structure is that undesirable materials can be passed along and concentrated within the ecosystem. From 1949 to 1957, the insecticide DDD (closely related to the better-known DDT) was applied in Clear Lake, California, to kill the gnats that swarmed around the lake in huge numbers every summer. It was sprayed onto the water at a concentration of 0.02 ppm.

The DDD density was 5 ppm in the phytoplankton, 15 ppm in the herbivores feeding on the phytoplankton, 100 ppm in the fish, and 1,600 ppm in the grebes (birds) that ate the fish. The grebes died, an unexpected consequence that demonstrates how important it is to understand the nature of the food chain.

Plant Succession

The flow of energy through a food chain illustrates the interconnectivity and dynamic nature of the biosphere. The vegetation layer, too, changes continuously over short and long periods of time. An important kind of vegetation change that occurs over decades to centuries is called **plant succession**, of which there are two basic types.

Primary succession occurs when plants become established on bare ground or in a lake, where soils are not yet present. Good examples can be seen where bare active dunes on a coastline become stabilized by vegetation (see the opening photo of Unit 20) or where plants colonize the debris left behind by a retreating glacier. The first plants to

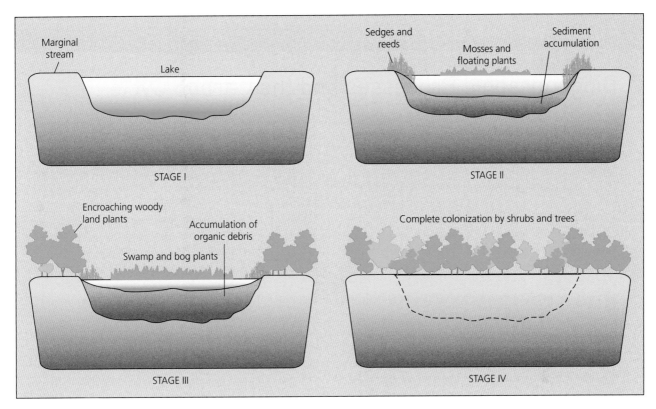

FIGURE 24.7 Idealized sequence of a type of primary succession in which a lake is filled and eventually colonized by shrubs and trees.

become established often have special adaptations that allow them to survive in the absence of nutrient-rich soil. For example, they may be able to survive periods of water stress or have bacteria associated with their roots that can fix, or capture, nitrogen from the atmosphere (see Unit 20). As more plants cover the bare ground, a humus-enriched **A** soil horizon forms, and nutrient availability increases. This process in turn allows a variety of new plant species to become established. Often, herbaceous plants dominate the early stages of succession, followed by shrubs and trees. Trees that thrive in full sun are followed by those that can begin their lives in the shade of older trees. This sequence of vegetation change illustrates the great importance of both soil processes and solar radiation in ecosystems.

Figure 24.7 shows another example of primary succession—the growth of vegetation in an area that has been a lake (Stage I). As the lake gradually fills with sediments, the water becomes chemically enriched; mosses and sedges as well as floating rafts of vegetation build up (Stage II). Plant productivity increases in the lake, and other plants encroach around its edges (Stage III). After the lake has completely filled with organic debris, allowing formation of Histosols (see Unit 23), dry-land plants and trees finally take over (Stage IV).

Secondary succession follows some form of vegetation **disturbance**, an event that disrupts an ecosystem, such as fire, an outbreak of tree-killing insects, or strong winds that blow down a forest. The soil is still present with its store of nutrients and the capacity to hold moisture, so conditions for plants in the first stage of secondary succession

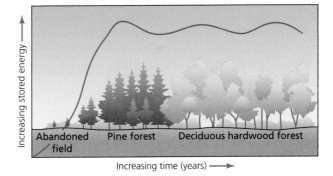

FIGURE 24.8 Increase in stored energy of the biomass in a typical secondary succession. In this case, a deciduous hardwood forest takes over an abandoned field over a period lasting about 175 years.

are not as harsh as those experienced in primary succession. Nevertheless, there is still a sequence of changes in vegetation—for example, from early stages dominated by herbaceous plants to a developing forest in the later stages. In all types of plant succession, the stages are often fairly predictable within a particular region, as illustrated in Figure 24.8. Certain plants that typically occur in one stage of succession can often be identified. For example, quaking aspen (*Populus tremuloides*) often forms the first forest canopy after fire or other disturbance in the Great Lakes region of the United States. More shade-tolerant trees grow up under the aspen and become dominant as the short-lived aspen die (see Focus on the Science: Biogeographic

Biogeographic Processes and the Carbon Cycle

How much has the destruction of tropical forests affected the level of carbon dioxide in Earth's atmosphere? Are there other changes in the biosphere that could be affecting this important greenhouse gas? These are pressing questions today, given the rapid increase of carbon dioxide in the atmosphere and its warming effect on Earth's climate. Many of the concepts and tools needed to answer these questions have already been developed through decades of research on the processes described in this unit. The global cycle of carbon was discussed in Unit 20. Here we consider in more detail the vital role played by the biosphere in this cycle.

Worldwide, photosynthesis removes vast quantities of carbon dioxide from the atmosphere each year. Yet, in mature tropical forests or other ecosystems where the rates of photosynthesis are high, respiration is also very active (we are including here the respiration by animals and by microorganisms that decompose organic matter, not just that carried out by the plants). If total biomass changes little over time, which is thought to be true of many mature forests, then photosynthesis and respiration are roughly balanced. Carbon dioxide taken up from the atmosphere by photosynthesis is returned by respiration, so there is little overall effect. If a mature forest is cut down and burned, this balance is upset. Burning of the wood and foliage abruptly releases much of the carbon they hold to the atmosphere as carbon dioxide, an oxidation process similar to respiration. Cultivation of the soils after the forest is destroyed often speeds up decomposition and releases even more carbon dioxide to the air.

Regrowth of forest on farm fields or pastures has the opposite effect. Where this happens naturally, there is an initial stage of growth of herbaceous plants and shrubs, which are soon replaced by rapidly growing young trees; this is typical secondary succession (see Fig. 24.8). The young trees rapidly take up carbon dioxide through photosynthesis, more than they release with respiration, so on balance, carbon dioxide is removed from the atmosphere. At the same time, the soils may also recover the organic matter that they lost in cultivation. In the later stages of succession, the increase in total biomass slows and eventually ends altogether.

Tropical forests contain a large part of all biomass on the continents, so when widespread destruction of these forests accelerated in the twentieth century, we would expect that a significant amount of carbon

dioxide was added to the atmosphere in the process. Expensive instruments are needed to directly measure the carbon dioxide released when a forest is destroyed, and that kind of study is rare. On the other hand, the total biomass in various tropical forests has been measured by researchers who have tediously collected, dried, and weighed vegetation. Geographers have also mapped the patterns of tropical deforestation from the 1970s to the present using satellite images. From these two types of information, it is possible to estimate the carbon dioxide released to the atmosphere as tropical forests have been cut and burned, also taking into account the new biomass in pastures or crops that replace the forest. Over the last century, the amount of carbon released by burning fossil fuels increased much more rapidly than the amount released by loss of forests. Today, fossil fuel use adds seven to eight times as much carbon to the atmosphere as deforestation or other kinds of land-use change. As new carbon dioxide from all sources is added to the atmosphere each year, part of this addition is offset by carbon that moves from the atmosphere into other parts of the environment, so-called **carbon sinks**. Overall, the total amount of carbon in the atmosphere has continued to rise.

Now the story takes a surprising turn. There is growing evidence that parts of the biosphere may now be taking up more carbon dioxide from the atmosphere than they did in the past, partly counteracting the effects of tropical deforestation. It is not certain where in the biosphere that carbon is being taken up and stored, however. One possibility is secondary succession taking place on former agricultural fields that were carved out of forest. It is clear this has happened in some temperate regions of the world: a drive through rural New England will reveal many old stone fences marking fields that have now returned to forest. Areas of secondary succession are also common within some tropical forests. However, the best estimates of the carbon stored through forest regrowth indicate that it is still outweighed by deforestation and cannot account for the net uptake of carbon by the biosphere. Another more likely possibility is that biomass is now actually increasing significantly in some mature tropical and higher-latitude forests because the trees are growing faster in response to higher levels of carbon dioxide in the atmosphere. This explanation is increasingly accepted, though ongoing research may turn up new evidence on this scientific puzzle.

Processes and the Carbon Cycle for more on the role of succession in the global cycle of carbon).

When ecologists and biogeographers first adopted the concept of succession in the early twentieth century, there was a great deal of emphasis on the final stage: the **climax community** of vegetation. It was believed that when this final stage was reached, the vegetation and its ecosystem would be in complete harmony with the soil, the climate, and other parts of the environment. It is certainly true that vegetation changes slow down in the later stages of succession, and the plant community may appear to change little over time. However, it is now known that slow fluctuations in composition are common even in mature forests of long-lived trees. The climate and other environmental conditions continually change as well, at least enough to prevent a completely stable balance between plants and their environment.

Geographic Dispersal

Geographers are interested in the spatial distribution of plant and animal species, a topic we will address in more detail in Units 25 and 26. Terrestrial and aquatic (freshwater and marine) plants, animals, and other *biota* (living organisms) are naturally present in specific geographic locations because of their adaptation to the physical and biological aspects of their environments. In the remaining part of this unit, we investigate this adaptation to environments and how it influences the geographic dispersal of particular groups of organisms.

Both physical and biotic factors determine the spread and geographic limits of any particular group of organisms. When we look closely at a single physical factor, we often find that a species has an *optimum range* for that factor—for example, an optimum range of temperature or moisture. Within that range, the species not only survives but also reproduces and maintains a large, healthy population, as shown in Figure 24.9. Outside this range, a species increasingly encounters *zones of physiological stress*. Although it can still survive in these zones, the population remains small. When conditions are even more extreme, in the *zones of intolerance*, the species is absent altogether, except possibly for short, intermittent periods. The concept illustrated in Figure 24.9 does not always apply as well to biotic factors. The conditions we call *biotic factors* are often really interactions between the species of interest and other animals or plants, but like temperature or moisture, those interactions with other organisms strongly influence whether a species is abundant, rare, or not present at all. A combination of several physical and biotic factors thus determines the overall distribution of the species.

PHYSICAL FACTORS

Temperature A common limiting factor for both plants and animals is temperature. Recall that in Unit 14 we mentioned the correspondence, suggested by Köppen, of the northern limit of forests in Eurasia and North America and certain temperature conditions. Another example of temperature's effect on spatial distributions was discovered by British ecologist Sir Edward Salisbury (1886–1978) in the 1920s. He studied a creeping woody plant known as the wild madder (*Rubia peregrina*) and found that the northern boundary of the wild madder in Europe coincided closely with the January 4.5°C (40°F) isotherm (line of equal temperature, see Unit 7). This temperature is critical, because in January the plant forms new shoots, and lower temperatures inhibit their development and subsequent growth. It is easy to see the value of this kind of information in predicting how plants or animals will respond to climate change.

Availability of Water The availability of water is another vital factor that limits the spread of plants and animals throughout the physical world. Water is essential in photosynthesis and in other functions of plants and animals, but plants vary widely in their adaptation to dry or wet climates. Plants that are adapted to dry areas are called **xerophytes**. The cactus family is a good example. Cacti are *succulent* plants, storing water in their swelling stems to survive long

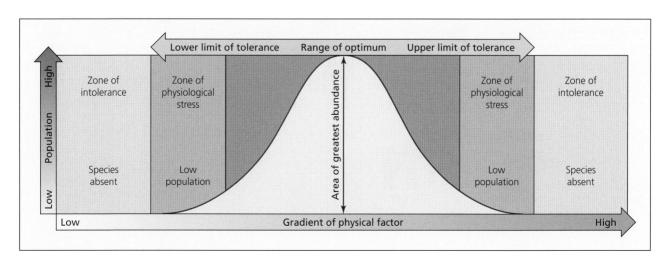

FIGURE 24.9 Model of population abundance of a species in relation to the physical factors in its environment.

dry periods, and they use a distinctive form of photosynthesis that reduces water loss as carbon dioxide is brought in through the stomata. Plants that require abundant water are *hygrophytes*; rainforests, swamps, marshes, lakes, and bogs are their typical habitats. *Hydrophytes* (note the single letter differentiating this term from hygrophytes!) grow in water or saturated soils. Roots immersed in water often suffer from a limited oxygen supply for respiration, but some hydrophytes such as water lilies actually pump oxygen into their root systems. Finally, the many plants that thrive in areas of neither extreme moisture nor extreme aridity are *mesophytes*.

In tropical climates that experience a dry season, flowering trees and plants drop their leaves to reduce water loss during the dry season. This phenomenon spread to plants of higher latitudes, where plants drop their leaves to avoid water shortages in winter, when roots cannot take up water from frozen soils. Trees and other plants that drop their leaves seasonally are called **deciduous**, and those that keep their leaves year-round are called **evergreen**.

Other Climatic Factors   Several other factors related to climate play a role in the dispersal of plants. These include the availability of light, the action of winds, and the duration of snow cover. Competition between plants for light is discussed later in this chapter as a biotic factor, but the maximum amount of light available is determined by latitude and the associated length of daylight. Growth in the short warm season of humid microthermal (**D**) climates is enhanced by the long daylight hours of summer, and plants can mature

surprisingly fast (see Fig. 17.4) in these climates. Wind influences the spread of plants in several ways. It can limit growth or even destroy plants and trees in extreme situations, but it also spreads pollen and the seeds of some species.

Distribution of Soils   The distribution of soils also affects plant dispersal. Factors concerned with the soil are known as *edaphic factors*, the most important of which are the presence of nutrients, soil structure and texture, and salinity. Soil structure and texture affect a plant's ability to root, but most important, they influence the soil's capacity to take up and retain water that can be used by plants (discussed in Unit 22). Effects of nutrient availability on plant distribution are common and take many forms. For example, certain plants have adaptations that allow them to flourish in alkaline soils with high calcium carbonate content, in which nutrients such as phosphorus and iron are bound in mineral forms that make them inaccessible to most plants. A number of plants are adapted to soils in which the salt content is high enough to be toxic to most vegetation.

Landforms   Landforms control vegetation distribution in many ways. On a large scale, as we noted in Unit 18, the local climate changes with altitude, and vegetation changes in response. On the flanks of Mt. Kenya in East Africa (Fig. 24.10), there is a transition from savanna grassland below 1,650 m (5,400 ft) to alpine vegetation above 3,650 m (12,000 ft). Landforms have local-scale effects as well. Steep slopes foster rapid drainage and may lead to a lack of soil water. Furthermore,

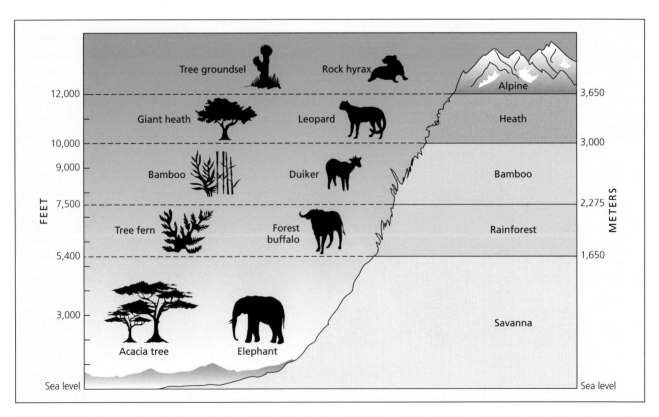

FIGURE 24.10   Zones of vegetation and animal life on the flanks of Mt. Kenya in East Africa, which lies directly on the equator.

the aspect of a mountain slope (the direction it faces) controls the amount of incoming radiation and determines the degree of shelter from the wind, thereby influencing the local distribution of plant and animal species.

BIOTIC FACTORS

Biotic effects on the spatial distribution of plants and animals are mainly those that involve interaction between species.

Competition *Competition* for resources such as food, space, light, or soil nutrients is one of the most important biotic factors. Competition can occur between individuals of the same species, often limiting the overall population. This occurs when animals defend their territories against others of the same kind. However, for understanding the distribution of plants or animal species, competition between individuals of different species is more important. For example, timber wolves are known to kill coyotes that compete for some of the same kinds of prey, reducing the population of coyotes in areas where wolves are present. Although competition among plants for sunlight, soil nutrients, or water is not as obvious, it is just as important. A plant is able to use these resources at

the expense of neighboring plants if it grows faster and shades them or extends its roots into the soil more effectively. Sometimes new species compete for resources so well that they eliminate old species. For example, in parts of the intermountain western United States, native grasses have been almost totally replaced by European cheatgrass (*Bromus tectorum*). One reason for the success of the cheatgrass is that it produces 65 to 200 times as many seeds as the native species does. Plants and animals also evolve adaptations that essentially allow them to avoid direct competition for a resource. In deciduous forests of the middle latitudes, for example, many shrubs and herbaceous plants grow intensely in spring before the leaves of taller trees filter out light.

Amensalism Another form of biological interaction is **amensalism**, in which growth of one species is inhibited by another. In the coastal hills of Southern California, sage shrubs grow on slopes, and grasses inhabit the valleys. Sometimes, however, the sage shrubs occur in the grassland zone. As Figure 24.11 shows, the shrubs are usually surrounded by grass-free bare ground. Foraging by birds and small animals clears some of the ground, but a major cause of this bareness

FROM THE FIELDNOTES

FIGURE 24.11 "Along the northern California coast you can see the competition between sagebrush and grass in progress, the grass losing out over time. The Mediterranean climatic regime, high relief, and thin soils give the better-adapted sage a durable advantage."

FIGURE 24.12 After their introduction and following the elimination of natural predators, the rabbit population of Australia exploded. States and individual landowners built fences (as shown here) and hunted or poisoned rabbits to protect pastures and farmlands. The introduction of a rabbit-killing disease initially seemed effective, but eventually the rabbits began to develop immunity to it.

is the cineole (a liquid with a camphor-like odor) and camphor oil emitted by the sage. Both are toxic to grass seedlings; the sage, therefore, suppresses the growth of grass.

Predation We use the term **predation** for any case in which individuals of one species feed on another species. This includes not only lions killing and eating antelopes but also cattle grazing on pasture grasses. Anyone who saw the landscapes of Australia when they were ravaged by rabbits (Fig. 24.12) cannot doubt the efficiency of predation as a factor in the distribution of vegetation. But examples of one species eating all members of another species tend to be rare. In a more natural situation, predation affects the plant distribution mainly by reducing the pressure of competition among prey species. The presence of predators may increase the number of species in a given ecosystem. Charles Darwin suggested that ungrazed pasture in southern England was dominated by fast-growing tall grasses that kept out light. Consequently, the ungrazed areas contained only about 11 species, whereas the grazed lands possessed as many as 20.

Mutualism In yet another biological interaction, termed **mutualism**, individuals of two species interact to their mutual benefit. Many flowering plants depend on pollinators—bees, other insects, birds, bats, and other mammals—to carry pollen from one plant to another so that they can produce seeds and reproduce. In an extreme case, the "traveler's tree" of Madagascar (*Ravenala madagascarensis*) is pollinated by lemurs, mammals that weigh about 5 kg (10 lb). The lemurs and other pollinators are drawn to flowers to feed on nectar, so they also benefit from this interaction. How these cases of mutualism evolved is a fascinating scientific problem, but a more immediate concern is that some pollinators are endangered species, also threatening the survival of their plant hosts.

DISTURBANCE

As we mentioned earlier, disturbance is the starting point of secondary succession. Disturbance can also be considered one of the factors that influences plant dispersal. It is often closely related to physical factors, especially climate, but can be a biotic factor as well when it is caused by living organisms, especially humans. Consider the role of disturbance by fire in the prairies and forest of the Midwestern United States before much of the landscape became fields of corn and soybeans. In prehistoric times, prairies grew in regions where the climate could have supported forest, such as eastern Illinois and Indiana. Most trees could not become established in the open prairies because they were killed or injured by frequent fires. The prairie grasses were adapted to fire and could draw on their well-developed root systems to quickly regenerate after fires. Certain trees were especially common in the prairie-forest transition zone—for example, bur oak (*Quercus macrocarpa*), which has a thick corky bark that protects it from fire. Fire in this environment was influenced strongly by physical factors, especially climate and weather. In forests near the prairie border, fire was more common in dry periods, but in the prairies, a wet year produced abundant grass that could fuel large fires when it dried. At the same time, as many geographers have emphasized, humans may have played an important role, setting fires for a variety of reasons, such as creating habitat favored by deer or other game. Humans certainly play a major role in other kinds of disturbance that influence plant distribution. Think of the common weeds found mainly in gardens and fields where humans often disturb the soil.

SPECIES DISPERSAL AND ENDEMIC SPECIES

We have just considered how the presence of particular organisms in a specific location is closely related to the physical and biological environment in that setting. This is only part of the story, however. Ancestral species from which modern species evolved arrived in a given area by movement over land; swimming, rafting, or flying (**dispersal**); or being carried along as landmasses drifted apart over tens of millions of years (**vicariance**).

An often-cited example of vicariance is the flora, fauna, and other biota on the island of Madagascar, which split off and moved away from the African continent with the breakup of the Gondwana supercontinent (see Fig. 30.1) and has been an isolated island for at least 70 million years. As parts of the populations of ancestral species were isolated, they evolved into new species better adapted to their local environment. As a result, Madagascar, other oceanic islands, and certain isolated (or once-isolated) land areas contain a significant percentage of species of plants, animals, and other life-forms that exist nowhere else on Earth. Biologists call such species **endemic**. Endemic species are particularly vulnerable to extinction because of environmental changes caused by human modification of the landscape (e.g., deforestation) or the intended or unintended

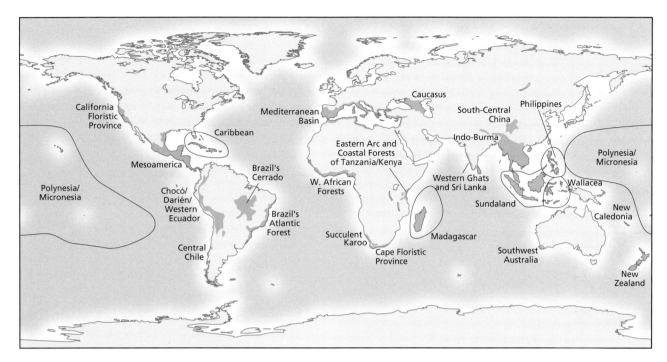

FIGURE 24.13 Global distribution of biodiversity "hot spots." These areas contain high concentrations of endemic species that are threatened by human activities.

introduction of other organisms (Fig. 24.13). It does not take long for even a fairly large island to be deforested or for an invasive species such as rats or pigs to occupy an entire island and outcompete endemic species.

Ever since humans began to travel beyond their home areas, they have brought plants, animals, and other biota with them. The process of dispersing species far from their natural areas accelerated with long-distance sea voyaging. In more recent times, ships have been joined by aircraft and motor vehicles as agents that transfer species from one region to another. These modes of transportation allow such natural barriers to the dispersal of species as oceans,

inhospitable climates, and mountain ranges to be bypassed. The rabbits brought to Australia by Europeans are a well-known example of the destruction some of these introduced species can cause (see Fig. 24.12).

Although researchers have not been able to fully explain every major geographic variation, the limiting physical and biotic factors we have discussed in this unit contribute a great deal to our understanding of the distribution of plant and animal species. In Unit 25, we explore how the biogeographic processes discussed in this unit are expressed spatially on the world map of vegetative associations.

Key Terms

amensalism *page 315*

biomass *page 308*

carbon sink *page 312*

carnivores *page 308*

climax community *page 313*

deciduous *page 314*

dispersal *page 316*

disturbance *page 311*

ecosystem *page 308*

endemic species *page 316*

evergreen *page 314*

food chain *page 308*

herbivores *page 308*

mutualism *page 316*

net primary productivity *page 308*

photosynthesis *page 306*

plant succession *page 310*

predation *page 316*

primary succession *page 310*

respiration *page 307*

secondary succession *page 311*

trophic level *page 308*

vicariance *page 316*

xerophyte *page 313*

Scan Here for a quick vocabulary review

ENHANCE

Review Questions

1. Describe the inputs and products of photosynthesis and respiration.

2. What are the major factors that influence the rate of photosynthesis, and why is there a trade-off between photosynthesis and the use of water by plants?

3. What are food chains? What is the importance of the primary producers in a food chain, and how much of the energy initially captured by photosynthesis is transmitted to higher trophic levels?

4. Explain the difference between primary and secondary succession.

5. Explain how the geographic distribution of a plant species is related to major physical factors using the concept illustrated in Figure 24.9.

6. *Amensalism*, *mutualism*, and *predation* all refer to interactions between animals or plants of different species. Explain the specific type of interactions referred to by each of these terms.

Scan Here for a quick concept review

www.oup.com/us/mason

UNIT
25

The Global Distribution of Plants

A tropical rainforest in Australia. The rich forests of the humid tropics hold much of the total biomass of land-based vegetation, and a large portion of all plant and animal species.

OBJECTIVES

- Explain the use of biomes to map global patterns of vegetation and cite the factors that differentiate major biomes

- List and describe the principal terrestrial biomes

In Unit 14, we discussed the geographical challenges of classifying and mapping Earth's climates. The map of world climates (see Fig. 14.3) shows the distribution of *macro*climates—climatic regions on a broad global scale. Embedded within those macroclimates, as we noted, are local *micro*climates, which have their own distinctive characteristics. The global map, therefore, is only a general guide to what we will find in particular places.

Mapping vegetation is an even more difficult problem. On one of your field trips, stop near any vegetated area and note the large number of plants you can identify, probably ranging from trees and grasses to ferns or mosses. Part of the local area you examine may be covered with trees; another part may be open grassland; some of it may be exposed rock, carrying mosses or lichens. How can any global map represent this intricate plant mosaic? The answer is to use a strategy similar to the method used for mapping climate: to look for criteria that allow us to distinguish the largest possible units of plant association. In the case of climate, this was done by analyzing temperatures, moisture, and seasonality. In classifying Earth's plant cover, biogeographers use the concept of the *biome* to derive their world map.

Biomes

A **biome** is the broadest justifiable subdivision of the plant and animal world, an assemblage of plants and animals that is of subcontinental dimensions. Biomes are distinguished mainly by the structure of their vegetation (the relative proportions of trees, shrubs, and grasses) and by important adaptations to climate that are shared by many of the predominant plant species. Biomes also differ in their relative abundance of different forms of animal life (e.g., grazing animals vs. those that browse on trees and shrubs). In addition, particular soil orders are commonly associated with specific biomes.

In this unit, we use the concept of the biome to explain the global distribution of plants primarily as a response to variations of climate and terrain. It is important to keep in mind, however, that a biome is a collection of many individual plant and animal species, each of which has a unique history of evolution and dispersal over time. We will discuss evolution and adaptation to particular niches in the environment in the next unit using animal distributions to illustrate those processes.

If you were to look at maps of biomes in biology textbooks, you would find quite a range of interpretations. Moreover, biologists include both *terrestrial* and *marine* biomes in their classifications; in biogeography, we tend to concentrate on the terrestrial biomes. Therefore, you should view Figure 25.1 as one justifiable representation of Earth's land biomes, but not the only possible one (see Focus on the Science: North America's Vegetation Regions box for a view of how vegetation maps change from global to continental scales).

We must also remember that biomes are not so sharply defined regionally that their limits are clearly demarcated on Earth's surface. The lines separating biomes on the world map, therefore, represent broad transition zones. If you were to walk from one biome to another, you would observe a gradual change as certain species thin out and vanish while others make their appearance and become established.

The distribution of biome regions results from two major factors: climate and terrain. Both are reflected on the map of global vegetation. The key *climatic factors* are (1) the atmosphere and its circulation systems, which determine where moisture-carrying air masses do (and do not) go; and (2) the energy source for those circulation systems, solar radiation. The Sun's energy not only drives atmospheric movements, but also sustains photosynthesis and propels the endless march of the seasons. The main *terrain factors* are (1) the distribution of the landmasses and ocean basins, and (2) the topography of the continents.

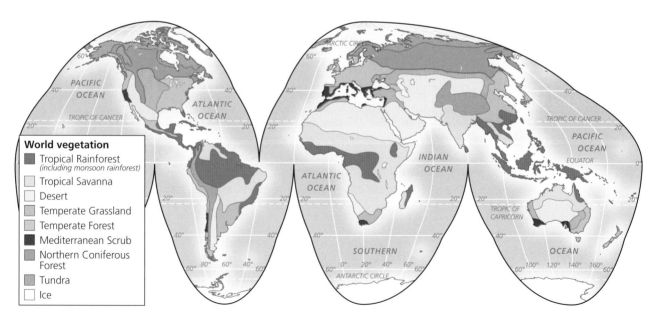

FIGURE 25.1   Global distribution of the principal terrestrial biomes.

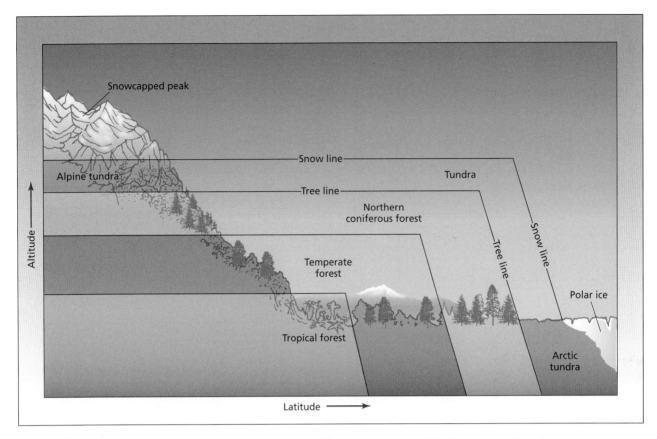

FIGURE 25.2 Vegetation changes with latitude and altitude. The temperature, which affects vegetation, decreases as one travels up a mountain or away from the equator so that, if there is plenty of moisture, the vegetation is similar at high altitudes and at high latitudes, as shown here.

When we examine the effects of terrain in more detail, we see that it affects vegetation mainly through its influence on local and regional patterns of climate. Mountain ranges produce rain shadows (Unit 12), and on the leeward side of mountains we often find vegetation reflecting drier conditions than are found at the same elevation on the windward side.

Furthermore, we know that temperature decreases and precipitation usually increases with altitude on mountain slopes (Unit 17). Therefore, we would expect that the latitudinal transition of biomes from the hot equatorial

regions to the cold polar zones can also be observed along the slope of a high mountain. Ecuador's Mt. Chimborazo in northwestern South America stands with its foot on the equator and is capped with snow. The great geographer Alexander von Humboldt climbed to just below the summit of this enormous volcanic mountain and later used it to illustrate a key concept that he first proposed: when we climb upward in altitude or move poleward in latitude, we cross similar zones of vegetation; in fact, we cross similar biomes (Fig. 25.2). Note that it is not just mountains that

FOCUS ON THE SCIENCE

North America's Vegetation Regions

At the continental scale, we can map the results of the forces and processes that shape vegetation in much more detail than can be included in the map of global biomes (see Fig. 25.1). The map of North America's vegetation displayed in Figure 25.3, constructed by biogeographer Thomas Vale, parallels our world map of biomes in some ways, most notably in its broad distinctions between forest, grassland, tundra, and

desert zones. However, Vale's map is a more detailed version of Figure 25.1, made possible by the larger scale of the North American map.

An example of the effects of shifting map scales is the forest–desert boundary in the U.S. Pacific Northwest, which results from the rain shadow effect east of the Cascade Mountains (see Unit 12). On the world map (see Fig. 25.1), that fairly rapid west–east change

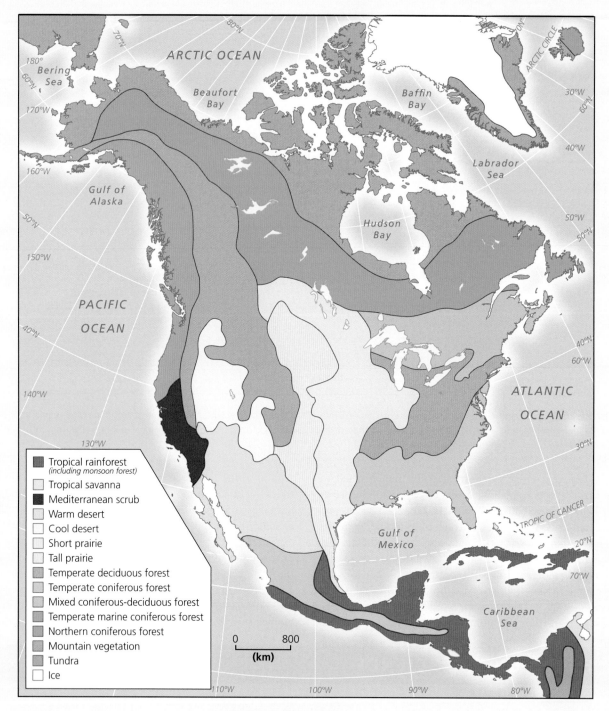

FIGURE 25.3 Distribution of natural vegetation in North America.

Legend:
- Tropical rainforest (including monsoon forest)
- Tropical savanna
- Mediterranean scrub
- Warm desert
- Cool desert
- Short prairie
- Tall prairie
- Temperate deciduous forest
- Temperate coniferous forest
- Mixed coniferous-deciduous forest
- Temperate marine coniferous forest
- Northern coniferous forest
- Mountain vegetation
- Tundra
- Ice

0   800
(km)

can be represented only by a line dividing two adjacent biomes. But on the North American map (see Fig. 25.3), that now-fuzzy "line" becomes a transition zone, and Vale found it necessary to introduce a narrow, intermediate *mountain vegetation* region to contain the distinctive communities of plants within the Cascade range itself.

The Cascades example also reminds us that the geography of vegetation at the continental scale results from the same factors that operate at the global biome level: climate and terrain. The influence of terrain is especially evident in Figure 25.3, with such major topographic features as the Cascades and the ranges of the Rocky Mountains in the western third of the United States not only visible but prominent. To the east, topography appears to play a less significant role, but the southwestward-pointing prong of the *mixed coniferous–deciduous forest* region near the central northeastern seaboard clearly reflects the presence of the Appalachian Mountains.

have a tree line, a zone above which trees will not grow; the entire Earth exhibits a tree line, poleward of which the northern coniferous forests are replaced by the stunted plants of the frigid tundra.

Some biomes that are widely distributed in the Northern Hemisphere hardly ever occur south of the equator because the Southern Hemisphere does not include large landmasses at comparable latitudes. The varied topography and elevation of the landmasses disrupt much of the regularity the map might have shown had all of the continents been flatter and lower. Accordingly, the orientations of Earth's great mountain ranges can clearly be seen on the world map of biomes (see Fig. 25.1).

Another perspective on the factors influencing the global distribution of biomes is represented in Figure 25.4. In this scheme, the latitude is increasing along the left side of the triangle, and the moisture is decreasing along the base. Notice that forest biomes are generated within three latitudinal zones: tropical, temperate, and subarctic. Also note that the tropical forests develop more than one distinct biome before giving way to savanna and ultimately desert.

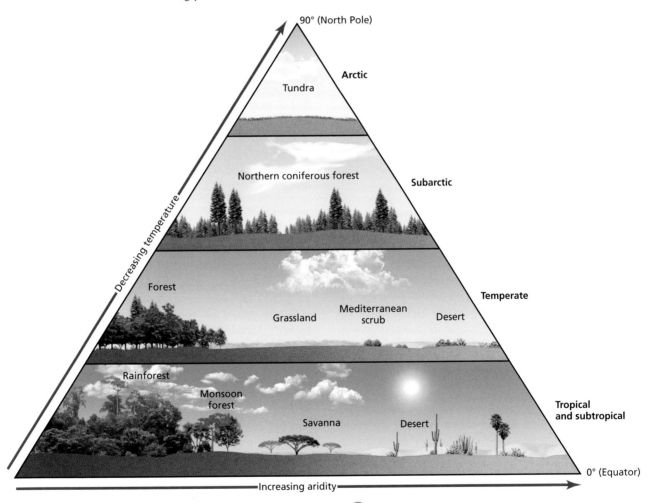

FIGURE 25.4 Simplified scheme of the major terrestrial biomes arranged along gradients of increasing aridity at different latitudes, illustrating the dominant influence of moisture and temperature on the structure of plant communities.

ENHANCE

Principal Terrestrial Biomes

Let us now examine Earth's major biomes, keeping in mind the tentative nature of any such regional scheme. Using Figure 25.1 as our frame of reference, together with the photographs that accompany our discussion, we can gain an impression of the location and character of each of the eight biomes treated in this unit.

TROPICAL RAINFOREST

The **tropical rainforest biome**, featuring vegetation dominated by tall, closely spaced evergreen trees, is a teeming arena of life that is home to a great number and diversity of both plant and animal species. The biogeographer Alfred Russel Wallace (1823–1913) wrote of the tropical rainforest in 1853: "What we may fairly allow of tropical vegetation is that there is a much greater number of species, and a greater variety of forms, than in temperate zones." After spending years in the tropical forests of South America and Asia, Wallace most likely knew this was an understatement. We now know that more species of plants and animals live in tropical rainforests than in all the other biomes combined. Botanists have found more than 200 species of trees per hectare (2.5 acres) in some rainforests of Southeast Asia and the Amazon (Phillips et al., "Dynamics and Species Richness of Tropical Forests," *Proceedings of the National Academy of Science* 91 [1994]: 2805–2809). In addition to this diversity of species, the sheer magnitude of the biomass contained in tropical forest is worth noting. This biome contains much of the total pool of carbon in terrestrial vegetation.

Until recently, tropical rainforests covered almost half the forested area of Earth (see Fig. 25.1). Much of this biome is in the humid equatorial climates with no wet season or a short dry season (see **Af** and **Am**, Fig. 14.3), but in some areas, it extends into the **Aw** climate, with a more pronounced dry winter. The largest expanses are found in the Amazon River Basin in northern South America and the Congo River Basin in west equatorial Africa. Soils of the tropical rainforest are predominantly Ultisols and Oxisols, with scattered Spodosols in sandy parent materials.

Solar radiation in the tropics is relatively high year-round, and together with high temperatures and moisture, this allows high net primary productivity (see Fig. 24.3). Most of the sunlight in this biome is captured by the tall trees. The crowns of the trees are so close together (see Fig. 24.2) that sometimes only 1 percent of the light above the forest reaches the ground. As a result, only a few shade-tolerant plants can live on the forest floor. The trees are large, often reaching heights of 40 to 60 m (130 to 200 ft). Because their roots are usually shallow, the bases of the trees are supported by buttresses. Another feature is the frequent presence of epiphytes and lianas. *Epiphytes* are plants that use the trees for support but are not parasites. *Lianas* are vines rooted in the ground with leaves and flowers in the *canopy*, the top parts of the trees.

Leaf litter that falls to the rainforest floor rapidly decomposes in this warm, humid environment. The shallow root systems of the trees are probably an adaptation that allows them to quickly capture nutrients released from the decaying litter. Many epiphytes actually draw nutrients and water from leaf litter trapped in hollows in the trunks or branches of large trees. Although it is easy to walk through a mature tropical rainforest, many areas contain thick, impenetrable undergrowth. This growth springs up where the original forest was disturbed, often where it was cut and burned for shifting agriculture. Natural disturbance is quite rare, although some tropical rainforests lie in the path of hurricanes and tropical storms and can be destroyed by the strong winds they generate. In recent decades, large areas of the tropical rainforests have been disturbed by logging or cleared to create agricultural fields or pastures. The clearing does not yield very productive soils, however. Despite the warm, humid climate, high crop yields are possible only with large, expensive fertilizer inputs. (The effects of tropical deforestation on the global carbon cycle are discussed in Unit 24, in Focus on the Science: Biogeographic Processes and the Carbon Cycle.)

Monsoon rainforests differ somewhat from the tropical rainforests that receive adequate moisture throughout the year, although for simplicity we include them in the tropical rainforest biome. Monsoon rainforests are established in areas with a dry season (often **Aw** climates) and therefore exhibit less variety in plant species. Where the dry season is more pronounced, many trees are deciduous, losing their leaves to save water during the dry winters. Vegetation is lower and less dense, and it grows in layers or tiers composed of species adjusted to various light intensities.

TROPICAL SAVANNA

The **savanna biome** encompasses a transitional environment between the tropical rainforest and the desert, consisting of tropical grassland with widely spaced trees (Fig. 25.5). Savanna vegetation has developed primarily because of the seasonally wet and dry climate (**Aw**) found in large areas of Africa, South America, northern Australia, and India as well as in parts of Southeast Asia. However, periodic burning plays a significant role in limiting tree growth, along with climate. In some places, the grasses form a highly inflammable straw mat in the dry season, which is ignited by lightning or by people seeking to maintain open vegetation. Both Alfisols and Ultisols can be found in this biome, with Vertisols in areas where clay-rich sediments have accumulated.

Thorn forests, characterized by dense, spiny, low trees, predominate, and bulbous plants are abundant in the savanna. The most common trees are deciduous, dropping their leaves in the dry season. These include the acacia and the curious, water-storing, fat-trunked baobab tree. Grasses in the savanna are usually tall and have stiff, coarse blades. Typical of the food chain of this biome are large herds of grazing animals, and at higher trophic levels, predators such as lions that prey on the grazers.

FIGURE 25.5 "The East African savanna is sometimes called a 'parkland' savanna because trees are widely spaced and give the landscape a regularity that seems cultivated, not wild. The umbrella-like acacia tree is an ally of wildlife and human traveler alike, as I can attest—its shade made many a hot day bearable. The savanna feeds grazing animals as well as browsers because it offers grasses as well as leaves to its migrants; the flat-topped acacia tends to be trimmed at around 5 m (17 ft) by giraffes, which are able to strip leaves from the thorniest of branches. And, as this photo shows, the acacia supports other species as well, as the large and occupied eagle's nest confirms."

DESERT

The **desert biome** is characterized by sparse vegetation, to the point that plants are not present at all in some areas (see Figs. 15.10 and 23.2). The desert biome coincides mainly with areas of arid (**BW**) climates and extends into semiarid (**BS**) climates in some regions. Aridisols and Entisols are the most common soil orders.

The sparse vegetation helps explain why the land areas with the lowest net primary productivity tend to fall within the desert biome (compare Fig. 25.1 with Fig. 24.3). Another reason for the low productivity is that desert plants grow slowly or for only brief parts of the year because soil moisture is in such limited supply. Whereas the grasses of the savanna are *perennials*, persisting from year to year, many desert plants are *ephemerals*, completing their entire life cycle during a single growing season. Ephemerals often grow quickly after the short but intense seasonal rains, covering open sandy or rocky areas in a spectacular display (Fig. 25.6). The seeds of ephemerals can lie in the soils for many years and germinate rapidly after a rainstorm. The perennial plants in the desert biome, such as cacti and euphorbias (spurges), are dormant much of the year.

FIGURE 25.6 "A field trip through Arizona desert country reminded us that the desert biome encompasses rich and diversified vegetation. Despite the often thin and unproductive Aridisols prevailing under desert climatic conditions, the infrequent and scant rainfall recorded here is enough to sustain a wide range of plant species as well as a varied fauna."

Those that have fleshy, water-storing leaves or stems are known as *succulents*. Others have small, leathery leaves or are deciduous. Woody plants have extremely long roots or are restricted to localized areas of water.

The vegetation of desert ecosystems mainly supports small animals that are well adapted to the arid conditions. Rodents live in cool burrows; insects and reptiles have waterproof skins that help them retain water. Larger grazing animals or browsers (animals that eat tree or shrub leaves and twigs) are relatively scarce, in contrast to their abundance in the savanna or grassland biomes.

TEMPERATE GRASSLAND

The **temperate grassland biome** generally occurs over large areas of continental interiors. Climate is the most important factor in the distribution of this biome, which falls mainly in the semiarid steppe (**BS**) climate region, between the arid desert biome and regions humid enough to support forest. Other factors can also be important in establishing the boundary

between temperate grassland and forest, however. In the Midwestern United States, the tall-grass prairie occupied a climatic zone with relatively high rainfall, in part because fire limited the encroachment of forest (see Unit 24). Mollisols are the characteristic soils of the temperate grasslands. Perennial and sod-forming grasses such as those shown in Figure 25.7 are dominant. The temperate grassland biome, like the savanna, is inhabited by herds of grazing animals and their predators (think of the vast herds of bison that once roamed the U.S. Great Plains and the wolves that hunted them).

This biome has been highly susceptible to human influence. As the U.S. example suggests, large areas have been turned over to crop and/or livestock farming. This is true of the interior areas of North America, the Pampas grasslands of Argentina, and the steppes of southern Russia, where the biome is most widespread (see Fig. 25.1). The tall-grass prairie of the Midwestern United States today exists only in scattered remnants, and even in the dry short-grass steppes of the western Great Plains, large areas of natural vegetation have been replaced by farm fields.

FIGURE 25.7 Oklahoma's Tallgrass Prairie Preserve contains diverse prairie grasses, among which four predominate: big and little bluestem, switchgrass, and indiangrass.

FIGURE 25.8 Deciduous broadleaf trees near the peak of their fall foliage ring a small lake in Vermont's Green Mountains National Forest.

TEMPERATE FOREST

There are several varieties of temperate forest. The **temperate deciduous forest biome** occurs in the eastern United States (Fig. 25.8), Europe, and eastern China, mainly in humid temperate climates with no dry season (**Cf**). Alfisols dominate the northern parts of this biome, with Ultisols more common to the south and Inceptisols on steeper slopes. These forests of broadleaf trees are shared by herbaceous plants, which are most abundant in spring before the growth of new leaves on the trees. An outstanding characteristic of temperate deciduous forests is the similarity of plants found in their three locations in the Northern Hemisphere. Oak, beech, birch, walnut, maple, ash, and chestnut trees are all common (each location has its own species of these trees, but they are closely related). Although deciduous trees are predominant, pines and other conifers occur in this biome as well. Browsing animals such as deer are common, as are omnivores like bears, pandas, and raccoons. As with the temperate grasslands, large areas of this forest type have been cleared and converted to agricultural production.

Temperate evergreen forest biomes are found on western coasts in temperate latitudes where abundant precipitation is the norm (**Cf** and **Cs** climates). The soils supporting these forests are varied and include Alfisols, Ultisols, Inceptisols, and Andisols. In the Northern Hemisphere, they take the form of needle-leaf forests, with the coastal redwoods and Douglas firs of the northwestern coast of North America being representative (Fig. 25.9). Some of the tallest and oldest trees on Earth are located in coastal northern California (redwoods) or in the interior just to the east (sequoias). The temperate evergreen rainforest on the western coast of New Zealand exemplifies the much less extensive broadleaf and small-leaf evergreen forests of the Southern Hemisphere.

FIGURE 25.9 Lush, old-growth Douglas fir forest in the U.S. Pacific Northwest.

MEDITERRANEAN SCRUB

The Mediterranean climate (see Unit 16) is characterized by hot dry summers and cool moist winters. Such climates prevail along the shores of the Mediterranean Sea, along the coast of California, in central Chile, in South Africa's Cape Province, and in southern and southwestern Australia (see Fig. 14.3). The **Mediterranean scrub biome** corresponds to these **Csa** and **Csb** climates. Alfisols and Mollisols are common in this biome, with Inceptisols and Entisols found on steep rocky hillsides. Vegetation consists of widely spaced evergreen or deciduous trees (pine and oak) and often-dense, hard-leaf evergreen scrub. Thick waxy leaves are well adapted to the long, hot, dry summers.

Mediterranean vegetation creates a distinctive natural landscape (Fig. 25.10). Even though this biome's regions are widely separated and isolated from each other, their

FIGURE 25.10 "Mediterranean physical and cultural landscapes coexist in a distinctive way. Where human activity has encroached onto even the steepest slopes, Mediterranean vegetation somehow survives. Italy's Amalfi coast, south of Naples, provides an example."

appearances are quite similar. In coastal California, the Mediterranean landscape is called *chaparral*; in the Mediterranean region of southern Europe, it is referred to as *maquis* (French) or *macchia* (Italian); in Chile, it is known as *mattoral*; and in South Africa, it is called *fynbos*. Unless you can identify species that are unique to each of these locales, you can easily mistake one regional landscape for another. Wildfire is common in many of these landscapes.

Mediterranean regions are among the world's most densely populated and most intensively cultivated. Human activity has profoundly altered the Mediterranean biome through the use of fire, grazing, and cultivation. Today, vineyards and olive groves have replaced countless hectares of natural chaparral and maquis. Before the rise of ancient Greece, the hills of the Greek peninsula were covered with oaks and pines that had adapted to this climatic regime. Only a few of the legendary cedars of Lebanon now survive. The cork oak, another example of adaptation, still stands in certain corners of the Mediterranean lands. The oaks of Mediterranean landscapes in California now stand above grasses that are almost entirely invasive species, introduced by Euroamerican settlers since the mid-1800s.

NORTHERN CONIFEROUS FOREST

The upper midlatitude **northern coniferous forest biome** takes a Latin name in North America, where it is usually referred to as *boreal forest*. In Russia, it is called the *snowforest* or *taiga*. Regardless of the local name, these forests are found in humid cold climates (**Df** and **Dw**). Spodosols are the characteristic soils of this biome, except in the cold northern reaches where permafrost is present and Gelisols have formed. The most common coniferous (cone-bearing) trees in this biome are spruce, fir, pine, and larch. These needle-leaf trees can withstand periodic drought resulting from long periods of freezing conditions (larch also sheds its needles in winter). The trees are slender and grow to heights of 12 to 18 m (40 to 60 ft); they generally live less than 300 years but grow quite densely (Fig. 25.11). Depressions, bogs, and lakes hide among the trees. In such areas, low-growing bushes with leathery leaves, mosses, and grasses rise out of the waterlogged soil. This type of growth assemblage, combined with stunted and peculiarly shaped trees, is known as *muskeg*. The moose (*Alces alces*) is a large browser in this biome, with a very wide range from North America through Siberia to Scandinavia.

All the biomes discussed here could be differentiated even more precisely. For instance, the boreal forest of Canada can be divided into three subzones: (1) the main boreal forest, which is characterized by the meeting of the crowns of the trees; (2) the open boreal woodland, marked by patches where the trees are broken up by open spaces of grass or muskeg; and (3) a mixture of woodland in the valleys and tundra vegetation on the ridges, called forest tundra, which is found along the polar margins of this biome.

FIGURE 25.11 A river meanders across a taiga (snowforest) landscape on the eastern side of the Ural Mountains in Siberia. In this Russian midsummer scene, the water level is relatively low.

TUNDRA

The **tundra biome** is the most continuous of all the biomes; it occurs almost unbroken along the poleward margins of the northern continents (see Fig. 25.1). It is also found on the islands near Antarctica and in alpine environments above the tree line on mountains at every latitude. The distribution of the tundra biome corresponds to the cold polar climate (**E**), for the most part, and Gelisols are the typical soils.

Only cold-tolerant plants can survive harsh tundra conditions. The most common are mosses, lichens, sedges, and sometimes, near the forest border, dwarf trees. Ephemeral plants are rare; the perennial shrubs are pruned back by the icy winter winds and seldom reach their maximum height. Plant roots cannot be extensive in this biome because the top of the permafrost is seldom more than 1 m (3.3 ft) below the surface. The permafrost also prevents good surface drainage. With so many severe limitations on plant growth, it should not be surprising that net primary productivity in the tundra biome is almost as low as in deserts (see Fig. 24.3). However, the fauna is surprisingly varied considering the small biomass available. It includes large herbivores, such as reindeer, caribou, and musk ox, and a variety of carnivores. During the short summer, shallow pools of water at the surface become the home of large insect populations. In the Northern Hemisphere, birds migrate from the south to feed on these insects. Projections of global warming over the next century suggest that the greatest temperature increase will be at high latitudes. If so, the area now occupied by the tundra biome will shrink, threatening the tundra fauna, including the birds that arrive as seasonal migrants.

Key Terms

biome *page 320*

desert biome *page 325*

Mediterranean scrub biome *page 327*

northern coniferous forest biome *page 329*

savanna biome *page 324*

temperate deciduous forest biome *page 327*

temperate evergreen forest biome *page 327*

temperate grassland biome *page 326*

tropical rainforest biome *page 324*

tundra biome *page 329*

Scan Here for a quick vocabulary review

Review Questions

1. Referring to the map of global biomes (see Fig. 25.1), find several places where you could fly over three or more adjacent biomes on the same continent. For each of these hypothetical flights, explain why one biome changes into another, focusing on the role of climate.

2. Describe the differences between northern coniferous forest, temperate deciduous forest, and tropical rainforest biomes and explain these differences in terms of climate and solar radiation.

3. Describe the differences between desert, temperate grassland, savanna, and Mediterranean scrub biomes and explain these differences in terms of climate.

Scan Here for a quick concept review

www.oup.com/us/mason

UNIT 26

Zoogeography: Spatial Aspects of Animal Populations

The vast Serengeti Plain is one of Africa's most extensive and effective wildlife refuges. Zebras, wildebeests, topi, and other herbivores there still number in the millions.

OBJECTIVES

- Briefly summarize the history of zoogeography and how it led to the concept of evolution through natural selection
- Define the concept of the ecological niche
- Describe the modern theory of evolution and related principles such as natural selection, and understand their importance

in explaining the present-day spatial distribution of animals

- Understand the concept of zoogeographic realms
- Relate zoogeography to the larger context of environmental conservation

The field of zoogeography (pronounced *ZOH-oh-ge-og-ra-phy*; you can review the definition in Unit 20) deals primarily with the geographic distribution of animal **species** (a species is a group of organisms with similar characteristics that can interbreed and produce fertile offspring). The biomes discussed in Unit 25 are defined mainly in terms of their dominant vegetation. Each biome also has characteristic animal species, such as the musk ox in the tundra biome. Therefore, one approach to studying zoogeography would be to focus on relating animal distribution to environmental factors such as climate, the same approach we took in discussing the geography of plants and biomes in Units 24 and 25. When we study global patterns of zoogeography, however, we find very different groups of animals predominating in quite similar environments on various continents. One of the most striking examples is the dominance of the major biomes of Australia by marsupials like kangaroos and wombats, which are only very distantly related to the most important mammals found in the same biomes on other continents.

These observations are best explained as results of the dispersal and evolution of animal species. *Dispersal* is the spread of animals across Earth's surface, which is limited by barriers such as oceans or deserts. The particular species that reached various parts of the world became the ancestors of new species that adapted to specific environments through **evolution**, defined as change in the characteristics of a population that is heritable, or capable of being passed on from one generation to the next. It is important to note that, along with the environmental factors such as climate and soils that were emphasized in Units 24 and 25, evolution and dispersal are also important in understanding many aspects of plant geography. Therefore, many of the concepts discussed in this unit can be applied to plants as well. To understand why the emphasis on dispersal and evolution has become so important in zoogeography, we must first take a brief look at the historical development of zoogeography.

Emergence of Zoogeography

The field of zoogeography began to develop following the publication of works by the great naturalist and explorer Alexander von Humboldt (1769–1859) and by Charles Darwin (1809–1882), who initially focused on geology but became increasingly interested in biological problems. Humboldt traveled much of the world, driven by his wide-ranging interests in geography of all kinds, and published widely read books on what he found. When he reached South America's Andes, he recognized that altitude, temperature, and natural vegetation are interrelated. Our current understanding of how plant distribution is influenced by environmental factors, as discussed in Unit 24, originated with this observation. Humboldt's work was an inspiration for Darwin, Alfred Russel Wallace (mentioned earlier and whom we will encounter again shortly), and many other scientists of the nineteenth century.

Darwin's momentous 1859 work *The Origin of Species* presented his case for the process of evolution through **natural selection**, the process whereby organisms better adapted to their environment have a greater tendency to survive and produce offspring. Darwin developed that case over decades, starting with his own long ocean voyage of exploration on the H.M.S. *Beagle*, and continuing through many years of mulling over his own observations and information gathered from the work of other scientists. In the Galapagos Islands off Ecuador in the Pacific Ocean, Darwin collected a number of bird specimens. When he had these examined by specialists after returning to England, he was impressed to find that distinct species occurred on various islands of the Galapagos. In particular, a group of finch species differed in the form of their beaks, with various forms corresponding to different food sources on each island. Ultimately, Darwin realized that natural selection provides an explanation for how these species could have diverged from each other after they colonized each island. We explore that concept in detail later in this unit.

As more became known about animals and plants in places distant from the centers of learning, fascinating zoogeographic questions continued to present themselves. One of the most interesting observations involved the transition of fauna from Southeast Asia to Australia. Somewhere in the intervening Indonesian archipelago, the animals of Southeast Asia give way to the very different animals typical of Australia. Australia is Earth's last major refuge of *marsupials*, animals whose young are born early in their development and are then carried in a pouch on the abdomen. Kangaroos, wallabies, wombats, and koalas (Fig. 26.1) are among Australia's many marsupials. Australia also is the home of the only two remaining egg-laying mammals, the platypus and the spiny anteater (*echidna*). Most interestingly, as noted earlier, marsupial species in Australia dominate environments similar to those occupied by very different groups of animals elsewhere in the world. Marsupials are found in New Guinea and on islands in the eastern part of Indonesia, but nonmarsupial animals (e.g., tigers, rhinoceroses, elephants, and primates) prevail in western Indonesia. What is the origin of the zoogeographic boundary between these sharply contrasting faunas?

One answer to this still-debated question was provided in an important work published by Wallace in 1876, *The Geographical Distribution of Animals*. (Wallace had earlier arrived at the idea of natural selection independently from Darwin.) Based in part on his own extensive exploration of animal life in what is now Indonesia, Wallace showed how far various species had progressed eastward along the island stepping-stones between mainland Southeast Asia and continental Australia. He also mapped the westward extent of marsupials from Australia. On the basis of these and other data, Wallace drew a line across the Indonesian archipelago, a zoogeographic boundary between the fauna of Southeast Asia and that of Australia now referred to as **Wallace's Line** (Fig. 26.2). Later researchers argued that the boundary should be placed somewhat farther east. An example of an alternative, Weber's Line, is also shown in Figure 26.2. Regardless of its exact location, it is clear that the transition between the

FIGURE 26.1 "Moving slowly and quietly through a eucalyptus forest in New South Wales, Australia, I was rewarded with this extraordinary sight, a koala resting in a tangle of branches. The koala 'bear' is part of Australia's unique fauna; it is a marsupial and carries its offspring in its pouch for as long as seven months. It eats about 1.3 kg (3 lb) of eucalyptus leaves daily, but only a particular kind and quality of leaf; in the wild its life span averages 20 years. The koala has diminished in number from an estimated several million to perhaps 150,000, and the population continues to decline as humans encroach on its natural habitat and diseases take their toll."

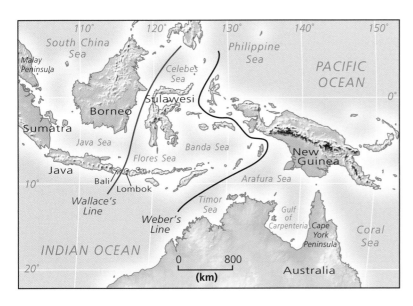

FIGURE 26.2 Wallace's Line across the Indonesian archipelago separates the faunal assemblages of Southeast Asia and Australia. Weber's line is an alternative boundary that has also been proposed.

typical animal species of Southeast Asia and those of Australia generally corresponds to a region where deep ocean waters separated islands even at times in the geologic past when sea level was lower. Those deep waters formed a barrier to animal migration so that the animals in Australia and Southeast Asia evolved independently. We now know this separation has lasted millions of years. In both regions, new species arose as animals occupied and became isolated in new environments, and as existing environments were altered by climatic change, volcanic activity, and other processes.

Range, Habitat, and Niche

With that historical background, we now turn to basic concepts of zoogeography that have emerged directly from the original works of Humboldt, Darwin, and Wallace. Humboldt's observations in South America and other regions led him to recognize that most species occur in a particular set of environmental conditions, involving temperature, moisture, and many other factors (a point that was elaborated on in Unit 24). More recently, biogeographers have developed a more specific set of concepts to describe this relationship between organisms and their environment, including the ideas of range, habitat, and niche.

When an **animal range** is described in zoogeography, the focus is usually on the animal's spatial distribution—the area in which it occurs. The ranges of four North American mammals are mapped in Figure 26.3. A relationship with climate and related factors is often apparent in such maps; for example, the black bear clearly occurs in a colder

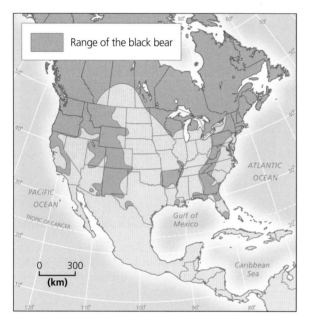

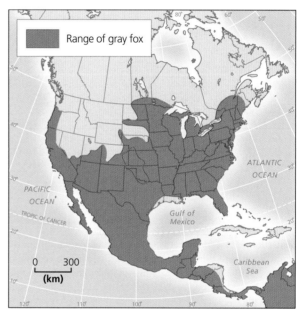

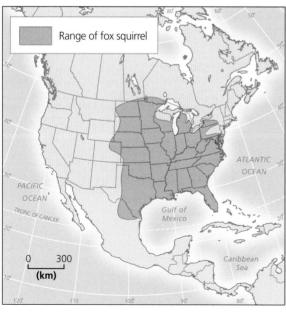

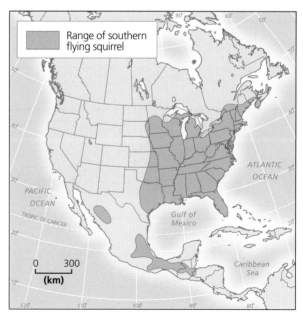

FIGURE 26.3   Ranges of some North American mammals.

range of climates than the gray fox. However, vegetation is also important, because the black bear range extends far south in forested areas (even into Florida) but does not include the mostly treeless Great Plains states of Nebraska and Kansas. The **habitat** of a species is the particular environment it normally occupies within its geographic range. A habitat usually is described in quite general terms, such as grassland or seashore or alpine. For example, within its range shown in Figure 26.3, the fox squirrel is largely found in a habitat with trees, whether it is a forest or a city park.

A more subtle but very important concept is the **ecological niche**, the environmental space within which an organism operates most efficiently or to which it is most effectively adapted. The concept of environmental space not only refers to an area on a map but also includes a range of climatic conditions, vegetation, availability of resources like food or light, and interactions with other plants and animals. Some niches extend over large areas, perhaps coinciding with entire biomes or continental parts of biomes. The ecological niche of the South American jaguar includes a substantial part of the Amazonian rainforest.

Other niches are quite small or at least have become small because of human impacts or other types of environmental change. The Kirtland's warbler (*Dendroica kirtlandii*) is a songbird that breeds only in dense young stands of a tree called the jack pine (*Pinus banksiana*) in the Great Lakes region, where the songbird spends the summer (Fig. 26.4). In that region, the jack pine is confined mainly to sandy soils where wildfire is frequent. Jack pine is more widespread in the subarctic region of Canada (see Fig. 26.4), but the warbler apparently cannot tolerate the lower temperatures there. Thus, in summer, the ecological niche of the Kirtland's warbler is limited to quite small patches of jack pine forest, primarily in Michigan (it migrates to the Bahamas in winter). When suitable areas of jack pine shrank because of logging and fire control in the twentieth century, the warbler came close to extinction.

Processes of Evolution

But how does this kind of fit between an animal or plant and its niche develop in the first place? Evolutionary theory answers this question using the concept of natural selection, proposed by Darwin and Wallace in a pair of papers in 1858 and greatly refined since that time. Natural selection stems from reproductive processes. The word *reproduction* itself accurately conveys the fact that to a large degree, individual plants or animals are faithful copies of their parents because of the genetic information (*genes*) passed from one generation to the next. However, there is always a certain degree of genetic variation within a population of a particular species, which can be expressed as observable variation in animal or plant traits, such as eye color or leaf shape. Darwin was not aware of the genetic basis for inheritance, but he was a keen observer of visible natural variation within species, such as domestic pigeons, which he studied closely. In fact, some of the true extent of genetic variation

A

B

FIGURE 26.4 Kirtland's warbler perched in a jack pine tree in its summer range (A). The jack pine has a large range in Canada, but the Kirtland's warbler spends its summer only in the southernmost, warmest parts of the jack pine's range (B).

is hidden because particular traits are only expressed when passed on from both parents. **Mutations** (inheritable changes in the DNA of a gene) continually add to the existing genetic variation within populations of organisms.

Now we come to a crucial point for understanding natural selection. Not all individuals pass on their genetic information to the next generation. A population of animals or plants always produces more offspring than survive to reproduce, because many immature individuals die by accident, disease, or predation. Within a particular population of organisms, only some pass on their genes because, for example, they are better able to survive cold weather, attract pollinators, or find a mate of the same species. Those individuals that survive and reproduce have been naturally selected.

This process of natural selection is responsible for all of the adaptations to environments we have already discussed and readily explains the fit between animals or plants and their ecological niches. Naturalists long before Darwin had noted many obvious examples of adaptation. Then, as now, the giraffe was a favorite example: its long neck allows it to obtain food from the tops of trees that other animals cannot reach (Fig. 26.5). The concept of natural selection now allows us to explain that long neck as the end-product of a process in which individuals with longer necks—a random occurrence reflecting genetic variability—were more likely to survive and reproduce.

A good example of more complex adaptations to specific niches in a shared habitat is the Serengeti Plain of East Africa, an area of savanna grassland grazed and browsed by many species. The herbivores there are so finely adapted that they use different portions of the Serengeti's vegetation at different times. During part of the year, mixed herds graze on the short grass, satisfying their protein requirements without being required to use too much energy in respiration while obtaining the food. Eventually, the short grass becomes overgrazed. Then the largest animals, the zebras and buffalo, move into areas of mixed vegetation comprising tall grasses, short grasses, and herbs. Zebras and buffalo eat the stems and tops of the taller grasses and trample and soften the lower vegetation. Wildebeests can then graze the middle level of vegetation, trampling the level below. Finally the Thomson's gazelles move into the softened area to eat the low leaves, herbs, and fallen fruit at ground level.

Thinking about this example more carefully leads us to a final important problem of evolution: the development of new species. Each of the herbivore species of the Serengeti is related, often closely, to other species with quite different niches. In fact, they share a common ancestor with those related species and have diverged from them over time. On other continents with extensive areas of savanna, mammals from quite different groups occupy niches similar to those inhabited by the Serengeti's wildebeest and gazelle. In Australia, for example, savanna herbivores are kangaroos or other marsupials. As changing global climate and tectonics created new environments over time, new species

emerged to fill niches in those environments. Those new species were descendants of whatever existing species were already present in particular parts of the world or whichever species managed to disperse into the new environments from elsewhere.

The title of Darwin's best-known book foreshadowed how important the problem of the *origin of species* would become for modern biogeography. Not only are biogeographers concerned with how climate and other factors influence the distribution of particular plant and animal species, but we also seek to understand how the ancestors of those present-day organisms dispersed over Earth's surface and diverged into new species. The *phylogeny* of bears—the diverging lines of descent that connect various bear species—illustrates this point well (Fig. 26.6). We know that the thick white fur of the polar bear is an adaptation to life on Arctic sea ice, and that American black bears have a very different niche in forests farther south (see Fig. 26.3). However, it is also important to know that these species diverged from common ancestors in fairly recent geologic time, probably within the past 2 million years.

As it turns out, geography plays a crucial role in the origin of new species through natural selection. The animals or plants that make up an existing species are generally divided into distinct populations inhabiting particular areas and interbreeding, effectively mixing their pool of genetic information. Some of these populations may live in environments where natural selection favors a specific part of the total range of variation within the species. Plants with unusual resistance to cold may be favored to survive and reproduce in a particularly cold portion of the species' range, for example.

Often, however, there is significant *gene flow* between populations that limits the effectiveness of selection. Think about birds that can easily wander from their home island to another one nearby where they can breed with others of the same species. The optimum situation for new species to emerge is a population isolated enough to limit gene flow from other populations. When this happens, natural selection can act to change the typical characteristics of the plants or animals in that population until they are distinct enough to form a new species, which cannot interbreed with its parent species even if geographic isolation ends.

While this general process is widely agreed upon, the details of how new species emerge are still the subject of extensive, ongoing research. In the Galapagos Islands, the finches that Darwin observed had diverged into about 15 different species adapted to contrasting environments on various islands. The biologists Peter and Rosemary Grant have studied these birds for decades, uncovering many interesting new insights on how species emerge. As the techniques of molecular biology have become more sophisticated, they often reveal more complexity in the descent of animal species than was recognized in the past. For example, the phylogeny of bears (see Fig. 26.6) turns out to be complicated by gene flow between apparently distinct species, such as brown bears and American black bears.

FIGURE 26.5 "Safari in Tanzania, March 1984. Spent a morning watching a small herd of giraffes as they browsed together, dispersed in the bush, reconvened, were briefly joined by a small herd of zebras, then separated again. In the competitive evolution of East Africa's herbivores, these longer-necked animals had advantages, notably their access to food beyond the reach of others, that led to larger numbers of offspring—who passed the genetic information involving the long neck to later generations."

Earth's Zoogeographic Realms

Let's revisit Wallace's Line (see Fig. 26.2), which marks the boundary between two **zoogeographic realms**, broad regions of the world in which animals tend to share common evolutionary origins. Now we can understand why the faunas on either side of Wallace's Line are so different, even in the case of animals occupying similar niches on one side or the other. East of the line, where few other groups of mammals arrived across the deep ocean waters, marsupials evolved into species adapted to the great variety of niches available in Australia and adjacent islands. West of the line, animals with quite different ancestries filled the same niches.

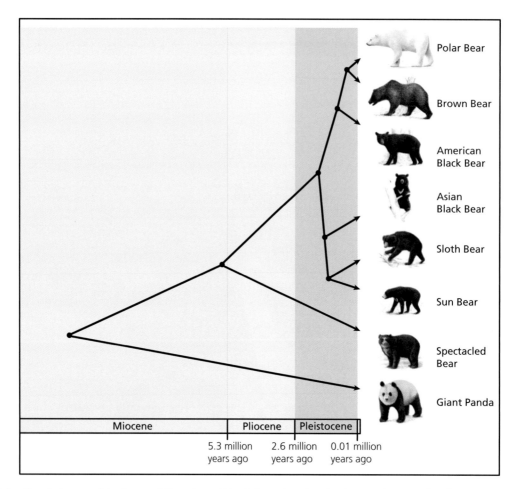

FIGURE 26.6 The phylogeny of the bears of Eurasia and North America, showing how various species diverged from each other at different points in geologic time (the horizontal scale). This tree of species is based on nuclear DNA and other evidence, but some techniques of genetic analysis suggest slightly different trees, as well as gene flow between some species after they diverged.

As with Wallace's Line, zoogeographic realm boundaries often coincide with natural barriers to animal migration such as high mountains (Himalayas), broad deserts (Sahara, Arabian), deep marine channels (Indonesia), and narrow land bridges (Central America). Some groups of animals are found only in one realm—for example, kangaroos only appear in the Australian realm. In both plants and animals across the various realms, biogeographers can discern evidence of the theory of **convergent evolution**, which holds that organisms in widely separated biogeographic realms, though descended from diverse ancestors, develop similar adaptations to similar ecological niches. To some extent, a global map of zoogeographic realms is an exercise in generalization, but there is general agreement on the boundaries of the major realms (Fig. 26.7).

By many measures, the *Paleotropic (Ethiopian) realm* contains Earth's most varied fauna, an enormously rich assemblage of animals, many of which are unique to this realm. Some species in this realm, however, have relatives in other realms. For example, the lion and the elephant also occur in the *Indomalayan (Oriental) realm*. Nevertheless, the Indomalayan realm has less diverse fauna. The fauna of the island of *Madagascar*, shown on our map as a discrete

realm, differs quite strongly from that of nearby Africa. Though relatively small compared to other realms, Madagascar has a distinct fauna because it has been isolated by deep ocean waters for millions of years. Madagascar is home to such unique fauna as the lemur, a small primate, but the island has nothing to match East Africa's herd animals, lions, or even poisonous snakes.

The *Australian realm*'s faunal assemblage exhibits the consequences of prolonged isolation and separate evolution. This is the realm of marsupials, described in the early pages of this unit, such as the kangaroo and the wombat, of the platypus and the Tasmanian "tiger." Although some biogeographers include New Zealand in the Australian realm, others map *New Zealand*, like Madagascar, as a discrete realm. There is ample reason for doing so: New Zealand's fauna includes no mammals other than bats, few terrestrial vertebrates of any kind, and nothing to match the assemblage of Australia—except when it comes to birds. New Zealand has a rich variety of bird life, with an unusual number of flightless species.

The *Neotropic realm* also has a rich and varied faunal assemblage, which includes such species as the pig-like tapir, the jaguar, and the boa constrictor. The two remaining realms, the *Nearctic* and the *Palearctic*, are much less

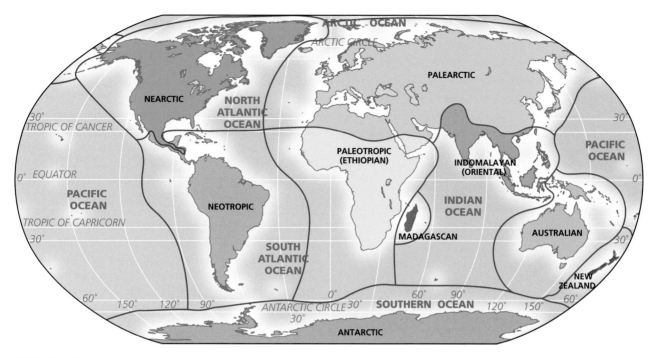

FIGURE 26.7 World zoogeographic realms.

rich and far less diverse than the other major zoogeographic realms. Some biogeographers prefer to map these together as a single realm, but evidence of long-term isolation (the Bering land bridge notwithstanding) is much stronger in the North American Nearctic than in the Eurasian Palearctic realm. Remarkable adaptations exist in both realms, such as those of the polar bear (both realms), the Siberian tiger and giant panda (Palearctic), and the bison and mountain lion (Nearctic).

Figure 26.7 presents the global zoogeographic map at a high level of generalization, and we should remember that this is a map of *realms*. Embedded within each realm are numerous subdivisions, or zoogeographic *regions*. Consider the Australian realm as an example. The fauna of mainland Australia, despite commonalities with the nearby large Pacific island of New Guinea, nonetheless differs sufficiently to make New Guinea and Australia separate zoogeographic regions within the Australian realm.

Ecological Zoogeography and Conservation

As interest in zoogeography grew in the twentieth century, new kinds of studies were undertaken. Once the broad outlines of evolution and animal phylogeny were established and zoogeographic realms had been mapped, new questions emerged. The most important of these can be considered problems of **ecological zoogeography**, the study of animals' interactions with their physical environment and other animal or plant species. The problem of explaining biological diversity became especially pressing as human impacts on the environment led to widespread extinction of species. Biologists believe that there may be as many as 30 million

species of organisms on our planet, of which only 1.7 million have been identified and classified. Obviously, many species live in the same habitat. What determines how many species can be accommodated in a specific, measured region? This is a central question in modern zoogeography. Studies of biological diversity and its causes on islands have proved especially fruitful in providing answers to this question.

ISLAND ZOOGEOGRAPHY

Studying biological diversity on islands presents unique opportunities. Small islands can be inventoried completely so that almost every species living there is accounted for. When zoogeographers began doing this work, their research had some expected and some unexpected results. In 1967, R. H. MacArthur (b. 1930) and E. O. Wilson (b. 1929) determined that the number of species living on an island is related to the size of that island. The larger the island, the larger the number of species it can accommodate, a generalization that turns out to be true for birds, insects, and many other organisms (Fig. 26.8). This is a specific case of a broader pattern widely recognized in biogeography, which is that as we look at larger areas, we find more species.

One explanation for this pattern might be that on larger islands (or in larger areas in general), the landscape is more diverse, and there are more potential ecological niches to be occupied. MacArthur and Wilson concluded that this was not the case with the islands they studied, and they proposed a remarkably simple alternative explanation. They argued that the number of species on an island is related to the balance between arrivals of new species on the island and extinctions of those already present. The number of species tends to increase or decrease until it

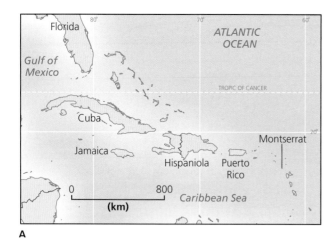

A

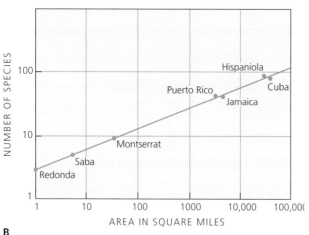

B

FIGURE 26.8 A basic concept of island biogeography concerns the increase in numbers of species with increasing island size. Islands in the Caribbean vary widely in size (A), and as their size increases, they hold greater numbers of amphibian and reptile species (B). The islands of Saba and Redonda in the northeastern Caribbean are too small to be visible on the map shown in (A).

reaches the stable value at which new arrivals are balanced by extinctions. New arrivals could be birds ranging out over the ocean from the mainland or a nearby island or insects and plant seeds carried on floating debris. The stable number of species tends to be larger for islands near the mainland because more species arrive there over time. Moreover, the stable number of species is lower on small islands because extinctions happen more often there. That is because, on small islands, populations of each species are smaller, so it is easier for all of the individuals of a species to die out quickly, leading to extinction.

This conclusion, that the number of species on an island has a predictable stable level, was tested on some small islands off the coast of Florida. Researchers counted the species of insects, spiders, crabs, and other arthropods living on each island. Then they sprayed the islands with pesticides, wiping out the entire population. After a few years, the islands were reinhabited naturally, as insects and other animals found their way back to them. Approximately the

same number of species had returned, although the number of individuals was proportionally different. That is, there were crabs and spiders before the extinction, but now there were more crabs and fewer spiders. This indicates not only that a certain number of species is characteristic of an island in a certain environment but also that the kinds of species making up this number vary and may depend on the order in which they arrived.

The ideas of MacArthur and Wilson had a powerful impact on the field of biogeography. In particular, they suggested that if we take into account a few simple processes such as dispersal and extinction, we can explain—or even predict—the diversity of species on islands of different sizes and with different degrees of isolation from other landmasses. It was quickly recognized that these ideas might apply not only to real islands, but to other isolated patches of habitat as well, such as mountain ranges surrounded by desert. Later work has shown that the story is more complicated than MacArthur and Wilson assumed. For example, large islands actually do have more diverse habitats in many cases, so that factor is more important in explaining species diversity than originally thought. Nevertheless, it is still clear that the biological diversity of islands or other patches of habitat depends on their size and isolation, because those factors influence animal and plant dispersal and extinction. In other words, geography matters! In the next section of this unit, we will see how these concepts have turned out to be quite important for practical problems of species conservation.

CONSERVATION OF ANIMAL SPECIES

The explosive growth of Earth's human population and the destructive patterns of consumption by prospering peoples have combined to render many species of animals extinct, endangered, or threatened. Awareness of the many threats to the remaining fauna has increased in recent decades, and some species have been salvaged from the brink of extinction. But for many others, hope is fading. To understand the crucial insights that ecological zoogeography can provide into the problem of species conservation, we must first consider the kinds of pressures faced by many species today and the history of efforts to preserve endangered species.

Threats to species survival often involve human encroachment on animal habitats. In many parts of the world—in the foothills of the Himalayas, on the plains of East Africa, and in the forests of tropical South America, for example—people and animals compete for land. Not long ago, substantial parts of India teemed with herds of wild buffalo; many kinds of antelope and deer, as well as their predators, lions and tigers; and large numbers of elephants and rhinoceroses. As the population of South Asia rose dramatically, and as the area experienced rapid economic growth, the habitat in which these species can thrive rapidly declined. Today, the Indian lion is virtually extinct; the great tiger is endangered and survives in only small numbers in remote forest areas.

FIGURE 26.9 "Safari, Tanzania, February 1989. A herd of buffalo stops at a small waterhole in Manyara National Park. The rift valley wall that marks the western limit of the Lake Manyara Rift is in the background; if the photographer turned around, Lake Manyara would be visible. The park constitutes a narrow strip of land, but it has a rich fauna. Such richness of animal life prevailed across much of tropical Africa before the arrival of the European colonialists, who upset the long-established balance among indigenous peoples, their livestock, and the realm's wildlife. Hunting for 'sport' was not an African custom; neither was killing for fashion."

In East Africa, similar pressures affect the region's famous wildlife reserves (Fig. 26.9). In the Amazon Basin, rainforest habitat destruction threatens the survival of many species of animals as well as plants (see Perspectives on the Human Environment: Biodiversity Under Siege). All this is reminiscent of what happened to the wildlife of North America following the arrival of the Europeans. The vast herds of bison grazing on the Great Plains were decimated by hunting within a few decades. Huge flocks of passenger pigeons once darkened the skies of eastern North America, but loss of forest habitat and hunting led quickly to their extinction.

In the United States, conservation efforts over the last century have had some notable successes, including the revival of the bison, the wild turkey, and the wolf as well as the survival of the grizzly bear and the bald eagle. There have also been failures, such as the extinction of birds like the ivory-billed woodpecker and Carolina parakeet. Despite this mixed record of success and failure in preservation, conservationists have now learned some important lessons. Most endangered animals cannot be protected simply through bans on hunting or by fencing off preserves. It is essential to actively manage the habitat of species that are at risk and ensure that it is adequate for survival. This is where ecological zoogeography can provide great insights.

Knowledge of the numbers, ranges, habits, and reproductive success of species is crucial to their effective protection. And the most important information may be the specific ecological niche occupied by animal species at risk.

The Kirtland's warbler, introduced earlier in this unit, is a good example. Efforts to preserve this species would have failed without knowledge of its need for jack pine stands of a certain age and density. With that knowledge, stands of this kind have been preserved or newly created through the controlled use of fire by land managers. The population of Kirtland's warbler is now believed to be expanding. The northern spotted owl in the U.S. Pacific Northwest is another well-known example. To make its nests, this owl needs old trees with soft, rotting wood, not the healthy young trees of new forestation. The spotted owl is therefore restricted to the dwindling stands of old, mature forest, so efforts to preserve this endangered species have focused on maintaining those patches of older forests. In general, the more information available about an animal's niche, the more effective conservation efforts can be.

Today, though, we should also take into account the potential effects of rapid climate change in coming decades. Patches of habitat that are suitable for an animal today may be too warm or too dry in 50 or 100 years. The niches of other animals may actually expand to new areas because of climate change. The Kirtland's warbler is a good example, since it could possibly expand into areas of jack pine in Canada if the climate there becomes warmer (see Fig. 26.4). This kind of shift in an animal's range will only occur if it can successfully migrate into the newly suitable areas. Conservation planning will then have to ensure there are no barriers that prevent such migration.

The studies of island zoogeography discussed earlier may be helpful in this kind of planning—even if there are no real islands involved. Consider the patches of jack pine used by the Kirtland's warbler or the old-growth forest used by the northern spotted owl. For these two birds, these are much like islands of favorable habitat, surrounded by other landscapes that they cannot use. Based on studies of real islands, we would expect that a species would be less likely to disappear from a large patch of habitat that can sustain a larger population. Also recall the finding that new species are more likely to arrive at islands closer to the mainland. With that knowledge, we might expect that Kirtland's warblers or spotted owls would be more likely to make their way to unoccupied patches of suitable habitat if they are not too far from patches where they now live or, better yet, if they are connected by corridors of useful habitat. In fact, some Kirtland's warblers have recently moved into new patches of suitable habitat in northern Michigan and Wisconsin, so in this case, the "islands" were not too far apart. Many recent species conservation efforts have used concepts of island zoogeography to design wildlife preserves.

Zoogeography therefore has many theoretical and practical dimensions. The spatial aspects of ranges, habitats, and niches require research and analysis. The results of such investigations are directly relevant to those who seek to protect wildlife and who make policy to ensure species survival.

Biodiversity Under Siege

Biodiversity, shorthand for *biological diversity*, refers to the variety of living organisms within a local area, within an ecosystem or biome, or for the planet as a whole. One important component of biodiversity is the *richness of species*, which refers to the total number of species in a particular area of Earth. How many living species does the global environment contain? Biologists have made widely varying estimates of the number of species yet to be discovered, but all agree that it is far greater than the number that have been identified and described. A recent study suggests that there are 7.7 million animal species on Earth, of which fewer than 1 million have been identified (Mora et al., "How Many Species Are There on Earth and in the Ocean?" *PLoS Biology* 9 (2011): 1–8). Most of those undiscovered animals are likely to be insects or other invertebrates, as mammals, birds, and fish have been so well studied. Besides this diversity of animals, the same study estimates that there are more than 600,000 species of fungi, of which only 43,271 have been cataloged, and that there are 8.7 million discovered and undiscovered species of all kinds of organisms, including plants.

Biodiversity may be regarded as one of our planet's most important resources and is a rising concern today because it is threatened in so many parts of the world as species disappear from local areas or become extinct. Many people place great value on the opportunity to visit places with high biodiversity, such as tropical forests, coral reefs, or just local woodlands full of birds and wildflowers. Yet the loss of biodiversity can also have much more practical economic or environmental impacts. For example, a major part of the biodiversity in forest ecosystems is due to the many animals and fungi that live out of sight in the soil. When some of these below-ground species are lost because of pollution or soil disturbance, trees may not be able to draw the nutrients and water they need from the soil, reducing tree growth and lowering the economic value of the forest. In many rural areas of the world, wild plants and animals still provide important sources of food or income, which can be lost as species disappear.

These and many other effects of biodiversity loss were discussed in the Millennium Ecosystem Assessment (Mace et al., "Biodiversity," in *Ecosystems and Human Well-Being: Current State and Trends*, ed. R. Hassan, R. Scholes, and N. Ash [Washington, D.C.: Island Press, 2005], pp. 77–122), which brought together a large team of scientists to evaluate current trends in the biosphere. This assessment also estimated the rate of species loss by extinction and how much it has increased in recent times because of human impacts. At least five times in Earth's past history, there have been periods of rapid extinction, when large numbers of species disappeared. The most recent was 65 million years ago, when the dinosaurs disappeared. Over most of geologic time, however, fewer than five species have gone extinct per year. Today, it is estimated that 100 to 1,000 times this number of species are lost per year. This estimate includes species known to scientists as well as many that have not even been identified and cataloged before they disappear (many of which are insects or other invertebrates).

Until recently, the main reason for such rapid extinction has been loss of habitat caused by deforestation, agriculture, urbanization, and air and water pollution, all the result of expanding human activities. Today, there is growing concern that climate change will soon become a leading cause of extinction, especially for species found only in limited areas that will become inhospitable because of rising temperature, decreasing moisture, or rising sea level.

The number of species now considered to be threatened with extinction is alarmingly high. The Millennium Ecosystem Assessment reports that 12 percent of all bird species, 23 percent of all mammal species, and 32 percent of all amphibian species are threatened. The levels are much higher for more specific groups of organisms or parts of the world; for example, 33 percent of flowering plant species in the United States are threatened, as are 57 percent of the world's species of penguins. A large majority of all threatened species are found in the tropical forests, in part because so many of the world's species live there, but also because disturbance of these forests by logging and conversion to agriculture has increased greatly in the past several decades (see Unit 25). Islands with many **endemic species** (highly localized species) host many threatened plants and animals.

Although biodiversity is one of the life sciences' newest research arenas, we already know enough to draw the following conclusion: as human technology continues to advance, it is increasingly accompanied by the most rapid extinction of natural species the world has undergone since the mass extinctions that occurred at the time the dinosaurs disappeared. We also recognize that the loss of biodiversity brought about by this wave of extinctions will have significant environmental and economic costs.

Key Terms

animal ranges *page 334*

biodiversity *page 342*

convergent evolution *page 338*

ecological niche *page 335*

ecological zoogeography *page 339*

endemic species *page 343*

evolution *page 332*

habitat *page 335*

mutation *page 336*

natural selection *page 332*

species *page 332*

Wallace's Line *page 332*

zoogeographic realms *page 337*

Scan Here for a quick vocabulary review

ENHANCE

Review Questions

1. How were Alexander von Humboldt and Alfred Russel Wallace instrumental in developing the field of zoogeography?

2. What is the *ecological niche* of an animal, and how does it differ from the animal's range?

3. Why do animals that live in a particular habitat in one zoogeographic realm often differ from those living in the same habitat in another realm?

4. Why are islands good places to study questions such as how to explain the number of species in a geographic area?

5. Why is understanding a rare animal's ecological niche important for conservation of that animal?

6. What can the zoogeography of islands tell us about successful strategies for species conservation?

Scan Here for a quick concept review

ENHANCE

 www.oup.com/us/mason

The Restless Crust

Restless Crust

The lithosphere, oldest of Earth's spheres, began to form as soon as molten rock first solidified in patches on the planet's turbulent surface more than 4 billion years ago. Since then, Earth's crust has thickened and matured, but it has not stabilized. New igneous rock forms from molten magma, while forces of weathering and erosion break down existing rocks exposed at Earth's surface. Loose material accumulates and becomes compressed into sedimentary rock. Both igneous and sedimentary rocks may be heated and transformed (metamorphosed) into harder metamorphic rock types. Rocks are uplifted and exposed to erosion at the Earth surface, or else they plunge downward into Earth's interior. Either way, the rock cycle, a system in which rock is continually formed and altered into new rock types, moves forward.

The rock cycle operates largely because of rock mobility within the worldwide system of *plate tectonics*. Great slabs of solid crust move, driven by mobile sub-surface rock. When two of these slabs converge, the less dense (continental) slab overrides the denser (oceanic) one, starting a process called *subduction* in which the oceanic crust is forced downward. At the surface, subduction creates spectacular scenery, causes earthquakes, and generates volcanic activity. Below, the sinking crustal slabs become part of the slowly moving mass called the *asthenosphere*, deep beneath the crust. At the same time, new crustal material forms where hot rock flows upward from great depths along ridges beneath the oceans. The movement and interaction of plates explains much of the topography we see on Earth, including great continental mountain ranges like the Himalayas as well as the midoceanic ridges and deep trenches of the ocean floor.

Minerals and Igneous Rocks

The rough surface of granite in Joshua Tree National Park, California reveals large crystals that formed as magma slowly cooled, deep beneath Earth's surface. Weathering and erosion have now exposed this igneous rock at the land surface, shaping it into striking pinnacles and rounded boulders.

OBJECTIVES

• Understand the properties and major groups of minerals

• Briefly outline the three types of rocks and the processes that produce them

• Discuss important aspects of igneous rocks and their influence on landscape forms

Earth's outermost solid layer of rock is the **crust** (in Unit 29 we discuss how the crust forms the upper part of a thicker layer called the *lithosphere*). The crust consists of many different types of rocks, which range from concrete-like hardness to soap-like softness. When heated to high temperatures, some melt and flow; others remain solid. When exposed to the forces of weathering and erosion, some withstand these conditions better than others. To understand why rocks of the crust behave in such widely varied ways, we need to study their basic properties and how they form, starting with the minerals that are the building blocks of most rocks.

Minerals

In chemistry, we learn that all known matter is made up of 92 naturally occurring chemical elements arranged in a periodic table according to their *atomic numbers*. An element cannot be broken down further by heating or by chemical reaction. Every element is identified by a one- or two-letter symbol. Aluminum, for example, is Al; iron is Fe. Elements exist independently or in combination with other elements; each element consists of atoms containing *protons*, *neutrons*, and *electrons*. Minerals can consist of a single element, such as diamond (which is pure carbon) or gold (a metallic element), or they can be combinations of elements, such as quartz (a compound of silicon and oxygen).

Minerals are **crystalline**; that is, their atoms are arranged in regular, repeating structures (Fig. 27.1). It is often impossible to see these structures in hand-held samples, but X-ray diffraction instruments reveal them. Sometimes, however, nature displays these crystalline structures in spectacular fashion (Fig. 27.2). A crystal as large as those shown in Figure 27.2 contains vast numbers of atoms arranged in the mineral's characteristic structure, which is repeated many times. Thus, minerals have distinct properties resulting from the strength and stability of the atomic bonds in their crystal lattices.

Minerals are *in*organic. That is, they do *not* fit the definition of *organic* used in chemistry, which refers to materials

FIGURE 27.2 A magnified view of unusually well-developed crystals of two minerals: quartz (the whitish and clear grains) and iron pyrite (the dark or gold-colored grains) in a rock sample from a locale near Lark, Utah. The shapes of these crystals are closely related to the crystalline structures of the two minerals.

made up mainly of carbon, hydrogen, and oxygen, with the carbon bonded directly to hydrogen. Although solid organic compounds can also have a crystalline structure (sugar is a good example), these are not minerals. In summary, a **mineral** is a naturally occurring inorganic element or compound having a definite chemical composition, physical properties, and crystalline structure.

Many minerals can be recognized quickly by the shape of their crystals as well as by their color and hardness. For crystals of a given mineral to form, there must be time for their atoms to arrange themselves into the proper pattern. As the formation time increases, so does the potential size of the mineral crystals. Imagine a reservoir of molten rock contained somewhere deep inside the crust. This mass of molten rock cools slowly, and therefore its vast numbers of atoms have the opportunity to arrange themselves into large individual crystals of various minerals. When the

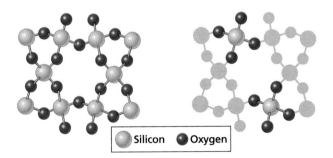

Silicon   Oxygen

FIGURE 27.1 Atoms of silicon and oxygen combine to form the crystalline structure of the common mineral quartz (on the left). The same elements arranged in a different structure would not form the same mineral. As shown on the right, the quartz structure is built by repeating a simple unit, the tetrahedron formed by one silicon bonded to four oxygens.

mass finally hardens, these crystals are readily recognizable to the naked eye.

But what happens if that mass of molten rock does not cool slowly deep inside the crust but instead pours out through a fissure or vent onto the surface of the crust? Now cooling takes place rapidly, and there is less time for the atoms to arrange themselves into large orderly crystals. When cooling is extremely rapid, as when blobs of molten rock are blown into the air from an erupting volcano, the result is a glass-like rock in which the atoms are arranged almost randomly, and in a manner very different from the orderly structure found in minerals. Obsidian, a glass-like rock that can be chipped into sharp stone tools, results from extremely rapid cooling of a lava flow (Fig. 27.3). Although obsidian has the same chemical composition as the minerals that make up granitic rocks, it is not a true mineral because of its lack of well-defined crystalline structure.

Rocks are composed primarily of minerals, though they can also contain other materials, such as coal (a mixture of organic compounds) or volcanic glass (like a mineral in composition but noncrystalline). A few rocks, such as quartzite, consist of only one mineral (in this case, mainly quartz).

FIGURE 27.3 Obsidian, the shiny volcanic glass, and rhyolite form when lava cools very rapidly under certain circumstances. Dark bands of obsidian can be seen in this extrusion in Inyo National Forest, California.

But most rocks contain several minerals, and the way rocks break or bend, weather, and erode has much to do with the particular mixture of minerals they contain.

MINERAL PROPERTIES

As just discussed, all minerals have characteristic crystalline structures, which can be studied using instruments designed for this purpose. Other specific properties often help in identifying and differentiating minerals, including chemical composition, hardness, cleavage or fracture, color and streak, and luster.

Chemical Composition   One of the first methods scientists used to study minerals was to determine the chemical elements they contained. It was discovered fairly quickly that the manner in which those elements were combined to form a crystalline structure was most important. To take the simplest example, the mineral *diamond*, one of the hardest-known substances, is pure carbon (C). But so is *graphite*, the soft "lead" in a pencil. Chemically, diamond and graphite are the same, but their crystalline structures differ. In a diamond, all atoms are bonded strongly to each other. In graphite, certain bonds are weaker, creating sheets that are easily split apart. Note, too, that while diamond is clear, transparent, and hard, graphite is opaque, black, and soft—the very opposite qualities (Fig. 27.4).

Hardness   As we saw with diamond and graphite, hardness is another important property of minerals. This quality can be useful in identifying minerals in the field, and it can tell us much about the overall hardness of the rock in which minerals occur. As long ago as 1822, a German mineralogist named Friedrich Mohs (1773–1839) noticed that certain minerals could put a scratch into other minerals, but not vice versa. Diamond, the hardest mineral of all, will scratch all other natural mineral surfaces but cannot be scratched by any of the others. Mohs established a hardness scale ranging from 1 to 10, with diamond (10) being the hardest. Mohs determined that talc (the base mineral of talcum powder) was the softest naturally occurring mineral, and this he numbered 1. Mohs' Hardness Scale (Table 27.1) continues to be used to this day. Notice that quartz, a commonly appearing mineral, is ranked 7 and is quite hard. By way of comparison, here are some everyday items ranked according to their approximate hardness: pocketknife blade, 5–6; glass, 5; copper penny, 3.5; fingernail, 2.5.

Cleavage/Fracture   The crystal form of minerals quickly identifies them in some cases, but not many crystals can grow unimpeded to the fully developed form shown in Figure 27.2. More useful is the property of *cleavage*, the tendency of minerals to break in certain directions to form bright plane surfaces, revealing the zones of weakness in the crystalline structure. When you break a rock sample across a large crystal of a certain mineral, the way that crystal breaks may help identify it.

A

B

FIGURE 27.4 Diamonds reflect this valuable mineral's hardness, clarity, and transparency (A). By contrast, the graphite form of the same carbon exhibits opposite characteristics (B) and is so inexpensive that it is commonly used as pencil "lead."

TABLE 27.1 Mohs Hardness Scale

Mineral	Hardness
Diamond	10
Corundum	9
Topaz	8
Quartz	7
Potassium feldspar	6
Apatite	5
Fluorite	4
Calcite	3
Gypsum	2
Talc	1

Color/Streak A mineral's color is its most easily observable property. Some minerals have distinct colors, such as the yellow of sulfur and the deep blue of azurite. Other minerals have identical colors or occur in numerous colors and therefore cannot be differentiated according to this property. However, the color of a mineral's *streak* (its powdered form when rubbed against a porcelain plate) can also sometimes help identify it. For example, although both galena and graphite are metallic gray in color, their streaks are gray and black, respectively. Like color, streak is often unhelpful in that most minerals have white or colorless streaks.

Luster A mineral also displays a surface sheen, or *luster*, which, along with color, can be a useful identifying quality. For instance, the difference between real gold and a similar-looking but much less valuable mineral, pyrite (FeS_2), can be detected by their comparative lusters. Not surprisingly, pyrite is called fool's gold!

MINERAL TYPES

Geologists divide the minerals into two major groups: the *silicates* and the *nonsilicates*. Each group is in turn subdivided. This classification is a central concern of the field of mineralogy. The importance of the silicates is easy to appreciate when one considers the composition of rocks in Earth's crust. Although at least 92 chemical elements are known in nature, only eight make up more than 98 percent of the crust by weight (Table 27.2). Moreover, the two most common elements in the crust—silicon and oxygen—constitute almost 75 percent of it. These two elements are the essential components of **silicate minerals**.

The silicates always contain silicon (Si) and oxygen (O) and, in most cases, other elements as well. All silicates have crystalline structures that share a basic building block (see Fig. 27.1): one silicon atom bonded to four oxygen atoms to form a tetrahedron (a three-dimensional shape with four triangular faces; each apex is an oxygen atom). The oxygen atoms can be shared with other tetrahedra to form various structures that distinguish the different groups of silicates. **Quartz** is the silicate with the simplest structure, as shown in Figure 27.1, a three-dimensional framework containing only silicon and oxygen atoms (its formula is SiO_2). Because the silicon-to-oxygen bonds are quite strong, quartz is highly resistant to breakdown through the chemical reactions of weathering. **Feldspars** also have silica tetrahedra arranged in a three-dimensional structure, but they contain aluminum and a variety of other elements and are a bit less resistant to weathering than quartz. *Micas* and *clay*

TABLE 27.2 Composition of Earth's Crust

Element	Percentage (by weight)
Oxygen (O)	46.6
Silicon (Si)	27.7
Aluminum (Al)	8.1
Iron (Fe)	5.0
Calcium (Ca)	3.6
Sodium (Na)	2.8
Potassium (K)	2.6
Magnesium (Mg)	2.1
TOTAL	98.5

The World's Oldest Rocks

Geologists have known for decades that Earth was formed about 4.6 billion years ago, condensing from a rotating, gaseous mass along with the other terrestrial planets of the inner solar system. Yet the oldest known rocks were considerably younger than Earth, leaving a significant gap in our understanding of the rock cycle that began soon after Earth formed. Recent discoveries of extremely old rock formations in the continental shields of Canada and Australia have now begun to shed much more light on the earliest stages of Earth's history.

In 1999, dating of gneiss (a metamorphic rock) on the Canadian Shield indicated that it is 4.03 billion years old (Bowring and Williams, "Priscoan (4.00–4.03 Ga)

Orthogneisses from Northwestern Canada," *Contributions to Mineralogy and Petrology* 134 [1999]: 3-16). Even older mineral grains, one with an age of 4.4 billion years, have been found in sedimentary rocks in Australia (Wilde et al., "Evidence from Detrital Zircons for the Existence of Continental Crust and Oceans on the Earth 4.4 Gyr Ago," *Nature* 409 [2001]: 175-178). This age records the time when these grains originally crystallized, as magma cooled to form an igneous rock. Later, they were released through weathering and erosion and were carried by water before becoming part of a sedimentary rock; therefore, these grains provide the earliest record of the rock cycle.

---

minerals are characterized by silica tetrahedra combined in sheets and, like the feldspars, always contain aluminum and other elements as well as silica. A number of other silicate structures are also possible.

The *nonsilicates* include the carbonates, sulfates, sulfides, and halides. Among these, the **carbonate minerals** are of greatest interest in physical geography. All carbonates contain a basic building block made up of one carbon and three oxygens (CO_3). This structure combined with calcium forms calcite ($CaCO_3$), the mineral of which limestone is made. Add magnesium (Mg) to the formula, and the mineral dolomite is formed. Both calcite and dolomite dissolve in water much more rapidly than the silicates. Therefore, rocks made up of these carbonate minerals can be weathered by rainfall and groundwater, creating remarkable landforms (which we discuss in Unit 42).

The *sulfates* all contain sulfur and oxygen combined to form SO_4. In some places, the calcium sulfate gypsum lies exposed over sufficiently large areas to influence the development of landforms. The *sulfides* (S, sulfur, without oxygen), on the other hand, occur in veins and ores and do not build or sustain landforms themselves. Pyrite (FeS_2) is such a mineral. The *halides* consist of metals combined with such elements as chlorine, fluorine, and iodine. The most common is halite, a compound of sodium (Na) and chloride (Cl), the substance that makes ocean water salty.

The final two categories of nonsilicates are the oxides and natural elements. The *oxides* are formed by a combination of metal and oxygen. Oxides of various metals can form economically important deposits, such as iron ores of hematite (Fe_2O_3) or magnetite (Fe_3O_4). However, oxide minerals can also form in many other environments, including soils; iron oxides often give soil horizons red or

brown colors. The *natural elements* are those rare and prized commodities that are among the most valuable on Earth: gold (Au), silver (Ag), platinum (Pt), and sometimes copper (Cu), tin (Sn), and antimony (Sb).

Classification of Rock Types

From what has been said about the minerals that make up the crustal rocks, it is evident that the diversity of rock types is almost unlimited. Still, when rocks are classified according to their mode of origin, they all fall into one of three categories. One class of rocks forms as a result of the cooling and solidification of **magma** (molten rock), a process that produces **igneous rocks**. The deposition, compression, and cementation of rock and mineral fragments produce **sedimentary rocks**, the second class of rocks. And when existing rocks are modified by heat or pressure or both, they are transformed into **metamorphic rocks**, the third class. We discuss igneous rocks in the remainder of this unit; sedimentary and metamorphic rocks are the subject of Unit 28.

Igneous Rocks

The term *igneous* means "origin by fire," from the Latin word *ignis*, meaning fire. The ancient Romans, upon seeing flaming lava erupt from Italy's Mt. Vesuvius and Mt. Etna, undoubtedly concluded that fire stoked the rock-forming ovens inside Earth. But igneous rocks actually form as a result of cooling—the lowering of the temperature of molten magma or **lava** (magma that reaches Earth's surface) until it solidifies. This can happen deep inside the crust or on the surface. Magma is not only a complex melt of many minerals; it also contains gases, including water vapor, and sometimes solid grains of certain minerals that crystallize at high temperatures. Magma flows outward and upward, forcing itself into and through existing layers of rocks in the crust. If its upward thrust ceases before it reaches the surface, the resulting rocks formed from the cooled magma are **intrusive igneous rocks**. If it penetrates all the way to the surface and spills out as lava or ash, the rocks formed from these materials are **extrusive igneous rocks**.

As we noted previously, slowly cooled intrusive igneous rocks tend to have larger mineral crystals than faster-cooling extrusive ones. Intrusive rocks are generally coarse grained, with mineral crystals as much as 1 cm (0.4 in) or more in diameter. Coarse-grained intrusive rocks that form at especially great depth are called *plutonic* igneous rocks, another term of Roman origin (Pluto was the Roman god of the underworld).

The color of igneous rocks can tell us much about their origins. When the original magma is rich in silica, it yields **felsic** rocks, both intrusive and extrusive. Such rocks are light colored, with pink or light gray feldspar and clear, almost-colorless quartz dominating. Magma that is poorer in silica yields darker intrusive or extrusive **mafic** rocks. We know that light-colored, coarse-grained **granite** forms as a felsic intrusive rock; if the same magma were to spill out onto the surface, it would yield a light-colored but much finer-grained felsic extrusive rock called *rhyolite*. Dark-colored, coarse-grained *gabbro* comes from relatively low-silica magma; if that same magma were to reach the surface before cooling to form an extrusive rock, it would be a fine-grained dark-colored **basalt**.

INTRUSIVE FORMS

Igneous intrusions (masses of intrusive rock) play a role in forming some of the world's most striking landscapes and landforms. In the analysis of these landscapes, the form taken by an intrusion becomes an important factor.

A **batholith** is a massive body of intrusive rock that has melted and assimilated many of the existing rock structures it has invaded (Fig. 27.5). A **stock** is similar to a batholith but smaller. Sometimes, magma inserts itself as a thin layer between strata of existing rocks without disturbing these older layers to any great extent; such an intrusion is a **sill**. Alternatively, magma can cut vertically across existing layers and form a kind of barrier wall called a **dike**. An especially interesting intrusive form is the **laccolith**. In this case, a magma pipe leads to a chamber, which grows, dome-like, pushing the overlying strata into a gentle bulge without destroying them (see Fig. 27.5). The form of an intrusion depends partly on the *viscosity* (fluidness) of the magma. Thick, viscous masses remain compact. In contrast, if magma is quite fluid, it can penetrate narrow cracks in existing rock strata and inject itself between layers. Viscosity is related mainly to the composition of the magma: low-silica mafic magma tends to have lower viscosity and can flow much more readily.

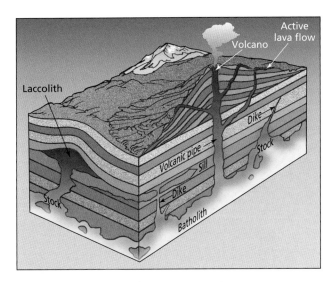

FIGURE 27.5 Diagrammatic cross-section through the uppermost crust, showing the various forms of igneous intrusions.

JOINTING AND EXFOLIATION

Igneous rocks such as granite and basalt display a property that is of great importance in their breakdown under weathering and erosion. **Jointing** is the tendency of rocks to develop parallel sets of fractures without any obvious movement along the plane of separation (Fig. 27.6). Granite often exhibits a rectangular joint pattern so that it breaks naturally into blocks. Basalt, on the other hand, usually possesses a columnar joint system that produces hexagonal forms. This jointing pattern in basalt is closely related to the rapid cooling of this extrusive rock near the ground surface (Fig. 27.7). The contraction of the rock as it cools produces planes of weakness and separation—the *joint planes*.

Not all jointing, however, is related to contraction during the initial cooling of igneous rocks. A special kind of jointing, found in certain kinds of granite, produces a joint pattern resembling a series of concentric shells not unlike the layers of an onion (Fig. 27.8). The outer layers, or shells, peel away progressively, leaving the lower layers exposed. This phenomenon, called **exfoliation**, is caused by the release of confining pressure. Large granite domes that display striking patterns of exfoliation today were at one time buried deep inside the crust and under enormous pressure. As erosion removed the overlying rocks, the pressure was reduced, and these rock masses expanded. The outer shells, unable to resist this expansion force, cracked and peeled along hidden (concentric) joint planes. Joints also form in sedimentary and metamorphic rocks, in part because of the stresses created by tectonic plate movement (see Unit 30).

IGNEOUS ROCKS IN THE LANDSCAPE

Igneous rocks tend to be more resistant to weathering and erosion than common sedimentary rocks. Therefore, intrusive igneous structures often form characteristic landforms when their overburden of sedimentary rock is

FIGURE 27.6 "With a faculty member of the University of Tasmania's Department of Geography we drove from Hobart, the capital, to Port Arthur, the former penal colony. The Tasman Peninsula presents numerous sites of physical geographic interest, and among the most fascinating is this wave-cut platform, where marine erosion has exploited joints in igneous rock. The joint pattern now looks like a roughly tiled floor, planed down by waves rolling over it and exposed by subsequent uplift. At high tide, waves still wash over the platform, filling the joint planes and enhancing the pattern."

FIGURE 27.7 Devil's Postpile, an example of the columns produced by hexagonal jointing in basalt, now preserved in a U.S. national monument. The extrusive igneous rock here cooled from lava that erupted within the last million years in the Sierra Nevada of California. The length and almost-perfect form of the columns resulted from the unusual thickness of the flow, which allowed it to cool a little more slowly than most lava flows.

FIGURE 27.8 "The Devil's Gate historic site in Wyoming is off the beaten track today, but in the 1840s and 1850s, thousands of emigrants passed this way along the Oregon Trail. The trail follows the well-named Sweetwater River, which the travelers relied on for fresh water from Rocky Mountain snows. The stark, barren domes of the Granite Mountains rise from the river valley. As erosion has removed overlying rock, the remaining granite has expanded, eventually peeling off in huge sheets through the process of exfoliation. Emigrants on the trail may have occasionally heard the crash of another sheet sliding off the mountains, a reminder that they were no longer in Illinois or Ohio."

removed through weathering and erosion. For example, when the strata overlying a laccolith (see Fig. 27.5) are eroded away, the granitic core stands as a mound above the landscape, sometimes encircled by low ridges representing remnants of the softer sedimentary cover. Batholiths exposed by erosion of the overlying rock can form mountainous terrain, with slopes of bare granite eroding through the process of exfoliation (see Fig. 27.8).

The intrusive sill, which long ago squeezed between sedimentary layers (see Fig. 27.5), resists erosion longer than the softer sedimentary rocks do. Eventually, such a sill is likely to cap a table-like landform called a *mesa* (see Fig. 40.8), a remnant of the intrusion. A dike (see Fig. 27.5), which is also more resistant than its surroundings, will

stand out above the countryside as a serpentine ridge (see Fig. 40.7 of New Mexico's Ship Rock on p. 488).

Although these landforms associated with igneous intrusion are impressive, the most spectacular landforms associated with igneous rocks, especially the world's great volcanoes, undoubtedly are shaped by extrusive igneous activity. The violent eruption of Mt. St. Helens in the U.S. Pacific Northwest in 1980 provided physical geographers and other scientists with an opportunity to witness volcanic processes and their consequences. Vesuvius, the great volcano that looms over the Italian city of Naples, is the most legendary of all such mountains. The processes and landforms of volcanism are investigated in Unit 32. In the following unit, we continue our survey of the two remaining major rock types.

Key Terms

basalt *page 351*

batholith *page 351*

carbonate mineral *page 350*

crust *page 347*

crystalline *page 347*

dike *page 351*

exfoliation *page 351*

extrusive igneous rock *page 351*

feldspar *page 349*

felsic *page 351*

granite *page 351*

igneous rock *page 350*

intrusive igneous rock *page 351*

jointing *page 351*

laccolith *page 351*

lava *page 350*

mafic *page 351*

magma *page 350*

metamorphic rock *page 350*

mineral *page 347*

quartz *page 349*

rock *page 348*

sedimentary rock *page 350*

silicate mineral *page 349*

sill *page 351*

stock *page 351*

Scan Here for a quick vocabulary review

ENHANCE

Review Questions

1. Explain why crystalline sugar and obsidian are not minerals.

2. Why is it not surprising that quartz (SiO_2) is such a common mineral in the continental crust?

3. If we identify a specimen as *mafic extrusive igneous rock*, what does this name tell us about the rock's composition and how it formed? What about a *felsic intrusive igneous rock*?

4. If you were looking at an igneous intrusion that was uncovered by erosion, how would you distinguish a *laccolith* from a *sill* or a *dike*?

5. Where and why does *exfoliation* occur?

Scan Here for a quick concept review

ENHANCE

www.oup.com/us/mason

UNIT
28
Sedimentary and Metamorphic Rocks

Dinosaurs walked the slopes of sand dunes during the Jurassic in what is now southern Utah. Today, the cross-bedded dune sand has become sedimentary rock, sculpted by weathering and erosion into a landscape of remarkable beauty.

OBJECTIVES

- Describe the major types of sedimentary rocks, how they form, and what they tell us about past environments
- Explain the formation of metamorphic rocks
- Summarize the concept of the rock cycle

Igneous rocks, which we described in detail in Unit 27, have been called Earth's primary rocks—the first solidified material derived from the molten mass that once was the primeval crust. The other two great classes of rocks could therefore be called secondary, because they are derived from preexisting rocks. These are the sedimentary and metamorphic rocks.

Sedimentary Rocks

Sedimentary rocks result from the deposition, compaction, and cementation of rock fragments, mineral grains, and dissolved material derived from other rocks. These materials are broken away or weathered from existing rocks by water, wind, and ice, through processes we explore in Parts Six and Seven. Ancient Roman scholars understood what they saw: *sedimentum* is the Latin word for settling. Many sedimentary rocks begin their existence as loose deposits of sand or gravel that come to rest at the bottom of a sea or lake, on a beach, or in a desert (Fig. 28.1).

Eventually, the loose sediment is transformed into rock, a process called **lithification**. Lithification has two distinctly different components. First, as successive layers of sediment accumulate, the weight of the sediment expels

FROM THE FIELDNOTES

FIGURE 28.1 "In western Inner Mongolia we followed the new highway connecting the isolated salt-producing towns of Jartai and Yabirai, stopping where poorly sorted debris—a mixture of sizes from boulders to fine sand—covers the surface on both sides of the road. Flash floods carried this mixture of boulders, gravel, and sand down the dry canyons of the mountain range to the northwest and deposited it on the alluvial fan where we stopped, perhaps as recently as the last few decades. As we watched, the wind winnowed sand from the surface sediments, leaving some of it in mounds around shrubs but carrying most of it toward sand dunes a few kilometers away. All of this sediment could eventually be carried away by streams and the wind, leaving no trace behind, but this is a subsiding basin, sinking along faults as the mountains rise on either side. Therefore, the bouldery debris and dune sand might also be trapped here, buried, and cemented to form sedimentary rock, preserving a record of this desert environment."

most of the water between the grains. Pressure caused by the weight of the overlying materials compacts and consolidates lower layers. The rock fragments and grains are squeezed tightly together. This is the process of **compaction** (Fig. 28.2A). Compaction rarely takes place alone, however. Most sedimentary material has some water in the pore spaces between the grains, and this fluid contains dissolved minerals. This mineral matter is deposited in thin films on the grain surfaces and has the effect of gluing them together, a process called **cementation** (Fig. 28.2B). The most common cements in sedimentary rocks are calcite and silica (in this case, the silica is material that has the composition of quartz but a poorly developed crystalline structure). Together, compaction and cementation can transform a bed of loose sand into a layer of cohesive sedimentary rock.

CLASTIC AND NONCLASTIC SEDIMENTARY ROCKS

The range of agents and materials that combine to produce sedimentary rocks is wide, and as a result the structure and texture of these rocks also vary greatly. Even the finest windblown dust can become lithified. The same is true for a mixture of boulders, cobbles, pebbles, and sand swept down by a river and subsequently compacted and cemented. Sedimentary rocks made from particles of other rocks are referred to as **clastic**, from the ancient Greek *klastos*, meaning broken. The majority of sedimentary rocks are clastic. **Nonclastic** sedimentary rocks form from chemical solution or from organic deposition.

FIGURE 28.3 A conglomerate in the Rocky Mountains. Before lithification, this was loose gravel deposited by a stream, but it is now cemented and can stand in a vertical cliff.

Clastic sediments are most conveniently classified according to the size of their fragments, which range from boulders to fine clay particles. The coarsest-grained sedimentary rock is **conglomerate**, a composite rock made of gravels, pebbles, and sometimes even boulders (Fig. 28.3). An important property of conglomerates is that the pebbles or boulders tend to be quite well rounded. This characteristic is evidence that sharp edges were worn off as the pebbles were transported by water for some distance, perhaps rolled down a river valley or washed back and forth across a beach. A large pebble may reveal the source area from where it was removed, possibly giving us information about ancient drainage courses. Sometimes pebbles are elliptical in shape, and in the conglomerate, a significant number of them are cemented with their long axes in the same direction. Such information helps reveal the direction of stream flow or the orientation of the coastline where the sediment accumulated.

When pebble-sized fragments in a conglomerate are not rounded but angular and jagged, we call it a *breccia*. The rough shape of the pebbles indicates that they were not carried far by water. When compaction and cementation occur after a landslide, for instance, the result is a breccia. Again, the properties of the fragments constitute a key to the past.

Another common and important sedimentary rock is **sandstone**. In a sandstone, the grains, as the name implies, are sand-sized, and they often are quartz grains. Some sandstones are quite hard and resist erosion even in humid climates, because their cementing material is silica. But other sandstones are less compacted and are cemented by calcite or dolomite. Such sandstones are softer and more susceptible to weathering and erosion. Sandstones can also serve as a key to the past. They may have rounded or angular grains depending on the distance they have traveled.

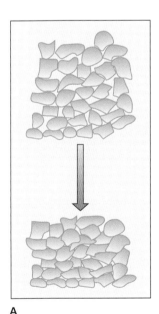
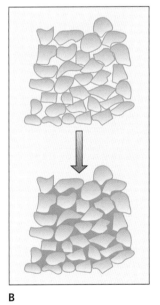

A　　　　　　　**B**

FIGURE 28.2 Compaction and cementation in sedimentary rocks. (A) In *compaction*, the grains are packed tightly together by weight from above. (B) In *cementation*, the spaces between the grains are filled through the deposition of a cement, such as silica or calcium carbonate.

Oilfield Formation

Petroleum, one of the world's leading fuels, became a full-fledged energy resource in the second half of the nineteenth century when oil-drilling technology was pioneered. Drilling began with a few barrels a day drawn from a single well in northwestern Pennsylvania in 1859. From that humble beginning, global production rose to more than *77 million* barrels a day at the start of the twenty-first century—a staggering output of approximately 28 billion barrels per year. Because oil-producing countries can count on a substantial income, the search for additional petroleum deposits continues all over the world.

Petroleum occurs in sedimentary rocks. Its formation started with large, shallow bodies of water where, scientists believe, microscopic algae produced the organic matter that eventually became petroleum. At their death, these tiny organisms became part of the sediments accumulating on the seabed.

Over millions of years, thick sediments containing a large quantity of organic matter accumulated and were transformed into sedimentary rock layers. Furthermore, deep within those rocks, the original organic material slowly changed under the pressure of overlying rocks and high temperatures, some of it becoming petroleum or natural gas. Both are organic compounds, like the algae they originated from, but with considerably different composition and structure.

Gas and oil can migrate through rocks, following the pores (open spaces). As they migrate, they may arrive at a location where the rock layers were compressed and bent into arching structures called *folds* (see Unit 34) and there become trapped. Where trapped oil accumulates in an upfold in the rock layers or a dome capped by a bed of impermeable rock, an oil reservoir forms (Fig. 28.4). Natural gas often forms above such an oil pool—the two energy

resources frequently occur together—and oil floats above any groundwater that may lie below the upfolded rock layers. The petroleum deposit remains under pressure until its existence is discovered by exploration. Then a well can be drilled to pump the black liquid to the surface.

Until recently, almost all oil and gas was pumped from traps like this, especially in rocks such as sandstone, where the relatively large pores allowed fossil fuels to flow fairly quickly toward the well. Now, the rapidly expanding use of hydraulic fracturing ("fracking") allows the pumping of fossil fuels—especially gas—from much less porous rocks such as shale. Hydraulic fracturing involves injecting fluids into the rock to create fractures—large new pores—through which gas or oil can flow. This allows natural gas production where it was never possible before but also stirs up intense controversy over its environmental impacts.

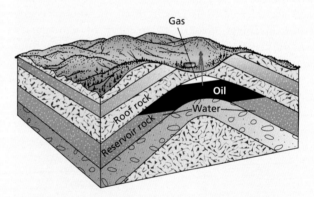

FIGURE 28.4 An oil pool (a body of rock in which oil occupies all the pore spaces) trapped in an upward-arching layer of reservoir rock. These curving rock structures are known as folds; they constitute the most important of all oil traps.

The bedding, or *sedimentary structures*, in sandstones often indicate to the trained eye whether the sand was deposited in a stream, on a shallow ocean floor, or by wind in a dune field. Sandstones that formed in ancient dune fields are an especially interesting example (see the photo at the beginning of this unit). In these rocks, **cross-bedding** (a sedimentary structure with beds inclined at an angle to the overall layering of the rock) formed on steep dune faces, where sand avalanched down after being blown over the

crest of the dune. Sometimes, the tracks of animals that walked over the dunes, including dinosaurs, are still preserved.

As with conglomerates, the size and shape of sandstone grains may vary, providing information on their history. Even-sized sands indicate effective sorting of particle sizes by wind or water. Rounded grains have probably been carried long distances or have gone through several cycles of erosion and deposition. Sandstones also have economic

FIGURE 28.5 Outcropping of shale sandwiched between two sandstone layers in Zion National Park, Utah. This soft, thin-layered sedimentary rock is easily weathered and eroded.

importance: because they are porous, they can contain substantial amounts of water or oil. Under certain structural circumstances, such water or oil can form a reservoir suitable for exploitation (see Perspectives on the Human Environment: Oilfield Formation).

Another common sedimentary rock is **shale**, the finest-grained clastic sedimentary rock type. Shale is compacted mud. Whereas sandstone contains quartz grains that are often visible to the naked eye, shale is made primarily from tiny clay mineral particles, which can only be seen under the most powerful microscopes. Shale also has a tendency to split into thin layers, making this already-soft rock even more susceptible to weathering and erosion (Fig. 28.5). In many places (e.g., the Appalachian Mountains in the eastern United States), the low valleys are often underlain by soft shale and the higher ridges by other rocks, including hard sandstone. A broader term, *mudstone*, refers to all clastic sedimentary rocks dominated by clay or silt particles, rather than sand.

One of the most interesting sedimentary rocks, because of both the way it forms and its response to weathering and erosion, is **limestone**. Limestone can form from the accumulation of marine shell fragments on a beach or on the ocean floor, which qualifies it as a special kind of clastic sedimentary rock. Most limestone, however, results from the respiration and photosynthesis of marine organisms in which calcium carbonate is crystalized from seawater. This calcium carbonate ($CaCO_3$) then settles on the ocean floor, and accumulations may reach hundreds of meters in thickness. Limestone varies in composition and texture, but much of it is finely textured and, when exposed on the continental landscape, hard and resistant to erosion. Limestone, however, is susceptible to solution (dissolving in water), and under certain environmental conditions, it creates a unique landscape both above and below the ground (see Unit 42).

Evaporites form from the deposits left behind as water evaporates. Such evaporites as halite (salt), gypsum, and anhydrite have some economic importance, but their areal extent is small. Biological sediments include the carbonate rocks formed by coral reefs, cherts formed from silica skeletons of diatoms and radiolarians (marine microorganisms), and the various forms of coal.

SEDIMENTARY ROCKS AS RECORDS OF PAST ENVIRONMENTS

A sequence of sedimentary rocks in the landscape is unmistakable because it displays variations in texture, color, and thickness of the various layers (Fig. 28.6). When they originally form, most sedimentary rocks are layered more or less horizontally. Cross-bedding, mentioned earlier, is a special case in which beds of rock originally form at a fairly steep angle, although those sloping beds are contained within thick horizontal layers. Depending on the size and shape of the cross-beds, they may provide important indications about where and how the sediment was deposited. The large cross-beds produced by deposition on dunes, mentioned earlier, provide a good example.

Layered sedimentary rocks tend to have planes of weakness between individual beds, which are often attacked by weathering and erosion. These rocks usually also have joints at angles to the bedding, which are produced by a variety of processes ranging from desiccation (drying) in

FIGURE 28.6 The Colorado River has cut a canyon in the Colorado Plateau that exposes hundreds of millions of years of rock accumulation. Here, at the Marble Canyon segment of the Grand Canyon, sedimentary strata display diverse colors reflecting the environmental circumstances of their deposition. Some are weakly cemented and erode quickly, forming relatively gentle inclines on the canyon wall. Others, like the uppermost layers, are more resistant and form scarps that retain their vertical configuration for a very long time. The exposure seen here is about 900 m (3,000 ft) high, the last phase of a geologic sequence that began here shortly after the planet's formation and continued intermittently for about 4 billion years.

FIGURE 28.7 An unconformity in sedimentary rocks in Utah. The lower, reddish-brown beds were originally horizontal but were folded after deposition. They were then eroded to a nearly flat surface before deposition of the overlying light-colored beds.

sedimentary rocks to tectonic stresses (related to the tectonic plate movement discussed in Unit 29). Furthermore, over time, sedimentary rocks may be folded, faulted, and otherwise deformed (Unit 34).

The layering, or **stratification**, of sedimentary rocks can provide important clues to the conditions that changed as a succession of rock beds, referred to as **strata,** was being deposited. Often, distinct surfaces are evident, separating strata of different grain size or with different types of cross-bedding or other sedimentary structures. Sometimes, it is apparent that the sequence was interrupted and that a period of deposition was broken by a period of tectonic tilting or folding and erosion before the deposition resumed. Where such an interruption is evident in the **stratigraphy** of sedimentary rocks—the order and arrangement of strata—the contact between the eroded strata and the strata of resumed deposition is called an **unconformity** (Fig. 28.7).

The texture and color of sedimentary layers allow us to deduce the kinds of environments under which they were deposited. Sedimentary rocks therefore are crucial in the reconstruction of past environments. Even more important, sedimentary rocks contain fossils (Fig. 28.8). Much of what is known about Earth's history is based on the fossil record (see Focus on the Science: Development of the Geologic Timescale). Interpretations from the fossil record, as well as conclusions drawn from sedimentary rock sequences far removed from one another, make possible correlations, which provide further evidence for reconstructions of the past.

FIGURE 28.8 Fossilized fish in sedimentary rocks provide valuable clues to the geologic past. This easily recognized school of fish, found in the Green River Formation in Wyoming, has been preserved for about 40 million years.

Sedimentary rocks can be observed as they accumulate today, which provides further insight into similar conditions in the distant geologic past. You may have seen ripples on a riverbed or a desert sand dune. These ripples can be created by shallow water flowing in a stream or by the persistent blowing of wind. Often they are erased, only to reform later. But sometimes they are buried, cemented, and preserved in lithifying rock. Millions of years ago, ripples were formed that have become exposed by erosion today, providing evidence of wind or stream flow in the distant past.

Metamorphic Rocks

Metamorphic rocks have been altered by varying degrees of heat and pressure, but without melting into magma. The term *metamorphic* comes from an ancient Greek word meaning change, but the complex processes involved in rock metamorphism have only begun to be understood in modern times. Metamorphism can include both recrystallization of existing minerals and formation of new minerals through rearrangement of atoms, both of which occur while the rock remains a solid. Rocks can also be compressed or stretched as metamorphosis takes place, often aligning mineral grains into a pattern of parallel bands called **foliation**. All rock types may be subject to metamorphism.

Metamorphism usually results from tectonic movement of the crust (see Unit 30) or, less often, from volcanic activity (see Unit 32). Where the crust is compressed and thickened by tectonic processes, some of the rocks within the crust are forced to greater depths, where temperature and pressure cause metamorphism. In particular, widespread regional metamorphism occurs deep in the crustal "roots" of great mountain ranges. Where metamorphic rocks occur at the ground surface, such as in the Canadian Shield or the Piedmont region of the southeastern United States, they often represent the exposed roots of ancient mountain ranges now utterly destroyed by erosion. When magma rises and intrudes into the crust, rocks in the zone near batholiths or dikes are affected (Fig. 28.9); this process is known as **contact metamorphism**. When a sheet of lava flows out over the surface, its heat also changes the rocks it covers.

METAMORPHIC ROCK TYPES

Some metamorphic rocks have quite familiar names. Sandstone, made of quartz grains and a silica cement,

FIGURE 28.10   A sample of schist, a metamorphic rock, from the Superstition Mountains of southern Arizona. The shiny appearance results from the alignment of small, elongated flakes of mica oriented in the same direction through the foliation process.

becomes **quartzite**, a hard rock that resists weathering. Limestone is converted into much denser and harder **marble**, used by sculptors for statues that can sometimes withstand exposure to the elements for thousands of years. Shale may be metamorphosed into **slate**, a popular building material; slate retains shale's quality of breaking along parallel planes.

Sometimes, metamorphism alters the preexisting rocks so totally that it is not possible to determine what the previous form of the rock may have been. A common metamorphic rock is **schist**. This rock is fine grained, and it breaks along roughly parallel planes (but unevenly, unlike slate). If schist was, at least in part, shale in its premetamorphic form, there is little resemblance left (see Fig. 28.5). When schist is seen outcropping on the landscape, it displays wavy bands, such as those shown in Figure 28.10. This is the foliation described earlier, a good visual clue in identifying common metamorphic rocks. Another frequently seen example of foliated metamorphic rocks is **gneiss**, the metamorphic rock derived from granite (Fig. 28.12).

METAMORPHIC ROCKS IN THE LANDSCAPE

Because metamorphic rocks have been subjected to heat and pressure, we might conclude that they would be the most resistant of all rocks to weathering and erosion. But even metamorphic rocks have their weak points and planes (see Fig. 28.12). Slate, for example, is weak at the surfaces along which it breaks, because water can penetrate along these planes and loosen the rock slabs. Schist often occurs

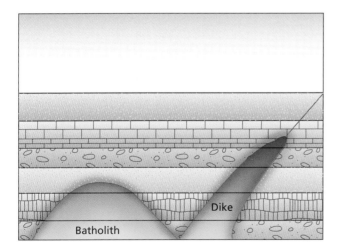

FIGURE 28.9   An example of contact metamorphism, caused by intrusion of magma into sedimentary rock strata. The "halo" of reddish color around the intruded magma marks the zone where sedimentary rock will be metamorphosed by the heat of the magma.

Development of the Geologic Timescale

Charles Lyell (1797–1875), the great British geologist, first visited North America in 1841. He and his wife, Mary, spent a great deal of their visit studying rocks in the field, often clambering over rugged hillsides and trudging along muddy shorelines. Lyell was especially eager to see the thick rocks of Silurian and Devonian age that stretched across the state of New York and the Eocene sediments of the coastal plain that extends from New Jersey to Alabama. *Silurian*, *Devonian*, and *Eocene* are all terms geologists still use to refer to the age of rocks today. You can find them and the times they represent on the *geologic timescale* in Figure 28.11. Each falls within a much longer unit called an era (e.g., Silurian falls in the Paleozoic era). The eras are grouped into even longer intervals called *eons*, the most recent of which is the Phanerozoic.

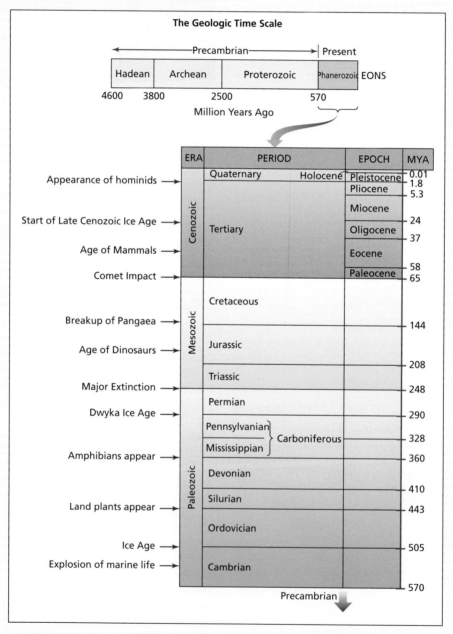

FIGURE 28.11 The geologic timescale. In the right column, showing ages, MYA stands for millions of years ago.

But how could Lyell judge which rocks were Silurian or Devonian or that the Devonian rocks were younger? Methods for radiometric dating of rocks—determining their age in years using radioactive decay—did not emerge until the twentieth century. However, by 1841 geologists had painstakingly placed many of the sedimentary rocks of Europe and North America in order by relative age, from oldest to youngest. Fossils were often the key; if a certain type of fossil occurred only in certain rocks in England, then rocks containing that fossil in Russia were probably about the same age. Rocks of similar age, with diagnostic groups of fossils, were assigned to geologic periods, often with names related to localities where they were initially identified. For example, the Silurian draws its name from the Silures, an ancient Celtic group that lived in Wales, where Silurian rocks were first distinguished. Fossils of fish with bones and the earliest land plants appear in Silurian rocks but not in the underlying older rocks. The younger Devonian rocks are named for Devon, a county of England, and contain fossils showing that both fish and land plants had become much more diverse and abundant, including the first fossils of seed-producing plants.

Fossils differentiate the eras quite clearly because of the dramatic changes in predominant forms of life over geologic time. The Silurian and Devonian both fall within the Paleozoic era, which began with an explosion of diversity in marine life and was also when plants first colonized the continents. Next came the Mesozoic era, sometimes called the Age of Reptiles, when dinosaurs flourished. Only long after Lyell's death was it discovered that a comet struck Earth at the end of the Mesozoic (the *K-T boundary impact*), contributing to the wave of mass extinctions that occurred at that time.

Lyell's own specialty was the Cenozoic era, when mammals came to dominate the fauna of the continents. In the late Cenozoic, climates seesawed between cold and warm, with glaciers and ice sheets expanding over large areas during the cold periods. Meanwhile, adaptable hominids had made their appearance, and more than 100,000 years ago, the human species, *Homo sapiens*, rose to dominate the planet. We have not said much about the three earliest eons, the Hadean, Archaean, and Proterozoic (often grouped together as the Precambrian). Though they span much of the history of Earth, much less is known about these eons than about the Phanerozoic.

To better understand the magnitudes of the time periods involved in the geologic timescale, let us relate Earth's life span to your age. If you are 20 years old and we take as the age of the planet the time of formation of the oldest rocks we can find (more than 4 billion years), consider the following:

1. One *year* of your life equals 230 million years of Earth's. That puts you in the early Mesozoic just one year ago.
2. One *month* of your life equals just over 19 million years of Earth's. The Rocky Mountains formed just three and a half months ago.
3. One *week* of your life equals nearly 5 million years of Earth's. The Pleistocene glaciations began three days ago.
4. One *day* of your life equals about 630,000 years of Earth's. Human evolution was still in its early stages just yesterday at this time.
5. One *hour* of your life equals over 26,000 years of Earth's. In that single hour, the human population grew from less than 100,000 to more than 7 billion, and the major civilizations developed.

Where will we be one hour from now?

---

in huge masses and therefore seems to resist weathering and erosion quite effectively. But schist is weak along its foliation bands and breaks down quite rapidly. Even gneiss is weakest along its foliation planes, especially where dark minerals such as micas have collected. In some areas where gneiss is extensive, the dark minerals have been weathered so effectively that vegetation has taken hold in these bands. From the air, one can follow the foliation bands by noting the vegetation growing in the weakest ones.

The Rock Cycle

Earth's first rocks were igneous rocks—rocks solidified from the still-molten outer sphere more than 4 billion years ago. Ever since, existing rocks have been modified and remodified, and there are few remains of these original ancient-shield rocks. What we see in the landscape today are only the most recent forms in which rocks have been cast by the processes acting upon them.

Over time, vast quantities of rock are broken down by weathering and carried away by erosion. The resulting sediments accumulate and become rocks, which are eventually pushed up into new mountains, or alternatively pushed downward to depths where they are metamorphosed or even melted. Magma rises from the depths where melting occurs, cooling to form igneous rocks; uplift pushes all types of rocks upward toward the surface; erosion uncovers them and wears them down. This cycle of transformation, which affects all rocks, is conceptualized as the **rock cycle** (Fig. 28.13).

FIGURE 28.12 "We were looking for a place to cross the rain-swollen Tana River in eastern Kenya. The normally placid stream was a raging torrent in places, and this site, usually a series of rapids, showed no promise. But the physical geography here was notable because of the rocks that cause the rapids. Foliation in metamorphic rocks lines up the minerals in parallel bands so that the recrystallized rocks appear to be streaked with alternating light- and dark-colored stripes. Metamorphic rocks tend to be quite resistant to erosion, as this gneiss outcrop, under constant attack by the river, illustrates."

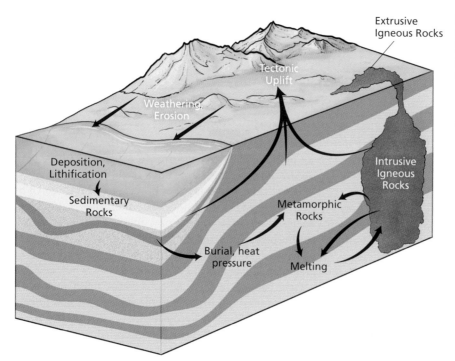

FIGURE 28.13 The rock cycle. The flow of materials within, above, and below Earth's crust continually forms and destroys igneous, sedimentary, and metamorphic rocks.

ENHANCE

The rock cycle has neither beginning nor end, so you can start following it anywhere on the diagram. It is also not a simple circle, but has multiple pathways. For example, igneous rocks may be uplifted and exposed to weathering and erosion, eventually contributing sand or silt grains to new sedimentary rocks. Alternatively, igneous rocks can be melted again or altered by pressure and temperature into metamorphic rocks. It is important to remember that the movement of matter through the rock cycle occurs over almost unimaginably long periods of geologic time. In our lifetimes, we are witnesses to just a brief instant in a cycle that affects the entire planet in space and its whole history in time.

Now that we have become familiar with the characteristics and cycling of Earth materials, we are ready to consider the forces of the restless crust. Our examination of these processes, which contribute importantly to the shaping of surface landscapes, begins in Unit 29 with an overview of Earth's internal structure.

Key Terms

cementation *page 357*

clastic sedimentary rocks *page 357*

compaction *page 357*

conglomerate *page 357*

contact metamorphism *page 361*

cross-bedding *page 358*

foliation *page 361*

gneiss *page 361*

limestone *page 359*

lithification *page 356*

marble *page 361*

nonclastic sedimentary rocks *page 357*

quartzite *page 361*

rock cycle *page 363*

sandstone *page 357*

schist *page 361*

shale *page 359*

slate *page 361*

strata *page 360*

stratification *page 360*

stratigraphy *page 360*

unconformity *page 360*

Scan Here for a quick vocabulary review

ENHANCE

Review Questions

1. Under what conditions can loose sediment be lithified?

2. If you were studying a sedimentary rock in the field, what properties would you look at to classify this rock and determine the environment in which it originated?

3. How do clastic and nonclastic sedimentary rocks differ?

4. How does metamorphism differ from melting and igneous rock formation?

5. What are the metamorphic equivalents of sandstone, limestone, and shale?

6. Starting with the formation of an igneous rock from magma, describe the possible pathways material from that rock can follow within the overall rock cycle. Then do the same, starting with the release of sand grains from a metamorphic rock and moving through weathering and erosion.

Scan Here for a quick concept review

ENHANCE

www.oup.com/us/mason

UNIT 29

Planet Earth in Profile: The Layered Interior

Lava flows into the sea along the coast of Hawai'i. The surface of the flow is already cooling to form basalt, a dense, dark-colored igneous rock. Similar rocks make up the bedrock floor of the deep ocean basins.

OBJECTIVES

- Describe the evidence that helps explain Earth's internal structure

- Outline Earth's internal layers and discuss some of the evidence leading to their discovery

- State the salient properties of Earth's outer layers, its crust, and the underlying mantle

- Describe the processes at work that modify Earth's surface and create physical landscapes of great diversity

In places like the Grand Canyon or the slopes of the Rocky Mountains, we can study vast exposures of the sedimentary, metamorphic, and igneous rocks discussed in Units 27 and 28. We can hold a piece of granite and interpret its large crystals of quartz and feldspar as the result of magma with high silica content cooling slowly deep beneath the surface. We can identify fossils in limestone that record its origin on a shallow ocean floor. Such direct observations are not possible for rocks much farther below Earth's surface. More than a third of a century after people first set foot on the Moon, the deepest boreholes on Earth have penetrated barely 12 km (7.5 mi) below the surface. Since the radius of the planet is 6,370 km (3,959 mi), we have penetrated less than 1/500th of the distance to the center of Earth.

Nonetheless, scientists have established the fact that the interior of the planet is layered like the atmosphere, and they have deduced the chemical composition and physical properties of the chief layers below the crust. This research is of importance in physical geography because the crust is affected by processes that take place in the layers below it. Consequently it is important to understand what is known about Earth's internal structure and how this information has been acquired.

Evidence of Earth's Internal Structure

Evidence that supports the concept of an internally layered Earth comes from several sources. By studying the wavelengths of light emanating from the Sun, it is possible to determine the chemical elements the Sun contains and their proportions, because certain specific wavelengths correspond to particular elements. Geophysicists have concluded that these proportions should be the same, on average, for Earth as a whole as they are on the Sun. However, compared to the average composition deduced from solar radiation, such abundant elements as iron, nickel, and magnesium are relatively depleted in the Earth's crust. This suggests that those heavy elements must be concentrated in deeper layers of the planet, to compensate for their lower abundance in the crust.

Rocks that formed deep below the surface, even below the crust, occasionally have been uplifted to levels where they can be sampled from drill cores or outcrops. These samples of rock from great depths do contain higher concentrations of iron and magnesium than "average" crustal rocks, as predicted. In addition, when the mass and size of the planet are used to measure its density, the resulting figure is 5.5 g/cm³. This is about double the density of rocks found in the continental crust. Again, the heavier, denser part of Earth must be in the deep interior.

EARTHQUAKES AND SEISMIC WAVES

Further evidence for the internal structure of Earth comes from the planet's magnetic field and from the high temperatures and pressures known to prevail at deeper levels. But the most convincing body of evidence is derived from the analysis of **earthquakes**—the shaking and trembling of Earth's surface caused by sudden releases of stress within the crust.

Earthquakes occur in many areas of the crust, and we discuss their causes in Unit 33. Earthquakes generate pulses of energy called **seismic waves** that can propagate, or spread out from their source, through the entire Earth. A strong earthquake in the Northern Hemisphere will be recorded by *seismographs* in the Southern Hemisphere. Today, thousands of seismographs continuously record the shocks and tremors in the crust (Fig. 29.1), and computers help interpret these earthquake data. This source has given us a picture of the interior structure of our planet.

Seismic waves take time to travel through the planet. The speed of an earthquake wave increases as the rock it travels through becomes more rigid. With greater depth below Earth's surface, the density of rock increases, sometimes abruptly. Most of these jumps in density coincide with abrupt increases in rigidity and seismic wave speed. When seismic waves reach such an abrupt change in density and rigidity, they may be bounced back (Fig. 29.2A); this is known as *seismic reflection*. Alternatively, waves may for the most part be bent rather than reflected, especially if the contrast in rock properties is less severe. In this case, *seismic refraction* changes the course of the seismic wave (Fig. 29.2B).

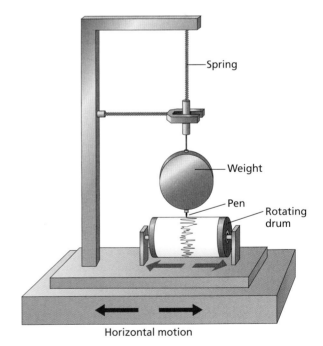

FIGURE 29.1 Diagram of a simple seismograph. Seismic waves cause the base of the seismograph to shake along with the ground it rests on, but the weight limits movement of the pen. Therefore, even very slight shaking is recorded as a wiggly line on the paper wrapped around the rotating drum. The line plotted on the paper is called a seismogram.

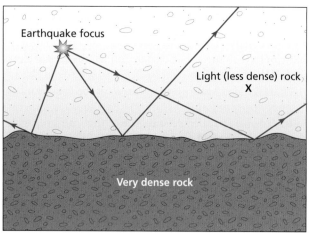

A

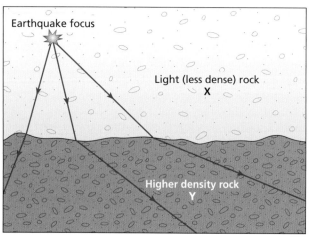

B

FIGURE 29.2 When seismic waves travel through the interior of Earth, several things happen. When they reach a plane where the rock material becomes much denser and more rigid, they may be *reflected* back (A). If the contrast in rock properties is less, they may be *refracted* (B). Their velocities are also affected. Speeds are less in the layers marked **X** and greater in the layer marked **Y**.

Seismic waves behave differently as they propagate through Earth. One type of wave travels along the surface of the crust and is termed a *surface wave*. Two other types of waves travel through the interior of Earth and are referred to as *body waves*. The body waves are of two kinds, known as **P** (*primary* or *pressure*) waves and **S** (*secondary* or *shear*) waves. The **P** waves are compressional waves. As they propagate, the material they are passing through vibrates back and forth parallel to the direction of **P**-wave movement. They even travel through liquids, although their velocity is then reduced. As **S** waves propagate, vibration moves material at right angles to the direction the waves are traveling. Secondary waves do not propagate through liquids. This is of great importance because, if **S** waves fail to reach a seismograph in an opposite hemisphere, it may be concluded that liquid material inside Earth halted their progress.

When an earthquake occurs, seismographs nearest its point of origin begin to record the passing of a sequence of waves. The seismogram (see Fig. 29.1) reveals the passage of the **P**, **S**, and surface waves with little separation in time. At seismograph stations farther away, waves traveling at various speeds arrive at different times and are more easily distinguished. However, some stations will not register any **P** or **S** waves, and others will record only **P** waves. From these initially puzzling observations, significant conclusions about the interior of Earth can be drawn.

USING SEISMIC WAVES TO DETERMINE EARTH'S INTERNAL STRUCTURE

The paths of seismic waves reveal the existence of a layer deep within Earth that **S** waves (shown by white arrows) cannot travel through. This process is illustrated in Figure 29.3. If an earthquake occurs at 0 degrees, **P** as well as **S** waves are recorded by seismographs everywhere from the epicenter to 103 degrees from the earthquake's source (a distance of 11,270 km [7,000 mi]). From 103 to 142 degrees, the next 4,150 km (2,600 mi), neither **P** nor **S** waves are recorded (except for **P** waves propagated along the crust). However, from 142 to 180 degrees (15,420 to 19,470 km [9,600 to 12,140 mi] distant from the quake), **P** waves—always shown by black arrows—reappear. From this evidence, it is concluded that Earth possesses a liquid layer that begins about 2,900 km (1,800 mi) below the surface. At the contact between this liquid layer and the layer above it, **S** waves cease to be propagated, so they do not appear in a large *shadow zone* opposite the point where the earthquake occurred. The **P** waves waves are refracted, or bent, where they encounter the liquid layer, directing them away from smaller shadow zones (see Fig. 29.3).

But some **P** waves that arrive on the far side of the planet, between 142 and 180 degrees, are refracted not just once or twice, but four times! Moreover, their speed has increased. This means that the **P** waves that reach the seismographs located at the point on Earth exactly opposite to the earthquake source must have traveled through an extremely dense, rigid mass inside the liquid layer. Confirmation of the existence of such a dense mass at the core of Earth comes from the fact that many **P** waves are reflected back at its outer edge (Fig. 29.4). From the travel times of these reflected **P** waves and the seismograms inside 142 degrees (see Fig. 29.3), scientists have concluded that Earth has a solid inner core—a ball of heavy, dense material. This is where the iron and nickel, depleted from the upper layers, are concentrated.

Earth's Internal Layers

On the basis of the seismic-wave travel we just discussed and other evidence, Earth is believed to have four major layers: a solid inner core, a liquid outer core, a mantle, and the thin crust (Fig. 29.5). The mantle can be subdivided

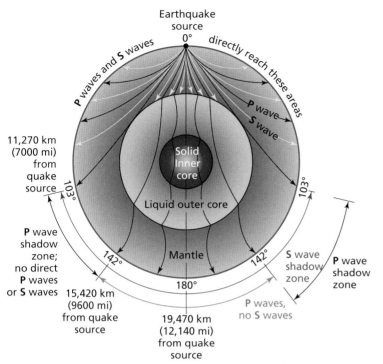

FIGURE 29.3 Imagine that a strong earthquake occurs at the North Pole (0 degrees on the drawing). This diagram shows the paths of the radiating **P** and **S** waves as they travel through the planet (waves that travel along the surface are not shown). **P** waves are black and **S** waves are white in the figure. Note that refraction of **P** waves by the liquid outer core diverts these waves away from a large *shadow zone* in the Southern Hemisphere between 103 and approximately 142 degrees from the quake's source at 0 degrees. This shadow zone and the larger one for **S** waves, which do not travel through liquids, confirm the presence of the liquid outer core and allow us to calculate its depth below Earth's surface.

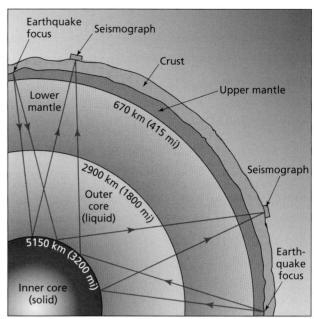

FIGURE 29.4 Certain **P** waves are reflected back toward the crust when they reach the outer edge of the solid inner core. From their travel times, the position of the contact between solid inner core and liquid outer core can be deduced. Note that the refraction of these waves as they traverse the interior of Earth is not shown.

into two parts: the much thicker lower mantle and the upper mantle, which can be subdivided even further, as we discuss in the following material.

INNER AND OUTER CORE

The solid **inner core** has a radius of only 1,220 km (760 mi). Its surface lies 5,150 km (3,200 mi) below sea level. Iron and nickel exist here in a solid state, scientists believe, because pressures are enormous. Such high pressures raise the melting-point temperature for these metals, so they do not melt despite the intense heat prevailing in the inner core.

The liquid **outer core** forms a layer 2,250 km (1,400 mi) thick. Its outer surface is around 2,900 km (1,800 mi) below sea level, just slightly less than halfway to the center of the planet. The liquid outer core may consist of essentially the same materials as the solid inner core, but because pressures

in the outer core are lower, the melting-point temperature is lower, and a molten state prevails. The density of the inner and outer cores combined has been calculated as 12.5 g/cm³, which compensates for the lightness of the crust (2.8 g/cm³) and accounts for the density of the planet as a whole (5.5 g/cm³).

MANTLE

Above the liquid outer core lies the solid **lower mantle** (see Fig. 29.5). The lower mantle has a thickness of about 2,230 km (1,385 mi). Seismic data show that the lower mantle is in a solid state. Geologists believe that this layer is composed of iron, magnesium, and silicon compounds.

The **upper mantle**, which extends from the base of the crust to a depth of just 670 km (415 mi), is of great interest to geologists as well as to physical geographers because it interacts with the overlying crust in many ways. The upper mantle is less dense overall than the lower mantle. We know this because **P** and **S** waves travel more slowly in the upper mantle than in the lower mantle. The upper mantle itself has two distinctly different layers. The lower of these is a zone that is mostly solid but is quite *plastic*; that is, it can readily change shape and flow under stress. Some commonplace materials display this behavior. For example, a steel bar that is heated can be easily bent or reshaped by hammering, even if it does not actually melt. This plastic part of the upper mantle probably contains a small percentage of molten rock, which reduces the velocity of **P** and

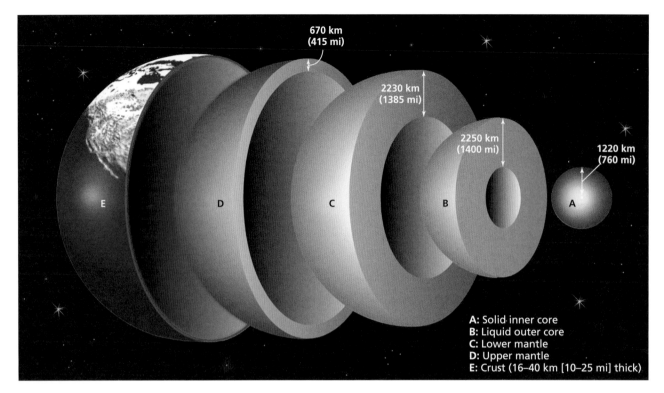

670 km
(415 mi)

2230 km
(1385 mi)

2250 km
(1400 mi)

1220 km
(760 mi)

A: Solid inner core
B: Liquid outer core
C: Lower mantle
D: Upper mantle
E: Crust (16–40 km [10–25 mi] thick)

FIGURE 29.5 Principal layers of the inner Earth.

S waves. The uppermost part of the mantle just beneath the crust is quite different and is strong enough to remain rigid under stress. This zone near the top of the upper mantle is part of the **lithosphere**, the relatively strong and brittle outer sphere of Earth that also includes the crust.

Earth's Outer Layers

Now that you are familiar with Earth's interior, let's turn to the rocks closer to Earth's surface. Those rocks can actually be subdivided in two ways, depending on the properties we are most interested in. First, we can classify them as forming either part of the upper mantle or part of the crust (see Fig. 29.5). After discussing the nature of the crust and how it is distinguished from the mantle through the use of seismic waves, we will turn to the second classification of Earth's outer layers, based on their behavior under stress.

The uppermost parts of the crust are the only portions of the solid Earth about which scientists have direct first-hand knowledge. The rocks that make up the outer shell of our planet have been analyzed from the surface, from mine shafts, and from boreholes. Even the deepest boreholes, however, only begin to shed light on what lies below.

STRUCTURAL PROPERTIES OF THE CRUST

One of the most significant discoveries relating to Earth's crust occurred in 1909. In that year, Croatian scientist Andrija Mohorovičić (1857–1936) concluded from his study of earthquake waves that the density of Earth materials changes markedly at the contact between crust and mantle. This

contact plane has been named the **Mohorovičić discontinuity**, or **Moho**, an abbreviation of his name. Despite more than a century of far more sophisticated analyses and interpretations, Mohorovičić's conclusion has proven correct: a density discontinuity does indeed mark the base of Earth's crust.

Mohorovičić's discovery made it possible to calculate the thickness of the crust. Earthquake waves speed up at the Moho discontinuity, clearly marking the boundary between the crust and mantle. In some places, this happens a mere 5 km (3 mi) down from the surface; elsewhere, the change in earthquake-wave velocity does not come until a depth of 40 km (25 mi) or even deeper. This proves that the crust is not of even thickness. It also shows that the crust is thinner than the shell of an egg relative to the planet's diameter.

When the Moho was mapped, it was found to lie much closer to the surface under the ocean floors than under the continental landmasses (Fig. 29.6). This discovery confirmed a conclusion also drawn from gravity measurements: the continents have crustal "roots" that create, in a rough way, a reverse image of the topography at the surface (a matter explored in Unit 32). Under the oceans, the crust averages only 8 km (5 mi) in thickness; under the exposed continental surfaces, the average depth is about 40 km (25 mi).

CONTINENTAL AND OCEANIC CRUST

For many years, it was not realized that a fundamental difference between continental and oceanic crust might account for these differences in thickness. In part, this was because rocks under the shallow ocean floor near continents are often quite similar to those found on land. But as more

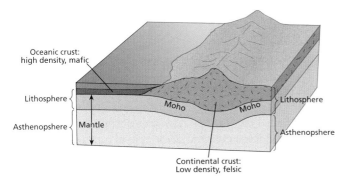

Oceanic crust:
high density, mafic

Lithosphere

Asthenopshere

Moho   Moho

Mantle

Lithosphere

Asthenopshere

Continental crust:
Low density, felsic

FIGURE 29.6 The crust is separated from the underlying mantle by the Mohorovičić discontinuity (Moho); below the continents the crust is much thicker and less dense than the crust underlying the ocean basins. The relatively rigid, brittle lithosphere includes both the crust and the uppermost part of the mantle, while the underlying asthenosphere is made up of more plastic rock deeper within the mantle.

information was collected on the rocks of the deep ocean floor, it became clear that they were quite different than the rocks making up most of the continental landmasses.

In fact, fundamental differences exist between continental crust and oceanic crust. The rocks that make up most of the continental landmasses have the lowest density of all, so the continents are sometimes described as "rafts" that float on denser material below. These low-density rocks are *felsic*; as discussed in Unit 27, this means that they are distinguished by being enriched with silica and typically have light colors (Fig. 29.7A). Granite is the most common

felsic igneous rock, and its density is about the same as the density of the continental landmasses as a whole (2.8 g/cm³). As a result, you will also see felsic continental crust referred to as "granitic" or "granitoid" crust, although the landmasses are made up of many other rocks as well.

The crust beneath the deep oceans, on the other hand, consists mainly of higher-density *mafic* igneous rocks (also defined in Unit 27). Mafic rocks have a relatively low content of silicon and other light elements, and are both dense and generally dark colored (Fig. 29.7B). The most common mafic rock in the crust is heavy, dark-colored basalt. Oceanic crust therefore is often referred to as basaltic crust, although many other rocks also form part of it. In combination, rocks of the oceanic crust have a density of about 3.0 g/cm³. Deep beneath the surface in the upper mantle, *ultramafic* rocks with even lower silica content predominate.

Although the continental crust is dominated by granitic rocks, especially below the surface, it also contains occasional masses of mafic rocks like basalt. These record locations where magma with a mafic composition emerged at the Earth surface because of local geologic conditions. Some especially large areas of basalt on the continents have been explained as the products of *mantle plumes*, which we will discuss in Unit 32, but this interpretation is still debated.

THE LITHOSPHERE

The crust terminates at the Moho, but the uppermost segment of the mantle is also fairly strong and brittle, like the crust just above it. As noted earlier, we can group the

FROM THE FIELDNOTES

A

B

FIGURE 29.7 "Baking in the African sun is some of the Earth's oldest rock: ancient granite, felsic, lightweight rock that has been part of the African shield for 3 billion years (A). Whenever you travel across the African landscape, you note how light colored these crystallines are, colored white, beige, or pink according to their composition (dominant quartz creates the lightest color; feldspars form light shades of red). But when new rock emerges from the Earth's interior, as on Pacific islands or along mid-ocean ridges, it is dominated by basalt, dark colored and heavier (B). This mafic rock forms much of the crust beneath the oceans."

crust and upper mantle together as the lithosphere. This is the second approach to categorizing layers of rocks near Earth's surface, and it is based on their response to stress. The lithosphere tends to remain rigid and break under stress rather than flowing as a plastic material. The part of the upper mantle below the lithosphere, where the hot rock is plastic and flows readily under stress, is often called the **asthenosphere**. Both the lithosphere and the asthenosphere are shown in Figure 29.6, along with the crust and the mantle. Like the Moho, the boundary between the lithosphere and the asthenosphere is at a much deeper level below the continental landmasses than under the ocean floor. Beneath the landmasses, the boundary averages 80 km (50 mi) below the surface. Beneath the oceans, it lies only about 40 km (25 mi) below the seafloor. Unlike the Moho, the contact zone between the strong lithosphere and the soft asthenosphere is gradual rather than abrupt.

We now know that both the asthenosphere and the lithosphere are mostly solid, but bodies of partly or completely molten rock can occur in both layers. This rock generally melts below the crust but can rise through it and even flow out onto the surface as lava if it has positive buoyancy—that is, if the molten rock is less dense than the surrounding solid rock. As magma rises through the crust, some of the felsic rock it is moving through may also melt, changing the composition of the magma. A body of magma can also separate into a less dense and more felsic upper portion and a denser and more mafic lower portion. As more has become known about the lithosphere and the asthenosphere, it has become easier to explain exactly where the conditions of temperature and pressure have allowed rock to melt, and whether that molten rock could rise to the surface.

The discovery of the asthenosphere was important to our understanding of what happens at much shallower depths in the crust, and even at Earth's surface. Because the asthenosphere is in a hot plastic state, the lithosphere can move over it. This movement of the lithosphere is related to the formation of mountains and even the movement of whole continents.

LITHOSPHERIC PLATES

The crust varies in thickness and is also a discontinuous layer. To us, living on the landmasses, the idea that the crust is not a continuous, unbroken shell can be difficult to grasp. In fact, the crust and the rest of the lithosphere are fragmented into a number of segments called **lithospheric plates** (or sometimes *tectonic plates*). These plates move in response to the plastic flow in the hot asthenosphere. Many of Earth's mountain ranges, including the mightiest of them all, the Himalayas, are zones where the moving plates have come together in gigantic collisions. This aspect of the lithosphere is so important to our study of landscapes and landforms that it is treated in a pair of units (30 and 31).

The Crustal Surface

Earth's crust is subject to tectonic forces from below. The rocks that form the crust are pushed together, stretched, fractured, and bent by the movement of the lithospheric plates. These forces tend to create great contrasts at the crustal surface—jagged peaks and sharp crests, steep slopes and escarpments, huge domes and vast depressions. Before beginning an in-depth examination of those forces in the remaining units of Part Five, it is useful to take another look at the surface of the continents and their varied relief.

TOPOGRAPHIC RELIEF

The term **relief** refers to the vertical difference between the highest and lowest elevations in a given area. A range of tall mountains and deep valleys, such as the Rocky Mountains, is an area of *high relief* (Fig. 29.8). Flat plains are areas of *low relief* (Fig. 29.9). An area of low relief can lie at a high elevation: a nearly flat plateau with an elevation of 3,000 m (10,000 ft) has lower relief than a mountainous area with peaks no higher than 2,000 m (6,600 ft) and valleys at 500 m (1,650 ft).

When we view the continental landmasses even at a small scale (see Fig. 2.5), it is evident that they have areas of high relief and other areas of low relief. North America, for example, has large areas of low relief, especially in central and eastern Canada, the interior United States, and the coastal plain bordering the Atlantic Ocean and the Gulf of Mexico. High relief prevails in the western third of the continent from the Rocky Mountains westward. In the east lies an area of moderate relief in the Appalachian Mountains.

The two types of relief just identified represent two kinds of continental geology. Earth's landmasses consist of two basic geologic components: *continental shields* and *orogenic belts*. In North America, the region centered on Hudson Bay is a continental shield, expressed topographically as a plain of low relief. The Rocky Mountains represent the topographic results of a period of mountain building and constitute an orogenic belt.

CONTINENTAL SHIELDS

All the continental landmasses contain shields as well as orogenic belts. The **continental shields** are large, stable, relatively flat expanses of extremely old metamorphic and igneous rocks. They have formed the nuclei of larger continental landmasses since the Precambrian, more than 540 million years ago. Another term, *craton*, refers to a continental shield plus the areas around it where younger sedimentary rocks cover the same stable base of ancient crystalline rock.

FIGURE 29.8 "Not one square foot of flat land in this dissected, high-relief area on the North Island of New Zealand. High relief prevails throughout most of New Zealand, which is positioned in the boundary zone between the Australian and Pacific Plates and is subject to volcanism and earthquakes. Glaciation, as well as stream erosion, further modify its topography, so that this is one of the most scenic locales on the planet. But this vista is not the result of nature's work alone. Before human settlers arrived less than 1,000 years ago, dense forests covered this area (like much of the rest of the islands). The Maori burned significant portions of it, but it was the Europeans and their livestock who had the greater impact. They converted forest into pasture for millions of sheep, in some areas sparing not a single tree for as far as the eye can see. In this area, the pastures are seeded annually from airplanes, producing a verdant countryside—but one that is a cultural, not a natural, landscape."

FIGURE 29.9 "Few places on Earth seem flatter than the High Plains, which stretch from Nebraska south to the Texas Panhandle. Yet we are standing more than 1,100 m (3,600 ft) above sea level, higher than many low mountains of the eastern U.S. More importantly at the moment, not even the slightest rise blocks our view of a menacing storm rolling in just as we finish fieldwork. Two tornadoes touched down nearby yesterday, so it's best to find shelter."

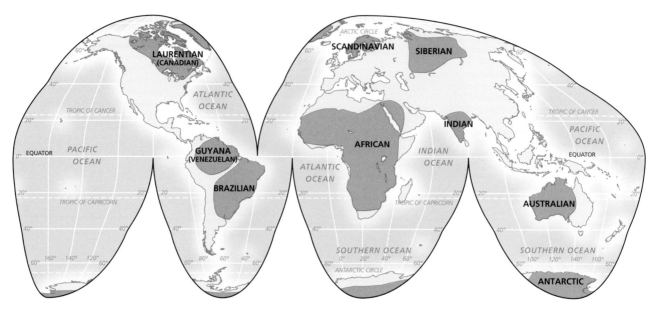

FIGURE 29.10 Continental shields of the world, representing ancient igneous and metamorphic rocks that have been exposed to weathering and erosion for hundreds of millions of years.

The shield in northern North America is called the *Canadian* (*Laurentian*) *Shield* (Fig. 29.10). In South America, there are two major shield zones: the *Guyana* (*Venezuelan*) *Shield* and the *Brazilian Shield*. These shield areas, unlike the Canadian Shield, are uplands today and present the aspect of low-relief plateaus rather than plains.

Eurasia has three major shields: the *Scandinavian Shield* in the northwest, the *Siberian Shield* in the north, and the *Indian Shield* in the south. The world's largest presently exposed shield is the *African Shield*, a vast region of ancient rocks that extends into the Arabian Peninsula at its northeastern extremity. Some of the oldest known rocks have been found in the *Australian Shield*, which occupies the western two-thirds of that continent. And under the ice in eastern Antarctica lies the *Antarctic Shield*. Wherever these shield zones form the exposed landscape, they exhibit expanses of low relief (Fig. 29.11).

FIGURE 29.12 Inca-built structures of Machu Picchu, high in the Andes of Peru, part of the orogenic belt that follows the west coast of South America. This is high relief in the extreme, but the Inca managed to construct large stone buildings and terrace even very steep slopes there. The purposes for which Machu Picchu was built are still uncertain. It may have served as a fortress and/or as a ceremonial center.

FIGURE 29.11 A stream flowing slowly across the vast tundra east of Yellowknife in Canada's Northwest Territories. This is the Canadian Shield, its crystalline rocks scoured by ice but now ice-free, its depressions and valleys filled with lakes and rivers.

OROGENIC BELTS

In contrast to shields, the **orogenic belts**—series of linear mountain chains—are zones of high relief. The term *orogenic* derives from the ancient Greek word *oros*, meaning mountain. As we discuss later, episodes of mountain building, called *orogenies*, have occurred in one region or another throughout Earth's 4.6-billion-year lifetime. The outcomes of these episodes are recorded on the topographic map by linear mountain chains, such as the old Appalachians and the much younger Sierra Nevada in North America. The Andes Mountains in South America, the Alps and Himalayas in Eurasia, and the Great Dividing Range in Australia all represent orogenic activity, when rocks were uplifted and often crushed into folds like a giant accordion (Fig. 29.12).

If Earth had no atmosphere and no moisture, these orogenic belts would stand unchallenged, destroyed only by new tectonic forces. But Earth does have an atmosphere, and as a result, the geologic buildup is attacked by a set of processes that work to wear it down. These are the various processes of *weathering*—the chemical, physical, and biological breakdown of rock—and erosion that will be the focus of many of the units in Parts Six and Seven. As a consequence of those processes, mountain belts are ultimately worn down and destroyed. In fact, the low continental shields contain rocks that once formed the roots of high mountain ranges. Uplift and downwearing of orogenic belts are essential drivers of the rock cycle discussed in Unit 28, as well as the tectonic and volcanic processes we will discuss in the next few units.

Key Terms

asthenosphere *page 372*
continental shield *page 372*
earthquake *page 367*
inner core *page 369*
lithosphere *page 370*

lithospheric plate *page 372*
lower mantle *page 369*
Mohorovičić discontinuity
 (Moho) *page 370*
orogenic belt *page 375*

outer core *page 369*
relief *page 372*
seismic wave *page 367*
upper mantle *page 369*

Scan Here for a quick vocabulary review

Review Questions

1. What are seismic waves, and how do they help us understand the layering of Earth's interior?
2. In addition to the behavior of seismic waves, what other evidence do we have that deep layers such as the outer core are much denser than rocks in Earth's crust?
3. What is the Mohorovičić discontinuity, and what is its significance?
4. What are the differences between oceanic and continental crust?
5. How does the lithosphere differ from the asthenosphere, and how does this difference relate to the concept of lithospheric plates?

Scan Here for a quick concept review

www.oup.com/us/mason

Plates of the Lithosphere

Where plates diverge. New land arises from submarine eruptions related to the Mid-Atlantic Ridge south of Iceland.

OBJECTIVES

- Describe Alfred Wegener's hypothesis of continental drift
- Explain how studies of the ocean floor led to the modern theory of plate tectonics, and how the major lithospheric plates and their motion have been mapped
- Describe the motion of lithospheric plates, the types of plate boundaries that result, and the processes that occur at each kind of boundary

When Christopher Columbus reached America in 1492, his discovery was recorded in his ship's log—and on the first map to be based on the Atlantic Ocean's western shores. When Columbus returned on his next three voyages, and as others followed him, the Atlantic coastline of the Americas became more familiar to Europeans. During the sixteenth century, Portuguese navigators and cartographers mapped Africa's Atlantic coasts all the way to the Cape of Good Hope at the continent's southern tip.

As a result, by the early 1600s, the general configuration of the Atlantic Ocean was fairly well known. In 1619, the great English naturalist Francis Bacon (1561–1626), looking at the evolving map of the Atlantic Ocean, made an observation containing the kernel of a momentous concept. He said that the opposite coasts of South America and Africa seemed to fit so well that it looked as though the two continents might at one time have been joined. It was a notion soon forgotten and not revived until nearly three centuries later. And even then, the idea that whole continents could move relative to each other was greeted with skepticism and, in some circles, derision.

Wegener's Hypothesis

In 1915, German Earth scientist Alfred Wegener (1880–1930) published a book that contained a bold new hypothesis. Not just Africa and South America, Wegener suggested, but all the landmasses on Earth once were united in a giant supercontinent during the Paleozoic and Mesozoic eras of geologic time (see Fig. 28.11). This primeval landmass, which he named **Pangaea** (meaning All-Earth), ultimately broke apart, forming the continents and oceans as we know them today. Wegener theorized that Pangaea consisted of

two major parts: Laurasia in the north and Gondwana in the south. Today, Eurasia and North America are the remnants of Laurasia; South America, Africa, India, Australia, and Antarctica form the principal fragments of Gondwana (Fig. 30.1).

Wegener's book, *The Origin of Continents and Oceans* (1915), was not translated into English until the end of the 1920s. By then, Wegener's notion of **continental drift**—the fragmentation of Pangaea and the slow movement of the continents away from this supercontinent—was already a topic of debate among geologists and physical geographers in many parts of the world. American geologist F. B. Taylor (1860–1938) had written a long article in support of continental drift. A South African geologist, Alexander du Toit (1878–1948), also supported Wegener's hypothesis and busied himself with gathering evidence from opposite sides of the South Atlantic Ocean. But most geologists, especially in Britain and the United States, could not conceive of the possibility that whole continents might be mobile, behaving like giant rafts.

Wegener's hypothesis was based on much more than the match between continental outlines noticed by Francis Bacon; in fact, Wegener marshaled a good deal of circumstantial evidence. Closely related fossil plants and animals that seemed likely to have evolved in the same geographic region were scattered across South Africa, South America, India, Australia, and even Antarctica. Paleozoic and Mesozoic rock formations in which those fossils were found also seemed closely related, though they were separated by the Atlantic and Indian Oceans. Plausible as continental drift was to those who believed this evidence, there was one major problem: a process that could move continents like rafts over the solid Earth was unknown. Wegener suggested

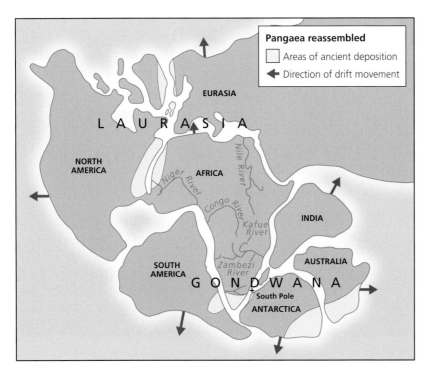

FIGURE 30.1 The breakup of the supercontinent Pangaea began more than 100 million years ago. Note the rotation of its remnants as they move away from Africa and how areas of ancient deposition help us understand where today's landmasses were once joined together.

that Earth's gravitational force, which is slightly lower at the equator, was over time strong enough to pull the continents apart. Taylor proposed that the Moon was torn from Earth in what is today the Pacific Basin and that the continents have been steadily moving into the gap thereby created. None of these ideas won much support.

Some scientists, however, kept working on the problem. It was fully accepted by this time that the crust can move vertically, producing mountain ranges and deep ocean trenches, but explaining horizontal motion over long distances was the problem. British geologist Arthur Holmes (1890–1965), one of the first to use radioactive decay to determine the age of rocks, was also interested in continental drift. He proposed as early as 1939 that heat-sustained convection cells might be in the interior of Earth and that these gigantic cells could be responsible for dragging the landmasses along (see Unit 31).

Plate Tectonics

Since Wegener proposed that the Atlantic Ocean opened up as continents drifted apart, the logical place to find more convincing evidence for his hypothesis was on the deep ocean floor, but this was the part of Earth's surface least known to geologists and geographers. The existence of a Mid-Atlantic Ridge—a mostly submarine mountain range extending from Iceland south to the Antarctic latitudes approximately in the center of the Atlantic Ocean—had been known for many years. For decades, it was believed that this feature was unique to the Atlantic Ocean. But during the 1950s and 1960s, evidence from deep-sea observations made by a growing number of research ships carrying new sonar equipment began to reveal the global map of the ocean floors in unprecedented detail. The emerging map clearly revealed that midoceanic ridges are present in all the ocean basins (Fig. 30.2).

Another important discovery involved the ages of rocks brought up from the ocean floor. These ages can be determined partly by using the decay of radioisotopes but also by relying on past changes in Earth's magnetic field recorded by mineral grains in ocean-floor basalt. Surprisingly, the basaltic rocks beneath the deep ocean floors were found to be no older than the Jurassic period, much younger than many continental rocks. Rocks older than the Jurassic are found only on small areas of the seafloor near continents. Moreover, the deep ocean-floor lithosphere is youngest near the midoceanic ridges and becomes older with increasing distance from the ridges in a fairly orderly pattern (Fig. 30.3).

These were momentous developments, and as often happens in science, a rush of exciting new information and ideas followed the initial breakthrough. Within a few years, a new theory of continental drift had taken hold, which finally resolved many of the issues that had prevented Earth scientists

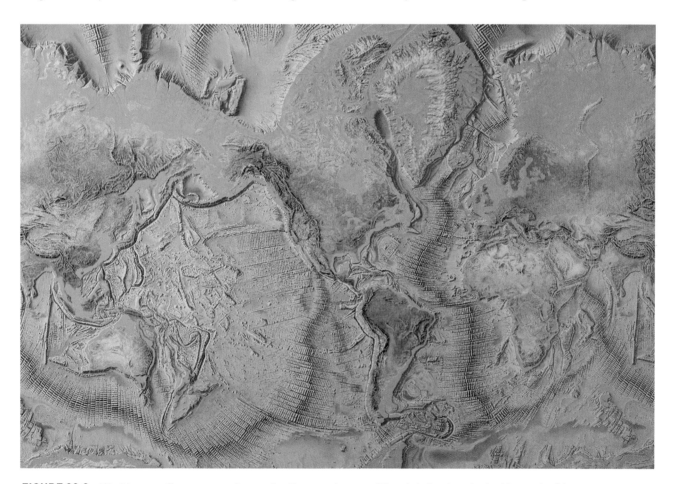

FIGURE 30.2 World ocean-floor map, underscoring the prominence of the global network of midoceanic ridges.

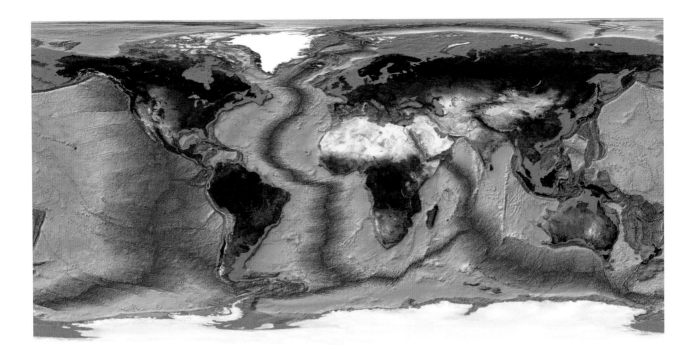

0 20 40 60 80 100 120 140 160 180 200 220 240 260 280
Age of oceanic lithosphere (millions of years)

FIGURE 30.3 Age of the lithosphere beneath the oceans. The youngest lithosphere is near the midoceanic ridges and becomes older with increasing distance from those ridges.

from accepting Wegener's ideas. It was recognized that the midoceanic ridges are not just submarine mountain ranges like those on the landmasses. They are also locations where hot rock continually rises from deep in the mantle, melting and then solidifying. Through this process, new igneous rock forms and is quickly pushed away horizontally by still-newer rock forcing its way up from below all along the midoceanic ridge. This process came to be called **seafloor spreading**; it involved the creation of new lithospheric material and its continuous movement away from its source. The new lithospheric material is relatively thin and dense, and has a mafic (low-silica) composition, as is typical of lithosphere beneath all of the major ocean basins (see Unit 29). As shown in Figure 30.4, seafloor spreading originates as a **rift** develops in a continental landmass. The rift widens to become a narrow sea and, eventually, an open ocean—exactly the sequence of events suggested by Wegener's reconstruction of the breakup of Pangaea. Seafloor spreading clearly explains why the age of the ocean-floor rocks increases away from the midoceanic ridges in such an orderly fashion.

If the midoceanic ridges are zones where new lithosphere forms and diverges, it is clear that Earth's lithosphere is divided into segments—large fragments separated along the ridges. These segments are called **lithospheric plates** (also known as *tectonic plates*). That concept led fairly quickly to the full theory of **plate tectonics**, which explains continental drift—along with earthquakes, volcanism, and mountain building—as the result of the formation, movement, and destruction of lithospheric plates.

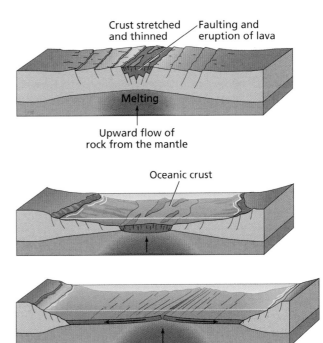

FIGURE 30.4 Concept of seafloor spreading, in which upwelling of rock from the mantle leads to the formation of new lithosphere, which then moves away to either side of the ridge. The process starts with stretching and faulting of the crust on a continent (top), followed by rift development and development of a narrow sea (middle), and finally seafloor spreading from a ridge below an open ocean (bottom).

Where ocean floors appear to spread apart, we are actually seeing two plates moving away from each other. The continents move along with the plates they are parts of (Fig. 30.4). Moreover, because the plates beneath the ocean floor spread outward from the midoceanic ridges, the lithosphere must be pushed together elsewhere. Where plates are pushed together, one plate plunges downward to be destroyed at the same time as the lithosphere is being formed at the midoceanic ridges. In this way, the lithosphere is formed in one zone and destroyed in another. In other cases, the converging plates become highly compressed, an important process of mountain building

DISTRIBUTION OF PLATES

When seafloor spreading was first recognized and the midoceanic ridges were mapped, it appeared that Earth's lithosphere was divided into seven major plates, all but one carrying a major landmass. Today, one of the original seven is generally thought to be split into two plates (Indian and Australian), and many researchers believe the African Plate is actually two plates (Nubian and Somali) separated by the East African rift. Therefore, either eight or nine large plates exist (Fig. 30.5):

1. The *Pacific Plate* extends over most of the Pacific Ocean floor from south of Alaska to the Antarctic Plate.
2. The *North American Plate* meets the Pacific Plate along California's San Andreas Fault and related structures. It carries the North American landmass as well as a

portion of easternmost Eurasia and the northern segment of Japan.

3. The *Eurasian Plate* forms the boundary with the North American Plate at the Mid-Atlantic Ridge north of 35 degrees N. It carries much of the Eurasian landmass north of the Himalayas, except for eastern Siberia, which is part of the North American Plate.
4. The *African Plate* on most maps extends eastward from the Mid-Atlantic Ridge between 35 degrees N and 55 degrees S. It carries Africa and the island of Madagascar, and it meets the Antarctic Plate under the Southern Ocean and the Australian and Indian Plates beneath the Indian Ocean. Many researchers now split this plate into the Nubian and Somali Plates, separated by the East African rift.
5. The *South American Plate* extends westward from the Mid-Atlantic Ridge south of 15 degrees N. It carries the South American landmass.
6. The *Australian Plate* carries Australia and meets the Pacific Plate in New Zealand.
7. The *Indian Plate* carries the Indian subcontinent and meets the Eurasian Plate at the Himalayas.
8. The *Antarctic Plate* encircles the Antarctic landmass (which it carries) under the Southern Ocean.

In addition, numerous smaller plates exist, most of them recognized by all researchers, but some still a matter of dispute. Several good examples of smaller plates lie just west of the South American and North American

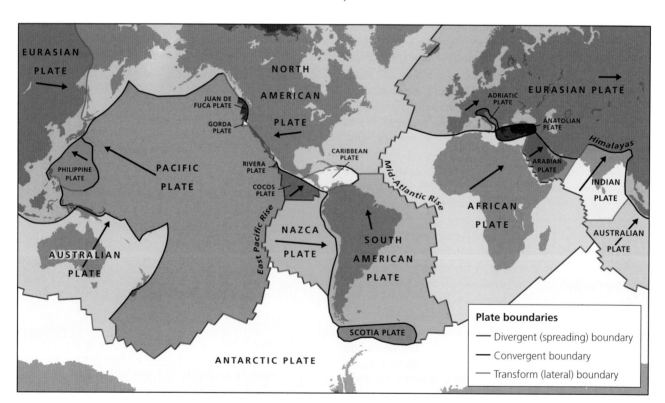

FIGURE 30.5 Lithospheric plates of Earth. The *relative* movement of the two plates at each boundary falls into one of three categories: divergent (spreading), convergent, or transform (lateral) motion. The arrows on the plates show motion relative to a single fixed reference point (maps of plate motion differ greatly depending on the reference point that is chosen).

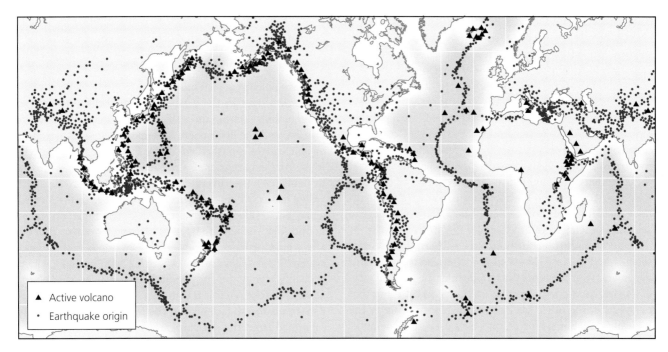

FIGURE 30.6 Global distribution of recent earthquakes and active volcanoes.

Plates: from north to south, they include the Juan de Fuca, Rivera, Cocos, and Nazca Plates (see Fig. 30.5). Because plate tectonics is still an area of active research, it should not be surprising that various maps of the lithospheric plates differ somewhat, with some maps showing small plates that are not recognized by other maps. It is also important to understand that although we often assume that the plates are perfectly solid and rigid, this is not completely true. That is, one part of a plate can actually move relative to another, and the boundaries between the plates are sometimes fuzzier than they appear on maps.

LOCATION OF PLATE BOUNDARIES

You may now wonder how geologists originally identified the boundaries between lithospheric plates once they were recognized in the 1960s. Some evidence collected before that time was quite useful, such as maps showing the global distribution of earthquakes (Fig. 30.6). Note that the map in Figure 30.6 shows that earthquakes often originate in the midoceanic ridges, in island arcs, and in mountain belts such as South America's Andes and South Asia's Himalayas. A few well-known faults, including the San Andreas Fault in California, are not associated with high mountains or midoceanic ridges but are still marked by bands of frequent earthquakes. It was concluded that all of these linear earthquake zones represented plate boundaries, where the crust was stressed as plates converged, pulled apart, or moved side by side, causing frequent earthquakes.

The discovery of seafloor spreading, discussed earlier, led quickly to the recognition that the midoceanic ridges mark the boundaries between tectonic plates. A careful look at the midoceanic ridges in Figure 30.2 and the plate boundaries that follow those ridges in Figure 30.5 reveals

that the ridges are not smooth and unbroken. Instead, there are abrupt steps where right-angled fractures in the crust (*faults*) offset segments of the ridges. Without these offsetting faults, the two large rigid plates could not move steadily apart from each other along an irregular boundary.

The world distribution of active volcanoes was also used to define plate boundaries (see Fig. 30.6). Of course, continental-surface patterns of volcanic activity were much better known than submarine volcanism. Volcanic activity is so common in western South and North America, in Asia's offshore *archipelagoes* (island chains), and in New Zealand that this circum-Pacific belt had long been known as the **Pacific Ring of Fire**. It is now recognized that this ring marks the edges of the Pacific Plate and several smaller plates. The high Andes Mountains follow one segment of the Ring of Fire. When Charles Darwin visited South America on the voyage of the *Beagle*, he witnessed a strong earthquake that caused uplift of 1 m (3 ft) or more along the coast of Chile. From this observation, Darwin concluded that the Andes are actively rising through crustal movement during earthquakes. The same is true of other high mountain ranges, including the Alps and the Himalayas, although they do not contain active volcanoes. The new theory of plate tectonics finally offered a simple explanation for the uplift of these mountains: they were located at the boundary between two converging plates.

Movement of Plates

The discovery of seafloor spreading also made it possible to determine the direction and even the speed of motion for many of the larger plates. Recall that the rocks forming the crust on either side of a midoceanic ridge get older with distance away from the ridge (see Fig. 30.4). If we know the age of the crust at some distance from the midoceanic ridge, then we can calculate how fast it has moved away from the ridge where it formed. Today, the direction and speed of plate motion can be much more directly measured using GPS measurements repeated at exactly the same point over a period of years. Although GPS measurements are highly accurate in recording movements of the crust, they can be affected by local movements that are not related to overall plate motion. Still, these measurements directly demonstrate that the continents are in motion, as proposed many years ago by Wegener. According to both long-term estimates from seafloor spreading and short-term GPS data, the plates move at velocities ranging from 1 to 12 cm (0.4 to 4.7 in) per year.

Understanding plate motion is trickier than it may seem at first because the speed and direction of movement depend on your reference point. Consider the point of view of a driver passing another car on the highway. From that perspective, the other car looks as if it is traveling in the opposite direction as the driver's own car. From the point of view of a police officer measuring speeds at the side of the road, the two cars are going in the same direction but at different speeds.

The same concept applies to plate motion. We can describe the motion of one plate relative to another or the motion of both plates in relation to the planet beneath them. This can be illustrated using the map of the distribution of plates shown in Figure 30.5. The arrow symbols on this map portray both the direction and the speed of plate motion in relation to the solid Earth below the mobile plates (longer arrows mean faster motion). Using that reference, the African Plate is moving northeastward, while the South American Plate moves northward. These paths of motion will gradually move the two plates farther apart. Therefore, if we could stand on the African Plate and somehow observe the movement of the South American Plate, it would appear to be moving directly away from us. We would not notice the northward component of the South American Plate's motion because we would be moving in somewhat the same direction as well. Keep this problem in mind whenever you study maps of tectonic plates and their movements, because it is not always clear what reference point was used in constructing those maps.

It is the relative motion of two plates (i.e., the motion of one plate as viewed from the other) that determines what happens at the boundary between them. Because the South American and African Plates are moving apart from each other, in a relative sense, new lithosphere for both these plates must be created along the Mid-Atlantic Ridge that forms the boundary between them. This type of relative motion at a boundary is called *divergence*, or spreading. *Convergence* describes the situation in which two plates are moving toward each other, and *transform* motion occurs when the plates are moving side by side. The movement of lithospheric plates is directly responsible for many of Earth's major landscapes and landforms. All our studies of geomorphic processes and features must take into account the effect of crustal mobility. For this purpose, understanding the details of what happens at each kind of plate boundary is crucial.

DIVERGENT PLATE BOUNDARIES

We have noted that plates *diverge*, or separate, along the midoceanic ridges in the process called seafloor spreading. Magma wells up from the asthenosphere, new lithosphere is created, and the lithosphere on opposing sides of the midoceanic ridges is pushed apart (see Fig. 30.4). Here, the tectonic forces are tensional, and the crust is thin and broken by faults. At the **divergent plate boundary** itself, a **rift valley** forms, where the crust drops between faults located along each side of the valley. Although most divergent boundaries are beneath oceans, early stages of this process affect continental crust. A notable example is the major system of rift valleys in eastern Africa (see Fig. 34.7), which probably signal the formation of a new plate boundary along this zone of crustal thinning. The Red Sea represents a more advanced stage of this process: the Arabian Plate has separated from the African Plate, and between them now lies a basalt-floored sea. As time goes on, the Red Sea is likely to widen and become a new ocean (forming in stages similar to those shown in Fig. 30.4). The Atlantic Ocean formed through a similar process as part of the breakup of Pangaea. Ancient lava flows along the rocky shores of Lake Superior record a much older rift that began to tear apart the North American Plate more than 1 billion years ago but, for reasons yet to be determined, stopped developing.

CONVERGENT PLATE BOUNDARIES

Because plates form and spread outward in certain areas of the crust, they must *converge* in other zones. The processes that occur at **convergent plate boundaries** depend on the type of lithosphere involved on each side. Recall from Unit 29 that continental crust has a relatively lower density, and in these terms it is "light" compared to oceanic crust, which is denser, heavier, and more prone to sink or be forced downward where plates converge. When an oceanic plate meets a plate carrying a continental landmass at its leading edge, the less dense continental plate overrides the denser oceanic plate, which sinks downward. This process is **subduction**, and the area where it occurs is called a *subduction zone* (Fig. 30.7).

A subduction zone is a place of intense tectonic and volcanic activity. The movement of the plates is comparatively slow, averaging 2 to 3 cm (about 1 in) per year. But this motion is enough to generate enormous energy, which can be released in the greatest earthquakes experienced anywhere on Earth. At the leading edge of the continental plate, a wedge-shaped mass of rock accumulates from material scraped off the descending oceanic plate as well as from sediments eroded from the continent. This *accretionary wedge* (see Fig. 30.7) is intensely faulted and folded by compression between the converging plates, and older parts of it rise to form low mountains at the continent's edge.

Even more important processes of the subduction zone originate deep below the continental plate. The rocks of the

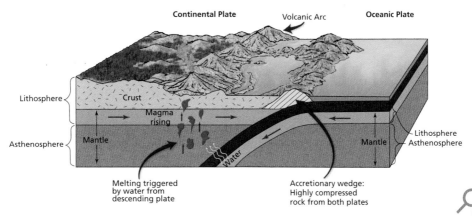

Continental Plate Volcanic Arc Oceanic Plate

Lithosphere — Crust

Magma rising

Asthenosphere — Mantle

Lithosphere
Asthenosphere

Mantle

Water

Melting triggered
by water from
descending plate

Accretionary wedge:
Highly compressed
rock from both plates

FIGURE 30.7 When a continental plate converges with an oceanic plate, the process of subduction carries the denser, less buoyant oceanic plate downward beneath the thicker but less dense continental plate. An accretionary wedge of material from both plates accumulates at the leading edge of the continent, and a volcanic arc (a chain of volcanoes) forms farther inland. The volcanoes arise because water from the sinking ocean plate triggers melting in the mantle above it, and the resulting magma rises through the continental crust.

ENHANCE

descending oceanic plate contain water. As that plate sinks into the mantle, the increasing heat eventually drives that water into solid mantle rock above the descending plate. The addition of water lowers the melting point of the rock in the mantle, so blobs of molten rock form and rise into the crust of the continent. Some of the magma reaches the surface and produces eruptions that form a chain of volcanoes, referred to as a **volcanic arc**. Many of the best-known volcanoes on Earth occur in arcs above subduction zones.

Three types of convergent plate boundaries exist, but only two result in subduction. We describe all three types in the material that follows.

Oceanic–Continental Plate Convergence One of the best examples of the kind of oceanic–continental subduction zone just described (see Fig. 30.7) lies along the northwestern coast of the United States, in northernmost California, Oregon, and Washington. The small oceanic Juan de Fuca Plate is being subducted below the continental North American Plate. Older parts of the accretionary wedge have risen to form the Olympic and Coast Range Mountains. Farther east, a volcanic arc demonstrates that melting is being triggered at depth by water carried downward with the Juan de Fuca Plate (we will discuss these volcanoes in more detail in Unit 32). South America's western margin, where the oceanic Nazca Plate is being subducted beneath the continental South American Plate, is another good example of oceanic–continental plate convergence.

Oceanic–Oceanic Plate Convergence When a convergent plate boundary involves two oceanic plates with thin, dense lithosphere, neither plate is very buoyant or resistant to subduction. However, one of them is a little less buoyant and ends up plunging below the other. A deep oceanic trench and a volcanic **island arc** mark this boundary (Fig. 30.8). The Marianas and other island chains around the Pacific Ring of Fire are the best examples of this type of subduction zone.

Continental–Continental Plate Convergence Where convergence involves two continental plates, both plates have relatively low density, and neither has any tendency to sink below the other. The best example is the convergence between the Eurasian Plate and the Indian Plate. Such continental convergence creates massive compression and thickening of the continental crust; it involves the formation of large faults and the folding of rock strata but little or no volcanic activity (Fig. 30.9). As we will see in Unit 31, this process is directly responsible for the development of some of Earth's greatest mountain ranges.

TRANSFORM PLATE BOUNDARIES

For many years, California's San Andreas Fault was known to be a place of crustal instability. This became especially apparent as scientists studied the devastating earthquake of 1906 in San Francisco and found that it resulted from movement of rocks along this great fault line. The real significance of the San Andreas Fault could not be recognized

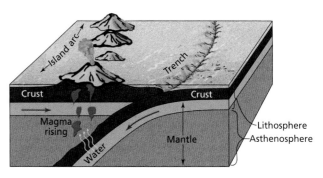

Island arc

Trench

Crust   Crust

Magma
rising

Mantle

Water

Lithosphere
Asthenosphere

FIGURE 30.8 Convergent plate boundary involving two oceanic plates. Where one oceanic plate is subducted beneath the other, a deep oceanic trench forms. Just above the descending plate, water triggers the melting of mantle rock, and magma rises to form an arc of volcanic islands.

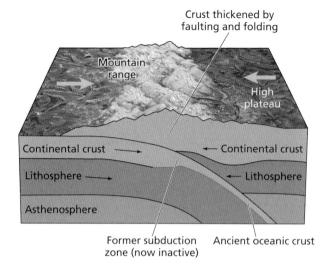

FIGURE 30.9 Simplified cross-section of a collision between two continental landmasses. At an earlier time than is shown here, the oceanic lithosphere of the plate to the left was being subducted below the continental lithosphere of the right plate. The left plate carried a continent, however, and when it arrived at the subduction zone, its low-density rocks would not sink into the asthenosphere, so subduction came to an end. Much compression and deformation of the crust occurred through faulting and folding, and high mountains developed above the now-inactive subduction zone.

FIGURE 30.10 The San Andreas Fault in action. Rows of trees in a Southern California orange grove were offset 4.5 m (15 ft) along a branch of the fault during the May 1940 Imperial Valley earthquake.

until plate tectonics became understood, however. The fault marks a plate boundary—not a boundary of divergence or convergence, but a boundary along which two plates are sliding past each other (Fig. 30.10). This is referred to as a **transform plate boundary**. Transform (or *lateral*) movement along such plate boundaries may not produce the dramatic topographic consequences displayed

by convergent movement, but it, too, is accompanied by earthquakes and crustal deformation.

The area to the west of the San Andreas Fault is part of the Pacific Plate, whereas the area to the east is part of the North American Plate (Fig. 30.11). The fault extends

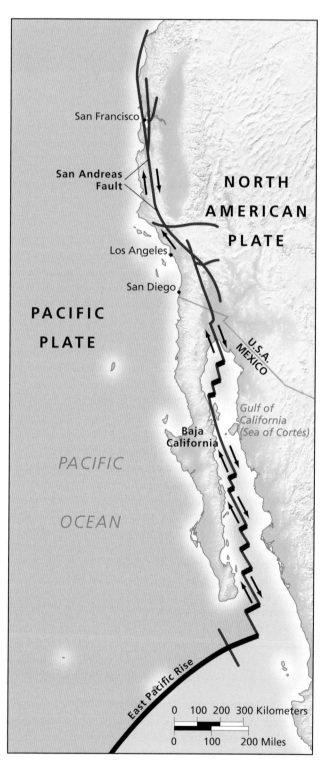

FIGURE 30.11 California's San Andreas Fault in its regional context. This fault separates the Pacific Plate from the North American Plate, and marks where the two plates slide past each other at a transform boundary.

The 1994 Northridge Earthquake and Earthquake Prediction

In the predawn darkness of January 17, 1994, the Los Angeles area was rocked by a destructive earthquake centered in Northridge in the San Fernando Valley, approximately 35 km (22 mi) northwest of downtown Los Angeles. The quake's magnitude was not particularly severe, but it struck a heavily populated urban sector, killing 61, injuring 8,500, and damaging more than 26,000 buildings as well as three vital freeways (Fig. 30.12). When everything was added up weeks later, the destruction totaled a staggering $30 billion.

As geologists descended on the valley to explain the event, they immediately discovered the earthquake had occurred along a then-unknown fault—but one that was well within the corridor containing the San Andreas Fault system (see Fig. 30.11). Ominously, this earthquake confirmed the existence of yet another important fault lying directly beneath the nation's second-largest metropolis.

Plate tectonics explains why both the Los Angeles area and the San Francisco Bay region are susceptible to damaging earthquakes: they both lie directly along the boundary between two actively moving lithospheric plates. Yet the 1994 quake also highlights the difficulty in predicting the timing and strength of earthquakes, even now that we understand their causes much better than in 1906, when a great earthquake devastated San Francisco. For many years, researchers pursued the goal of earthquake prediction using the idea that after a big earthquake on a fault of the San Andreas system, the moving plates steadily accumulate new stresses, which finally build up again to the breaking point. If this buildup can be measured, then in theory, it might be possible to accurately estimate the probability of an earthquake in various parts of California within the next 5 to 10 years.

Unfortunately, earthquake experts are increasingly pessimistic that predictions like this will be possible any time soon, and those that have been attempted have often failed. Although we can say with some certainty that California will experience earthquakes larger and more damaging than the 1994 quake in the future, when they will occur cannot be predicted. Even their exact location may be unexpected, like the Northridge earthquake that occurred on its previously unknown fault.

FIGURE 30.12 Even a moderate earthquake can produce major disruptions in a populous metropolitan area. This Los Angeles freeway interchange was put out of commission by the San Fernando Valley quake on January 17, 1994.

southward into the Gulf of California, thereby separating Mexico's Baja California from the Mexican mainland, which is part of the North American Plate. At its northern end, the San Andreas Fault enters the Pacific Ocean north of San Francisco. Baja California and Southern California (including metropolitan Los Angeles) are sliding north-northwestward past the North American Plate because of the northward motion of the Pacific Plate as a whole. This process is going on at a fairly high rate of speed, estimated to average about 5 cm (2 in) per year.

In the case of the San Andreas Fault, earthquakes attend this movement. Moreover, the San Andreas is the major fault in a larger and complex system of faults that have developed because of the side-by-side motion of the North American and Pacific Plates (see Perspectives on the Human Environment: The 1994 Northridge Earthquake and Earthquake Prediction). No volcanic activity is associated with this type of plate boundary.

Alfred Wegener's original vision of continents in motion over geologic time, stimulated by the observation of geographic patterns, has been proven essentially correct. We continue the discussion of plate dynamics in Unit 31 by asking more questions about the breakup of Pangaea, the supercontinent originally proposed by Wegener.

Key Terms

continental drift *page 377*

convergent plate boundary *page 382*

divergent plate boundary *page 382*

island arc *page 383*

lithospheric plates *page 379*

Pacific Ring of Fire *page 381*

Pangaea *page 377*

plate tectonics *page 379*

rift *page 379*

rift valley *page 382*

seafloor spreading *page 379*

subduction *page 382*

transform plate boundary *page 384*

volcanic arc *page 383*

Scan Here for a quick vocabulary review

Review Questions

1. Briefly describe some of the evidence that existed in support of the notion of continental drift at the time that Wegener proposed it.
2. What discoveries prompted the hypothesis of seafloor spreading?
3. Explain how major plate boundaries can be identified.
4. How can the rate of plate motion be determined using the age of rocks on the ocean floor?
5. Why does subduction occur at some convergent boundaries but not at others?
6. How are earthquake hazards in California related to plate boundaries?

Scan Here for a quick concept review

www.oup.com/us/mason

UNIT 31

Plate Movement: Causes and Effects

Monte Fitz Roy, in Patagonia near the southern end of the Andes, is famous as one of Earth's most challenging mountains for climbers. Such rugged topography is typical of mountains that are actively rising where lithospheric plates converge.

OBJECTIVES

- Outline the mechanisms and processes that move lithospheric plates
- Discuss the concept of isostasy and relate it to the topography of the continents
- Describe the evolution of Earth's continental landmasses

Unit 30 traced scientists' growing understanding of plate tectonics, from Alfred Wegener's initial hypothesis of continental drift to the modern view of lithospheric plate movement. In this unit, we return to an issue that slowed the acceptance of Wegener's ideas: What drives the motion of the plates? We then describe how Earth's ocean basins and high mountain ranges can be explained as effects of plate tectonics. Finally, we examine the origin and growth of continents, starting with a look back in time before the formation of Pangaea, the supercontinent first envisaged by Wegener.

The Mechanism of Plate Motion

The movement of the plates that form Earth's crust has been established beyond a reasonable doubt. Although we can now directly measure plate motion, the mechanism propelling the plates, in the past as well as the present, is still not completely understood. Most explanations include a set of internal convection cells in Earth's mantle (Fig. 31.1); this was first proposed by Arthur Holmes before the development of modern plate tectonics. According to this model, in some parts of the world rock rises toward the surface from deep in the mantle, while in other regions, it descends back to great depths, forming a cycle of continuous movement.

This cycling of rock clearly would explain the processes that are active at divergent plate boundaries. Hot rock rises toward the surface there, driving the plates apart and continually adding new material to them through igneous activity. This process, pushing plates away from divergent boundaries, is an important contributor to the overall motion of the plates. In recent years, researchers have increasingly emphasized another reason for plate motion: the plates are *pulled* as well as pushed. Far from the midoceanic ridges, on the other side of the typical plate, oceanic lithosphere is

plunging downward into the mantle at a subduction zone. Because this lithosphere is so dense compared to the surrounding rock, it tends to sink and pull the rest of the plate along behind it. It appears that where a greater portion of the edge of a plate is being subducted, the plate moves faster, suggesting that the pull of the sinking slab is crucial to plate motion. The sinking plate continues the overall cycle of convection, which is completed as rock deep in the mantle flows back toward the divergent boundaries.

Basic questions about the convection cells remain unanswered. How deep are the cells? Some researchers have argued that they include the whole thickness of the mantle, but others believe the lowest point reached by the circulating rock is much shallower. How do new areas of upward flow develop, where a new divergent boundary forms and splits a continent in two? As our understanding of plate tectonics continues to evolve, clearer answers to these questions will be possible, but no doubt new questions will emerge as well.

Isostasy

The upper surfaces of the continents display a high degree of topographic variation. Mountain ranges rise high above surrounding plains; plateaus and hills alternately dominate the landscape elsewhere. Mountain ranges have mass. Because of the law of gravity, they exert a certain attraction on other objects. This attraction even acts on a plumb line, a weight hanging on a wire or string and used by surveyors and carpenters to mark a vertical line. If we were to hang a plumb line somewhere on the flank of a mountain range, the mountains would attract the weight, drawing the plumb line very slightly away from the vertical, not enough to be visible but enough to be important for highly accurate surveys.

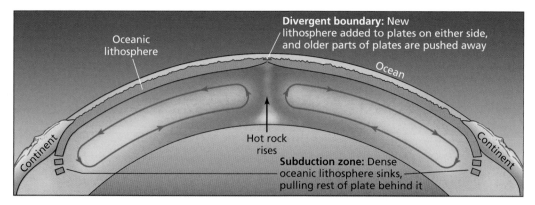

FIGURE 31.1 A highly simplified view of convection cells in the mantle. Hot rock rises to a divergent boundary, or spreading center, marked by a midoceanic ridge. There, new oceanic lithosphere is added to the plates on each side, which are pushed away in the process. The opposite edge of each plate plunges downward in a subduction zone, pulling the rest of the plate with it. The upwelling rock at divergent boundaries, the movement away from those boundaries, and the return flow of rock deep in the mantle together form cycles of slowly moving rock.

This may seem like a curiosity of little importance, but the behavior of a plumb line near mountains led scientists to some important ideas about how the topography of Earth develops and is maintained. The story of how this happened is worth reviewing in some detail as a good example of the twists and turns often followed by science.

More than a century ago, the head of the British survey of India, George Everest, discovered a discrepancy between the vertical lines determined by his surveyors' plumb lines and the true vertical calculated from sightings on stars and the Sun. Everest (after whom Mt. Everest is named) explained this discrepancy as the effect of many small errors in surveying. John Pratt, a mathematician working in India at the time, was not satisfied with that answer and suggested that gravitational attraction from the great mass of the Himalayas pulled the plumb lines away from the vertical (Fig. 31.2). To his surprise, however, his calculations showed that the mountains' mass should have pulled surveyors' plumb lines much *farther* from the vertical than was actually observed.

Soon afterward, astronomer George Airy offered an explanation. The mountains must have "roots" formed from rocks of relatively low density. These roots extend far below the Himalayas and displace heavier rock that would have caused greater attraction. We now know the mountain roots are granitic rock, which has considerably lower density than the ultramafic rocks of the upper mantle (see Unit 29).

Airy used the analogy of ice floating in water, as shown in Figure 31.3A. Ice floats because it is less dense than water, and where a block of ice protrudes high above the water surface, this height must be balanced by a greater thickness of ice extending below water level. If a block of ice is pulled up out of the water or pushed down into it, it quickly returns to the level where it is in balance. Airy proposed a similar idea: Earth's thin crust essentially floats on denser rocks below it. The thicker the crust, the deeper its roots extend into the mantle and also the higher it stands above surrounding areas with thinner crust (Fig. 31.3B).

After Airy proposed his hypothesis, Pratt added one more twist to the story. He argued that lower density, not greater thickness, causes some parts of the crust to stand much higher above sea level than others. To use the analogy of ice again, ice full of air bubbles would have lower

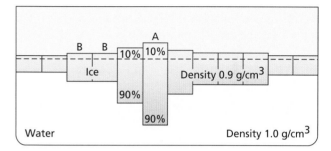

A

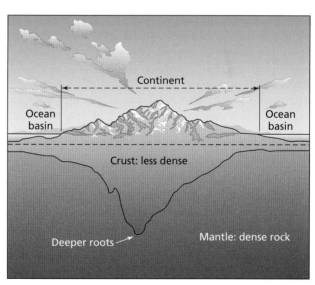

B

FIGURE 31.3 Isostasy. The distribution and behavior of crust and mantle are analogous to blocks of ice floating in water (A). Note that no matter how thick the block, the same percentages (10 percent/90 percent) float above and below the surface. Each block is therefore in balance. The ratio for mountains and their "roots" of crustal rock is a little higher, but the principle is the same (B).

ENHANCE

FIGURE 31.2 The Himalayas form an awesome mountain wall when seen from the south. The gravitational pull of these massive mountains caused errors in the British survey of India. This photo is of central Nepal (a small country wedged between India's Ganges Plain and the Himalayan massif) in the vicinity of Annapurna, the world's eleventh-tallest peak.

density than pure ice and would float higher in water. Today, we recognize that both Airy and Pratt were right, depending on the situation. Their ideas are combined in the modern concept of **isostasy**: *iso* means the same or equal, and *stasy* comes from the ancient Greek word meaning to stand. Isostasy is a condition of equilibrium between floating landmasses and the denser rock beneath them.

It is now believed that the entire lithosphere—not just the crust—is involved in isostatic adjustment. Both the thickness and the density of the lithosphere influence the level at which it is in balance and thus how high or low the surface topography is relative to sea level. The crust—and therefore the lithosphere as a whole—is often unusually thick below high mountains such as the Himalayas, sustaining their great elevation. The ocean basins are low parts of Earth's surface topography because of the thin, dense oceanic lithosphere (Unit 29). We also recognize today that as the lithosphere adjusts to an isostatic balance, it does not simply move vertically like a floating block of ice. Instead, it is somewhat flexible, bending downward in some areas and flexing upward in other locations.

PLATE TECTONICS, ISOSTASY, AND MOUNTAIN BUILDING

The deep roots of the Himalayas were involved in the discovery of isostasy, and they provide an excellent example of the close connections among plate motion, crustal thickness, and the uplift of high mountain ranges. The collision of the Indian and Eurasian Plates has led to intense compression of the crust, forming folds and giant faults (which we will discuss in detail in Unit 34). As a result, the crust squeezed between these plates has been greatly thickened, and in response, the lithosphere has risen, lifting the Himalayan peaks and the Tibetan Plateau just to the north high above sea level. That is, the thicker lithosphere has adjusted to a new isostatic balance, with its deepened roots allowing it to float somewhat higher in the asthenosphere. (See Unit 29 for a review of Earth's outer layer.)

The Alps and the Caucasus Mountains are also relatively young mountain ranges that formed at convergent plate boundaries, and the Appalachians mark a former boundary where continents collided during the assembly of Pangaea. As you might expect, compression and thickening of the crust are usually greatest in collisions, but these processes can also occur where an oceanic plate is subducted (Unit 30). This is clearly the case with the western edge of South America, where the Andes have risen above an active subduction zone.

A careful look at maps of Earth's topography reveals more complicated cases, however. In Asia, many high mountain ranges are found at some distance from where the Eurasian and Indian Plates meet. The Tian Shan, Kunlun Shan, and Qilian Shan of northwestern China (*shan* means mountains in Chinese) all have high snow-capped peaks that are still rising today, yet they are separated from the convergent plate boundary by the large Tibetan Plateau. Geologists believe that the uplift of these peaks is related to stress from the same collision that produced the Himalayas. The stress was transmitted away from the collision zone through the lithosphere. The Rocky Mountains of the western United States and Canada are also far from the nearest plate boundary, and explaining their formation through plate tectonics remains a challenge.

ISOSTASY AND EROSION

We can also use the concept of isostasy to envision what would happen if a high mountain range were subjected to a lengthy period of erosion. Consider the analogy with ice shown in Figure 31.3A. If we were to remove the upper 10 percent of the block of ice marked **A**, we would expect that block to rise slightly—not quite to the height it was before, but nearly so. If we were to place the removed portion of **A** on the two blocks marked **B**, they would sink slightly, and their upper surface would adjust to a slightly higher elevation than before. Thus, column **a** would have a lower height and a shallower underwater "root," whereas the blocks marked **B** would have a greater height and a deeper root. The ratio of ice thickness above and below water would remain the same for all blocks.

This analogy helps explain why erosional forces in the real world have not completely flattened all mountain ranges. Based on the amount of eroded rock carried downstream each year by the world's major rivers, mountains formed hundreds of millions of years ago, such as the Appalachians of the eastern United States, should have been worn down to sea level long ago. Yet parts of the Appalachians still stand high above their surroundings (Fig. 31.4). What seems to happen is that as erosion removes the load from the ridges, isostatic adjustment raises the remaining mountain mass in compensation for the loss. Rocks formed deep below the surface, tens of thousands of meters down, are thereby gradually raised as the mountain continually readjusts to erosion. Eventually, these rocks that formed deep in mountain roots are exposed to our view and to weathering and erosion.

Isostasy also helps us to answer some puzzling questions about the deposition of enormous thicknesses of sedimentary rocks. These often form where rivers carried sediment into an ocean or lake basin over a long period of geologic time. A good place to see this process at work today is where a large river delivers its sediment load, such as the Yellow River delta shown in Figure 31.5. In these locations, sedimentary rocks often contain indications that they were initially deposited in shallow water, yet they now lie within a sequence of rocks thick enough to fill deep ocean basins. One sequence of sedimentary rocks found in Africa involved the accumulation of nearly 6.5 km (more than 20,000 ft) of various sediments followed by an outpouring of great quantities of lava. Other parts of the world contain even thicker deposits. Although thousands of

FIGURE 31.4 View from Pilot Mountain, North Carolina, overlooking the Piedmont, which yields eastward to the coastal plain. A vista like this suggests that the mountains are being eroded into lowland topography flanking them, but as the text points out, things are not that simple.

meters of sediment collected over millions of years, the depth of the water somehow remained about the same. In the accumulating sediments near the Bahamas, for instance, rocks formed in shallow or intertidal flats are now 5,500 m (18,000 ft) thick.

Isostasy can readily explain these observations. Each centimeter of sediment that is deposited adds that much more thickness and mass to the crust, and over time the lithosphere gradually shifts downward toward a new position of balance. This downward movement of the lithosphere allows great thicknesses of sediment to accumulate even if deposition is always near sea level.

ISOSTASY AND REGIONAL LANDSCAPES

In studying the effects of isostasy, we tend to be preoccupied with mountain ranges, mountain building, plate compression, and associated phenomena. But we should not lose sight of the consequences of isostasy in areas of less prominent, less dramatic relief. Erosion is active on the continents' plains, too, and millions of tons of material are carried away by streams, rivers, and other erosional agents. Even moving ice and wind denude and reduce land surfaces. In contrast to the mountainous zones, the plains are

vast in area and slopes are gentler. Rivers erode less spectacularly on the plains than in the mountains because they flow down gentler slopes and their valleys are wider, among other factors. These circumstances mean that eroded material is removed from the plains at a slower rate. And in response, the eroding lithosphere slowly rises so that the elevation of the plains decreases even more slowly than the rate of erosion. This helps to explain why extensive areas of low relief can persist for long periods of time in the continental interiors. Not only is the rate of erosion slow in these landscapes, but it is also partly offset by slow isostatic uplift. This is much the same explanation as we discussed earlier for the persistence of old mountain ranges like the Appalachians.

Ice Sheets and Isostatic Rebound Several other manifestations of isostasy also are of interest. When ice sheets spread over continental areas during glaciations, the weight of the ice (which can reach a thickness of several thousand meters) causes isostatic subsidence (sinking) of the crust below. The resulting subsidence is similar to that produced by sediment accumulation in other regions. This is what happened in northern North America and Europe during the most recent glaciation. When these continental

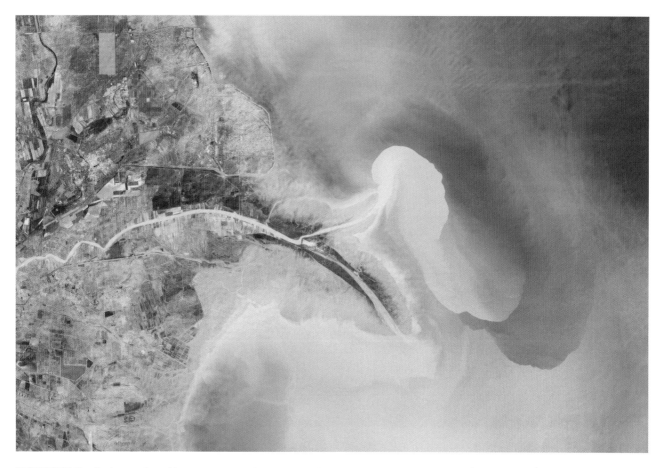

FIGURE 31.5 A plume of muddy water enters the ocean at the mouth of China's Huang He (Yellow River). This river carries vast quantities of material eroded from the windblown silt deposits of the Loess Plateau and landslide-prone mountain valleys in western China. Much of the sediment is deposited in shallow water near the river mouth, and it accumulates so rapidly that the coastline has shifted tens of kilometers out into the ocean in the last 200 years alone. As sediment piles up, the lithosphere flexes downward so that the river delta always remains near sea level.

glaciers reached their maximum extent, the crust below was depressed by their weight (1,000 m [3,300 ft] of ice depresses the underlying crust by about 300 m [1,000 ft]). When the ice sheets retreated less than 15,000 years ago, isostatic readjustment caused the crust to rebound. The upward readjustment, however, could not keep pace with the relatively rapid melting of the ice. In geologic terms, the melting removed the enormous load of the ice sheets almost instantaneously.

Studies show that the ensuing isostatic rebound is still going on; Earth's crust is still not in equilibrium. In the heart of the Scandinavian Peninsula in northern Europe, the site of a huge ice sheet, the crust is rising at more than 1 m (3.3 ft) per century, a very fast rate. In some coastal areas of Norway, metal rings placed in rocks centuries ago to tie up boats are now much too high above sea level to be of use. In the U.S. Great Lakes region, isostatic rebound is also ongoing. As the crust rises there, it is raising the bed of the St. Mary's River, the outlet from Lake Superior into Lake Huron; this, in turn, raises the water level of Lake Superior relative to Lake Huron.

Dams and Crustal Equilibrium Even human works on the surface of Earth can produce isostatic reaction. When a river dam is constructed, the weight of the impounded water behind it may be enough to produce isostatic accommodation in the crust. Measurable readjustment of this kind has taken place in the area of Kariba Lake, formed upstream of the great dam on the Zambezi River in southern Africa, and around Lake Mead behind Hoover Dam on the Colorado River in Nevada. We cannot see these changes with the naked eye, but scientific instruments detect them. To us in our everyday existence, the crust may seem permanent, unchanging, and solid, but even our own comparatively minuscule works can disturb its equilibrium.

Evolution of the Continents

The most recent phase of lithospheric plate movement started with the breaking up of Pangaea, the supercontinent first envisaged by Alfred Wegener nearly a century ago (see Fig. 30.1). Continental drift proved to be one manifestation of plate tectonics, partial evidence for a process of planetary proportions involving immense amounts of energy. Confirmation of the breakup of first Pangaea and then Laurasia and Gondwana raises new questions, however.

The fragmentation of Pangaea—the collision between India and Asia, and the separation of Africa and South America—took place within the last 160 million years of Earth's history. This is a relatively brief period of geologic time, accounting for only 4 to 5 percent of Earth's total existence as a planet (see Focus on the Science: Development of the Geologic Timescale in Unit 28). What happened during previous phases of plate movement? Before Pangaea, were there earlier supercontinents that formed through a coalescence of landmasses, only to be pulled apart again?

CONTINENTAL SHIELDS

Certainly, rocks older than 160 million years exist. As we have already noted, Earth formed more than 4.5 billion years ago (see Focus on the Science: The World's Oldest Rocks, in Unit 27). The oldest known rocks are from the geologic era known as the Precambrian, and they date from more than 4 billion years ago (see Fig. 28.11). These are mostly igneous and metamorphic rocks, which today form part of the continental shields. The comparatively inactive continental shields are of interest in physical geography because they carry landscapes that also are quite old and have, in some areas, changed little over many millions of years. The shield areas of Africa, South America, and Australia all provide unique insight into the effects of long-term weathering and slow erosion. The Canadian

(Laurentian) Shield is just as old but has been altered by recent glacial erosion.

Once reliable methods were developed for determining the age of the ancient rocks of the continental shields, it became clear that various parts of each shield formed at quite different times during the Precambrian. In fact, the shields were originally assembled out of smaller pieces of continental crust in somewhat the same way as Pangaea. Therefore, they still bear the geological traces of numerous Precambrian subduction zones and collisions between continental plates. Since the Precambrian, the shields have been the stable cores of continents, while the plates they are on have collided to form larger continents or in other cases rifted apart to form new oceans. The details of these earlier plate movements are still hotly debated, but the collisions that built the supercontinent of Pangaea are fairly well known. One of the more important results of these collisions was the formation of the Appalachian Mountains in eastern North America and other mountain ranges in Europe.

TERRANES

Another part of the story leading up to the formation and breakup of Pangaea has been especially intriguing to geologists and is also important for understanding the physical geography of the continents today. Even before the concept of plate tectonics emerged, it was known that certain parts of landmasses do not, geologically speaking, seem to belong where they are located. That is, their rocks and geologic histories are completely different from those of their surroundings, as if they initially developed in some other

---

PERSPECTIVES ON THE HUMAN ENVIRONMENT

California's Gold and Plate Tectonics

The discovery of gold in 1848 at Sutter's Mill in the Sierra Nevada foothills of California is a well-known story, as is the gold rush that followed. The origin of California's gold and its connections with plate tectonics are much less widely known.

Gold is an exceedingly rare element, and it can only be mined profitably where it has been concentrated by a unique set of geologic processes. Hydrothermal water—deep groundwater that is heated by magma and can dissolve gold—often plays a role. For California's gold, the earliest stages of concentration may have occurred where gold was deposited by hydrothermal waters spewing out from vents on the ocean floor. Some of the gold-bearing ocean-floor

sediments eventually ended up in terranes that were attached to the growing continent of North America.

In the Cretaceous period (see Fig. 28.11), what is now California was an active subduction zone, as oceanic lithosphere moving in from the west plunged below the North American continent. As usual, melting occurred above the descending plate, and huge masses of magma rose into the crust, forming the granite batholith we see exposed today in places like Yosemite National Park. Again, hydrothermal waters formed—heated by the intruding magma—and dissolved gold from the ancient seafloor deposits and other sources, depositing the gold in veins of quartz (Fig. 31.6). As the granitic rock of the batholith was

continued

continued

exposed to weathering and erosion over tens of millions of years, gold nuggets and flakes were carried away with the sand and gravel transported by streams.

Gold was first discovered in 1848 in the sandy beds of modern streams flowing out of the Sierra Nevada. Within a few years, miners began to seek gold in the older stream gravels that underlie much of the landscape of the goldfield region. Hydraulic mining, in which these gravels were blasted away with high-pressure water to separate the denser gold particles, devastated large areas until it was outlawed by court order in 1884. By then, miners had moved on to hard-rock mining, tunneling directly into the gold-bearing quartz veins.

Long before 1884, the gold rush had permanently changed the human geography of California and the western United States, as hundreds of thousands of gold seekers and other immigrants poured into California from the eastern states, Europe, Latin America, and China. The new city of San Francisco sprang up overnight and soon large, productive farms were established in California's valleys. For Native Americans of the goldfield region, the gold rush was an unmitigated disaster, however. Their way of life was destroyed, and many were killed by gold seekers or died of introduced diseases.

FIGURE 31.6 Specimen of gold from a quartz vein. Weathering and erosion of the rock that contained this vein would eventually have released nuggets and flakes of gold to be carried by streams.

part of the world. That is essentially the conclusion that was reached once the reality of continental drift was accepted: these chunks of continent must have moved from faraway locales to their current foreign positions. Once there, they were attached to an existing continent in a process called **accretion**. These fragments can be regional in extent; consistent in age, rock type, and structure; and either continental or oceanic in origin. Earth scientists refer to them as **terranes**. The term **suspect terrane** (or *exotic terrane*) has sometimes been used to convey how out of place some of these fragments are and how difficult it is to identify their origin.

One of Earth's most complex mosaics involving suspect terranes extends along western North America from Alaska to California, where long-term accretion has expanded the landmass far beyond its original western margins at the end of the Precambrian. Numerous terranes have been mapped here. One of the best known is the Wrangellia terrane, parts of which occur in mainland and peninsular Alaska, in the Canadian province of British Columbia, and south of the Canada–United States border. Wrangellia is of volcanic, island-arc origin; its exposed fragments stand in sharp contrast to the surrounding regional geology (Fig. 31.7). In fact, much of California, most of Alaska, and large parts of the Piedmont region in the southeastern United States are all made up of terranes accreted to the older core of North America (see Perspectives on the Human Environment: California's Gold and Plate Tectonics). Similar suspect terranes are known to exist in the mountain belts of other landmasses.

Exactly how most of these terranes reached their present locations and where they originated are still matters of speculation. Some are pieces of continental crust that were broken off larger landmasses through rifting, migrated with a tectonic plate, and then collided with the continent to which they are now attached. Other suspect terranes may be former island arcs, pushed into continental margins and enveloped by mountain building. Clearly, this is one way a continental landmass may grow in present times, even if the Precambrian phase of shield formation has long passed.

The dynamics of plate movement produce much of the restlessness that characterizes Earth's crust. One of the most spectacular surface manifestations of this geologic activity is volcanism, a topic that is the focus of Unit 32.

FIGURE 31.7 "Traversing Glacier Bay in southeast Alaska, we were given a seminar by a National Park Service guide who enjoyed asking challenging questions. Knowing some physical geography helped, but here she had me stumped. 'Look at that outcrop,' she said. 'What can you tell me about it?' I said that the rocks looked darker than the regional gray-granite masses rising steeply from the bay, but in the absence of any knowledge of volcanic activity here, I could not do any better. 'You're looking at lava,' she explained; 'this is a fragment of one of those suspect terranes, an old basaltic island arc welded onto the local regional geology. It's called Wrangellia, and pieces of it can be identified from mainland Alaska all the way down the coast to British Columbia and the U.S. Northwest. It's out of place and we don't know how it got here, but it sure stands out in this landscape.' That is the kind of field experience you don't forget."

Key Terms

accretion *page 394*

isostasy *page 390*

suspect terrane *page 394*

terrane *page 394*

Scan Here for a quick vocabulary review

ENHANCE

Review Questions

1. Why does an old mountain range like the Appalachians still persist despite massive erosion?

2. Why is Earth's crust continually rising in elevation in Scandinavia and the Lake Superior region?

3. How do we know that the processes of plate tectonics were active before the breakup of Pangaea?

4. How does thickening of the crust where two plates converge lead to mountain uplift?

5. What kind of evidence might you look for in trying to identify a terrane that was accreted to North America?

Scan Here for a quick concept review

ENHANCE

www.oup.com/us/mason

Volcanism and Its Landforms

Mt. Shasta towers 3000 m (10,000 ft) above the surrounding landscape of northern California. This potentially active volcano last erupted in 1786 and is one of a chain of great volcanic peaks that formed above a subduction zone in the Pacific Northwest.

OBJECTIVES

- Describe the global distribution of volcanic activity and explain its relationship with plate boundaries
- Explain how the composition of magma influences volcanic eruptions
- Summarize the major types of volcanic landforms and the hazards associated with them
- Cite some dramatic historical examples of human interaction with volcanic environments
- Describe the landscapes that result from volcanism

Volcanism is the eruption of molten rock at Earth's surface, which is often accompanied by rock fragments and explosive gases. The process takes various forms, one of which is the creation of new lithosphere at the midoceanic ridges (see Fig. 30.4). Along some 50,000 km (31,000 mi) of ocean-floor fissures, molten rock penetrates to the surface and begins its divergent movement. It is a dramatic process involving huge quantities of magma, the formation of bizarre submarine topography, the heating and boiling of seawater, and the clustering of unique forms of deep-sea oceanic life along the spreading ridges. This spectacle is all hidden from us by the ocean water above. What we know of it comes from the reports of scientist-explorers who have approached the turbulent scene in specially constructed submarines capable of withstanding the pressure at great depths and the high temperatures near the emerging magma.

Islands situated on midoceanic ridges afford a glimpse of this process that is mostly concealed from view. Iceland and its smaller neighboring islands along the Mid-Atlantic Ridge between Greenland and Norway, where the ridge rises above the ocean surface (see Fig. 2.6), are totally of volcanic origin, and there is continuing volcanic activity there. In 1973, a small but populated and economically important island off Iceland's southwest coast, Heimaey, experienced a devastating episode of volcanic activity. First, Heimaey was cut by fissures, and all 5,300 of its inhabitants were quickly evacuated. In the months that followed, lava poured from these new gashes in the island, and volcanic explosions rained fiery pieces of ejected magma onto homes and commercial buildings (Fig. 32.1).

Heimaey actually increased in size, but the lava flows threatened to fill and destroy its important fishing port. This threat led to an amazing confrontation between people and nature. The islanders quickly built a network of plastic pipes at the leading edge of the advancing lava. They pumped seawater over and into the lava to cool it more quickly than nature could. Their hope was that the lava would form a solid dam, which might stop the advance and restrain the lava coming behind it. This daring scheme worked: part of the harbor was lost to the lava, but a critical part of it was saved and actually improved. When this volcanic episode ended, life returned to Heimaey. But Iceland and its neighbors are on an active midoceanic ridge, and volcanism will surely attack them again. What happened above the surface at Heimaey is happening, continuously, all along the 50,000 km (31,000 mi) of submerged spreading ridges.

Distribution of Volcanic Activity

In addition to being common near midoceanic ridges, volcanism also occurs on the continents in the vicinity of plate boundaries (see Figs. 30.5 and 30.6) and leaves a characteristic signature in the form of volcanic landscapes. Most volcanism not associated with seafloor spreading is related to subduction zones at convergent plate boundaries (see Figs. 30.7 and 30.8). Not surprisingly, a majority of the

FIGURE 32.1 The 1973 eruption on the Icelandic island of Heimaey generated lava flows and ashfalls that forced the evacuation of the 5,300 inhabitants and caused considerable destruction. It also produced a heroic reaction in which local citizens, assisted by the U.S. military, fought back.

ENHANCE

world's active volcanoes lie along the Pacific Ring of Fire. But note that some volcanic activity is associated neither with midoceanic ridges nor with subduction zones. The island of Hawai'i, for example, is near the middle of the Pacific Plate. Lava has poured from one of its volcanic mountains, Kilauea, almost continuously since 1983. On the African Plate, where West Africa and equatorial Africa meet, lies Mt. Cameroon, another active volcano far from spreading ridges and subduction zones. A map of the world reveals a number of similar examples both on ocean-floor crust and on continental crust. Later in this unit, we discuss an explanation for these puzzling volcanoes and how they may help track the motion of tectonic plates.

Volcanoes are often classified as active, dormant, or extinct, in part because these categories may help predict where dangerous eruptions could occur in the near future. An *active* volcano has often been defined as one that has erupted in recorded history. Written records extend much farther back in some parts of the world than others, however, and many volcano researchers prefer to classify as active any volcano that has erupted in the past 10,000 years. A *dormant* volcano has not erupted in recent geologic time, but an eruption is still considered possible. Magma often gives evidence of its presence underground and its potential to produce future eruptions. A good example is the area of Yellowstone National Park, where the last volcanic activity involved minor lava flows tens of thousands of years ago. The famous geysers and hot springs record the presence of magma at a fairly shallow depth, however, warning us of the possibility of future activity at this volcanic center where cataclysmic eruptions occurred in the geologic past. In fact, movement of this shallow pool of magma has caused the ground surface in central Yellowstone National Park to rise and fall slightly over the past several decades. When there is no such clear sign of magma below, and especially if the volcano has been worn down and greatly modified through weathering and erosion, it is tentatively identified as *extinct*. Such a designation is always risky; some volcanoes have erupted after being classified as extinct.

As noted earlier, the great majority of continental and island-arc volcanoes are in or near subduction zones. Volcanic activity there is concentrated, and parts of the landscape are dominated by the unmistakable topography of eruptive volcanism. Many of the world's most famous mountains are volcanic peaks marking convergent plate boundaries: Mt. Fuji (Japan), Mt. Vesuvius (Italy), Mt. Rainier (U.S. Pacific Northwest), Mt. Chimborazo (Ecuador), and many others. Frequently, such mountains stand tall enough to be capped by snow, their craters emitting a plume of smoke. It is one of the natural landscape's most dramatic spectacles.

Properties of Magma

Compared to common liquids such as water, all magma has quite high viscosity (a fluid's resistance to flowing). However, some magmas have higher viscosity than others because of their composition. Basaltic lavas, such as those flowing from the midoceanic ridges, are relatively low in silica and are therefore fairly fluid when they erupt. They flow freely and often (on land) quite rapidly—up to 50 km/h (30 mph) on steep slopes. Other lavas are richer in silica and tend to be more viscous; as a result, they flow more slowly.

Magma also contains steam and other gases under pressure. These gases cannot easily escape from silica-rich magma because of its high viscosity, so pressure builds up and eruptions tend to be explosive. Gobs of lava are thrown high into the air and solidify as they fall back to the mountain's flanks. Such projectiles, not unreasonably, are called volcanic *bombs*. Smaller fragments may fall through the air as volcanic *cinders* or volcanic *ash*. After the explosive 1980 eruption of Mt. St. Helens in Washington State, lighter volcanic ash fell over a wide area downwind from the mountain. Geologists use the term **pyroclastics** for all such solidified fragments that erupt explosively from a volcano (*pyroclastic* is ancient Greek for broken by fire). *Pumice* is a rock that forms from pyroclastic material under certain

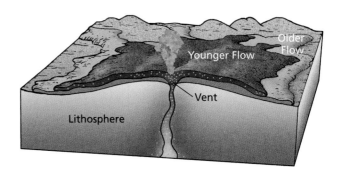

FIGURE 32.2 During a typical flood basalt eruption, lava flows onto the surface from a vent and spreads out in a sheet rather than forming a dome. Younger flows cover all or parts of older flows. As eruptions continue over time, a thick stack of cooled flows builds up.

conditions and has such low density that it actually floats (rafts of pumice are sometimes seen on rivers or the ocean surface after explosive volcanic eruptions).

Gases can escape from low-silica magma much more easily than they can from silica-rich magma, so explosive eruptions are uncommon. Often, such basaltic magma erupts to produce lava flows rather than great clouds of pyroclastic material. Lava erupts from a **vent**, an opening in Earth's crust, or longer and more linear openings called **fissures**, and it may form large sheets across the countryside rather than volcanic peaks (Fig. 32.2). Enormous amounts of basaltic lava—referred to as **flood basalts**—can accumulate in this fashion. For example, just before Gondwana broke up, fissure eruptions produced a vast lava plateau of which parts still exist in India, South Africa, South America, and Antarctica. In the United States, the most extensive flood basalts occur on the Columbia Plateau of Washington, Oregon, and Idaho (Fig. 32.3).

When basaltic magma erupts, a common byproduct is sulfur dioxide, released to the atmosphere, where it can become sulfuric acid. In 1783 and 1784, eruptions of basaltic magma at the Laki volcano in Iceland produced sulfuric acid fogs that drifted over much of northwestern Europe, peeling paint from buildings and killing crops.

Volcanic Landforms

The most characteristic product of a volcanic eruption is the towering mountain form, represented by such peaks as Fuji (Japan), Rainier (United States), Popocatépetl (Mexico), Vesuvius (Italy), and Kilimanjaro (Tanzania). But not all volcanic action produces such impressive landforms. Some volcanoes are larger in volume but less prominent in shape than the well-known volcanic peaks. Others are smaller and less durable. Although a great variety of volcanic landforms can be identified, five major types are especially common: calderas, composite volcanoes, cinder cones, shield volcanoes, and plateaus or plains formed by extensive flood basalts. These landform types can often be related to the silica content of the magma that produced them.

A

B

FIGURE 32.3 The flood basalts of the Columbia Plateau and Snake River Plain cover much of the interior Pacific Northwest of the United States (A). The horizontal layering of these successive lava flows is strikingly visible in the landscape (B).

CALDERAS

A volcano's lavas and pyroclastics come from a subterranean magma chamber, a reservoir of active molten rock material that forces its way upward through one or more volcanic vents. When that magma reservoir is emptied rapidly in an eruption, it ceases to support the volcano, and the ground surface can collapse over a large area, creating a basin called a **caldera** (Fig. 32.4). Calderas generally result from a particularly violent eruption. Relatively small calderas form through the collapse of a composite volcano, the landform type discussed next. Oregon's Crater Lake is

A

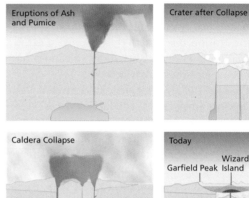

B

FIGURE 32.4 Crater Lake (A) is a relatively small caldera in Oregon that formed about 7700 years ago through the collapse of a composite volcano (Mt. Mazama) during a violent eruption. The lake's development is shown here as a sequence of four stages (B). Eruption of a large volume of pyroclastics (ash and pumice) led to collapse of the volcano, forming the present crater, which then filled with water. Wizard Island is a cone formed by a minor later eruption.

a classic example of a caldera formed in this way. The lake, almost 600 m (2,000 ft) deep, fills a circular crater 10 km (6 mi) across, with walls more than 1,200 m (4,000 ft) high. This remarkably beautiful caldera appears quite large when viewed from the rim, but it is far from the largest caldera on Earth.

Three enormous calderas were produced by eruptions in the Yellowstone area of Wyoming, Montana, and Idaho over the last 2.5 million years. The youngest of these, which formed about 640,000 years ago, had an oval shape about 50 to 70 km (30 to 42 mi) across in the center of today's Yellowstone National Park. Its form is no longer clearly visible because of erosion and later volcanic activity. As discussed earlier, a large body of magma remains below Yellowstone today and could produce similar catastrophic eruptions in the future. An even larger caldera was produced about 28 million years ago by the so-called Fish Canyon Tuff eruption in Colorado, when up to 5,000 km³ (1,200 mi³) of ash and other material was spewed out by one of the largest known volcanic eruptions in Earth history. Large calderas such as those of Yellowstone and the one that formed from eruption of the Fish Canyon Tuff are produced by relatively high-silica magma.

COMPOSITE VOLCANOES

Most of the great volcanoes formed over subduction zones are **composite volcanoes**, which disgorge a succession of lavas and pyroclastics. These have the shape most people associate with volcanoes, the high steep-sided peak exemplified by Mt. Fuji, Mt. Rainier, and Popocatépetl. In cross-section, such volcanoes look layered, with lavas of various thicknesses and textures interspersed with strata formed by compacted pyroclastics (Fig. 32.5). Neither the heavier pyroclastics nor the rather viscous lava travel far from the crater. Thus, the evolving volcano soon takes on its fairly steep-sided, often quite symmetrical appearance. Many composite volcanoes are long-lived and rise to elevations of thousands of meters. Appropriately, they also are called *stratovolcanoes*.

Composite volcanoes are usually formed by magma with intermediate silica content, often called *andesitic* or *dacitic* because of the extrusive igneous rock types that form from it. This magma is viscous enough to trap gases and make their eruptions explosive and, in some cases, notoriously dangerous. Molten lava flows from these volcanoes but is not the main threat to life in their surroundings. Pyroclastics can be hurled far from the crater, and volcanic ash creates hazards near the volcano as well as farther afield. In 2010, ash from the eruption of the Eyjafjallajökull volcano in Iceland drifted over Europe, bringing air travel to a halt in many areas until the ash dispersed. The most serious threats to life near a composite volcano are lahars and pyroclastic flows, however.

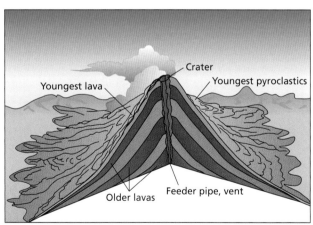

FIGURE 32.5 Simplified cutaway view of a composite volcano showing a sequence of lava flows interbedded with compacted pyroclastics.

ENHANCE

FIGURE 32.6 Aftermath of the 1985 mudflow caused by a nearby erupting volcano, which late at night buried the Colombian town of Armero without warning. This lahar killed at least 20,000 people in the immediate area, making it one of the worst natural disasters of the late twentieth century.

Lahars Tall composite volcanoes are capped by snow and may shelter glaciers on their slopes. The hot ash from a major eruption melts the snow and ice, which forms a flood of ash, mud, and water rushing downslope. Such a mudflow of volcanic origin is called a **lahar** (pronounced luh-*harr*), from a word used for these flows in Indonesia, a nation with many active and dangerous composite volcanoes. Such a mudflow can be extremely destructive. In 1985, a volcano named Nevado del Ruiz in the Andes of central Colombia erupted, and much of its snowcap melted. In the mudflow that swiftly followed, more than 20,000 people perished. After it was over, the scene at a town in its path at the base of the mountain range was one of utter devastation, a mass of mud containing bodies of people and animals, houses and vehicles, trees and boulders (Fig. 32.6). Lahars are mainly triggered by eruptions but occasionally may also result from intensive, warm-season orographic rainfall on the snowpack of a volcano's uppermost slopes.

Pyroclastic Flows Perhaps even more dangerous than a lahar is the outburst of hot gas and glowing volcanic ash particles that often accompanies explosive eruptions of composite volcanoes. Such an outburst produces a **pyroclastic flow**, also often referred to as a *nuée ardente* (French for glowing cloud; pronounced noo-*ay* ahr-*dahnt*). This is a fluid mass of cooling bits of magma suspended in hot gases that races downslope at speeds exceeding 100 km/h (65 mph). Everything in its path is incinerated; the force of the descending ash cloud, which is heavier than the surrounding air, and its searing temperatures ensure total destruction. In 1902, a pyroclastic flow burst from Mt. Pelée on the eastern Caribbean island of Martinique. It descended on the nearby port town of St. Pierre located at the mountain's base in a matter of minutes, killing an estimated 30,000 persons (the only survivor was a prisoner in a cell in the town's thick-walled prison). The Paris of the Caribbean, as St. Pierre had been described, never recovered and still carries the scars of its fate more than a century later.

Predicting Risks from Composite Volcanoes Although eruptions of composite volcanoes are often explosive, they do not occur without warning. The 1980 eruption of Mt. St. Helens in southern Washington State was predicted by volcanologists who measured the telltale signs of resumed activity, including frequent earthquakes and a growing bulge on the north side of the peak. The area around Mt. St. Helens was evacuated, and few lives were lost considering the dramatic nature of the eruption. (One fatality was a young scientist at an observation post on a nearby ridge, reminding us that even today volcanology remains one of the most dangerous scientific professions.) Today many volcanoes are actively monitored. Recent experience has shown that many thousands of lives can be saved by warning systems and planned evacuations. In 1991, after intense debate among volcanologists and government officials, the decision was made to evacuate tens of thousands of people from the area near Mt. Pinatubo in the Philippines because of activity indicating a possible eruption. Soon after, a major eruption produced pyroclastic flows and devastating lahars (Fig. 32.7). Plans for similar evacuations are in place for parts of Washington State that are threatened by lahars from Mt. Rainier.

CINDER CONES

Some volcanic landforms consist not of lava but almost entirely of pyroclastics. Normally, such **cinder cones** remain quite small, frequently forming during a brief period of explosive activity. They are often produced by magma with intermediate to low silica content. An extensive area of cinder cone development lies in East Africa, associated with the rift valley system of that region (Fig. 32.8). In North America, Sunset Crater (Arizona) and the Craters of the Moon (Idaho)

FIGURE 32.7 A huge cloud of volcanic ash and gas rises above Mt. Pinatubo in the Philippines on June 12, 1991. Three days later, an explosive eruption, one of the largest in the past century, spread a layer of ash over a vast area around the mountain. There were immediate environmental as well as political consequences: a dust cloud orbited Earth and reduced global warming, and a nearby U.S. military base on Pinatubo's island (Luzon) was damaged beyond repair.

ENHANCE

are examples of this volcanic landform. Geologists were able to observe closely the growth of a cinder cone in Mexico from 1943, when it was born in a cornfield in Michoacán State (about 320 km [200 mi] west of Mexico City), until 1952. This cinder cone, named Paricutín, grew to a height of 400 m (1,300 ft) in its first eight months of activity and remained intermittently active for nearly a decade.

SHIELD VOLCANOES

Shield volcanoes are built up mainly from lava flows produced by fluid basaltic, low-silica magmas. The eruptions of shield volcanoes are sometimes vigorous enough to form dramatic "fountains" of molten rock and some cinders, but they are much less violent than the eruptions of composite volcanoes. The basaltic lava is very hot, however, and it flows in sheets that—over many eruptions—gradually build up a broad mountain with gently sloping sides (Fig. 32.9). Compared to their horizontal dimensions, which are quite large, the tops of such volcanoes are rather unspectacular and seem rounded rather than peaked. This low-profile appearance has given them the name shield volcanoes.

The most intensively studied shield volcanoes undoubtedly are those on the main island of Hawai'i, where the U.S. Geological Survey operates an observatory. Mauna Loa is the largest active shield volcano on Hawai'i. It stands on the ocean floor and from there rises 10,000 m (32,000 ft), with the uppermost 3,500 m (13,680 ft) protruding above sea level. To the north, Mauna Kea's crest is slightly higher. And to the east lies Kilauea (pronounced kill-uh-*way*-uh), one of the most active volcanoes on Earth, which began its current eruption in 1983. Kilauea's lavas have even flowed across a housing subdivision, obliterating homes and streets, and reached the ocean, thereby adding a small amount of land to the island (Fig. 32.10).

The volcanic landscape in this area displays some interesting shapes and forms. Lava that is especially fluid when it emerges from the crater develops a smooth "skin" upon hardening. As the lava continues to move, this slightly hardened surface is then wrinkled into a ropy pattern called **pahoehoe** (Fig. 32.11). Less fluid lava hardens into angular, blocky forms called **aa**, so named, according to Hawai'ian tradition, because of the shouts of people trying to walk barefoot on this jagged terrain!

BASALT PLATEAUS AND PLAINS

As mentioned earlier, flood basalts represent the accumulation of many lava flows that spread out over large areas, often from numerous vents across an area. In this case, the result is not a single volcanic mountain but extensive lava-covered plains or plateaus. Two outstanding examples of these landforms can be found in the northwestern United States: the Columbia Plateau and the nearby Snake River Plain (see Fig. 32.3). The Columbia Plateau was formed by eruptions of highly fluid lava between 17 and 15 million years ago, which eventually covered 50,000 km² (20,000 mi²). The surface of

FIGURE 32.8 "East Africa's volcanic landscape displays landforms ranging from great peaks such as Mounts Kilimanjaro and Kenya to small cinder cones and local fissure eruptions. Driving across this area you are reminded that volcanic activity continues; some of the cinder cones have not yet developed much vegetation. Elsewhere plants are just beginning to establish themselves on the lava. In places the landscape poses challenging questions. Many cinder cones, for example, have small craterlike depressions near their crests, but those 'craters' tend to lie to one side, as in this photo taken in Kenya. As it happens, a majority of them lie in the same compass direction from the top of the hill. What might be the cause of this pattern?"

ENHANCE

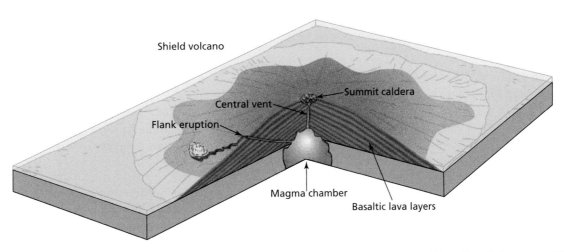

FIGURE 32.9 Simplified cutaway view of a shield volcano, modeled after shield volcanoes of Hawai'i, which are over 320 km (200 mi) wide at the base and rise 13 km (8 mi) above the ocean floor.

FIGURE 32.10 Unearthly landscape of Kilauea's upper slopes on the island of Hawai'i. Kilauea is a classic shield volcano. This photo portrays an eruption in progress, one of a still continuing series that began more than 30 years ago. Lava emanating from here can flow for many kilometers and reach the sea.

the Snake River Plain is covered with younger lavas, in some cases only a few hundred years old. Also on the Snake River Plain are the remnants of a chain of huge calderas that erupted over the last 17 million years, which form a curious pattern: the oldest calderas are the farthest southwest, and they become progressively younger toward the northeast. The youngest caldera in this group is the one that erupted 640,000 years ago in what is now Yellowstone National Park.

Hot Spots

The past and present activity of volcanoes in Hawai'i, Yellowstone, and the Snake River Plain raise important questions about the connection between volcanic activity and plate tectonics. The most basic question is: Why are there volcanoes in these locations at all? None are near subduction zones or divergent plate boundaries, the locations where we expect to find most volcanic activity. Hawai'i is in the middle of a plate, almost as far from a plate boundary as one can get on Earth's surface. Mt. Cameroon in Africa, mentioned earlier, is another volcano far from plate boundaries.

There are some notable similarities between these locations of unexpected volcanism. The calderas of the Snake

River Plain form a chain in which the most recent activity has been at one end, in the Yellowstone region. Mt. Cameroon is also part of a volcanic chain. The Hawai'ian Islands provide an even more striking example. The Big Island—Hawai'i—displays the currently active volcanism. But all the islands in this archipelago originally formed as shield volcanoes, and they bear a chain of mostly extinct volcanoes that become older as one moves from southeast to northwest, that is, from Hawai'i to Kauai (Fig. 32.12A). Long-term erosion after cessation of volcanic activity has reduced the size of the islands northwest of Hawai'i, especially Kauai and Oahu, and obscured the original shape of their shield volcanoes. Furthermore, if we travel northwest from Kauai we first reach Midway Island, and then a chain of now-submerged remnants of volcanoes called *seamounts* that are clearly visible on maps of ocean-floor topography (Fig. 32.12B). The seamount chain eventually bends north-northwestward, where the submerged peaks are known as the Emperor Seamounts.

It is remarkable that the only pronounced volcanic activity along this entire corridor is on Hawai'i—at the end of the chain. Along the rest of the chain, the age of volcanic rocks (i.e., the time since the volcanoes were active) increases toward the northwest. Volcanic rocks of Kauai are up to 5.6 million years old, those of Midway 25 million years old, and those at the northern end of the Emperor Seamounts 75 million years old.

The concept that can explain volcanic activity in Hawai'i and probably in Yellowstone as well was first proposed by Canadian geologist J. Tuzo Wilson (1908–1993). He theorized that the Pacific Plate has been moving over a **hot spot** in the mantle. A hot spot is a location where hot rock persistently flows upward from the lower mantle (called a *mantle plume*), triggering frequent volcanic activity. Because hot spots are rooted deep in the mantle, they remain fixed as lithospheric plates move over them. As the Pacific Plate moved over the Hawai'ian hot spot, shield volcanoes formed (see Fig. 32.12A). Thus, one location after another experienced volcanic activity, acquiring volcanic landforms as the plate moved. As each island moved off the hot spot, volcanic activity there ceased.

Today, Hawai'i is active, but eventually the plate carrying that island will have moved far enough to the northwest that Hawai'i's volcanoes will be extinct, and a new island will be the center of volcanic eruptions. Already, a large undersea volcano named Loihi is rising about 35 km (22 mi) southeast of Hawai'i. That seamount, which is 975 m (3,000 ft) beneath the waves, is expected to emerge above the ocean surface between 10,000 and 100,000 years from now.

If the hot spot theory is correct and subcrustal hot spots are indeed stationary, then it is possible to calculate the speed and direction of plate movement. As the map (see Fig. 32.12A) shows, the Emperor Seamounts extend in a more northerly direction than the Midway–Hawai'i chain. If the same hot spot is responsible for both the Emperor Seamounts and the Midway–Hawai'i chain, this indicates

FIGURE 32.11 "Spending a semester teaching at the University of Hawai'i gave me the opportunity to spend much time on the Big Island, where the great volcanoes are active today. Here on Hawai'i, the product of very recent (and current) shield volcanism, snow dusts the mountaintops even as palm trees grace the tropical beaches. To experience an area where the night sky reflects the glow of molten magma and the day reveals clouds of superheated gases emanating from caldera walls is memorable. Shown here is a mass of recently erupted pahoehoe lava, its smooth 'skin' wrinkled into ropy patterns by continued movement of molten rock inside."

that the moving Pacific Plate, which today travels toward the northwest (see Fig. 30.5), changed direction about 40 million years ago. Given the age of Midway's lavas (25 million years), we may conclude that the plate traveled some 2,700 km (1,700 mi) over this period. This works out to about 11 cm (4.5 in) per year, a rate of movement consistent with average rates for other plates (see Unit 30).

The volcanic effects of hot spots under comparatively thin oceanic lithosphere can be discerned rather easily and are revealed by the ocean-floor topography (see Fig. 32.12A). But it is likely that hot spots are also active under thicker continental crust. Many researchers interpret the chain of calderas ending in Yellowstone as the track of a hot spot that the North American Plate has moved over. The giant flood basalts of the Columbia Plateau may be related to the same hot spot. This example illustrates the point that the effects of hot spots under continental plates are quite complex: they may produce eruptions of basaltic magma in some locations and high-silica rhyolitic caldera eruptions in other places. Some geologists have recently cautioned that the hot spot concept may have been applied too widely and that volcanic activity far from plate margins could have a variety of other causes.

Volcanoes and Human History

The explosive eruptions of composite volcanoes can destroy cities almost instantly. The example of St. Pierre in the Caribbean was mentioned earlier. Similarly, an eruption of Mt. Vesuvius in 79 CE buried the Roman cities of Pompeii and Herculaneum, preserving them as two of the most remarkable archaeological sites ever discovered. Other volcanoes are believed to have changed the course of human history, but in more complicated ways.

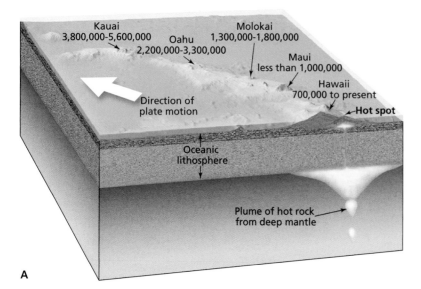

Kauai
3,800,000-5,600,000
Oahu
2,200,000-3,300,000
Molokai
1,300,000-1,800,000
Maui
less than 1,000,000
Hawaii
700,000 to present
Hot spot
Direction of
plate motion
Oceanic
lithosphere
Plume of hot rock
from deep mantle

A

FIGURE 32.12 The Hawai'ian islands are a chain of volcanoes formed by the seafloor moving over a hot spot (A). The hot spot is beneath the big island of Hawai'i today, and volcanoes are active there. In the past each of the other islands was over the hot spot and was built up by volcanic activity (the age of the volcanoes, in years, is shown for each island). These islands are part of a much longer chain of former volcanoes that reveal the track of the hot spot, including Midway Island and the Emperor Seamounts, which once stood above sea level but are now submerged (B).

ENHANCE

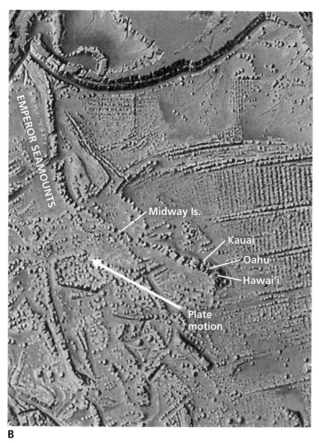

EMPEROR SEAMOUNTS

Midway Is.

Kauai

Oahu

Hawai'i

Plate
motion

B

KRAKATAU AND TAMBORA

In 1883, Indonesia's Krakatau volcano blew up with a roar heard in Australia, 3,000 km (1,900 mi) away. The blast had a force estimated to be equivalent to 100 million tons of dynamite. It rained pyroclastics over an area of 750,000 km² (300,000 mi²) and propelled volcanic dust through the troposphere and stratosphere to an altitude of 80 km (50 mi). For years following Krakatau's eruptive explosion, this dust orbited Earth, affected solar radiation, and colored sunsets brilliant red. The eruption and its aftermath generated great, destructive ocean waves. First, the explosion itself set in motion a giant *tsunami*, a sea wave set off by crustal disturbance. Next, water rushed into the newly opened caldera, setting off further explosions and successive tsunamis. When these waves reached the more heavily populated coasts of Indonesia, an estimated 40,000 people were killed.

Explosive eruptions of Krakatauan dimensions are rare, but Krakatau was actually the second-largest eruption of the nineteenth century in its corner of the world. The eruption of the Indonesian volcano Tambora in 1815 was even larger, and its effects on humans reached far beyond Indonesia itself. The volcanic dust it generated interfered so strongly with incoming solar radiation that the weather turned bitterly cold. In Europe and North America, the year 1816 was widely described as the "year without a summer." A high percentage of crops failed, food and fuel shortages developed, and problems lasted into the ensuing winter and beyond.

SANTORINI

The volcanic island of Santorini (Thera) stood in the Mediterranean about 110 km (70 mi) north of Crete, where the Minoan civilization thrived. On a fateful day sometime around 1645 BCE, Santorini exploded. So much volcanic ash was blasted into the air that skies were darkened for days. It has been suggested that this was the event described in the Bible's Old Testament as the act of God in retribution against the Pharaoh: "Thick darkness in all the land of Egypt for three days." The Mediterranean Sea churned a sequence of tsunamis that lashed the coasts of other islands.

A

B

FIGURE 32.13 Thera (Santorini), in the eastern Mediterranean, bears witness to a cataclysmic volcanic eruption more than 3,600 years ago that may have changed the course of human history. What was left after that eruption was an enormous caldera, 60 km (37 mi) in circumference and open to the sea, on whose inner wall homes and businesses now perch precariously (A). But Santorini's activity has not ended, and earthquakes continue to destroy lives and property. A new cone is slowly rising in the center of the caldera, where fresh lava signals the rise of a new mountain (B). In the background, atop the caldera wall, is the much-damaged town of Oja.

As for Santorini itself, when daylight reappeared, its core was gone and had been replaced by a vast caldera, now filled with seawater (Fig. 32.13). The loss of life in the eruption must have been devastating. Minoan civilization was the Mediterranean's most advanced, and great palaces graced Santorini as well as Crete. Fertile, productive, well-located Santorini was once host to towns and villages, ports and farms; fleets of boats carried trade between it and prosperous Crete. Many archaeologists and historical geographers have speculated that the eruption of Santorini triggered the end of the Minoan culture. Whatever the cause, Minoan civilization went into decline at about the time of Santorini's explosion. It has also been suggested that the legend of a drowned city of Atlantis was born from Santorini's destruction.

Landscapes of Volcanism

Volcanic activity, especially the mountainous type, creates unique and distinct landscapes. Volcanic landscapes are limited in their geographic extent, but they do dominate certain areas. A single composite cone, by its sheer size or threat, can dominate physical and mental landscapes over a much wider area.

Mt. Vesuvius still towers over the southern Italian city of Naples. Active and monitored anxiously, Vesuvius has erupted disastrously some eighteen times since the first century CE. During the twentieth century alone, it erupted in 1906, 1929, and 1944, causing destruction and death in each instance. Farther south, on Sicily, the island just off the toe of the Italian peninsula, stands Mt. Etna. Again, "the mountain," as such dominating volcanic landforms seem to be called wherever they stand, pervades the physical landscape of the entire area. East Africa's Kilimanjaro, 5,861 m (19,340 ft) tall, carries its snowcap within sight of the equator (see Fig. 17.16). It has not erupted in recorded history, but its presence is a constant reminder of what could happen.

It may initially be hard to understand why so many people have always lived in the shadow of these great, dangerous volcanoes. The answer often lies in the fertile soils that develop there, formed from volcanic ash deposits and with an unusual capacity for holding the water and nutrients needed for productive farming. Especially in the tropics, the intensively cultivated farmland covering the ridges and valleys of volcanic regions contrasts strongly with nearby nonvolcanic landscapes where forest cover dominated until recent times.

Volcanic landscapes are most prevalent, as noted earlier, along the Pacific Rim of subduction zones that run from southern Chile counterclockwise to New Zealand (see Fig. 30.6). On the Asian side of the Pacific, many of these landscapes are island arcs. The Aleutian Islands form an especially long arc reaching from North America almost to Asia. In other parts of the Pacific Rim, snow-capped composite volcanoes tower above lowlands or low, eroded mountain ranges. This is the case in Japan, Kamchatka (part of eastern Russia), and the Pacific Northwest of the United States and Canada, where the Cascade Range includes a line of volcanoes from Mt. Garibaldi in British Columbia to Lassen Peak in California (Fig. 32.14). Two of these Cascade volcanoes, Mt. Shasta and Mt. Rainier, reach elevations almost as high as the tallest peaks in the Rocky Mountains and Sierra Nevada. In Chile, Peru, Ecuador, and Colombia, South America's Andes Mountains are studded with great volcanic cones, but the overall height of this mountain range and the high plateau, or *altiplano*, to the east is related more to compression and thickening of the crust than to volcanic activity.

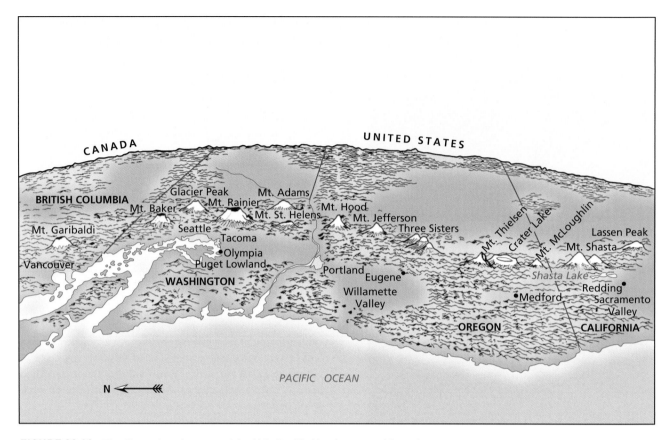

FIGURE 32.14 The Cascade volcanoes of the U.S. Pacific Northwest and Canada.

The Cascades volcanoes of the United States (see Fig. 32.14) provide an excellent example of how rapidly and dramatically large composite volcanoes can build up and then decay as they become inactive. The towering peaks of Mt. Rainier and Mt. Shasta have formed within the last 500,000 to 600,000 years. These two volcanoes continue to be active (Rainier erupted in the 1840s and Shasta in 1786), which explains their present height. As soon as a volcano of this type ceases to erupt, it is rapidly worn down by water and glacier ice in the moist climate of the Pacific Northwest. Inactive volcanoes that have not been completely destroyed are small jagged peaks that are only a fraction of their former height. Mt. Thielsen, which is visible in Figure 32.14, is a good example.

Key Terms

aa *page 402*

caldera *page 400*

cinder cone *page 402*

composite volcano *page 401*

fissure *page 399*

flood basalts *page 399*

hot spot *page 404*

lahar *page 401*

pahoehoe *page 402*

pyroclastic flow *page 401*

pyroclastics *page 399*

shield volcano *page 402*

vent *page 399*

Scan Here for a quick vocabulary review

ENHANCE

Review Questions

1. In what three geologic settings do volcanoes occur?

2. What type(s) of magma are associated with particularly hazardous volcanic eruptions?

3. How and why do the following three volcanic landforms differ from each other: shield volcanoes, composite volcanoes, and calderas?

4. What are hot spots, and how do they help explain current volcanic activity in the Hawai'ian Islands?

5. What type of eruptions produced the rocks of the Columbia Plateau, and what is a possible explanation for why they occurred where they did?

Scan Here for a quick concept review

ENHANCE

www.oup.com/us/mason

UNIT 33

Earthquakes and Landscapes

Pavement cracked by severe shaking of soft ground in the Marina District, San Francisco, during the Loma Prieta earthquake of 1989.

OBJECTIVES

- Describe how earthquakes are produced by fault movement and how earthquake strength is measured
- Explain the cause and behavior of tsunamis
- Summarize the types of damage that earthquakes produce
- Relate the spatial pattern of earthquakes to plate tectonics
- Describe the landscapes and landforms that bear the signature of fault movement and earthquakes

In Unit 29, we described how earthquakes in the crust and upper mantle generate seismic waves that travel through the lithosphere as well as the interior of Earth (see Fig. 29.4). These waves yield key evidence that furthers our understanding of the internal structure of the planet. As the seismic database expands and methods of analysis improve, the inferred properties of Earth's internal structure and composition are becoming better known.

Earthquakes have an obvious impact on cultural as well as physical landscapes. Following a major earthquake, physical evidence of its occurrence can be seen on the ground in the form of open fractures, new scarps, areas where the ground surface has risen or subsided, and, often, numerous landslides and mudslides triggered by shaking. Moreover, the shocks and aftershocks of an earthquake can do major damage to buildings and other components of the cultural landscape. Add to this the fact that certain areas of the world are much more susceptible to earthquake damage than others, and no doubt remains that we should study this environmental hazard in a geographic context.

What Produces Earthquakes?

As lithospheric plates converge, spread apart, or move side by side, stress is produced in rocks near the plate boundaries, often producing faults. A **fault** is a fracture in crustal rock involving the displacement of rock on one side of the fracture with respect to rock on the other side. Some faults, such as the San Andreas Fault, are giant breaks that continue for hundreds of kilometers (see Fig. 30.11); others are shorter. Along the contact zone between the North American and Pacific Plates, the upper crust is riddled with faults, all resulting from the stresses imposed by plate movement. Like volcanoes, certain faults are active, whereas others are no longer subject to enough stress to cause movement. In the great continental shields that form the cores of continents lie many inactive faults that bear witness to earlier ages of crustal instability. Today, those faults appear on geologic maps, but they do not pose earthquake hazards.

All rocks have a certain rupture strength, that is, they have the strength to resist breaking until stress reaches a certain level. Below that level, they will deform or bend rather than fracturing (Fig. 33.1). When the stress finally exceeds the rupture strength, rocks break suddenly to form a fault with one side slipping past the other. Even after the rock breaks to form a fault, the two sides are usually locked together by friction so no motion occurs. Once again, the rock on either side of the fault deforms as stress builds up. Now the fault is a plane of weakness, and as the friction is finally overcome, the two sides of the fault slip past each other again. In the process, the rock on either side rebounds to its original shape, and the cycle begins again. Each sudden movement along the fault produces an **earthquake**, the sudden release of energy that has been slowly built up during phases when the fault is locked and there is increasing rock deformation.

Before faulting:
Horizontal layers of rock

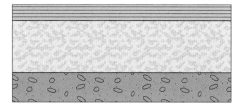

Stress causes deformation (bending)

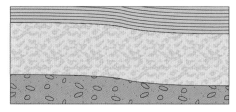

Rock breaks forming fault, and rebounds from deformation → **Earthquake**

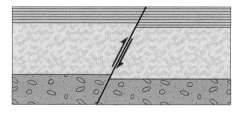

Fault locked by friction, continued stress causes deformation

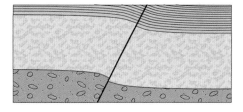

Friction overcome, fault slips, rock rebounds from deformation → **Earthquake**

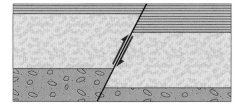

FIGURE 33.1   Development and continued movement of a fault. Stress (in this case, compression) causes horizontal rock layers to deform, or bend slightly, until their rupture strength is exceeded and the rock breaks to form a fault. Continued stress causes new deformation of the rock, followed by sudden fault movement when friction is overcome. With each slippage of rock along the fault, the deformed (bent) rock on either side rebounds to its original shape and seismic waves are generated, which we experience as an earthquake.

This energy release takes the form of seismic waves that radiate in all directions from the place of fault movement. Earthquakes can originate at or near the surface of Earth, deep inside the crust, or even in the upper mantle. The place of origin is the **focus**, and the point directly above the focus on Earth's surface is the **epicenter** (Fig. 33.2). Earthquakes range from tremors so small that they are barely detectable to great shocks that can destroy entire cities. The strength of an earthquake can be described in two ways, which sound similar but are distinctly different concepts. An earthquake's **magnitude** is the energy released as the fault ruptures; its **intensity** refers to the severity of the shaking at a particular location.

Measuring Earthquake Strength

You have probably heard of the *Richter Scale*, developed in 1935 by geophysicist Charles Richter (1900–1985) to describe earthquake magnitude, which assigns a number to an earthquake based on the measurement of the physical force of the ground motion. This scale is logarithmic, so that each increase of 1 unit represents an increase of 10 times in magnitude. For example, an earthquake of magnitude 4 releases 10 times as much energy as one of magnitude 3, and 100 times as much as a quake of magnitude 2. The original Richter Scale had shortcomings, and has largely been replaced by the *Moment Magnitude Scale*. This new scale is also logarithmic, and the moment magnitude for most earthquakes is similar to the Richter magnitude. When the U.S. Geological Survey announces the magnitude of a major earthquake, the Moment Magnitude Scale is used, even

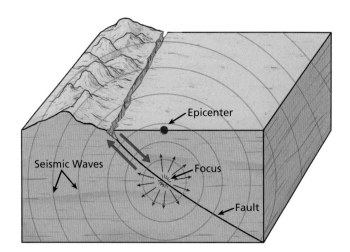

FIGURE 33.2 The focus of an earthquake is the point at which the fault begins to rupture, which may be deep below ground. The epicenter is the point at Earth's surface vertically above the point of focus. The energy released at the earthquake's focus radiates outward to other parts of Earth in the form of seismic waves, represented by the rings spreading from the focus.

though the news media often still report this number as being "on the Richter Scale."

The strongest earthquakes recorded in the past century have had magnitudes of 9.0 to 9.5, and they all occurred at subduction zones around the Pacific Rim. We will look at these subduction-zone quakes later in the unit. Other major earthquakes tend to have magnitudes between 7 and 8, as illustrated by the list of well-known earthquakes in Table 33.1.

TABLE 33.1 Noteworthy Earthquakes of the Twentieth and Twenty-First Centuries

Year	Place	Moment Magnitude	Estimated Death Toll
1906	San Francisco, California	7.8	3,000
1908	Messina, Italy	7.2	72,000
1920	Ningxia, China	7.8	200,000
1923	Tokyo-Yokohama, Japan	7.9	143,000
1960	Temuco-Valdivia, Chile	9.5	1,655
1964	Prince William Sound, Alaska	9.2	131
1970	Chimbote, Peru	7.9	70,000
1976	Tangshan, China	7.5	242,000*
1976	Guatemala	7.5	23,000
1985	West-Central Mexico	8.0	9,500
1988	Armenia, (then) U.S.S.R.	6.8	25,000
1990	Northwestern Iran	7.4	50,000+
2001	Gujarat, India	7.6	20,000
2004	Sumatra, Indonesia	9.1	228,000
2005	Northern Pakistan	7.6	86,000
2008	Sichuan, China	7.9	87,600
2010	Port-au-Prince, Haiti	7.0	316,000
2011	Northern Japan	9.0	20,986

*Official death toll. Reliable reports persist that as many as 700,000 died in this earthquake. Moment magnitudes listed are those calculated by the U.S. Geological Survey.

In the seconds and minutes after each of the great earthquakes listed in Table 33.1, seismic waves traveled rapidly away from the ruptured fault. During the 1906 earthquake, which had a focus near San Francisco, shaking was soon felt from Oregon to Los Angeles; however, the shaking was much stronger near the ruptured segment of the San Andreas Fault than it was farther away. This effect is called **attenuation**: overall, the ground shaking produced by an earthquake becomes weaker with increasing distance from the focus.

On the other hand, local areas of unusually strong shaking often exist, even at some distance from the focus. These often turn out to be places where the ground is soft, such as a river valley filled with loose, muddy, water-saturated sediment. In such places, the shaking motion increases, an effect called **amplification**. That is, the seismic waves become stronger, somewhat like the way an amplifier in a sound system strengthens weak electrical signals. When a major earthquake struck Mexico City in 1985, little impact occurred where the bedrock was solid. But where the buildings stood on the soft ground of an old lakebed on which part of Mexico City is built, the damage was great, because the earthquake waves were significantly amplified. Many high-rise structures collapsed or simply fell over as the ground wobbled like a bowl of gelatin. Where cities were built along ocean or lake shorelines, new ground for building was often created by filling shallow bays with garbage or construction debris. These are also places where amplification is common, and severe damage can result.

An earthquake's intensity, rather than its magnitude, is the best indicator of its impact on the cultural landscape—on people, their activities, and structures. Although an earthquake has only one magnitude, its intensity can vary from place to place because of attenuation, amplification, and other effects on the strength of shaking. Intensity is reported on the *Mercalli Scale*, which was first developed by Italian geologist Giuseppe Mercalli in 1905 and modernized in 1931. Now called the *Modified Mercalli Scale* (Table 33.2), it assigns a number ranging from I to XII to an earthquake (Roman numerals are always used). For instance, an earthquake of intensity IV is felt indoors, and hanging objects swing. Intensity V produces broken windows and dishes, awakens many sleepers, and cracks plaster. Intensity IX damages building foundations and breaks in-ground pipes. At intensity XII, damage is total, and even heavy objects are thrown into the air.

Tsunamis

The Japanese word for a great sea wave, **tsunami**, has come into general use to identify a seismic sea wave. Though they are sometimes mistakenly called tidal waves, tsunamis have nothing to do with ordinary ocean tides. Instead, they are produced when ocean water is suddenly displaced by movement of the crust in an earthquake or by a large landslide (see Fig. 33.9 for a specific example of how this can occur). This sudden displacement results in one or more large waves that rapidly radiate outward from the point of origin. In the open ocean, they look like very broad swells on the water surface, and ships and boats can ride them out quite easily. But a tsunami's height increases when it reaches shallow water near shore, and as it reaches the shore, it can create huge breakers that smash into coastal

TABLE 33.2 Modified Mercalli Scale of Earthquake Intensities

Intensity	Qualitative Title	Description of Effects
I	Negligible	Detected by instruments only.
II	Feeble	Felt by sensitive people. Suspended objects swing.
III	Slight	Vibration like passing truck. Standing cars may rock.
IV	Moderate	Felt indoors. Some sleepers awakened. Hanging objects swing. Sensation like a heavy truck striking building. Windows and dishes rattle. Standing cars rock.
V	Rather strong	Felt by most people; many awakened. Some plaster falls. Dishes and windows broken. Pendulum clocks may stop.
VI	Strong	Felt by all; many are frightened. Chimneys topple. Furniture moves.
VII	Very strong	Alarm; most people run outdoors. Weak structures damaged moderately. Felt in moving cars.
VIII	Destructive	General alarm; everyone runs outdoors. Weak structures severely damaged; slight damage to strong structures. Monuments toppled. Heavy furniture overturned.
IX	Ruinous	Panic. Total destruction of weak structures; considerable damage to specially designed structures. Foundations damaged. Underground pipes broken. Ground fissured.
X	Disastrous	Panic. Only the best-constructed buildings survive. Foundations ruined. Rails bent. Ground badly cracked. Large landslides.
XI	Very disastrous	Panic. Few masonry structures remain standing. Broad fissures in ground.
XII	Catastrophic	Extreme panic. Total destruction. Waves are seen on the ground. Objects are thrown into the air.

Source: After P. J. Wyllie, *The Way the Earth Works* (New York: Wiley, 1976), p. 45.

FIGURE 33.3 Aftermath of the tsunami triggered by a severe earthquake in northern Japan, 2011. In Ishinomaki, shown here, large fishing boats were swept inland. The replica of the Statue of Liberty, built the year before the quake, surprisingly remained standing.

FIGURE 33.4 A ferocious earthquake (magnitude 9.2) devastated parts of Anchorage, Alaska, on March 27, 1964. This is Fourth Avenue the day after the quake; it shows only part of the destruction that extended from the harbor to the interior.

towns and villages, reaching heights far above even the largest wind-driven waves (Fig. 33.3). A tsunami triggered by the 1964 Prince William Sound earthquake in Alaska (see Table 33.1) reached a height of 67 m (218 ft) above sea level along the shores of a narrow inlet.

Tsunami waves have been known to travel all the way across the Pacific Ocean at speeds of more than 800 km/h (500 mph). A major earthquake in Chile in 1960 generated a tsunami that reached Hawai'i about 15 hours later, smashing into coastal lowlands with breakers 7 m (23 ft) high. Ten hours later, the still-advancing wave was strong enough to cause damage in Japan. Situated in the mid-Pacific, the Hawai'ian Islands are especially vulnerable to tsunamis. In 1946, a tsunami caused by an earthquake near Alaska struck the town of Hilo on the island of Hawai'i and killed 156 persons.

The most devastating tsunami in recent years was produced by a strong earthquake off the coast of Sumatra in 2004. This wave traveled across the Indian Ocean, striking coastlines from Sri Lanka to Thailand and causing many of the more than 200,000 deaths related to this earthquake. At the time of this disaster, an effective tsunami warning system had not yet been developed for the Indian Ocean, although one was in place for the Pacific. When an almost equally strong earthquake produced a tsunami off the coast of Japan in 2011 (see Fig. 33.3), the wave was so large that it overwhelmed sea walls and other structures designed to protect coastal settlements and nuclear power stations. However, as the tsunami raced across the Pacific, accurate warnings of its arrival time and height limited the damage.

Earthquake Damage

One of the most important points revealed by Table 33.1 is that the loss of life and destruction during earthquakes is not directly related to their magnitude. Only 131 people died in the 9.2-magnitude earthquake of 1964 centered near Prince William Sound, Alaska, even though this was the second-strongest earthquake recorded anywhere in the world since 1900. The limited loss of life was largely because of low population density near the epicenter and along the coastlines that were directly affected by the large tsunami triggered by the quake. There was significant damage in the city of Anchorage, however, where soft clay sediments were turned into flowing mush, carrying away entire neighborhoods and destroying hundreds of homes as they flowed downhill (Fig. 33.4).

Other earthquakes with much lower magnitudes have been far deadlier, partly because they affected more populated areas and completely destroyed buildings that were not able to withstand strong shaking (Fig. 33.5). The potential for loss of life in poor urban areas where many buildings are not well constructed and may collapse was made tragically clear by the loss of more than 300,000 lives during the 2010 earthquake in Haiti, even though its magnitude was only 7.0.

The direct effects of shaking only account for a fraction of the damage in many earthquakes, however. The 1906 San Francisco quake (magnitude 7.8) again provides a good example. This earthquake leveled many of the buildings in the young city of San Francisco, but in addition, fires started by ruptured gas lines raged out of control and could not be fought because water supply systems were also disrupted or destroyed (Fig. 33.6). Landslides, mudslides, and snow avalanches can all be triggered by earthquakes as well (Fig. 33.7).

FIGURE 33.5 In January 2001, a major earthquake devastated parts of India's Gujarat state. As many as 20,000 people were killed and many more injured, and entire towns were obliterated. As Figure 30.5 shows, Gujarat lies in a vulnerable area, where the Indian, Arabian, and Eurasian tectonic plates converge.

FIGURE 33.6 Earthquakes affecting major urban areas often start fires from broken gas mains, kitchen stoves, and other sources. The damage from the 1906 earthquake that struck San Francisco was worsened enormously by such fires, which raged out of control for days afterward.

FIGURE 33.7 Landslides are a major hazard arising from earthquakes in high-relief landscapes. In January 2001, an earthquake (magnitude 7.6) struck near San Salvador, the capital of El Salvador in Central America. The town of Santa Tecla would have been little affected except for this massive landslide, which brought death and destruction.

Earthquake Distribution

The vast majority of earthquakes occur near plate boundaries, which fortunately include many locations on the ocean floors and other remote areas less vulnerable than some of the heavily populated places listed in Table 33.1.

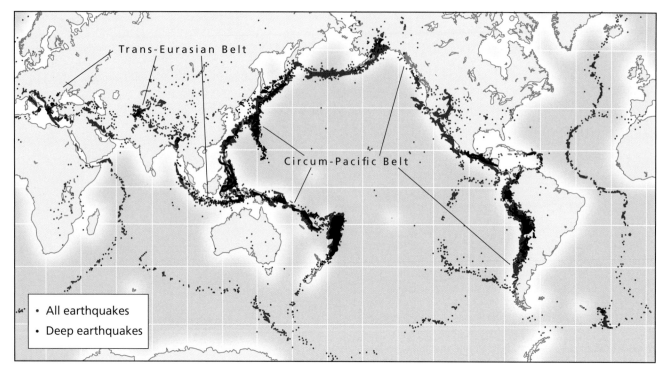

FIGURE 33.8 Global distribution of recent earthquakes. The deep earthquakes, shown by black dots, originated more than 100 km (63 mi) below the surface.

The greatest concentration of quakes is along the *Circum-Pacific Belt* of subduction zones associated with the Pacific and Nazca Plates and their neighboring plates (Fig. 33.8). About 80 percent of all shallow-focus earthquakes (earthquakes that occur at depths of less than 100 km [63 mi]) originate in this belt. As the map indicates, earthquakes with deeper foci in the lower lithosphere are even more heavily concentrated in this zone.

When we compare Figure 33.8 with a map of the world distribution of population (see Fig. 2.9), we quickly see where large numbers of people are at risk from earthquakes. All of Japan's 127 million people live in an area of high earthquake incidence and therefore high risk. The Philippines and Indonesia, home to almost 400 million people, also are earthquake-prone. But south of these island nations, the earthquake-affected zone extends into a less populated area of the Pacific Ocean, lessening in intensity just to the north of New Zealand.

In the north, the Circum-Pacific Belt affects sparsely populated parts of eastern Russia and the Aleutian Islands, but also penetrates the somewhat more populated areas of southern Alaska. Population is also rather sparse along the Pacific coast of Canada until the vicinity of Vancouver Island is reached. New insights on earthquake risks for the cities of Vancouver and Seattle will be discussed in the next section, and it has long been known that much of coastal California is prone to strong earthquakes. Most of western Mexico and Central America and virtually all of the west coast of South America are active earthquake zones, including many major population centers. The strongest earthquake in recent history occurred in 1960 along the coast of Chile, though loss of life was fortunately limited, in part because Chileans had already adopted building construction techniques designed to reduce earthquake damage.

Another zone of high earthquake incidence is the *Trans-Eurasian Belt*, which extends generally eastward from the Mediterranean Sea through Southwest Asia and the Himalayas into Southeast Asia, where it meets the Circum-Pacific Belt. The incidence of major earthquakes in this corridor is not as high as it is in parts of the Circum-Pacific Belt, but some important population centers are at risk. These include Italy, the countries of the former Yugoslavia, Greece, Turkey, Iraq, Iran, and the highlands of Afghanistan, northern Pakistan, northernmost India, and Nepal.

A third zone of earthquakes is associated with the global system of *midoceanic ridges* (compare Figs. 33.8 and 30.2). The earthquakes there are generally less frequent and less severe than those in the Circum-Pacific Belt, and except for population clusters on islands formed by these ridges, this earthquake zone does not endanger large numbers of people.

GREAT SUBDUCTION-ZONE EARTHQUAKES

One of the world's most recent serious earthquakes occurred on March 11, 2011, in northern Japan. This extremely powerful quake (now known as the Tohoku Earthquake) is believed to have been the fourth-strongest in the world since 1900. At least 20,986 people were confirmed dead or were missing after the quake. The worst damage and loss of life were caused by a tsunami (see Fig. 33.3), which reached heights almost 30 m (100 ft) above average sea level in parts of Japan and even caused damage and a few deaths far across the Pacific in California and Indonesia. The tsunami also severely damaged several nuclear reactors and their safety systems, leading to a desperate fight to limit leakage of radioactive material, which went on for many months.

The focus of the 2011 earthquake was located at a subduction zone where the Pacific Plate is plunging under a continental plate, just to the east of northern Japan (see Fig. 30.5). This fits a pattern that has become increasingly clear since the development of plate tectonics theory in the 1960s. Almost all of the strongest earthquakes since 1900 occurred at subduction zones, including the earthquake of 1964 in Alaska, the 1960 earthquake and several other strong quakes in Chile, and the great earthquake of 2004 off the coast of Sumatra. In effect, the plate boundary itself acts as a giant fault, and sudden movement of the rocks on each side produces extremely powerful shaking. Great **subduction-zone earthquakes** of this type almost always produce large tsunamis.

A distinctive type of ground movement occurs in many subduction zones, which can also help explain why earthquakes there usually produce tsunamis (Fig. 33.9). Before the earthquake, the edge of the overlying continental plate tends to buckle upward under pressure from the convergence of the two plates. When the earthquake occurs, the overlying continental plate slips upward along the contact surface between the plates and flattens out again. As this happens, the buckled-up area abruptly drops by as much as 2 m (6 ft). Because the thin outer margin of the continental plate is below sea level, its motion can easily generate a tsunami. In addition, the rapid subsidence of the plate surface affects coastal lands where small changes in elevation above or below sea level have major impacts for vegetation, soils, and people. Coastal areas of southern Alaska sank abruptly in the 1964 earthquake, often killing trees by immersing their roots in salt water (Fig. 33.10). These drowned trees provided an important clue to researchers studying earthquake risk in another subduction zone off the coast of Washington and Oregon (see Focus on the Science: Great Earthquakes in the Pacific Northwest).

INTRAPLATE EARTHQUAKES

Contact zones between tectonic plates clearly form Earth's most active earthquake belts. But earthquakes, some of them severe, can and do occur in other areas of the world.

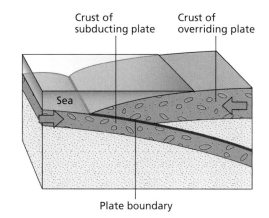

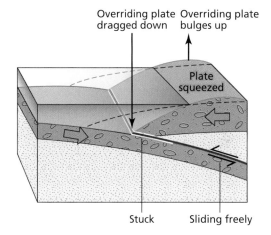

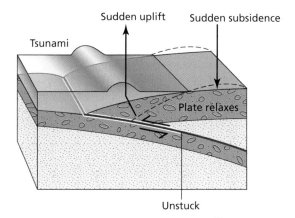

FIGURE 33.9 Explanation for uplift of the land surface above a subduction zone, followed by rapid subsidence and tsunami formation during major earthquakes.

Note in Figure 33.8 the scattered pattern of epicenters in interior Asia, in eastern Africa from Ethiopia to South Africa, and in North America east of the Pacific coast subduction zone. The causes of these **intraplate earthquakes**

FIGURE 33.10 Coastal area in southern Alaska that abruptly subsided 2 m (6 ft) during a subduction-zone earthquake in 1964 to an elevation below sea level at high tide.

are still not well understood, but they can produce severe damage because people in the affected areas are less well prepared for earthquakes than their counterparts in areas with greater activity.

For instance, municipal ordinances governing high-rise construction in California cities take seismic hazards into consideration; in the eastern United States,

local building codes do not. Yet the most severe earthquake of the last 300 years in the United States outside Alaska did not occur in California, but in the central Mississippi Valley.

In 1811 and 1812, three great earthquakes, all with epicenters near New Madrid in the "boot heel" of extreme southeastern Missouri, changed the course of the Mississippi River and created a 7,300-hectare (18,000-acre) lake in a former stream channel. The severity of these quakes has been estimated to have been between 7.0 and 8.0 on the Moment Magnitude Scale. Had nearby St. Louis and Memphis been major cities at that time, the death toll could have been enormous. In 1886, an earthquake of magnitude 7.0 devastated Charleston, South Carolina, and the ground shook as far away as New York City and Chicago. Thus, parts of the Midwestern and eastern United States must also be considered potentially hazardous seismic zones.

Earthquakes and Landscapes

Earthquakes are not comparable to volcanic action as builders of a distinct landscape. But earthquakes do *modify* physical and cultural landscapes, and some landforms actually are created by earthquake movements. When movement along a fault generates an earthquake or when a fault is newly created following the long-term deformation of rock strata, the result may be visible at the surface in the form of a **fault scarp**. As

Great Earthquakes in the Pacific Northwest

The basic concepts of plate tectonics were developed and hotly debated by geologists when memories of the great earthquakes of 1960 and 1964 in Chile and Alaska were still fresh, and the connection between those events and subduction was quickly recognized. The earthquakes of 2004 and 2011 in two other subduction zones near Sumatra and Japan re-emphasized the potential for extremely powerful quakes in that setting. Yet at least one well-studied subduction zone has no historical record of such devastating earthquakes: the area just off the coast of Oregon, Washington, and British Columbia in the Pacific Northwest, where the small Juan de Fuca

Plate plunges beneath the continental North American Plate (see Fig. 30.5). Native Americans in the region did, however, preserve the memory of a terrible earthquake that occurred on a winter night before the arrival of Europeans.

In the 1970s, geologists began an intensive search for evidence of large prehistoric earthquakes in the Pacific Northwest. They studied the effects of great subduction-zone earthquakes in Alaska, Chile, and other regions and soon recognized that rapid subsidence of coastal lands could leave important traces that might be preserved for centuries. Along the coasts of Oregon and

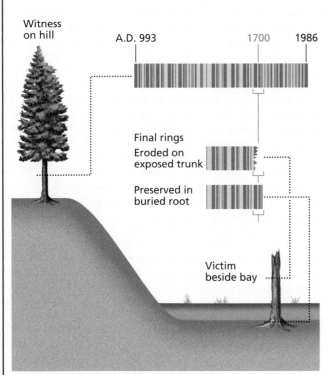

Washington, these researchers painstakingly recorded evidence: trees killed by submersion of their roots in salt water, soils in which plants or other organisms indicate rapid "drowning," and sand and gravel layers similar to those deposited by modern tsunamis. Eventually, it was determined that more than one earthquake must have produced these indicators, with the most recent one occurring around the year 1700 CE. More detailed dating using tree-growth rings indicated that the quake occurred early in 1700 (Fig. 33.11). Finally, historical records were found in Japan of a tsunami from an unknown source that struck on January 27–28, 1700. Based on its timing, the tsunami could have been produced by an earthquake that took place at about 9 p.m., January 26, 1700, off the coast of Washington—that is, a great earthquake that occurred on a winter night, as recalled by Native Americans.

Little doubt now exists that great earthquakes have occurred off the coast of this region in the past and that they could recur at any time in the future. With this knowledge, there is now much greater urgency in preparing for serious earthquakes in densely populated coastal areas of the region, including the cities of Seattle and Vancouver (based on Atwater et al., "The Orphan Tsunami of 1700—Japanese Clues to a Parent Earthquake in North America," U.S. Geological Survey Professional Paper 1707 [2005]).

FIGURE 33.11 Abrupt subsidence of the land during past earthquakes along the coast of Oregon and Washington killed numerous trees, some of which are still present as dead snags. The pattern of wide and narrow tree rings (indicating years of fast and slow growth) can be matched between the killed trees (victim trees) and others still living nearby (witness trees), allowing the year of the earthquake to be determined.

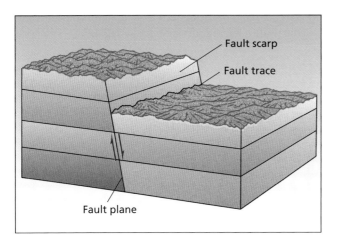

FIGURE 33.12 Fault scarp resulting from the vertical movement of one block with respect to another. The blocks are in contact along the fault plane. The exposed face of the fault plane is the fault scarp. If the upper block were eroded down to the level of the lower block, the only surface evidence of the fault plane would be the fault trace.

Figure 33.12 illustrates, a fault scarp is the exposed cliff-like face of the **fault plane**, the surface of contact along which blocks on either side of a fault move.

Not every earthquake produces a fault scarp, and in some cases, the movement is more horizontal than vertical.

But if one block is raised with respect to another, a fault scarp is produced. Such a scarp may have a vertical extent (or *face*) of less than 1 m (3.3 ft); others may exceed 100 m (330 ft). Repeated vertical movement along the fault, by raising the height of the exposed face in stages, produces higher scarps. As Figure 33.12 shows, the lower edge of the fault scarp is called the **fault trace**. Imagine that erosion wears down the raised block to the left so that the scarp is no longer visible. The existence of the fault is then only revealed at the surface by the trace. Along the trace may lie a band of crushed, jagged rock fragments called *fault breccia*, evidence of the powerful forces that created the fault.

This unit has focused on the sudden, dramatic movements of rocks that produce earthquakes and their accompanying impacts on overlying landscapes. But the surfaces of Earth's landmasses are also affected by the consequences of less spectacular stresses on their underlying rocks. Unit 34 examines those types of stresses and the rock structures they shape.

Key Terms

amplification *page 413*

attenuation *page 413*

earthquake *page 411*

epicenter *page 412*

fault *page 411*

fault plane *page 419*

fault scarp *page 418*

fault trace *page 420*

focus *page 412*

intensity *page 412*

intraplate earthquake *page 417*

magnitude *page 412*

subduction-zone
   earthquake *page 417*

tsunami *page 413*

Scan Here for a quick vocabulary review

Review Questions

1. Explain why an earthquake has only one *magnitude* but its *intensity* can vary from place to place.
2. Explain why earthquakes of very high magnitude do not always cause great loss of life.
3. Where on Earth are earthquakes most frequent? Why?
4. All of the earthquakes listed in Table 33.1 with a magnitude greater than 9.0 occurred in the Circum-Pacific Belt of seismic activity. How can you explain this pattern?
5. Why do coastal areas suddenly subside during great subduction-zone earthquakes?
6. How are fault scarps produced?
7. Why are parts of the eastern United States considered to constitute a hazardous earthquake region?

Scan Here for a quick concept review

www.oup.com/us/mason

Faults, Folds, and Landscapes

The eastern face of the Sierra Nevada rises above Owens Valley, California, in one of the most dramatic landscapes of the western United States. The mountains have risen through movement on a major fault, which parallels the mountain front.

OBJECTIVES

- Define basic terminology used in describing rock structure
- Compare and contrast types of fault movements and the landforms they produce
- Explain the folding of rocks and the landforms produced when folded rocks are eroded
- Describe the occurrence of regional deformation of the crust in addition to more localized faults and folds

The landscapes of the continental landmasses are sculpted from rocks with diverse properties. In previous units we studied some of the properties of rocks that result from their origin: the hard, resistant, crystalline batholiths formed deep in the crust and exposed by erosion (Unit 27); the lava flows (Units 27 and 32); the foliated schists produced by metamorphism (Unit 28); and the layers of sedimentary rock strata (Unit 28).

Rocks also have structure that reflects the way they have reacted to *stress*, and that structure strongly influences the formation of landscapes through long-term erosion. Stress results from force applied to rock, which acts to compress it, stretch it out, or cause masses to slide side by side. In zones where plates converge, rock strata are compressed, and divergent plate movement tends to produce tension that acts to stretch rocks out horizontally. Where sediments accumulate, their growing weight pushes the underlying rocks downward. Where erosion removes rock, the crust has a tendency to "rebound."

In response to the many forms of stress, some rocks are brittle and fracture, especially when they are not far below the ground surface and are relatively cool. Others bend slightly but spring back when the stress is removed. Deeper beneath the surface, under higher temperatures and pressures, rocks can exhibit plastic behavior that allows them to fold and flow under stress. When the stress is removed, these structures remain and influence the shape of the landscape as they are uncovered by erosion.

Rocks are deformed in so many ways that the resulting forms seem endlessly complicated. But in fact, certain geologic structures occur many times over and can be recognized even on topographic maps that do not contain any geologic information. We have said elsewhere that our understanding of the surface must begin with what lies below, and in this unit we study the relationships between geologic structure and visible landscape.

Terminology of Structure

To describe the nature and orientation of structures below the surface accurately, we must use consistent terminology. Imagine a layer of quartzite rising above the surface to form a ridge that extends from northeast to southwest according to your compass (Fig. 34.1). This orientation is the **strike** of that ridge; defined more technically, it is the compass direction of the line of intersection between a rock layer and a horizontal plane. In your field notes, you would report the strike of the ridge as N 45 degrees E (45 degrees east of due north). The strike of any linear feature is always recorded to range from 0 degrees to 90 degrees east or west of north.

Note how the quartzite layer illustrated in Figure 34.1 angles downward between softer strata. This is referred to as the **dip** of that layer, which is the angle at which it tilts from the horizontal. If the quartzite layer tilts 30 degrees from the horizontal, its dip is recorded as 30 degrees. In addition, the direction of the dip must be established. If you were to stand on the ridge and see the quartzite layer

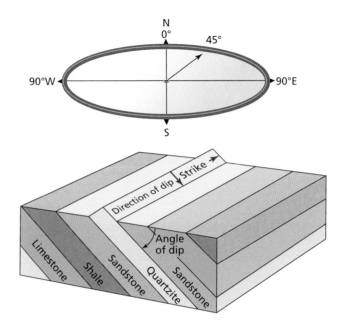

FIGURE 34.1 Strike, direction of dip, and angle of dip.

tilting downward toward the southeast, its dip would be 30 degrees SE. As the diagram indicates, the direction of dip is always at a right angle to the strike.

A prominent feature such as our quartzite ridge is called an *outcrop*, a locality where rock is exposed at the ground surface. Here, the hardness of this metamorphic rock and its resistance to erosion have combined to create a prominent topographic feature whose length and straightness indicate tilting but little or no other deformation. Where rock strata are bent or folded, the determination of strike and dip can become more complicated, and the field map may record many variations in strike and dip along segments of the outcrop.

Faults

Recall from Unit 33 that a **fault** involves the displacement of rock on one side of a fracture with respect to the rock on the other side. In that unit, our focus was on earthquakes, which result from sudden slippage along a fault plane. In this unit we consider how faulting can play a major role in shaping the landscape of many regions. Faulting results when brittle rocks come under stress, cannot bend or fold, and therefore break. Rocks can break under stress without displacement of the rock on either side. In this case, the fracture that results is called a *joint* rather than a fault. Slippage of rock along faults is the key to distinguishing them from joints, and the direction of displacement is used to identify various types of faults, as we discuss later in this unit.

The displacement of rock along a fault during a single earthquake, even a severe one, is usually smaller than about 3 m (10 ft); however, repeated fault movements over a long period of time can displace the rock on either side by hundreds of meters. At the surface, a fault may be barely visible, or it may be marked by a tall fault scarp, a "step" in the landscape resulting from uplift of rock on one side of the fault. Often, when a sequence of sedimentary layers is faulted, it is possible to locate the same bed on opposite sides of the fault plane, and the amount of vertical displacement can be determined. However, if the blocks on either side of the fault consist of the same rock (say, a mass of granite), it may not be easy to measure displacement.

Some areas, such as the zones near convergent boundaries between tectonic plates, are more strongly affected by faulting than other locales. We have already encountered faults in our study of convergent plate boundaries and in connection with earthquakes. In this unit, we identify several types of faults and see how they are expressed at the surface of the crust.

An understanding of plate tectonics is very helpful in comprehending why various types of faults occur where they do. Where plates converge and collide, *compressional* stresses are strong, and rocks are crushed tightly together. The lithosphere is forced to occupy less horizontal space, and rocks respond by breaking, bending, folding, sliding, and squeezing upward and downward. Where plates diverge, and in other settings where the crust is subjected to spreading processes, the stress is *tensional*, and rocks are pulled apart. Especially at a shallow depth below the surface, rocks can break under tension, and faults result. And where plates slide past each other or where forces below the crust are lateral, the stress is *transverse*. This type of stress also produces faulting and associated earthquakes. Although active faulting is most common near the boundaries of tectonic plates, numerous faults can be found even in the "stable" shields of the continental cores. These may be legacies of earlier periods of stress or the products of stresses transmitted long distances from plate boundaries. Stress that is unrelated to plate motion can also cause faulting. Let us now examine the basic types of faults.

COMPRESSIONAL FAULTS

Where crustal rock is compressed into a smaller horizontal space, shortening of the crust is achieved by the rock on one side of the fault moving both upward and over the rock on the other side (Fig. 34.2). Such a structure is a **reverse fault**. Note that there is some horizontal movement of the two sides toward each other because the fault is tilted rather than perfectly vertical. Whenever a fault is inclined like this, the side of the fault that overhangs the other side is called the **hanging wall**. The other side of the fault is the **footwall**. When vertical movement occurs during faulting, blocks that move upward (with respect to adjacent blocks) are referred to as *upthrown*, whereas blocks that move downward are termed *downthrown*.

In the case of reverse faults, the upthrown block is on the hanging wall. Therefore, each time the upthrown block moves upward along the fault, it creates an initial scarp that overhangs the downthrown block (Fig. 34.2A). Such an overhanging scarp soon collapses under the effects of gravity, weathering, and erosion, and produces a slope at

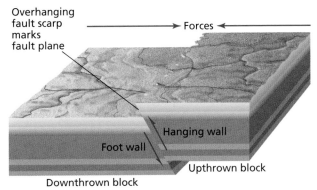

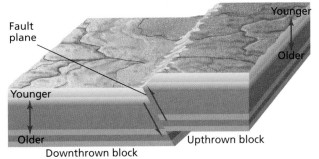

FIGURE 34.2 In a reverse fault, the upthrown block is on the hanging wall, so fault movement produces an initially overhanging scarp (A). This collapses almost as soon as it forms, however, and erosional forces produce a new slope, which lies at an angle to the original (B).

an angle to the original fault plane (Fig. 34.2B). As a result, we usually cannot identify a reverse fault in the field by looking for this kind of overhanging scarp. On the other hand, identification is often easy if we can find outcrops that reveal a cross-section of the rocks on both sides of the fault plane. Along a reverse fault, older rock on the hanging wall will have slipped *up* the fault plane so that it is in contact with *younger* rock on the footwall (see Fig. 34.2B).

When the angle of a fault plane in a compressional fault is very low, the structure is referred to as a **thrust fault** (Fig. 34.3). Note that the overriding block slides almost horizontally over the downthrown block, covering much more of it than is the case in a reverse fault. It is not unusual to find the same rock strata repeated above and below large, nearly horizontal thrust faults; in effect, the thickness of these rocks has been doubled. Huge thrust faults are often found on one or both sides of great mountain ranges produced by compression of the crust. The highest peaks on Earth, in the Himalayas, are formed from rock that pushed southward over other rocks along giant thrust faults (Fig. 34.4). In the process, the crust was thickened, producing the deep root needed to support such high mountains according to the principles of isostasy (Unit 31).

A

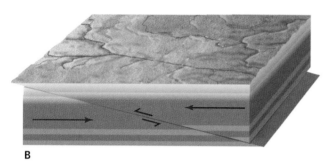

B

FIGURE 34.3 Development of a thrust fault. Rock strata before faulting (A) have less total thickness than they do after fault movement (B).

Many of the central Rocky Mountain ranges, including the Front Range just west of Denver, Colorado, are large blocks that were uplifted by movement on thrust faults located on one or both sides of the range. More generally, an uplifted block of rock bounded by faults can be called a *horst*.

TENSIONAL FAULTS

Tensional stresses pull crustal rock apart so that there is more horizontal space for crustal material—the opposite of the horizontal shortening produced by compression. The result is one or more **normal faults** (Fig. 34.5). The typical normal fault has a steep but somewhat inclined fault plane, like many reverse faults. However, the downthrown block forms the hanging wall of a normal fault. That is, the hanging wall slips downward relative to the footwall. As this happens,

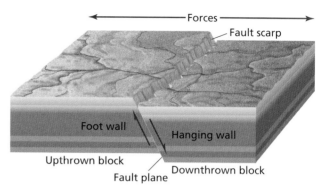

FIGURE 34.5 Unlike compressional stresses, tensional stresses pull the crust apart, resulting in normal faults. The hanging wall slips downward relative to the footwall.

the rocks on either side also move laterally away from each other, extending the crust horizontally. A cross-section of the rocks on either side of the fault is crucial to identifying a normal fault. Younger rocks on the hanging wall will have moved downward so that they are in contact with older rocks on the footwall. As with a reverse fault, the scarp of a normal fault is modified by erosion so that its slope does not exactly correspond to the dip angle of the fault plane. Sometimes the fault scarp has been eroded for so long that it has retreated from its original position. In such instances, the fault trace (see Fig. 33.12) reveals the original location of the scarp.

Normal Faults and Landscapes   Large normal faults are responsible for some of the most spectacular scenery of the western United States. The eastern front of the Sierra Nevada in California towers above the desert valleys to the east, forming a classic scene that has appeared as a backdrop in hundreds of Westerns. This mountain front formed through long-term movement of a normal fault, with the Sierras on the upthrown block. The western front of the

FIGURE 34.4 Block diagram of major rock structures under the Himalayas and Tibetan Plateau, showing the large thrust faults that have formed at this convergent plate boundary. The two largest thrust faults are known as the Main Boundary Thrust and the Main Central Thrust, both of which involve huge masses of rock thrust southward for tens of kilometers.

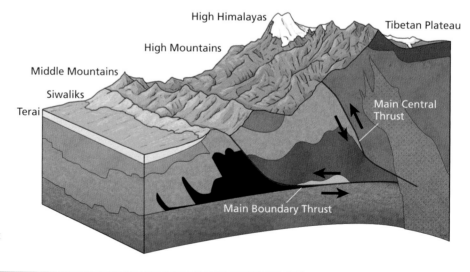

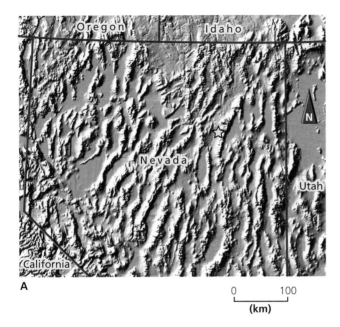

A

0   100
(km)

B

FIGURE 34.6 The Basin and Range region of the western United States. A shaded relief map of northern Nevada reveals the characteristic topography: alternating basins and ranges resulting from tension on the crust and normal fault development (A). The Ruby Mountains in Nevada, marked on the map with a star, are the upthrown block on the footwall of a normal fault. The fault trace follows the foot of the steep front of the range, rising above the Ruby Marshes on the basin floor (B).

ENHANCE

Wasatch Mountains behind Salt Lake City, Utah, formed in the same manner.

Between the Sierra Nevada and the Wasatch Mountains is the Basin and Range region (Fig. 34.6). The name comes from its distinctive topography: dozens of narrow mountain ranges separated by desert basins. In almost every case, the ranges are the upthrown blocks of normal faults; some are active and others are no longer prone to movement. Therefore, the Basin and Range region records

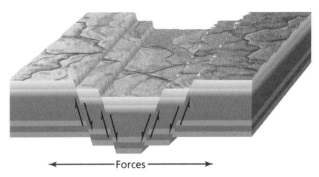

—— Forces ——

FIGURE 34.7 Formation of a rift valley (a graben) as tensional forces generate parallel normal faults between which crustal blocks slide downward.

tension acting on the crust over a large area, thinning the crust and breaking it into numerous ranges and basins. The origin of this tension is not certain, but it is similar in some respects to the rifting apart of continents that ultimately leads to the opening of a new ocean basin (see Fig. 30.4).

Rift Valleys   In other cases, tension on the crust is focused in a somewhat narrower band and creates conditions favorable to the development of rift valley topography (Fig. 34.7). The midoceanic ridge system, as noted in Unit 30, is essentially a vast rift system in which tensional forces are pulling the crust apart. On the continental landmasses, the best example of rift valley topography lies in eastern Africa (Fig. 34.8). We do not need a geologic map to see this magnificent example of crustal rifting because East Africa's Great Lakes fill the rift valley over much of its length. The system actually extends northward through the Red Sea and into the Gulf of Aqaba and the Jordan River Valley beyond. In eastern Africa, the rift system crosses Ethiopia and reaches Lake Turkana, the northernmost large lake in the rifts. South of the latitude of Lake Turkana, the system splits into two gigantic arcs, one to the east of Lake Victoria and the other to the west. These arcs come together just north of Lake Malawi. The system continues southward through the valley in which that lake lies, and it does not end until it has crossed Swaziland. As long ago as 1921, British geomorphologist J. W. Gregory (1864–1942) concluded that the floors of the rift valleys, in places thousands of meters below the adjacent uplands, are **grabens** (sunken blocks; see Fig. 34.7) between usually parallel normal faults (Fig. 34.9).

Many geologists now believe the East African rifts are located where there is intense upward flow of hot rock from deep in the mantle—like the mantle plumes that underlie hot spots such as Hawai'i (Unit 32), but distributed over the length of the rifts. This upward flow is what causes tension in the crust so that it breaks along normal faults. If this view is correct, we would expect that widespread volcanism would occur along the East Africa rift valleys, as it does in hot spots. This is indeed the case: composite cones, lava domes, cinder cones (see Fig. 32.8), and fissure eruptions all mark the region. As recently as January 2002, a major eruption sent lava flows into the heart of the Congolese city of Goma north of Lake Tanganyika. The East African rifts most likely represent the early stages of development of a divergent plate boundary. In fact, the rift zone there is already considered a plate boundary by many researchers (see Unit 31).

TRANSVERSE FAULTS

Where blocks of crustal rock move laterally, motion along the fault plane is horizontal, not vertical. Thus, there are no up-thrown or downthrown blocks, and movement is parallel to the fault. This is a **strike-slip fault**, so called because movement occurs along the strike of the fault plane (Fig. 34.10).

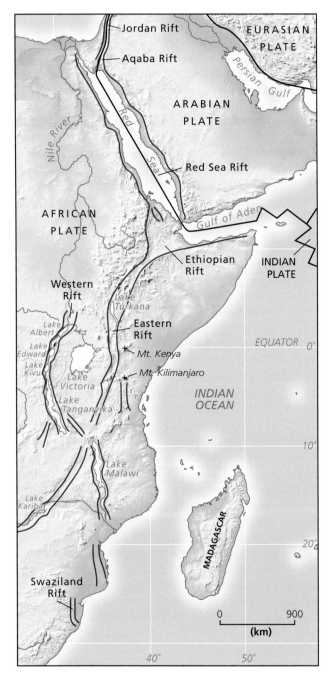

FIGURE 34.8 African rift valley system, which extends from beyond Africa in the north (the Jordan–Aqaba segment) to Swaziland in the south.

FIGURE 34.9 Aerial view of the East African rift valley in southern Ethiopia. The tensional stresses pulling the crust apart here are evident in the fault scarps and the down-thrown crustal strips at their base. This scene provides a striking example of the landscape mapped in Fig. 34.8.

When we discussed transform plate boundaries in Units 30 and 33, we identified the San Andreas Fault as the contact plane between the Pacific and North American Plates. The San Andreas Fault is a special case of a strike-slip fault that marks a plate boundary. To describe a strike-slip fault completely, the direction of movement is also given. This is determined by looking *across* the fault and stating the direction in which the opposite block is moving. For example, consider the San Andreas Fault again. If you were to stand on the North American Plate side and look across to the west, you would see the Pacific Plate sliding past to the right (see Fig. 30.11). If instead you were to look across the fault from the Pacific Plate, North America would appear to be moving toward the right. Accordingly, the San Andreas Fault is a *right-lateral* strike-slip fault. If you were to look across a *left-lateral* strike-slip fault, the far side from your viewpoint would appear to be moving to the left.

Fold Structures

When rocks are compressed, they respond to the stress by **folding** as well as by faulting. Even the hardest rocks have some capacity to bend before fracturing. Folding commonly occurs at depths where the combination of temperature and

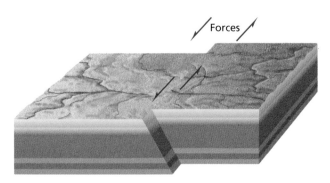

FIGURE 34.10 A strike-slip fault.

FIGURE 34.11 "The Mescal Mountains in south-central Arizona display some complex structures. The sedimentary strata shown here were laid down horizontally, but were later folded into the anticline of which both limbs are visible in the lower half of the photo and one in the upper part."

pressure favors plastic behavior of the rock, making it more likely that it will fold than break in faults. Fold structures are most easily recognized in layered sedimentary rocks. The folded layers themselves may be visible in outcrops, but more often we see a pattern of ridges and valleys marking more or less erosion-resistant layers within the folded mass. Folds, like faults, come in all dimensions. Some are too small to see, others are road-cut size, and still others are the size of entire mountain ridges (Fig. 34.11).

Folds are primarily compressional features. Not surprisingly, folding and thrust faults produced by compression often occur together. The accordion-like folds in the Andes Mountains, the Appalachians, and parts of Eurasia's Alpine Mountain system (e.g., the Zagros Mountains of Iran) are the result of intense compression where lithospheric plates converge.

ANTICLINES AND SYNCLINES

Folds are rarely simple symmetrical structures. Often, they form a jumble of upfolds and downfolds that make it difficult to discern even a general design in the field. But when we map the distribution of the rock layers and analyze the topography, we discover that recognizable and recurrent structures

do exist. The most obvious of these structures are upfolds, or **anticlines**, and downfolds, or **synclines**. As illustrated in Figure 34.12, an anticline is an arch-like fold, with the limbs dipping away on either side of the axis, or center line, of the fold. A syncline, on the other hand, is trough-like, and its limbs dip toward its central axis (see Fig. 34.12).

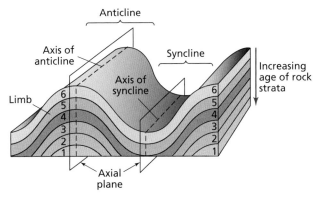

FIGURE 34.12 Anticlines are arching upfolds; synclines are trough-like downfolds. Note the relative age of rock layers within each type of fold indicated by the numbers (layer 6 is the youngest; layer 1 is the oldest).

When originally flat-lying sedimentary strata are folded, the cores of the synclines are made up of rocks that are younger than those at the same elevation in adjacent anticlines. For example, at the axis of the syncline in Figure 34.12, the elevation of the youngest rock layer (6) is about the same as that of older strata (2 and 3) in the anticlines on either side of the syncline. Now imagine that erosion removes the upper parts of both anticlines and synclines (Fig. 34.13A). If we walk across this landscape and produce a geologic map showing where various strata are at the surface, we can easily identify anticlines and synclines (Fig. 34.13B). The synclines will be cored by younger strata, and the eroded anticlines will have cores of older rock. Measurements of strike and dip, also recorded on the map in Figure 34.13B, confirm these interpretations: strata dip toward the center of synclines and away from the center of anticlines.

PLUNGING FOLDS AND ASSOCIATED LANDSCAPES

Folds are rarely as symmetrical as those shown in Figure 34.12, however, and their axes are not usually horizontal. Anticlines and synclines often *plunge*, which means that

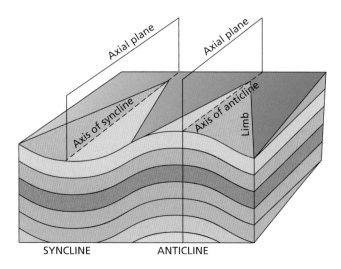

FIGURE 34.14 Anticline–syncline structure plunging in opposite directions.

their axes dip. You can demonstrate the effect of this by taking a cardboard tube and cutting it in half lengthwise. Hold it horizontally and you have a symmetrical anticline (and a syncline, if you place the other half adjacent to it). Tilt the tube about, say, 30 degrees from the horizontal by the edge of a table or other flat surface. Draw the horizontal line and cut the tube along it. The second cut represents the outcrop of the youngest rocks of the anticline. Note how they form a "nose" at one end and open progressively toward the other end. In the field, such an outcrop signals an eroded, plunging anticline; the adjacent syncline plunges in the other direction (Fig. 34.14).

In the intensely folded central Appalachian Mountains, plunging anticlines and synclines adjoin each other

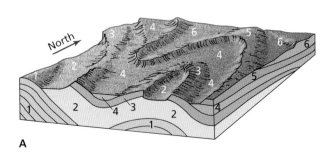

A

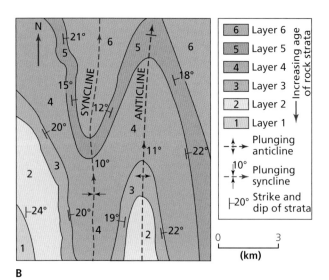

B

FIGURE 34.13 Eroded anticline–syncline landscape of surface outcrops with parallel strikes (A). The age sequence of these rock formations can be interpreted by referring to the geologic map (B).

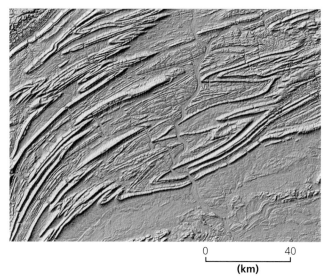

FIGURE 34.15 A shaded relief map reveals the complex pattern of ridges and valleys produced by erosion of folded rock in the Appalachian Mountains of central Pennsylvania. The ridges represent the more resistant rock strata cropping out on the limbs of plunging anticlines and synclines.

in wide belts eroded into attractive scenery by streams over many millions of years. But as the Appalachians' topography suggests, there is considerable regularity to the pattern (Fig. 34.15). The folded rocks of Europe's Alps are much more severely distorted. Many of the anticlines have become *recumbent* (lying on their side), and some are completely overturned. Add to this the presence of intense faulting and erosion by streams and glaciers, and complex, spectacular alpine landscapes are produced.

Regional Deformation

Not all types of crustal deformation are encompassed by the examples of faulting and folding discussed in this unit. A less easily detected but important type of deformation involves *upwarping* or *downwarping* of the crust over large areas. The resulting structures can be thought of as very large anticlines or synclines, but with such gentle sloping sides that they are not apparent at a local scale.

Key Terms

anticline *page 427*
dip *page 422*
fault *page 422*
folding *page 426*
footwall *page 423*

graben *page 425*
hanging wall *page 423*
normal fault *page 424*
reverse fault *page 423*
strike *page 422*

strike-slip fault *page 426*
syncline *page 427*
thrust fault *page 423*

Scan Here for a quick vocabulary review

Review Questions

1. What type of evidence could help you to distinguish between a normal and a reverse fault when looking at an isolated rock outcrop?

2. Explain how the types of faults that are common in the Basin and Range and East African rift regions allow us to determine the predominant type of stress acting on the crust in those areas.

3. What type of rock structures help to identify areas where the crust has been compressed by the convergence of lithospheric plates?

4. In Figure 34.13, which direction are the anticlines and synclines plunging?

5. Differentiate between the concepts of strike and dip in describing a ridge on Earth's surface.

6. What kind of plate boundary may be present where there is a large strike-slip fault?

Scan Here for a quick concept review

www.oup.com/us/mason

PART SIX

Sculpting the Surface: Weathering, Mass Movements, and Flowing Water

Sculpting the Surface: Weathering, Mass Movements, and Flowing Water

The natural landscapes we observe and study are the products of *weathering*, *erosion*, and *deposition*, which combine to create the infinite variety that marks the terrestrial surface of our planet. Weathering is the breakdown and alteration of rocks and minerals by mechanical, chemical, and biological processes. The particles produced by weathering—from boulders to microscopic flakes of clay—are eroded by mass movements, water, wind, and glacier ice; carried some distance; and then deposited. Weathering and stream erosion, joined by mass movements on slopes, are the predominant processes of land sculpture across large areas of the continents. Those processes interact with the mountain-building and volcanic activity that result from movement of the lithospheric plates. Tectonic processes rapidly uplift mountain ranges, dramatically changing the landscape over time. Erosion by mass movements and flowing water is especially rapid on the steep, high slopes of actively rising mountains, however, and the rate of erosion can eventually increase enough to keep pace with uplift. Ultimately, mountain ranges are worn down by erosion when active uplift ends, although this process may extend over long periods of geologic time.

UNIT 35

The Formation of Landforms and Landscapes

These strange rock forms in Goblin Valley, Utah were sculpted by weathering and erosion, like other landforms ranging from massive mountain ranges to small hillside gullies. Weathering breaks down the rock strata and erosion by water, and wind removes the weathered debris.

OBJECTIVES

- Describe the major processes that shape landforms and landscapes
- Explain the importance of gravity as a driving force for erosion and how gravity-driven erosion is influenced by the slope and relief of the land surface
- Summarize the interactions among tectonic activity, weathering, and erosion

What we see in Earth's landscapes is the temporary result of many processes—*temporary* because all landscapes change over time, even if the change is often far too slow to be perceivable within a human life span. In Parts Six and Seven of this book, we focus on processes that shape landforms and landscapes by breaking down rock or by eroding, transporting, and depositing the products of rock disintegration. In this unit, we start with an overview of these varied land-shaping processes and consider the broad similarities among them and how these processes interact with tectonic and igneous activity. The rest of Part Six deals with three specific groups of processes, all of which involve liquid water at or below Earth's surface. Part Seven takes up the work of ice, wind, waves, and currents in shaping Earth's landscapes.

Processes That Shape Landforms and Landscapes

A **landform** is a feature of Earth's topography that can be distinguished and studied as a single unit. A solitary mountain (e.g., a composite volcano), a hill, a single valley, a dune, and a sinkhole all are landforms. In everyday language, we might use the word *landscape* for the area of land we can see from a high point. In physical geography, we define a **landscape** as an aggregation of landforms. As we look out over a landscape, we focus on the individual landforms it is made up of and how they are arranged (Fig. 35.1). A volcanic landscape, for example, may consist of a region of composite volcanoes, lava flows, and other features resulting from

volcanic activity, along with scattered stream channels and other landforms that result from other processes. Dunes may be part of a desert or coastal landscape.

The processes that form the landscapes of the continents range from the quiet disintegration of rocks to the roaring cascade of boulder-filled mountain streams. The most basic distinction among these processes is between weathering and all of the other processes, which involve erosion and movement of sediment.

WEATHERING

Weathering is the breakdown of rocks in place; other agents such as running water or wind are needed to carry the products of weathering away from the location where it occurs (Fig. 35.2). The various processes of **erosion** begin with the detachment of pieces of solid rock, or the setting in motion of loose material on the land surface. However, all erosion also involves *transportation* of the eroded material some distance, whether it is down a hillside or thousands of kilometers.

Weathering often sets the stage for erosion by bringing about the disintegration of rock that is easily eroded by water or wind. Weathering attacks both in-place bedrock and rock fragments in soils and sediments through chemical action or a great variety of mechanical stresses. As we will discuss in Unit 36, rocks can be weathered by exposure to temperature extremes, by frost action, and even by the growing roots of plants. A special case of weathering, which produces the unique landscapes referred to as *karst*, is discussed in detail in Unit 42.

FIGURE 35.1 A landscape in the Rocky Mountains of Colorado, shaped by glaciers, mass movements, and flowing water. The mountains in the background are landforms, as is the ridge in the foreground, which is a *moraine* (a ridge of debris deposited at the edge of a glacier that existed here in the past).

FIGURE 35.2 Where water stands on this bare granite surface after rain, weathering breaks the rock down into gravelly debris and dissolved material. The weathering products remain here unless removed by erosion, in this case by water washing the debris off the rock through the small channel in the foreground.

MASS MOVEMENT

The spontaneous downslope movement of Earth materials under the force of gravity is called **mass movement**. Anyone who has witnessed or seen a picture of a landslide or debris flow knows what such slope failure involves. Houses, roads, and even entire hillsides tumble downslope, often after the weathered surface material has become saturated with water (Fig. 35.3). In Unit 32, we discussed the great mudflow (or *lahar*) that occurred on Colombia's volcano, Nevado del Ruiz, when it erupted in 1985. Saturated by abruptly melted snow from the volcano's peak, this mass movement traveled more than 50 km (30 mi) down valleys from the base of the mountain and buried at least 20,000 people living in its path (see Fig. 32.6). As examples like this show, mass movements can involve large volumes of material, sometimes with disastrous consequences for people. But other forms of mass movement are slow and almost imperceptible. We discuss mass movement in more detail in Unit 37.

RUNNING WATER

We devote most of our attention in this part of the text to the fascinating processes of erosion and transportation by flowing water, wind, glacial ice, and shoreline waves. Of these, running water is by far the most effective and significant agent of erosion worldwide, as you will learn in Units 38 to 41.

Through the action of stream flow, the particles resulting from weathering, including some brought downhill by mass movement, are carried away over long distances. During this process of transportation, additional breakdown occurs that affects both the transported particles *and* the rocks of the valleys through which they travel. For example, a boulder first loosened in a mass movement is removed by a stream and rolls or bounces along the streambed as the water transports it. It breaks into smaller pieces, which become rounded into pebbles, and the parts that are broken off during this process form smaller particles. In this way, streams not only transport the products of weathering but also continue the process of breaking them down.

In fact, when we consider erosion at the continental scale, it is hard to overemphasize the importance of great rivers that collect eroded sediment from large areas and convey it to lowland basins or the ocean. Rivers have played this key role in the rock cycle since early in Earth's history. Many of them have also eroded enormous quantities of rock as they carved their own valleys into the landscape (Fig. 35.4). Few places on Earth are unaffected by the erosional power of water. Even in deserts, water plays a major role in shaping the landscape.

When we study maps of streams, it is clear that many river systems consist of a major artery that is joined by *tributaries*, smaller streams that feed into the main river (Fig. 35.5). The spatial pattern of these systems can reveal much about the rock structures below. Certain patterns are associated with particular structures and landscapes. Often, the configuration of a regional stream system can yield insights about the underlying geology and geomorphic history of an area, as we will explain in Unit 40.

GLACIERS

Streams are not the only agents of erosion. Ice in the form of glaciers is also important in certain landscapes today and has affected much larger areas in the past. As recently as

FIGURE 35.4 The Chang Jiang (Yangtze River) in China originates deep in the Asian interior and flows across the heart of the country to the Pacific Ocean. In the process, it has sculpted numerous dramatic landforms, but perhaps none as scenic as the famous Three Gorges. The Chinese government has since built a massive dam and flood control system that partially drowned the gorges seen here, forever changing the local (and regional) hydrography and altering human–environment relationships that have prevailed here for millennia.

ENHANCE

FIGURE 35.3 A landslide closed the American River Bike Trail near Sacramento, California, in 1985. Heavy rains and floods waterlogged the soil, resulting in the collapse of slopes throughout the area.

FIGURE 35.5 Stream system in its drainage basin. The main river is built up by the smaller tributary streams that empty into it.

15,000 to 20,000 years ago, great glaciers covered much of northern North America and Europe. That was only the most recent of many glacial periods of the past 2 to 3 million years. During Earth's glacial periods, sheets of ice modified the landscape by scouring the surface in some areas and depositing the eroded rock elsewhere. At the same time, valleys in high mountain areas (e.g., the Alps and the Rockies) were filled by glaciers that snaked slowly downslope. These mountain glaciers widened and deepened their containing valleys, carrying away huge loads of rock in the process. When the global climate warmed up near the end of each glacial period and the glaciers melted back, the landscape that emerged from beneath the melting ice had taken on an unmistakable and distinctive character. We examine the evidence of these events in Units 43 to 46.

WIND

Wind, too, is an erosional agent. Globally, compared to running water and moving ice, wind is a minor factor in landscape genesis. In certain areas of the world, however, wind *is* an effective modifier of the surface. Wind can propel sand at high velocities and wear down exposed rock surfaces in its path. Wind is also capable of moving large volumes of sand from one place to another, creating characteristic dune landscapes (Fig. 35.6). Even larger volumes of *loess*, or windblown dust deposits, have accumulated in certain areas like northern China and the central Great Plains of the United States. The processes and landscapes of wind action are treated in Unit 47.

COASTAL WAVES AND CURRENTS

Coastal landscapes are of special interest in physical geography. Where land and sea make contact, waves erode and rocks resist, and the resulting scenery often is spectacular as well as scientifically intriguing. As in the case of wind action, it is not the waves alone that erode so effectively. Waves can move pieces of loosened rock that grind against rock outcrops along the shoreline. The erosional results are cliffs and other high-relief landforms (Fig. 35.7). Coastal landscapes and landforms are the focus of Units 48 and 49.

SEDIMENT DEPOSITION AND LANDFORMS

Streams, glaciers, wind, and waves are erosional agents, but they also have the capacity to deposit sediment in distinctive spatial patterns. The **deposition** of sediment occurs whenever solid material that has been transported some distance is dropped out of flowing water or wind, melts out of a glacier, or simply comes to rest at the foot of a slope. Deposition, like erosion, creates characteristic landforms and landscapes. The processes of erosion and deposition are closely linked. The key is transportation: the removal of weathered rock, its long-distance conveyance, and often its reduction in size by crushing and grinding in the process. For example, materials removed from interior highlands by rivers may be laid down as sand or mud in a coastal delta. Those Pleistocene glaciers that scoured the surface of much of Canada deposited their ground-up load of sediments to the south, much of it in the U.S. Midwest. Mountain glaciers excavated the upper valleys, but they filled their lower valleys with deposits. Wind action wears rocks down in some places but builds dunes or loess deposits in others. Waves that cut cliffs also contribute to the movement of sand along shorelines to create beaches in other locales. All of this is part of the rock cycle (see Fig. 28.13) sustained by the hydrologic cycle (see Fig. 11.3).

FIGURE 35.6 In desert environments, the wind often forms dunes from loose sand released by rock weathering or deposited by streams. These massive dunes form part of the coastal Namib Desert between Namibia's mountains (seen in the background) and the South Atlantic Ocean behind the photographer.

ENHANCE

Gravity, Slopes, and Relief

The most important force driving erosion and the transportation of eroded rock material to a new site of deposition is Earth's *gravity*. When soil and loose rock on a hillside become saturated with water, gravity is able to overcome the friction holding them in place and pull them downslope, where stream water (also responding to gravity) removes them (see Fig. 35.3). When a mineral grain on a rock face is loosened, gravity causes it to fall to the ground below, and the wind may carry it away. Streams flow and glaciers move down their valleys under the force of gravity. That force is resisted by friction within the flowing water or ice as well as where water or ice contact the bed of the stream or glacier. Steeper slopes increase the gravitational force, allowing faster flow in some cases, but more importantly, increasing the ability of the water or ice to erode rock or sediment it flows over. Gravity-driven movement is less central to the work of wind and waves but still plays a part. For example, wave action on beaches includes not only the upward rush of the breaking wave but also the gravity-driven backwash of water down the beach slope.

Because erosional processes driven by gravity generally work faster on steeper slopes, rugged landscapes with many steep slopes tend to have higher erosion rates than those with only gentle slopes—as long as other factors such as climate or bedrock type do not differ too much. The frequency of steep slopes in a landscape is often closely related to **relief**, the local range in elevation. Areas with high relief generally have many high, steep slopes and high rates of erosion.

FROM THE FIELDNOTES

FIGURE 35.7 "Oregon's coastal Route 101 provides some magnificent scenery and many superb field examples of coastal landforms. Drive it southward, so you are on the ocean side! Ample rainfall in this **Csb** environment sustains luxuriant vegetation. Below, the waves do their work even as tectonic forces modify the geology. Here, two natural bridges have formed following wave penetration of a fault-weakened section of the coastline, the waters rushing in through one and out through the other."

Over geologic time, however, erosional processes wear landscapes down toward sea level. The eventual result should be a reduction in relief—a lowering of the mountains, widening of the valleys, and so forth. Tectonic uplift near plate margins and isostatic rebound of the crust as erosion proceeds offsets this tendency (we discuss the interaction between these processes in more detail in the next section). But as long as erosion exceeds uplift, relief will eventually be reduced, and erosion rates will decrease accordingly. (Measuring those rates is increasingly feasible, as discussed in Focus on the Science: Measuring Erosion.) The ancient continental shields (Unit 29) are in an advanced stage of this process, and rates of erosion by mass movements and streams in such landscapes are much lower than in high-relief mountain ranges. Where shields occur in warm, humid climates, such as in parts of Africa and South America, rock is weathered to great depths, and the weathering products accumulate because of slow erosion. The Canadian Shield of North America does not have a climate that favors weathering, but it has been extensively eroded by ice sheets during past glacial periods.

Erosion and Tectonics

We often speak of erosional processes *sculpting* Earth's surface. It is worth considering that an actual sculptor carving a statue must start with a particular rough block of rock, and the nature of that block will influence what can be carved out of it. This is even truer for the sculpting of landscapes. As the units in Part Five detail, the continents are affected by the horizontal and vertical movements of the lithosphere. These movements result in the folding, faulting, and tilting of the crust; mountain uplift; subsidence of basins; igneous intrusions; and volcanic eruptions. The rock structures are the rough blocks that are carved by water, wind, glaciers, and waves, and these structures strongly influence the ultimate product.

FOCUS ON THE SCIENCE

Measuring Erosion

We often speak of rapid or slow erosion, but can we actually measure the rates of erosion in various settings and compare them? This is a more daunting task than it may seem. Sometimes it is possible to trap eroded sediment washing off a hillside or blowing off the soil surface in the wind for a few minutes or hours and calculate the rate directly. However, setting up equipment to make these measurements is expensive and time consuming, and erosion is often too slow or intermittent to study in that fashion. The sediment load carried by a river tells us how much eroded material is leaving the drainage basin, but we rarely know exactly where in the basin it is coming from. A simple but often effective way to measure erosion at a single point is to insert a metal pin in rock or soil and then come back periodically over time to see if the surface has been lowered by erosion relative to the pin. Still, this method only works if erosion is fast enough to produce measurable surface lowering over a few years.

Within the past two decades, geomorphologists have discovered a powerful new technique to measure erosion, even when it occurs at very slow rates.

Though not perceptible to us, cosmic rays continually bombard Earth's surface. Within certain common minerals, such as quartz, this bombardment turns existing atoms into *cosmogenic radionuclides*—special forms of elements such as beryllium (Be), aluminum (Al), or chlorine (Cl). Over long periods of time, the cosmogenic radionuclides build up near the surface of a rock, but some of these atoms are continually removed by erosion. Where erosion is slow, the buildup is great, but where a surface is rapidly eroding, only low levels of these radionuclides accumulate.

Using a rock hammer or drill, a sample can be collected from a rock surface and analyzed to determine the content of a particular cosmogenic radionuclide. Beryllium-10 is a common choice. Measurement of the content allows calculation of a long-term erosion rate, though, as you might expect, there are many complicating factors (including snow cover, which shields the rock from cosmic rays). Variations on this method can also be applied to soil-covered landscapes. As this new technique is applied around the world, we are learning a great deal about how fast landforms and landscapes change over time (Fig. 35.8).

continued

continued

FIGURE 35.8 A gently sloping mountain summit in the Wind River Range, Wyoming. Using cosmogenic radionu-clides, geomorphologists Eric Small and Robert Anderson estimated that summits like this are lowered by erosion at rates of 10 m (3 ft) per million years. The erosion rates of valleys below this summit are probably quite a bit higher because they are periodically eroded by glaciers (Small and Anderson, "Pleistocene Relief Production in Laramide Mountain Ranges, Western United States," *Geology* 26 [1998]: 123–126).

This analogy is not perfect, however, because erosion does not wait until rock structures or volcanoes are fully developed before it starts to work. Mountains are eroded as they are uplifted by tectonic forces. This kind of interaction is even more dramatic in the case of large composite volcanoes (Unit 32). As these mountains are built up by repeated volcanic eruptions, they are also eroded by glaciers and mass movements. The addition of new rock mass through eruptions keeps up with or outpaces erosion until the volcano is no longer active. Then erosion wins out and quickly reduces the size of the volcanic peak. Mt. Rainier, an active volcano in Washington State, is an interesting case study. Now one of the highest peaks in the continental United States, it is quite a young mountain in geologic terms, but it will also be short-lived. When volcanic activity ceases to build up this peak, its layers of volcanic ash and lava will be rapidly eroded from its steep slopes, probably reducing it to a small remnant of its former height within a few hundred thousand years.

Therefore, we cannot identify many landforms produced entirely by tectonic or volcanic processes; almost all are modified to some extent by erosion, even in early stages of their development. At the same time, erosion also actively shapes landforms and landscapes long after active tectonic uplift or volcanic eruptions have ended. The dramatic folding and thrust faults that we see in the Appalachian Mountains today were produced by convergence between lithospheric plates that ended more than 250 million years ago. Yet the Appalachians are still being eroded by streams and mass movements.

Although there is no active tectonic uplift in old mountain ranges like the Appalachians, the crust still slowly rises as rock is removed by erosion (see the discussion of isostasy in Unit 31). This partially offsets the effects of erosion and has allowed the Appalachians to persist as a mountain range much longer than they otherwise would have. Because this isostatic uplift maintains relatively steep slopes, it also allows relatively high rates of erosion to persist in parts of the Appalachian landscape. We focus on erosion and deposition in the following units, but it is important to keep in mind that past or present tectonic activity and crustal movements always exert some influence on landscapes as well.

Key Terms

deposition *page 435*

erosion *page 433*

landform *page 433*

landscape *page 433*

mass movement *page 434*

relief *page 436*

weathering *page 433*

Scan Here for a quick vocabulary review

Review Questions

1. Describe a landscape you are familiar with. What are the predominant landforms, at least as you can identify them at this point? Can you determine what major processes may have shaped them? Revisit this question after reading more of Parts Six and Seven.

2. Explain why weathering, by itself, cannot reduce mountain ranges to low-relief plains.

3. Based on the brief descriptions of weathering and erosion processes in this unit, explain why and how the hydrologic cycle (the global cycle of water) plays a vital role in producing most landforms and landscapes.

4. Why is erosion often rapid in areas of active tectonic uplift?

Scan Here for a quick concept review

www.oup.com/us/mason

Weathering Processes

Weathering pits on the surface of a rock outcrop in Tennessee. Small depressions in the rock held water after rain, and the rock in contact with the water was chemically weathered. Over time, weathering deepened the depressions into distinct pits, which hold water longer, increasing the rate of weathering. Solid products of weathering must be splashed or blown out of the pits when they have been reduced to fine enough particle sizes.

OBJECTIVES

- Explain the major processes of mechanical weathering
- Describe the reactions involved in chemical weathering and the nature of the products of weathering
- Summarize the concept of biological weathering
- List the general environmental controls over weathering processes and discuss how they influence the geography of weathering

On the northeast coast of Brazil, about 1,300 km (800 mi) north of Rio de Janeiro, lies the city of Salvador, capital of the state of Bahía. Two centuries ago, Salvador was a wealthy, thriving place. Agricultural products from nearby plantations and whales from offshore waters yielded huge profits. The Salvadorans, under the tropical Sun, built themselves one of the world's most beautiful cities. Ornate churches, magnificent mansions, cobblestoned streets, and manicured plazas graced the townscape. Sidewalks were laid in small tiles in intricate patterns so that they became works of art. And speaking of art, the townspeople commissioned the creation of many statues to honor their heroes.

But times eventually changed. Competition in world markets lowered incomes from farm products, and over-exploitation eliminated the whale population. There was no money to maintain the old city; people moved away, and homes stood abandoned. Salvador had to wait for a new era of prosperity. When things improved, the better-off residents preferred to live in high-rise luxury buildings on the coast or in modern houses away from the old town. Old Salvador was left unattended, exposed to the elements (Fig. 36.1). The equatorial Sun beat down on roofs and walls, moisture seeped behind plaster, and mold and mildew formed as bright pastel colors changed to gray and black. Bricks were loosened and fell to the ground, and statues began to lose their features as noses, ears, and fingers wore away. People who knew the old city in its heyday were amazed at the rapid rate of decay that was evident everywhere.

The fate of Old Salvador reminds us that buildings of brick and stone can be destroyed by the quiet, persistent work of atmospheric moisture and heat as well as by swirling rivers and thundering waves. The quiet destruction of buildings proceeds in many ways. Moisture dissolves mortar and limestone blocks and causes other building materials to swell and peel. In coastal cities, the air may carry sea salts, which form crystals that can grow in crevices and exert enough force as they grow to break apart stone. Bricks expand in the heat of the Sun and cool at night, loosening the mortared joints between them.

The rock underlying natural landscapes is affected by these processes just as much as buildings and statues are. The same silent, unspectacular processes affect rocks as well as mortar and brick. Just as water seeps behind plaster, it can invade rocks along cracks and joints. Once there, it can dissolve minerals or freeze and expand to break apart rock. Salt crystals can break apart rocks as well as buildings.

In combination, these processes are called *weathering*—an appropriate term because it signifies the impact of the elements of weather examined in Part Two of this book. Weathering goes on continuously not only at exposed surfaces but also beneath the ground and within rock strata. Do not be misled by the analogy just made with decaying buildings and assume that weathering is a largely destructive process. In fact, it is the crucial first step in the formation of Earth's fascinating and often beautiful natural landscapes. Weathering is also essential for the development of soils that support diverse plant communities.

Several kinds of weathering processes occur. The three principal types are (1) *mechanical weathering*, (2) *chemical weathering*, and (3) *biological weathering*. In this unit, we examine each of these types of weathering in turn. But we cannot make the mistake of assuming that these processes are mutually exclusive, or that if one occurs the others do not. In fact, weathering processes usually operate in some combination, and it is often difficult to separate the effects of each process.

Mechanical Weathering

Mechanical weathering, also called *physical weathering*, involves the destruction of rocks through the imposition of stress. A prominent example is **frost action**. We are all aware of the power of ice to damage roads and sidewalks and split open water pipes as a result of water increasing in volume as it freezes. Similarly, the water contained by rocks—in cracks, joints, and even smaller pores—can freeze into crystals that shatter the strongest igneous or metamorphic rocks. This ice can produce about 21,000 metric tons of pressure for every 1 m² (30,000 lb per in²). Of course, frost action operates only where winter brings subfreezing temperatures. Where rock is exposed in a region with cold winters, water that penetrates into the rocks' joint planes can freeze and thaw repeatedly during a single season, wedging apart blocks and slowly breaking down boulders and rock outcrops (Fig. 36.2). At high altitudes in mountain ranges, freezing and frost action can occur throughout the year.

When the fragments of rock accumulate near their original location, they can form a **blockfield** (Fig. 36.3), also sometimes called *rock sea* or the German word *felsenmeer*. Below bedrock cliffs, boulders released by mechanical weathering form steep slopes of **talus** or *scree* (Fig. 36.4). Although the boulders in blockfields and talus slopes are near their bedrock source, they have often been carried some distance downslope by various kinds of mass movement (which we will discuss in Unit 37).

In arid regions and in coastal areas, the development and growth of salt crystals have an effect similar to that of ice. When water in the pores of rocks such as sandstone evaporates, small residual salt crystals form. The growth of those tiny crystals pries the rocks apart (in a process called *salt wedging*) and weakens their internal structure. As a result, caves and hollows form on the face of scarps. The wind may remove the loosened grains and reinforce the process.

In Unit 27, we discussed the process of *exfoliation*, when sheets of rock peel off outcrops and boulders. Exfoliation is a common form of mechanical weathering in granitic rocks, and it typically occurs when expansion of rock produces enough stress for a surface layer to break loose from the larger mass. Exfoliation of great sheets of rock

FIGURE 36.1 "A first visit to the coastal city of Salvador, Bahía (Brazil), in the early 1980s was a depressing experience. In the old part of the city, the historic Pelourinho District, hundreds of centuries-old buildings ranging from ornate villas to elaborate churches lay in dreadful disrepair. Weathering—mechanical, chemical, and biological—was destroying one of civilization's great legacies. The top floor of this mansion revealed evidence of all three, and it, like many others, appeared beyond salvation. But the story has a happy ending. United Nations World Heritage designation and international assistance brought on a massive restoration program, and during the 1990s much of the Pelourinho's heritage was saved."

FIGURE 36.2 This large boulder in a Wisconsin forest was deposited by a glacier and is now undergoing mechanical weathering. Expansion of freezing water in fractures broke loose a small fragment on the top right side of the boulder (turned over for this photo to show the newly broken surfaces).

from granite domes (see Fig. 27.8) results from expansion after overlying rock has been removed by erosion.

Stress is also produced as rocks expand and contract in response to temperature changes. During forest fires, thin sheets of rock peel off boulders exposed to the heat. Rocks exposed on the ground surface in deserts are often broken by cracks, clearly demonstrating some kind of mechanical weathering. Early geomorphologists assumed that these fractures could form through expansion of rocks heated in the daytime sun followed by contraction during the cool nights that are common in deserts. Lab experiments carried out in the 1930s seemed to rule out this possibility. However, within the last decade, new research has shown that enough stress can develop to fracture a rock when solar radiation heats first one side of a rock and then the other over the course of a day. This example shows how much remains to be learned about even fairly simple processes such as mechanical weathering.

Chemical Weathering

The minerals that rocks are made of are subject to alteration by **chemical weathering**. Most chemical weathering takes place out of sight, below ground in soils or fractured rock. Weathering pits, like the one shown in the opening photo of this unit, are an exception. Some minerals, such as quartz, resist chemical alteration quite successfully, but others, such as the calcium carbonate in limestone, dissolve easily. In rock composed of a combination of minerals, some minerals usually break down more rapidly than others. In granite, for instance, the feldspars are chemically more reactive than the quartz and are destroyed more quickly by chemical weathering. The resistance of the quartz is related to the strong bonds between silicon and oxygen atoms that form the entire structure of quartz crystals, as discussed in Unit 27.

You may have seen a heavily pitted granite surface. In such cases, the most readily weathered minerals have been broken down by chemical weathering, often leaving behind clay as one of the weathering products. The once-solid rock disintegrates when enough of the less resistant mineral grains have been destroyed. The quartz grains still stand up but are now loose and can be blown or washed away along with the clay produced by weathering. Even a hard rock that contains abundant quartz cannot withstand the weathering process forever.

Contact of water with rock is essential for most chemical weathering because of the role water plays in chemical weathering reactions. Fractures that can convey water deep into solid rock have striking effects on the patterns of chemical weathering. In a variety of rocks, but especially in granite, weathering works its way into the rock from fractures. This process leaves rounded masses of unweathered

FROM THE FIELDNOTES

FIGURE 36.3 "Looking up this steep slope formed by loose, angular boulders, I wondered whether there was a risk of sudden collapse. But this *felsenmeer* (rock sea), developed from mechanical weathering of quartzite, was stable. After having been pried apart, the angular fragments of various sizes have moved very little, creating a distinct element in this Montana landscape."

FIGURE 36.4   Talus below the summit of Wheeler Peak, Great Basin National Park, Nevada.

solid rock in areas between major fractures. The result is **spheroidal weathering** (Fig. 36.5). Rounded boulders produced in this way can be exposed by erosion, in which case they could be mistaken for boulders that became rounded by rolling along a streambed (Fig. 36.6).

PROCESSES OF CHEMICAL WEATHERING

Several kinds of chemical reactions are important in chemical weathering. Table 36.1 lists examples of the major types of reactions. **Carbonation** occurs when carbon dioxide (CO_2) gas dissolves in water and reacts with it to form a mildly acidic solution. More specifically, the reaction of CO_2 and water (H_2O) forms carbonic acid (H_2CO_3). Some of the carbonic acid then separates into bicarbonate (HCO_3^-) and hydrogen (H^+) ions (recall that *ions* are electrically charged atoms or groups of atoms; see Unit 22).

The hydrogen ions represent acidity and can react with minerals, breaking down their crystalline structure. The simplest example is the weathering of limestone and dolomite, rocks formed from carbonate minerals (see Unit 27). The common carbonate minerals dissolve completely, and

fairly rapidly, when they come in contact with the mildly acidic water produced by carbonation (see Table 36.1). This kind of weathering even attacks limestone underground and contributes to the formation of caves and subterranean corridors (Unit 42).

Carbon dioxide dissolves in rainfall, so rainwater is already mildly acidic when it reaches the ground surface. That acidity is often enhanced within the soil. Levels of carbon dioxide gas can be much higher in the air within soil pores than in the atmosphere above. This is because plant roots and microorganisms continually release carbon dioxide through respiration, and it does not escape quickly to the air above the soil. Therefore, more carbon dioxide will dissolve into rainwater once it has seeped into the soil, adding to its acidity. Fallen leaves and other organic matter release organic acids into the water as well. These natural processes combine to make water in the soil quite effective in attacking minerals. And when coal and other fossil fuels are burned, oxides of nitrogen and sulfur are released into the atmosphere. Those oxides then react with rainfall in areas downwind to make the rainwater much more acidic than it would be naturally, which also enhances weathering rates.

FIGURE 36.5 A rounded boulder of solid granite produced by spheroidal weathering. The boulder was once part of a solid rock mass, which has disintegrated through chemical weathering, forming the soft material now surrounding the boulder.

Silicate minerals are also broken down by weathering when they come in contact with acidic water, though much more slowly than the carbonates. Silicate mineral weathering primarily involves various forms of **hydrolysis**, in which water (H_2O) molecules separate to release hydrogen (H^+) ions, and those ions then react with minerals (see Table 36.1). As we just discussed, the reaction of carbon dioxide with rainwater or soil water results in the release of hydrogen ions, and this process is the crucial starting point of much silicate mineral weathering.

When weathering breaks down iron-bearing silicate minerals, some of the iron that is released reacts with oxygen to form *iron oxides*, reddish-brown to orange- or yellow-colored minerals that form a distinctive rust-like coating on weathering rocks and also give many soil horizons their color. The formation of iron oxides is an example of **oxidation**, a chemical process involving oxygen and iron (or numerous other elements that can be oxidized like iron). Unlike hydrolysis, which consumes acidity, oxidation produces acidity, releasing H^+ ions into solution.

Many kinds of oxidation can play a role in chemical weathering. An especially interesting example is the oxidation of *pyrite* (iron sulfide) and other sulfide minerals (see Table 36.1). Sulfides are common in shales and in the muds that accumulate in coastal marshes and bays. They are stable when deeply buried below the groundwater table or submerged by seawater. However, when exposed to oxygen at the ground surface or in well-drained soils, sulfides rapidly weather by oxidation, generating a great deal of acidity. This happens when coastal marshes are drained to create farmland or are naturally uplifted above sea level, producing acidic soils with unique problems for agricultural use. Shales and other fine-grained rocks containing sulfides are often dumped at the surface in coal-mining operations, and wastes left from mining some important metal ores also contain sulfides. As a result, mine wastes in many parts of the world contain sulfides that weather rapidly by oxidation, and the acidic waters that result have contaminated and even killed most life in streams draining mining districts.

PRODUCTS OF CHEMICAL WEATHERING

Weathering of pure carbonate minerals produces only dissolved ions, but weathering of common silicate minerals through hydrolysis often results in solid as well as dissolved products. Feldspars, one of the most important groups of minerals, provide a good example. As the original feldspar is attacked by acidic water, some of the atoms released from the feldspar crystal structure are dissolved, especially any sodium (Na) and calcium (Ca) that are present, along with some of the silicon (Si) and

TABLE 36.1 Reactions Involved in Chemical Weathering

Carbonation and Carbonate Mineral Weathering

1.	CO_2	+	H_2O	→	H_2CO_3	→	H^+	+	HCO_3^-
	carbon dioxide (gas)		water		carbonic acid		hydrogen ion		bicarbonate ion
2.	$CaCO_3$	+	H^+	→	Ca^{2+}	+	HCO_3^-		
	calcite (a carbonate, solid)		hydrogen ion		calcium ion		bicarbonate ion		

Hydrolysis and Silicate Mineral Weathering (a simple example; more complicated silicate weathering reactions also involve formation of clays or other secondary minerals)

1.	$4H_2O$	→	$4H^+$	+	$4OH^-$		
	water		hydrogen ion		hydroxide ion		
2.	$4H^+$	+	Mg_2SiO_4	→	$2Mg^{2+}$	+	H_4SiO_4
	hydrogen ions		olivine (a silicate, solid)		magnesium ions		silicic acid (dissolved)

Oxidation (example of pyrite)

	FeS_2	+	$2H_2O$	+	$4O_2$	→	$FeO(OH)$	+	$4H^+$	+	$2SO_4^{2-}$
	pyrite (a sulfide mineral)		water		oxygen		goethite (an oxide mineral)		hydrogen ions		sulfate ions

FIGURE 36.6 "Walking along the east-facing slope of the Swaziland Lowveld, checking the geological map against rock exposures. Light-colored granite, rich in silica, suddenly changes to what the map suggests is dolerite (but another map shows it as 'dark, large-grained granite'). Whatever the analysis, the exposure of this darker rock is marked by an expanse of boulders, most of them rounded and many undergoing spheroidal weathering. Peeled the shells off the boulder on the right until I reached the yet-unaffected core; the one at the left is for contrast."

other elements. At the same time, clay minerals of various kinds form out of the remaining feldspar components.

In fact, most clay minerals form through chemical weathering of feldspars and other silicate minerals, including some clays that are important mineral resources. *Kaolinite* is a clay mineral often produced by weathering of feldspars. It is the major ingredient in the clay used to produce porcelain ceramics, a process discovered more than 2,000 years ago in China, which explains why the common name "china" is applied to high-quality porcelain. Kaolinite-rich clay is also used to coat glossy paper. Some rich kaolinite deposits formed through weathering by hydrothermal water (groundwater heated by magma), but others formed through more ordinary weathering by acidic water percolating through soils. Aluminum, one of our most important industrial materials, is refined from *bauxite*, a mixture of minerals produced by the intense chemical weathering of silicate minerals.

Biological Weathering

Living organisms are important agents of weathering, including both physical and chemical weathering processes. Therefore, many physical geographers distinguish **biological weathering** from nonbiological processes that also break down rocks. Of course, most chemical weathering in soils is influenced by the biological process of respiration, which contributes to the acidity of soil water, as discussed earlier. Burrows made by animals and pores created by root growth can also allow acidic water or oxygen to reach greater depths, increasing the effectiveness of weathering by hydrolysis and oxidation. In other cases, plants, animals, and microorganisms play more direct and active roles. Tree roots grow into fractures in rock and can create stress that eventually breaks loose rock fragments, a kind of mechanical weathering enhanced by plant growth. Plant roots also release chemical compounds that may increase the rate of chemical weathering in the area near the root. Lichens, a combination of algae and fungi that live on bare rock, draw nutrients from the rock by ion exchange (discussed in Unit 22). This process is likely to speed up the overall disintegration of the rock beneath the lichen, and lichens also keep moisture in contact with the rock surface. The rock surface shown in the opening photo of this unit may be affected by biological processes such as these. Some evidence also shows that certain bacteria can enhance chemical weathering.

Geography of Weathering

Mechanical weathering can be effective in almost any climate, though certain kinds of mechanical weathering have a more limited range. Frost action weathers rock only where temperatures drop low enough to freeze water in fractures, and the wedging effect of salt crystal growth is mostly limited to coastal areas and deserts. The work of mechanical weathering is most evident where blocks of rock weathered from cliffs accumulate on talus slopes. We can find such landscapes most easily in rugged mountain ranges, but they are not limited to that setting or to a narrow range of climates. In the gently rolling farm country of the Midwestern United States and Ontario, Canada, rocky talus slopes occur below cliffs of especially hard, erosion-resistant dolomite bedrock that is mechanically weathered.

As we noted earlier, chemical weathering often goes on out of sight within the soil, but globally it is more important than mechanical weathering in transforming rock into

FIGURE 36.7 Outcrops of granite in southeastern Wyoming. These rounded forms developed from a larger mass of rock through spheroidal weathering. The soft weathered material around them was then removed by erosion. They continue to weather today.

sediment and dissolved material carried by stream systems. Water is important for chemical weathering, and most chemical reactions occur faster at higher temperatures; therefore, it is not surprising that climate influences the geography of chemical weathering. Everything else being equal, the most intense chemical weathering is found in warm, humid tropical (**A**) climates. In many deep tropical soil profiles, only quartz and a few other resistant minerals remain from the original parent material. Clays and iron and aluminum oxides have replaced the feldspars and other common silicate minerals. In the United States, the best place to find deeply weathered landscapes is in the Southeast. The "red clay roads" of the Piedmont region in South Carolina and Georgia expose soils produced through long-term chemical weathering of metamorphic rock, with their color reflecting the oxidation of iron released from weathered silicate minerals.

Yet the effects of chemical weathering can be dramatic even in much cooler and drier climates depending on the local bedrock types and the geologic history of the landscape. Limestone and dolomite weather rapidly in all but the coldest humid climates, though they are more resistant in dry regions. Weathered granite terrain can be found in climates ranging from humid to fairly dry. Earlier, we noted that weathering of granite occurs near fractures that extend deep below the surface, but between those fractures rounded masses of unweathered solid granite remain. The weathered, disintegrated granite near the fractures forms a loose mass of feldspar and quartz grains known as *grus*. Water erosion then strips away some of the grus and leaves solid rock exposed as rounded towers or mushroom-like forms (Fig. 36.7). The exposed rock does not weather quickly without soil cover, but sometimes rounded channels form where water flows down the side of rock pinnacles (some are visible in Fig. 36.7). Why do such landscapes shaped by chemical weathering occur in places with low rainfall and low temperature today? One answer is that little weathering is required to cause near-total disintegration of the granite. On the other hand, many landscapes of deeply weathered granite are probably quite old and have been exposed to weathering for millions of years. Given all that time, deep weathering can probably occur even in a cool, dry climate.

In our study of the mass removal of loose Earth materials in Unit 40, we focus on great rivers that sweep vast amounts of rock fragments and particles downstream. Much of that material was first loosened by the quiet, relentless processes of weathering. The intermediate force that acts on weathered materials prior to their removal is gravity, and its important influence on slope stability is explored in Unit 37.

Key Terms

biological weathering *page 446*
blockfield *page 441*
carbonation *page 444*
chemical weathering *page 443*

frost action *page 441*
hydrolysis *page 445*
mechanical weathering *page 441*
oxidation *page 445*

spheroidal weathering *page 444*
talus *page 441*

Scan Here for a quick vocabulary review

Review Questions

1. Describe how freezing of water and salt crystal formation cause mechanical weathering.
2. Explain why water coming into contact with rock plays a critical role in both mechanical and chemical weathering and give specific examples.
3. How and why does spheroidal weathering occur?
4. Under what climatic conditions would you expect chemical weathering to be most rapid and effective? Explain your answer.

5. How is chemical weathering affected by the interaction of carbon dioxide in the air and water falling as rain or seeping through the soil?
6. How are clay minerals related to weathering processes?
7. Choose three examples of biological weathering and explain why they can also be considered special cases of either mechanical or chemical weathering.

Scan Here for a quick concept review

www.oup.com/us/mason

Mass Movements

Talus slope in the southern Rocky Mountains of Colorado.

OBJECTIVES

- Explain why slopes fail under the force of gravity and the factors that influence slope failure

- Describe the various types of mass movements and the circumstances under which they usually occur

- Discuss the importance of mass movement processes in shaping landscapes and the hazards associated with them

In 1998, as Hurricane Mitch devastated the hills and mountains of Honduras, the world was again reminded of the dangers posed by unstable slopes. As this ferocious storm slowly made its way across the heart of this Central American country over the course of two days, it unleashed almost 200 cm (75 in) of wind-driven rain, which triggered massive flooding, landslides, and rapid flows of debris. Almost 10,000 people died, 1.5 million were left homeless, and more than 70 percent of the country's transport facilities were washed away; agricultural losses alone resulted in an economic crisis.

The landslides and flows triggered by Hurricane Mitch are examples of **mass movement**, the downhill movement of earth materials by the force of gravity. The Hurricane Mitch disaster is, admittedly, a rather spectacular example of the hazards posed by mass movement processes. But in all mountain zones, rocks, soil, and other unconsolidated materials continuously move downslope in quantities ranging from individual boulders to entire hillsides. This movement happens at different rates of speed: some rocks actually fall downslope, bounding along and occasionally even landing on highways, while other material moves slowly, almost imperceptibly.

There are times when the signs of an imminent collapse are unmistakable, and people at risk can be warned. At other times, a mass movement event occurs in an instant. In 1970, an earthquake loosened a small mass of rock material high atop Mt. Huascarán in the Peruvian Andes, and this material fell about 650 m (2,000 ft) down the uppermost slopes. It normally would have done little damage, but at an elevation of about 6,000 m (20,000 ft), it happened to land on a steep slope of loose material, snow, and ice on the mountainside. The falling mass and the loose material combined and roared downslope. According to later calculations, the mass of debris moved at a speed of more than 435 km/h (270 mph). By the time it stopped, it had traveled 18 km (11 mi). The loss of life is estimated to have exceeded 25,000 persons. No traces were found of entire communities, in particular Yungay (Fig. 37.1).

Slope Stability and Mass Movement

The force of gravity plays a major role in the modification of landscapes. Gravity drives the flow of water in streams, of course, but in this unit we are concerned with the downhill movement of solid material mixed with varying amounts of water. Gravity pulls downward on all materials. Slopes consist of many different kinds of material. Some are solid bedrock; others consist of bedrock as well as loose rock fragments; still others are covered by soil and soft sediment or weathered rock. Solid bedrock can withstand the force of gravity because it is strong and tightly bonded. But if weathering loosens a fragment of bedrock, that particle's first motion is likely to be caused by gravity.

Why do landslides occur on some slopes while others remain stable for long periods of time? You can do a simple

FIGURE 37.1 The 1970 flow that destroyed the Peruvian town of Yungay (left of center). A rockslide from the mountain in the distance disintegrated into a flow that moved rapidly down the valley.

ENHANCE

experiment that helps answer this question. Take a flat surface the size of a cafeteria tray and cover it with a layer of sand about 2.5 cm (1 in) thick. Now tilt the tray somewhat, say to an angle of 10 degrees from the horizontal. The sand is unlikely to move. Tilt it slowly to a steeper angle, and at a given moment, movement will begin. At first, you may be able to observe some small particles moving individually, but soon the whole mass will move downslope because you have exceeded the maximum angle at which the sand remains at rest.

All slopes have such a threshold of steepness that separates stability from **slope failure**, the point at which the slope materials begin to slide or flow away downslope. The **angle of repose** is defined as the steepest angle at which a given kind of slope material will remain stable and not fail. For loose sand, the angle of repose is generally around 33 to 35 degrees. Clay or silt-rich soils typically have a lower angle of repose than sand. Coarse boulder debris may have

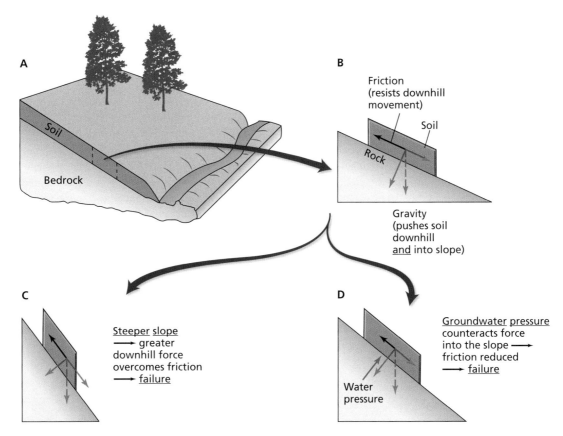

FIGURE 37.2 Causes of slope failure using the example of a layer of soil resting on bedrock (A). We can identify forces acting on a block of this soil (B). The force of gravity tends to push the block downslope but also into the slope (perpendicular to the surface). Friction resists downhill motion and increases along with the component of the gravitational force pushing the soil block *into* the slope. On a low-angle slope, friction is great enough to resist the force pushing the block downslope, and the hillside is stable. On a steeper slope (C), the component of gravity driving the block downslope is greater, while the component pushing it into the slope is less. Therefore, friction is overcome, and the block starts sliding away downhill. When water (often from heavy rainfall) saturates the soil, the water pressure in the soil pores counteracts the force pushing the block into the slope (D). This also reduces friction, allowing the block to slide and slope failure to occur.

an angle of repose as steep as 45 degrees, and cliffs of strong bedrock remain stable at much steeper angles than that.

To understand why failure is so closely related to the steepness of the slope, we turn from our experiment with a sand-covered tray to another simple example: a layer of soil resting on an inclined surface of underlying bedrock (Fig. 37.2A). Various forces act on this layer of soil, some tending to drive it down the slope in a mass movement, and others working to keep the soil in place. We can simplify our analysis by considering the forces acting on a particular part of the soil layer, represented by a block resting on an inclined surface (Fig. 37.2B). The weight of the block—that is, the force of gravity acting on it—tends to push the block both down the slope and into the slope, perpendicular to the bedrock surface. The component of the gravitational force pushing the block downhill creates *shear stress* where the block rests on the underlying surface (*stress* is force applied to an area). Downslope movement of the block of soil in response to this shear stress is resisted by *friction* at the base of the block.

The ability of geologic materials to develop such resistance to movement is their **frictional strength**. If the base of the block and the surface it rests on are rough and irregular, the friction will be greater. However, the friction also increases along with the component of the block's weight that pushes it into the slope. The greater the force pushing the block into the slope, the more the bumps and hollows in the block and the underlying surface interlock with each other and resist motion. Therefore, the friction resisting motion is greater for a heavier block.

Now consider what happens as the slope becomes steeper, as the slope of the sand-covered tray did as it was tilted (Fig. 37.2C). More and more of the block's weight will tend to push it *down* the slope rather than *into* the slope. That is, there will be more and more shear stress and less friction. Eventually, this stress will overcome the friction, and the block will slide; that is, slope failure will occur. Another important change in stability will occur if the soil is saturated with water (Fig. 37.2D). Then the pressure of water in the soil pores opposes the force pushing the material into the slope. The friction resisting movement is reduced, and slope failure can occur at a much lower angle than when the soil or rock is dry.

In this simple example, only friction resists motion, but for many slopes, we have to consider another source of resistance called **cohesion**. Cohesion refers to a tendency of slope materials to stick together and resist failure that is not due to friction and doesn't depend on the weight of overlying materials. Dry sand has no cohesion, and only friction between the sand grains resists motion. In clay-rich soils, there may be little friction to resist motion, but cohesion is created by electrical attractions between clay particles. In our example of a block on a slope, cohesion could be added by some kind of cementing material between the block and underlying surface or by tree roots binding the block in place. Tree roots are an important source of cohesion on many natural slopes.

Not all mass movements involve a layer of soil sliding over bedrock. In many cases the surface where failure occurs forms within soft sediments or in fractured or weathered bedrock. The causes of slope failure in the simple example we have just examined apply to most cases of natural landslides, however. Slopes can become over-steepened (see Fig. 37.2C), increasing shear stress, through a variety of processes. When a stream deepens its valley, the valley sides will become steeper, and eventually slope failure will occur (Fig. 37.3). Roads cut into steep slopes

FIGURE 37.3 "From Carmel along G-16 toward the Soledad loop. Vineyards along the river, deeply incised. Mediterranean vegetation on the steeper hillslopes. Noted a small tributary that had oversteepened a slope and caused a landslide not even the dense natural vegetation could prevent."

have the same effect, and road building in mountainous areas often runs the risk of triggering landslides. Earthquake shaking is also a common trigger of slope failure.

The saturation of slope materials by water is the crucial factor for many landslides and other mass movements, however. By saturating slope materials and decreasing the friction that resists failure, large rainstorms can produce dozens or even hundreds of landslides as they move across mountainous terrain (see Fig. 37.2D). This is precisely what happened when Hurricane Mitch crawled across Honduras and triggered countless slope failures.

Forms of Mass Movement

From the term *mass movement*, it is clear that the materials involved are moved *en masse*—in bulk. The solid material in motion can be dry, but more often it contains some water. We refer to such movements in our everyday language by several terms. A *landslide*, for instance, is a form of mass movement. So is an **avalanche**, involving not rock debris but snow and ice (Fig. 37.4). Another form of mass movement is the almost imperceptible downslope creep of soil. The term *mass wasting* is also employed to describe these various gravity-induced movements.

It is difficult to find a completely satisfactory system for classifying mass movements. One way to distinguish them is by the type of downhill motion they involve. Accordingly, we recognize (1) *creep* movements, the slow, imperceptible motion of the soil layer downslope; (2) *slide* movements, in which more or less intact blocks of material slide along a planar or curved surface; (3) *flow* movements, in which the material in motion has disintegrated into a fluid mass; and (4) *fall* movements, in which blocks of rock or sediment fall or topple through the air.

CREEP AND SOLIFLUCTION

Mass movements produce characteristic marks and scars on the landscape. Sometimes, this evidence is so slight that we need to assume our detective role to prove that mass movement is indeed taking place. The slowest mass movement—**creep**—cannot be observed as it actually happens because it involves such slow downhill motion of the soil on slopes. The upper layer of the soil moves faster than the layers below, and vertical objects are tilted downslope in the process. As a result, creep sometimes reveals itself in trees, fence posts, and even gravestones that have been tilted slightly, as shown in Figure 37.5.

Soil creep is not a slope failure of the type we discussed earlier. Instead, it is a more complicated type of motion that appears to result from the alternate freezing and thawing of soil particles or from alternate periods of wetting and drying. The particles are slightly raised or expanded during freezing and wetting and then tend to settle slightly lower along the slope when they are thawed or dried. In reality, there is considerable randomness in the path followed by soil particles that are raised and lowered in this fashion, but the net movement is

FIGURE 37.4 Start of a snow avalanche on Mt. Dickey, Alaska. As the snow thunders downslope, its volume increases and the area it affects widens.

ENHANCE

FIGURE 37.5 Effects of soil creep in the cultural and physical landscape—objects tilting downslope.

downhill. The total change during a single freeze–thaw or wetting–drying sequence is minuscule, but over periods of years the results are substantial. Animal burrowing can have a somewhat similar effect, because soil pushed out of a burrow on a steep slope will tend to end up a little farther downhill than where it originated. The same thing happens when trees tip over and pull up a mass of soil in their roots. Slow downhill soil movement can occur in many landscapes under a wide variety of climates, and it is often hard to judge whether most of this movement is related to freezing and thawing, wetting and drying, or burrowing and tree tipping.

A special kind of slow mass movement is called **solifluction**, a term from Latin that literally means soil flow. Solifluction occurs mainly in cold regions of the high latitudes or in high-mountain landscapes, especially where permafrost is present. As the name suggests, solifluction does include a true downslope flow of soil. However, it often also includes a component of soil creep as a result of the raising and lowering action of freeze–thaw cycles (see Unit 46).

SLIDES

Although soil creep is quite difficult to detect in the landscape, slides are usually obvious—sometimes for thousands of years after they occur. Slides start with a slope failure in which a block of rock and soil breaks loose and slides over a distinct surface. The initial block often separates into two or more smaller blocks soon after motion begins and eventually may disintegrate into a flow. **Landslide** is a general term for this type of mass movement. The surface on which the blocks slide downhill may be fairly planar or distinctly curved. A **slump** involves sliding motion on a curved surface (Fig. 37.6). The blocks that

form a slump rotate distinctly, with the top of each block tilted toward the slope, which makes this type of mass movement easy to recognize. Slumping is especially common in shales and other soft, clay-rich rocks and sediment with high cohesion. A drive through terrain underlain by shale will often reveal dozens of small slumps on hillsides along the road. Although these cause damage to roads or structures, they are far from the most hazardous mass movements.

Large landslides, and especially **rockslides**, are quite a different matter. Rockslides are distinguished by the fact that they involve a large mass of bedrock that breaks loose and moves rapidly downslope (Fig. 37.7). Rockslides in mountainous areas can roar hundreds or even thousands of meters downslope with a thunderous sound and great force. The destructiveness of large rockslides makes them one of the most dangerous forms of mass movement, especially when they occur in inhabited areas.

Groundwater pressure after heavy rainfall, by reducing the friction that resists slope failure, can trigger both large and small landslides and rockslides. This is the most common cause of landslides worldwide, but they are produced in other ways as well. Earthquakes can provide the vibrations necessary to set the downward slide in motion (see Fig. 33.7). In the devastating Sichuan earthquake of 2008 in central China, landslides added to the destruction caused directly by seismic shaking.

Oversteepening of slopes by the erosion of streams or road construction can also be a factor (see Perspectives on Human Environment: The Human Factor). In a study conducted in 1976, the Japanese Ministry of Construction located 35,000 sites that were potentially dangerous because of land development. Torrential summer rains that year on Japan's Izu Peninsula, near Tokyo, had caused almost 500 landslides, killing nine people in a popular holiday resort and leaving nearly 10,000 homeless.

Both slumps and large rockslides start out with the motion of one or more large intact blocks or slabs. Once in motion, however, the blocks often disintegrate into a flowing mass. An impressive example is shown in Figure 37.8. The 1970 Huascarán disaster in Peru (see Fig. 37.1) began with a large earthquake-triggered slide of rock and ice but became a flow that involved more than 50 million m³ (1.7 billion ft³) of rock debris, snow, and ice. Therefore, many mass movements that we might classify as slides or flows based on the predominant form of motion could more accurately be considered hybrids of the two types.

FLOWS

Flows involve relatively fluid downhill movement of weathered rock, loose sediments, or soil. They can be dry or involve a mixture of debris and water. In both cases, the flowing mass is denser and much more viscous (i.e., it flows less readily) than water flowing in a stream. Dry sand often flows down the steep downwind face of sand dunes, a good example of dry flows. Wet flows are more common in most

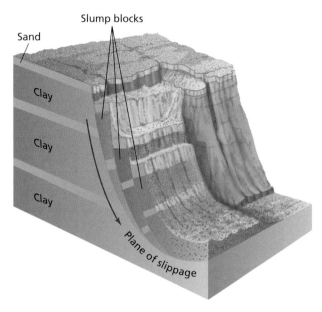

FIGURE 37.6 Example of slumping in which three slump blocks have moved downslope. Note the backward rotation of each slump block, the scarp at the head of the slump cavity, and the forward flow at the toe of the mass.

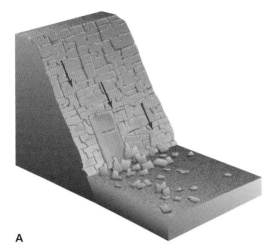

A

FIGURE 37.7 Rockslide. Note the steep slope and the parallel planes of weakness in the rock face (shown with red arrows) (A). Such breakaway mass movements are a constant danger in high-relief terrain. The photo (B) shows a rockslide that closed the Karakoram Highway in northern Pakistan's Hunza Valley—except to intrepid truck drivers who risk their lives by negotiating the rocky obstruction.

B

settings and can move at rates ranging from barely noticeable to extremely rapid.

The term **earth flow** is generally used for slow-moving flows, especially if they involve fairly fine-grained and clay-rich soil or sediment. Often beginning with a slump, an earth flow moves downhill so slowly that measurements over a period of years may be needed to determine the rate of motion (Fig. 37.9). Many earth flows are long lived, and some have been monitored for decades.

Another form of flow is the **debris flow**—a fluid, fast-moving slurry of sediment and water (Fig. 37.10). Debris flows may contain weathered rock material of all sizes, including huge boulders. A debris flow originates when heavy rainfall comes to an area where large quantities of weathered rock or other loose debris are available on steep slopes (the torrential rains of Hurricane Mitch fell on steep slopes covered with soil and weathered rock, for example). The term *mudflow* is used for rapid flows of this type that are dominated by fine muddy sediment. In fact, the news media often describe all kinds of debris flows as mudflows.

The debris flow may start with a slope failure high on a slope in which the sliding blocks collapse into a flowing mass. Sometimes an accumulation of debris in a steep mountain stream channel is set in motion as water rushes downhill. The debris flow typically rushes down an existing small stream channel, eroding it in the process, but sometimes it creates a new channel on a mountainside. Upon reaching the floor of a larger valley, the flow can continue downstream for some distance, even several

FIGURE 37.8 A dramatic photograph of a slump that collapsed into a flow, halting traffic on a major road in the San Francisco Bay area in 1952.

FIGURE 37.10 Aftermath of a debris flow that raced down a valley in Sichuan Province, China, in 2010. This flow and numerous others at the same time were triggered by heavy monsoon rains and occurred in an area already devastated by a major earthquake in 2008.

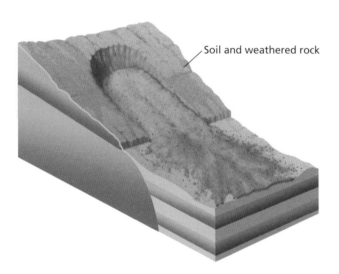

Soil and weathered rock

FIGURE 37.9 Earth flow. This lobe-shaped mass leaves a small scarp at its upper end and pushes a tongue of debris onto the valley floor in the foreground.

kilometers (as shown in Fig. 37.10). Eventually, the flow comes to a halt, because the slope it is flowing down has decreased, and in some cases because water has been lost from the flow into underlying deposits.

Debris flows cause great destruction when they affect populated areas, and they are a leading natural hazard in mountainous regions. For example, in the rugged terrain of interior China, debris flows may travel down certain small valleys every few years. Towns are built on valley floors because these are the only sites with much level ground, but this commonly places them in the path of debris flows (see Fig. 37.10). Artificial channels to divert flows and warning systems have been installed to reduce the risk. Debris flows are also one of the most important processes supplying sediment to streams in this region, as in many other mountainous landscapes.

FALL

The last of the mass movements we discuss is the fastest of all—**fall**—in which pieces of rock loosened by weathering literally fall through the air, often then rolling some distance downslope. Such rock fragments and boulders usually do not go far, ending at the base of the slope or cliff from which they broke away (Fig. 37.11). They form a *talus,*

FIGURE 37.11 Rockfall—free falling of detached bodies of bedrock from a cliff or steep slope. The loosened boulders usually come to rest at the base of the slope from which they fell.

The Importance of Mass Movements

It is easy to underestimate the importance of mass movements among the processes of erosion. Creep occurs in many landscapes but is almost imperceptible. Large, rapid slides and flows are associated with particular kinds of slopes and geological materials. But creep is the most important form of erosion that occurs on many vegetation-covered slopes, and small slope failures frequently take place along the banks of most streams. Furthermore, rapid mass movements are of great importance in mountain ranges that are experiencing active tectonic uplift, especially if they also receive abundant orographic rainfall. In these cases, powerful streams in deep canyon-like valleys cause much oversteepening and constant collapse along their valley walls and sweep away eroded material brought down from higher slopes by debris flows. Such rapidly eroding, tectonically active landscapes contribute a large portion of all the sediment carried toward the oceans by rivers. Much of that sediment starts on its path through mass movements.

Mass movements also perform the important job of exposing new bedrock to the forces of weathering. Each time mass movement removes a slab of soil and weathered rock from a bedrock slope, a fresh rock surface is exposed to weathering. If mass movements did not occur or were less effective, the entire system of weathering and erosion would be slowed down.

an accumulation of rock fragments, large and small, at the foot of the slope, as we saw in Figure 36.4. Such an accumulation, also known as a *scree slope*, lies at a surprisingly high angle—more than 35 degrees for large fragments. When an especially large boulder falls onto the slope, the talus may slide some distance before attaining a new adjustment.

PERSPECTIVES ON THE HUMAN ENVIRONMENT

The Human Factor

Mass movements most frequently are caused by natural forces and conditions, but human activities can also contribute to their occurrence. Sometimes people fail to heed nature's warnings, as in the case of the disastrous Langarone landslide in northern Italy in 1963. The materials on the slopes of the Langarone Valley were known to have low shear strengths. Nevertheless, a dam was constructed across the valley, and a large artificial reservoir (Lake Vaiont) was impounded behind it. The water rose and saturated the rocks on the valley sides, adding to the risk of landslides. When heavy rains pounded the upper slopes in the summer and early autumn of 1963, major landslides seemed to be inevitable.

In early October, the severity of the hazard was finally realized, and it was decided to drain the reservoir. But it was too late, and on October 9, a mass of rock debris with a surface area of 3 km² (1.2 mi²) slid into Lake Vaiont, causing a huge splash wave in the reservoir. The wave spilled over the top of the dam and swept without warning down the lower Langarone Valley, killing more than 2,600 persons.

Yet despite this tragic lesson, the building of houses and even larger structures on slide-prone slopes continues in many countries. This includes the United States, which contains numerous areas that are susceptible to landslide risk (Fig. 37.12). Most of these areas are not likely to experience disasters on the scale of the one that struck the Langarone Valley. Yet mass movements in all of these areas can do expensive and avoidable damage to roads and buildings and even occasionally cause the loss of lives.

continued

continued

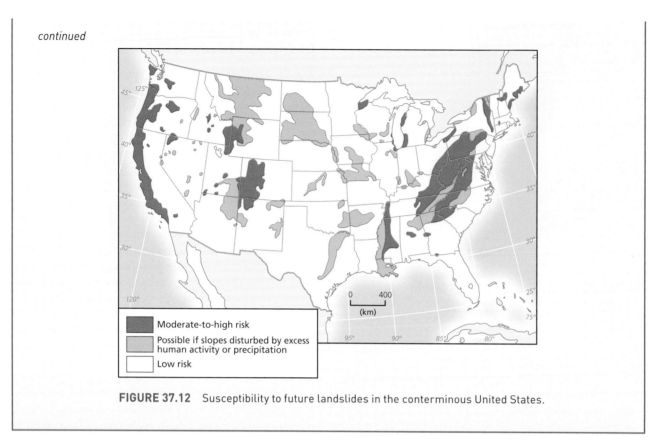

FIGURE 37.12 Susceptibility to future landslides in the conterminous United States.

Moderate-to-high risk

Possible if slopes disturbed by excess human activity or precipitation

Low risk

FIGURE 37.13 Eroding cliffs along the coast of Chesapeake Bay at Governor's Run, Maryland, are susceptible to slumping, threatening waterfront homes. These houses had already been condemned at the time this photo was taken.

Another reason to spend time examining the sliding and flowing of earth materials is that knowledge of mass movement can help in planning and policymaking. It is quite costly to use public money to rebuild highways in unstable areas (see Fig. 37.8). Moreover, many people build houses in locations chosen for their spectacular scenic views, with little or no regard for the stability of the land. Some of these residents have had rude awakenings, as Figure 37.13 illustrates.

Key Terms

angle of repose *page 450*

avalanche *page 453*

cohesion *page 452*

creep *page 453*

debris flow *page 455*

earth flow *page 455*

fall *page 456*

frictional strength *page 451*

landslide *page 454*

mass movement *page 450*

rockslide *page 454*

slope failure *page 450*

slump *page 454*

solifluction *page 454*

Scan Here for a quick vocabulary review

ENHANCE

Review Questions

1. Why does slope failure occur when a slope is oversteepened?

2. What are the two types of resistance to slope failure and mass movement, and how do they work?

3. How do creep, slides, and flows differ as specific forms of mass movement?

4. Describe the nature of debris flow behavior and some of the natural hazards associated with it.

5. Why does heavy rainfall trigger landslides in many landscapes with steep slopes?

6. In addition to rainfall, what are some other causes of mass movements?

7. What is solifluction? In what environment is it most common?

Scan Here for a quick concept review

ENHANCE

www.oup.com/us/mason

UNIT 38

Water in the Lithosphere

Groundwater heated by magma emerges from hot springs in Yellowstone National Park. Many of the exotic colors are produced by microorganisms adapted to life in hot spring waters.

OBJECTIVES

- Describe the various paths taken by precipitation that falls to the surface of the lithosphere

- Explain the fundamental aspects of stream flow

- Summarize basic concepts related to groundwater flow and human uses of groundwater

Water is the essence of life on Earth. Humanity's earliest civilizations arose in the valleys of great rivers. Our ancestors learned to control the seasonal floods of these streams, and irrigation made agriculture possible in dry regions. Today, we refer to those cultures as *hydraulic civilizations* in recognition of their ability to control and exploit water. Our current dependence on water is no less fundamental. Society's technological progress notwithstanding, water remains Earth's most critical resource. We could conceivably do without oil or coal, but we cannot survive without water. Thus, the historical geography of human settlement on this planet is in no small part the history of the search for, and use of, water.

The operations of the hydrologic cycle were introduced in Unit 11, which treats the part of the hydrosphere found in the atmosphere (water vapor). In this unit, the focus is on the part of the hydrosphere contained in the lithosphere. In Units 39 to 41, we study rivers and streams, the primary agents of landscape formation. But before water collects in streams, it must travel over or through the surface layer. Arriving as precipitation (mostly as rainfall), some water evaporates. Some falls on plants; some moistens the upper soil layers; some seeps deeper down and collects underground in the pores of rocks and sediments. And some runs downslope and collects, first in small channels and then in larger creeks, to become part of the volume of permanent streams.

FIGURE 38.1 Vegetation intercepts a percentage of rainfall. These drops on a few blades of grass do not seem to amount to much, but over a large area, such interception can amount to a substantial volume of water that does not reach or saturate the soil.

Water at the Surface

Our inquiry starts with raindrops that arrive at Earth's surface. What happens there depends on the characteristics of the rainfall, and on the nature and condition of the land surface. Rainfall varies in intensity, duration, and total amount. The *intensity* is the amount of water that falls in a given time. A rainfall intensity of 12.5 mm (0.5 in) per hour is not especially heavy for summer storms in the Midwestern United States, but it could still produce abundant runoff and even flooding if it had a long *duration*, the length of time that rainfall continues at a given intensity. For example, rainfall of this intensity that kept up for a duration of 10 hours would produce a total *amount* of 125 mm (5 in) of rain.

In some cases, raindrops never reach the ground; they fall on vegetation and evaporate before they can penetrate the soil. Figure 38.1 shows how this **interception** occurs. Especially for rainfall of low intensity or short duration, interception can substantially reduce the proportion of the total rainfall that reaches the soil surface. The amount of water intercepted by vegetation depends on the structure of the plants involved. In Australia, for example, the fairly open leaf canopies of eucalyptus trees intercept only 2 to 3 percent of the rain, and the rest falls directly to the ground or quickly drips off the leaves. Hemlock and Douglas fir trees in the northwestern United States have small needles rather than leaves such as those of the eucalyptus, so we might expect them to intercept little rainfall. Each tall

hemlock or Douglas fir tree holds a vast number of needles, however, and they form dense foliage that intercepts as much as 40 percent of the rain.

INFILTRATION AND RUNOFF

If rainwater reaches the ground, it can be absorbed by the surface. Most natural surfaces absorb a portion of the water that falls on them and are considered **permeable**, although the degree of permeability depends on many factors. Materials such as concrete roads or granite outcrops that do not permit water to pass through them are said to be **impermeable**.

Infiltration is the flow of water into Earth's surface through the pores (open spaces between particles) in the soil mass. The amount of water that infiltrates in a single rainstorm depends on the soil's **infiltration capacity**, which is the maximum rate of infiltration into that soil under a particular set of conditions. The last part of that definition is important, because the infiltration capacity of the same soil can change greatly over time, depending on several factors: (1) the physical characteristics of the soil, (2) the type and extent of the vegetation cover, (3) how much moisture is already in the soil, and (4) how much the structure of the soil surface has been altered by rain that has already fallen.

The most evident soil characteristics that affect infiltration are texture and structure. For the most part, water infiltrates more rapidly into coarse, sandy soils than into clay soils. The pores between sand grains are much larger than those between closely packed silt or clay particles, and those larger pores allow water to flow more rapidly into the soil. A soil with fairly abundant silt and clay can still allow rapid infiltration, however, if it has a well-developed structure. In that case, water readily enters the soil through large pores between the aggregates or peds (see Unit 22), each of which is made up of many mineral grains clumped together. Of course, in cold regions soils freeze in winter. The formation of ice in soil pores greatly reduces the infiltration rate, even in soils that can absorb water readily when they are not frozen.

Rain that falls as a light drizzle will all readily soak into the soil, because the rainfall rate is much less than the soil's infiltration capacity. Researchers have focused their attention on infiltration during rains of higher intensity and duration, repeatedly measuring the rate of infiltration over the course of individual rainstorms on experimental plots around the world. These measurements clearly reveal how the infiltration capacity of a soil can change dramatically as a storm proceeds.

Figure 38.2 shows the typical shape of plots representing infiltration rate over time; the rapid decrease at the start is clear. There are two important reasons for these trends. First, a soil generally takes up more water when it is dry than when it is already wet from previous rain. In part, this is because the water infiltrating into a dry soil has many pores to fill. As pores near the ground surface are filled, rainfall can only infiltrate as fast as the water already in the soil flows downward into deeper soil. Therefore, the infiltration rate usually drops off quickly after the start of a rainstorm. This is a good reason to pay attention to the duration of a storm as well as its intensity; a heavy but short-lived rainstorm might not even fill the soil pores to the point at which the infiltration rate declines.

Vegetation and human land use also have important influences, as you can see by comparing the curves for an uncultivated meadow and a tilled cornfield in Figure 38.2. A soil's infiltration capacity will almost always decline less during storms when it is under natural vegetation than

when it is cultivated and is free of vegetation at some times of the year. The most important role of vegetation is to shield the soil surface from rainfall. Where there is dense vegetation, leaves and stems absorb the impacts of raindrops, but where there is little or no vegetation cover, raindrops strike the ground with surprising force (see Fig. 39.1). Soil aggregates are destroyed, and small particles are scattered in all directions and washed into pores. A thin, dense layer of soil with few large pores soon forms, sealing the surface and greatly reducing the infiltration capacity. After the rain, this dense layer dries to form a hard crust, which will persist until it is broken up by burrowing animals or cultivation.

Certain land uses can have a significant effect on infiltration, even if they do not expose bare soil to rainfall. In a much-studied valley in Switzerland, forest soil absorbs 100 mm (4 in) of water in two minutes, but a pasture where cattle graze requires three hours to take in the same amount of water. The cattle most likely compacted the soil, closing up large pores and reducing permeability. People walking on a hiking trail or logging equipment driven over the forest floor can have similar effects.

If a rainstorm has a great enough intensity and duration, it will eventually exceed the infiltration capacity. Water will then begin to accumulate into small puddles and pools. It will collect in any hollow on a rough ground surface, detained behind millions of little natural dams. This situation is called *surface detention*. When there is more rainwater than the small detention hollows can hold, the water flows over the land as surface **runoff** (or *overland flow*). After a short period of detention, any rainfall that does not infiltrate the soil runs off it. Thus, as precipitation continues and infiltration rates decrease, runoff rates increase.

Where raindrop impacts can form a surface seal on bare soil, runoff is common. In the hilly landscape of southwestern Wisconsin, a long-term study found that cultivated fields produced runoff almost every year. A nearby forest plot on similar soils produced runoff only twice in 13 years. This study provides a good indication of how the response to rainfall changed across southwestern Wisconsin and many other parts of the United States when former forests or prairies were converted to farmland in the 1800s. The frequency of runoff and the rate of soil erosion both increased dramatically as a result.

Once surface runoff has started, the water continues to flow until it reaches a stream or river or an area of permeable soil or rock. An area of land on a slope receives all the water that runs off the higher elevations above it. The longer the total flow path, assuming that the runoff flows across uniform material, the greater the amount of runoff across the area. Therefore, the largest runoff rates occur at the base of a slope, just before the runoff enters a river. This emphasizes the need for sound land management practices. If water is needed for raising crops, then the less that runs off, the better. Because infiltration and runoff rates are inversely related, runoff depends on all the factors that

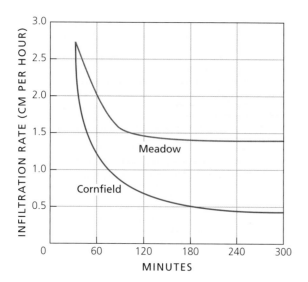

FIGURE 38.2 Infiltration rates for two different vegetated areas. Both show higher rates at the beginning than at the end of the rainfall, but the meadow has higher rates throughout.

FIGURE 38.3 Downward rush of water in a high-gradient mountain stream at Roaring Fork, Great Smoky Mountains National Park, Tennessee.

affect infiltration. By employing appropriate farming techniques, high infiltration rates can be maintained and runoff rates reduced.

WATER FLOW IN STREAMS

For some, looking down at a flowing river holds the same kind of fascination as looking into the flames of a fire. The physical geographer's interest in stream flow is less romantic but just as compelling. A study of river flow helps us make predictions about such matters as pollution and floods. From the smallest flow in a tiny rivulet to the largest flow in a major river, certain general rules apply.

One fundamental property of a stream channel is its **slope** (also called *gradient*). Slope is calculated by dividing the change in elevation between two points along the stream course by the length of the stream between those points. (Dividing vertical distance by horizontal distance is a common way to calculate any slope, but in this case those distances are measured along a stream course that may be winding rather than straight.) A mountain stream that has rapids and falls in its valley has a steep slope (Fig. 38.3). By contrast, the lower portions of such rivers as the Mississippi and the Amazon have very low slopes. The Amazon River, for instance, falls only about 6 m (20 ft) over its final 800 km (500 mi).

The slope of a stream is important because water flow in an open channel is driven by gravity. The force of gravity driving water downstream is related to both the slope of the water surface and the depth of flowing water; steeper, deeper streams are driven downstream with greater force. This force is resisted, however, in various ways. The water itself has some *viscosity* (internal resistance to flow). Rocks or even sand grains on the streambed, tree roots in the channel, and stream bends all add resistance. The average speed, or **velocity**, of water in a stream depends on the balance between the gravitational force driving the water downstream and the resistance to flow.

It is important to refer to the *average* velocity in making this generalization because the actual velocity of water is not the same in all parts of a stream channel. The stream tends to move fastest in the center of its channel, just below the surface of the water, and slowest along the bottom and sides of the channel, where rocks, tree roots, or other features resist the flow (Fig. 38.4). As illustrated in Figure 38.4C, the distribution of velocity is even more complicated when we look at how it changes through bends in a stream channel. The highest velocity is associated with the *thalweg*, the deepest part of the channel, which swings toward the outside of bends.

Measurements of stream velocity produce some surprises. For example, some sparkling mountain "torrents" have lower average velocities than the placid lower segments of the Mississippi River! This can be explained by the shallow depth of the small mountain streams and the resistance to flow in their rocky channels. Only careful velocity measurements can prove the eye wrong.

The **discharge** of a river is the volume of water passing a given cross-section of its channel within a certain amount of time. Discharge (Q) is calculated from the average velocity (V) and the cross-sectional area of the flowing water (A) using the following equation:

$$Q = V \times A$$

The cross-sectional area is equal to the average depth of water times the width of the water surface. Long-term records of discharge are very important in planning dam construction and flood control strategies. Discharge records also give us some indication of how much sediment the river can potentially carry downstream. This reflects the river's effectiveness as an erosional agent.

Discharge varies—by day, by season, by year, and over longer periods. A graph of a river's discharge over time is called a **hydrograph**. The hydrograph in Figure 38.5 records a 10-day period during which the discharge was affected by a storm in Nova Scotia's St. Mary River drainage basin. This is only one event in that river's life. The longer and more detailed the hydrographic record, the more accurate our knowledge of a river basin's hydrology.

Hydrologists have devised standard methods to measure the processes of stream flow. Current meters that measure velocity are designed to be lowered into streams

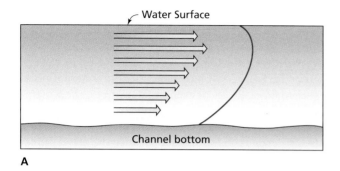

A

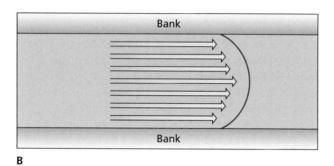

B

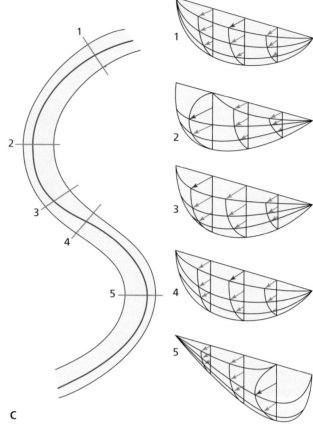

C

FIGURE 38.4 Velocity variation in a river; longer arrows indicate faster flow. The basic pattern is shown from the side (A) and from above (B). In river bends (C), the zone of highest velocity (red arrows) is associated with the thalweg (red line). The fastest flow swings toward the outer bank and moves deeper below the surface *within* bends (cross-sections 2 and 5) rather than *between* bends (cross-sections 1, 3, and 4).

from bridges or from a cable car (Fig. 38.6). Because the velocity of the flowing water varies with distance from the banks and streambed, an accurate calculation of discharge requires measurements of depth and velocity at numerous locations across a large stream, including measurement of velocity at two or more depths in each location. Automated gauging stations continuously measure water level along tens of thousands of rivers around the world. With careful calibration, these measurements can be used to make accurate estimates of discharge and study its change over time.

Water Beneath the Surface

Despite the beauty and importance of rivers and lakes, together they contain less than one percent of all the freshwater in the world (see Figure 11.4). By far, the greatest proportion of the world's freshwater (about three-fourths) is locked in glaciers and ice sheets. But most of the remaining one-fourth of the total freshwater supply is hidden beneath the ground, within the lithosphere, and is called **groundwater**. A raindrop that falls to Earth may remain above ground, as we have seen, but it may also infiltrate into the soil. Two zones within the ground may hold this

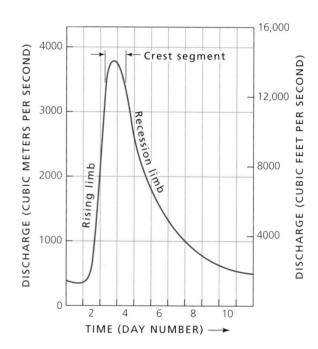

FIGURE 38.5 Hydrograph for the St. Mary River, Nova Scotia, Canada, showing the discharge associated with an individual storm.

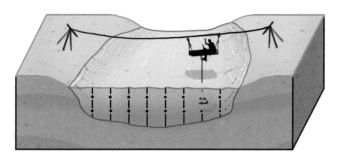

FIGURE 38.6 An accurate measurement of stream discharge requires measurements of depth and velocity of the flowing water at multiple locations across the channel. Here, they are taken using a small car hanging from a cable stretched across a river.

water, distinguished by whether or not they are saturated—that is, whether all open space in the rock or sediment is filled with water. The upper zone, which is usually unsaturated except at times of heavy rain, is the **vadose zone**, or the *zone of aeration*. Below this is the **saturated zone**, sometimes called the *phreatic zone* (Fig. 38.7). The vadose zone includes the soil, so as a raindrop infiltrates into the soil surface, it enters the vadose zone.

SOIL MOISTURE IN THE VADOSE ZONE

Once rainwater has infiltrated into the soil, it is called *soil moisture*, and any further movement is by processes other than infiltration. The movement of water, under the influence of gravity, from the upper part of the soil down through the pores to deeper horizons or underlying rocks is called *percolation*. Percolation is most rapid when the soil is quite wet and many of the largest pores are filled with water. Once infiltration has ended, eventually water drains out of the larger pores and percolates downward below the soil. As discussed in Unit 22, water is held within the small pores quite strongly against the force of gravity. As the soil drains, therefore, percolation under the force of gravity becomes much less important.

Water is also drawn from wetter to drier soil, a process sometimes called *capillary action*. This is the same process

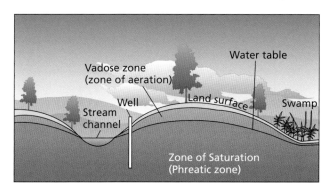

FIGURE 38.7 Two subsurface water-holding zones. The term *groundwater* applies to water lying below the water table in the saturated zone.

we observe when water is drawn up into the pores of a dry paper towel—water will move some distance into small pores even against the force of gravity. Capillary action can move water downward or upward, depending on moisture conditions in the soil; in all cases this process draws water from moister to drier parts of the soil. Moisture can also move within the soil through evaporation, movement of water vapor, and recondensation onto new surfaces.

Field capacity is the maximum amount of water that a soil can hold by *capillary tension*, the force that holds water in small pores against the force of gravity. If we wet a paper towel and then hold it up in the air, quite a bit of water will remain in the paper's pores. Soils retain water even more effectively in their smaller pores. Essentially, field capacity is the water still held in those smaller pores just after water has rapidly drained from the largest pores. Plant roots can extract much of the water that remains at field capacity and carry it up through their stems to the leaves, where it evaporates. Eventually, the only remaining water is in the tiniest pores or in thin films coating mineral grains. This *hygroscopic* water cannot be taken up by plant roots. When the remaining water in a soil is so tightly held that plants can no longer take up enough water to keep from wilting, the soil is said to be at the *wilting point*, which varies for different soils and different plant species.

The texture and structure of the soil both help determine the amount of water it holds at field capacity and at the wilting point. The *moisture retention capacity* of the soil is the water content at field capacity minus the content at the wilting point. This is the water typically available for use by plants, so this is an important quantity for agricultural purposes and for understanding how water stress affects natural ecosystems.

GROUNDWATER IN THE SATURATED ZONE

If you dig or drill a hole far enough into Earth, chances are that sooner or later you will come to a layer that is permanently saturated with water. This is the top of the saturated zone, a surface called the **water table** (see Fig. 38.7). Instead of lying horizontally, the water table tends to follow the topography of the land surface, though with less relief, as Figure 38.7 indicates.

We can classify the materials of the lithosphere below the water table according to their water-holding properties. These are also illustrated in Figures 38.8 and 38.9. Porous and permeable layers that can be at least partially saturated are **aquifers**. Sandstone and limestone often are good aquifers. Other rock layers, such as those of mudstone and shale, consist of tightly packed particles, and groundwater flow through them is so slow that they are effectively impermeable. These are known as **aquicludes** (sometimes called *aquitards*). You can imagine aquifers and aquicludes in the earth as layers of sponge-like material separated by layers of plastic that prevent water flow. An *unconfined*

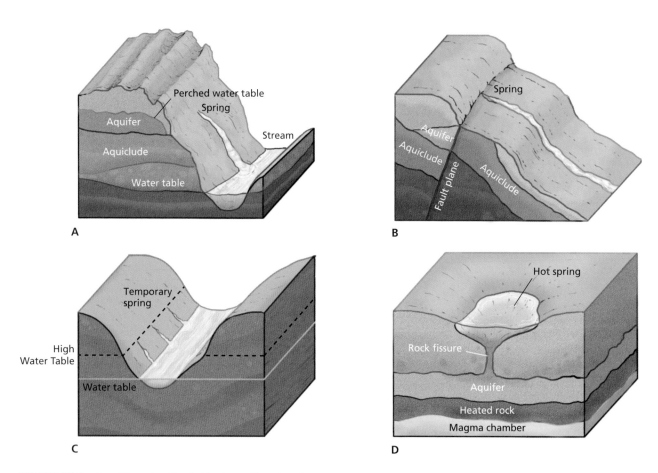

FIGURE 38.8 Conditions leading to the formation of springs.

aquifer obtains its water from local infiltration, as if the sponge were at the surface (Fig. 38.9). The water table is the top of an unconfined aquifer. A *confined aquifer* exists between aquicludes.

In an unconfined aquifer, groundwater flows from where the water table is high to where it is low. Since the water table follows the overlying topography, this means

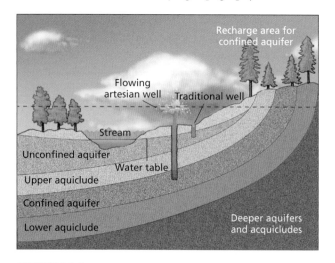

FIGURE 38.9 Aquifers, aquicludes, and their relationship to the water table and wells. The dashed red line is the level to which water from the confined aquifer will rise in a well.

that groundwater flows from parts of the aquifer underlying hills or other high terrain toward valleys. Where the water table intersects the ground surface, usually in low areas, it emerges from the ground in a process called **groundwater discharge**. Groundwater discharge often involves water seeping out of the beds of streams and lakes, where it tends to go unnoticed.

A **spring** occurs where discharge forms a stream of flowing water that emerges from the ground. The distribution of springs in a particular landscape was often valuable information in the past, when well-drilling technology did not provide easy access to groundwater. Take a look at a map of the region you live in and note how many place names record the location of springs.

Springs can occur in a number of settings. Most commonly, they form along the edges of valleys where the land surface meets the water table. A slightly more complicated setting involves an aquiclude that halts the downward percolation of water, which is then forced to flow from a hillside, as shown in Figure 38.8A. Occasionally, as this diagram also shows, an aquiclude leads to the formation of a separate water table, called a **perched water table**, at a higher elevation than the main water table. Sometimes water finds its way through joints in otherwise impermeable rocks, such as granite, and springs form where it reaches the surface. Faulting can affect groundwater flow paths so that springs form, as shown in Figure 38.8B. Furthermore, after long periods of heavy rainfall exceptionally high water tables can occasionally raise the water level high enough to cause temporary springs, as shown in Figure 38.8C.

Water lost from the aquifer through discharge is replaced by **groundwater recharge**, most of which is water that has infiltrated and percolated downward from the land surface above the aquifer. Recharge is more complicated for a confined aquifer, which often obtains its water from a distant area where the rock layer of the aquifer eventually reaches the surface. This is illustrated in the upper right portion of Figure 38.9.

An interesting variation on ordinary groundwater systems involves **hydrothermal** water. This is groundwater heated by magma that is relatively close to the surface (Fig. 38.8D). Many quite distinctive minerals and valuable ores of lead, zinc, copper, gold, and other metals form when material dissolved in hydrothermal water precipitates to form solid crystalline material.

Hydrothermal water is also responsible for the formation of *hot springs*. The water flowing from these has a temperature above that of the human body (37°C [98.6°F]). Hot springs are common in the mountains of Idaho and in nearby Yellowstone National Park, where magma rises at a hot spot or mantle plume (Fig. 38.10 and the photo that opens this unit). Near Rotorua, New Zealand, hot springs are used for cooking and laundry, and steam from them is used for generating electricity. Similarly, *geothermal* steam powers turbines in the hills north of San Francisco, which produce electricity for more than 1.5 million homes. Virtually all of Reykjavik, the capital of Iceland, is heated by water from hot springs, as are parts of Boise, Idaho.

Human Use of Groundwater

In many parts of the world, settlements are located near wells and springs. In such cases, geology directly influences human locational decision-making. In theory, wells can be dug or drilled wherever an aquifer lies below the surface, though in some places the aquifer is so deep that a well is too expensive and impractical.

The *traditional well* is simply a circular opening or drill hole in the ground that penetrates the water table. Water is then drawn or pumped to the surface. Most modern wells are drilled, with a casing (a large-diameter pipe) installed from the surface to some depth below the water table to prevent contamination from surface runoff (e.g., road salt) and infiltration by other chemicals. A submersible pump is then installed to raise water to the surface.

The French region of Artois is blessed with a confined aquifer whose water supply is recharged from a remote location. Wells sunk into this aquifer produce water that rises to the surface under its own natural pressure (see Fig. 38.9). Artois has lent its name to this type of *artesian well* ("artesian" means from Artois), but such wells are common in many other parts of the world as well. The pressure in artesian wells can sometimes be quite strong. One dug in Belle Plaine, Iowa, spouted water over 30 m (100 ft) into the air and had to be plugged with 15 wagonloads of rock to bring it under control.

Once a well is installed, it becomes part of the groundwater flow system. Water drawn from wells must be replaced by groundwater recharge, or the water stored in the aquifer will decrease. In an unconfined aquifer, this means the water table will drop to a lower level, sometimes below the level of existing wells or springs. Aquifers all over the world have been affected by this kind of *drawdown* when pumping from wells exceeds recharge.

In parts of the western Great Plains and California's Central Valley, heavy groundwater use for irrigation has lowered the water table dramatically. In the Central Valley, this drawdown has even caused the ground surface to subside, by more than 8 m (26 ft) in some places, as groundwater extraction allowed the sediments under the valley floor to become more compact. In 2015, after years of drought depleted the water supply from rivers, Central Valley farmers relied increasingly on groundwater, raising concerns that drawdown and subsidence could become even more severe.

Individual wells can also cause local drawdown; that is, the local water table directly surrounding the well drops, forming a *cone of depression* (Fig. 38.11A). Excessive pumping from wells in coastal areas can cause another problem:

FIGURE 38.10 Old Faithful geyser in Yellowstone National Park, which erupts approximately once an hour. A *geyser* is a hot spring that periodically expels jets of heated water and steam.

ENHANCE

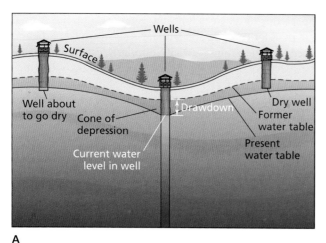

A

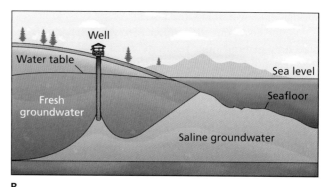

B

FIGURE 38.11 Potential disadvantages of wells. When water is pumped from a well, the local water table is lowered in the form of a cone of depression (A). Excessive pumping will make the cone so large that the water table drops below neighboring wells, leaving them dry. Near a coastline, excessive pumping can lead to the intrusion of seawater into the well (B).

ENHANCE

groundwater that is saline because of its connection with seawater can intrude into an aquifer beneath adjacent land areas, replacing freshwater. Because of the greater density of saltwater, it will always tend to intrude somewhat below the freshwater part of a coastal aquifer. As more and more freshwater is pumped from wells, the intrusion of saltwater into the aquifer increases, as diagrammed in Figure 38.11B. Eventually, many wells start to yield saltwater and become useless.

PERSPECTIVES ON THE HUMAN ENVIRONMENT

Groundwater Contamination

Residents of highly developed, industrialized countries have long taken it for granted that when they turn on a faucet, safe, potable water will flow from the tap. Throughout most of the rest of the world, however, the available water—including groundwater—is often unfit for human consumption because of contamination by various types of pollution.

A very common source of water pollution in wells and springs is human and animal waste. Drainage from septic tanks, malfunctioning sewers, privies, barnyards, and pastures widely contaminates groundwater. If water contaminated with pathogens passes through soil and/or rock with sizable openings, such as coarse gravel or cavity-pocked limestone, it can travel considerable distances while remaining polluted.

Large quantities of refuse and industrial waste products are also deposited in shallow basins at the land surface, including landfills for solid waste and lagoons for material dissolved or suspended in water (Fig. 38.12). Before the 1970s, little attention was paid anywhere in the world to making sure wastes did not leak from these basins downward into the groundwater system. Such leakage is possible even for solid waste because rainwater that percolates downward through the site can carry away soluble—and sometimes toxic—substances. Other sources of contamination were often ignored as well, including chemicals spilled or deliberately dumped on the ground surface and leaking fuel storage tanks. In this manner, harmful chemicals slowly seeped into aquifers and contaminated them. The pollutants migrated from landfill sites as plumes of contaminated water following the regional groundwater flow paths.

In the United States and other developed countries today, much stricter rules require liners and other measures to keep contaminants from leaking out of landfills, along with long-term monitoring of groundwater quality near the landfill. A great deal of effort and expense has gone into cleaning up earlier contamination, though without complete success. Many of the practices that formerly led to contamination in the United States continue in developing countries, where limited resources and ineffective enforcement hamper efforts to control them.

Now a new set of groundwater quality issues has come to the fore, closely linked with agricultural land use. High levels of nitrogen fertilizer are often used to ensure high crop yields in modern commercial agriculture, but as a result nitrogen in the form of nitrate ions escapes uptake by plants and leaks below the soil profile into groundwater systems. High nitrate levels—a health hazard for infants, in particular—have been detected in aquifers underlying areas of intensive farming in the United States and other countries. Certain pesticides have also been detected in groundwater at unacceptable levels. The solution here is more complex than designing safe landfills. Many agricultural practices need to be adjusted to limit the loss of fertilizers or pesticides below the soil profile or to reduce their use in other ways, and farmers must be convinced to adopt the new practices.

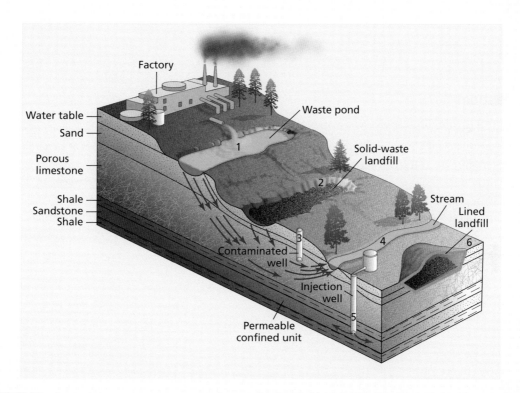

FIGURE 38.12 Groundwater system contamination resulting from improper disposal or storage of waste. Hazardous materials or pathogens from waste in an open waste pond (1) and an unlined landfill (2) percolate downward and contaminate an underlying aquifer. Also contaminated are a well downslope (3) and a stream (4) at the base of the hill. Injection into a deep, confined rock layer that lies well below aquifers used for water supplies (5) was often recommended in the past, but the safety of this procedure is now questioned. A carefully constructed surface landfill that is fully lined to prevent downward seepage of wastes (6) is an accepted practice today, but it must be carefully monitored. Reduction of waste production and recycling of waste products are often the best options where they are feasible.

The movement of water on and through the surface of Earth has important consequences for the erosion, transportation, and deposition of surface materials. In the following units, we highlight the various pathways this water takes, the environmental controls determining their relative magnitudes, and the resultant geomorphic work that is being done. We begin our survey in Unit 39, which explores the processes of stream erosion.

Key Terms

aquiclude *page 465*
aquifer *page 465*
discharge *page 463*
field capacity *page 465*
groundwater *page 464*
groundwater discharge *page 466*
groundwater recharge *page 467*
hydrograph *page 463*

hydrothermal *page 467*
impermeable *page 461*
infiltration *page 461*
infiltration capacity *page 461*
interception *page 461*
perched water table *page 466*
permeable *page 461*
runoff *page 462*

saturated zone *page 465*
slope *page 463*
spring *page 466*
vadose zone *page 465*
velocity *page 463*
water table *page 465*

Scan Here for a quick vocabulary review

ENHANCE

Review Questions

1. Why does the infiltration rate generally decrease over the course of a rainstorm? Cite at least two reasons.

2. How do vegetation cover, soil texture, and soil structure influence infiltration?

3. What is a hydrograph, and how is the information shown on a hydrograph collected?

4. Explain why field capacity is an important property of soils from the standpoint of plant growth.

5. How does the water table relate to the saturated and vadose zones, and how does it generally relate to the surface topography?

6. Describe the difference between an aquifer and an aquiclude.

7. What are some of the effects on aquifers when excessive amounts of water are pumped from wells?

Scan Here for a quick concept review

ENHANCE

 www.oup.com/us/mason

UNIT
39 Slopes and Streams

With each thunderstorm, water flows down these highly dissected slopes in Badlands National Park, South Dakota, carrying away more of the soft, easily eroded bedrock. Over time, this rapid erosion is destroying the gently sloping, grass-covered tableland in the background.

OBJECTIVES

- Describe the processes responsible for the erosion of slopes by water
- Explain the concept of sediment yield from a stream's drainage basin
- Analyze the movement of sediment through river systems by processes of erosion, transportation, and deposition
- Discuss the river as a system

Streams are the most important sculptors of terrestrial landscapes. In Part Seven, we study the landscapes carved by glaciers, molded by wind, and shaped by waves. As impressive as they are, none of these erosional agents come close to flowing water as the principal creator of landforms and landscapes on our planet. Even where ice sheets advanced and receded and where deserts exist today, channelized flowing water has played a major role in modifying the surface.

The Latin word for river is *fluvius*, from which the term *fluvial* is derived, denoting running water. **Fluvial processes** are the geomorphic processes associated with flowing water, and fluvial landforms and landscapes are produced by streams. Recall from Unit 35 that rivers erode sediment, transport it over short or long distances, and then deposit it. As a result, the landscape contains fluvial landforms that are *erosional*, created primarily by the removal of rock, and fluvial landforms that are *depositional*, resulting mainly from the accumulation of sediment. The Grand Canyon of Arizona is an erosional landscape; the delta of the Mississippi River is an assemblage of depositional landforms. Of course, stream sediment is temporarily deposited on sandbars in the Grand Canyon, and erosion reshapes parts of the Mississippi Delta, but our concern here is to make a basic distinction between landforms based on the predominant process that created them. After we study fluvial processes in this unit, we will examine the landscapes they create in later units.

Erosion of Slopes by Water

In Unit 11, we examined the hydrologic cycle, the global system that carries water from sea to land and back again. This unceasing circulation of water ensures the continuation of fluvial erosion because the water that falls on the elevated landmasses always flows back toward sea level. As it does so, it carries the products of weathering with it. In Unit 37, we focused on the processes of mass movement, often the first step in moving weathered rock toward the stream system. In the current unit we turn to erosion of slopes by water falling as raindrops and flowing downslope to streams.

Rainfall comes in many different forms, from the steady, gentle rain of a cloudy autumn day to the violent, heavy downpour associated with a midsummer afternoon thunderstorm. During a misty drizzle, the soil generally is able to absorb all or most of the water because the infiltration rate remains higher than the rate of rainfall. This means that no water collects at the surface. But as we discussed in Unit 38, when rain falls at higher intensities, or for long durations, it may sooner or later exceed the soil's infiltration capacity, resulting in *runoff*. Runoff generally begins as *sheet flow* (sometimes called *sheet wash*). Sheet flow is defined as a thin layer of water that moves downslope without being confined to channels; however,

FIGURE 39.1 Rainsplash erosion. The impact of a drop of water is shown here on a lake surface, but if the drop hits the ground, soil particles are thrown into the air. When this occurs on a slope, the particles thrown downhill strike the ground a little farther from their starting point than those thrown uphill. Cumulatively, this process contributes to downslope movement of the uppermost soil surface.

it is often made up of many small strands of flowing water rather than a uniform sheet.

Raindrops dislodge and scatter soil particles, as we saw in Unit 38. In that unit, our focus was on the formation of a dense, impermeable surface layer of soil that limits infiltration, but raindrop impacts also drive downhill movement of soil in a process called **rainsplash erosion** (Fig. 39.1). If the surface is flat, much of the scattered soil may be deposited nearby, and the area suffers little net loss of soil. But if the exposed soil lies on a slope, rainsplash erosion results in a downslope transfer of soil. The steeper the slope, the faster downslope movement proceeds. Furthermore, many of the dislodged soil particles are quickly carried away by sheet flow.

Besides carrying soil dislodged by raindrop impacts, sheet flow can cause considerable erosion by detaching the soil it flows across, a process referred to as **sheet erosion**. As runoff continues, small short-lived channels called *rills* emerge and grow. Rills may merge into larger and longer-lasting *gullies* (Fig. 39.2). These in turn feed water and eroded soil into permanent streams. Both water and sediment become part of the river system that is shaping the regional landscape.

Vegetation plays an important role in restraining the erosional forces of raindrop impact and runoff. Leaves and branches break the fall of raindrops and reduce their erosive impact, and the roots of plants bind the soil and help it resist removal. Where natural vegetation is removed to make way for agriculture or perhaps decimated by overgrazing, accelerated erosion often results.

FIGURE 39.2 Sheet erosion is common in the barren, semiarid landscape of the Bighorn Basin, Wyoming. Rills have formed in the slope at left, feeding eroded sediment into the larger gully in the center of the photo.

Drainage Basins and Sediment Yields

A river system consists of a *trunk river* joined by a number of *tributary streams*, which are themselves fed by smaller tributaries. These branch streams diminish in size the farther they are upstream from the trunk river. The complete system of the trunk and its tributaries forms a network that occupies a region known as a **drainage basin**. One well-known drainage basin is the Mississippi Basin in North America: the Mississippi River is the trunk river in that basin, and the Missouri and Ohio rivers are its chief tributaries (Fig. 39.3). A great number of smaller drainage basins are nested within a large unit like the Mississippi Basin.

A drainage basin is defined by the organization and orientation of water flow. Within a drainage basin, all streams flow into other streams, which ultimately join the trunk river. In the United States, a drainage basin is also sometimes referred to as a *watershed*. Adjacent drainage basins are separated from each other by topographic rises called *drainage divides*. One especially prominent drainage divide is the Continental Divide in the western United States, where it follows the spine of the Rocky Mountains (see the red line in Fig. 39.3). Water from the eastern slope of the central and northern Rocky Mountains flows into the Mississippi drainage basin and reaches the Atlantic Ocean via the Gulf of Mexico. Water falling on the western slope becomes part of the Colorado or Columbia river system and eventually flows into the Pacific Ocean.

To assess the erosional activity of a river system in a defined basin, physical geographers take a number of measurements. An obvious one is the quantity of sediment that passes a cross-section of the trunk river at or near its mouth. The mass of sediment leaving a stream basin per year is the **sediment yield**. We might expect that the larger a drainage basin is, the larger the sediment load carried out of it will be. But this is not always the case. The vast Amazon Basin of equatorial South America, for instance, is more than six times as large as the basin of India's Ganges River. But the Amazon's annual sediment load, measured at its mouth, is less than one-fifth that of the Ganges. What this means is that the sediment yield per square kilometer of the Amazon Basin is less than one-thirtieth of the yield per square kilometer from the Ganges Basin.

By dividing the total sediment yield by the area of the basin, we can begin to compare the effectiveness of erosion in each of these great river basins. But determining the rate of erosion is more complicated because the Amazon is supplied with sediment mainly from the Andes Mountains, which make up only a small portion of the entire drainage basin. Similarly, much of the sediment load of the Ganges comes from the Himalayas, not from the lowlands it flows through near its mouth. Furthermore, the sediment yield may not reveal all the erosional work that goes on inside the drainage basin because some material is deposited as **alluvium** (sediment laid down by a stream on its valley floor) and thus does not reach the lowest course of the trunk river.

What, then, are the conditions that influence the rate of erosion? The amount of *precipitation* is an obvious factor; more available water means that more erosion occurs. But the *vegetation* cover also plays an important role because it inhibits erosion. The *relief* in the drainage basin is important as well. In a basin where the relief is generally low, slopes are less steep and erosion is less active than in a basin where the relief is high. A large portion of the Amazon Basin has dense forest cover and low relief, a combination that reduces the erosion rate significantly. Another factor is the underlying *lithology* (rock type). Soft sedimentary rocks erode much more rapidly than hard crystalline rocks. We must also consider *human impact* as a factor influencing stream basin erosion rates. Human activity can affect the fluvial system in many ways. The most common human impact is an increased rate of regional erosion (Fig. 39.4); however, when we build dams, sediment is trapped in the reservoirs that result, reducing the sediment yield that major rivers carry all the way to the ocean.

Some rivers—for example, rivers that drain areas of easily weathered limestone bedrock—carry large quantities of dissolved material. We take this dissolved load into account along with sediment load when we calculate the rate of **denudation**, the overall lowering of a landscape that results from the combined processes of weathering and erosion.

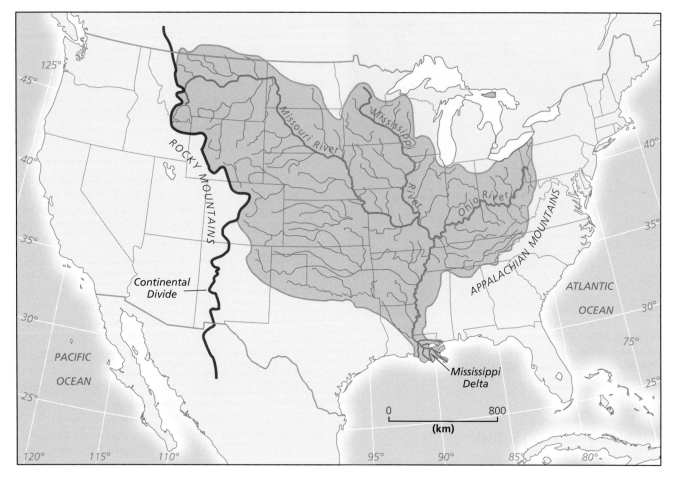

FIGURE 39.3 Drainage basin of the Mississippi River. It includes all the areas of the central United States drained by the Mississippi (trunk river) and its chief tributaries, the Missouri and Ohio rivers.

The Work of Streams

As streams modify the landscape, they perform numerous functions, which can be grouped under three headings: *erosion*, *transportation*, and *deposition*. Before we examine these processes, let us consider the dynamic nature of the river system as a whole. The Greek philosopher Heraclitus is famous for saying that we can never step in the same river twice. Although he may have been considering the constant renewal of water in the stream, a river's channel, valley, and drainage basin are continually changing as well.

Bends in the river channel migrate and change shape over time, as do sandbars. Where the stream channel erodes a slope on one side of the valley or another, the valley is widened. In the area where the river has its origin, streamlets merge to form the main valley. Over time, the "head," or source, of the valley is extended upslope in a process called *headward erosion*. At the other end of the valley, the stream reaches the coast of a sea or lake and may deposit its load in a **delta** (Fig. 39.5). The delta gradually extends out into the water body until the river abruptly changes course—often as it spills out of its banks during a flood and takes a shorter path to the sea. Then the river begins to build its delta in a somewhat different direction.

FLOODS

In Unit 38, we introduced the concept of stream *discharge*—the volume of water flowing past a point along the stream channel in a given period of time. Recall that stream discharge often varies dramatically over time as a result of changing water inputs from adjacent upland areas and tributaries. The simplest definition of a **flood** is a distinct peak of discharge, though many floods defined in this way would barely be noticed. *Overbank* floods, when a stream overflows its channel, draw more attention, in part because of potential hazards to people and property.

Floods are also of great interest to physical geographers because they do much of the total work of rivers in shaping the landscape. Interestingly, though, for many streams of humid regions it is the much more frequent small floods that are most important in this role. Geomorphologists M. Gordon "Reds" Wolman and John P. Miller proposed that the most effective stream flows, those most important in moving sediment and shaping the channel, are about the size needed to fill the channel to the top of its banks. Such *bankfull* floods typically occur every one to three years on average. Although geomorphologists have subsequently found numerous exceptions to this generalization, we still find that many humid-region streams

FIGURE 39.4 The Loess Plateau of north-central China is a landscape carved by water erosion out of thick deposits of wind-blown dust. The sediment yield from this easily eroded landscape to the Huang He (Yellow River) dramatically increased over the past 2,000 years as almost all available land was put into use as agricultural fields.

are shaped mainly by relatively small, frequent floods. Extreme floods can have dramatic effects such as the aggressive scouring of streambeds and banks, the reshaping of the channel, and the undercutting of valley sides, but they occur so rarely that their effects are limited compared to the work of many smaller floods.

At this point, you might logically ask how we know that a certain size of flood occurs every one to three years on average. In fact, we have a remarkable set of data to work with: measurements of stream flow collected continuously for many decades on thousands of large and small streams. In the United States, the U.S. Geological Survey collects this information through automated stations and field measurements by technicians. One way to summarize this information is to plot the largest flood in each year over the period of record (or the largest floods above a certain threshold discharge). A typical example of this sort of plot is shown in Figure 39.6; note the frequent smaller floods and the rare large ones.

From data like these, hydrologists can use statistical methods to estimate the **recurrence interval** of a particular stream discharge. This is the average time between

floods that reach or exceed a given magnitude. For example, for the river shown in Figure 39.6, a discharge of 7,000 cubic feet per second (cfs) has a recurrence interval of about 10 years. That is, the average time between floods of 7,000 cfs or larger is 10 years. Alternatively, we could say the "10-year flood" is 7,000 cfs, but that wording may suggest that such floods occur every 10 years in a predictable fashion. In reality, the actual time between floods of 7,000 cfs or larger has varied widely; in fact, there was no flood of this size in the 23 years from 1979 to 2002. This is the highly variable nature of stream flow. Knowing the recurrence interval for a flood of 7,000 cfs is still useful, however, since it tells us that floods this large don't occur almost every year and are also not extremely rare events.

EROSION BY RIVERS

As we have noted, streams are the most important and efficient of all agents of erosion, continuously modifying their valleys and changing the landscape. Stream erosion takes place in three ways: *hydraulic action*, *abrasion*, and *corrosion*.

FIGURE 39.5 An astronaut's view of the Nile Delta, looking southward from a position over the Mediterranean Sea (bottom). The city of Cairo, Egypt's capital, lies near the junction of the apex of the delta's triangle and the Nile River channel. A small part of the El Faiyum Depression is visible just west (to the right) of the Nile River at the top of the photograph. A portion of the Gulf of Suez, Great Bitter Lake, and the Suez Canal can be seen to the east (left); some cirrus clouds hover close to the ancient city of Alexandria near the western point of the delta (far right). The dark color of the delta itself reflects intensive agriculture in contrast to the nearly empty desert flanking it. A huge population is concentrated here and, as the image shows, in the lower valley of the Nile beyond.

Abrasion **Abrasion** refers to the mechanical erosive action of boulders, pebbles, and smaller grains of sediment as they are carried along within the river. These fragments dislodge other particles along the streambed and banks, thereby contributing to the process of deepening and widening. Gravel- and sand-sized particles tend to scour rock exposed in the streambed, wearing it down while evening out the rough edges. One striking form of abrasion is the erosion of potholes in rock streambeds, in which rounded holes up to several meters deep are "drilled" into bedrock by swirling sand or gravel. Abrasion and hydraulic action most often function in combination.

Hydraulic Action The work of the water itself as it dislodges and drags away rock material from the valley floor and sides is referred to as its **hydraulic action**. Large volumes of fast-moving water can break loose sizable boulders and move them downstream, a process sometimes called *plucking*.

Corrosion In terms of the volume of rock removed, **corrosion** is usually the least important form of erosion by streams. It is the process by which certain rocks and minerals are dissolved by water. Limestone, for instance, is eroded not only by hydraulic action and abrasion but also

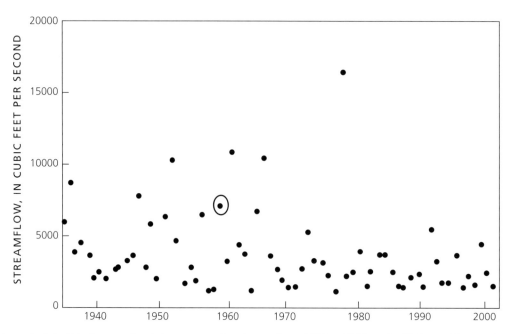

FIGURE 39.6 A plot of the largest flood that occurred each year from 1934 to 2002 on the Kickapoo River at Steuben, Wisconsin, based on U.S. Geological Survey measurements. The circled flood has an estimated recurrence interval of 10 years.

through corrosion. In Unit 42, we discuss the special landscapes formed when corrosion by water flowing underground is the dominant form of erosion.

TRANSPORTATION BY RIVERS

Erosion and transportation go hand in hand. The materials loosened by hydraulic action and abrasion are carried downstream together with dissolved minerals. Along the way, different transportation processes are dominant. In the mountains, streams carry boulders and pebbles down their steep channels; near the coast, there are few if any pebbles on the beds of large rivers, but the water is brown or gray with finer sediment. The three major categories of material transported by streams are *bedload*, *suspended load*, and *dissolved load*.

Bedload **Bedload** refers to particles that remain fairly close to the streambed as they are transported. Bedload usually consists of sand, gravel, or boulders, and it is moved downstream because of force applied to it by the flowing water. In all but the deepest flows, the swirling eddies of water cannot lift the largest particles above the bed against the force of gravity; therefore, they slide or roll along the riverbed, a process known as **traction**. Somewhat smaller particles hop along the bed, a process called **saltation** (Fig. 39.7).

Suspended Load Fine sediment, including almost all of the clay and silt and part of the sand, is carried within the stream by a process known as **suspension** (see Fig. 39.7). The grains that make up this **suspended load** are denser than water, so they do not actually float and instead settle toward the bed under the force of gravity. However, this

settling is a slow process for such small grains, and they are often dragged upward by eddies of water. Thus, once a suspended grain is set in motion, it may not return to the bed for a long distance downstream, and suspended particles become scattered throughout the depth of the stream. When suspension dominates, stream water becomes muddy, and a large river can carry huge amounts of fine suspended sediment toward its mouth.

Dissolved Load Some rock material dissolves in stream water and is carried downriver as **dissolved load**. Numerous minerals can be partially dissolved, and even a clear mountain stream contains ions of calcium, sodium, potassium, and other dissolved materials.

STREAM POWER

When we compare a small mountain stream to a great river like the lower Mississippi (Fig. 39.8), it seems obvious that the large river can move greater quantities of sediment downstream per year. Yet there may be large boulders that are clearly in motion during floods on the mountain stream that we would not find moving on the bed of the lower Mississippi River. The term **transport capacity** denotes the maximum load of sediment that a stream can carry at a given discharge. **Competence** refers to the largest particle size that a stream can set in motion. The Mississippi clearly has a large transport capacity, but under most conditions it does not have the competence to move boulders. The mountain stream in contrast has impressive competence. Conditions of flow such as stream depth, slope of the flowing water surface, and discharge influence both transport capacity and competence.

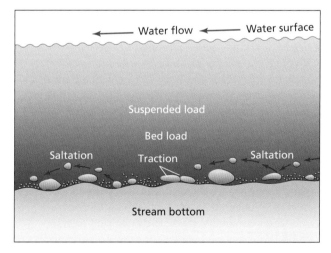

FIGURE 39.7 Larger sand and gravel fragments roll and slide along the streambed in traction, while smaller fragments in the bedload advance downstream by saltation. Even smaller silt and clay particles are carried above in the suspended load, maintained there by the stream's turbulence.

ENHANCE

FIGURE 39.8 The lower Mississippi River and its floodplain in northeastern Louisiana, looking downstream toward the state capital, Baton Rouge. Sediment eroded from the entire Mississippi Basin (shown in Fig. 39.3) passes through this reach of the river, although much sediment is stored along its way to the ocean for periods ranging from decades up to thousands of years.

A river's transport capacity is closely related to its **stream power**, which is a general indicator of the stream's ability to do work. The total stream power of a river increases with *discharge* and *slope* of the water surface. Therefore, the larger the discharge and the steeper the slope, the more sediment the river can transport downstream. Given that information, we can conclude that the Mississippi River does indeed have greater transport capacity than a small mountain stream. Although the slope of the Mississippi is much less, its discharge is vastly greater.

It is important to remember that a river's transport capacity is the *maximum* sediment load that it can carry. The actual sediment load can be less if the supply of material available to the stream is limited. Most rivers have the capacity to carry more suspended silt and clay than they normally do if they were supplied with this added material by soil erosion in the drainage basin. The amount of bedload that is transported downstream is often closer to the river's full capacity, but there are interesting cases in which a stream has excess capacity. Where a river flows over bedrock in a narrow canyon, some of its power is used to erode the rock beneath it rather than simply carrying gravel or boulders supplied from upstream.

The competence of a stream is related to the force it can apply to particles on the streambed; much more force is required to set a boulder in motion than a grain of sand. This force increases with the depth of water and the stream's slope. This is why mountain streams can sometimes move large boulders that could not be transported by much larger rivers. The mountain streams are very steep, and when melting snow produces an early summer flood, the combination of depth and slope is enough to move boulders and all smaller particles on the bed. The shape of the stream channel is clearly important here. A wide, shallow channel has less competence than a narrow, deep channel, even if the discharge is the same.

FOCUS ON THE SCIENCE

Measuring Stream Sediment

The Grand Canyon and other canyons of the Colorado River system are not only striking landforms, but also outstanding case studies of the power of streams to erode rock. When we see the rivers that carved out these canyons, it can be hard to comprehend the amount of time required for them to accomplish that task.

Scientists have long been interested in how fast rivers can erode rock. To answer this question, we clearly need information on how much eroded rock material is carried away toward the oceans each year by streams such as the Colorado and its tributaries. Because a stream's sediment load varies greatly with discharge, many field measurements are required to make even a rough estimate of the total material transported per year.

Fortunately, there are also good practical reasons for such an expensive and time-consuming measurement program. Dams have been built on the Colorado River, as on many streams, to control the flow and store water in reservoirs for use by farms, households, and industries. Whenever a large dam is planned, one of the important considerations is the time it will take for the reservoir behind it to fill with sediment carried in from upstream. Lake Mead is a large reservoir downstream from the Grand Canyon that stores water used by millions of people and by crops on millions of acres of farmland. Since this reservoir was first planned, sediment input from muddy rivers like the Little Colorado (Fig. 39.9) has been a major concern.

Mainly because of such practical issues, the U.S. Geological Survey carries out numerous measurements of sediment load on the Little Colorado and many other streams. These measurements are also of great value to scientists studying stream processes and long-term development of the landscape, however. Measuring suspended load is relatively simple. Bottles are lowered into the stream at a series of locations across the channel to take up muddy water at a carefully controlled rate. The bottle must be moved vertically during sampling because sediment concentration can change greatly from depths near the bed to the surface. Suspended sediment is then filtered from the water trapped in the bottle and dried to determine the mass of sediment in a given volume of stream water. Measuring bedload is much more difficult and is rarely done on most streams. The rounded boulders visible in Figure 39.9 (lower right) have moved as bedload during past floods but would be difficult to trap in any practical sampling device.

FIGURE 39.9 A technician from the U.S. Geological Survey measures suspended sediment in the Little Colorado River, just upstream from where it joins the much larger Colorado River within the Grand Canyon. He is using a bottle held on a pole to collect a sample of sediment-laden water; other bottles are visible in his basket.

DEPOSITION BY RIVERS

Stream power also helps explain when and where rivers deposit sediment (see Focus on the Science: Measuring Stream Sediment). If we observe a river over a period of years, we can see many cycles of erosion and deposition. When the river deepens during a flood, the resulting increased discharge adds stream power, and sand and gravel on the bed begin to move downstream in larger and larger quantities. As the flood recedes and stream power drops, the bedload is deposited again to wait for the next flood. In this case, the sediment is only stored temporarily.

Over the longer term, the sediment supply from the drainage basin can increase, overwhelming the capacity of the stream to transport it far downstream. Then, along at least some parts of the stream's course, more sediment will be deposited than is eroded. Eventually, this sediment accumulation will fill the river's valley and raise the level of its bed, a process called **aggradation**. We will return to this process and other long-term adjustments of the stream system later in this unit.

Finally, there are locations in the landscape where rivers continue to deposit sediment over long periods of time. In Unit 41, we discuss landforms produced in such settings, including deltas, where a stream flows into the ocean or a lake, and alluvial fans, where streams emerge from mountains onto lowlands.

BASE LEVEL

John Wesley Powell, one of the best-known American geologists of the nineteenth century, is famous for his trips by wooden boat down the canyons of the Colorado River. As he and his party floated through the Grand Canyon, he must have pondered the limits of erosion by streams: Why do some carve deep canyons, whereas others flow across plateaus and plains in shallow valleys? In trying to answer this question, Powell proposed the term **base level**—the level below which a stream cannot erode its bed. He recognized that for a river like the Colorado, which flows to the ocean, sea level is the **ultimate base level**. No stream in the Colorado River Basin can erode below that ultimate limit. This is true of all river systems that reach the ocean. The lower Mississippi River channel shown in Figure 39.8 is not far above sea level, but the Gulf of Mexico is still downhill from the location of that photo. A continuous downhill grade is required to move sediment eroded from the Mississippi River Basin all the way to the ocean.

When we consider this need for a channel that continuously drops in elevation as it winds its way to the ocean, we see that there is another important kind of base level. For every reach or segment of a stream, we can assume that it will not erode below the level of the next reach downstream. Smaller streams cannot erode below the level of larger rivers or lakes that they flow into. These points are especially evident in the

FIGURE 39.10 View toward the canyon of the Colorado River in Utah (the large stream in the middle of the photo), looking down the side canyon eroded by a small tributary stream. The Colorado River is the local base level for the tributary, which cannot erode its canyon below the level of the Colorado.

Colorado River system, where Powell probably first pondered the idea of base level. The Colorado has eroded canyons all along its course through Utah and northern Arizona, including the Grand Canyon and a series of canyons farther upstream. Through each of these canyons, the river slopes downward toward the upper end of the next canyon. Large and small tributaries of the Colorado have eroded their own canyons but have not eroded below the level of the Colorado itself (Fig. 39.10). Therefore, Powell also recognized that there are many **local base levels** within a river system. The local base level for tributary streams is the elevation of the Colorado River where they join it. Similarly, the elevation of the river channel at the upper end of a particular canyon forms the local base level for the reach of the river just upstream.

Over the past several million years, the Colorado has eroded downward into its bed, deepening its canyons and continually lowering the local base level for its tributaries. The tributaries have responded by eroding deeper canyons themselves, lowering the local base level for many smaller streams in the process. The long-term evolution of most other river systems can also be understood by applying the concept of base level. In many cases, however, we need to consider tectonic uplift of streams relative to their base levels, which essentially has the same effect as lowering the base level.

The River as a System

The river is one of the simplest and most easily understood examples of a system that we have on Earth's surface. It is an open system, with both matter and energy flowing through it, as diagrammed in Figure 39.11. The most obvious material flowing through the system is water, entering

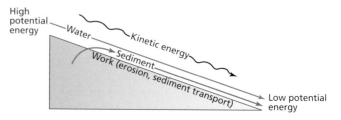

FIGURE 39.11 A river as an open system.

as precipitation in any part of the drainage basin and leaving by evaporation and flow into the ocean.

ENERGY AND WORK IN A RIVER SYSTEM

Through the action of the hydrologic cycle (see Fig. 11.3)—powered by energy from the Sun—water gains potential energy. Any object possesses *potential energy* by virtue of being raised above Earth's surface—work done against the force of gravity. Water vapor evaporates from the oceans, is raised and carried long distances through the atmosphere, and then condenses and falls as precipitation, with some falling on high parts of the continents where it has greater potential energy. The motion of this water as it runs off over the surface and then downstream in a river represents the transformation of potential energy into kinetic energy, the energy of movement (see Fig. 39.11). The water uses kinetic energy to carry its load and move itself. By the time the water has reached its ultimate base level, which is usually the sea, there is no more potential energy available. Thus, no kinetic energy can be generated, and the river is unable to perform further work.

In addition to water, another kind of material, sediment, enters and leaves the fluvial system. This sediment is produced by weathering processes, soil erosion, mass movement, and scouring of the streambed and banks. Its movement through the fluvial system is facilitated by the kinetic energy of the flowing water. The two "flows," water and sediment, define the behavior of the river system.

A GRADED RIVER SYSTEM

We can learn much about the behavior of streams by studying their longitudinal *profiles*, a plot of elevation against distance along the stream (Fig. 39.12). Longitudinal profiles show where the stream channel drops rapidly or slowly in the downstream direction. Many rivers have a distinctly concave profile, dropping quickly in their upper reaches, but with a much gentler slope downstream, where discharge is greater. Geomorphologists have often scrutinized the longitudinal profiles of rivers to understand why they developed their characteristic shape and why they change over time.

The *graded stream* is a concept often used to explain adjustments of the longitudinal profile and other stream characteristics. As with quite a few other important scientific concepts, it has proven difficult to define the graded stream in a way that satisfies all the scientists that use the term. The most common view is that a **graded stream** is one in which stream slope and channel characteristics are adjusted to provide only the capacity needed for transportation of the sediment supplied from the drainage basin. The graded river has enough stream power to convey sediment downstream, so the stream does not accumulate in parts of the stream's valley; that is, *aggradation* does not occur. At the same time, all of this stream power is consumed in transporting sediment, so the stream does not actively erode the bedrock beneath its channel. A graded stream is a *steady-state system*, and its longitudinal profile should be stable over time.

How easy is it to decide whether a stream is graded, then? We should keep in mind the constant change in the stream

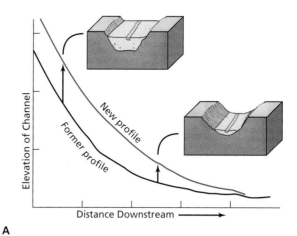

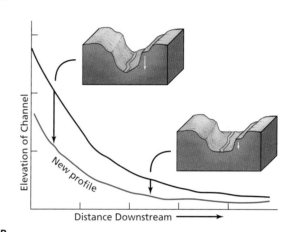

A

B

FIGURE 39.12 Longitudinal profiles of streams (plots showing how stream channel elevation drops with distance downstream), illustrating responses to changing transport capacity and sediment yield from the drainage basin. Aggradation that is greater upstream steepens the profile (A), while incision that is greater upstream reduces the slope (B).

ENHANCE

system as each flood shifts sediment from one location to another and reshapes the channel somewhat. Therefore, it can be quite difficult to discern whether the stream is actually conveying all of the sediment it is supplied with, or whether it is instead storing it somewhere along the channel or valley floor. It is probably more useful to focus on cases where sediment supply and capacity to transport sediment are clearly thrown out of balance. In those cases, when we examine the river's response, it often appears that it involves adjustment toward a graded condition—that is, toward a new balance between sediment supply and the river's transport capacity.

A dramatic example occurred in the California goldfields during the late 1800s, when miners began to use large hoses to blast loose gold-bearing gravels (see Perspectives on the Human Environment: California's Gold and Plate Tectonics, Unit 31). This *hydraulic mining* choked rivers with sediment washed out of the mines, causing dramatic

FIGURE 39.13 Aggradation of sediment in a stream valley of the California goldfields as a result of hydraulic mining, in a historic photo taken by geomorphologist G. K. Gilbert. Thick layered gravel on the left represents the initial accumulation of sediment washing into the stream, which buried and killed trees on the right. Some incision has occurred since the end of mining, exposing the dead trees.

aggradation near the goldfields because the rivers did not have adequate stream power to move all the sediment downstream (Fig. 39.13). When the mining was halted by a court decision, the rivers quickly used their power to erode away the sediment that had filled their valleys only a few decades earlier. This kind of downward erosion by streams, lowering the level of their bed, is called **incision**.

The same sequence of aggradation and incision occurs naturally where a glacier or ice sheet advances into a drainage basin. Glaciers are highly effective at eroding the rock beneath them, and the resulting debris is fed into streams that drain the glacier, causing aggradation. When the glacier retreats out of the drainage basin and sediment supply is reduced, incision usually follows.

To understand these responses of the fluvial system to changed conditions, consider what happens if the upstream part of a river experiences aggradation because it is near a glacier margin or an area of hydraulic mining. When downstream reaches of the river do not aggrade as much (or at all), the longitudinal profile is steepened (see Fig. 39.12A). This increases stream power and the capacity to transport the extra sediment being supplied to the stream, so the stream is adjusting toward a new balance—that is, toward a graded condition. When the sediment supply from mining or a glacier is cut off, incision concentrated in upstream areas makes the longitudinal profile *less* steep (see Fig. 39.12B). This is also an adjustment toward a new balance, because less stream power is needed to handle the reduced sediment supply.

Adjustment also occurs when a stream's profile is steepened by tectonic activity or when large floods become more frequent because of climate change or input of meltwater from glaciers. In all of these situations, the response will likely be incision, because of the excess stream power produced by a steeper channel or greater flood discharge. Incision that is concentrated upstream, however, will ultimately reduce the channel slope, lowering the stream power and bringing it back toward balance with sediment supply.

Finally, we should note that not all adjustments of a stream to changing sediment load or discharge involve aggradation or incision. Sometimes a stream can adequately adjust simply by flowing more directly down its valley or by forming a more winding, sinuous channel. These changes effectively increase or decrease the stream's slope and adjust its capacity to move sediment.

In Units 40 and 41, we examine the products of rivers' work in the landscape. These products range from spectacular canyons and deep gorges to extensive deltaic plains. As we study these landforms, we will learn still more about the functions of rivers, those great sculptors of scenery.

Key Terms

abrasion *page 476*

aggradation *page 479*

alluvium *page 473*

base level *page 479*

bedload *page 477*

competence *page 477*

corrosion *page 476*

delta *page 474*

denudation *page 473*

dissolved load *page 477*

drainage basin *page 473*

flood *page 474*

fluvial processes *page 472*

graded stream *page 481*

hydraulic action *page 476*

incision *page 482*

local base level *page 480*

rainsplash erosion *page 472*

recurrence interval *page 475*

saltation *page 477*

sediment yield *page 473*

sheet erosion *page 472*

stream power *page 478*

suspended load *page 471*

suspension *page 477*

traction *page 477*

transport capacity *page 477*

ultimate base level *page 479*

Scan Here for a quick vocabulary review

ENHANCE

Review Questions

1. What is the difference between rainsplash erosion and sheet erosion, and how do they work together?

2. Define the sediment yield of a drainage basin and describe some of the factors influencing it.

3. Compare and contrast the types of downstream motion and particle sizes associated with bedload and suspended load in a stream.

4. Explain the difference between the competence and transport capacity of a stream and describe the factors that influence each of these characteristics.

5. What is the difference between ultimate and local base level?

6. What are some possible causes of stream aggradation and incision, and how might these processes represent adjustment of the stream toward a graded condition?

Scan Here for a quick concept review

www.oup.com/us/mason

Landscapes Shaped by Stream Erosion

The Zambezi River exploits a combination of fault and joint structures to form the gigantic Victoria Falls.

OBJECTIVES

- Outline the effects of geologic structure, lithology, tectonics, and climate on the patterns of stream erosion
- Explain the important types of drainage networks and their origin
- Briefly describe cases in which stream erosion overcomes the control of geologic structure
- Summarize the major theories of long-term landscape development

When rivers erode the landscape, many processes occur simultaneously. Rock material is removed, transported, pulverized, and deposited. Stream valleys are widened and deepened. Mass movements are activated. Slopes are flattened in some places and steepened in others. Long-buried rocks are exhumed, exposed, and eroded. Endlessly, the work of streams modifies the topography.

The regional landscape and the individual landforms that comprise it reveal the erosional process or processes that dominate in a particular area. For example, large areas of North America and Europe are dominated by glacial landforms. Most of these landforms date back to the last glacial period, though they were modified to varying degrees by slope or stream erosion after the ice sheets retreated. Coastlines feature landforms shaped by waves, wind, and tides, and large fields of windblown sand dunes are found in certain arid and semiarid landscapes.

Landscapes shaped by stream erosion and deposition—along with mass movements and water erosion on the slopes of stream-eroded valleys—are far more extensive worldwide than those dominated by other geomorphic processes. It is important to note, however, that tectonic uplift, subsidence, faulting, and folding of crustal rocks also play critical roles in providing the potential energy that drives stream erosion and creating rock structures that strongly influence the patterns of erosion.

Factors Affecting the Patterns of Stream Erosion

Among the many factors influencing the spatial patterns of fluvial erosion, four have special significance: climate, bedrock type (or *lithology*), geologic structure, and tectonic activity.

CLIMATE

Climate plays a role in the evolution of landscapes, but this role remains open to debate. Landscapes of humid areas are often thought to be rounded and dominated by convex slopes, whereas we expect landscapes of desert environments to be stark, angular, and dominated by concave slopes. But detailed geomorphological research has shown that there are many exceptions to these generalizations, or perhaps they are not really valid generalizations at all. Part of our impression, geomorphologists say, is related to the cloak of vegetation in humid areas as opposed to the barrenness of arid zones. Strip away this vegetation and the contrast between humid and arid landscapes may be less pronounced than we expect.

On the other hand, there is a good reason to believe that many landscapes still bear the imprint of *past* climates, especially those of the last glacial period. At that time, large ice sheets covered parts of North America and Europe. When the ice sheets were at their maximum size, around 21,000 years ago, the climate of much of the world was quite different from today—colder overall, but also

drier in many places and wetter in others. Rivers that drained the ice sheets had large sediment loads and aggraded (see Unit 39), filling valleys with debris that is still partially preserved in the current landscape. Even rivers that had no connection to the ice sheets had very different discharges or sediment loads than they do in the present; some eroded more rapidly into bedrock, while others aggraded. Permafrost developed in midlatitude landscapes, and many slopes eroded more rapidly. When trying to interpret the factors influencing stream erosion of a particular landscape, we need to consider the geologic past as well as the present.

BEDROCK TYPE (LITHOLOGY)

Mount Monadnock is a well-known landform of southern New Hampshire, said to be climbed by more hikers each year than any other mountain in the United States. The word **monadnock** has come to be applied to many mountains and large hills that stand high above the surrounding landscape, like the original Mt. Monadnock (Fig. 40.1). The geology of Mt. Monadnock explains why it has not been reduced by erosion. It is composed mainly of interbedded schist and quartzite, both hard metamorphic rocks. Quartzite in particular is almost immune to chemical weathering in the New England climate. Although it can be mechanically weathered by frost action, it tends to break into large durable blocks that are not easily moved by surface runoff.

Mount Monadnock offers a clear example of the importance of rock type as a factor affecting the pattern of landscape erosion. Geologists and physical geographers use the term **lithology** to refer to the bedrock type of a local area. Different rock types have very different properties of hardness and resistance against weathering and erosion, and they also exhibit varying capacities to form

FIGURE 40.1 Mount Monadnock, southwestern New Hampshire.

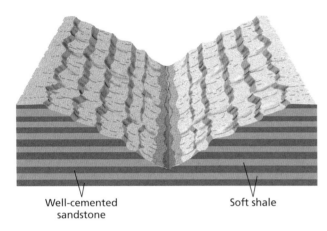

FIGURE 40.2 Alternating cliffs and gentle slopes produced by the differential erosion of resistant sandstone and weak shale.

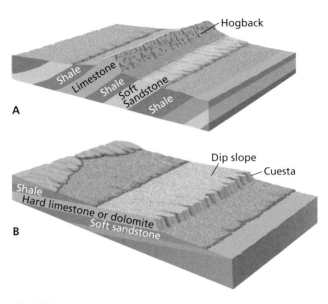

FIGURE 40.3 Hogback (A) and cuesta (B) landforms. Note that hogbacks dip at a high angle, whereas cuestas dip at a low angle. (These drawings represent a dry climatic setting, where limestone and dolomite are resistant rocks and often form hogbacks or cuestas.)

steep slopes. Strong crystalline rock masses, for example, can stand in vertical cliffs, but certain weak shales cannot form stable slopes of more than about 35 degrees. Figure 40.2 shows a common pattern visible, with many local variations, in landscapes throughout the world. Weak shales lie between more resistant sandstone layers, creating a segmented slope in which the sandstone forms small cliffs while the shale forms low-angle slopes. If the bedrock shown in this diagram had been uniform, no such pattern of alternation would have developed, and the valley would exhibit an unbroken V shape, with the angle of the sides depending on the nature of the rock.

It is hard to generalize about which rock types are more or less resistant, except in the case of the extremes such as quartzite (highly resistant) or shale (almost always nonresistant). Climate is a factor in resistance as well. For example, limestone is often resistant to weathering and erosion in dry climates but nonresistant in humid areas. The widely differing resistance of rock types to weathering and erosion is essential in explaining how geologic structure influences the pattern of stream erosion, our next topic.

GEOLOGIC STRUCTURE

Geologic structure refers to features such as synclines, anticlines, domes, and faults that were formed originally by tectonic processes such as the folding and faulting of rock (see Unit 34). Streams sculpt these geologic structures into characteristic landforms. One of the simplest examples is a landform that results from sedimentary rock strata that dip at high or low angles as they do on the flanks of an anticline or syncline. Some of these strata are much more resistant than others to erosion.

Hogbacks and Cuestas Where the rock strata are steeply tilted, the most resistant layers form prominent, steep-sided ridges known as **hogbacks** (Fig. 40.3A). Erosion by streams and slope processes has removed the less resistant, softer rock on either side of the resistant unit. In contrast, where the sedimentary strata dip at a lower angle,

the resistant rock forms a low ridge with one fairly steep slope and another very gentle slope (Fig. 40.3B). This landform, known as a *cuesta*, is not usually as pronounced in the landscape as a hogback, but cuestas can be hundreds of kilometers long. The gentle slope on one side of a cuesta is often called a *dip slope* because it parallels the dipping strata.

Ridges and Valleys When geologic structures become more complicated, so do the resulting landforms. The erosion of synclines and adjacent anticlines produces a series of parallel ridges and valleys that reveal the rock structures. If the axes of the folds plunge, as is frequently the case, the result is a terrain of zigzag ridges (see Fig. 34.13). Note that the steep face of the resistant layer (numbered 3 on that diagram) faces *inward* on the anticline and *outward* on the syncline, providing evidence of the properties of the folds below the surface. The Appalachian Mountains exhibit many examples of this topography (see Fig. 34.15).

Domes In areas where sedimentary strata have been pushed upward to form a dome, stream erosion also produces a characteristic landscape in which the affected layers form a circular pattern, as shown in Figure 40.4. One of the world's best examples of this type of landscape lies in the Black Hills in the area of the South Dakota–Wyoming border. There the crystalline core of the dome has already been exposed by erosion, but the overlying sedimentary layers still cover the dome's flanks in all directions. As a result, the regional topography consists of a series of concentric cuestas of considerable prominence separated by persistent valleys. Not only physical geographic configurations but also the human geographic

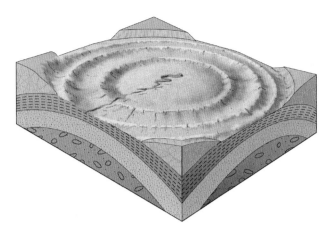

FIGURE 40.4 Circular cuestas produced by stream erosion of sedimentary rock layers that have been pushed upward to form a structural dome.

features in the area reveal the dominance of this concentric pattern: ridges, rivers, roads, and towns all exhibit semicircular patterns.

Faults Faults, too, are exposed and sometimes given relief by stream erosion. Normal and reverse faults create topography by themselves because one block moves upward or downward with respect to the other. After each episode of faulting, mass movements and stream erosion soon begin to modify the newly formed topography. In fact, physical geographers distinguish between a *fault scarp* (Fig. 40.5A), a scarp (cliff) created by geologic action

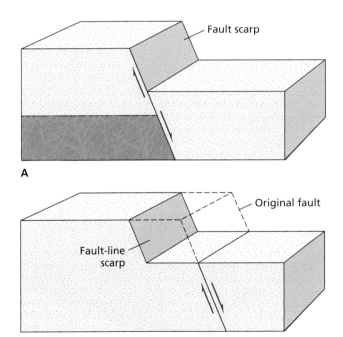

FIGURE 40.5 A fault scarp (A) originates as a result of geologic activity without much erosional modification. A fault-line scarp (B) originates as a fault scarp but then undergoes significant erosional change.

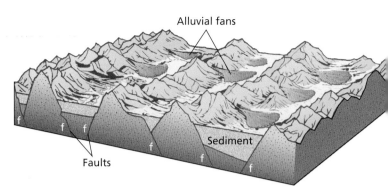

FIGURE 40.6 The erosion of upthrown blocks leads to the filling of their intervening valleys with sediment. In the southwestern United States, parallel faults create a so-called basin-and-range topography in which the fault-formed valleys are filling up with sediment from the ranges. Short streams flowing off the uplands produce alluvial fans—fan-shaped sedimentary landforms (see Unit 42).

without significant erosional change, and a *fault-line scarp*, a scarp that originated as a fault scarp but then was modified, or even displaced, by erosion (Fig. 40.5B). Where the geologic structures are formed by numerous parallel faults, as happens in regions where lithospheric plates are affected by convergent or divergent movements, the fault-generated terrain is also modified by stream erosion. Even prominent upthrust blocks may be worn down to a reduced relief, and valleys fill with sediment derived from these uplands (Fig. 40.6).

Extrusive Igneous Structures In Unit 32, we discussed volcanoes as geologic phenomena and as landforms. Created by volcanic action, volcanic landforms are quickly modified by stream erosion. Extinct and long-dormant volcanic cones reveal their inactivity through numerous and often-deep stream channels carved into their slopes. Eventually, the entire volcanic mountain may be worn down to a plug of cooled magma and radiating dikes (Fig. 40.7). Basalt and other extrusive igneous rocks cooled from lava can become the *caprocks* of plateaus or smaller landforms known as *mesas* and *buttes* (Fig. 40.8). A caprock is a layer of particularly erosion-resistant rock that forms the broad, nearly flat summits of these landforms.

Intrusive Igneous Structures The sedimentary, fault-dominated, and volcanic structures we have discussed so far have quite characteristic erosional forms. From topographic and drainage patterns, we can often deduce what lies beneath the surface. But vast areas of the landmasses are underlain by granitic and metamorphosed rocks that do not display such regularity. We noted in Part Five that large batholiths formed within the crust have been uplifted and exposed by erosion and that large regions of

FIGURE 40.7 "Flying to the west coast from Miami I had a superb view of Ship Rock in New Mexico, a famous geologic as well as cultural landmark. It is the 420-m (1400-ft)-high remnant of a large volcano that was once active in this now stable area. Dikes radiate outward from the eroding core, marking the dimensions of this extinct giant."

ENHANCE

metamorphosed crystalline rocks form the landscapes of the ancient shields. Some batholiths now stand above the surface as dome-shaped mountains smoothed by weathering and erosion (see Fig. 27.8). The great domes that rise above the urban landscape of Brazil's Rio de Janeiro are such products of deep-seated intrusion and subsequent erosion, as is Stone Mountain near Atlanta, Georgia, a dome of granite and related rocks now rising nearly 200 m (650 ft) above the surrounding plain (Fig. 40.9).

Geologic structure, therefore, is an important factor in stream erosion, a relationship that cannot be overlooked when using human technology to influence river courses (see Perspectives on the Human Environment: Controlling

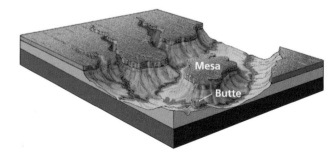

FIGURE 40.8 Mesas and buttes created because of resistant caprock that forms the smooth top of these landforms. Here, the caprock is basalt formed from lava flows, but it can be formed by resistant sedimentary rock as well.

FIGURE 40.9 Stone Mountain, Georgia, now all but engulfed by the burgeoning eastern suburbs of Atlanta, is one of the best-known granitic domes in the United States.

Rome's River). It also affects the development of drainage systems. Later in this unit, we will discover how the actual pattern of drainage lines can reveal the nature of the underlying structural geology.

TECTONIC ACTIVITY

In Unit 39, we noted the importance of the potential energy possessed by water as it begins to flow downhill from some high part of the landscape. This potential energy, converted to kinetic energy, is essential for the fluvial system to do its work of erosion. Where tectonic processes have produced high mountains and plateaus, the potential energy available is much greater than on a gently sloping plain near sea level. Without tectonic uplift, erosion eventually reduces any landscape to a low-relief plain where erosion is very slow; some of the ancient continental shields are good examples of this pattern.

Because of plate motion and tectonic uplift at plate margins, however, there are many parts of the continents where stream erosion is much more rapidly shaping the landscape. In swiftly rising mountains like the Himalayas, tectonic uplift continuously increases the slope of streams, which respond by incising their valleys deeper into bedrock. The slopes of peaks around them are then continuously steepened, driving more rapid erosion. In fact, erosion across the landscape may become fast enough to balance tectonic uplift, keeping the relief from increasing or decreasing much over time (although this is a matter of debate among geomorphologists working in such tectonically active mountain ranges). Even old mountains like the Appalachians, which are now far from plate margins, still stand far above sea level and retain steep slopes and

actively eroding streams. In part, this is because of isostatic uplift of the crust in response to erosion (see Unit 31). Thus, even very modest uplift or subsidence can change the slope of streams enough to alter the spatial pattern of erosion.

Drainage Patterns

Rock structures strongly influence the patterns of stream channel networks that develop on them. The drainage pattern often reveals much about the geology that lies below. By examining the way a river system has evolved in a certain area, we can begin to unravel the origins of the regional landscape. But before we study actual drainage patterns, we should acquaint ourselves with the concept of drainage density.

Drainage density is the combined total length of the stream channels that exist in a unit area of a drainage basin, such as one square kilometer. Areas of high drainage density are often carved into intricate networks of large and small valleys, each containing the channel of the stream that produced it. The higher the drainage density, all other things being equal, the more rapid and efficient is the movement of water and eroded soil or rock from the slopes into stream channels. As a result, high drainage density is also associated with high rates of erosion.

The highest drainage densities worldwide are in *badlands*, areas of rapidly eroding slopes dissected by a vast number of channels of all sizes. Badlands National Park in South Dakota preserves an especially good example (see photo that opens Unit 39). Badlands are common in semiarid climates, but climate is not the only factor influencing drainage density; bedrock and soil type, local relief, and recent base-level changes are also important.

Besides drainage density, the geometric pattern of stream networks can also vary widely. A **radial drainage** pattern, for instance, is typical of the drainage of a conical mountain flowing in all directions (Fig. 40.11A). We can tell that the drainage in this example flows outward in all directions from the way tributaries join larger streams. In most cases, tributaries join larger streams at angles of less than 90 degrees, and often at much smaller angles. Radial patterns like the one shown in Figure 40.11A develop most often on volcanic cones.

Another highly distinctive drainage type is the **annular drainage** pattern, the kind that develops on domes like South Dakota's Black Hills structure. Here, the concentric pattern of valleys is reflected by the positioning of the stream segments, which drain the interior of the excavated dome (Fig. 40.11B).

One of the most distinctive drainage patterns is the **trellis drainage** pattern, in which streams seem to flow in only two orientations (Fig. 40.11C). This pattern is very regular and orderly, and it often develops on parallel-folded or dipping sedimentary rocks of alternating degrees of hardness. The main courses are persistent, but tributaries are short and join the larger streams at right angles.

Controlling Rome's River

Rivers flow through most of the world's large cities, and trying to make these watercourses behave has been a priority since people first clustered alongside them. Europe's great cities in particular have had long experience in attempting to control local river channels. No European city, however, has been engaged in this battle longer than Rome (J. Tagliabue, "They Still Try to Make a Turbulent Tiber Behave," *New York Times*, September 16, 1993, p. A6).

Taming the Tiber is a struggle that dates back more than 2,000 years to the time of Julius Caesar, when stone bridges (still in use today) were built across the river and "improvements" were made to its banks. Nonetheless, because the Tiber is a relatively short stream that emanates from the nearby rugged Apennine Mountains, serious floods for years continued to bedevil its valley regularly. In the late nineteenth century, following modern Italy's unification and the restoration of Rome as the country's capital, the new government was spurred to action and ordered that the Tiber be corseted by stone walls to protect the city's treasured riverside monuments (Fig. 40.10). Unfortunately, to save money, the construction program avoided erecting heavy structures that could have significantly enhanced the walls' stability. Not surprisingly, parts of the stonework collapsed a few years later during an especially bad flood.

Upstream from Rome, hydroelectric dams were installed in the second half of the last century. As a result, sand and silt that would have replenished the riverbed in Rome are now trapped behind the dams. Now, instead of carrying these sediments through the city, the Tiber's power is concentrated on cutting into bedrock, which threatens to undermine bridges, stone walls, and buildings adjacent to them.

Finally realizing that their quick fixes would invariably be negated by natural fluvial processes, Rome's engineers have more recently tried to achieve a long-term solution. To meet the challenge, the most vulnerable bridges were reanchored to the river bottom, and huge concrete slabs were inserted below the waters to trap silt and gravel to reinforce bridge foundations in the 1980s and 1990s. But the problem persists, as the upstream dams are still in place and new sediments are still unavailable to replenish the riverbed.

Whatever happens to the latest river-control structures, human ingenuity and technology will never completely succeed in controlling the Tiber or any other major river. That realization has been a long time in coming to the so-called Eternal City, which after only two millennia has begun to work more harmoniously with nature to do what is possible to mitigate the Tiber's hazards.

FIGURE 40.10 The Castel Sant'Angelo (Hadrian's Tomb) overlooks the walled embankments of the Tiber River as it flows through the heart of Rome.

The **rectangular drainage** pattern also reveals right-angle contacts between main rivers and tributaries, but this pattern is less well developed than in the case of trellis drainage (Fig. 40.11D). What the diagram does not show is that rectangular patterns tend to be confined to smaller areas, where a joint or a fault system dominates the structural geology. Trellis patterns, on the other hand, usually extend over wider areas.

The tree-limb pattern shown in Figure 40.11E is appropriately termed the **dendritic drainage** pattern because it

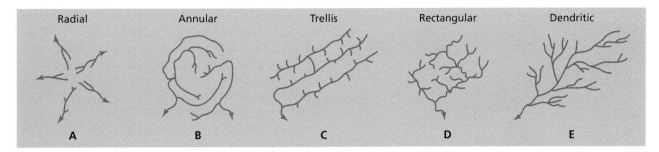

FIGURE 40.11 Five distinctive drainage patterns. Each provides clues about its underlying geologic structures.

resembles the branches of a tree. It is the most commonly observed drainage pattern, and it is typical on extensive batholiths of generally uniform hardness or on flat-lying sedimentary rocks. The entire drainage basin is likely to slope gently in the direction of the flow of the trunk river.

Overcoming Geologic Structure

The drainage patterns we have just discussed would suggest that structure exercises powerful control over river systems. Certainly, the underlying geology influences the development of the patterns in Figure 40.11, but there are places where rivers seem to ignore structural trends. In some areas, rivers cut across mountain ranges when they could easily have flowed around them. What is behind these discordant relationships?

SUPERIMPOSED AND ANTECEDENT STREAMS

In many such instances when rivers cut across mountain ranges, the river channel first developed on a surface that lay *above* the still-buried mountain range. In Figure 40.12A, note that the stream has developed on sedimentary rocks that covered the underlying ranges. If the drawing showed the entire drainage pattern, it probably would be dendritic, but for simplicity we show only the trunk river. As the stream erodes downward, it reaches the crests of the mountain ranges. In Figure 40.12B, the emerging range probably causes some waterfalls to develop where the stream cuts across its harder rocks, but a **water gap** (a pass in a ridge or mountain range through which a stream flows) is developing, and the stream stays in place. As more of the overlying sedimentary material is removed by the river and its tributaries, the ranges rise ever higher above the stream valley. This process, illustrated in Figure 40.12C, is referred to as stream *superimposition*. The river is called a **superimposed stream** because it maintained its course regardless of the changing lithologies and structures it encountered.

When we see a river slicing across a mountain range, however, we should not conclude without further research that it is superimposed. There are other possible explanations. The ridge through which the river now flows may not have been buried but may have been pushed up tectonically. A river flowing across an area of uplift (Fig. 40.13A)

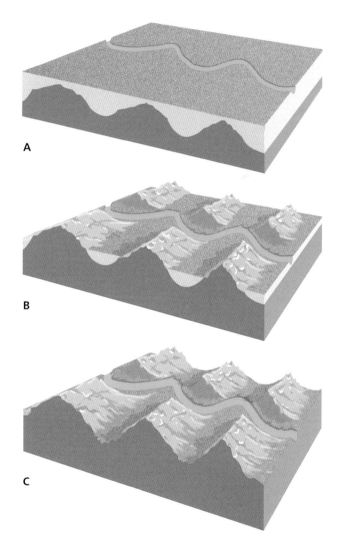

FIGURE 40.12 Evolution of a superimposed stream, which carves water gaps where it slices through three emerging mountain ridges.

may have been able to keep pace, eroding downward as rapidly as the ridge formed (Figs. 40.13B and 40.13C). This river would therefore predate the ridge and would be referred to as an **antecedent stream**.

STREAM CAPTURE

One of the most intriguing stream processes involves the "capture" of a segment of one stream by another river.

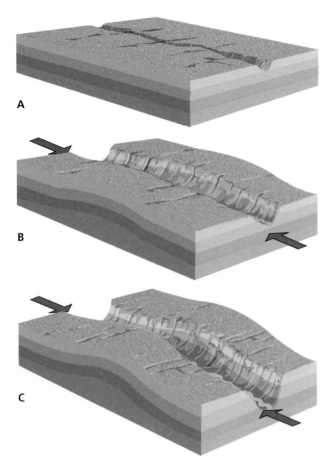

A

B

C

FIGURE 40.13 Evolution of an antecedent stream, which kept flowing (and eroding downward) as the mountain ridge was being tectonically uplifted across its path. Tectonic compression causing the uplift is shown by the purple arrows.

Also called **stream piracy**, this process diverts streams from one channel into another, weakening or even reversing some river courses while strengthening others. Stream capture is not just a theoretical notion, and it does not affect only small streams.

Looking at a map of south-central Africa, you will see the great Zambezi River. Flowing toward the Zambezi from the north is the Upper Kafue River. But at Lake Iteshi, the Kafue River makes an elbow turn eastward and joins the Zambezi, as shown in Figure 40.14. What happened here is a classic case of stream capture. The Upper Kafue River once flowed south-southwestward, reaching the Zambezi as shown by the dashed blue line in the figure. But another river, the Lower Kafue, was lengthening its valley by headward erosion (Unit 39) toward the west. When the upper channel of the Lower Kafue intersected the Upper Kafue in the area of Lake Iteshi, the flow of water from the Upper Kafue began to divert into the Lower Kafue, and its original connecting channel to the Zambezi was abandoned. Today, a small stream (appropriately termed an *underfit*) occupies the large valley once cut by the Kafue between Lake Iteshi and the Zambezi. In sum, the Upper Kafue's waters have been pirated.

This is a large-dimension case of capture; many smaller instances can be found on maps of drainage systems. It is important, however, to realize the effect of capture. The capturing stream, by diverting into its channel the waters of another river, increases its capacity to erode and transport. In the case shown in Figure 40.14, the power of the Lower Kafue River to erode and transport sediment was increased after its piracy was successful.

Long-Term Evolution of the Landscape

Physical geographers perform research on landforms and drainage systems, erosional processes, and stream histories. Like other scientists, however, they also want to understand the "grand design"—the overall shaping of the landscape, the sculpting of regional geomorphology, and the processes that achieve this. At first, it would seem that this is merely a matter of the sum of all the parts. If we

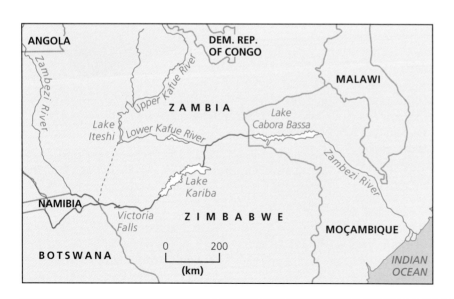

FIGURE 40.14 Stream piracy enabled the headward-eroding Lower Kafue River to capture the Upper Kafue in the vicinity of Lake Iteshi. Previously, the Upper Kafue River drained into the Zambezi via the (now-abandoned) connecting channel shown by the dashed blue line.

FIGURE 40.15 William Morris Davis, Professor of Geography at Harvard University in the days when Harvard still had a distinguished Geography Department, was founder of the Association of American Geographers (1904) and produced numerous substantive and theoretical works in the field of physical geography.

know about the factors affecting erosional efficiency (e.g., those discussed in Units 38 and 39), then shouldn't the evolution of the landscape surely be understood? The answer is, not yet. Physical geographers today still debate some very basic issues concerning regional geomorphology.

One particular debate has been going on for more than a century. It was started by William Morris Davis (1850–1934), a pioneering physical geographer and professor at Harvard University from 1878 to 1912 (Fig. 40.15). Davis proposed that a **cycle of erosion** affects most of Earth's landscapes. This cycle has three elements: geologic *structure*, geographic *process*, and time or *stage*. Every landscape, Davis argued, has an underlying geologic structure; that landscape is being acted upon by streams or other erosional processes, and it is at a certain stage of its long-term evolution.

Davis envisioned tectonic uplift of an extensive area occurring fairly rapidly, followed by long periods of time when the uplifted areas were shaped by erosion and experienced little further tectonic activity. The sequence of events that Davis thought would follow uplift is illustrated in a simplified form in Figure 40.16. Essentially, an area of low relief and gentle slopes is raised above its base level; this in turn accelerates erosion by the network of streams draining the uplifted landscape (Fig. 40.16A). In this "young" stage, the river valleys are deep and narrow and are separated by broad ridgetops, remnants of the old low-relief surface. Over time, the stream network extends into these ridgetops,

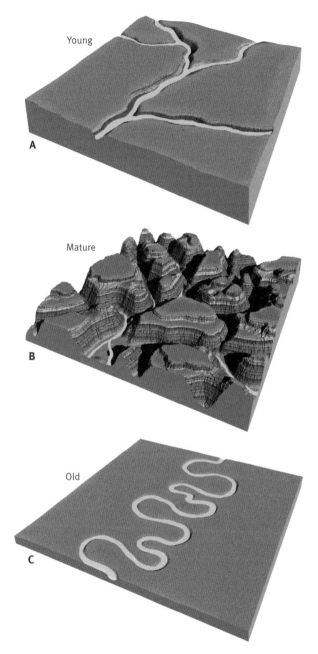

FIGURE 40.16 Simplified version of the cycle of erosion proposed by William Morris Davis.

and eventually, the "mature" landscape is entirely occupied by stream valleys and bordering steep slopes (Fig. 40.16B). Following even more protracted erosion, the steep slopes are worn down to gentle slopes. Finally, most hills are eroded away, leaving a nearly flat plain (Fig. 40.16C) that Davis called a **peneplain** (a "near plain"). Though not shown in Figure 40.16C, the peneplain might still have a few prominent, not-yet-eroded remnants on it. The remnants, which we discussed at the outset of this unit, are called *monadnocks*, after Mt. Monadnock (see Fig. 40.1).

Note that in Davis's view, the slopes that dominate the landscape in Figure 40.16B are worn *down*; that is, they become more convex in appearance and then are flattened, a process called **slope decline**. For many years, the cycle

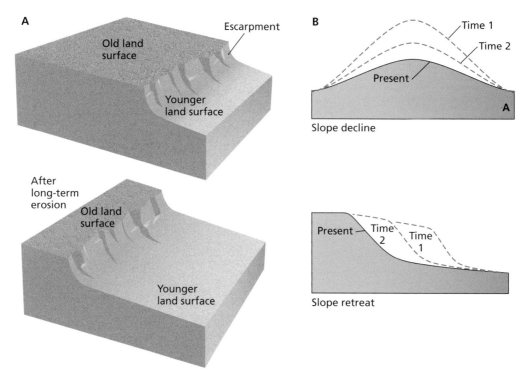

FIGURE 40.17 Concept of landscape evolution by retreat of escarpments (A). This concept assumes slope retreat as opposed to slope decline, as generally assumed by William Morris Davis. Cross-sections in (B) contrast slope decline and slope retreat.

proposed by Davis was generally accepted as the normal mode of landscape evolution, although some geomorphologists—and, to some extent, even Davis himself—noted various observations that were hard to reconcile with it.

German geographer Walther Penck (1988–1923) published an alternative theory in the 1920s, which included the important and now widely accepted concept that uplift and erosion occur simultaneously, not in sequence. He also argued that some slopes do not wear downward, but *backward* through **slope retreat**. That is, slopes can erode and shift horizontally without becoming much gentler over time (Fig. 40.17). Penck's ideas were poorly understood, however (he died young and his book was not translated into English until the 1950s). South African geomorphologist Lester King (1907–1989), who studied old landscapes of southern Africa, India, Australia, and South America, focused on the retreat of steep slopes or escarpments as the dominant process of long-term landscape development.

Perhaps the most striking alternative to the Davisian cycle was proposed by John Hack (1913–1991) in 1960, following ideas suggested in the nineteenth century by G. K. Gilbert (1843–1918). Hack argued that landscapes may be in **dynamic equilibrium**, in which the forms of slopes and

valleys or the general ruggedness of the topography do not change over time, even though large quantities of rock are removed by erosion. Essentially, slopes have adjusted to differences in rock resistance so that erosion rates are similar across the landscape as a whole; steeper slopes correspond to more resistant rock, and gentler slopes correspond to easily eroded material. With fairly uniform erosion rates, the entire landscape is lowered over time without significant changes in form.

The Davisian cycle and the concept of dynamic equilibrium are both testable, in theory, with enough information on long-term erosion rates in various parts of the landscape. For many years, it seemed quite difficult to obtain this information in most landscapes, and long-term landscape evolution was not seen as a very promising line of research. Now there is renewed interest in the topic, in part because erosion rates can be estimated using *cosmogenic radionuclides* (see Focus on the Science: Measuring Erosion, Unit 35). So far, this new research has produced results that seem consistent with more than one hypothesis of landscape development—or, in a few cases, with no existing hypothesis. Stay tuned for further developments in this long-standing problem of physical geography.

Key Terms

annular drainage *page 489*

antecedent stream *page 491*

cycle of erosion *page 493*

dendritic drainage *page 490*

drainage density *page 489*

dynamic equilibrium *page 494*

geologic structure *page 486*

hogback *page 486*

lithology *page 485*

monadnock *page 485*

peneplain *page 493*

radial drainage *page 489*

rectangular drainage *page 490*

slope decline *page 493*

slope retreat *page 494*

stream piracy *page 492*

superimposed stream *page 491*

trellis drainage *page 489*

water gap *page 491*

Scan Here for a quick vocabulary review

Review Questions

1. How does lithology influence the erosion of stream valleys and slopes and the drainage patterns that result?

2. How are rock structures, including folds, faults, and igneous intrusions, expressed by specific landforms or drainage patterns?

3. Consider a mountain range undergoing rapid tectonic uplift. Explain how the local relief (the height of mountain peaks relative to adjacent valleys) and the typical steepness of slopes may change little over time as a result of the interaction of uplift and erosion.

4. What is the difference between antecedent and superimposed streams?

5. Compare and contrast the cycle of erosion as envisioned by William Morris Davis and the concept of dynamic equilibrium.

Scan Here for a quick concept review

www.oup.com/us/mason

Landforms of the Fluvial System

A bar has developed across the mouth of this river in northern New Zealand. While sediment may be stored only temporarily in this bar, it is part of a larger and more persistent system of fluvial landforms, including the river channel, floodplain, terraces, and delta.

OBJECTIVES

- Describe the formation and characteristics of alluvial fans

- Discuss the common landforms produced by fluvial erosion and deposition in river valleys

- Summarize the evolution of river deltas

In Unit 40, we examined the work of streams on the landscape as sculptors, carvers, and cutters, and explained how that work is related to rock structure and other factors. In this unit, we focus on characteristic landforms found along stream valleys and where streams enter lowlands or water bodies. Many of these landforms are built by the deposition of stream sediment, but stream erosion is often involved as well. For example, the *floodplain* is a setting where we clearly see both erosion and deposition by a stream, and each of these processes helps shape floodplain landforms.

Alluvial Fans

Let us begin with a special case. In certain areas, especially in arid zones of the world, precipitation is infrequent and stream flow is discontinuous. Rainstorms in the mountains produce a rush of water in the valleys, and turbulent, sediment-laden streams flow toward adjacent plains. Emerging from the highlands, the water spreads out in a wider, shallower channel, and some of the stream flow may be lost by seepage into underlying sediments. As a result, the stream rapidly deposits much of its sedimentary load. Meanwhile, it has stopped raining in the mountains, and soon the stream runs dry. A stream that flows intermittently like this is called an **ephemeral stream** (as opposed to a permanent, continuously flowing stream). In many desert mountain ranges, debris flows—the rapidly moving slurries of sediment and water discussed in Unit 37—also flow down the same ephemeral stream channels from time to time. Where those channels emerge from the mountains, the debris flows soon come to a halt.

Such a sequence of events leads to the construction of an alluvial fan where an ephemeral stream emerges from desert mountains. As the term suggests, an **alluvial fan** is a fan-shaped deposit consisting of alluvial material (and often debris flow deposits as well), located where a mountain stream emerges onto a plain (Fig. 41.1). Streams that form alluvial fans are unlikely to flow far beyond the edge of the fan and may not be part of the regional drainage basin that ultimately leads to the ocean. When the mountain rains are below average, the stream may not even reach the outer margin of its own deposits.

Alluvial fans are not unique to arid regions, although many of the best-known examples have developed there. A careful look usually reveals small fans distributed along the foot of steep valley-side slopes in humid regions as well. But the large well-developed alluvial fans of certain desert landscapes are the best examples to consider in explaining how these landforms develop.

The alluvial fan attains its conical or semicircular shape because the stream that emerges from a canyon-shaped mountain valley does not develop a permanent course on the fan surface but instead follows various paths across the fan over time. When the sediment-laden water surges from the mouth of the canyon, it is no longer confined by bedrock canyon walls. As the water flows onto the fan, it spreads out into a wider, shallower channel. The shallower water cannot

FIGURE 41.1 Stream action shapes arid landscapes through deposition as well as erosion. Streams emanate from mountains such as the Last Chance Range in Death Valley National Park, California, laden with sediment eroded from the barren mountain slopes. Emerging from the mountain front, the streams become wider and shallower, and water seeps into their bed so that sediment is rapidly deposited to form huge alluvial fans. Larger fragments are dropped first (center), and smaller pebbles and sand travel somewhat farther (foreground). The wind has mobilized some of the sand and formed it into dunes (right foreground).

ENHANCE

continue to transport many of the coarser particles brought out of the canyon and quickly deposits them. So much deposition occurs along the stream channel that it is actually raised above adjacent areas of the fan. The next time rains generate a flash flood, the water can spill out of its previous channel and find a different path across the fan. As the stream flow shifts back and forth across the fan over the course of many floods, the entire landform is built up. Debris flow deposition also shifts from one part of the fan to another over time.

An alluvial fan therefore consists of a series of poorly stratified layers, thickest near the mountain front and progressively thinning outward. The coarsest sediments are normally found nearest the apex of the fan, where the stream emerges from the mountains. Finer-grained material is located toward the outer edges, although debris flows can sometimes carry large boulders far out onto the fan. Because the fan lies on bedrock and has layers of greater and lesser permeability, infiltrating water can be contained within it. Many alluvial fans in the southwestern United States and throughout the world are sources of groundwater for permanent settlements.

When environmental conditions are right for alluvial fan development, a mountain front may have not just one or two but dozens of larger and smaller alluvial fans. When

FIGURE 41.2 An extensive area of desert pavement covers the surface of a section of Signal Park in Arizona's Kofa National Wildlife Refuge.

this happens, the cone shapes of individual adjacent fans coalesce and are difficult to distinguish, and the landform is called an *alluvial apron,* or a **bajada** (from a Spanish word for slope). Large fans develop over tens of thousands to millions of years, and older, higher-standing parts of the surface of such fans are no longer subject to the shifting stream process we described earlier. Instead, the older areas become stable and often develop a surface layer of densely spaced gravel, referred to as *desert pavement* (Fig. 41.2). Well-developed soils—often Aridisols (see Unit 23)—also form on these older fan surfaces.

Landforms of River Valleys

Streams in narrow mountain valleys frequently flow over the bedrock surface, eroding it when discharge is high. The same is true of some streams in areas of hard crystalline rock, like the Canadian Shield, and even for stretches of large rivers such as the Mississippi or Chang Jiang (Yangtze). But most lowland rivers of any size flow mainly over **alluvium** (loose, uncemented stream sediment), which fills the valley bottom and varies widely in its thickness. By eroding and depositing alluvium, these streams develop characteristic channel patterns.

Most people, if asked to draw a typical river, would probably sketch a single sinuous (winding) channel with distinct bends called **meanders** (Fig. 41.3). This meandering form is the most common channel pattern worldwide, and the meanders themselves change over time in predictable ways. Erosion occurs on the outside of the meanders, while sediment is deposited on the inside, so the meanders grow larger over time, as Figure 41.3 illustrates. Meanders often also migrate slowly in the downstream direction, because there tends to be more erosion on the downstream side of the bend. The area of most recent deposition on the inside of the bend—called the **point bar**—is usually bare sand or gravel and slopes gently into the channel. This is a specific type of *bar,* a term used more widely for deposits of sand or gravel within a stream channel or along the edge of the stream that emerge above water at low discharge.

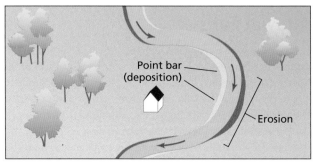

A

B

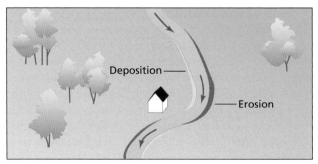

C

FIGURE 41.3 The development of meanders in a stream involves both deposition and erosion. (A) A stream bend displays evidence of deposition on the inside of the bend and erosion on the outside curve. (B) The river's bends are becoming larger through erosion on the outside of the bend and deposition on the inside of the bed, building the point bar. There is a bit more erosion on the downstream side of the outer bend, so the bend as a whole shifts downstream over time (note the shift relative to the small building to the left of the stream). (C) The process has advanced to produce tighter bends extending farther in both directions from a straight line down the stream valley, and the bend has moved somewhat farther downstream.

As a meander grows larger, its neck becomes narrow, and eventually the stream can break through and cut off the meander entirely (Fig. 41.4). The cutoff channel often retains water and forms a crescent-shaped **oxbow lake**. Meanwhile, new meanders develop and grow over time, restoring the overall sinuosity of the stream.

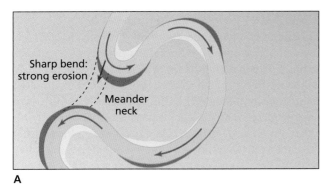

Sharp bend: strong erosion

Meander neck

A

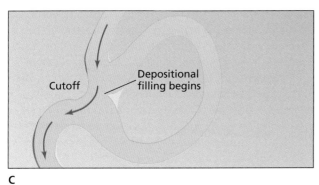

Near cutoff

B

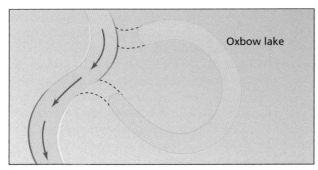

Cutoff

Depositional filling begins

C

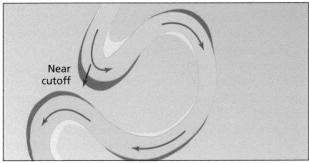

Oxbow lake

D

FIGURE 41.4 Process of meander cutoff, forming an oxbow lake.

ENHANCE

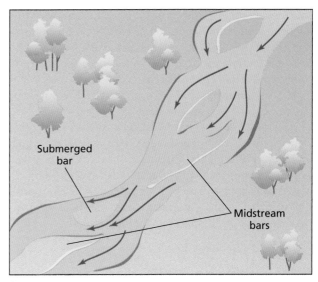

Submerged bar

Midstream bars

A

B

FIGURE 41.5 One process by which braided streams develop: formation of midstream bars, diverting flow around each side of the bar (A). The Platte River is a classic example of a braided stream in the central Great Plains of Nebraska (B). Multiple channels divide and flow around submerged bars, bars that are above water but still bare sand, and bars that have been covered with vegetation.

Not all rivers meander, and **braided streams** are common in some settings (Fig. 41.5). Braided channel patterns involve multiple strands of flowing water separated by sand, gravel bars, or vegetated islands. Frequent shifts in channel position are common. One way this happens is through deposition of sediment in a channel, forming a midstream bar (Fig. 41.5A). Water flows around this obstruction, eroding the banks on either side, and soon two distinct channels replace the single one. Water may also abruptly spill across bars and erode new channels.

Geomorphologists have long debated the reasons that some channels are braided rather than meandering; high sediment load, highly variable discharge, steep valley slopes, and loose sandy bank materials have all been suggested as factors. It is not uncommon to find rivers with intermediate channel patterns, such as a meandering channel that occasionally splits into multiple strands.

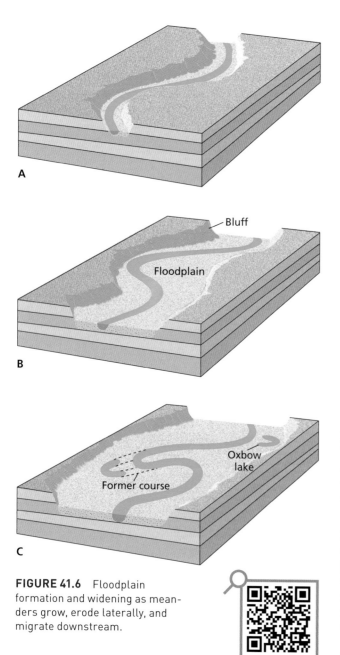

FIGURE 41.6 Floodplain formation and widening as meanders grow, erode laterally, and migrate downstream.

THE FLOODPLAIN

The migration of meanders or braided stream channels has two important effects: widening the river's valley and creating an extensive **floodplain**, which is an area of relatively flat, low-lying ground near the stream channel that is often inundated by floods (Fig. 41.6). Consider a stream in a narrow valley with bedrock slopes that rise abruptly on either side (Fig. 41.6A). In this narrow space, the shifting channel will often erode the slope on one side of the valley or the other. *Bluffs*, steep slopes that border the edge of a floodplain, are created in this manner (Fig. 41.6B), and the valley is gradually widened, allowing greater space for the floodplain to develop (Fig. 41.6C).

A floodplain is made up of alluvium that is continuously recycled through stream erosion and deposition. As meanders migrate, the stream erodes an existing part of the floodplain, but at the same time the floodplain is rebuilt through deposition on the inside of the meanders. When the stream spills over its banks in a flood, more sediment is added to the floodplain. Braided streams also continuously recycle floodplain deposits through the shifting of their multiple channels. Over time, the floodplain may become many kilometers wide, and *meander belts* themselves may form giant meanders, as shown in Figure 41.7. Also observe that the bottom of the stream channel, swinging back and forth across the valley, can erode a fairly level surface on the bedrock underlying the floodplain if the rock is not too deeply buried by alluvium.

When we study floodplains in more detail, we can often find a variety of landforms related to channel migration or sediment deposition by the stream, such as the oxbow lakes discussed previously (see Fig. 41.4). On aerial photos of a floodplain, former channel positions can often be seen even if they are not filled with water, because a darker tone indicates wetter soils than in adjacent areas.

Recall that floods in which the river overflows its banks occur every one to three years on many streams (see Unit 39). These floods deposit sediment that builds the river's **natural levees**, broad ridges that run along both sides of the channel (Fig. 41.8). As the river spills out of its channel, the coarsest material it carries is deposited closest to the overflow; hence, it ends up along the levees. These self-generated natural levees look somewhat like **artificial levees**, which are ridges or berms constructed along most large rivers in settled areas to confine the floods to the channel and protect farmland or towns from flooding. The natural levees are not continuous, though, so floodwaters can still easily spill onto the floodplain.

Extreme floods can have dramatic effects on floodplains, eroding artificial or natural levees and scouring deep holes in the floodplain surface. Large fan-shaped deposits of sand carried out of the channel by floodwaters record the power of these floods. Yet such large floods occur rarely enough that the floodplain can recover between them, and the floodplain that we normally see is largely the work of smaller, much more frequent floods.

Human Use of Floodplains The fertility of the soils and the flatness of the land on floodplains have attracted

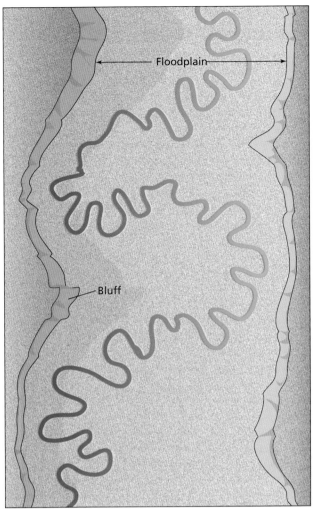

FIGURE 41.7 Segment of a meander belt.

ENHANCE

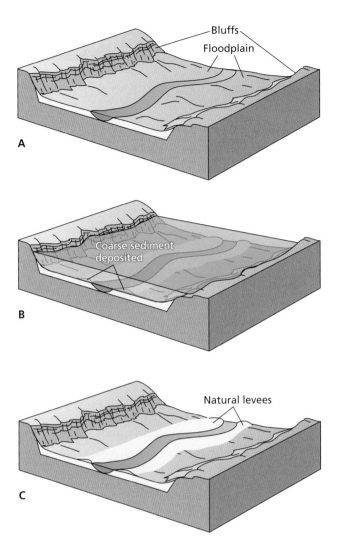

Bluffs
Floodplain

A

Coarse sediment deposited

B

Natural levees

C

FIGURE 41.8 Relationship between floods and natural levee development. Under most conditions the river's flow is confined to the channel (A). During large floods, water covers much of the floodplain, sometimes reaching from bluff to bluff. As water spills out of the channel, coarse deposits are laid down closest to the channel banks (B). After repeated flooding, these deposits are built up as natural levees that line the stream channel (C).

human settlement for thousands of years. Crops grow bountifully there, and the land is easy to farm and develop. The risks, however, are great, and they are not only financial. Floods in the United States have caused billions of dollars in damage and the loss of several dozen lives in many recent years (see Perspectives on the Human Environment: The Floods of 2011).

The decision to live on or avoid a floodplain is not strictly rational. Geographers Gilbert White, Robert Kates, and Ian Burton have shown that human adjustments to the known dangers of flooding do not increase consistently as the risk becomes greater, because people tend to make an optimistic rationalization for continuing to live in a potential flood zone.

Ironically, flood-control measures often encourage an unrealistic view of floodplain hazards. Artificial levees and dams that store floodwaters in reservoirs have been built

along many major rivers. Although dams and levees can protect floodplain settlements from most smaller floods, this may mislead floodplain residents into underestimating their risk. Eventually, a large enough flood will fill the reservoirs or overtop the levees. The most realistic attitude is to recognize that floodwaters and the sediment they carry have built the floodplain and will return to it sooner or later.

TERRACES

In Unit 39, we discussed the processes of aggradation and incision, in which a river responds to changes in its capacity to transport sediment or in the sediment supply from its drainage basin. The signature landform of these processes is the **terrace**, a bench that stands above the modern floodplain and represents an older floodplain that has now been

The Floods of 2011

In the spring of 2011, massive floods moved down the lower Mississippi River, setting new records for flood discharge and causing billions of dollars in damage. In the summer of 2011, far upstream in the same river system, floodwaters covered vast areas of the Missouri River floodplain. Then, near the end of August 2011, devastating floods destroyed roads, bridges, and farmland in rural areas of Vermont, New Hampshire, and New York. New stream flow records were set at 11 of 20 long-term monitoring stations in Vermont and New Hampshire.

Each of these episodes of flooding was caused by a unique combination of weather conditions and river-system behavior, but they all illustrate that the hazards of using land on floodplains are still very real despite great progress in weather and flood forecasting. During the winter of 2010–2011, unusually heavy snowfall on the northern Great Plains and in the northern Rocky Mountains set the stage for the lower Mississippi and Missouri River floods. Between April 15 and May 5, 2011, heavy rain fell as a series of midlatitude cyclones (see Unit 13) moved through the Midwest. As that water drained down the Ohio and Mississippi rivers, river levels rose. Eventually, the rivers overtopped many artificial levees.

A series of large dams have been built on the Missouri River, and the reservoirs behind them can store large volumes of water and reduce flooding downstream. By early summer, however, these reservoirs were full, and floodwater had to be released, covering the floor of the Missouri Valley in South Dakota, Iowa, Nebraska, and Missouri.

The Missouri and Mississippi floods were predicted long before they reached their peak. Though there was plenty of time to prepare, only some of the potential damage could be avoided. The floodplains of the Missouri and lower Mississippi rivers are used mainly as productive farmland, but they are also dotted with small towns, industrial facilities, and even nuclear power plants. The nuclear plant shown in Figure 41.9 was successfully protected from flooding, but many less protected structures were not, and the loss of most agricultural crops was unavoidable.

Flooding in the northeastern United States in August 2011 was caused by torrential rains associated with a tropical cyclone (see Unit 13), Hurricane Irene. Based on remote sensing, computer models, and decades of careful observations, forecasters accurately predicted Irene's track up the U.S. East Coast. Fortunately, the storm weakened and serious damage was avoided in major coastal cities like New York. Even the weakened storm still produced enough rain to cause rapidly rising floods where runoff was concentrated in narrow valleys like those of Vermont's Green Mountains. Roads, towns, and fields all tend to be located along streams in those same mountain valleys. Although there was enough warning to allow most people to reach higher ground, damage to roads isolated many towns and rural homes for days or even weeks.

FIGURE 41.9 Missouri River floodwaters surround the Ft. Calhoun nuclear power plant just north of Omaha, Nebraska, in June 2011. Barriers kept water out of the facility.

FIGURE 41.10 Terraces of the Snake River in Grand Teton National Park, Wyoming. Terraces are visible as flat benches in the middle distance, between the river and the mountains, and in the left foreground.

abandoned (Fig. 41.10). Terraces can occur at several levels above a stream, forming a staircase-like set of landforms.

Let us first consider an instance in which a stream incises, eroding the bedrock under the channel (Fig. 41.11). This could happen for a variety of reasons—for example, because tectonic uplift in the headwaters of the stream has steepened the channel or because climate change has increased vegetation density and reduced sediment yield from the drainage basin. In both cases, the stream has more power than it needs to transport the sediment with which it is supplied, so it uses that power to incise into bedrock.

A stream may incise for another reason as well if it joins a larger river downstream. That larger river could have incised and lowered the base level for all its tributaries (see Unit 39). Near the larger stream, the tributaries will now drop steeply, increasing stream power and causing rapid incision. That incision will then work its way up each tributary over time.

Whatever the cause of incision, the stream cuts downward from its original level on the floodplain (Fig. 41.11A) to a new level, where it stabilizes (Fig. 41.11B). Why does it stabilize rather than continuing to use its excess stream power to erode? By incising more in its upstream reaches, it may have lowered its slope and thus no longer has the power to erode downward while still carrying the sediment load supplied by its drainage basin (see Fig. 39.12). Or if the downcutting was caused by lowered base level, each segment of the stream will stabilize once its channel has adjusted to a new lower level.

Soon a new lower floodplain develops (Fig. 41.11C), complete with a new set of bounding bluffs. Remnants of the older floodplain now stand above these newer bluffs as terraces. These terraces reveal the evolution of the river system by showing that it must have incised in response to past climate change, tectonic activity, or lowered base level.

The sequence of events just described produces terraces that are essentially benches cut into the local bedrock,

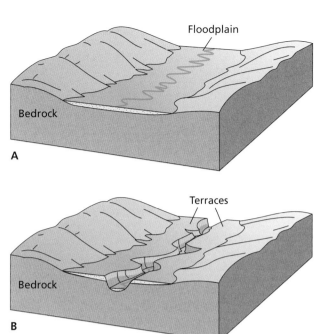

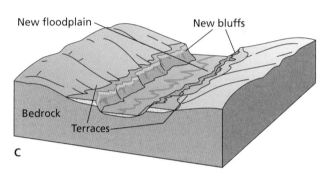

FIGURE 41.11 Strath terraces form when a stream incises and leaves parts of its former floodplain as benches above the modern floodplain.

covered with a thin layer of stream deposits laid down when the terrace was an active floodplain before incision. This type of terrace is often referred to as a *strath terrace*. The terraces shown in Figure 41.11 are *paired terraces*; that is, they lie at the same elevation on each side of the incised stream. Often, terraces are not paired because lateral (sideways) erosion destroys the terrace on one side of the valley.

A more complicated type of terrace formation involves aggradation before incision. That is, the stream initially cannot carry all the sediment with which it is supplied and deposits some of it. As discussed in Unit 39, this deposited sediment accumulates on the streambed and across the floodplain, building them up over time and filling the valley with thick stream sediment. The most likely cause for this behavior is an increase in sediment supply to the stream, possibly from a glacier upstream or from more rapid erosion of hillsides in the drainage basin. Recall that in Unit 39, we discussed the impact of hydraulic mining in California. That example shows that human activity like mining can also cause aggradation and valley filling.

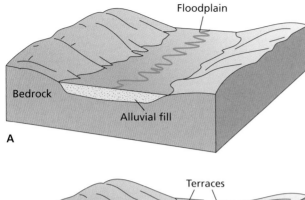

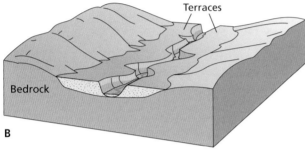

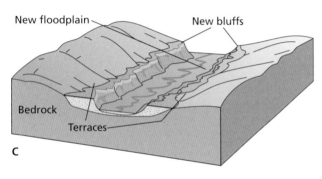

FIGURE 41.12 An alternative sequence of events can produce fill terraces. First, the stream aggrades and fills its valley with alluvium (A). Then, incision into the valley-filling alluvium produces terraces (B and C).

Greater aggradation in the upper reaches of a river steepens its slope and eventually gives it more capacity to transport the added sediment (see Fig. 39.12). Now consider what happens if the sediment supply to an aggrading stream is reduced, as happened when hydraulic mining was outlawed in California or when a glacier retreats and is no longer pouring sediment into a river system. The steepened stream then not only has capacity to move all of the reduced sediment supply but also can incise into the material it previously deposited. As it erodes downward, it creates terraces (Fig. 41.12). These are *fill terraces*, formed out of stream sediment that filled the valley when aggradation was taking place. The sequence of events that produces a set of strath or fill terraces can be repeated over time and can produce several terraces at different levels.

Deltas

About 2,500 years ago, the ancient Greek scholar Herodotus studied the mouth of the Nile River and found that this great river of northeastern Africa forms a giant fan-shaped deposit where it reaches the Mediterranean Sea. Noting the triangular shape of this area of sedimentation, he called it a **delta**, after the fourth letter of the Greek alphabet, which is written as Δ (Fig. 41.13A). Ever since, river-mouth deposits have been called deltas, even when they have a different shape. In this unit, we focus on deltas where streams enter the ocean, but it is important to note that deltas also form along lakeshores.

As a river enters the ocean, its bedload of sand or coarser particles is rapidly deposited. The finer material the river carries in suspension forms a plume of sediment drifting out into the ocean, often visible from the air (see Fig. 31.5). In the ocean, the salt water causes clay particles to clump together, so that even they start to settle fairly close to the river mouth. It is these processes of deposition that build the delta.

As the river flows onto the delta, stream flow separates into multiple channels called **distributaries** that carry the river's water in several directions over the surface. Thus, a trunk river receives *tributaries* in its drainage basin and develops *distributaries* where it forms a delta. Each distributary may divide into several channels before reaching the ocean. One major distributary and its branches often carry much of the flow for a time, but the area around that distributary system is built up rapidly by sediment deposition. This area of deposition extends seaward as a *lobe* of the delta. However, water and sediment are eventually diverted to a lower part of the delta surface, and the process begins anew.

As the map of the Nile Delta (see Fig. 41.13A) shows, the Nile forms a few prominent distributaries and many smaller ones. When Herodotus did his fieldwork, the Nile Delta was an uninhabited swampy area. Today, it is an area of dense rural settlement and intensive cultivation (Fig. 41.14). Control of the Nile's distributaries and land reclamation have made this transformation possible.

Three factors determine the form of a delta: (1) the discharge of the stream and the amount of sediment it carries; (2) the configuration of the offshore continental shelf near the river mouth; and (3) the strength of currents, waves, and tides. The Nile and Mississippi rivers have large deltas, but the Congo River of west equatorial Africa does not. This is because the Nile and the Mississippi flow into relatively quiet waters, and offshore depths increase gradually. The Congo River, in contrast, flows into deeper water immediately offshore, coastal currents are strong, and the Congo's sediment load is less than either the Nile's or the Mississippi's.

The Nile Delta has a much smoother outline than the Mississippi Delta, however (Figs. 41.13A and 41.13B). Waves and coastal currents sweep the Nile River sediments into bars and barrier islands that enclose coastal lagoons, creating a smooth shoreline. As the delta grows seaward, the *delta plain* (the flat landward portion of the delta) stabilizes. On the Mississippi Delta, delivery and deposition of sediment by the river overwhelm the less effective work

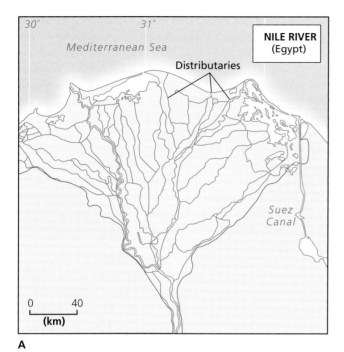

A

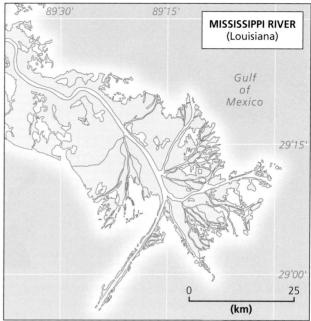

B

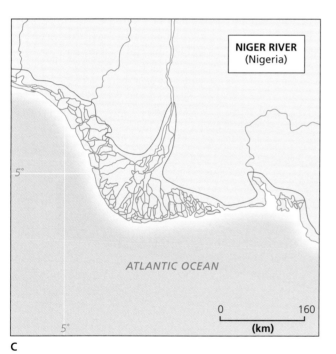

C

FIGURE 41.13 The spatial form of a delta depends on the quantity of sediment carried by the river, the configuration of the continental shelf beyond the river mouth, and the power of waves and currents in the sea. The Nile Delta (A) exhibits the classic triangular shape. The Mississippi Delta (B), exhibiting a birdfoot shape, results from large quantities of sediment carried into the ocean too fast for extensive reworking by the local currents and waves. The Niger Delta (C) is shaped by strong waves and currents that sweep sediment along the coast.

of waves and currents, creating a *birdfoot delta* (Fig. 41.13B). The birdfoot shape differs considerably from that of the Niger Delta on the coast of West Africa shown in Figure 41.13C. There, currents and waves sweep sediment along the shoreline even more than they do at the mouth of the Nile.

No two deltas form in exactly the same way, and the process is complicated. A simplified version of delta formation is illustrated in Figure 41.15. The finest deposits are the *bottomset beds*. These are composed mainly of the fine suspended particles carried out beyond the main body of the delta before they settle to the bottom. The sloping *foreset*

beds form where the river's bedload is deposited most rapidly, cascading down the front of the delta. As the foreset beds accumulate, the delta grows out into the ocean. In the meantime, the river adds material to the *topset beds*, the horizontal layers that are gradually built out over the foreset beds and form the surface sediments of the delta plain. The topset beds include a mix of coarser and finer sediment, often deposited where water spills out of distributaries during floods.

The thickness and resulting weight of deltaic sediments can depress the coastal crust isostatically, complicating the process of delta development still further. Loose sediments

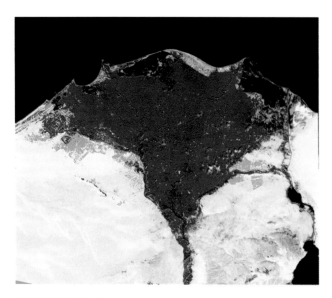

FIGURE 41.14 Remotely sensed image of the Nile Delta. Intensively cultivated croplands, actually bright green, appear red in this image.

forming the delta are compacted over time and cause additional subsidence, or sinking of the land surface. In actively growing parts of the delta, the land surface is continually built upward by sediment deposition, so subsidence is not evident. In other parts of the delta and surrounding coastal areas, the effects of subsidence are seen where land areas that were once above sea level are submerged. In parts of the Mississippi Delta where it enters the Gulf of Mexico in southeastern Louisiana, much land has been submerged over the past few decades (at a rate of approximately 80 km² [30 mi²] per year).

Deltas are among the largest depositional features of the fluvial system. Those boulders and pebbles that were dragged down mountain valleys now exist as fine grains on the coast or offshore on the ocean bottom. They were transported and deposited many times over and pulverized and weathered into fine material in the process—a vital part of the work of rivers, those great sculptors and builders of the landscape.

We have now completed our survey of the geomorphic processes associated with running water and the fluvial landscapes and landforms they shape. But we are not yet ready to turn our attention away from water as a geomorphic agent. A special case of denudation involves the removal of rock not by physical breakdown but by chemical weathering. This process produces the highly distinctive landscapes of karst regions, which we examine in Unit 42.

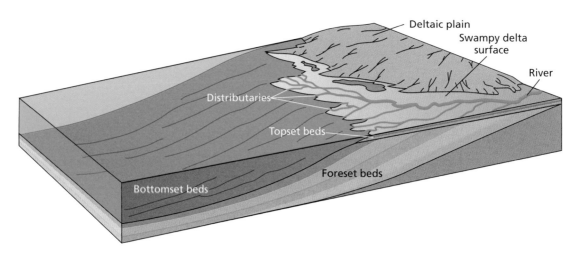

FIGURE 41.15 Internal structure of a small delta. Large deltas, such as those mapped in Figure 41.13, are far more complex in their composition.

Key Terms

alluvial fan *page 497*
alluvium *page 498*
artificial levee *page 500*
bajada *page 498*
braided stream *page 499*

delta *page 504*
distributaries *page 504*
ephemeral stream *page 497*
floodplain *page 497*
meander *page 498*

natural levee *page 500*
oxbow lake *page 498*
point bar *page 498*
terrace *page 501*

Scan Here for a quick vocabulary review

Review Questions

1. Why and how do alluvial fans form where streams emerge from desert mountain ranges?

2. Compare and contrast meandering and braided channel patterns.

3. How is the process of meander migration related to the formation of floodplains?

4. Describe the process of meander migration; include the formation of a point bar and an oxbow lake.

5. Describe at least two sequences of events that can produce stream terraces and explain how those events are related to the balance between sediment supply and transport capacity.

6. Describe the similarities and differences between alluvial fans and river deltas.

7. Explain why large deltas can have such varied forms.

Scan Here for a quick vocabulary review

www.oup.com/us/mason

UNIT
42

Karst Processes and Landforms

Limestone towers dominate this classical landscape of tropical karst in Ha Long Bay, Vietnam, the work of long-term weathering in a warm, humid climate.

OBJECTIVES

- Discuss the processes that produce karst landscapes and the environmental factors that influence them

- List the landforms that characterize karst landscapes and describe their formation

- Relate karst processes to the development of extensive underground cave systems

FIGURE 42.1 The depression-pocked surface of the Pennyroyal Plain, lying to the east of Bowling Green, Kentucky, is a classic temperate karst landscape.

As we note elsewhere in Part Six, water erodes rocks of all kinds, sculpting the surface into many distinctive landscapes. Chemical or mechanical weathering sets the stage for water erosion by breaking down rock into erodible particles, and it also dissolves some rock components, which are then carried away in solution. Certain types of rocks, however, can be dissolved almost completely by groundwater, leaving little residual solid material to be removed by stream flow. This process of **dissolution** is a form of chemical weathering that allows corrosion of rock by flowing water (see Unit 39). Such corrosion by underground waters produces landforms and entire landscapes that look very different from those formed when stream erosion carves away solid material from the land surface. The word **karst** is used for these landscapes and landforms as well as the processes that produce them.

A unique characteristic of karst landforms is that many of the most important examples occur below ground. The most important of these is the **cave**, technically any substantial opening in bedrock that leads to an interior open space. The word *substantial* here has a human connotation: it is generally agreed that, to be called a cave, an opening must be large enough for an average-sized adult to enter. Caves formed in limestone or dolomite occur in many areas of the world, and some are so large and spectacular that they have become quite famous. Mammoth Cave, located in west-central Kentucky, is a network of underground chambers and passages totaling over 500 km (310 mi) in length. Carlsbad Caverns in New Mexico has more than 37 km (23 mi) of explored chambers and tunnels. Other major cave systems are found in the Appalachians, Indiana, Missouri, Texas, Utah, and elsewhere in the United States. They also can be found in many parts of Europe, China, Australia, Africa, and South America. Not all caves occur in karst landscapes; for example, caves are also carved by waves along shorelines. Such caves are not karst features, however.

Archaeologists have discovered that many caves in karst landscapes were occupied by humans hundreds of thousands of years ago, and some caves contain valuable evidence about their occupants. In Lascaux Cave in southwestern France, the cave walls were decorated by artists; in these drawings, we can easily identify a wide variety of animals with which these people were familiar, some now extinct. Other caves, such as the Sterkfontein Caves in South Africa, have yielded skeletons that have helped anthropologists unlock the secrets of human evolution. Caves therefore are more than mere curiosities to Earth scientists and archaeologists.

Other karst landforms are visible at the surface, creating unique landscapes (Fig. 42.1). The term *karst* has its origin in a landscape of this type in east-central Europe where Slovenia, Croatia, and Italy meet. There, in a zone bordering the Adriatic Sea, an arm of the Mediterranean, lie some of the most spectacular and characteristic of all karst landscapes. Surface streams disappear into subsurface channels, steep-sided and closed depressions dot the countryside, and stark limestone hills crown a seemingly chaotic topography. This terrain extends to the Adriatic coast itself, and where sea and limestone meet the shore becomes a monument of natural sculpture.

Karst landscapes are not always as spectacular as this, but some karst areas are world famous for their angular beauty. Perhaps the most remarkable of all is in southeastern China, centered on the city of Guilin (see Fig. 16.6B). Here, the Li River winds its way through a landscape that has for millennia inspired artists and writers. And to the west, in the province of Yunnan, lies another unique manifestation of karst processes, the fantasy-like Stone Forest (Fig. 42.2).

Karst terrain is widely distributed across Earth, occurring on all the continents in hundreds of localities. More than a century ago, in 1893, Serbian scholar Jovan Cvijic (pronounced yoh-*vahn svee*-itch; 1865–1927) produced the first comprehensive study of karst processes and landscapes under a title that can be translated as *The Karst Phenomenon*. Ever since, the term *karst* has been in use. Cvijic also described and gave names to many landforms resulting from karst processes. However, he was not aware of all the numerous places where karst topography existed, and later additional karst phenomena were identified and named. As these studies progressed, karst geomorphology became an important part of physical geography.

Karst Development

Karst landforms and landscapes are the products of a complex set of processes influenced by bedrock type, local relief, surface drainage, and climate. The most basic requirement is rock made mainly of minerals that are relatively *soluble* in water under the prevailing climate and other conditions. The best-known karst landscapes have developed in limestone bedrock. Calcite ($CaCO_3$), the predominant mineral in limestone, is fairly soluble and dissolves with no solid residue. Karst is also fairly common in *dolomite*, a rock similar to limestone but with high magnesium content. Karst features can be found in other

FIGURE 42.2 "It was a long road trip from Kunming in China's Yunnan Province to the so-called Stone Forest, through heavily eroded lands and on some badly neglected roads. But the destination was worth it: an aptly named expanse of limestone columns, interspersed with small lakes whose rise and fall are marked on the rock walls. Here a once-continuous, thick layer of jointed limestone is attacked by solution from above (rainwater) as well as below (groundwater). The cumulative result is a jagged landscape as far as the eye can see."

rocks, however, such as the highly soluble rock *gypsum*. Karst has even developed in quartz-rich sandstones or quartzites in certain tropical settings such as the Roraima region of Venezuela, Brazil, and Guyana. Although quartz is not considered a very soluble mineral at all, it will apparently dissolve to form karst, given enough time in a warm, humid tropical climate.

Even where thick limestones are present, karst does not always develop. Karst is rare in limestones that contain more than 20 to 30 percent of minerals other than calcite that will not completely dissolve in water. Thin beds of shale or sandstone within thick limestone may prevent karst features from forming as well. In both cases, material that does not dissolve readily plugs the developing conduits and caves and limits karst development. The porosity of the limestone also influences the dissolution process. Limestones have a great variety of pores of various shapes and sizes depending on the environment in which they formed. Some limestones have pores allowing water to reach much of the rock mass and dissolve it quickly, but other limestones are dense and therefore dissolve slowly. When there are

many joints or faults in the limestone or dolomite strata, subsurface water can flow more easily. This in turn increases the likelihood of karst landscape development.

WATER, CLIMATE, AND KARST DEVELOPMENT

In karst areas, water shapes landforms by dissolving limestone or other rock both above and below ground. Much of the work of creating karst features is done below ground by water flowing through pores and fractures in the rock. This flow includes water percolating through the vadose zone (where pores are partly filled with air; see Unit 38) as well as groundwater flowing through water-saturated rock. Gradually, the water dissolves rock surrounding the original pores and fractures, enlarging them to form **conduits** (pipe-like openings) and caves. Surface runoff flowing across bare limestone can dissolve some of the rock, but surface streams are usually poorly developed and often sink into the underground flow system.

Given the key role water plays in the process, it is not surprising that karst landscapes develop more slowly in arid climates than in humid regions. Seasonal ground frost or permafrost limit percolation of water from the surface and therefore limit karst development as well. Still, karst features do occur in deserts and permafrost regions, although they may form very slowly there.

Soil and Vegetation The soil and vegetation in a karst area are also important. In Unit 36, we discussed the acidity formed by the reaction of rainwater or soil water with carbon dioxide (CO_2). This process (*carbonation*) makes the water much more effective as a solvent for limestone. Carbon dioxide is released through respiration by roots and microorganisms in the soil, and the air in soil pores is usually enriched in CO_2. Therefore, water that has percolated through soils that are rich in roots and organic matter has a greater capacity to dissolve limestone than the water that initially fell as rain. This helps to explain why more fully developed karst topography is found in warmer as well as moister climatic regions. Higher temperatures often promote biological activity, and this in turn enhances the effectiveness of the available water as an agent of dissolution.

RELIEF

The formation of karst landscapes is further promoted when the area of limestone or other soluble bedrock has at least moderate relief. Where the surface is flat or nearly so and where surface streams have not succeeded in creating some local relief, underground drainage and dissolution are slowed, and karst formation is inhibited. In this setting, sluggishly moving water in underground conduits soon becomes *saturated* with dissolved limestone and thereby loses its capacity to dissolve more of it.

Where local relief is moderate or high, this groundwater flows down a steeper gradient, so it moves more rapidly. Freshwater replaces older saturated water in the system more rapidly, allowing accelerated dissolution. Therefore, karst development is more effective in a landscape dissected by deep valleys. Erosion of deep valleys has another effect: as valleys get deeper, more of the rock high above the valleys is above the water table and unsaturated, allowing older, higher parts of cave systems to drain and fill with air. Meanwhile, the cave system expands at lower levels. Many large cave networks are on several levels, indicating that dropping water tables resulting from incision and valley deepening played a role in their evolution (Fig. 42.3).

Groundwater in Karst Regions

Below the water table, porous rock is saturated, its pores and other open spaces occupied by slow-flowing groundwater. This is true for karst aquifers as well. But in karst, some of the pores are greatly enlarged by dissolution of the rock around them, providing pathways for much more

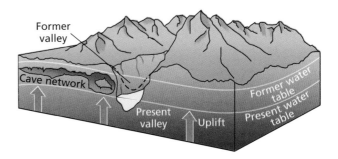

FIGURE 42.3 Cave network formed by water that entered limestone fractures and enlarged them below the (former) water table. When that water table dropped—partly as a result of uplift and partly as a result of the nearby stream's deepening of its valley—the cave system filled with air.

rapid groundwater flow. Therefore, in karst regions, some of the groundwater may be slowly creeping through small pores, while another portion is flowing through large openings almost as rapidly as a surface stream. Think of a plumbing system in a building in which most of the pipes are connected but some are thousands of times larger in diameter than others.

One method that researchers have found to study the complicated flow paths of groundwater in karst is to put colored dyes in surface water where it disappears below the ground and then check for the appearance of the colored water in certain accessible cave locations and at springs. In some cases, the dye will quickly emerge, even miles away. However, if the dye was added at another point in the system, it might move too slowly to trace for any distance. From a practical point of view, it is wise to assume that any contaminant added to groundwater in a karst landscape *can* move quite rapidly, possibly to a spring or well that is used for drinking water or that feeds a trout stream.

Karst Landforms and Landscapes

A topographic map of a karst area quickly reveals the unusual character of relief and drainage (Fig. 42.4). It seems that all the rules we have learned so far are broken. Surface streams are interrupted and stop (or shrink) in mid-valley. Contour lines reveal basins without outlets and without streams entering or leaving them. Steep-sided hills rise from flat plains without displaying slope characteristics familiar to us.

When we examine maps more closely, it becomes clear that the signature of karst topography is not the same everywhere. Geomorphologists identify at least three types of karst landscape: *temperate*, *tropical*, and *Caribbean*.

1. **Temperate karst**, of which the type area in Slovenia and Croatia offers an outstanding example, forms more slowly than tropical karst. Disappearing streams, jagged rock masses, solution depressions, and extensive cave networks mark temperate karst (Fig. 42.5).
2. **Tropical karst** develops rapidly as a result of the higher amounts of rainfall and humidity, biological activity,

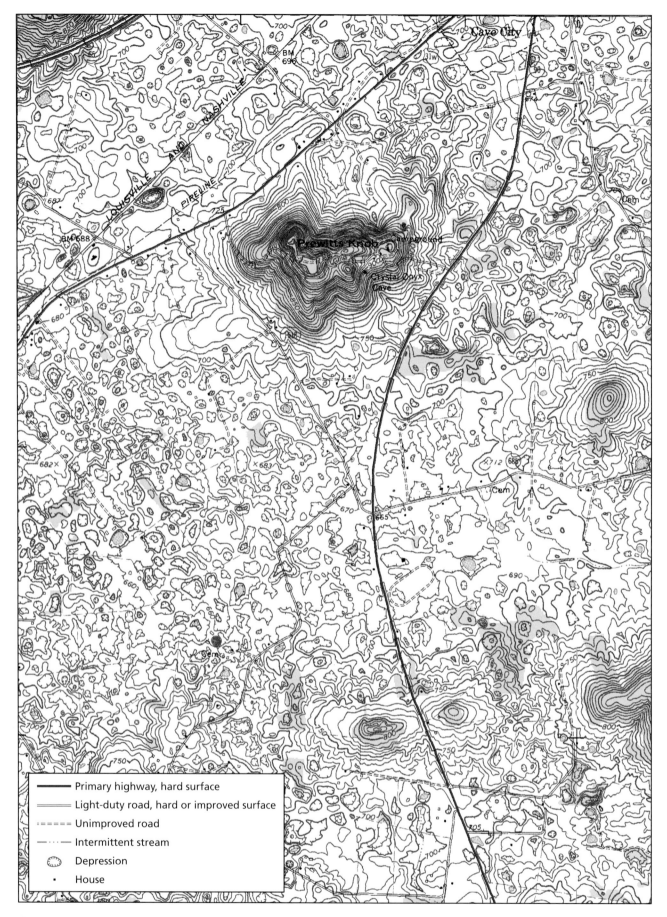

FIGURE 42.4 Karst landscapes exhibit unusual relief and drainage characteristics. This topographic map shows the surface near Mammoth Cave in west-central Kentucky, one of the most prominent areas of temperate karst in the United States. Note (1) the myriad small depressions lacking outlets; (2) the pair of intermittent, disappearing streams; and (3) the steep slopes of Prewitts Knob rising from the nearly flat surrounding plain.

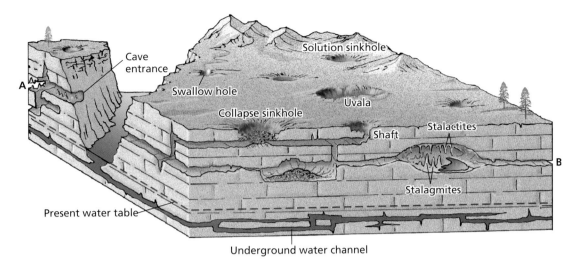

FIGURE 42.5 Surface and underground features of temperate karst. When the groundwater table was higher, the solution features at levels A and B formed. Now the water table is lower and solution proceeds. A new cave network will form when the next drop in the water table occurs.

FIGURE 42.6 Satellite image of east-central Florida, a classic Caribbean karst region, centered on the Orlando metropolitan area (NASA's rocket-launch complex at Cape Canaveral juts out into the Atlantic at the far right). The large number of round lakes, formed in sinkholes, is characteristic of this landscape. Lake Apopka, lying just to the west of the city's western suburbs at the latitude of Cape Canaveral, is the largest in the Orlando area; its sprawling urban development is revealed by the pale whitish color. This image also shows that lakes have coalesced in a number of places (e.g., in the extreme south), and the topography continues to evolve.

and organic acids in the soil. Steep-sided hills tend to be covered with vegetation, and solution features are larger than in temperate karst landscapes (see photo opening this unit).

3. **Caribbean karst** is a special case found only in a few locations. In the type area of central Florida (Fig. 42.6), nearly flat-lying limestone is eroded underground, although it lies barely above sea level. Water comes from the hill country to the north, seeps through the limestone, and leaves the system through offshore submarine springs. A similar situation exists in Mexico's Yucatán Peninsula. There, as in Florida, the roofs of the subsurface conduits have collapsed in many places, creating those characteristic depressions in the ground (see Fig. 42.1).

DISAPPEARING STREAMS AND SINKHOLES

Let us now examine some landforms we might encounter in any karst landscape. Where a surface stream "disappears," flowing into an underground channel, the place of descent is sometimes called a *swallow hole* (see Fig. 42.5, top left). This may occur at a fault or an enlarged joint that has been widened by dissolution and leads to a subsurface conduit system. Interrupted drainage of this sort is a general indicator of karst conditions. Another common karst landform is the surface depression, ranging in size from small hollows to large basins. The dominant process of formation, as we have noted, is dissolution. Depressions also can form from the collapse of part of the roof of an underground stream conduit. Logically, the former are referred to as *solution sinkholes* and the latter as *collapse sinkholes* (see Fig. 42.5).

Solution sinkholes (see Fig. 42.5, top right) range widely in size; some are as small as a bathtub and some larger than a football stadium. A single area of karst topography may contain tens of thousands of them, some old and established, others just starting to form. Their relative location is probably related to the original configuration of the terrain when karstification began. Low-lying places became solution hollows first. These subsequently expanded, and others formed as the surface was lowered overall. In cross-section (see the sinkhole at the front edge

Florida's Winter Park Sinkhole

Cultural landscapes often experience sudden change in areas of collapse and/or suffosion sinkholes. A well-publicized disruption of this kind occurred in 1981 when, without warning, a large sinkhole materialized near the center of Winter Park, an elegant suburb of Orlando in central Florida's karst region. Within moments of its appearance, the yawning abyss swallowed a three-bedroom house, half a municipal swimming pool, several motor vehicles (including five new Porsches parked at the rear of a dealership), and part of a nearby street. In all, the damage amounted to more than $2 million in property losses—as well as the disappearance of some prime real estate.

This scene is documented in the photo in Figure 42.7, taken on the day of the collapse. The new sinkhole was longer than a football field (107 m [350 ft]), with a depth of at least 40 m (130 ft). The sinkhole soon stabilized and remains essentially unchanged today. The only difference is that it has mostly filled with water to become one of the thousands of small round lakes that pockmark this part of peninsular Florida (see Fig. 42.6).

Thanks to high-profile television news coverage at the time, the Winter Park sinkhole remained an object of curiosity for years after its formation. Within a week, local authorities had surrounded the collapse with a chain-link fence, and the public just kept coming to see it. It became a landmark, complete with souvenir shops, and is a popular stop for many of the tourists who stream into the Orlando area to visit its theme parks and related attractions. Was it perhaps on your itinerary during a trip to Walt Disney World?

FIGURE 42.7 One of North America's most famous sinkholes lies in the middle of Winter Park, Florida. Appearing quite suddenly in the spring of 1981, it swallowed parts of an automobile dealership and a city pool (this photo was taken only a few hours after the collapse).

center of Fig. 42.5), solution hollows resemble funnels; water seeps down the sides to the approximate center of the basin. There a *shaft* leads downward, allowing the water to join the regional groundwater system.

Collapse sinkholes (see Fig. 42.5, left front corner), as we noted earlier, are created by the collapse or failure of the roof or overlying material of a cave, cavern, or underground channel. Because the dissolution process also created these subsurface tunnels, even a collapse sinkhole ultimately owes its origin to the dissolution process. Physical geographers distinguish between a *collapse sink*, in which the rock ceiling of the sinkhole collapses into the underground solution hole, and a *suffosion sink*, created when an overlying layer of unconsolidated material is left unsupported. Where such loose material lies on top of limestone strata affected by the karst process, lower parts of this material are drawn downward into the enlarging karst joints. This creates a void in the lower stratum of this unconsolidated layer. When the void becomes large enough, it will collapse. Unlike the slow development of solution sinkholes, that can happen quite suddenly (see Perspectives on the Human Environment: Florida's Winter Park Sinkhole). Obviously, building in a zone of collapse or suffosion sinkholes can be a hazardous proposition.

Occasionally, two or more neighboring sinkholes join to become an even larger depression called an **uvala** (see Fig. 42.5, center). Uvalas can reach a diameter of more than 1.5 km (1 mi). Some have rough, uneven floors that are dry and vegetated; others are filled with water and form scenic lakes. Studies of sinkhole terrain indicate that this kind of karst landscape develops best where the environment is humid, so that there is plenty of water to sustain the underground drainage system. It also appears that the rise and fall of the water table may have something to do with the distribution of collapse sinkholes. Where groundwater levels rise and fall rapidly and substantially, this process seems to trigger more frequent collapses.

KARST TOWERS

If sinkholes are the signature landforms of temperate karst areas, then the dominant feature in tropical karst regions is the *tower*. Sinkholes and towers have various other names, but there is no need to complicate our terminology. A **tower** is a cone-shaped, steep-sided hill that rises above a surface and may or may not be pocked with solution depressions. Even when many sinkholes are also present, the towers, which are sometimes hundreds of meters tall, dominate the landscape (see the photo opening this unit). In tropical karst zones, the contrast between towers and depressions is so sharp that the whole scene is sometimes referred to as **cockpit karst**, the term *cockpit* referring to the irregular, often steep-sided depressions between the towers.

Tropical karst studded with such towers is found in such locales as Puerto Rico, Jamaica, Cuba, and Vietnam (Fig. 42.8). It is uncertain exactly what determines the location and distribution of the towers. The towers are remnants of a thick bedrock sequence consisting of limestone and/or dolomite layers. Before the karst topography developed, and following regional uplift (or the lowering of sea level), the original surface presumably developed a soil cover and plants took hold. The original terrain likely determined the initial pattern of soil and vegetation (thick and well developed in some places, thin and sparse in others).

Where the surface was low, moisture collected and soil soon formed, but higher places remained barren. Eventually, these higher places became the tops of karst towers as the intervening hollows grew ever deeper. China's Guilin area (see Fig. 16.6B) and Yunnan Stone Forest (see Fig. 42.2) may have originated in this way, although fluctuating groundwater also was a factor. In fact, the limestone towers in the Stone Forest still rise from lakes with levels that vary seasonally.

FIGURE 42.8 A low aerial view of central Jamaica's "cockpit country," a classic, tower-studded tropical karst landscape.

Karst and Caves

We began this unit with a look at caves as part of the karst phenomenon. Now we are in a better position to examine these remarkable features in more detail. Our focus here is on caves that have formed through dissolution that enlarged fractures or other openings in soluble rock. Many caves were initially below the regional water table and were filled with water in earlier stages of their development. Some remain in that condition, but others have been drained, usually because nearby streams have eroded deeper valleys and the water table has declined in elevation. Of course, these are the caves that are easiest to enter and explore or to make into tourist attractions. Quite a few caves contain streams that have been diverted underground from the surface drainage network.

CAVE FEATURES

A fully developed cave consists of an entrance (*portal*) and one or more chambers, passages, and terminations. A *termination* marks the place beyond which a person cannot crawl any farther along an underground passage. Passages in a fully developed cave system form an interconnected network. The pattern of this network depends on the stratigraphy (the order and arrangement of rock layers; see Unit 28), faulting, and jointing of the bedrock sequence. It may consist of one major subsurface artery (the *linear form*); it may look like the branches of a tree (the *sinuous form*); or if block jointing is well developed, it may have an *angulate* (right-angle, stepped) form. Given the complexity of karst features, various other more detailed models have been developed to account for the overall structure of the maze of caves, caverns, and conduits.

Where passages grow exceptionally large, chambers or "rooms" develop. These chambers, some with the dimensions of a large hall, contain many fascinating forms. Lakes exist in some of them, and drops of water falling from the ceiling create eerie musical echoes in the dark void. Streams may even flow through them with the magnified sound of a waterfall.

Water dripping from the roof of caves often precipitates calcite ($CaCO_3$). This water, entering the cave, contains abundant dissolved calcium (Ca^{++}) and bicarbonate (HCO_3^-) ions, usually as a result of calcite or dolomite dissolution above the cave. Recall that water percolating downward from the soil often has abundant dissolved carbon dioxide because of the high levels of this gas in soil air. Cave air tends to contain somewhat less carbon dioxide, so some of it is released from the dripping water into the cave air. This makes calcite a little less soluble in the dripping water, and it precipitates to form distinctive accumulations of solid calcite. These include icicle-like **stalactites** hanging from the ceiling and **stalagmites** standing, like sentinels, on the floor (Fig. 42.9). These opposing pinnacles can become several meters tall and often coalesce to form **columns** (Fig. 42.9). Small wonder that caves attract so many visitors.

CAVE NETWORKS

To physical geographers, however, caves present other mysteries. In 1988, divers for the first time penetrated the water-filled tunnels of a cave system beneath northern Florida's Woodville Karst Plain near Tallahassee. They entered a sinkhole lake and followed a flooded passage using battery-powered motors and floodlights. The passage went 75 m (240 ft) below the surface, and they followed it for more than 2.5 km (1.5 mi) until they saw daylight above and returned to the surface through another sinkhole. The cave passages, they reported, were as much as 30 m (100 ft) wide but also narrowed considerably. They saw side passages joining the main conduit and realized that they were seeing only a fraction of a very large and unmapped network. These Florida cave systems, now below sea level and filled with water, were formed more than 35 million years ago during a period of lower sea level.

Karst terrain and associated caves, as we noted at the beginning of this unit, are widely distributed across Earth. Some of the world's most impressive karst regions are only now becoming known and understood (e.g., the karst structures of Australia's Kimberley region in the northern part of Western Australia). Karst topography has been submerged by coastal subsidence, and it also has been uplifted into high mountains (there is karst terrain under Canadian

FIGURE 42.9 Close-up view of some of the wonders of New Mexico's Carlsbad Caverns, showing myriad columns, stalagmites, and stalactites.

glaciers and on frigid Andean slopes in South America). Thus, the map of karst and cave distribution contains valuable evidence for climatic as well as geologic change. The geomorphic signature of karst is indeed one of the most distinctive elements in the global environmental mosaic.

Key Terms

Caribbean karst *page 513*
cave *page 509*
cockpit karst *page 515*
collapse sinkhole *page 514*
column *page 516*

conduit *page 510*
dissolution *page 509*
karst *page 509*
solution sinkhole *page 513*
stalactite *page 516*

stalagmite *page 516*
temperate karst *page 511*
tower *page 515*
tropical karst *page 511*
uvala *page 515*

Scan Here for a quick vocabulary review

Review Questions

1. Which rock types are prone to karst development and why?
2. Explain why dense vegetation and soil rich in organic matter may favor karst development.
3. How do relief and the incision of streams influence karst development?
4. Describe the processes by which underground conduits and caves form in karst regions.

5. Explain the difference between collapse and solution sinkholes.
6. What are some differences between the landforms of temperate and tropical karst?
7. How do towers form in karst regions?
8. Describe some of the major features of caves and how they form.

Scan Here for a quick concept review

www.oup.com/us/mason

PART SEVEN

Sculpting the Surface: Ice, Wind, and Coastal Processes

Sculpting the Surface: Ice, Wind, and Coastal Processes

When you drive or fly across the continents, you can readily recognize landscapes carved by flowing water into networks of stream valleys and ridges. Yet if you travel northward in Europe or North America you will eventually find quite different landscapes, made up of chaotic ridges, mounds of debris, and closed depressions, none of which can easily be explained as the work of streams. One of the most important early discoveries of Earth science was the recognition that these landscapes were shaped by glacier ice in a much colder climate of the past. At high latitudes where cold climates persist today, vast areas are affected by ice in another form, in seasonally or permanently frozen ground. If you venture into dry regions of the world, you will find the remarkably regular, repeating forms of sand dunes, landforms shaped by wind. Along the shores of the ocean or lakes, distinctive features formed through erosion and deposition by waves, currents, and tides are recognizable. Glaciers, frost action, wind, and coastal processes each leave their own distinctive signatures on the landscape, and learning to interpret those signatures is an important step in understanding and adapting to many of the environments in which we live.

UNIT 43

Glacial Erosion and Deposition

The Tonsina Glacier in Alaska flows down a mountain valley from the zone of ice accumulation at higher elevations. Loss of ice from the lower part of the glacier is evident from the meltwater flowing out of it at the lower left. The bare rock near the lower end of the glacier was covered by ice in the recent past, but now the glacier is retreating.

OBJECTIVES

- Explain how glaciers act as important agents of landscape formation and discuss the different categories of glaciers

- Give a brief overview of how past glaciation has influenced Earth's surface over large areas of the continents

- Describe the formation of glacier ice and its movement through the glacial system

- Summarize how glaciers shape the landscape through erosion and deposition

In this unit and Units 44 to 46, we address various aspects of the *cryosphere*. The **cryosphere**, one of the five major components of the Earth system (see Fig. 2.4), consists of all the forms of frozen water that exist above, on, and just below Earth's surface. It includes all types of *glaciers*, snow cover, ice floating atop water bodies, and permanently frozen ground (permafrost). Our focus here will be on glaciers as powerful agents of landscape modification that create an array of distinctive erosional and depositional landforms. Beyond the basic question of how glacial landforms develop, they also interest physical geographers because they provide valuable evidence in the study of environmental change. Glacial landforms in regions where no glaciers exist at present record past periods of time when the global climate was much colder than it is today (Fig. 43.1). We can also use glacial landforms to reconstruct the size of past glaciers and estimate how much the storage of water as ice in those glaciers affected global sea level.

A **glacier** is a body of ice formed on land that is in motion. A glacier's motion is not readily apparent over short time periods, and to a casual observer glaciers appear to be sedentary accumulations of ice, snow, and rock debris. Yet glaciers do move, and they steadily erode their valleys. Hunters and guides in the Alps pointed this out to visiting geographers and geologists more than two centuries ago, and the scientists confirmed the reality of glacial movement by putting stakes in the ice and measuring its annual downslope advance. But exactly how glaciers flow over and erode the rock or soft sediment beneath them is still not completely understood and continues to be a subject of ongoing research.

In many ways, glaciers are more difficult to understand than rivers. Although we can directly observe and measure the flow of water in a shallow river, and the movement of sediment that it brings about, it is much more difficult to observe what happens as ice flows over rock and erodes it. In most cases, there is only indirect evidence of what

FROM THE FIELDNOTES

A

B

FIGURE 43.1 "Flying over glaciated terrain suggests how the planet changes as climatic cycles run their course. When mountain glaciers develop over an area originally sculpted by rivers, they bury most of the terrain and fill the valleys with ice, leaving only the crests and peaks of ranges and mountains protruding above, as you can imagine from a view over a section of the Alaska Range (A). When the glaciers have melted away (and they may even melt from the Alaska Range), the exposed topography reveals their work in a variety of landforms, including their steep-sided valleys, jagged mountain peaks, and sharp-edged ridges (B). This photo was taken over a part of Colorado's southern Rocky Mountains, where most of the once-prevailing ice has melted away in the warmth of the current interval."

happens on the underside of glaciers. A rare exception occurred when a tunnel was dug beneath a glacier for a hydroelectric project in Norway. This allowed researchers to study the interaction of the glacier ice and the bed of rock or debris underneath it. They were able to observe processes such as the formation of thin water layers that allowed the ice to slide over its bed (Iverson et al., "Soft-bed Experiments Beneath Engabreen, Norway: Regelation Infiltration, Basal Slip and Bed Deformation," *Journal of Glaciology* 53 [2007]: 323–340).

The glaciers of Switzerland, Norway, and other mountainous regions are **mountain (alpine) glaciers**. We subdivide these glaciers according to their location. Some are small and confined to hollows below mountain peaks (called *cirques*); thus, they are called *cirque glaciers* (Fig. 43.2). Larger *valley glaciers* fill deep troughs carved into mountain ranges (see the opening photo of this unit) where the combination of cold temperatures and high snowfall allows more ice accumulation. Glaciers called

continental glaciers, or **ice sheets**, consist of huge masses of ice that are not confined to valleys but bury large regions beneath them. Antarctica, a continent almost twice as large as Australia, is almost completely covered by vast ice sheets and so, too, is Greenland, the world's largest island.

Glaciers of the Past

During the past several million years, Earth has periodically experienced **glacial periods**, when Earth's atmosphere became much cooler than it is at present (see Unit 18). During each glacial period, there was great expansion of ice sheets in high to middle latitudes and growth of large glaciers in mountains where they are small or nonexistent today. Glaciers even formed on the high volcanic mountains of Hawai'i during past glacial periods. After a glacial period has reached its peak, the climate begins to warm up. Then the glaciers start receding rapidly in a phase known as **deglaciation**. After deglaciation, the global climate

FIGURE 43.2 Dinwoody Glacier in the Wind River Mountains of Wyoming occupies a large cirque, or mountainside hollow. In this summer view, fresh snow is visible on the highest parts of the glacier, above the snowline. The gray ice surface lower on the glacier indicates the predominance of ice loss through melting (ablation).

ENHANCE

enters a relatively warm **interglacial period** that lasts tens of thousands of years until Earth enters a new cooling episode, and eventually the glaciers expand again.

At the peak of the last glacial period, 25,000 to 15,000 years ago, much of Canada was covered by continental glaciers, and these ice sheets reached south of the Great Lakes. Earth today is in an interglacial period, and in the comparatively warm climate, glaciers have withdrawn to the coldest of the polar (and mountainous) regions. Areas once covered by continental ice sheets are dominated by other geomorphic processes.

As we discussed in Unit 18, Earth's climate has shifted between glacial and interglacial conditions many times over the past 2 million years, like the back-and-forth motion of a pendulum. As a result, ice sheets expanded across North America and Europe and then contracted and disappeared many times. During some glacial periods, North American ice sheets reached as far south as southern Illinois and northern Missouri, depositing boulders from Canada and the Great Lakes region.

The more recent glacial periods occurred during the Pleistocene epoch of the Cenozoic era (see Fig. 28.11). In fact, the Pleistocene was marked by many pronounced glacial and interglacial periods. From averaging the global temperatures over glacial and interglacial periods of the Pleistocene, we know that Earth experienced its coldest climate in more than 65 million years during the Pleistocene. But geologists also now know that glacial periods occurred before the Pleistocene as well. Ice sheets first formed in Antarctica tens of millions of years ago, and glaciers were present at high latitudes and in some mountain ranges several million years ago during the Pliocene epoch. Great ice sheets first covered northern North America and Europe between 2 and 3 million years ago. Therefore, scientists use the label **Late Cenozoic Ice Age** to refer to the entire period of cooling climate and frequent glaciations that reached a peak in the Pleistocene.

The Late Cenozoic Ice Age is only the latest in a series of such events in Earth's environmental history. For example, the supercontinent of Gondwana (see Unit 30) experienced repeated formation of massive ice sheets during the Pennsyvanian and Permian periods of the Paleozoic era some 320 to 240 million years ago. When Gondwana split apart much later, the resulting fragments (Africa, South America, India, Australia, and Antarctica) all carried evidence of the activity of these glaciers in their landscapes and underlying rock strata. When geologists discovered this evidence, they had a major clue to the former existence of Gondwana. Moreover, much older rocks indicate several even earlier ice ages.

In one important respect, the present interglacial period is unlike any other this planet has witnessed. During the interglacial period now in progress, the world's human population has grown explosively. Geologically, these last 11,500 years constitute the Holocene epoch (see Fig. 28.11). Geographically, the Holocene has witnessed the transformation of the planet—not only by climatic change but also by human activity. The present interglacial period is unprecedented because, for the first time in Earth's history, humans have become agents of environmental change.

On the basis of what is known about the patterns of past glaciations and interglacial periods, we might assume that another cooling episode lies ahead and that the glaciers will once again expand and advance. But human interference in the composition of the atmosphere may affect the course of events. As discussed in Unit 19, climate scientists now believe that greenhouse gases added to the atmosphere through human use of fossil fuels and land-use change are already warming Earth's climate above its natural interglacial level. A warmer climate will necessarily cause additional melting of ice in polar and high-mountain regions. At least through the near future, then, we are more likely to experience further shrinkage of glaciers than we are new growth. When we study glaciers, present or past, it is important to remember their significant connections to global environments. Even today, when the extent of ice is comparatively limited, the glaciers of Greenland and Antarctica influence the radiation and heat balances of our planet.

Continental ice sheets contain huge volumes of freshwater and thereby affect the global water balance as well. When a glaciation begins, precipitation in the form of snow is compacted into glacial ice, some of which is not returned to the oceans. Therefore, sea level drops as glaciation proceeds. When deglaciation begins, the melting glaciers yield their large volumes of water, and sea level rises again. Thus, glaciation and deglaciation are accompanied, in turn, by falling and rising sea levels. The melting of large ice sheets during the last deglaciation caused a dramatic sea-level rise, which then tapered off as the present interglacial period continued and Earth's remaining glaciers stabilized. Given the current prediction that the remaining glaciers will rapidly shrink as climate warms over the next century, we can expect additional sea-level rise as a result.

The Glacier as a System

To understand the behavior of a glacier, we can visualize it as an open system, with both input and output of matter. Input of water to the glacial system occurs when snow falls on the glacier surface. On higher parts of the glacier, that snow persists through the summer rather than melting away, and eventually it becomes ice. Ice flows through the glacier to its lower margin and eventually melts to form water that drains away into streams, lakes, or the ocean. Ground-up rock also moves through the glacier system, embedded in the ice, as we will discuss later.

Recall that when we considered the river as a system in Unit 39, we noted that water falling as rain or snow on higher parts of the basin has potential energy because of its elevation, which is transformed into kinetic energy— the energy of motion—as the water flows downhill. Part of the kinetic energy that is continually generated as the river flows down its sloping channel is used to erode and carry sediment. Similarly, the flow of ice in a glacier represents the conversion of potential energy to kinetic energy, some of which does work on the landscape.

To understand the glacial system in more detail, in this section, we first focus on the accumulation of glacier ice. Ice accumulation does not occur uniformly across the whole glacier; instead, it continually builds up the thickness and height of one part of the glacier, creating a gradient from high to low potential energy. Ice flows down that gradient and in the process does work. After we look more closely at ice formation, we will examine the specific ways in which glaciers flow, and how they erode the rock beneath them and carry away the eroded debris.

FORMATION OF GLACIER ICE

Glacier ice forms from compacted, recrystallized snow. Not all snow, even in mountainous areas, becomes part of a glacier, of course. When you travel through (or over) a high-mountain area such as the Rocky Mountains, you may observe the *snow line*, a line above which snow remains on the ground throughout the year (see Fig. 43.2). Below this snow line, each winter's accumulation of snow melts during the following summer, and none of it is transformed into ice. But above the snow line, where it is colder and snowfall may also be greater, some permanent snow survives the summer and contributes to the growing thickness of the snowpack. Where summer snow loss is less than winter gain, conditions are favorable for the formation of glacial ice.

Snow is converted into ice in stages. Newly fallen snowflakes are light and delicately structured crystals. A layer of freshly fallen snow generally has a low density. Some melting of the outer "points" of the snowflake crystals may take place and change them into irregular but more spherical grains (Fig. 43.3A). A later snowfall may also cover and compress the layer of older snow, packing the crystals more tightly together and destroying their original structure. All this has the effect of increasing the density of the lower layer and reducing its open spaces, or porosity. In areas where periodic melting occurs, fluffy snow can be converted into dense, granular snow in a matter of days.

But this first stage does not yet yield glacial ice. The granular, compacted snow—called **firn**—undergoes further compression and recrystallization (Fig. 43.3B). That takes time, and it requires more time in cold polar areas than in warmer temperate zones. This is because in the temperate areas where melting occurs, percolating meltwater fills the remaining pore spaces, refreezes there, and adds to the weight of the snowpack. Glaciologists calculate that the transformation from firn to ice in temperate areas may require less than 50 years. In polar areas, it may take 10 times as long.

A glacier in coastal Alaska may need a firn layer less than 15 m (50 ft) deep for ice to form. On the other hand, in the colder and drier Antarctic, 100 m (330 ft) of firn is

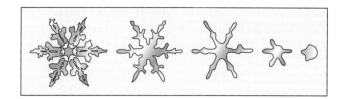

A

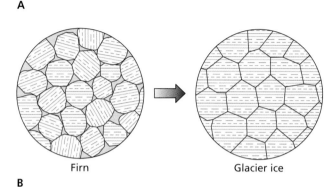

Firn → Glacier ice

B

FIGURE 43.3 The transformation of a snowflake into a granule of old snow (firn) can take several weeks as the outer "points" of the crystal melt (A). The conversion of firn into glacier ice takes decades, even centuries, in the coldest climates as recrystallization and further compression slowly squeeze out the open spaces between individual granules (B).

required to produce glacier ice of the same density. Snow accumulation on the Antarctic ice sheet is very slow, but the firn is of enormous depth.

ICE ACCUMULATION AND ABLATION

When we observe a mountain glacier in summer, there is often a striking contrast in appearance between the upper and lower parts of the ice mass (see Fig. 43.2). High on the glacier, light areas reflect the persistence of unmelted snow—and occasional new snowfall—through the summer. Lower down, the "dirty" ice surface reveals the debris concentrated as snow and ice melt.

This contrast approximately marks a very important boundary between the **zone of accumulation**, which is the lighter-colored zone higher on the glacier, and the **zone of ablation**, which is located at lower elevation. *Ablation* refers to loss of ice, which occurs largely through melting for most glaciers. *Sublimation* (conversion from solid ice directly to water vapor) also plays a small part in ablation, and glaciers that flow into the ocean lose ice as icebergs. In the zone of accumulation, where snow persists in summer, more frozen water is added by snowfall each year than is lost by ablation. In the zone of ablation, on the lower part of the glacier, more is lost than is added.

If there is a net loss of ice in the ablation zone, why doesn't that part of the glacier eventually disappear? This does not happen because the zone of ablation is continually resupplied with ice (Fig. 43.4). Ice moves continuously from where the glacier surface is higher (the zone of accumulation) to where it is lower (the zone of ablation). If the

A

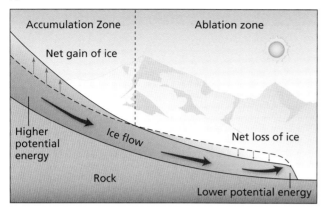

B

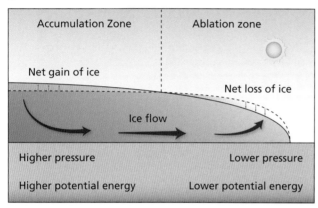

FIGURE 43.4 A valley glacier (A) and a large ice sheet (B) differ greatly in size and shape, but both are open systems, with new snow added in their upper zones of accumulation and material lost in their lower zones of ablation. In both cases the gradient of potential energy drives ice flow from the accumulation zone to the ablation zone.

glacier is in equilibrium, it gains as much ice in its accumulation zone as it loses in its ablation zone, and the glacier neither grows nor shrinks. But equilibrium conditions rarely persist over a long period of time, and glaciers tend to fluctuate in size. When the *mass balance* of ice is positive—that is, more ice accumulates than is lost by ablation for the glacier as a whole—the glacier thickens, and its lower edge advances downhill. When the balance is negative—that is, there is more loss than gain—the glacier thins, and its lower margin recedes. Tonsina Glacier, shown in the opening photo of this unit, provides an excellent example of a glacier that has a retreating margin because of a negative mass balance.

Even as a glacier shrinks, however, ice continues to flow from the accumulation zone to the ablation zone. The dark gray, especially "dirty" ice at the lowest end of a glacier (see Fig. 43.2) reminds us that the ice melting there has flowed from higher parts of the glacier and, in the process, has

picked up debris eroded from rock beneath the glacier. That debris is carried to the lower margin of the ice, where it accumulates to form the glacial landforms we discuss in Unit 44.

Let's return to the concept of a potential energy gradient within the glacier. In a steep mountain glacier, both the surface and the bed of the glacier slope distinctly downhill from the accumulation zone into the ablation zone. In this case it is not difficult to understand that the ice is driven downhill by gravity, losing potential energy as it drops in elevation (Fig. 43.4A). Now consider a large ice sheet, like the one that covered northern North America during past glacial periods (Fig. 43.4B). The surface of that huge ice sheet rose up from its edges to one or more high central domes, where the zone of accumulation was located. However, the ground beneath the ice did not drop much toward the ice sheet margins and actually rose in some cases. In this case it is easier to think about a gradient of pressure (also a kind of potential energy). The pressure deep beneath the thick ice of the accumulation zone was much greater than the pressure beneath the thinner ice of the ablation zone near the ice margins. That gradient drove the flow of the ice from the central domes toward the edge of the ice sheet.

Glacial Movement and Erosion

Not only the motion, but also the enormous erosional power of glaciers is ultimately related to those same energy gradients we just discussed. When mountain glaciers have melted away and vacated their valleys, they leave exposed some of the world's most spectacular scenery (Fig. 43.5). Valley sides are sheer and scarp-like; waterfalls plunge hundreds of meters onto flat, wide valley floors. The glaciers may be gone, but the landscape bears the dramatic imprint of their work.

THE MOVEMENT OF ICE

To understand how glaciers erode the landscape, we must first consider how solid glacier ice flows. A profile through the center of a mountain glacier reveals that its upper layer is often cut by large cracks called **crevasses**; this reveals that the ice in this layer is brittle and breaks easily when it is compressed or stretched. Below this rigid layer, however, the ice takes on the properties of a plastic material that can actually flow under stress, although it remains a solid (Fig. 43.6). The ability of the deeper ice to flow explains one of the two major types of glacier movement, called **glacial creep**. Glacial creep involves the internal deformation of the ice, with crystals slipping over one another in response to the stress within the sloping ice mass.

The second type of motion, called **glacial sliding**, involves the movement of the entire glacier over the rocks below it. Not all glaciers move by sliding, but this type of movement is especially interesting because it is responsible for much of the erosional work done by glaciers. Glacial

FIGURE 43.5 "The magnificent scenery of Yosemite National Park in California's Sierra Nevada was sculpted by streams and glaciers. Half Dome (right) seems to have been halved by a powerful glacier coming down the U-shaped valley it overlooks, but scientists are not unanimous on this point."

ENHANCE

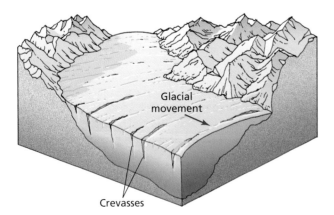

Glacial movement

Crevasses

FIGURE 43.6 A glacier's brittle upper layers are studded with large crevasses. At a greater depth, the ice flows instead of breaking under stress.

sliding is enhanced by the existence of a thin film of water between the bottom of the ice and the bedrock floor. This film of pressurized water is just millimeters thick and is probably discontinuous. It is enough, however, to reduce friction between glacier and bedrock and speed the glacier's movement downslope.

Most glaciers move slowly. The rate of motion measured on the glacier surface is only a few centimeters per day in some cold polar regions, but as much as 4 or 5 m (15 ft) or even more per day in rapidly moving alpine glaciers. Occasionally, a glacier develops a **surge**, a rapid movement of as much as 1 m per hour or more, sustained over a period of months and producing an advance of several kilometers in one season. A well-known glacial surge occurred in Alaska in 1937 when the Black Rapids Glacier suddenly advanced across the Delta River Valley. This "galloping glacier" threatened the Richardson Highway, at that time the only road into the interior of Alaska, before retreating almost as rapidly as it had advanced. Glacier surges result from especially rapid sliding, probably because of unusually high water pressure built up at the bottom of the ice.

The entire glacier does not move as fast as the ice at the surface (Fig. 43.7). Ice moves the slowest—and sometimes does not flow at all—near the bed of the glacier, or where it

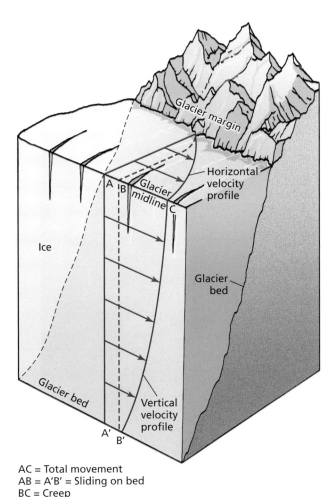

AC = Total movement
AB = A'B' = Sliding on bed
BC = Creep

FIGURE 43.7 Differential movement rates within an advancing glacier. The flow is slowest near the rock walls of the valley and the glacier's bed. The fastest flow is in the upper central part of the glacier.

is in contact with the rock on either side of it. In those locations, friction resists movement, and any motion that does occur is through sliding. Farther above the bed, or away from the valley sides, the ice moves faster, as motion by glacial creep is added to the movement through sliding. The fastest flow is in the upper central part of a glacier.

TEMPERATURE AND GLACIAL MOVEMENT

Temperature clearly influences the behavior of glaciers; for example, when the climate becomes warmer, glaciers generally retreat. Temperature is also a critical factor in glacial movement and erosion. Indeed, whether the **basal ice** at the bottom of a glacier is at or near its melting point may be the most important factor of all in determining a glacier's capacity to erode the rock beneath it.

Glaciers either gain or lose heat at the upper surface of the ice depending on how much solar radiation they absorb and the temperature of the air above. In summer, heightened solar radiation and warmer air can cause considerable warming of the uppermost ice compared to deeper layers. In winter, the glacier surface can become colder

than deeper ice. Heat also flows upward into the glacier from underlying Earth materials, warming the deeper layers of ice.

Overall, then, the temperature at various depths can vary quite a bit, but it follows two basic patterns. In cold polar climates, glacier ice tends to be below the freezing point of water from the top to the bottom of the glacier, except for a thin upper layer that may warm to the melting point in summer. In these glaciers, temperature rises gradually toward the base of the glacier because of the addition of heat from below. The basal ice never warms enough to melt, however. Therefore, these cold polar glaciers can move by glacial creep but not by glacial sliding. Movement by creep alone is quite slow, and, even more important, little erosion takes place.

In somewhat warmer climates, almost the entire thickness of the glacier is at or near the melting point. Liquid water can occur at various levels within the glacier, including at its base, and the glacier moves by sliding as well as by creep. A sliding glacier is much more effective at the kinds of erosion we will discuss in the next section of this unit. It should be noted that there are some glaciers that fall between the extremes just described. For example, they may have upper layers that are well below the melting point, but their deeper ice is warm, and there is melting of basal ice and motion by sliding.

GLACIAL EROSION

Erosion by glaciers can take place through **plucking** (Fig. 43.9A), also referred to as *quarrying*. In plucking, blocks or fragments of bedrock beneath the glacier are broken loose and frozen into the ice as it moves forward. Plucking is greatly facilitated when there is liquid water under pressure just below the glacier. This can occur where the glacier flows over a bump in the underlying rock surface, increasing pressure in a local area under the ice. The meltwater is forced into cracks in the rock and pries them open, breaking loose blocks of rock that may be frozen into the glacier and carried away.

Glacial erosion also can occur by **abrasion**, the scraping process produced by the impact of rock debris carried in the ice upon the bedrock below. Despite its appearance, ice is not a hard substance; on the Mohs Hardness scale (see Table 27.1), it would rate only about 1.5. Abrasion therefore must be performed by the rock fragments that the moving ice drags along the underlying bedrock. Some of these rock fragments are, of course, quite soft themselves and do not have much effect on glacial erosion. Harder fragments, however, do have a powerful impact on the bedrock floor beneath the glacier. The glacier pushes a boulder downward while dragging it along. This slight downward motion results from the melting away of basal ice and is necessary for effective abrasion to occur. Thus, both abrasion and plucking depend on the basal ice being warm enough to melt, at least in some parts of the glacier's underside.

FIGURE 43.8 "Our Northwestern University field camp was based in Platteville, Wisconsin, during the summer, and one of our tasks was to find evidence in the landscape of the action of Wisconsinan glaciers. It was not difficult (except in the Driftless Area, which the glaciers missed!), because these most recent glaciers left numerous manifestations of their former presence. This is hard bedrock planed by a glacier and carved with striations, scratches made when the ice dragged embedded rocks across the surface. From this evidence we can reconstruct the direction of movement, indicated by the compass."

Even where melting at the glacier bed allows erosion to occur, it may not be very rapid. Harder rocks are generally eroded much more slowly than soft sediment, of course. Where hard rocks are broken by many fractures they can erode more rapidly, however, especially through plucking. High rates of glacial abrasion generally require fast-moving ice. The greater the weight of overlying ice, the more rapidly the glacier moves. As a result, erosion by a glacier flowing through a mountain valley may be much more rapid toward the middle of the valley than at its edges.

Abrasion produces several telltale features in the landscape. When the abrading debris consists of fine but hard particles (e.g., quartz grains) and the underlying bedrock also is quite hard, abrasion produces a polished surface that looks as though the bedrock has been sandpapered. But when the rock fragments are larger, the underlying surface may be scratched quite deeply. These scratches, made as the boulder or pebble is dragged along

the floor, are called glacial **striations** (Fig. 43.8). Striations are often meters long and millimeters to centimeters deep. Because striations lie parallel to the direction of the moving ice, they can be useful indicators of the direction of ice movement where the topography provides few other clues.

Plucking also leaves evidence in the landscape. The most common landform associated with glacial plucking is the *roche moutonnée* (Fig. 43.9B), which actually involves the work of abrasion as well. This characteristic, asymmetrical mound appears to result from abrasion on one side (the side from which the ice advanced) and plucking on the opposite side. A complicated process allows the glacier to quarry this leeward side, lifting out and carrying away loosened parts of the *roche*. Studies suggest that jointing in the bedrock contributes to the glacier's ability to "pluck" the mound over which it passes. *Roches moutonnées*, like striations, help us reconstruct the path of the glacier.

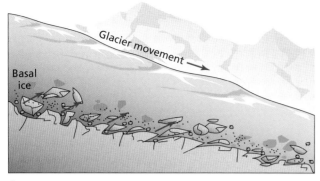

A

ENHANCE

FIGURE 43.9 Plucking occurs when glacial-bed rock fragments are torn loose and carried away downslope embedded in the basal ice flow (local variations in the pressure of water at the base of the ice play a key role in this) (A). Larger mound-like landforms, called *roches moutonnées*, are also created by this process, with the plucking found on the leeward side (B).

As in the case of rivers, glaciers deposit in some places as they erode in others. The erosion of their mountainous source areas results in deposition of debris in their lower valleys and at their terminal edges. Like rivers, glaciers grind up the sedimentary loads they carry. The debris carried downslope by alpine glaciers tends to concentrate in certain zones of the glacier and appears on the surface in the lower part of the ablation zone, where it is left behind by melting ice. Continental ice sheets, too, erode the landscapes they flow over and build up landforms out of debris deposited at their margins. Much of the local topography of the area bordering the Great Lakes as well as surrounding states is the result of erosion and deposition by continental ice sheets. We turn next to the landforms and landscapes created by the great ice sheets of the past.

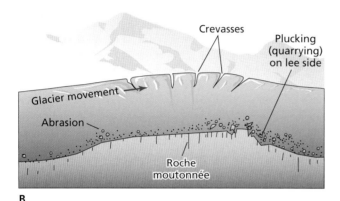

B

Key Terms

abrasion *page 527*
basal ice *page 527*
continental glacier *page 522*
crevasse *page 525*
cryosphere *page 521*
deglaciation *page 522*
firn *page 524*

glacial creep *page 525*
glacial period *page 522*
glacial sliding *page 525*
glacier *page 521*
ice sheet *page 522*
interglacial period *page 523*
Late Cenozoic Ice Age *page 523*

mountain (alpine) glacier *page 522*
plucking *page 527*
roche moutonnée *page 528*
striation *page 528*
surge (glacial) *page 526*
zone of ablation *page 524*
zone of accumulation *page 524*

ENHANCE

Scan Here for a quick vocabulary review

Review Questions

1. Where do continental glaciers presently exist?
2. What is meant by the term *Late Cenozoic Ice Age*?
3. How does snow become transformed into glacial ice?
4. Describe how the mass balance of a glacier controls the glacier's thickness and the area it covers.
5. Describe the difference between glacial creep and glacial sliding.

6. Explain how the profile of temperature within a glacier influences the rate of glacier movement and the amount of erosion that occurs as a result.
7. How is a *roche moutonnée* formed?

Scan Here for a quick vocabulary review

ENHANCE

www.oup.com/us/mason

UNIT
44

Landforms and Landscapes of Continental Glaciation

The Transantarctic Mountains are partly submerged by the Antarctic Ice Sheet, but the higher peaks rise above the ice. Large glaciers like the one in the center of this photo flow through this mountain range and drain ice from the high plateau of East Antarctica.

OBJECTIVES

- Compare and contrast contemporary continental ice sheets and the ice sheets of past glacial periods
- Summarize the history of continental glaciation in North America and how our understanding of that history has changed over time
- Identify typical landforms and landscapes produced by continental glaciers and discuss the glacial processes that formed them
- Describe the development of the Great Lakes
- Briefly discuss the pluvial lakes that formed during glacial periods in the western United States

During glacial periods, Earth's surface is transformed. Great ice sheets form over landmasses at high latitudes. Entire regions are submerged under ice—mountains, plateaus, plains, and all else. The weight of the ice, which may reach a thickness of more than 3,000 m (10,000 ft), pushes the underlying crustal bedrock downward isostatically (see Unit 31). So much water is converted into snow (and subsequently into glacial ice) that sea level can drop more than 100 m (330 ft). Large areas of continental shelf are exposed, coastlines are relocated accordingly, and continental outlines change shape. As the ice sheets expand, thereby expanding the region of polar-type temperatures, global climatic zones are compressed toward the lower latitudes. Midlatitude lands that were previously temperate become cold, barren, and subpolar in character; vegetation belts shift equatorward.

In this unit, we focus on the great continental glaciers, present and past, and on the landforms they create. Like mountain glaciers, the continental ice sheets migrate, erode by abrasion, and create characteristic landforms through erosion as well as deposition (see Unit 43). In terms of total area affected, continental glaciers have the larger impact by far. It is estimated that glacial deposits laid down by ice sheets cover nearly 9 percent of the North American continent. Add to this the vast continental shields scoured bare beneath the ice sheets, and the significant role of continental glaciers becomes even more evident.

Ice Sheets Present and Past

Earth's present climate is relatively warm compared to the atmospheric conditions that prevailed 21,000 years ago at the peak of the last glacial period, and glaciers have receded from many areas. But large ice sheets persist to this day in two places—Antarctica and Greenland—and they provide us with many insights regarding the past ice sheets that shaped landscapes of North America and Europe.

ANTARCTIC ICE SHEET

Ice sheets in Antarctica began forming about 40 million years ago and have existed in some form ever since. The *Antarctic Ice Sheet* today covers most of the continent of Antarctica (Fig. 44.1). Many researchers subdivide this massive ice sheet into the East and West Antarctic Ice Sheets, which behave independently of each other. Our focus here is on the general properties of the Antarctic Ice Sheet as a whole, which is of a size comparable to that of the ice sheets that covered Canada and the northern United States during past glacial periods. The Antarctic Ice Sheet covers an area of more than 13 million km^2 (some 5.2 million mi^2), constituting almost 9 percent of the "land" area of the globe. About 61 percent of all the freshwater on Earth is presently locked up in Antarctic ice. If this ice sheet were to melt completely, global sea level would rise by some 60 m (200 ft), and possibly more. This observation

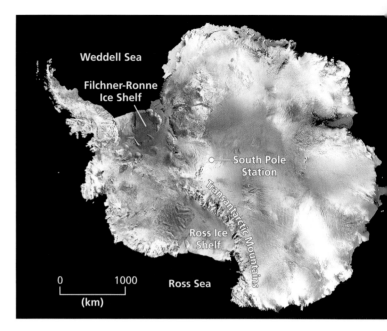

FIGURE 44.1 Infrared satellite image of Antarctica in its standard orientation, with the Antarctic Peninsula extending northwest toward South America. Snow is represented in white, ice in blue, and nunataks (mountain peaks) in black (the girdling Southern Ocean also appears in black). The mostly buried Transantarctic Mountains can be observed extending across the continent from the landward end of the Antarctic Peninsula. The huge bay to the east of the peninsula is the Weddell Sea, and the blue area at the apex of that sea is the Filchner-Ronne Ice Shelf. The Ross Ice Shelf directly across from the Filchner-Ronne, flanking the Transantarctic Mountains, is even larger.

helps us understand why global sea level was far lower during the last glacial period than it is today; at that time, two other large ice sheets were present on Earth, with sizes similar to that of the current Antarctic Ice Sheet.

Beneath this great glacier lies an entire continental landmass, including an Andes-sized mountain range and a vast plateau. There are even lakes of liquid water below the ice. In places on top of this plateau—particularly the region to the right of the South Pole in Figure 44.1—the ice is more than 4,000 m (13,200 ft) thick. In schematic profile, it looks like a giant dome resting on the landmass below (Fig. 44.2). The weight of the Antarctic Ice Sheet is so great that the landmass below it has sunk by hundreds of meters and would rebound upward if the ice melted. The same was true of the great ice sheets present in North America and Europe during past glacial periods.

At present, the Antarctic glacier system experiences very little mass input in the interior zone of accumulation. Average annual snowfall in the interior amounts to less than 10 cm (4 in) of water equivalent, which qualifies this region as a desert according to our climate classification system (see Unit 15). But because that meager amount of snow falls over a large area with a climate too cold for significant melting, it is sufficient to keep the great ice sheet flowing slowly outward.

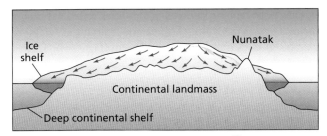

FIGURE 44.2 The Antarctic Ice Sheet forms a gigantic dome that depresses the landmass below. In a few places, nunataks (protruding mountain peaks) rise above the icecap. Vertical scale and the lateral extent of the ice shelves are both markedly exaggerated.

ENHANCE

The basic pattern of flow in the Antarctic Ice Sheet seems simple when seen at the scale of the map in Figure 44.3. Ice moves generally toward the coasts from the high *ice divides* in the interior of the continent. But in fact, it is now known that there are narrow zones of fast-flowing ice, especially toward the coasts. These zones are much like a network of converging streams of water, so they are referred to as **ice streams**. The surface of the ice in some of these streams moves faster than 1 km per year (0.6 mi per year), which is quite rapid for glacier motion.

Ice shelves are floating extensions of the main Antarctic Ice Sheet or smaller glaciers, which remain attached to the larger ice mass as they protrude from land into the frigid

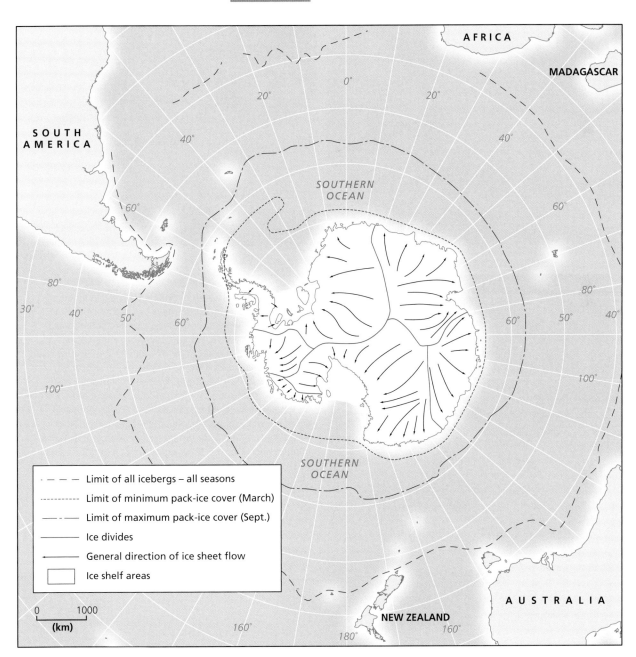

FIGURE 44.3 Flow lines of the Antarctic Ice Sheet and floating pack-ice limits in the surrounding Southern Ocean.

seawater (see far left edge of Fig. 44.2). Some of the ice shelves are large enough to be major features of the geography of Antarctica. At their outer edge, the ice shelves break up into huge tabular icebergs in a process called **calving**, a form of ablation. Calving also occurs from other *tidewater glaciers*, glaciers that flow into the ocean (Fig. 44.4). Several

FROM THE FIELDNOTES

A

B

C

FIGURE 44.4 "To watch (and hear) the calving of a tidewater glacier is one of the most memorable field experiences ever. We had sailed slowly north toward the head of Alaska's Glacier Bay, where the Grand Pacific Glacier marks the end of navigation. But it was a smaller nearby glacier entering from the left (west), the Margerie Glacier, that provided the action. Even before we reached it, we began to hear what sounded alternately like thunder, gunshots, loud groans, and gusts of wind. As its thick ice disgorged into Glacier Bay accompanied by booming and crashing noises that echoed up the valley, huge columns of it collapsed into the water, making large waves and leaving car-sized chunks of ice floating in widening semicircles. And not just ice, but also boulders (A), pebble-sized rocks (B), and surges of pent-up meltwater (C) crashed into the bay, roiling and muddying its waters and contributing to the glacial sediments accumulating below."

ENHANCE

ice shelves around the Antarctic Peninsula have collapsed in recent decades, probably in response to warmer conditions that increased melting from the base of the ice shelves, and also because of reduced ice flow to the coast.

THE GREENLAND ICE SHEET

The *Greenland Ice Sheet* covers most of the surface of Greenland, the largest of all the world's islands (Fig. 44.5), and is about one-eighth as large as Antarctica's continental glacier. Although in terms of volume this ice sheet is far less important than the Antarctic Ice Sheet, it still contains about 11 percent of the world's freshwater supply. Like the Antarctic Ice Sheet, Greenland's glacier exhibits the shape of a dome (see Fig. 44.2), reaching its highest elevation at

FIGURE 44.6   Floating pack ice fills most of the Arctic Ocean. The cracks and pressure ridges reflect the crushing and breaking that take place in this constantly churning surface, which at present is being monitored carefully in the search for evidence of global warming.

more than 3,000 m (10,000 ft) in the east-central part of the island. From there, the surface drops quite rapidly toward the coasts. The ice flows in the same direction, reaching the coast through many *outlet glaciers* that descend from the larger ice sheet through steep-walled valleys. Icebergs calved from these outlet glaciers drift south into the Arctic Ocean. The floating **pack ice** that covers much of the Arctic Ocean (Fig. 44.6) forms through a different process: the freezing of seawater.

ICE SHEETS OF PAST GLACIAL PERIODS

The Antarctic and Greenland continental glaciers are the two Holocene survivors of a larger group of Pleistocene ice sheets. During the glacial periods of the Late Cenozoic, larger and smaller continental glaciers formed in the middle latitudes of both hemispheres, but to a greater extent in the Northern Hemisphere, where there is far more land.

The most massive of all was the *Laurentide Ice Sheet*, the vast continental glacier that occupied Canada east of the Rocky Mountains and expanded repeatedly to cover the North American continent as far south as the Ohio River Valley and southern Illinois (Fig. 44.7). To the west of the Laurentide Ice Sheet, the northern Rocky Mountains were blanketed by a separate, smaller body of ice called the *Cordilleran Ice Sheet*, which extended just south of the present-day United States–Canada border. Much smaller **icecaps** blanketed mountainous areas of the western United States, such as the plateau where Yellowstone National Park is located today. Icecaps differ from ice sheets in that they cover much smaller areas, usually less than 50,000 km² (20,000 mi²) in size.

In Eurasia, the largest continental glacier was the *Scandinavian (Fennoscandian) Ice Sheet*. This glacier, centered on the Scandinavian Peninsula, extended southward into central Europe (see Fig. 44.7). At its southwestern margin, it coalesced with a smaller regional icecap situated atop the British Isles. The Scandinavian Ice Sheet was only slightly

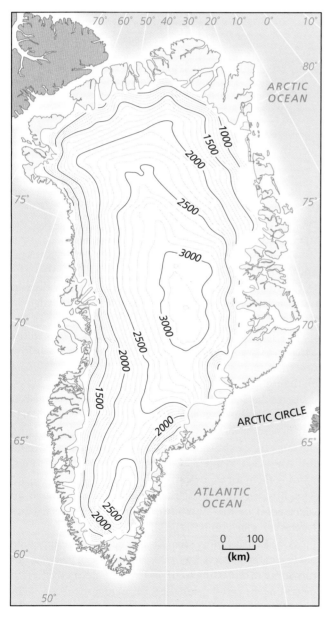

FIGURE 44.5   The Greenland Ice Sheet, which is one-eighth the size of the Antarctic Ice Sheet (see Fig. 44.1). Ice thickness contours are given in meters.

FIGURE 44.7 Late Cenozoic continental ice sheets and glaciers of the Northern Hemisphere, shown at the maximum size they reached during the last glacial period (not all reached this size at the same time). Arrows indicate the general direction of ice flow. Today's coastlines are shown as dashed lines. The Pleistocene coastlines on the map developed when global sea level was more than 100 m (330 ft) lower than at present. The pack ice covering the Arctic Ocean at that time extended much farther south, reaching the North Atlantic.

less extensive than the Laurentide Ice Sheet. Elsewhere in Europe, a separate icecap also developed over the Alps. Smaller icecaps covered many other mountain ranges of Eurasia as well.

In the Southern Hemisphere, only the Antarctic Ice Sheet reached continental proportions because of the small extent of other high-latitude landmasses. Otherwise, the largest glaciers south of the equator were icecaps in southern South America, which covered the southern Andes Mountains and much of Patagonia to the east. Another icecap developed on the South Island of New Zealand, where substantial mountain glaciers still exist.

North America's Glaciations

Before the days of deep-sea drilling, researchers had to rely on landforms and glacial sediments for evidence of the number and timing of Late Cenozoic glaciations. In the late nineteenth century, several persistent and highly capable glacial geologists made great strides toward this goal, mapping many of the basic glacial landforms we still recognize today and determining their relative age. It is hard to exaggerate the achievements of these scientists, who traveled by horse and buggy and worked without modern topographic maps. One of the best-known American geologists of his time, T. C. Chamberlin (1843–1828), named the most recent stage of glaciation the *Wisconsinan*, after the state of Wisconsin, where he thought the evidence for it was best

preserved. (Not coincidentally, perhaps, this was also the state where Chamberlin began his career.)

Older glaciations were named the Illinoian, Kansan, and Nebraskan, again for the states where their deposits were thought to be most widespread or most evident in outcrops. An ancient soil was found buried beneath deposits of the Wisconsinan glaciation, but above those of the earlier Illinoian glaciation. This buried soil was named the Sangamon Soil for the county in Illinois where it was first described in coalmines and wells. It clearly formed during an interglacial period named the Sangamonian, which occurred between the Illinoian and Wisconsinan glacial periods. Older buried soils corresponding to earlier interglacial periods were eventually identified as well. At around the same time in Europe, glacial geologists working around the Alps also mapped the deposits of four glaciations, naming them for four rivers of southern Germany: Würm, Riss, Mindel, and Günz. These were generally thought to correspond to the Wisconsinan, Illinoian, Kansan, and Nebraskan glaciations of the United States.

As with many other scientific concepts that are appealing because of their simplicity, the idea of four major glacial stages turned out to be *too* simple. In the United States, new efforts to determine the age of glacial deposits were made in the 1960s and 1970s using evidence such as volcanic ash layers from the great eruptions at the Yellowstone hot spot (see Unit 32). Where one of these ash layers overlies glacial sediments, they must represent a glaciation that occurred before that

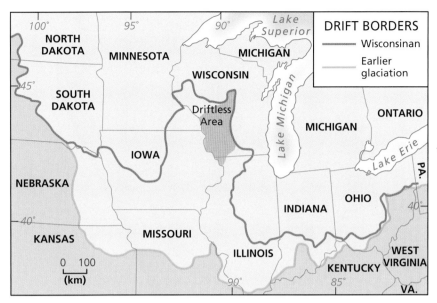

FIGURE 44.8 Areas covered by the Laurentide Ice Sheet during the Wisconsinan and earlier glacial periods in the north-central United States. This map is based on the distribution of glacial drift and the distribution of terminal moraines.

particular eruption. Glaciations after that eruption would have laid down sediment *on top* of the ash layer. Based on this new research, the older glacial sediments that had been called Kansan and Nebraskan turned out to represent five or more glacial periods, not just two, as previously thought. At the same time, studies of ice sheet and deep-sea cores (see Unit 18) revealed that dozens of glacial periods occurred in the Late Cenozoic, although not all of them may have resulted in ice advances into the Midwestern United States.

Today, we still recognize the Wisconsinan and Illinoian glacial periods in North America as well as the Sangamonian interglacial period between them. The ice sheet of the Wisconsinan glaciation reached its greatest size between about 25,000 and 16,000 years ago, although the ice sheet lingered on in Canada until less than 8,000 years ago. The peak of the Illinoian glaciation was around 150,000 to 130,000 years ago. We no longer use the terms *Kansan* and *Nebraskan*, because it is recognized that there were numerous glaciations before the Illinoian, and it is difficult to sort them out from the evidence that remains in the landscape today. Recently, glacial deposits in northern Missouri were dated to about 2.4 million years ago, and these may represent the advance of the very earliest continental ice sheet in North America.

The landscapes shaped by Wisconsinan ice sheets are very different from those affected by Illinoian and older glaciations. In particular, the Wisconsinan glacial landforms are much more distinct because they have not been worn down by stream and slope erosion. Figure 44.8 shows areas affected by Wisconsinan and older ice sheets; this map has not changed much since the original work of T. C. Chamberlin and other pioneering glacial geologists.

Landscapes Shaped by Continental Ice Sheets

In Unit 43, we studied the ways glaciers erode and described two kinds of erosional features: polished and striated surfaces and *roches moutonnées*. These features are quite common in glaciated mountain ranges, but they are harder to find in many of the vast areas covered by continental ice sheets. Nevertheless, erosion by such ice sheets has strongly influenced extensive landscapes. Much of the surface of Canada's Laurentian Shield (see Unit 29) is an ice-scoured plain marked by depressions now filled with water to form lakes. This plain has been eroded down to hard bedrock by each of the ice sheets that formed over northern North America. The Scandinavian Ice Sheet produced similar landscapes in northern Sweden, Finland, and northwestern Russia. Glacial erosion is also partly responsible for carving out the basins of some larger lakes, including the Great Lakes, as we will discuss later in this unit.

Many of the landscape features produced by continental ice sheets are depositional, not erosional, landforms, however. A continental glacier transports huge amounts of rock debris as the ice within it flows toward the ablation zone, scouring and sculpting the surface beneath it (see Unit 43). When the Laurentide Ice Sheet moved from the hard crystalline rocks of the Canadian Shield to the softer rocks of adjacent areas, its erosion became more rapid, and it increased its sedimentary load. If we were able to take a view in profile of such a sediment-charged glacier, we would note that virtually all of the sediment load is carried near the bed of the glacier. Some of the rock material is fine grained, but much of the load is pebble- and boulder-sized, or even larger. Laden with all this debris, the ice inches forward, eroding material it moves across and depositing it mostly near the edge of the ice. In the process, distinctive landforms are produced (Fig. 44.9).

GLACIAL SEDIMENT

Solid material carried in ice can be deposited directly out of the lower layers of the glacier as the ice holding it melts or as the debris is plastered onto underlying sediment. The result is a deposit called **till**, which is *unsorted*, meaning it

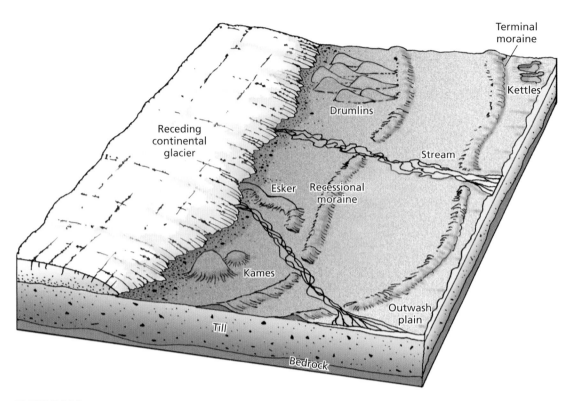

FIGURE 44.9 Landforms produced by a continental ice sheet. The ice advanced to the position marked by the terminal moraine and then retreated and re-advanced slightly to the position of the recessional moraine. The drumlins, esker, and kames all formed under the ice but were uncovered as the ice sheet retreated.

A

B

FIGURE 44.10 Two major types of glacial drift. Glacial till (A) is unsorted, with particles ranging in size from clay to boulders. Outwash (B) is predominantly sand and gravel and is stratified, with distinct layers sorted by particle size.

contains fragments ranging in size from fine clay particles to boulders (Fig. 44.10A). Other rock material carried by the ice is moved up to the glacier surface in the ablation zone, where this debris is exposed by melting. As a result, most glaciers have a noticeable accumulation of debris covering the ice near the glacier margin. As the glacier begins to retreat, it leaves behind this surface debris to form a jumble of mostly unsorted material. Not all glacial geologists

FIGURE 44.11 Hummocky topography with many erratic boulders scattered over the surface. Photo taken by W. C. Alden in 1908. He and other early glacial geologists studied topography like this to identify areas of Wisconsinan glaciation.

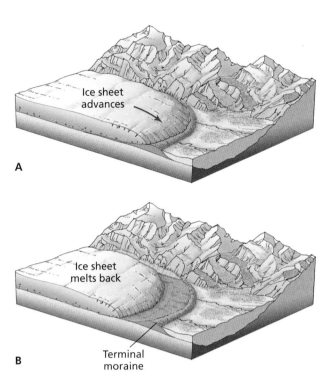

A

B

Terminal moraine

FIGURE 44.12 A terminal moraine is formed from debris deposited at an ice sheet's outer edge (A). After the ice sheet melts back, this debris remains as a ridge that records the former position of the ice margin (B).

consider this to be a type of till, but it has similar properties and is hard to distinguish from till that melts out of the base of the ice sheet. Finally, melting water often flows off the surface of the glacier or out from underneath it, especially in summer. This water carries sediment some distance and sorts it by particle size (as silt, sand, or gravel). Such sorted, water-lain material is generally known as **outwash**, and it is usually stratified, or layered, with distinct beds of varying particle size (Fig. 44.10B). Together, all of these kinds of sediment can be referred to as **glacial drift**.

Much of the Midwestern and northeast United States is covered by glacial drift of varying thickness. Sometimes, more than 30 m (100 ft) of glacial drift cover a very different buried landscape that was carved out of bedrock by streams before the earliest glacial periods. In other places, we find bedrock under less than 1 m (3 ft) of drift.

RECOGNIZING GLACIATED LANDSCAPES

Landscapes covered by drift from the Wisconsinan glaciation are often easy to identify (Fig. 44.11) because of their hummocky topography with many small depressions, lakes and wetlands, and common glacial **erratics** (boulders carried far from their source area by glaciers). In contrast, areas covered by ice only during earlier glaciations look more like typical stream-eroded landscapes. Often, only a few erratics piled at the edge of fields reveal the fact that the ice sheet was once present. Perhaps the best way to appreciate the nature of glacial landscapes is to compare them to areas nearby that were bypassed by the ice sheets, such as southwestern Wisconsin (see Focus on the Science: The Driftless Area).

MORAINES

One of the most important landforms studied by glacial geomorphologists is the **moraine**, a curving irregular ridge

of glacial drift that marks the former outline of a glacier or ice sheet. Continental glaciers carry large loads of rock debris and constantly bring this material to the margin of the ice (Fig. 44.12A). When the ice sheet is stable rather than shrinking or expanding, its margin remains in about the same position for a time, with ice flow continually bringing new debris to that margin. This process builds up the ridge we recognize as a moraine after the glacier has retreated away from it (Fig. 44.12B).

A **terminal moraine** forms at the outermost limit of the ice sheet during a major advance. As the ice sheet begins to retreat from its outermost limit, it forms additional moraines where the retreat halts for some time or where there is a minor re-advance of the ice. These are called **recessional moraines** (see Fig. 44.9). The major terminal moraines marking the outermost limit of the ice sheet during the Wisconsinan glaciation were mapped more than 100 years ago (see Fig. 44.8). For example, the Shelbyville Moraine marks the outer limit of Wisconsinan glaciation in central Illinois. To the southwest of that low ridge, the landscape has not been covered with ice since the Illinoian glaciation. North of the Shelbyville Moraine, there are numerous recessional moraines that developed as the ice sheet retreated later in the Wisconsinan glaciation.

DRUMLINS

A unique landform associated with ice sheet movement is the **drumlin**, a smooth, steep-sided hill that is elliptical in shape (Fig. 44.13). Drumlins almost always occur in

The Driftless Area

Time and again, the ice sheets of the Late Cenozoic moved southward in North America, covering older landscapes with glacial drift from Nebraska to New England. But one small area never was buried (see Fig. 44.8). In southwestern Wisconsin lies a 400-km (250-mi) stretch of scenic landscape of moderate relief supported by exposed bedrock and exhibiting typical landforms of stream erosion.

Surrounded on all sides by the characteristic scenery of glacial drift, Wisconsin's *Driftless Area* provides a unique glimpse of the landscape now concealed from view throughout the Midwest. Economically, this region's rugged topography has deflected the Corn Belt to the south (even though the climate is hospitable to agriculture), and local agriculture is dominated by less crop-intensive dairying.

The most common explanation for the absence of glaciation in the Driftless Area has to do with diversion of ice flow away from this region by topography to the north. The deep trough now occupied by Lake Superior rises steeply on the south to higher ground underlain by hard igneous and metamorphic rocks. Ice flow was diverted to the southwest by this trough so that the ice sheet did not cover the Driftless Area.

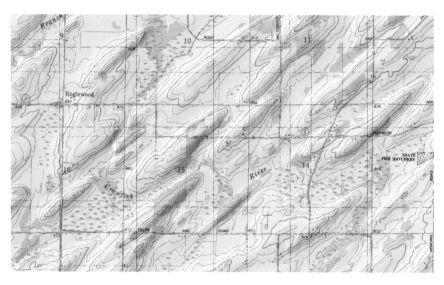

FIGURE 44.13 The unmistakable topography of drumlins is evident on this shaded-relief topographic map of a small part of a drumlin field in south-central Wisconsin. Ice flow was clearly from northeast (upper right) to southwest (lower left). The drumlins here dominate the landscape, influencing the location of wetlands, streams, and even roads.

groups, and there are *drumlin fields* containing thousands of these features. The long axis of a drumlin lies parallel to the direction of ice movement. Drumlins develop beneath glacier ice, and it is generally believed they are formed when the ice sheet erodes away material on either side or perhaps squeezes up preexisting sediment into the typical drumlin shape. The process of formation is still somewhat mysterious, in part because no one has ever seen a drumlin forming, hidden beneath a glacier. In contrast, the formation of moraines can be observed at the margin of many present-day glaciers.

LANDFORMS PRODUCED BY GLACIAL MELTWATER

As discussed earlier, meltwater flowing beneath, on top of, or out in front of glaciers deposits sorted sediment called outwash. A variety of distinctive landforms are built in the process. One of the most interesting of these is the *esker*. Water flowing at the base of the glacier can form a long tunnel under the ice. Rock debris collects in this tunnel and is sorted considerably by the water even as the tube is being clogged. Eventually, when all of the ice has melted away, the tunnel's outline is marked by a long, steep-sided, sinuous ridge, the **esker** (see Fig. 44.9, lower center).

Eskers are generally thought to form when a large mass of glacier ice has become *stagnant*—that is, no longer actively flowing. This may occur when the active part of an ice sheet retreats quickly, leaving large areas of stagnant ice behind. Another distinctive and common landform of stagnant ice is a **kame**, a rounded hill of gravel and sand deposited by meltwater (Fig. 44.14). Kames can probably form in a variety of ways. One especially interesting process involves water flowing down a funnel-shaped vertical opening in the glacier, carrying sand and gravel down to the bed, where it forms a large pile that becomes the kame.

FIGURE 44.14 A kame in Wisconsin. Photo taken by the glacial geologist W. C. Alden in 1908. Note the small gravel pit on the left side of the kame.

Many landforms of glacial meltwater form in front of, not beneath, the glacier. An **outwash plain** is the product of meltwater carrying (and to some extent sorting) rock debris from the wasting ice front. In some places, large blocks of unmelted ice are initially buried in the outwash plain. When these blocks melt, they create steep-sided, water-filled depressions known as **kettles** (see Fig. 44.9, upper right). Occasionally, there are so many of these kettles that the affected portion of the outwash plain is described as *pitted*. (A pitted outwash plain is visible in front of a valley glacier in the lower left of the photo that opens Unit 43.)

GLACIAL LAKES

Without glaciers, there would be far fewer lakes in the world. The "10,000 lakes" that the state of Minnesota claims on its license plates are almost all in basins produced by glacial erosion or deposition. The kettles in outwash plains we just mentioned are only one of the many kinds of basins that can hold lakes in glaciated landscapes. As another example, moraines that cross stream valleys can act as natural dams, creating lakes behind them.

However, another kind of glacial lake exists only while the glacier is still present or at least nearby. Lakes of this type develop where the glacier itself blocks the previous stream network, forming a closed basin where water collects. One of the largest lakes in recent geologic history formed this way in what is now the Red River Valley of North Dakota, Minnesota, and Manitoba. The natural drainage of this area is to the north, toward Hudson Bay, but that path was blocked by the ice sheet during the Wisconsinan glaciations. As a result, water was ponded and soon spread over a large area, far greater than the area of modern Lake Superior.

This water body has been named Lake Agassiz, after geologist Louis Agassiz (1807–1873), who was the best-known early advocate of the idea of continental glaciation. Water occasionally spilled out of Lake Agassiz into the Mississippi River system to the south, the Mackenzie River to the north, and the Great Lakes to the east. Finally, when the center of the remaining ice sheet collapsed a little more than 8,000 years ago, the lake drained out through the

newly uncovered Hudson Bay. The former bed of Lake Agassiz now forms the fertile farmland of the Red River Valley. Many smaller lakes also formed next to the ice sheet in various places; some later drained like Lake Agassiz, and others are still present in some form.

The Great Lakes and Their Evolution

The Great Lakes, too, owe their origin to the Late Cenozoic ice sheets. Even before glaciers first advanced over the area occupied by the present five Great Lakes (the largest cluster of freshwater lakes in the world), there were probably lowlands where each lake is today because of the underlying rock structures. An ancient rift zone underlies Lake Superior, and the rocks below Lake Michigan and Lake Huron include shales that are much more easily eroded than the dolomites forming higher ground on either side of these lakes. The ice sheets excavated these lowlands into closed basins that hold today's lakes. A good indication that the basins were formed by glacial erosion is the depth of Lake Superior. Its deepest point is 223 m (733 ft) below sea level. A river system could not erode this deep basin and carry the eroded rock out to the ocean (see the discussion of base level in Unit 39), but glaciers have carved out deep basins like this in several parts of the world.

After the ice receded, the Great Lakes formed, but their size and water levels have varied greatly over the last 14,000 years. At times in the past, each lake has had a level considerably higher than it does today. Around the edges of the modern water bodies are *lake plains*, areas of flat, dry land that were underwater when the nearby lake was at a higher level (Fig. 44.15). The Great Lakes also dropped to water levels that were lower than present levels at other times. The level of water in each lake basin depends on the elevation of the lowest outlet, where the water drains out toward the ocean. What makes the history of lake levels so complicated is that most of the lakes have more than one possible outlet (see Fig. 44.15), and some outlets were either blocked or at a different elevation in the past.

In fact, lakes formed in the modern Great Lakes basins while the ice sheet still filled part of each basin, blocking all outlets to the north and east. As a result, the lakes rose to much higher water levels than exist today and eventually spilled out to the south over high outlets. The earliest version of Lake Superior drained into the Mississippi River through its St. Croix outlet (see number 6, Fig. 44.15), and early versions of Lake Michigan drained south through its Chicago outlet (see number 5, Fig. 44.15). An early water body in the Lake Ontario Basin spilled into the Hudson River system (see number 14, Fig. 44.15).

Even when the ice had entirely retreated out of the Great Lakes region, lake levels kept changing. One major factor was isostatic rebound of the crust after the weight of the ice sheet was removed. Outlets to the north that had been under thicker ice rebounded more than those to the south. Thus, at one point, Lake Huron drained through the

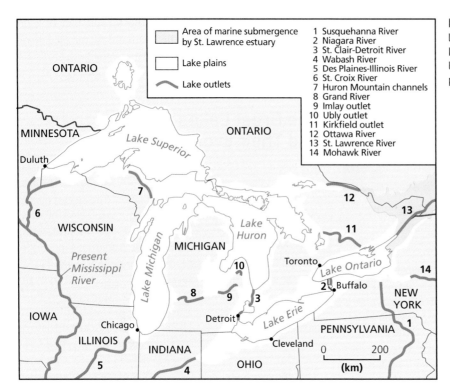

FIGURE 44.15 Lake plains (former lake bottoms) around the modern Great Lakes and major outlets used by these lakes during their evolution into the present-day Great Lakes.

Map legend:
- Area of marine submergence by St. Lawrence estuary
- Lake plains
- Lake outlets

1 Susquehanna River
2 Niagara River
3 St. Clair-Detroit River
4 Wabash River
5 Des Plaines-Illinois River
6 St. Croix River
7 Huron Mountain channels
8 Grand River
9 Imlay outlet
10 Ubly outlet
11 Kirkfield outlet
12 Ottawa River
13 St. Lawrence River
14 Mohawk River

Ottawa River (see number 12, Fig. 44.15), but that outlet rose through rebound and was abandoned. Levels in all of the upper lakes adjusted as a result. Finally, climate change over the last 10,000 years has also had an impact on water levels in the Great Lakes, although the details of that process are still debated.

The future of the Great Lakes is uncertain. Future climate change will affect the input of water from precipitation and its loss through evaporation. At the same time, rebound of the crust continues at a slow rate and is still affecting the water level in Lake Superior. The overall effect of such change on lake levels is quite difficult to predict.

Pluvial Lakes

As the map of the maximum extent of the Laurentide Ice Sheet and Rocky Mountain icecaps indicates, the continental glacier never reached as far as Utah, Nevada, Oregon, or California (see Fig. 44.7). Still, the climate of glacial periods had far-reaching effects in those areas. Today, the basins in this area of the American West are arid. During the glaciations, however, precipitation in this region was substantially higher than at present. Evaporation was also reduced in the colder glacial climate, and more than 100 lakes, known as **pluvial lakes**, developed as a result (Fig. 44.16). Although these pluvial lakes were not directly caused by glaciers, they are discussed here because they developed at about the same time the Laurentide Ice Sheet covered northern North America.

The largest pluvial lake in North America, Lake Bonneville, was the forerunner of Utah's Great Salt Lake. Lake Bonneville at one stage was about as large as Lake Michigan is now.

FIGURE 44.16 Pluvial lakes of the western United States during the last glaciation. Blue dashed lines represent overflow channels.

It reached a maximum depth of more than 300 m (1,000 ft) and overflowed northward in enormous floods through the Snake River of southern Idaho and hence into the Columbia River. Although Lake Bonneville has now shrunk into the Great Salt Lake, its former shorelines still can be seen on the slopes of the mountains that encircled it. Today, only a few of the pluvial lakes still contain water, and this water is saline; the Great Salt Lake is the largest of them. As for the other pluvial lakes, they have mostly evaporated away and left only geomorphic evidence of their former existence.

As you can imagine, many other features directly or indirectly connected with continental glaciations have been identified and their origins analyzed. By understanding the processes that formed these features, we are able to interpret the sequence of events that prevailed during the several glaciations of the Late Cenozoic Ice Age. The landforms of continental glaciers may not be as scenic as those associated with mountain glaciers (which we will examine in Unit 45), but they are valuable indicators of past processes and environments.

Key Terms

calving *page 534*

drumlin *page 539*

erratics *page 539*

esker *page 540*

glacial drift *page 539*

icecap *page 535*

ice shelf *page 533*

ice stream *page 533*

kame *page 540*

kettle *page 541*

moraine *page 539*

outwash *page 539*

outwash plain *page 541*

pack ice *page 535*

pluvial lake *page 542*

recessional moraine *page 539*

terminal moraine *page 539*

till *page 537*

Scan Here for a quick vocabulary review

Review Questions

1. Explain how studying the Antarctic Ice Sheet may help explain landscapes of the northern United States and Canada.

2. Name and cite the location of the two largest Northern Hemisphere Late Cenozoic ice sheets that are no longer present today.

3. Define *till* and *glacial drift* and explain the difference between them.

4. Define each of the following landforms and describe what information they provide about past locations or characteristics of an ice sheet: drumlin, esker, moraine, kame.

5. How many times may glaciation have affected the Midwestern United States during the Late Cenozoic Ice Age, and how has the answer to this question changed over time?

6. How might continental glaciation produce lakes on the landscape?

7. How do pluvial lakes relate to glaciation?

Scan Here for a quick vocabulary review

www.oup.com/us/mason

Landforms and Landscapes of Mountain Glaciers

Flying low over the Fox Glacier, New Zealand, reveals recent snow on older crevasses.

OBJECTIVES

- Summarize the current distribution of mountain glaciers
- Describe the characteristic landforms produced by mountain glacier erosion
- Discuss the landforms produced by glacial deposition in mountain landscapes

The global climate during the Cenozoic era was generally mild until the onset of the Late Cenozoic Ice Age. Before 40 million years ago, even Antarctica was mostly ice-free. High mountain ranges that host many of the world's glaciers today—including the Alps and the Himalayas—had not yet formed or were in the early stages of development. We do not have a clear picture of what the mountain ranges that were in existence then looked like, but they were probably dominated by landforms associated with river erosion and deposition.

When Earth's climate began to cool and the altitude of the snow line dropped, the formation of glaciers began. The Antarctic Ice Sheet was the first continental glacier to develop because of Antarctica's polar location and its high overall elevation. Gradually, on the other continents, permanent ice formed on higher mountain slopes, and *mountain glaciers* flowed down the high valleys. Along with the global trend toward a cooler climate, uplift of the Himalayas, the Alps, the Andes, the Sierra Nevada, and other mountain ranges raised them to elevations at which more and more of the mountainous mass could be glaciated. Glaciers occupied valleys first carved by rivers, and glacial erosion replaced stream erosion. Permanent ice appeared on high mountains even in equatorial locales, and Earth was indeed transformed.

In this unit, we focus on glaciers that form on mountains and erode and deposit in alpine settings. Mountain glaciers differ from the ice sheets of continental glaciation in that they are generally confined to cirques or valleys, and their behavior is influenced by the topography they inhabit (see Perspectives on the Human Environment: Mountain Glaciers: The View from Space). In other respects, they are quite similar to continental glaciers.

In Unit 44, we noted that two large ice sheets survive to the present (the Antarctic and Greenland Ice Sheets), and that several other major continental glaciers and icecaps wasted away with the onset of the Holocene interglacial period. Mountain glaciers, too, were larger and much more prevalent before the current interglacial period began. Many mountain glaciers melted away and vacated their valleys as their zones of accumulation diminished. But despite the warmth of the present global climate, numerous mountain glaciers endure. Some that have not disappeared have receded far up their formerly occupied valleys, leaving abundant evidence of their earlier advances.

Global Distribution of Mountain Glaciers Today

Every major landmass on Earth except Australia contains alpine glaciers. It has been estimated that there are as many as 150,000 individual mountain glaciers in the world today, ranging in size from huge bodies to small patches of ice. One of the largest mountain glaciers lies in Antarctica's Queen Maud Mountains, where it feeds the continental ice sheet. This is the Beardmore Glacier, and it remains a mountain valley glacier for more than 200 km (125 mi) before it merges into the Antarctic Ice Sheet. The Beardmore Glacier is as much as 40 km (25 mi) wide, larger than anything seen in the Alps or other midlatitude mountain ranges.

The remote Beardmore Glacier has been observed by comparatively few people, but other glaciers are in more accessible locales. Cruise ships visit the Alaskan glaciers where they enter the Gulf of Alaska, and tourists can see active mountain glaciers from cable-car gondolas in the French and Swiss Alps. Landscapes produced by alpine glaciation are among our most dramatic and underscore the significance of ice in shaping Earth's surface.

North America In North America, major clusters of mountain glaciers lie in the Arctic islands of Canada (Fig. 45.1), in southeastern Alaska, in Canada's Yukon and the Coast Mountains of British Columbia, and in the Canadian Rocky Mountains along the Alberta–British Columbia border (Fig. 45.2). Equatorward of 50 degrees N latitude, glacier formation becomes more exceptional. The South Cascade Glacier near Mt. Rainier in western Washington State is an example of a temperate glacier in a moist maritime environment, where abundant orographic snowfall sustains the glacier despite relatively high summer temperatures. In Colorado and Wyoming, where there is much less snowfall in the Rockies, winter snow accumulation is still enough to produce ice formation and maintain small remnant glaciers in a few locations, mostly on shaded northeastward-facing slopes.

South America In South America, there are large mountain glaciers in the southern Andes of Chile. These begin just south of latitude 45 degrees S (about the latitude of Minneapolis in the Northern Hemisphere) and become progressively larger at higher latitudes. At the Strait of Magellan, where the South American mainland ends, glacial ice reaches the sea.

Africa Almost all of Africa's glaciers are on two soaring mountains near the equator, Mt. Kilimanjaro (5,861 m [19,340 ft]) and Mt. Kenya (5,199 m [17,058 ft]), and today, most are rapidly receding or disappearing altogether. Glaciers descend from near the summits of both mountains. There are eleven small glaciers on Mt. Kenya, of which two remain substantial.

New Zealand The Southern Alps, as the mountain backbone of the South Island of New Zealand is called, reach their highest point on Mt. Cook (3,764 m [12,349 ft]). In the vicinity of Mt. Cook lie another 15 mountains more than 3,000 m (10,000 ft) high. The entire mountain range was covered by an icecap in glacial times, and several major glaciers survive in the area of Mt. Cook. Among the best known are the Franz Josef Glacier and the Fox Glacier. The Fox Glacier is receding quite rapidly, and its withdrawal is

Mountain Glaciers: The View from Space

The relationship between alpine glaciers and their surrounding landscapes is strikingly visible in satellite imagery. The Landsat satellite image in Figure 45.1 is a particularly fine example. The image is centered on Bylot Island, located off the northern coast of Canada's Baffin Island. In this image, red represents green tundra vegetation, and white or gray areas are snow and glacier ice, with a few patches of floating sea ice in the dark ocean water around the island.

In Figure 45.1, more than two-dozen large glaciers flowing through mountain valleys can readily be seen. There are several excellent examples of two or more individual valley glaciers joining as they flow downstream to form a single main (trunk) glacier. All of the valley glaciers originate in the mountains of the island's interior (which exceed 600 m [2,000 ft] in elevation), and they flow outward and downslope toward the sea. Because of Bylot's high latitude (73°N), the snow line is at a relatively low elevation, and many glaciers come close to reaching the coast (two in the northwest actually do). Overall, the island's area is 10,880 km² (4,200 mi²), its longest east–west dimension measures approximately 145 km (90 mi), and its widest north–south extent is about 115 km (70 mi). Surprisingly, given Bylot's remote location and inhospitable setting, humans have long made inroads here. Several thousand Inuit people occupy the wide, low-lying, ice-free southern peninsula each spring and summer to take advantage of local marine resources.

FIGURE 45.1 Vertical satellite view of the alpine glacier complex that blankets most of Bylot Island in the Canadian Arctic. Several valley glaciers are marked, but many others are visible.

marked by signposts in its lower valley along the road leading to the present glacial margin (Fig. 45.3).

Europe   Many of the world's mountain glaciers lie in two major clusters on the Eurasian landmass. Of these two, the European Alps are undoubtedly the most famous, and the south-central Asian zone is by far the largest. Much of what is known about the erosional and depositional work of glaciers was learned through research performed in the European Alps. Europe's Alps extend in a broad arc from southeastern France through the area of the Swiss–Italian border into central Austria (Fig. 45.4). Mont Blanc, 4,807 m

FIGURE 45.2 "My flight from Seattle to Tokyo afforded spectacular views of the glacial landscape of Alaska. Cirques are filled with snow and ice, U-shaped valleys are steep-sided, and moraine-streaked glaciers coalesce and flow downvalley. The narrow medial moraine near the bottom of the photo widens quickly as rock debris is added. Tributary glaciers thicken the lateral moraines, and by the time the glacier makes its turn in the distance, it is choked with rocky debris. When I took this photo, we were near Gulkana, Alaska."

(15,771 ft) high, is the tallest peak, but several other mountains exceed 4,000 m (13,200 ft). Active glaciers abound in the Alps, and virtually every erosional and depositional landform associated with glaciation is found there. Glaciers also exist in Norway, where high latitude rather than altitude produces the cold climate that sustains them. Mt. Elbrus, the highest peak in the Caucasus Mountains (between the Black and Caspian seas along Russia's southern flank), reaches 5,642 m (18,510 ft) and carries several small glaciers.

Asia The glacial topography of the European Alps is dwarfed, however, by the vast expanse of glacial landscape that extends across the soaring highlands of south-central Asia from Afghanistan to southwestern China (Fig. 45.5). Many of the world's highest mountains—including the tallest of all, Nepal's Mt. Everest (8,850 m [29,035 ft])—lie in this zone. Much of the region still remains buried under ice and snow, and many of the mountain glaciers here are hundreds of meters thick and many kilometers wide. Maximum development occurs in the Himalayan–Tibetan area (Fig. 45.5). Vast as the ice and snow cover is, however, there is abundant evidence that here, too, the glaciers have receded during the Holocene: glacial topography and glacial deposits extend well beyond the margins of the present ice.

FIGURE 45.3 "The Fox Glacier flows westward off New Zealand's Southern Alps. There was a time when this glacier surged into the forested lowlands near the coast, cutting down trees like matchsticks; but today you can walk up a vacated valley to its receding face. The glacier's recession has been monitored for decades, and signposts in the valley mark where it stood years ago. Global warming, reduced snowfall in its catchment area, and possibly other factors as well combine to stagnate this once-aggressive glacier. Today morainal deposits in a wide, U-shaped valley evince the Fox Glacier's former power, but the glacier itself is covered by rock rubble, and meltwater streams emanate from its base, feeding small lakes downvalley."

FIGURE 45.4 The Alps form a gigantic crescent of spectacular mountain ranges in central Europe. The large lowland they frame just to their south is northern Italy's Po Plain. The Italian Peninsula, with its Apennine Mountains backbone, extends seaward from the southern margin of the Po Valley.

FIGURE 45.5 A space-shuttle view of the Himalaya Mountains, looking westward from a point above the easternmost part of India's Ganges Plain. Together with the Tibetan Plateau to the north (right), the Himalayas constitute Earth's most prominent highland zone.

Erosional Landforms of Mountain Glaciers

There is no mistaking a landscape sculpted by mountain glaciers, even long after the glaciers have melted away. Mountains, ridges, and valleys all bear the stamp of the glaciers' work. As we noted in Unit 43, the zones of mass accumulation of these glaciers are on high mountain slopes. There, snow is compacted into ice, and the ice moves downhill by creep and sliding under the force of gravity to occupy valleys formed earlier by stream erosion.

GLACIAL TROUGHS

A mountain glacier that occupies a river valley immediately begins to change the cross-section and profile of that valley. River valleys in mountain areas tend to have V-shaped profiles signifying the active downward erosion of relatively narrow stream channels. Glaciers fill former stream valleys to a much higher level than any river flood and erode the lower parts of valley-side slopes, widening the valley bottoms until the valleys becomes U-shaped (Fig. 45.6A). Glacial erosion can also greatly deepen valleys if ice flows through them during several glacial periods. The U-shaped glacial valley, or **glacial trough**, is one of the most characteristic of the glacial landforms. Many mountain valleys famous for their scenery are glacial troughs. One of the best examples in the world is Yosemite Valley in California (see Fig. 43.5).

Maps and aerial photographs of glacial troughs reveal another property of these valleys: they do not turn or bend nearly as tightly as rivers do. Glaciers shear off many of the mountainside spurs around which rivers once flowed, thereby straightening the valley course. Such **truncated spurs** are further evidence of glacial action in the landscape after the glaciers have melted away (Fig. 45.6B). This makes sense when we compare a valley glacier to a narrow stream channel that can occasionally erode one side of the valley or another but is

A

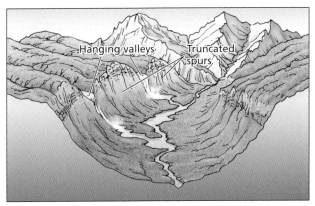

B

C

FIGURE 45.6 Evolution of a glacial trough. Diagram (A) shows the stage of peak glaciation, with the U-shaped trough gouged out by the trunk glacier that is advancing toward the viewer. When the valley glacier has melted away, truncated spurs and hanging valleys are apparent (B). If the glacial trough is near the coast and has been deepened below (rising) sea level, it will become inundated, and a fjord will form (C).

often in the middle, away from the side slopes. A glacier fills the entire valley, however, and actively erodes spurs that once protruded into the valley.

When a river is joined by a tributary, the water surface of both streams is at the same level, and the floor of the tributary valley tends to slope gently down to its junction with the larger valley, without a sharp break where they meet. But when a smaller glacier joins a larger one, the tributary glacier is much thinner than the larger trunk glacier, and its valley is not as deep. (Recall from Unit 43 that glacial erosion becomes much more effective as the ice thickness increases.) The ice surfaces of the tributary and trunk glaciers are at about the same level (see Fig. 45.6A), but their bedrock floors are separated by an abrupt downward step from the tributary to the trunk valley, sometimes hundreds of meters high. When both glaciers melt away, the valley of the tributary glacier, as viewed from the floor of the main glacier, seems to "hang" high above. Such a discordant junction is appropriately called a **hanging valley**, or hanging trough (see Fig. 45.6B), another definite sign of the landscape's glacial history. A hanging valley is often graced by a scenic waterfall where the stream now occupying the tributary glacier's valley joins the main trough.

HIGH-MOUNTAIN LANDFORMS

In the mountains above the glacial valleys, which are the source areas of the glaciers, the landscape also is transformed. The three block diagrams in Figure 45.7 illustrate the sequence of events. Initially, the landscape consists of rounded ridges and peaks (Fig. 45.7A). With the onset of glaciation, deep snow accumulates on the higher slopes. The snow persists through the summer and eventually forms large masses of flowing ice. The ice moves downslope under gravity, and glacial erosion begins (Fig. 45.7B).

In the upper area of continuous snow accumulation, the ice hollows out shallow basins, which become the glacier's source area. Not only does the ice excavate such basins, but frost wedging also breaks loose rock from the steep cliffs rising above the glacier, especially at its headward end (the uppermost part of the glacier). Continuing headward erosion combined with downward erosion beneath the glacier carve distinctive, amphitheater-like landforms known as **cirques** into the side of high peaks (Fig. 45.7C). A well-developed cirque is a bowl-shaped, steep-sided depression in the bedrock, with a gently sloping floor. Two, three, or even more cirques may develop near the top of a mountain. In time, these cirques, growing by headward erosion, intersect. Now nothing remains of the original rounded mountaintop in the center of the diagram in Figure 45.7C except a steep-sided, sharp-edged peak known as a **horn**. The Matterhorn in the Swiss Alps is the quintessential example. When the ice melts or the icecap thins out, these horns tower impressively above the landscape (Fig. 45.8).

Other dramatic elements of mountain glacier topography are shown in Figure 45.7C. One is the large number of

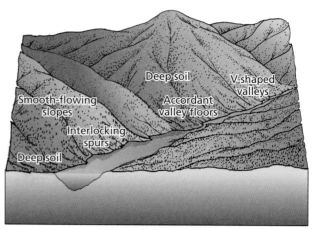

A BEFORE GLACIATION

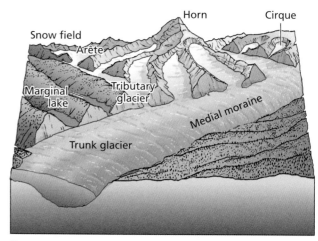

B DURING GLACIATION

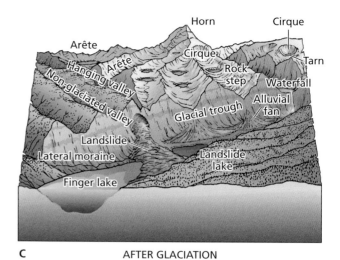

C AFTER GLACIATION

FIGURE 45.7 Transformation of a mountain landscape by alpine glaciation. Note how the initial rounded ridges and peaks are sculpted into a much sharper-edged topography by frost wedging and glacial erosion.

ENHANCE

FIGURE 45.8 "Boarded a two-seater airplane to fly from the airstrip at Whataroa (on New Zealand's South Island) up the Franz Josef Glacier toward the Mt. Cook area in the Southern Alps. This scenery seems even more spectacular than the 'real' Alps. How I wish that I could have my whole physical geography class with me here—what a matchless laboratory! This view of a basin-shaped, snow-accumulating, ice-generating cirque also reveals a developing horn (right background) and a sharp-edged arête (left foreground)."

ENHANCE

razor-sharp, often-jagged ridges that rise above the ice-containing troughs. These ridges frequently separate adjacent cirques or glacial troughs and are known as **arêtes** (see Fig. 45.8). They develop as glacial erosion enlarges the cirques or troughs on either side of a ridge, narrowing it to a knife's edge.

Another major feature appears as the glaciers move downslope and create step-like profiles, which are evident in the valleys in front of the horn in the center of Figure 45.7C. Such landforms result from a combination of factors. One relates to the differential resistance of various rocks over which the glacier passes. The jointing of the bedrock (and therefore its susceptibility to frost wedging) also affects this process. In the postglacial landscape, **rock steps** reveal the local effects of such glacial erosion.

LAKES

Depressions gouged by alpine glaciers are filled by water during interglacial periods, such as the present period. Where the climate is warm enough to melt the glacier ice in even high-altitude cirques, small circular lakes are found on the floors of the cirque basins. These lakes, dammed up behind the "lip" of the cirque, are known as **tarns** (one is labeled in the upper far right of Fig. 45.7C). Lakes also form on the rock steps (shown in the three valleys below the horn in the central portion of Fig. 45.7C). The largest lakes fill substantial parts of glacial troughs. Such lakes may be several kilometers wide and 50 km (30 mi) or more long. Some of the world's most scenic lakes, from the Alps of Switzerland to the Southern Alps of New Zealand, owe their origins to glacial erosion.

FJORDS

Among the most spectacular landforms associated with glacial erosion are fjords. A **fjord** is a narrow, steep-sided, elongated inlet of the ocean formed where a glacial trough has been inundated by seawater (Fig. 45.6C). During glacial periods, many glaciers reach the ocean. Recall that global sea level during past glacial periods was much lower than it is today. Furthermore, glaciers that reached the ocean at that lower sea level could continue to erode rock even many meters below it. Thus, vigorously eroding glaciers created troughs reaching far below the present higher sea level. When the ice melted, ocean water inundated the glacial valleys, creating a unique coastal landscape. Fjords developed mainly in places where glaciated mountains lie near a coastline, such as in southern Alaska and western Norway. Other famous and scenic fjords are found along the southwestern coast of Chile and the southwestern coast of New Zealand's South Island.

Depositional Landforms of Mountain Glaciers

Debris carried by an alpine glacier comes not only from the valley floor beneath the glacier but also from the valley sides above the glacial ice. *Frost wedging*, the repeated freezing and thawing of water in rock cracks and joints, loosens pieces of bedrock (see Unit 36). A vigorously eroding glacier also tends to undercut its valley sides so that mass movement contributes additional material to the glacial surface. Some of the debris carried downslope is ground into particles so fine that it is called **rock flour**, and when the glacier melts and deposits this rock flour, much of it can be blown away by the wind.

MORAINES

Larger rock fragments are often deposited at the margin of mountain glaciers in ridges of debris called *moraines*. Recall from our discussion of continental glaciers in Unit 44 that a *terminal moraine* is a ridge that builds up at the edge of the glacial ice where it has reached its maximum limit of advance (Fig. 45.9). The terminal moraine of a mountain valley glacier forms a ridge across the valley at the former location of the glacier's lower end. During a valley glacier's retreat, stationary periods or brief re-advances are marked by similar ridges called *recessional moraines*. Both terminal and recessional moraines tend to be curved, with the convex side of the moraine facing down the valley, corresponding to the shape of the glacier's curved snout. After the glacier retreats, these moraines can form dams that impound meltwater, creating temporary glacial lakes. Some moraine-dammed lakes persist in the landscape today.

Mountain glaciers also form **lateral moraines**, ridges of debris situated where the ice meets the valley-side slope along both sides of a glacier (see Fig. 45.9). When a trunk glacier is joined by a substantial tributary glacier, their lateral moraines join to become a **medial moraine**. Medial moraines are situated away from the valley-side slopes and mark the boundary between ice from each of the two glaciers (see Fig. 45.9; good examples are also visible in the opening photo of Unit 43). This pattern may be repeated

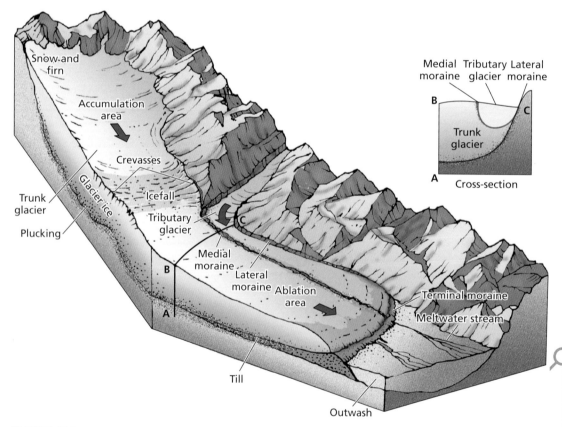

ENHANCE

FIGURE 45.9 Cutaway view of a valley glacier and one of its tributaries showing depositional features. Note the positions of the lateral, medial, and terminal moraines as well as outwash down the valley from the glacier.

several times as additional tributary glaciers enter the main glacier's channel.

POSTGLACIAL LANDSCAPE CHANGE

River action begins to modify the depositional landforms in glacial troughs as soon as the glaciers vacate them. Meltwater covers the newly exposed valley floor with outwash (see the front edge of Fig. 45.9). Modification of the glacial deposits (though not the bedrock topography) is usually quite rapid. Glacial lakes drain, material is sorted and redistributed, and vegetation recovers.

Glacial topography is scenic, and it attracts countless tourists to ski lodges and other highland resorts. But glaciated areas are not stable. Glaciers retreat from valleys, leaving behind oversteepened walls. Frost wedging loosens huge quantities of rock, much of it perched precariously on steep slopes. Snow accumulations can lead to avalanches that would be harmless in remote terrain but are often fatal when humans populate these landscapes. Mass movements of various kinds form a significant natural hazard in alpine-glaciated regions, as the large landslide and the lake it dammed suggest in the lower central portion of Figure 45.7C.

Key Terms

arête *page 551*
cirque *page 550*
fjord *page 551*
glacial trough *page 549*

hanging valley *page 550*
horn *page 550*
lateral moraine *page 552*
medial moraine *page 552*

rock flour *page 552*
rock step *page 551*
tarn *page 551*
truncated spur *page 549*

Scan Here for a quick vocabulary review

Review Questions

1. What two world regions contain the greatest concentrations of mountain glaciers?
2. How and why does a glacial trough differ in shape from a stream valley?
3. What is a hanging valley, and how does it form?
4. How are the processes that form cirques, arêtes, and horns related to each other?

5. Describe the types of moraines associated with mountain glaciers and explain where they are found and how they form.
6. How does mountain glaciation produce abundant lakes?
7. What is a fjord, and how is it formed?

Scan Here for a quick vocabulary review

www.oup.com/us/mason

UNIT
46

Periglacial Environments and Landscapes

In this summer view from the air, polygonal patterns on the Arctic Coastal Plain of Alaska reveal the presence of permafrost, or permanently frozen ground, below a thin thawed layer. Water ponds on the saturated soils because it cannot drain through the permafrost.

OBJECTIVES

- Discuss the nature and distribution of permafrost
- Explain the processes that shape periglacial landscapes
- Describe the major landforms of periglacial regions
- Briefly review the broader importance of periglacial environments

Earth is undergoing an interglacial period at present, yet large regions of the world are anything but warm, even during the summer. There are still large expanses of cold landscapes where **permafrost**—permanently frozen ground—is found at a shallow depth. The technical term for such environments is **periglacial**, which literally means near glacial, or on the perimeter of glaciation. This term is used today not only for cold landscapes near existing glaciers but also for regions with similarly cold climates where no glaciers are present. In this unit, we study the processes and landforms that characterize periglacial environments.

Periglacial zones today occupy high-polar and subpolar latitudes almost exclusively in the Northern Hemisphere. No true periglacial environments exist today in southern Africa or in Australia, although the highlands of Tasmania (off the southeastern coast of Australia) show evidence of periglacial conditions in the recent past. Small areas of South America (notably the island of Tierra del Fuego) exhibit periglacial conditions, as does the Antarctic Peninsula and other parts of Antarctica not covered by ice. Accordingly, this unit deals almost exclusively with the Northern Hemisphere. Nonetheless, it is estimated that as much as one-fifth of the entire land surface of Earth is dominated by periglacial conditions, and this alone should persuade us to learn more about these cold environments.

We can also assume that periglacial conditions migrate into the middle latitudes when ice sheets expand. When the Wisconsinan ice sheets covered much of northern North America, periglacial conditions extended far to the south, and the landscape still bears those imprints. This reminds us of the dramatic changes in climate and other conditions that have been experienced in many parts of the world, even over the last few tens of thousands of years.

Permafrost

It is difficult to define the environmental limits of periglacial regions precisely. Perhaps the most practical way to delimit periglacial conditions is based on a phenomenon unique to these regions: the presence of permafrost. In periglacial zones, the ground (soil as well as rock) below the surface layer is permanently frozen. What this means, of course, is that most of the water in this subsurface layer is frozen throughout the year. The permafrost layer (Fig. 46.1) normally begins between 15 cm (6 in) and 5 m (16.5 ft) below the surface.

The upper surface of the permafrost is called the *permafrost table*. The soil above the permafrost table is subject to annual freezing and thawing. This is the **active layer**; its thickness varies greatly, sometimes over short distances (Fig. 46.1). Many factors influence the thickness of this layer—not just local climate, but also soil texture, vegetation cover, depth of snow in winter, and whether the soil is in a low area and remains wet after thawing. Below the permafrost table, the frozen ground can be very deep. In high-latitude Siberia, permafrost depths of more than 1,200 m

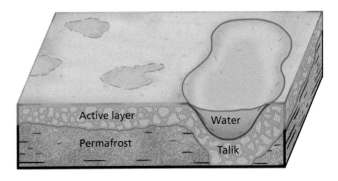

FIGURE 46.1 Subsurface layers characteristic of periglacial zones, showing the active layer that thaws in summer and the permafrost beneath it. Note the *talik*, a zone that remains unfrozen year-round, under the lake at right.

(4,000 ft) have been measured. The permafrost would keep thickening season after season except that heat from Earth's interior eventually limits this process.

In permafrost regions, special conditions occur under lakes or other water bodies that are deep enough that they do not freeze to the bottom in winter. Below the unfrozen water, the ground also remains unfrozen to some depth, even in winter, or permafrost may be absent altogether. Such an unfrozen zone above permafrost, or extending down through it, is called a *talik* (see the right side of Fig. 46.1).

The map in Figure 46.3 illustrates the distribution of permafrost in the Northern Hemisphere. Note that the map first differentiates between *continuous permafrost*, located in northernmost North America and in a broader zone in northeastern Eurasia, and *discontinuous permafrost*, found as far south as the latitude of northern China and the southern end of Canada's Hudson Bay. Continuous permafrost, as the term implies, is largely unbroken and often thick, although it may have gaps under lakes or wide rivers. Discontinuous permafrost is generally thin and contains unfrozen gaps. Both continuous and discontinuous permafrost zones were shifted much farther south during past glacial periods (the dashed-line boundary in Fig. 46.3 is a rough estimate of the southern extent of permafrost of any kind during past glacial periods). Also shown on the map are lower-latitude patches of permafrost. These patches, known as *alpine permafrost*, occur mainly at high elevations in mountains.

If you compare the map in Figure 46.3 with the map of world vegetation in Figure 25.1, you can see that a certain correspondence exists between the boundary separating continuous and discontinuous permafrost and the transition from tundra to forest vegetation. Although the relationship between vegetation and permafrost is complicated, one important connection involves snow cover. Thick snow cover insulates the underlying soil and limits the development of ground frost. Snow tends to be blown off higher ground in tundra landscapes, but thicker snow is retained where it is sheltered by forest, favoring discontinuous permafrost.

Humans and the Periglacial Environment

The periglacial world is a landscape of scoured bedrock, of basins and lakes, of thin and rocky soil, and of scattered glacial landforms such as kames, eskers, and drumlins. Winter is protracted and bitter; nights are frigid and long. Summer is short and cool, depending on latitude and exposure. The surface is frozen half the year or more, but when the accumulated snow melts, the ground is saturated.

As cold and inhospitable as periglacial environments are, people have lived in and migrated through these regions for many thousands of years. Those who stayed there adapted to the difficult conditions. Arctic peoples such as the Inuit of northern Canada and the Sami of Scandinavia skillfully exploited the environment's opportunities on both land and sea. Their numbers remained small, and their impact on the fragile periglacial domain was slight.

But beginning in the nineteenth century and continuing to the present, newcomers from warmer climates arrived in the periglacial world in growing numbers, driven by the search for resources or stationed at military installations. In Alaska and Canada, for example, the Klondike gold rush brought an initial wave of invasion, followed by mining and oil exploration in other areas, and the military activities of the Cold War. Frontier towns, highways, railroads, and pipelines were all constructed in permafrost regions.

The periglacial environment poses unique engineering, construction, and maintenance problems unknown in warmer regions. This quickly became apparent as structures not designed to avoid these problems failed in various ways (Fig. 46.2). More recent projects, such as the pipeline built to carry oil from the North Slope of Alaska, were designed to be stable on permafrost soils. Nonetheless, there are still many reasons for concern about the impacts of resource exploitation on fragile periglacial environments.

A

B

FIGURE 46.2   Two examples of structural failure in the periglacial environment of interior Alaska, typical of many examples observed in the early to middle twentieth century. A roadhouse on the Richardson Highway collapsed because heat from the house thawed permafrost beneath it (A). A road conducted heat into underlying ice wedges, and they began melting, causing deep troughs to form (B).

Geomorphic Processes in Periglacial Environments

The modification of the landscape in periglacial zones differs from the processes that occur in other regions, because freezing, thawing, and related mass movements play such dominant roles. Water, as it changes phase from the liquid to the frozen state, is an important force. When a permanently frozen layer exists below the surface, water cannot drain downward. Therefore, water often saturates the active layer in summer and forms abundant lenses of ice in winter. In the upper stratum of the permafrost, boulders are shattered by frost, and the fragments that result are constantly moved by freezing, thawing, and the force of gravity. This results in landforms that are unique to the periglacial domain.

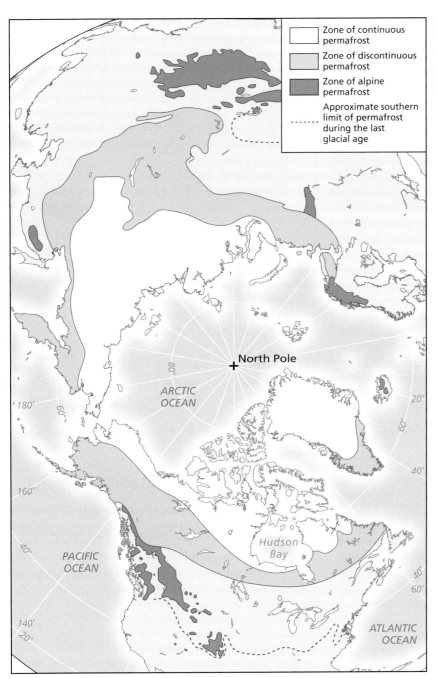

FIGURE 46.3 Distribution of permafrost in the Northern Hemisphere. The dotted line shows the possible southern extent of permafrost during glacial times. This boundary has been estimated from relict periglacial features.

Legend:
- Zone of continuous permafrost
- Zone of discontinuous permafrost
- Zone of alpine permafrost
- Approximate southern limit of permafrost during the last glacial age

FROST ACTION

The freezing and thawing of water in soil and rock is a key disintegrative force in periglacial environments. **Frost action** (also called *frost wedging*), a type of mechanical weathering discussed in Unit 36, is quite important for understanding the development of many periglacial landscapes. Recall that frost action occurs when the stress created by the freezing of water into ice overcomes the strength of the rock containing the water. Joints and cracks in nonporous crystalline rocks are zones of weakness that are exploited by frost action (see Fig. 36.2). Frost action is capable of dislodging boulders from cliffs, splintering boulders into angular pebbles, cracking pebbles into gravel-sized fragments, and reducing gravel to sand and

even finer particles. Thus, the active layer in periglacial regions (see Fig. 46.1) often consists of a mixture of ice-fractured fragments of all sizes.

The surface layer of periglacial terrain is often characterized by the sorting of fragments by size. This sorting, accomplished by repeated freezing and thawing of the active layer, produces a phenomenon called *patterned ground*, which we discuss in more detail later in the unit. Once rock fragments have been loosened by frost wedging, they are moved by frost heaving. **Frost heaving** involves vertical (upward) displacement of soils and the rock fragments they contain when the formation of ice in the ground expands the total mass. Later, when it thaws, the soil settles downward. Through repeated cycles of freezing and

FIGURE 46.4 The effects of frost heaving—in the form of boulders thrust upward out of the ground—can be seen on this alpine tundra slope overlooking Forest Canyon in the area of the Continental Divide Peaks of the Rocky Mountains in central Colorado.

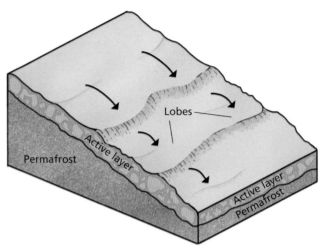

FIGURE 46.5 Even where slope angles are low, the whole active layer may move slowly downslope through the process of solifluction. The moving soil often moves in lobes that form rounded steps recognizable on many slopes of periglacial regions.

thawing, large rock fragments that are heaved upward do not settle back to their original position. Instead, they move upward differentially through the soil, eventually becoming concentrated at the surface. As a result, the ground in periglacial landscapes sometimes seems studded with boulders or gravel (Fig. 46.4). The same process also pushes concrete blocks, road segments, posts, poles, and other artificial fixtures out of the ground. As rocks or soil are pushed upward by frost, there also is a compensating horizontal movement of materials so that a kind of crude cycling of rocks or soil can occur.

In addition to frost heaving, another important process is **frost creep**, the downhill movement of particles in the active layer under the influence of gravity. This is a specific form of soil creep, a type of mass movement discussed in Unit 37. During frost creep, a piece of rock brought to the surface by frost heaving moves downslope during the thawing phase. Frost wedging, frost heaving, and frost creep combine to produce some remarkable landforms, which we discuss later in this unit.

SOLIFLUCTION

Solifluction is a form of slow mass movement that is a combination of frost creep and *gelifluction*—the slow flowage of saturated soil. Recall that we discussed this slope process, which is closely associated with periglacial conditions, in Unit 37. As we noted earlier, soils in permafrost areas are often saturated because of limited drainage. In the warm season, such saturated soil begins to move as a mass, even when the slope angle is low (Fig. 46.5). The texture of the soil is important because highly permeable materials such as gravel and sand are not likely to move by solifluction, whereas silt-laden soils will move quite freely.

Periglacial areas are cold, but they are not without vegetation, which significantly affects the temperature and thickness of the active layer and may stabilize it to some extent. Again, human intervention can have devastating effects. When the protective vegetative cover (whether tundra or forest) is removed, binding roots are destroyed, summer thawing reaches a greater depth, and more of the active layer is destabilized (see Fig. 46.2B). Recovery, in fact, may not occur at all, even after the damaged area is vacated. Periglacial ecosystems are particularly fragile.

Landforms of Periglacial Regions

Landforms in periglacial regions are not as dramatic or spectacular as those of mountain glaciated areas. They are nonetheless quite distinctive, resulting from a combination of frost action and mass movement. The geomorphic features thus produced often take the form of special patterns that look as though they were designed artificially.

ICE WEDGES

One of these remarkable patterns seen in periglacial regions is created by *ice wedges*. During the frigid Arctic winter (Fig. 46.6A), the ground in the active layer and the upper permafrost becomes cold and brittle. Then, when the temperature plunges quickly in an especially cold spell, the frozen ground contracts and cracks, much like the formation of cracks in mud as it shrinks while drying. You have probably noticed that drying mud fractures to form a polygonal network of cracks. Frozen ground fractures into much larger polygons. During the following summer (Fig. 46.6B), meltwater fills the cracks, eventually freezing to create thin wedges of ice. The next winter, the frozen ground cracks again, usually at the ice wedges, which are planes of weakness. More meltwater then enters the open cracks in the next summer, turns into ice, and widens the original wedges. This process is repeated over many seasons.

Some ice wedges reach a width of 3 m (10 ft) and a depth of 30 m (100 ft). The networks that they form are

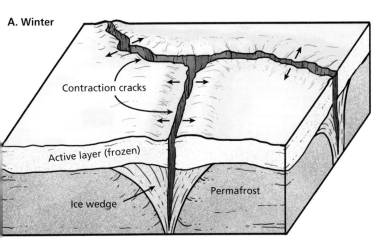

A. Winter

Contraction cracks

Active layer (frozen)

Ice wedge

Permafrost

B. Summer

Water

Water

Active layer (thawed)

New ice added to wedge

Permafrost

New ice added to wedge

FIGURE 46.6 Formation of ice-wedge polygons in a periglacial landscape with continuous permafrost. In winter (A), the frozen ground contracts and cracks during rapid drops in temperature. In summer (B), the active layer thaws and water fills the part of the crack extending into the permafrost, where it freezes. Repeated over many years, this produces a polygonal network of ice wedges, with a surface trough over the wedges and ridges on either side pushed up by wedge growth.

ENHANCE

referred to as **ice-wedge polygons** (Fig. 46.7). The cracks that develop in winter can be filled by windblown sand rather than ice, or by sand mixed with ice. The resulting features are called *sand wedges* or *composite wedges*. Ice-wedge polygons and sand-wedge polygons are typical of permafrost zones, and when climatic conditions change, they remain imprinted on the landscape. In fact, wedge-shaped structures forming polygons have been found in many places in Europe and the Midwestern United States. These structures originated as ice wedges during past glacial periods, when the cold climate allowed widespread permafrost formation beyond the ice sheet. Then, when the climate warmed and the permafrost thawed, overlying soil caved into and filled the cavity left by melting ice wedges, preserving the original polygons.

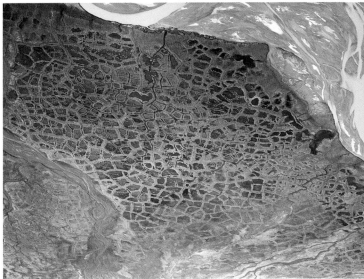

FIGURE 46.7 View from an airplane over a landscape of ice-wedge polygons in the Canadian Arctic.

PATTERNED GROUND

Another characteristic feature of periglacial regions, as we noted earlier, is patterned ground. **Patterned ground** consists of rock and soil debris shaped or sorted in such a manner that designs are formed on the surface resembling rings, polygons, lines, and other regular, repeated arrangements (see Fig. 46.8). Patterned ground is not just a local feature of limited spatial extent; the phenomenon can persist for kilometers, dominating the landscape and giving it an unmistakable appearance. Such forms are characteristic of periglacial regions, but unlike ice-wedge features, permafrost is not essential for patterned ground formation. Geomorphologists have argued that patterned ground results from interacting processes, including the lifting and sorting of fragments by frost heaving and downward settling driven by the force of gravity. Feedbacks between these processes are believed to maintain the regular patterns we observe, but much remains to be learned about these striking landforms.

PINGOS

Yet another highly distinctive periglacial landform is a mound called a **pingo**. Permafrost conditions are a prerequisite for the formation of these mounds, which are round or elliptical at the base and can grow quite large. Although many are comparatively small and occur in clusters, some isolated pingos are as large as 600 m (2,000 ft) in diameter and 60 m (200 ft) high (Fig. 46.9). The core of a pingo is ice, not rock or soil.

Pingos are believed to form where liquid water moves upward and freezes near the surface. Massive ice builds up and pushes the ground above it into a dome-like shape. Many pingos form on the bed of drained lakes where the permafrost table rises to the surface and pushes water

ENHANCE

FIGURE 46.9 A prominent pingo in the Mackenzie River Delta in north-westernmost Canada.

FIGURE 46.8 Ground-level view of a plain filled with rock debris sorted into myriad stone circles 3 to 5 m (10 to 17 ft) in diameter. This striking example of patterned ground was photographed in western Spitsbergen, part of the Arctic Ocean's Svalbard island chain, which belongs to Norway.

upward above it. In other places, groundwater flows upward to feed the pingo's growth. Pingos are unique to periglacial environments.

BLOCKFIELDS

We discussed **blockfields** in Unit 36, but we return to them here because of their importance in periglacial regions. Recall that these are large areas covered with blocky boulders (see Fig. 36.3) that have a variety of other names, such as *felsenmeer*, a German word meaning rock sea. The boulders or blocks (angular stones) that make up blockfields are often large, as much as 1 to 3 m (3.3 to 10 ft) in diameter. They are loosened from bedrock slopes, cliffs, or other exposed surfaces by frost action. Where the boulders or blocks form on slopes, they tend to move downhill as slow-moving expanses of rock as large as 100 m (330 ft) wide and more than 1 km (0.6 mi) long. Exactly how movement takes place is still uncertain.

Whatever the answer, boulder fields are known to occur in periglacial areas and in cold mountainous zones above the timberline. They clearly result from the combination of frost action and mass movement, and although the mechanisms may not be clearly understood, they do provide evidence that near-glacial conditions once prevailed where they exist. We might therefore expect that the

Driftless Area of Wisconsin (see Focus on the Science: The Driftless Area, Unit 44), the place the continental glaciers missed but surrounded, should be a likely place to find a boulder field. Not surprisingly, this type of landform does indeed occur there, where quartzite bedrock weathers into large blocky boulders.

So far, we have studied periglacial landforms that result mainly from frost action, the rearrangement of rock debris by freezing and thawing, and downslope movement driven mainly by gravity. You will note that we have not discussed river transportation. Although major rivers do flow through periglacial areas (e.g., the northward-flowing streams in Siberia, or the Mackenzie River in Canada), they do not create landforms distinctly different from those of other regions.

Understanding Periglacial Environments

Only a very small minority of Earth's population ever experiences periglacial environments and landscapes. These areas are the realm of migrating caribou and reindeer, of musk oxen and wolves, of huge flocks of birds, and of dense swarms of insects. The landscape is one of unfamiliar forms and plants, of mosses and lichens, of needle-leaf evergreen trees, and of large patches of barren ground.

Nevertheless, the periglacial realm has attracted outstanding scientific research over the years, and ingenious methods have been devised to study periglacial processes. In one study of the development of ice wedges, electrical wires that broke when the frozen ground cracked were used to detect exactly when this occurred and under what conditions. More recently, an extensive network of sites has been established for monitoring permafrost depth and temperatures around the Arctic.

One of the motivations for this research has been strictly practical. More information on periglacial environments was needed for effective planning of resource development in Alaska, Canada, and northern Russia. The effects of global warming on permafrost and the possibility that warming permafrost soils could release abundant carbon to the atmosphere (see Unit 24) are also important justifications for studying the periglacial environment. This environment is fascinating in itself, however, and many researchers working there are also motivated by curiosity about the unique set of geomorphic processes and landforms that can be found in the world's cold regions.

Key Terms

active layer *page 555*
blockfield *page 560*
frost action *page 557*
frost creep *page 558*

frost heaving *page 557*
ice-wedge polygon *page 559*
patterned ground *page 559*
periglacial *page 555*

permafrost *page 555*
pingo *page 559*
solifluction *page 558*

Scan Here for a quick vocabulary review

Review Questions

1. Define the term *periglacial* and describe the general spatial distribution of these environments.
2. Why are wet soils common in periglacial landscapes even though precipitation there is relatively low?
3. Describe permafrost and define the layers you would observe in an excavation that extended below the permafrost table.
4. How do patterned ground, ice-wedge polygons, and pingos form?
5. How do periglacial landforms such as ice-wedge polygons and blockfields help us reconstruct past climatic conditions?

Scan Here for a quick vocabulary review

www.oup.com/us/mason

UNIT 47

Wind as a Geomorphic Agent

An isolated dune migrates across the arid landscape of the Tibetan Plateau, shaped and re-shaped continually by the wind. Although the dune migrates mainly from right to left, the ripples on the dune surface were formed by the most recent wind blowing in the opposite direction.

OBJECTIVES

- Explain the process of wind erosion and how it is affected by vegetation cover and other factors

- Describe the transportation of sediment by the wind

- Summarize the common landforms produced by wind erosion and deposition and relate them to environmental factors

- Discuss the origin and environmental significance of loess

The effects of running water, flowing ice, and coastal wave action in shaping Earth's surface are generally obvious, but the role of **eolian** (wind-related) processes is usually subtler and more difficult to assess. Only in the striking and often-beautiful landscapes of great desert dune fields does the wind clearly emerge as the dominant geomorphic agent (Fig. 47.1).

In general, as conditions become more humid, the stabilization of the surface by vegetation diminishes the role of the wind, and other processes become more important in shaping the physical landscape. Complicating this simple assessment is the realization that much of Earth's surface bears the signature of processes that are no longer operating but were important in the past under somewhat different climates. There is no question that eolian processes have had an important influence on the landscapes of various regions during previous climatic regimes, even if they have little effect today. Furthermore, human activities often destabilize the surface vegetation, and in some areas, this has made eolian processes more significant (see Perspectives on the Human Environment: The Dust Bowl).

Wind Erosion

For wind to be an effective agent in shaping the landscape, it must be able to lift grains of sand or silt from the ground surface and move them downwind. As you might expect, the wind must be above a certain minimum speed to move loose particles from the surface. Geomorphologists have found that what is critical is actually the *shear stress* acting on the ground surface, that is, the force exerted on an area of the surface by the air flowing across it. If the wind is flowing across a smooth, bare surface, such as a bed of loose, fine sand, then an increase in wind speed will produce a corresponding increase in the shear stress that may set grains in motion. The connection between wind speed and wind erosion is more complicated where there is greater *surface roughness*, a term referring to many different features that impede the wind flow, from large trees to small pebbles. Vegetation is the most important source of surface roughness in most natural landscapes.

When vegetation is present, part of the shear stress exerted by the wind acts on the vegetation rather than on the ground surface. If the vegetation is dense enough, particles on the ground surface do not "feel" the wind at all. When researchers have made detailed measurements of wind speed at various heights above the ground, the impact of vegetation is clear. Even over a bare surface, friction results in a steep drop-off of wind speed near ground level. If enough vegetation is present, much of that steep decrease in wind speed is shifted upward above the plant canopy. Wind erosion is highly unlikely under tall, dense vegetation such as forest, and even a dense cover of short grasses can reduce erosion to an insignificant level. In fact, in northern China, where windblown sand dunes are burying roads or fields, the moving sands are effectively stabilized by setting rows of sticks in the ground to create roughness, mimicking the effect of live vegetation.

FIGURE 47.1 Some of the world's most striking and beautiful landscapes are shaped by wind, including desert dune fields such as this one in the Badain Jaran Desert of north-central China. The world's largest dunes occur in this region.

It is clear then, that there are several important factors that influence the effectiveness of the wind in eroding particular parts of the landscape. One is the frequency of strong winds that can generate sufficient shear stress to initiate the movement of sand or silt grains. Gradients of atmospheric pressure are the primary driver of winds (see Unit 8), so the spatial pattern of high and low pressure can influence where wind erosion is possible. For example, strong winds are often associated with steep pressure gradients around intense cyclones in the midlatitudes (see Unit 13). Even those strong winds cannot erode the soil surface if it is protected by enough vegetation. This is evident in the Great Plains of North America, where strong winds are a fact of life—as is quickly noticed by anyone who moves there from other regions—yet wind erosion does not occur except where drought or agricultural tillage have thinned the natural grass cover or removed it entirely. Thus, vegetation cover and the factors that influence it, such as precipitation and human land use, also have a strong influence on the effectiveness of eolian processes.

Finally, the size of particles on the ground surface plays a role. Sand grains with a diameter of about 0.1 mm (0.025 in) are the most easily set in motion. You might expect that smaller particles of silt or clay would be even more easily picked up by the wind, but these have a greater tendency to stick together and form surface crusts that resist erosion. Where there are few loose sand grains, small granular soil peds (aggregates) may be set in motion instead. Once the wind picks up sand grains or small aggregates, these moving particles strike and wear away the surface in a process called **wind abrasion**. Then any fine silt and clay particles on the soil surface are quickly launched into the wind, producing clouds of dust. Soil peds moving in the wind also tend to disintegrate into fine dust.

It is easy to understand how abrasion by sand grains can launch great clouds of dust into the air when we note that abrasion actually has enough force to wear away hard rock surfaces, fence posts, and walls. If you have ever walked along a dry beach on a windy day, you have probably felt sand particles striking your lower legs; if the wind is

FIGURE 47.2 A dust cloud bears down on a settlement in Arizona. Although much of the cloud is suspended dust, sand grains may also be moving near the surface in saltation (a hopping motion shown in Fig. 47.3).

strong enough, the experience can even be unpleasant. This illustrates the fact that wind abrasion is strongest near the ground surface and diminishes with height.

Wind Transportation

Once mineral grains are set in motion by the wind, their further movement takes place in three ways. The finest material, such as silt and clay, is carried high in the air in *suspension* and is usually referred to as dust. For the smallest dust particles, the force of gravity driving them to settle toward Earth is almost balanced by friction between the particles and surrounding air, so they can remain aloft almost indefinitely (see Focus on the Science: Why Don't Particulates Fall?, Unit 6). Larger dust particles settle more rapidly but can still be carried some distance before returning to the ground surface.

Near the source of the dust, the suspended material forms a thick cloud that extends upward from the ground surface (Fig. 47.2). The leading edge of the cloud is often well defined as it rolls across the landscape. Such dust storms are common in southern Arizona, Iraq, northern China, and many other dry regions of the world, especially where

agricultural fields, roads, or other disturbed areas are exposed to the wind. Fine mineral dust in suspension rises to such a height that it can be carried from one continent to another at high levels in the troposphere. For example, dust produced by wind erosion during spring storms in the dry regions of northern China is sometimes carried across the Pacific Ocean to the United States and Canada.

Unlike silt and clay particles, sand grains are too heavy to be picked up and carried very far above the ground. The wind can still pick them up off the surface, but they settle quickly back toward the ground under the force of gravity, so their overall motion involves short hops at low levels. This process is known as **saltation** (Fig. 47.3). As the hopping sand grains return to the surface, their momentum helps dislodge other grains and launch them into motion as well. Because they strike the ground with enough force to break loose material from hard surfaces, saltating grains are responsible for most wind abrasion.

The largest grains moving by saltation travel in the lowest 20 cm (8 in) or so above the surface. Almost all the rest of the saltating particles stay within 2 m (6.6 ft) of the ground, explaining why most abrasion occurs not far above the ground. Though looming clouds of dust like the one in Figure 47.2 are made up mainly of suspended particles, there often may be abundant sand moving in saltation near the base of the cloud. Wind also moves larger rock fragments by actually pushing them along the surface in a process called **surface creep**. Even pebbles of considerable weight can be moved by the strongest winds, found in places like the Dry Valleys of Antarctica.

Landforms Shaped by Wind

It is important to remember that wind action is not confined to deserts or semiarid steppelands. Wind also shapes parts of the landscape in glacial and periglacial zones, in savannas and humid midlatitude grasslands, and in other areas as well. Nevertheless, wind does its most effective work in dry environments. Deserts and semideserts are often strongly affected by eolian processes, and the landforms and landscapes of wind action are best developed in places like North Africa's Sahara, Southwest Asia's Arabian Desert, and Australia's Great Sandy Desert. In the material

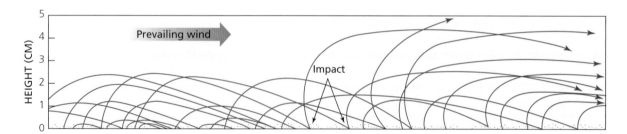

FIGURE 47.3 Wind causes movement of sand grains by saltation. Grains are lifted directly by the wind or launched through the impact of another grain; they are then carried only a short distance by the wind before gravity pulls them back to the loose sand surface, where they strike other particles, repeating the process.

FIGURE 47.4 The striking topography of a zone of yardangs carved out of tilted rocks dominates the foreground and center of this photo. This locale is near Minab, a coastal town on the Persian Gulf's Strait of Hormuz in southeastern Iran.

ENHANCE

that follows, we first examine the landforms produced primarily by wind erosion. We then move on to study landforms and sediments (most notably *dunes* and *loess* deposits) that result when the windblown sand or silt is deposited.

Erosional Landforms

When the wind sweeps along a surface and carries away the finer particles found there, this process is called **deflation**. Arid landscapes often include shallow basins without outlets. These basins can have a variety of shapes and origins, but many seem to have a preferred orientation relative to the prevailing wind direction and are most likely produced by deflation. They are believed to begin as small local hollows that are continuously enlarged as the wind removes freshly weathered particles. Eventually, some of these **deflation hollows** can grow quite large.

Not many other landforms can be attributed exclusively to wind erosion. One interesting surface feature often

thought to be a product of wind action is **desert pavement**, a surface concentration of closely packed pebbles (see Fig. 41.2). Traditionally, desert pavement has been explained as the result of deflation that carried away fine soil material, leaving only the pebbles too large to be moved by the wind. Recently, desert geomorphologists realized that desert pavement can form in quite a different manner, one that involves more deposition than erosion. Over time, windblown dust is trapped among the stones on an alluvial fan (see Unit 41) or some other gravel-covered landform. The fine-grained dust swells when it is wetted and shrinks when it dries, in the process continually pushing the stones upward to form a surface pavement (somewhat like a frost heave, discussed in Unit 46).

The most striking product of wind abrasion is the **yardang** (Fig. 47.4). Yardangs are low, narrow ridges that form parallel to the prevailing wind direction. They tend to develop in deserts where the wind carries loose sand across soft and easily eroded bedrock unprotected by vegetation. Wind abrasion scoops out troughs on either side of the yardang; the yardang itself is a remnant of uneroded rock.

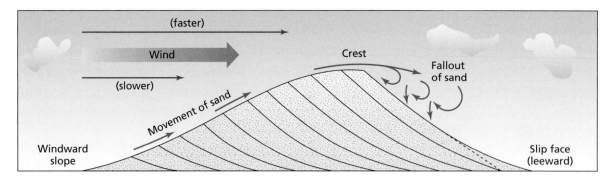

FIGURE 47.5 Cross-section of an active sand dune that is migrating from left to right. The lower-angle windward slope is the erosional side of the dune, with surface sand grains pushed upward toward the crest. The steeper leeward slip face is the depositional zone, where wind-driven sand is dropped by the wind and then flows downslope. The accumulation of sand grains on the advancing slip face produces strata inside the dune, much like the foreset beds in a delta (see Fig. 41.15).

Dunes

Dunes are shaped by both deposition and erosion of sediment by the wind, but we classify them as depositional landforms because they develop where, overall, there is more deposition than erosion of windblown sand. A large area covered by dunes is called an **erg**, or *sand sea* (see Fig. 47.1). Vast ergs occur in the Sahara and other great warm deserts of the world, but it is important to note that large portions of these deserts are not covered with dunes. Instead, the desert surface can be covered with a rolling, hummocky sheet of windblown sand (called a *sand sheet*, logically enough). More often, the ground surface is gravel-covered, bare rock, or even wind-eroded lake deposits (the latter are common in parts of the Sahara and record past periods of wetter climate). Windblown sand and dust pass over such landscapes without accumulating. Only where sand tends to accumulate over time will dunes develop. This may happen because winds converge in a low area between highlands and bring together sand from a large area. Sand also piles up at the foot of mountain ranges because the wind cannot move it up over steeply sloping mountains as fast as it arrives from areas upwind.

The profile of a typical dune has three elements: the *windward slope*, the *crest*, and the leeward slope, known as the **slip face** (Fig. 47.5). As this diagram shows, the windward slope has a lower angle than the leeward slope. The wind erodes the windward slope and drives the sand grains up that slope and over the crest. There, they are dropped on the upper part of the slip face. A distinct hump of sand builds up over time until it collapses into a flow of sand down the steep slip face (accounting for the term *slip face*). Thus, from the profile of a dune, we can often estimate the prevailing wind direction.

As sand flows are deposited on the slip face, that leeward face of the dune slowly shifts in the downwind direction. This process creates very distinctive large *cross-beds*, as mentioned in Unit 28. Such cross-beds can be used to identify windblown sands in sedimentary rocks that are hundreds of millions of years old, and even to determine prevailing wind directions at the time the ancient dunes were migrating. As

the leeward face shifts downwind, the windward face follows, as it is continually eroded. Dunes in arid deserts generally support little vegetation, so the wind acts on the entire surface of the dune continually, but the basic shape of the dune is unchanged as it migrates with the wind.

Ripples on the dune surface show that it is being actively worked by the wind, as in the photo that opens this unit. Ripples are small, regularly spaced ridges, usually less than 1 cm (0.4 in) high. They look like miniature dunes, and like dunes they have a steeper leeward slope and a gentler windward slope. When a dune's surface is bare sand, much of it formed into ripples, we can say that the dune as a whole is *active*. Over time, however, dunes may migrate into moister areas on the desert margin, or a climate change may affect a dune area. Then plants will take hold, and the vegetation will slow or even halt the dune's movement; in such instances, a dune is described as *stable*, or fixed.

DUNE FORMS

One of the most remarkable characteristics of dunes is that although they are quite varied in detail, there is a fairly small number of distinctive dune forms. From the form of a dune, we can often determine the prevailing direction of the strong winds that shape the dune and also learn about the supply of sand available to be formed into dunes.

Barchans The most easily recognized dune form is probably the **barchan**, a crescent-shaped dune. The best way to understand its form is to examine Figure 47.6A. The convex side of this dune is the windward side, so that the points of the crescent lie *downwind*, and the slip face of the dune is on the concave side. A cluster of barchans therefore immediately indicates the prevailing direction of strong winds in its locality. Each barchan is often surrounded by an area with little or no cover of windblown sand; therefore, the total amount of sand held in a group of barchans is relatively small. Geomorphologists believe that barchans form when there are strong prevailing winds from one direction, but there is a limited supply of sand that can accumulate into dunes.

The Dust Bowl

A dust storm blows up early in the novel *The Grapes of Wrath*. Author John Steinbeck (1902–1968) describes how the wind in the Dust Bowl swept away the soil and the livelihood of an Oklahoma farming community in the 1930s:

> The air and the sky darkened and through them the sun shone redly, and there was a raw sting in the air. During a night the wind raced faster over the air, dug cunningly among the rootlets of the corn, and the corn fought the wind with its weakened leaves. . . .
>
> The dawn came, but no day . . . and the wind cried and whimpered over the fallen corn.
>
> Men and women huddled in their houses, and they tied handkerchiefs over their noses when they went out, and wore goggles to protect their eyes.
>
> When the night came again it was black night, for the stars could not pierce the dust to get down . . . now the dust was evenly mixed with the air, an emulsion of dust and air. Houses were shut tight . . . but the dust came in so thinly that it could not be seen in the air, and it settled like pollen on the chairs and tables, on the dishes. The people brushed it from their shoulders. Little lines of dust lay at the door sills.*

In the novel, the "Okies" eventually pack up their families and head for California. As historians have pointed out, the migrants who left Oklahoma in the 1930s were often fleeing poverty in the eastern part of the state, not dust storms. In fact, many inhabitants of the areas hardest hit by dust storms stayed put, but the blowing dust and eroded fields were devastating—and frightening—all the same (Timothy Egan collected many firsthand accounts of these storms in his book *The Worst Hard Time* [Boston: Houghton Mifflin, 2006]).

How did this tragedy occur? After Congress passed the Homestead Act in 1862 and the Civil War ended in 1865, families poured onto the Great Plains, anxious to establish farms of their own. They plowed up the thick protective layer of vegetation to plant crops or put out cattle to graze on the dense grass cover. This region receives sparse rainfall, but enough falls to support a crop of wheat in most areas. Occasional severe droughts are a fact of life on the Plains, however. In fact, we know today that great expanses of this region are covered with sand dunes and thick dust deposits that formed during prehistoric droughts.

In the mid-1930s, the worst drought in decades afflicted much of the Great Plains. The crops did not survive, leaving bare soil unprotected against the wind; in pastures, the cattle devoured the sparse grass and exposed bare ground there as well. When strong winds blew, as they often do on the Plains, layers of topsoil were stripped off and whisked away in the form of dust storms. Patches of sand dunes began to migrate in the wind.

Though drought, dust storms, and active sand dunes have all occurred naturally on the Great Plains for thousands of years, the effects of drought were clearly aggravated this time by land use—especially agriculture that left soils bare. The dramatic dust clouds, some of which even drifted eastward as far as Washington, D.C., triggered major new soil conservation efforts by the federal government. By 1938, the Dust Bowl conditions finally began to subside, and heavier rainfall returned. Drought returned to the Plains in the 1950s and in 2000–2002, and dust storms did appear again, but not to the frightening extremes seen during the Dust Bowl.

What has changed since the 1930s? Farmers use better conservation practices, but probably even more important, groundwater is now pumped from the High Plains (Ogallala) Aquifer onto fields across much of the region hardest hit by the Dust Bowl. For now, this water can sustain crops even through dry years. The aquifer has been depleted over large areas, though, and has a limited life expectancy. Given predictions of warmer temperatures over the next century, higher evaporation will cause more rapid depletion of groundwater and greater susceptibility to droughts. Eventually, the great dark clouds of the Dust Bowl could return—or perhaps the land will be returned to protective grass cover when it can no longer be irrigated.

*Excerpts from Steinbeck, *The Grapes of Wrath* (New York: Viking, 1939), pp. 2–3.

Transverse Dunes A **transverse dune** is a more or less continuous ridge of sand positioned at a right angle to the prevailing wind (Fig. 47.6B). The leeward side of the ridge forms a steep slip face. The ridge can be straight, or it can have many small curves and look like a group of barchans joined together. Fields of transverse dunes are made up of many of these ridges, often spaced quite close together. Thus, fields of transverse dunes represent the accumulation

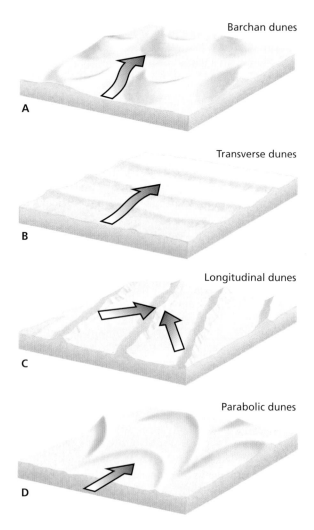

Barchan dunes

A

Transverse dunes

B

Longitudinal dunes

C

Parabolic dunes

D

FIGURE 47.6 The four most common types of sand dunes. In each diagram, the prevailing wind direction is indicated by the arrow(s).

of a significant amount of eolian sand. Not surprisingly, these dunes are believed to form when the sand supply is quite large and the prevailing strong winds blow from one direction.

Longitudinal Dunes Transverse dunes can be mistaken for *longitudinal dunes* (also known as *linear* or *seif dunes*). Like transverse dunes, **longitudinal dunes** form lengthy sand ridges, but they are not perpendicular to a single prevailing wind direction. The relationship of longitudinal dunes to wind directions has been controversial. It seems clear that longitudinal dunes will develop where strong winds blow from two distinctly different directions (Fig. 47.6C). The longitudinal dune is not perpendicular to winds from either direction but "splits the difference" between them. Slip faces develop on both sides of the dune ridge depending on which direction the wind happens to be blowing from.

On the other hand, there are some longitudinal dunes, especially those with some vegetation cover on their lower slopes, that seem to be aligned parallel to a single

prevailing wind direction. This issue is important because longitudinal dunes are a very prominent part of many arid landscapes. Flying from Adelaide on the coast of South Australia to Alice Springs in the heart of the continent, you can see an entire landscape that looks like corrugated cardboard. Thousands of longitudinal dunes more than 3 m (10 ft) high, with many as tall as 20 m (65 ft), extend continuously for as long as 100 km (63 mi).

Parabolic Dunes *Parabolic dunes* may superficially resemble barchans with their crescent shape, but they are in a class by themselves relative to the other dune forms just discussed. A **parabolic dune** has a crescent shape with the points directed *upwind* and a slip face on the convex side of the crescent (Fig. 47.6D). In addition, parabolic dunes are generally more elongated than barchans, and some have a very long and narrow hairpin shape. What makes these dunes different is their anchoring by vegetation. Generally, the two points or limbs of the parabolic dune are somewhat stabilized by vegetation. Only the central part of the parabola where the slip face is located is bare, active sand. A *blowout* (a deflation hollow in sand) usually lies between the two limbs of the crescent, and sand is blown out of this hollow onto the dune. Not surprisingly, parabolic dunes are common in semiarid landscapes and along coastlines, where there is enough precipitation to sustain the vegetation that partially anchors them. They probably often originate with the development of a blowout, and sand deflated from the blowout builds up the parabolic dune form over time.

DUNE LANDSCAPE RESEARCH

Present-day dune fields, both in deserts and along coasts, are unique landscapes, well worth studying for many reasons. Ecologists have found them to be very useful natural laboratories for investigating plant succession as vegetation gradually becomes established on initially bare sand. Dune field research also contributes to the interpretation of the past, because dunes offer clues about climate change. Certain areas where dunes now exist are no longer arid, and the dune landscape has become fixed. From the morphology of the dunes, however, conclusions may be drawn not only about earlier climates in general terms but, more specifically, about wind directions as well.

Another reason to know as much as we can about wind erosion and dune formation is immediate and practical. There are places along desert margins where advancing dunes are overtaking inhabited land today, and some dunes developed or were reactivated in the Dust Bowl of the 1930s. By understanding how dunes migrate and how wind action drives them, we are in a better position to develop ways to stabilize them and halt their progress.

Loess

Perhaps the most impressive example of the capacity of wind to modify the landscape comes not from desert dunes but from dust deposits that built up at almost imperceptible rates over hundreds of thousands to millions of years. Sedimentary deposits of dust are called **loess**. Loess deposits around the world are fairly similar in appearance: tan or light yellowish-brown silt without the boulders and pebbles that characterize glacial till or the distinct layers sorted by particle size that are seen in stream deposits. Loess was first identified in the Rhine Valley as long ago as 1821, which is how it got its German name (*loess* means loose in German and probably refers to how easily loess soils could be plowed). Loess also exists in northern France, along the Danube Valley of Eastern Europe, and in large areas of southern and central Russia.

The thickest accumulations of loess, however, are found in Asia, especially in north-central China. In the hilly middle basin of the Huang He (Yellow River) lie the thickest loess deposits in the world. In fact, the Chinese call this region the Loess Plateau, and here the loess averages 75 m (250 ft) in thickness and in places reaches a depth of as much as 380 m (1,250 ft) (Fig. 47.7). Loess also occurs over a sizable area of southern South America, including Argentina's productive Pampas.

In the United States, the most prominent loess deposits extend from the Great Plains to the lowlands of the Mississippi, Ohio, and Missouri river basins. As Figure 47.8 shows, some of the thickest deposits lie in Nebraska and Iowa, where as much as 60 m (200 ft) of loess has buried the underlying topography. Most of the other loess deposits of the central United States range from less than 1 m (3.3 ft) to about 30 m (100 ft) in thickness. Another major area of loess deposition occurs on the Columbia Plateau in the Pacific Northwest, near where the states of Washington, Oregon, and Idaho meet (see Fig. 47.8).

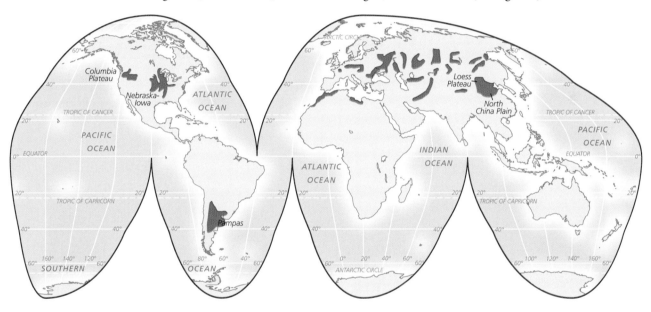

FIGURE 47.7 Major loess deposits of the world.

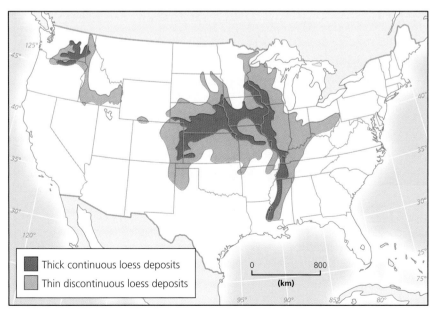

FIGURE 47.8 Loess regions of the United States, highlighting the major deposits of the east-central Great Plains, the Mississippi Valley, and the Pacific Northwest.

ORIGINS OF LOESS

The dust that becomes loess can come from various sources. The great loess deposits of China are made up mainly of dust blown from the deserts and semiarid regions to the north and west of the Loess Plateau. Rivers flow into these deserts from high, rapidly eroding mountain ranges, depositing silt on alluvial fans and in desert basins. That silt is then frequently eroded by the wind and is probably the main component of the Chinese loess. Dust is blown onto the Loess Plateau of China during both cold glacial and warm interglacial periods, although the rate of dust deposition is probably higher during the glacial periods.

In North America and Europe, most loess was deposited during glacial periods. In part, the loess particles were blown from glacial outwash deposits along major rivers, like the Mississippi or the Danube. Glacial erosion produces tremendous quantities of finely textured sediment (*rock flour*, see Unit 45), and such fine material became part of the outwash deposits formed during glacial recession. In addition, much of the landscape south of the ice sheet was a barren, windswept periglacial environment. As a consequence, wind erosion was widespread and not limited to outwash deposits. The thick loess of Nebraska, in particular, came primarily from nonglacial sources, areas of deflation on the western Great Plains. At times, in the landscapes south of the ice sheets, vast clouds of dust must have darkened the skies and obscured the Sun. The fine-grained dust was often carried long distances in suspension before it was deposited on the ground, sometimes hundreds of kilometers from its source.

LOESS LANDSCAPES

Loess-covered landscapes have a distinct and easily recognizable look related to both human land use and the topography created where water has eroded the loess. Just what makes

FROM THE FIELDNOTES

FIGURE 47.9 "We were on our way to fieldwork in the deserts to the north, but it was a beautiful day in the northwestern Loess Plateau, so we took our time and stopped at an overlook for lunch. The loess here is as much as 300 m (985 ft) thick, all of which has accumulated slowly over millions of years as layer upon layer of dust. As remarkable as the work of the wind has been here, few landscapes anywhere display such impressive geomorphic work by humans as well, carving the loess into terraces, trails, and dwellings. Human impacts have a long history here—the oldest evidence of agriculture in northern China was found nearby."

loess landscapes so distinctive? First, loess has a high capacity to retain water for plant use. It also contains a wide range of minerals derived from ground-up bedrock, which weather rapidly enough to supply nutrients needed by plants. Scrape off the soil horizons, and deeper layers of loess can still be used to grow crops, although the yields will be lower.

If we compare the maps of loess deposits, agricultural productivity, and population distribution, we can see that some of the world's most fertile areas—and some great human agglomerations—lie in loess regions. In the United States, the loess of Iowa and Illinois yields massive harvests of corn, soybeans, and other crops. In the drier Great Plains, fields of loess-derived soils in Kansas, Nebraska, and eastern Colorado are irrigated with groundwater to produce grain that feeds a large portion of the beef cattle raised in the United States. In terms of population size and density, however, nothing on Earth matches the landscapes of the North China Plain. The vast wheat fields there are primarily planted on loess-based soils, both where the loess was originally deposited by the wind and where it was reworked into floodplain deposits by streams. The most productive farmlands of Russia and Ukraine, too, are located on loess deposits.

Loess also has an unusually great capacity to stand in upright walls and columns (Fig. 47.9) and to resist collapse when it is excavated. In China's Loess Plateau, the local inhabitants have excavated the loess to create dwellings that are sometimes large and elaborate. As long as the region remains geologically stable, the millions of people dwelling underground are safe. But whereas loess withstands erosion and has vertical strength, it collapses when shaken. In 1920, a severe earthquake struck the Loess Plateau, and an estimated 180,000 people lost their lives, most of them buried inside their caved-in homes.

Finally, loess is susceptible to rapid erosion by water where natural plant cover is removed in fields or along roads and paths. Surface runoff quickly carves deep gullies in loess deposits. In regions like the Loess Plateau of China or the western plains of Nebraska, it is not unusual to find flat, fertile fields that suddenly drop away into a maze of narrow gullies and razor-edged ridges (see lower left corner, Fig. 47.9).

Key Terms

barchan *page 566*
deflation *page 565*
deflation hollow *page 565*
desert pavement *page 565*
dune *page 566*
eolian *page 563*

erg *page 566*
loess *page 568*
longitudinal dune *page 568*
parabolic dune *page 568*
ripples *page 566*
saltation *page 564*

slip face *page 566*
surface creep *page 564*
transverse dune *page 567*
wind abrasion *page 563*
yardang *page 565*

Scan Here for a quick vocabulary review

Review Questions

1. Describe how the ability of the wind to pick up and transport grains of sand or silt is influenced by vegetation or other surface roughness.

2. Which size of particles is most readily set in motion by the wind, and how does this affect the motion of particles of other sizes?

3. How are sand and fine silt transported by the wind and at what levels above the ground?

4. Describe the typical sand dune profile, how it migrates, and how the prevailing wind direction might be deduced from it.

5. Describe barchan, transverse, longitudinal, and parabolic dunes, and explain how they reflect wind directions, sand supply, and vegetation.

6. What is loess, and what types of sources does it come from? Describe its major physical properties.

Scan Here for a quick vocabulary review

www.oup.com/us/mason

Coastal Processes

High-relief rocky coast, Oregon. While this coastal landscape looks very different from the low sandy beaches and barrier islands of New Jersey or Georgia, they are all shaped by the action of wind-driven waves, currents, and tides.

OBJECTIVES

- Describe the physical properties of waves and how their behavior changes as they arrive at the shore

- Explain how wave action causes the erosion, transportation, and deposition of sediment in the coastal zone

- Discuss the nature of tides and their role in coastal erosion and deposition

- Summarize the effects of storms and crustal movement on coastal landscapes

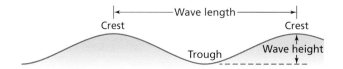

FIGURE 48.1 Wave height is the vertical distance between the wave crest and the wave trough. Wave length is the horizontal distance between two crests (or two troughs).

Some of the world's most spectacular scenery lies along the coasts of the continents. Sheer cliffs tower over surging waves. Curving beaches are fringed by steep-sided headlands. Magnificent bays are flanked by lofty mountains. Great rivers disgorge into the open ocean. Glaciers slide and calve into the water. Coastlines are shaped by many forces: by waves from the sea, by rivers emptying from land, by ice, and even by wind. Rising and falling sea levels, tides and currents, and tectonic forces all contribute to the development of coastal topography. The complexity and variety of coastal landforms and landscapes are the result, and these are the topics of this and the following unit.

Our interest in coastal processes stems from two considerations. First, we seek to understand the processes that shape the fascinating and often strikingly beautiful landscapes at the interface between land and sea. The second consideration is more practical: coastlines are probably the most intensively used landscapes for a variety of human activities. In such heavily developed areas, coastal processes can have very significant consequences. Our ability to manage these landscapes successfully rests on our knowledge of the environmental processes operating within them. Although this is also true in other regions of the world, change in coastal landscapes is often especially apparent to human observers. Beaches shrink or grow noticeably over only a few years, and eroding coastal bluffs collapse, taking expensive homes with them. Other changes may be subtler, but they can still have critical importance for wildlife habitats or for the economies of coastal regions.

In this and the following unit, we study the **littoral zone**, where the land meets the sea. In physical geography, the term **coast** refers in a general sense to the strip of land and sea where the various coastal processes combine to create characteristic landscapes ranging from dunes and beaches to islands and lagoons. The term **shore** has a more specific meaning and denotes the narrower belt of land bordering a body of water (the most seaward portion of a coast). A *shoreline* is the actual contact border between water and land. Thus, we often refer to a coastal landscape, of which the shoreline is but one part.

Waves

Many forces help shape coastal landforms, but the key agents are *waves*. Most waves (not only in oceans but also in lakes) are generated by wind. Waves form when energy is transferred from moving air to water. Large waves form in water when the wind *velocity* is high, the wind *direction* is persistent, the wind *duration* is protracted, and the *fetch* (the distance over which the wind blows) is long. When conditions are favorable, the ocean's upper layer is stirred into long rolling waves, or *swells*, which can travel thousands of kilometers before they break against a shore. When conditions are less favorable (e.g., changing wind directions), waves are generally smaller but may have a more complex pattern.

WAVES IN OPEN WATER

Once a series of swells are well developed and moving across open water, we can observe their properties (Fig. 48.1). The **wave height** is the vertical distance between the *crest*, or top, of a wave and its *trough*, or bottom. The height of a wave is important when it reaches the coast because a high wave does more erosional work than a low wave. The **wave length** is the horizontal distance from one crest to the next (or from trough to trough). A wave's *period* is the time interval between the passage of two successive crests past a fixed point.

In the open ocean, swells may not look large or high because they cannot be observed against a fixed point. But they often reach heights of up to 5 m (16.5 ft) and lengths ranging from 30 m (100 ft) to several hundred meters. Storm waves tend to be much higher, often exceeding 15 m (50 ft). The highest wind-driven wave ever measured reached 34 m (112 ft) during a Pacific storm in 1933.

When swells travel across the open ocean, they seem to move the water itself in the direction of their movement. But this is not the case. In reality, the passing wave throws the water into an orbital motion. As shown in Figure 48.2, a water parcel (or particle) affected by the passing wave takes a circular, vertical path. When the wave approaches, the parcel rises and reaches the height of the crest. Then the parcel drops to the level of the trough and comes back to where it started when the wave arrived. Waves that move water particles in this circular up-and-down motion are called **waves of oscillation**. As Figure 48.2 shows, the depth of water affected by a wave of oscillation corresponds to half the wave length; for example, if the wave length is 100 m (330 ft), the wave will stir the water to a depth of about 50 m (165 ft).

WAVES AGAINST THE SHORE

Once they have been formed by steady, strong winds, swells may move across vast reaches of open ocean without losing their strength or energy. As they approach the coastline, they usually enter shallower water and go through a series of changes referred to as **shoaling**. The free orbital motion of the water is disrupted when the wave begins to "feel" (be affected by) the ocean bottom. At this point, the swell becomes a *wave of translation*. No longer do water particles in orbital motion return to their original positions, and the wave can begin its erosional work.

Because we know that the depth of a wave is half its length, we can determine where it will feel bottom first.

Trough · Crest · Trough

Wave height

Orbital diameters
decrease downward

Negligible motion
(where depth equals
1/2 wave length)

Wave length

FIGURE 48.2 Orbital motion of water as a wave passes in deep water. Rather than moving with the wave, "parcels" of water follow the circular paths shown by the blue circles with arrows, with the greatest motion at the surface and less motion (smaller circles) as the depth increases. One full orbit around any circle corresponds to motion of the wave by one full wave length. To follow the changing position of a single parcel of water at the surface, note the changing position of the arrows in the largest circles, from right to left across the diagram. The parcel starts at the low point of the circle (right side of diagram) and rises clockwise around it to the top as the wave passes, then drops down again (left side). Follow the solid lines downward from each arrow to find the corresponding position of water parcels making smaller orbits below the surface. Dashed blue lines represent water-parcel positions one-eighth of an orbit later, and the dashed black line shows where the wave crest would have moved to at that point.

ENHANCE

The wave in our earlier example, with a length of 100 m (330 ft), will begin to interact with the ocean bottom when the water depth becomes less than 50 m (165 ft). Contact with the bottom slows the wave down so that the wave length is forced to decrease. The orbital motion of water particles continues but is now faster than the forward motion of the wave as a whole. As a result, water pushes upward, increasing the wave's height and steepness, and the circular path of water particles is compressed into an oval one (Fig. 48.3A). Eventually, the wave crest collapses (breaks) forward, creating a *breaker*. This happens along the breadth of the advancing wave, which is now capped by a foaming, turbulent mass of water.

From the beach, we can see a series of approaching waves developing breakers as they advance toward the shore (Fig. 48.3B); such a sequence of breaking waves is referred to as the **surf**. When a wave reaches the shore, it finally loses its form, and the water rushes up the beach in a thinning sheet called the **swash**. As anyone who has walked through the swash will remember, the dying wave still has some power, and the uprushing water carries sand and gravel landward. Then the last bit of wave energy is expended, and the water flows back down toward the sea as **backwash**, again carrying sand with it. Along the shore, therefore, sand is unceasingly moved landward and seaward by wave energy.

WAVE REFRACTION

When we look out across the surf from the top of a dune, it often seems as though the surf consists of waves arriving parallel, or very nearly parallel, to the coastline. In actuality, a truly parallel approach is quite rare, but the waves do change orientation toward that of the shoreline as they move into shallow water.

When a wave approaches a beach at an oblique angle, only part of it is slowed down at first—the part that first reaches shallow water. The rest of the wave continues to move at a higher velocity (Fig. 48.4). This process bends the wave as the faster end overtakes the end already slowed by shoaling. This bending is known as **wave refraction**. By the time the entire wave is in shallower water, its angle to the shoreline is much smaller than it was during its approach through deep water. Notice, too, that the angled approach of the waves relative to the beach sets up a *longshore current* flowing parallel to the shoreline.

When a coastline has prominent *headlands* (or *promontories*, points of land that jut into a body of water) and deep bays, wave refraction takes place as shown in Figure 48.5. The waves approach the indented coastline roughly parallel to its general orientation. They reach shallower water first in front of the headlands, so that they are slowed and their length is reduced. The segment of the wave headed for the bay has yet to reach shallow water, so it continues at open-water velocity. This has the effect of refracting or bending the wave, concentrating its erosional energy on the point of the headland. Rock material loosened from the promontories is transported toward the concave bends of the bays, where a beach forms. Wave action therefore has the effect of straightening a coastline, wearing back the promontories, and filling in the bays. Wave refraction is a crucial part of this process.

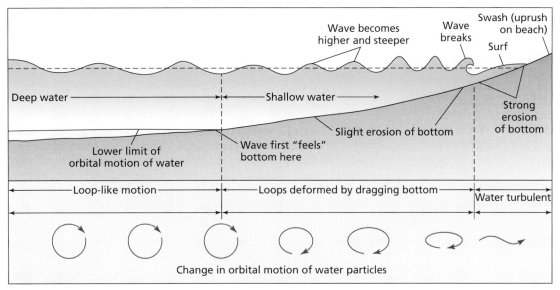

A

B

ENHANCE

FIGURE 48.3 Waves are transformed as they travel from deep water through shallow water to shore (A), producing evenly spaced breakers as they approach the beach (B). In the diagram, circles, ovals, and wave lengths are not drawn to scale with the waves on the surface.

Erosion and Transportation of Sediment by Wave Action

Waves are powerful erosional agents. Furthermore, although it may be hard at first to imagine how waves rushing up on a beach can actually move material parallel to the shoreline, they are highly effective at that process as well. In this unit, we first consider the work of waves in eroding rocky shorelines. Cliffs along a shoreline rarely extend straight down to great depths below the water level. If cliffs did take this shape, waves striking the cliff would be reflected back out into the ocean and would be less effective at erosion. Instead, there is usually at least a narrow bench or ramp cut into the rock at the foot of coastal cliffs, produced by wave erosion and often exposed at low tide.

As the waves break and rush across the bench to strike the cliff, the bench is enlarged through two processes of erosion. The first of these is *hydraulic action*, which is especially effective where rocks are strongly jointed or otherwise cracked (e.g., along bedding planes). Air enters the joints and cracks, and when the water pounds the rock face, this trapped air is compressed and exerts so much pressure that it can eventually break loose blocks of rock.

In the second process of erosion, waves break pieces of rock from the surface being attacked, and these fragments enhance the waves' erosional effectiveness. This process, known as **corrasion**, is similar in some ways to the abrasion of rock by particles carried by glacier ice or wind. A wave loaded with rock fragments, large and small, erodes much more rapidly than a wave of water alone. Some coastal bedrock is also susceptible to chemical action by

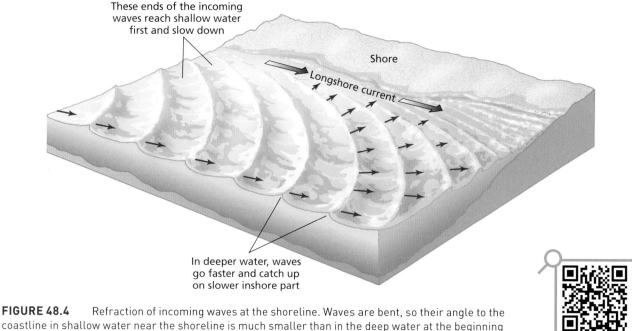

These ends of the incoming
waves reach shallow water
first and slow down

Shore

Longshore current

In deeper water, waves
go faster and catch up
on slower inshore part

FIGURE 48.4 Refraction of incoming waves at the shoreline. Waves are bent, so their angle to the coastline in shallow water near the shoreline is much smaller than in the deep water at the beginning of their approach. However, the inshore angle of the waves is still sufficient to produce a longshore current that flows parallel to the shoreline.

ENHANCE

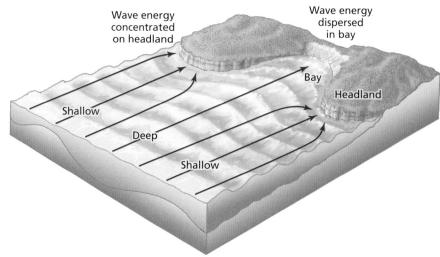

Wave energy
concentrated
on headland

Wave energy
dispersed
in bay

Shallow

Bay

Headland

Deep

Shallow

FIGURE 48.5 As the arrows indicate, the refraction of waves concentrates wave energy on headlands and dissipates it along bay shores. Note how the incoming waves on the sea surface are increasingly distorted as they approach the irregular coastline above a bottom that is deepest in front of the bay.

seawater. The breakdown of coastal bedrock by solution or other chemical means is referred to as *corrosion*.

Erosion of rock by waves is obvious and easy to observe along rocky points and headlands extending into the ocean (Fig. 48.6). This is exactly where we would expect the wave energy to be concentrated because of wave refraction, as discussed earlier (see Fig. 48.5). In bays between rocky headlands, the shoreline looks quite different and is often lined by sandy beaches (see unit-opening photo). This is also predictable because of the lower wave energy affecting a given length of shoreline in the bays. When we compare these two parts of the coastal landscape, it is clear that sediment eroded from the points must be carried away along the shoreline.

Furthermore, the sand accumulated in the bays was brought in from elsewhere; therefore, there must be an effective means of transporting sand and gravel parallel to the coast.

This is even more apparent when we consider the large quantities of sand carried to the coastal zone by rivers. Much of this sand is moved away from river mouths along the shoreline. In fact, there are entire coastal landscapes that lack rocky shorelines at all and are dominated by beaches and chains of sandy islands. All of these coastal landforms are created by sand carried from river mouths or eroding headlands located some distance away. To learn more about this, we next examine the processes that move sediment along the shoreline.

FIGURE 48.6 "On the advice of a Chilean colleague I took the coastal road north from Viña del Mar, and high relief evinces the submergence of this coastline as waves pound the promontories. The rocks themselves show ample evidence of the stresses imposed by tectonic-plate motion in the form of numerous fractures and faults, planes of weakness attacked by the waves. Much of South America's western coastline, from the cliffs of western Colombia to the fjords of southern Chile, reflects the forces of subduction; turn westward, and beneath those onrushing waves lies no gently sloping continental shelf. This is fast-changing scenery."

SHORE-ZONE CURRENTS

The rush of breaking waves toward the shoreline is the most obvious process of water movement through the shore zone. Water that rushes onto the beach cannot simply pile up there, however. Instead, there must be a return flow of some kind out into deeper water. The currents involved in this return flow are very important for movement of sediment. Water flows back out from the beach along the bottom, a kind of current called an *undertow*.

Along many shorelines, though, the flow of water back out to sea occurs in narrow zones with high velocity. Such a narrow, stream-like current flowing seaward from the shore is called a **rip current** (Fig. 48.7A). Standing on the beach, you can sometimes see the effect of a rip current on the surf. Because the rip current advances against the incoming surf, the breakers are interrupted where the rip current encounters them. From a higher vantage point, you may even be able to see the patches of mud carried by the rip current, as contrasted against the less muddy surf (Fig. 48.7B).

Rip currents must be fed by water flowing toward them along the shoreline (see Fig. 48.7A). This flow along the beach forms *feeder* currents, which are a kind of **longshore current**, the general term for flow parallel to the shoreline. Longshore currents may move water equally in both directions toward a rip current (see Fig. 48.7A). Often, though, the longshore currents tend to move water in one predominant direction. Thus, there is an overall direction of flow along the coast.

We can explain why water is driven in a single predominant direction along the shore if we consider what happens when the direction of wave movement is not perpendicular to the shoreline. We already have seen that wave refraction changes the orientation of wave crests so that they more closely parallel the shoreline and waves move more directly toward the shore. However, waves still often strike the shoreline at somewhat of an oblique angle, so that part of their motion is parallel to the shore. This motion tends to drive the flow of water along the coast, generating a longshore current, as shown in Figure 48.4.

Now it should be clear that there is a continual circulation of water in the shore zone as waves move toward the shore: longshore currents carry water along the shoreline,

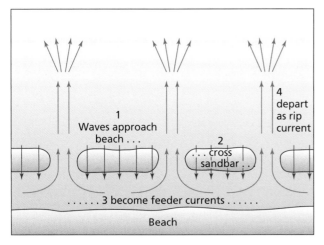

A

FIGURE 48.7 Rip currents are generated by small feeder currents (a kind of longshore current) in the shore zone, which at regular intervals rush seaward directly into the on-coming waves (A). They die out beyond the surf zone, and swimmers caught in them can easily exit to the side. Rip currents have an erosional function, moving the beach particles outward beyond the surf zone. They can be identified as bands of lighter-colored water—resulting from the sediment they carry—that extend offshore from the beach (B).

B

and rip currents return water out into the ocean or lake. Sometimes these circulating water flows only move sediment back and forth along the shore, with no overall transportation in one direction or another. The longshore currents, however, often move in a single predominant direction. When this is the case, sediment is also carried along the shore in that direction.

LONGSHORE DRIFT

The overall flow of water along the shore readily moves sand and other sediment along with it. Therefore, the nearshore currents that we just described contribute to **longshore drift**, the net movement of sediment along the shoreline through the action of waves and nearshore currents.

To understand another important process that contributes to longshore drift, let us return to the waves arriving on the beach at a slight angle. When a wave's swash rushes up the beach carrying sand and gravel, it does so at the angle of the arriving wave. But when the backwash carries sand back toward the sea, it flows straight downward at right angles to the shoreline. The combined effect of this is to move sand along the beach, as shown in Figure 48.8. This process, called **beach drift**, can easily be observed on a beach. Remember that the movement you see in one area of swash and backwash is continuously repeated along the entire length of the beach.

Longshore drift is a critical process in producing and continually reshaping many important landforms in the coastal zone. Along some stretches of the shoreline, more

sediment is carried away by longshore drift than is brought in. As a result, the shoreline is eroded, and beaches lose sand and shrink. When that happens, engineers may build *groins* (walls of rock or other erosion-resistant materials) and other structures at an angle to the beach and out into the surf, hoping to slow the drift of beach sand and slow the process of erosion (Fig. 48.9). Along other stretches of the same coast, longshore drift brings in more sediment than it carries away, building beaches, spits, and other landforms we will discuss in Unit 49.

Tides

So far in this unit, we have discussed the erosional and depositional action of waves that originate in the open ocean and are formed by strong and persistent winds. Certainly, these waves do the bulk of the work in shaping coastal landforms, but other kinds of water movements also play a role.

Tides, the cyclical rise and fall of sea level, occur along all ocean coasts. Their importance in shaping the shoreline varies widely from place to place. The sea level rises and falls twice each day (see Focus on the Science: Tides and Their Behavior). Again, the beach tells the story: what has been washed ashore during the *high tide* lies along the upper limit of the most recent swash, ready for beachcombing during *low tide*. Thus, the entire process of wave motion onto the beach, which we discussed previously, operates while the tides rise and fall. This has the effect of widening the sloping beach. During high tide, the uprushing swash reaches farther landward than it does during low tide;

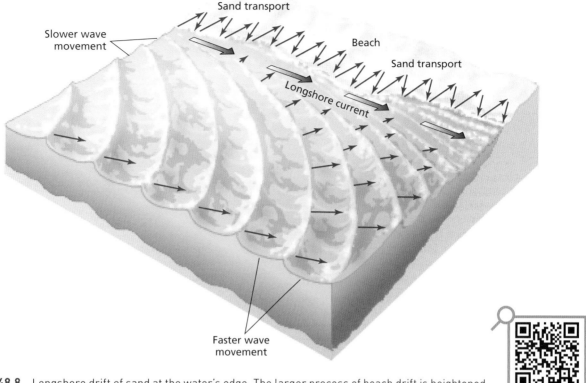

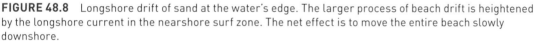

FIGURE 48.8 Longshore drift of sand at the water's edge. The larger process of beach drift is heightened by the longshore current in the nearshore surf zone. The net effect is to move the entire beach slowly downshore.

ENHANCE

during low tide, the backwash reaches farther seaward than it does during high tide.

Along a straight or nearly straight shoreline, the *tidal range* (the average vertical distance between sea levels at high tide and at low tide) may not be large, usually between about 1 and 4 m (3 and 12 ft). But in partially enclosed waters, such as a large bay, the tidal range is much larger. The morphology (shape) of the bay and its entrance affect the range of the tide. Probably the world's most famous tides occur in the Bay of Fundy on Canada's Atlantic Coast; there, the tidal range is as much as 15 m (50 ft), and it creates unusual problems for people living on the waterfront. A similar situation occurs in the bays that surround Moray Firth on Scotland's northeastern coast, as Figure 48.11 vividly demonstrates.

Tides have erosional and depositional functions. In rocky, narrow bays, where the tidal range is great and tides enter and depart with much energy and power, tidal waters erode the bedrock by hydraulic action and corrasion. The changing water level associated with tides sets in motion

FIGURE 48.9 Groins, extending into the surf at a 90-degree angle, mark the beaches of the resort hotels that line the Atlantic shore of Miami Beach. These structures prevent the excessive loss of sand by beach drift, which piles up on the upcurrent side of each groin. The narrowness of the beach, however, indicates that this method is not very effective.

Tides and Their Behavior

Earth's envelope of water—the hydrosphere—covers more than 70 percent of the planet. The surface of this layer of water unceasingly rises and falls in response to forces that affect its global distribution. This cyclical rise and fall of the sea level is known as the *tide* recorded at any given place in the world ocean. The *tidal range* is the vertical difference between sea levels at high tide and low tide.

Three principal forces control Earth's tides: (1) the rotation of the planet, (2) the gravitational pull of the Moon, and (3) the gravitational pull of the Sun. Earth's daily rotation has the effect of countering the gravitational pull of its own mass. The rotational velocity is greatest near the equator and lowest at the poles, so that the layer of water bulges slightly outward toward the equator. This is a permanent condition. But Earth orbits around the Sun and is in turn orbited by the Moon. This means that the gravitational pulls of Moon and Sun come from different directions at different times.

Tidal levels at a coastal location rise and fall rhythmically based on Earth's rotation and the 28-day lunar revolution, which produce two high tides and two low tides within a period slightly longer than 24 hours. When Earth, the Moon, and the Sun are aligned, as shown in Figure 48.10A, the effects of terrestrial rotation, lunar attraction, and the Sun's attraction are combined, and the result is an unusually high tide, or *spring tide*. In contrast, when the Moon's pull works at right angles against the Sun's attraction and the rotational bulge, the result is a *neap tide*, which has the least extreme range (Fig. 48.10B).

Tides play a major role in coastal erosion. They can generate strong tidal currents that rush into and out of river mouths. They also carry waves to higher coastal elevations during spring-tide extremes. When a severe storm attacks a coastline in conjunction with a spring

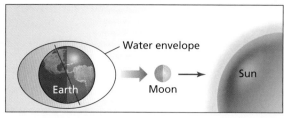

A

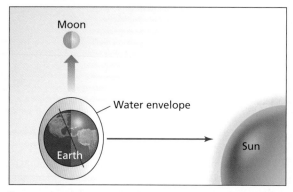

B

FIGURE 48.10 Schematic diagram of tides. (A) Spring tide: Earth's rotational bulge and the gravitational pulls of both the Sun and the Moon combine to produce an unusually high tide. (B) Neap tide: the Moon's gravitational pull is at a right angle to the Sun's pull and the terrestrial rotation bulge, resulting in the least extreme tides. Earth's water envelope and astronomical distances are strongly exaggerated.

tide, substantial erosion may occur. The contribution of tides to shoreline erosion is also influenced by the coastal topography both below the water (shallow and sloping or deep and steep) and above it (long and narrow or wide-curving bays). Compared to the constantly pounding waves, tides are not prominent as coastal modifiers, but their impact is still significant.

tidal currents, which rush through the sandy entrances of bays and lagoons, keeping those narrow thresholds clear of blockage. Where the tidal current slows down, the sediment it carries is deposited in a fan-shaped, delta-like formation, a feature we will examine further in Unit 49.

A *tidal bore* is created when a rapidly rising high tide creates a wave front that runs up a river or bay. A dramatic example is sometimes seen in the lower course of Brazil's Amazon River, where the tide is known to rush in like a foamy breaker that never collapses forward. This breaker is

reported to start as a wall of water as much as 8 m (25 ft) high, moving upstream at a rate of 20 km/h (13 mph). It loses its height as it advances upriver but has been observed more than 400 km (250 mi) inland from the Atlantic Coast.

Storms and Crustal Movements

During most of the year, the world's coasts are slowly modified by the erosional and depositional processes described in this unit. Except under special circumstances, these changes are slow to occur. But virtually all coastlines are vulnerable to unusual and rare events that can greatly transform them in a very short period of time. These events are storms. In Unit 13, we discussed the various kinds of storms that can develop over water. In the lower latitudes, tropical cyclones or hurricanes can generate enormous energy in ocean waves. At higher latitudes, storms are most often associated with weather fronts between contrasting air masses.

STORM SURGES

Whatever the source, the powerful winds whipped up in these storm systems spawn large waves that are propelled toward the shoreline, producing an unusually high water level called a **storm surge**. Low atmospheric pressure also causes the water surface to rise somewhat and contributes to the surge. During a storm surge, waves attack coastal-zone areas normally untouched by this kind of erosion.

A single severe storm surge can break through stable, vegetation-covered offshore islands, erode dunes lying well above the normal swash zone, and penetrate kilometers of coastal plain. When such a storm strikes the coast at the time of high tide or during a spring tide, its impact is all the more devastating (see Fig. 49.8). These dramatic events notwithstanding, physical geographers disagree as to the long-term geomorphic effect of severe storms. It is certainly true that much of the coast and shore eventually return to pre-storm conditions after a surge. But some effects of the storm may be long lasting, if not permanent.

CRUSTAL MOVEMENT

When we study coastal processes, we must be mindful not only of the many and complex marine processes discussed in this unit but also of the vertical mobility of Earth's crust. When we first observe a section of coastline, we classify it as a coast of erosion or deposition depending on the dominant landforms we identify there. These observations are based on the present appearance of the coast and the prevailing processes now at work.

Over the longer term, however, the coastal bedrock may be rising (relative to the sea level) or sinking. For instance, we have learned that the Scandinavian Peninsula is undergoing isostatic rebound following the melting of the heavy ice sheet that covered it until recently (see Unit 44). This means that its coasts are rising as well. Along the coastlines of Norway and Sweden, therefore, we should expect to find evidence of marine erosion now elevated above the zone where wave processes are currently taking place. In other areas, the coast is sinking. The coastal zone of Louisiana and Texas shows evidence of subsidence, in part because of the increasing weight of the sediments in the Mississippi Delta, but probably as a result of other causes as well.

Add to this the rising and falling of the sea level associated with glacial and interglacial conditions (see Unit 43),

FROM THE FIELDNOTES

A

B

FIGURE 48.11 "Sailing along the eastern coast of Scotland I noticed numerous signs of large tidal ranges in funnel-shaped bays, reminiscent of what I had seen in Canada's maritime provinces. We sailed toward Invergordon on the Moray Firth at high tide, when the causeway was barely above the water level (A). A field trip to Loch Ness and Fort William took about six hours, and when I walked back across the same causeway afterward, the tide was still going out, exposing a beach at the foot of the seawall (B) and the tall pillars on which the causeway rested. On that day in 1992, the tidal range here was 5.5 m (18 ft), about average for the site."

and we can see that it is impossible to generalize about coasts, even over short stretches. Ancient Greek port cities built on the waterfront just 2,500 years ago are now submerged deep below Mediterranean waters as a result of local coastal subsidence. But in the same region, there are places originally built on the waterfront that are now situated high above the highest waves. Unraveling the marine processes and Earth movements that combine to create the landscapes of coastlines is one of the most daunting challenges of physical geography.

Key Terms

backwash *page 574*
beach drift *page 578*
coast *page 573*
corrasion *page 575*
littoral zone *page 573*
longshore current *page 577*

longshore drift *page 578*
rip current *page 577*
shoaling *page 573*
shore *page 573*
storm surge *page 581*
surf *page 574*

swash *page 574*
tide *page 578*
wave height *page 573*
wave length *page 573*
wave refraction *page 574*
waves of oscillation *page 573*

Scan Here for a quick vocabulary review

Review Questions

1. Why is an understanding of coastal processes an important part of physical geography?
2. What conditions influence the height of waves?
3. Describe the motion of water parcels as a wave passes in the deep ocean. Contrast that motion with what happens as incoming waves enter shallow water.

4. Explain why wave energy is concentrated on headlands or points and why sediment tends to accumulate in bays.
5. What is a longshore current, and what is longshore drift? How are they generated?
6. What are tides? What forces control tides?

Scan Here for a quick concept review

www.oup.com/us/mason

UNIT 49

Coastal Landforms and Landscapes

Natural bridge on Aruba in the southern Caribbean.

OBJECTIVES

- Explain the formation of beaches, coastal dunes, barrier islands, and other depositional coastal landforms
- Identify landforms typical of erosional coastlines

- Relate erosional and depositional processes to a general classification of coastlines
- Briefly discuss coastal landforms produced by living organisms

In Unit 48, we reviewed the numerous processes that contribute to the formation of coastal landforms and landscapes. In this unit, we examine those landscapes themselves and identify the dominant processes at work there. When we see cliffs and caves, we can conclude that erosional activity is paramount. On the other hand, beaches and barrier islands indicate deposition. We therefore divide coastal landforms into two groups: depositional and erosional. However, we should keep in mind that most coastal landscapes display evidence of both deposition and erosion.

Depositional Landforms

Undoubtedly, the most characteristic depositional landform along the coastline is a **beach**, defined as a coastal zone of sediment that is shaped by the action of waves and nearshore currents. This definition indicates that a beach is wider than the part of it we can see above water. On the landward side, a beach begins at the foot of a line of dunes or some other feature, but on the seaward side, it continues beneath the surf. Beaches are constructed from sand and other material derived from both local and distant sources. Most of the beach material in the coastal environment comes from rivers. When streams enter the ocean, the sediments they carry are deposited and transported along the shore by waves and currents. In addition, beach material may be produced locally by the erosion of nearby sea cliffs and the physical breakdown of those particles as they are moved along the shore (see Unit 48).

BEACH DYNAMICS

The character of a beach reflects the nature of the material composing it. Most beaches along the U.S. Atlantic Coast are made of sand; their light color is a result of the quartz fragments that make up the sand grains. In areas where dark-colored igneous rocks serve as the source for beach material, as they do in parts of Hawai'i, beaches are dark colored. Along the coast of northern California, Oregon, and Washington, high-energy waves prevent sand from

accumulating, resulting in gravel-covered beaches (often called *shingle beaches*).

A beach profile has several parts (Fig. 49.1). The most seaward zone is the **nearshore**, which is submerged even during an average low tide. One or more **longshore bars**, submerged ridges of sand parallel to the beach with troughs between them, often develop in this zone. The **foreshore** is the zone that is alternately covered with water during high tide and exposed during low tide. This is the zone of beach drift and related processes. Landward of the foreshore lies the **backshore**, which extends inland from the high-tide waterline to the first line of windblown sand dunes. As Figure 49.1 shows, the backshore includes one or more benches called **berms**, which are beyond the reach of normal wave action but can be the site of erosion and deposition during major storms.

Beach profiles show a considerable amount of variation. Where seasonal contrasts in wave energy are strong, a beach will have one profile during the winter and another during the summer (Fig. 49.2). The summer's long, steady, low waves transport sand from the nearshore zone onto the beach and create a wide summer berm. The overall slope of the beach is relatively gentle. During the following winter, higher and more powerful storm waves erode much of the summer berm away, carrying the sand back to the nearshore zone and leaving a narrower winter berm. Overall, the beach becomes steeper. Thus, a beach displays evidence of erosion as well as deposition.

Beaches are best viewed as open systems characterized by inputs, outputs, and changes in storage. The size of the beach reflects the material in storage and is therefore a measure of the balance between the input and output of sediment moved by waves and nearshore currents. The input of sediment is derived from local erosion, from offshore, and from upshore. Outputs of sediments can occur offshore or downshore, and the width of the beach reflects the magnitude of the inputs and outputs.

Two scenarios illustrate the behavior of a beach. Temporary increases in wave energy associated with a storm, for example, can in certain cases promote temporary

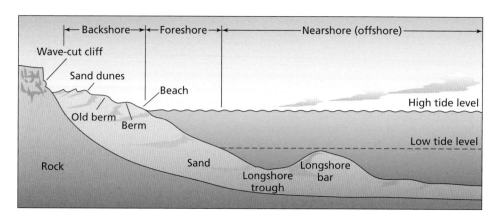

FIGURE 49.1 Parts of a beach in cross-sectional profile. The length of the profile is between 100 and 200 m (330 and 660 ft). Vertical exaggeration is about two to four.

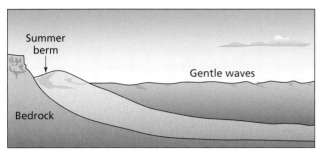

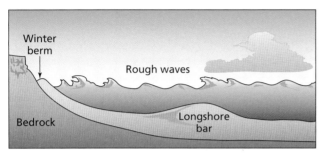

A

B

FIGURE 49.2 Seasonal variation in beach erosion. Gentle summer waves produce a wide berm that slopes gently landward (A). The rougher waves associated with winter storms produce a cold-season beach profile, which shows the summer berm eroded to a narrower, steeper-sloped remnant (B).

erosion (increased output to offshore zones). When wave energy conditions return to normal, offshore material is redeposited on the beach, and it rebuilds to its former configuration. Long-term changes in sediment supply, however, can disrupt this general balance and produce substantial, and rather permanent, changes in the beach. Often, coastlines experience a decline in sediment supply from rivers because of upstream dams and reservoirs. Since the wave energy is not affected by such construction, the decreased sediment supply results in heightened beach erosion. Accordingly, the beach gets smaller as the storage of sediment decreases to reflect the lower inputs from rivers.

The location and distribution of beaches is therefore related to both sediment availability and wave energy. Where the coastal topography is shaped by convergent lithospheric plate movement, coasts are steep, the wave energy is generally high, and beaches are comparatively few. On the mainland of the United States, for example, beaches on the coast of the Pacific Northwest generally are discontinuous, short, and narrow (see Fig. 49.10), but the Gulf and Atlantic coasts, from the Mexican border to New York's Long Island, are almost continuously beach-fringed, and beaches tend to be wide. Where continental shelves are wide, beaches usually are more continuous and well developed. Along high-relief coastlines, where shelves are frequently narrow, beach forms are generally restricted to more sheltered locations, and the material composing them tends to be larger.

COASTAL DUNES

Wind is an important geomorphic agent in coastal landscapes, and many beaches are fringed by sand dunes that are primarily the product of eolian deposition (Fig. 49.3). Coastal zones are subject to strong sea breezes, and sand on the beach berm is constantly being moved landward by these winds. Even when an offshore storm does not generate waves high enough to affect the berm, its winds can move large amounts of sand landward.

In this way, coastal dunes are nourished from the beach. These dunes are often partly stabilized by vegetation (see Fig. 49.3). The ability of plants to quickly stabilize dunes in a humid coastal region is an important difference between beach dunes and desert dunes. Locally, blowouts (wind-eroded hollows) often develop, leading to formation of parabolic dunes (see Unit 47). Overall, though, only the large supply of sand blowing off the beach keeps coastal dunes active in humid regions, and they become completely stabilized at some distance inland.

SANDSPITS AND SANDBARS

We turn now to the depositional landforms of the surf zone and beyond. One of the most characteristic of these is the **sandspit**. When longshore drift occurs and the shifting sediment reaches a bay or a bend in the shoreline, it may form an extension into open water as shown (twice) in the central portion of Figure 49.4. In effect, a spit is an extension of the beach. It begins as a small tongue of sand and grows larger over time. It may reach many kilometers in length and grow hundreds of meters wide, although most spits have more modest dimensions. One well-known sandspit is New Jersey's Sandy Hook, which guards the southernmost entrance to New York City's harbor (Fig. 49.5).

FIGURE 49.3 Vegetation anchors sand dunes at Asilomar State Beach in California.

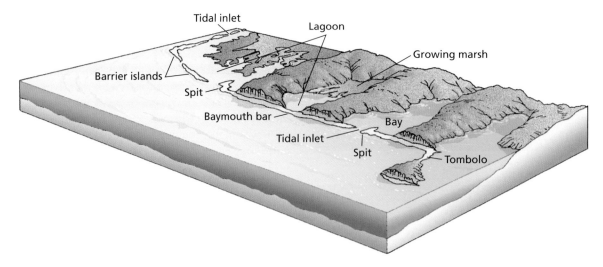

FIGURE 49.4 Common depositional landforms along a stretch of coastline.

Some sandspits continue to grow all the way across the mouth of a bay and become **baymouth bars** (see Fig. 49.4). The bay may simply be an indentation in the coastline, or it may be a river's **estuary** (the lowermost part of a river valley flooded by a rise in sea level). At first, tidal currents will breach the growing bar and keep the bay open (note the tidal inlet located to the right in Fig. 49.4). However, if the longshore drift is strong and sediment plentiful, the bay can be closed off. This has important consequences, because if the tidal action ceases, the bay is no longer supplied with cleansing ocean water. Behind the baymouth bar, the ecology of a bay then changes to that of a swamp or marshland.

A growing sandspit also may form a link between an offshore island and the mainland, creating a landform called a **tombolo** (see Fig. 49.4). Spits, bars, and tombolos take on many different shapes as they bend, curve, and shift during their evolution. Another landform of deposition is the *sandbar*, or **offshore bar**, which is found some distance from the beach and is not connected to land. Recall that we referred to longshore bars in connection with the beach profile (see Fig. 49.1). Such offshore bars expand and contract depending on wave action and sediment supply. Once formed, offshore bars interfere with the very waves that built them. Waves will break against the seaward side of a bar, then regenerate, and break a second time against the shore itself (see Fig. 49.2B).

BARRIER ISLANDS

Along certain stretches of Earth's coasts lie large and permanent offshore bars appropriately called **barrier islands**. These islands are made of sand, but they reach heights of 6 m (20 ft) above sea level and average from 2 to 5 km (1 to 3 mi) in width. They can lie up to 20 km (13 mi) from the coast but more commonly are about half that distance from shore, and they often stretch for dozens of kilometers, unbroken except by tidal inlets. The offshore barrier island

FIGURE 49.5 Sandy Hook (center), which protrudes for nearly 8 km (5 mi) into Lower New York Bay, marks the northern terminus of the New Jersey shore. In the hazy distance lies New York City, whose outer borough of Staten Island forms the coastline visible at the center left.

strip (known locally as the Outer Banks) that forms North Carolina's Cape Hatteras is a classic example (Fig. 49.6).

Geomorphologists have debated the origin of barrier islands for more than 100 years. The simplest view is that they formed as spits that grew across large bays and were eventually broken up by inlets to form chains of islands. One strength of this hypothesis is that we can see spits actively forming today. As sea level rose at the end of the last glacial period (see Unit 43), a great deal of glacially deposited sand and gravel was exposed to wave erosion in areas

FIGURE 49.6 Satellite view of North Carolina's slender Outer Banks, separated from the mainland by Pamlico Sound (left of center). The easternmost point of land near the center is Cape Hatteras, a popular tourist destination highly vulnerable to the vagaries of winds and currents in its often-turbulent marine environment.

A sheltered body of water called a *lagoon* almost always separates the barrier island from the mainland (see Fig. 49.4).

Because of the recreational and other opportunities they provide, which often attract intensive development (see Fig. 48.9), barrier islands are of more than geomorphic interest. Several cities, including Miami Beach, Galveston, and Atlantic City, and many smaller towns have developed on these strips of sand. Numerous long stretches of barrier islands extend from southern Texas to New York along the Gulf of Mexico and lower Atlantic Coast of the United States. Between these islands and the mainland lies the Intracoastal Waterway, an important artery for coastwise shipping.

Being low and exposed, barrier islands are vulnerable to hurricanes, and severe storm waves can temporarily erase parts of them. The vulnerability of an island to severe erosion depends in part on the height of the storm surge (see Unit 48) relative to the elevation of the island's dune ridges above sea level. Sea levels are expected to rise over the next century as warming global temperatures lead to melting of high-latitude glacier ice (Unit 43). As a consequence, storm surges will increasingly be able to overtop parts of barrier islands. In heavily developed areas, the hazards are particularly obvious (see Perspectives on the Human Environment: Hazards of Barrier Island Development).

Erosional Landforms

Where wave erosion (rather than deposition) is the dominant coastal process, a very different set of landforms develops. Exposed bedrock, high relief, steep slopes, and deep water are key features of this terrain. If there are islands, they are likely to be rocky remnants of the retreating coast, not sandy embankments built in shallow water. In Unit 48, we discussed the processes of erosion by waves, most importantly by hydraulic action and corrasion. Somewhat like a river adjusting toward a graded profile, wave erosion works to straighten an indented coastline. Wave refraction concentrates erosional energy on the headlands that stick out into the water, while sedimentary material collects in the concave bends of bays; we can observe the beginning of this process in Figure 48.5.

The sequence of events that follows is depicted in Figure 49.9. When headlands (Fig. 49.9A) are eroded by

such as the Atlantic coasts of the northeastern United States and northern Europe. As a result, barrier island formation could have occurred fairly rapidly as the eroded sediment moved along shorelines and formed swiftly growing spits. An alternative interpretation is that barrier islands had their origins as offshore bars during the last glaciation, when the sea level was much lower than it is today. As sea level rose, these offshore bars migrated coastward, growing as they shifted.

Regardless of how barrier islands formed, they have developed a distinctive profile. They are characterized by a gently sloping beach on the seaward side, a wind-built ridge of dunes in the middle, and a zone of natural vegetation (e.g., shrubs, grasses, mangroves) on the landward side (Fig. 49.7).

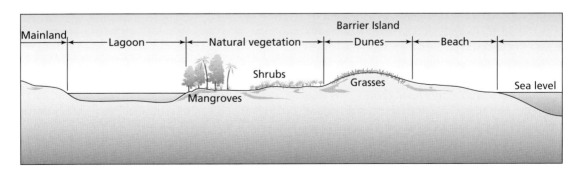

FIGURE 49.7 Cross-sectional profile of a barrier island and adjoining lagoon. Vertical exaggeration is about two times.

Hazards of Barrier Island Development

From Bar Harbor, Maine, to the mouth of the Rio Grande at Brownsville, Texas, more than 4,000 km (2,500 mi) away, 295 barrier islands lie along much of the U.S. Atlantic and Gulf of Mexico coastlines. Given Americans' love of the seashore—and the fact that more than half of them reside within an hour's drive of a coast—it should come as no surprise that most of these islands have witnessed the development of long oceanside strips of summer homes, resorts, high-rise condominiums, commercial and tourist facilities, fishing piers, and public beaches.

Barrier islands are dynamic landforms, affected by longshore drift, storm surges, and rising sea level. The myriad structures built on loose sand atop these islands therefore are often threatened by erosional processes, which are strongest on the island's seaward side, directly facing the most desirable beachfront development sites.

To protect themselves, barrier island communities construct seawalls parallel to the beach and groins perpendicular to it, stretching outward into the surf (see Fig. 48.9). Some communities also spend lavishly to replenish beaches by pumping in massive amounts of new sand from offshore, deeper-water sources. These measures, unfortunately, buy only a few years' time and may worsen erosion in the longer run. Seawalls block the onrushing surf but also prevent much new sand deposition—leading the beaches to soon disappear. Groins are more successful at trapping incoming sand, but studies have shown that saving one beach usually occurs at the expense of destroying another one nearby. As for replenishing shrinking beaches by pumping in sand, all the evidence shows that this, too, is only a stopgap measure. For the past quarter-century, the U.S. Army Corps of Engineers has shifted massive amounts of beach sand along 800 km (500 mi) of the eastern seaboard, with little more than a $10-billion expenditure to show for its herculean efforts.

The greatest hazard to these low-lying offshore islands is the 30 or so cyclonic storms that annually move over the East Coast. Even a moderate-strength midlatitude cyclone can accentuate coastal erosion processes to the point at which major property losses occur. The gravest of such threats to these vulnerable sand strips, of course, are the occasional bigger storms, which are long remembered by local residents. A particularly severe late-winter cyclone in 1962 smashed its way northward from Cape Hatteras

FIGURE 49.8 The aftermath of the March 1962 storm on a barrier island just off the central south shore of New York's Long Island. For 50 years this was widely remembered as one of the Mid-Atlantic region's worst coastal storms—that is, until October 29, 2012, when Hurricane Sandy unleashed an unprecedented scale of devastation centered on the New Jersey–New York coastal zone.

to New England, leaving in its wake a reconfigured coastline (as the flooding ocean created new inlets across barrier islands) and nearly $1 billion worth of storm damage in today's dollars (Fig. 49.8).

Even worse devastation is associated with hurricanes. After Hurricane Camille (one of the most powerful of the past century) attacked the Louisiana and Mississippi Gulf Coast in 1969, the United States was spared this kind of awesome damage for 20 years. Most important, the uncharacteristic lull of the 1970s and 1980s was accompanied by the largest coastal construction boom in history. Then Hurricane Hugo roared across the barrier islands of the South Carolina shoreline in 1989. Farther north, in 2012, the barrier islands of New Jersey received some of the worst damage inflicted as a result of Sandy (which was originally a tropical cyclone but was strengthened by merging with a midlatitude

At least among those who have studied the problem from an objective, scientific perspective, the realization is dawning that nature is certain to win the battle of the barrier islands in the end—despite the best efforts of policymakers, planners, and coastal engineers to manage the precarious human presence that has cost so many billions of dollars to put into (temporary) place.

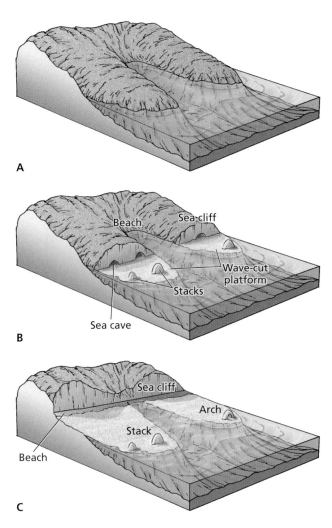

FIGURE 49.9 Straightening an indented coastline by wave erosion. When erosion is the dominant coastal process, a very different set of landforms develops (compare to the depositional landforms in Fig. 49.4).

waves, steep **sea cliffs** develop (Fig. 49.9B). Waves cannot reach the upper parts of these cliffs, but they vigorously erode their bottom part, especially along joints, layers of softer strata, and other weaknesses. The waves often create **sea caves** near the base of the cliff by undercutting it. Soon the overhanging part of the cliff collapses, so that wave action combines with mass movement to erode the coastal bedrock.

The cliff continues to retreat. A **wave-cut platform** develops at the foot of the cliff (see Fig. 49.9B), marking its recession. These platforms are bedrock benches that slope very gently seaward. At low tide, we can see boulders and cobbles broken from the cliff lying on this platform. Soon the waves of a high tide (or a storm) will hurl these fragments back against the cliff face. As the headlands retreat, certain parts will invariably prove to be more resistant than others.

Sections of the headlands survive as small islands, and as wave erosion continues, these islands are sometimes penetrated at their base and become **sea arches** (see Fig. 49.9C). Other remnants of the headlands stand alone as columns called **stacks** (see Figs. 49.9B and C). Arches (see unit-opening photo) and stacks are typical of coastlines that are being actively eroded, but they are temporary features. Soon they, too, will be eroded down to the level of the wave-cut platform. Eventually, the headlands are completely removed, as are the beaches at the heads of the bays. A nearly straight, retreating cliff eventually marks the entire coastal segment (see Fig. 49.9C), with a portion of the wave-cut platform, covered by sediment, at its base.

The speed of cliff retreat depends on a number of conditions, including the power of the waves and, importantly, the resistance of the coastal rocks. Cliffs can form in any coastal rock, ranging from hard crystalline rocks to soft and loose glacial deposits. England's famous White Cliffs of Dover are cut from chalk. Tall cliffs in the Hawai'ian Islands are carved from volcanic rocks. In parts of New England, particularly on Cape Cod, the coast is formed by glacial deposits, and this loose material retreats as much as 1 m (3.3 ft) per year. Along parts of the U.S. West Coast, deeper water and soft bedrock combine to produce an even faster rate of retreat, but in other segments hard crystalline rocks wear away at a much slower rate (Fig. 49.10).

Coastal Landscapes

The events we just described constitute a model for the evolution of an erosional shoreline from an irregular shape to a straight, uniform line of cliffs. The reason so few coastal landscapes resemble this model is that many of them have experienced marked fluctuations in the conditions that shape shoreline evolution. Recent sea-level fluctuations, as well as tectonic movements along plate margins, continually disrupt the operation of coastal processes. With this in mind, it is convenient to distinguish between two general types of coastlines: *emergent* and *submergent* coasts.

FIGURE 49.10 "A short climb to an overlook near the Oregon–California border provides an instructive perspective over an active segment of the Pacific coastline here. Deep water, unobstructed fetch, and pounding waves drive back the shore along steep cliffs. Note the virtual absence of stacks, suggesting that the rocks being attacked here are not as resistant to erosion as the harder crystallines seen farther north. The owners of those apartments must have dramatic ocean views, but future generations will face the reality of nature's onslaught."

EMERGENT COASTS

The landscapes of *uplifted* or **emergent coasts** carry the imprints of elevation by tectonic forces. Some coastal zones have been uplifted faster than postglacial sea levels have risen. The net effect of this is that such features as cliffs and wave-cut platforms are raised above today's sea level, sometimes by tens of meters (Fig. 49.11). When raised this way, a wave-cut platform is termed a **marine terrace**. Occasionally, landforms such as stacks still stand on the uplifted marine terrace, confirming the origin of this landform.

Depositional coastal landforms—dunes, beaches, bars, and spits—may also be uplifted above the level of wave action. These features are more rapidly erased by erosional processes than are bedrock cliffs and wave-cut platforms. Sometimes cultural features reveal recent uplift. Stone structures of coastal settlements (including docks) may survive longer than soft sedimentary landforms. When such settlements are well above the water, we can conclude that uplift has occurred. Some Maya buildings, constructed on the waterfront more than 1,000 years ago, now lie elevated on uplifted segments of the Mexican and Central American coasts.

SUBMERGENT COASTS

Other coastlines are **submergent coasts**—that is, these coasts are *drowned* rather than uplifted. This submergence was caused in large part by the rise of sea level over the past 10,000 years as water that had been held in large ice sheets during the last glacial period returned to the oceans. At the peak of the last glacial period, sea level stood more than 120 m (400 ft) below its present average mark. This exposed large parts of the continental shelves that are now under water. Rivers flowed across these areas of dry land just as they do today across the coastal plain, eroding their valleys to the edge of the ocean.

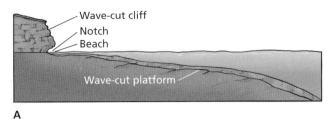

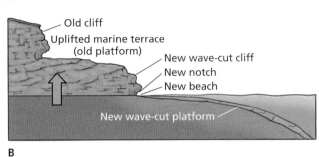

FIGURE 49.11 A wave-cut platform (A) is transformed into an uplifted marine terrace (B) when tectonic uplift elevates the coastal zone above the existing sea level.

this process can create an extensive network of coast-fringing ridges. To grow, corals need clear water, warm temperatures, and vigorous cleansing wave action. This wave action does erode the reef, but it also washes the growing coral.

Where conditions are favorable, wide, flat-topped coral reefs develop. Sometimes, they are attached to the shore; others lie offshore and create lagoons between the shore and the reef. Still other corals create **atolls**, roughly circular reefs that surround a lagoon without any land in the center. The round shape of atolls, many of which are found in the lower latitudes of the Pacific Ocean, was studied more than 150 years ago by Charles Darwin. He concluded that the corals probably grew on the rims of eroded volcanic cones. As the flattened volcano subsided (or sea level rose), the corals continued to build upward (Fig. 49.12).

Vegetation also influences the evolution of shorelines. In parts of West Africa, Southeast Asia, and the southeastern United States, *mangroves* and their elaborate root systems have become builders of shorelines. Once these unique plants have taken hold, the erosional power of

The courses of many such rivers, in fact, can be traced from their present mouths across the continental shelf to their former outlets. When glacial melting raised global sea levels by many meters, these marginal areas were submerged, and the river valleys became submarine canyons. The water rose quite rapidly until about 7,000 years ago. Over the past seven millennia, sea level has continued to rise in many parts of the world, but slowly enough for well-developed coastal landforms to develop.

In the drowned river mouths, we continue to witness the effects of submergence. Rivers often flow into these estuaries and leave no doubt regarding their origins. In some places, the tops of nearly submerged hills rise as small islands above the water within the estuary. If sea levels were still rising rapidly, this invasion of river valleys by advancing ocean water would continue today. However, the rise of sea level has slowed sufficiently to permit longshore drift to form large spits and bars across many estuaries.

Evidence of submergence also can be seen on more rugged, mountainous coastlines, especially where U-shaped troughs were carved by glaciers reaching the ocean during past glacial periods. Since deglaciation, rising water has filled the troughs to form fjords, or drowned glacial valleys, with deep waters and sheer and often-spectacular valley sides (see Fig. 45.6C). Submergent coastlines therefore display varied landscapes.

Living Shorelines

Living organisms, such as corals, algae, and mangroves, can shape or affect the development of shores and coasts. A **coral reef** is built by tiny marine organisms that discharge calcium carbonate. New colonies of organisms build on the marine limestone deposits left by their predecessors, and

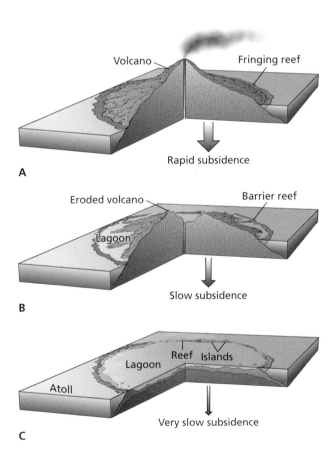

FIGURE 49.12 Relationship between coral atoll reefs and marine volcanoes. The coral reef originally develops around the rim of a subsiding volcanic cone (A). As the cone erodes, the corals continue to build upward, leaving a lagoon surrounded by a ring-like reef (B). When the cone has disappeared below the water, the circular atoll is the only feature remaining at the ocean surface (C).

waves and currents is harnessed by their root systems. As a result, a zone of densely vegetated mud flats develops, creating a unique ecological niche.

As these and the other examples in this unit illustrate, the landforms and landscapes of coastlines are formed and modified by many processes and conditions. No two stretches of shoreline are exactly alike because the history of a coastal landscape involves a unique combination of waves, tides, currents, wind, sea-level change, and crustal movement.

Key Terms

atoll *page 591*

backshore *page 584*

barrier island *page 586*

baymouth bar *page 586*

beach *page 584*

berm *page 584*

coral reef *page 591*

emergent coast *page 590*

estuary *page 586*

foreshore *page 584*

longshore bar *page 584*

marine terrace *page 590*

nearshore *page 584*

offshore bar *page 586*

sandspit *page 585*

sea arch *page 589*

sea cave *page 589*

sea cliff *page 589*

stack *page 589*

submergent coast *page 590*

tombolo *page 586*

wave-cut platform *page 589*

Scan Here for a quick vocabulary review

Review Questions

1. Describe a typical beach profile. Why do different beach profiles develop in summer and winter?

2. How do sandspits and baymouth bars form?

3. Describe the parts of a typical barrier island and explain the problems associated with development on these landforms.

4. Describe the processes by which irregular, embayed coastlines become straightened.

5. What is the fundamental reason for the existence of many submergent coastlines throughout the world?

Scan Here for a quick concept review

www.oup.com/us/mason

APPENDIX SI and Customary Units and Their Conversions

This appendix provides a table of units and their conversion from older units to Standard International (SI) units.

Length

Metric Measure

1 kilometer (km)	= 1000 meters (m)
1 meter (m)	= 100 centimeters (cm)
1 centimeter (cm)	= 10 millimeters (mm)

Nonmetric Measure

1 mile (mi)	= 5,280 feet (ft)
	= 1760 yards (yd)
1 yard (yd)	= 3 feet (ft)
1 foot (ft)	= 12 inches (in)
1 fathom (fath)	= 6 feet (ft)

Conversions

1 kilometer (km)	= 0.6214 mile (mi)
1 meter (m)	= 3.281 feet (ft)
	= 1.094 yards (yd)
1 centimeter (cm)	= 0.3937 inch (in)
1 millimeter (mm)	= 0.0394 inch (in)
1 mile (mi)	= 1.609 kilometers (km)
1 foot (ft)	= 0.3048 meter (m)
1 inch (in)	= 2.54 centimeters (cm)
	= 25.4 millimeters (mm)

Area

Metric Measure

1 square kilometer (km^2)	= 1,000,000 square meters (m^2)
	= 100 hectares (ha)
1 square meter (m^2)	= 10,000 square centimeters (cm^2)
1 hectare (ha)	= 10,000 square meters (m^2)

Nonmetric Measure

1 square mile (mi^2)	= 640 acres (ac)
1 acre (ac)	= 4840 square yards (yd^2)
1 square foot (ft^2)	= 144 square inches (in^2)

Conversions

1 square kilometer (km^2)	= 0.386 square mile (mi^2)
1 hectare (ha)	= 2.471 acres (ac)
1 square meter (m^2)	= 10.764 square feet (ft^2)
	= 1.196 square yards (yd^2)
1 square centimeter (cm^2)	= 0.155 square inch (in^2)
1 square mile (mi^2)	= 2.59 square kilometers (km^2)
1 acre (ac)	= 0.4047 hectare (ha)
1 square foot (ft^2)	= 0.0929 square meter (m^2)
1 square inch (in^2)	= 6.4516 square centimeters (cm^2)

Volume

Metric Measure

1 cubic meter (m^3)	= 1,000,000 cubic centimeters (cm^3)
1 liter (l)	= 1000 milliliters (ml)
	= 0.001 cubic meter (m^3)
1 milliliter (ml)	= 1 cubic centimeter (cm^3)

Nonmetric Measure

1 cubic foot (ft^3)	= 1728 cubic inches (in^3)
1 cubic yard (yd^3)	= 27 cubic feet (ft^3)

Conversions

1 cubic meter (m^3)	= 264.2 gallons (U.S.) (gal)
	= 35.314 cubic feet (ft^3)
1 liter (l)	= 1.057 quarts (U.S.) (qt)
	= 33.815 fluid ounces (U.S.) (fl oz)
1 cubic centimeter (cm^3)	= 0.0610 cubic inch (in^3)
1 cubic mile (mi^3)	= 4.168 cubic kilometers (km^3)
1 cubic foot (ft^3)	= 0.0283 cubic meter (m^3)
1 cubic inch (in^3)	= 16.39 cubic centimeters (cm^3)
1 gallon (gal)	= 3.784 liters (l)

Mass

Metric Measure

1000 kilograms (kg)	= 1 metric ton (t)
1 kilogram (kg)	= 1000 grams (g)

Nonmetric Measure

1 short ton (ton)	= 2000 pounds (lb)
1 long ton	= 2240 pounds (lb)
1 pound (lb)	= 16 ounces (oz)

Conversions

1 metric ton (t)	= 2205 pounds (lb)
1 kilogram (kg)	= 2.205 pounds (lb)
1 gram (g)	= 0.03527 ounce (oz)
1 pound (lb)	= 0.4536 kilogram (kg)
1 ounce (oz)	= 28.35 grams (g)

Pressure

standard sea-level air pressure	= 1013.25 millibars (mb)
	= 14.7 lb/in^2

Temperature

To change from Fahrenheit (F) to Celsius (C)

$$°C = °F - 32/1.8$$

To change from Celsius (C) to Fahrenheit (F)

$$°F = °C \times 1.8 + 32$$

Energy and Power

1 calorie (cal)	= the amount of heat that will raise the temperature of 1 g of water 1°C (1.8°F)
1 joule (J)	= 0.239 calorie (cal)
1 watt (W)	= 1 joule per second (J/s)
	= 14.34 calories per minute (cal/min)

PRONUNCIATION GUIDE

Aa (AH-AH)
Ablation (uh-BLAY-shunn)
Abyssal (uh-BISSLE)
Acacia (uh-KAY-shuh)
Aconcagua (ah-konn-KAH-gwah)
Adiabatic (addie-uh-BATTICK)
Adriatic (ay-dree-ATTIK)
Agassiz (AG-uh-see)
Albedo (al-BEE-doh)
Albuquerque (ALBA-ker-kee)
Aleutian (uh-LOOH-shun)
Algae (AL-jee)
Algeria (al-JEERY-uh)
Amalfi (ah-MAHL-fee)
Ameliorate (uh-MEEL-yer-ate)
Amensalism (uh-MEN-sull-ism)
Amino (uh-MEE-noh)
Anion (AN-eye-on)
Antarctica (ant-ARK-tick-uh)
Antimony (ANN-tuh-moh-nee)
Antipodally (an-TIPPUD-lee)
Aphelion (ap-HEELY-un)
Appalachia (appa-LAY-chee-uh)
Aqaba (AH-kuh-buh)
Aquiclude (AK-kwuh-cloode)
Aquifer (AK-kwuh-fer)
Archipelago (ark-uh-PELL-uh-goh)
Arête (uh-RETT)
Aridisol (uh-RID-ih-sol)
Armero (ahr-MAIR-roh)
Arroyo (uh-ROY-yoh)
Artois (ahr-TWAH)
Aruba (ah-ROO-bah)
Asthenosphere (ass-THENNO-sfeer)
Aswan (as-SWAHN)
Atacama (ah-tah-KAH-mah)
Atoll (AT-tole)
Aurora borealis (aw-ROAR-ruh baw-ree-ALLIS)
Autotroph (AW-doh-troaf)
Azores (AY-zoars)
Azoric (ay-ZOAR-rick)
Azurite (AZH-uh-rite)

Bahamas (buh-HAH-muzz)
Bahía (buh-HEE-uh)
Baja (BAH-hah)
Bajada (buh-HAH-dah)
Bali (BAH-lee)
Bangladesh (bang-gluh-DESH)
Baobab (BAY-oh-bab)
Baotou (bau-TOH)
Barchan (BAR-kan)
Barents (BARRENS)
Barnaul (bar-nuh-OOL)
Basal (BASE-ull)

Basalt (buh-SALT)
Baykal (bye-KAHL)
Beijing (bay-ZHING)
Bengal (BENG-gahl)
Benguela (ben-GWAY-luh)
Bergeron (BEAR-guh-roan)
Bering (BEH-ring)
Berkshire (BERK-sheer)
Bhuj (BOODGE)
Biome (BYE-ohm)
Biota (bye-OH-tuh)
Boise (BOY-zee)
Bonneville (BON-uh-vill)
Borneo (BOAR-nee-oh)
Brahmaputra (brahm-uh-POOH-truh)
Breccia (BRETCH-ee-uh)
Buoy (BOO-ee)
Butte (BYOOT)
Bylot (BYE-lot)

Cairo (KYE-roh)
Caldera (kal-DERRA)
Calve (KALV)
Calving (KAL-ving)
Camille (kuh-MEEL)
Capri (kuh-PREE)
Caribbean (kuh-RIB-ee-un/karra-BEE-un)
Caribou (KARRA-boo)
Castel Sant'Angelo (kuh-STELL sunt-UN-jeh-loh)
Catena (kuh-TEENA)
Cation (CAT-eye-on)
Caucasus (KAW-kuh-zuss)
Celebes (SELL-uh-beeze)
Celsius (SELL-see-us)
Cenozoic (senno-ZOH-ik)
Chang Jiang (chung-jee-AHNG)
Chaparral (SHAP-uh-RAL)
Chengdu (chung-DOO)
Cherrapunji (cherra-POON-jee)
Chesapeake (CHESSA-peek)
Chile (CHILLI/CHEE-lay)
Chimborazo (chim-buh-RAH-zoh)
Chinook (shin-NOOK)
Chugach (CHEW-gash)
Chukotskiy (chuh-KAHT-skee)
Cirque (SERK)
Cirrus (SIRRUS)
Clayey (CLAY-ee)
Cocos (KOH-kuss)
Conduit (KONN-doo-it)
Coniferous (kuh-NIFF-uh-russ)
Contentious (kon-TEN-shuss)
Coriolis, Gustave-Gaspard (kaw-ree-OH-liss, goo-STAHV gus-SPAH)
Corrasion (kaw-RAY-zhunn)

Costa Rica   (koss-tuh REE-kuh)
Cretacious   (kreh-TAY-shuss)
Crete   (KREET)
Crevasse   (kruh-VASS)
Croatia   (kroh-AY-shuh)
Crystalline   (KRISS-tuh-leen)
Cuesta   (KWESTA)
Cumulonimbus   (kyoo-myoo-loh-NIMBUS)
Cumulus   (KYOO-myoo-luss)
Cvijic, Jovan   (SVEE-itch, yoh-VAHN)

Dakar   (duh-KAHR)
Davisian   (duh-VISSY-un)
Debris   (deh-BREE)
Deciduous   (deh-SID-yoo-uss)
Deltaic   (del-TAY-ik)
Denali   (deh-NAH-lee)
Deoxyribonucleic   (day-oxie-RYBO-noo-CLAY-ik)
Desertification   (deh-ZERT-iff-uh-KAY-shun)
Diurnal   (dye-ERR-nul)
Dokuchayev, Vasily   (dock-koo-CHYE-eff, VAH-zilly)
Drakensberg   (DRAHK-unz-berg)
du Toit   (doo-TWAH)

Echidna   (eh-KID-nuh)
Ecliptic   (ee-KLIP-tick)
Ecuador   (ECK-wah-dor)
Ecumene   (ECK-yoo-mean)
Edaphic   (ee-DAFFIK)
Eemian   (EE-mee-un)
El Faiyum   (el-fye-YOOM)
El Niño   (el-NEEN-yoh)
Ellesmere   (ELZ-meer)
Ely   (EELY)
Emu   (EE-myoo)
Endemism   (en-DEM-izzum)
Eolian   (ee-OH-lee-un)
Ephemeral   (ee-FEMMA-rull)
Epiphyte   (EPPY-fite)
Epoch   (EH-pok)
Equinox   (EE-kwuh-nox)
Eratosthenes   (eh-ruh-TOS-thuh-neeze)
Estuary   (ESS-tyoo-erry)
Eucalyptus   (yoo-kuh-LIP-tuss)
Euphorbia   (yoo-FOR-bee-uh)

Fahrenheit   (FARREN-hite)
Falkland   (FAWK-lund)
Felsenmeer   (FEL-zen-mair)
Fenneman   (FEN-uh-mun)
Fiji   (FEE-jee)
Filchner-Ronne   (FILK-ner ROH-nuh)
Findeisen   (FIN-dyzen)
Firn   (FERN)
Fissure   (FISHER)
Fjord   (FYORD)
Foehn   (FERN ["r" silent])
Franz Josef   (frahnss YOH-zeff)
Fuji   (FOO-jee)
Fungi   (FUN-jye)

Galapagos   (guh-LAH-pah-guss)
Galena   (guh-LEENA)
Ganges   (GAN-jeeze)
Gaseous   (GASH-uss)
Geiger   (GHYE-guh)
Gelisol   (JELLY-sol)
Geodesy   (jee-ODD-uh-see)
Geostrophic   (jee-oh-STROFFIK)
Geosynchronous   (jee-oh-SIN-krunn-uss)
Geyser   (GUY-zer)
Gibraltar   (jih-BRAWL-tuh)
Gila   (HEE-luh)
Giza   (GHEE-zah)
Gneiss   (NICE)
Gnomonic   (no-MONIC)
Gondwana   (gond-WAHNA)
Goudie   (GOWDY)
Graben   (GRAH-ben)
Granule   (GRAN-yule)
Grebe   (GREEB)
Greenwich   (GREN-itch)
Gros Piton   (groh-peet-TAW)
Guangzi-Zhuang   (GWAHNG-zee JWAHNG)
Guatemala   (gwut-uh-MAH-lah)
Guilin   (gway-LIN)
Guinea   (GHINNY)
Gujarat   (GOO-jah-raht)
Gulkana   (gull-KAN-uh)
Gunz   (GOONZ)
Guyana   (guy-ANNA)
Guyot   (GHEE-oh)
Gypsum   (JIP-sum)
Gyre   (JYER)

Halide   (HAY-lyde)
Halong   (hah-LONG)
Hatteras   (HATTA-russ)
Heimaey   (HAY-may)
Hematite   (HEE-muh-tite)
Herbaceous   (her-BAY-shuss)
Herodotus   (heh-RODDA-tuss)
Hierarchical   (hyer-ARK-ik-kull)
Hilo   (HEE-loh)
Himalayas   (him-AHL-yuzz/himma-LAY-uzz)
Hindu Kush   (HIN-doo KOOSH)
Holocene   (HOLLO-seen)
Huang He   (HWAHNG-HUH)
Huascarán   (wahss-kuh-RAHN)
Humus   (HYOO-muss)
Hygroscopic   (hye-gruh-SKOPPIK)

Icarus   (ICK-uh-russ)
Igneous   (IGG-nee-us)
Indigenous   (in-DIDGE-uh-nuss)
Indonesia   (indo-NEE-zhuh)
In situ   (in-SYE-too)
Inuit   (IN-yoo-it)
Ion   (EYE-on)
Iquitos   (ih-KEE-tohss)
Irrawaddy   (ih-ruh-WODDY)
Isarithmic   (eye-suh-RITH-mik)
Isohyet   (EYE-so-hyatt)

Isostasy   (eye-SOSS-tuh-see)
Iteshi   (ih-TESHI)

Jawa   (JAH-vuh)
Juan de Fuca   (WAHN duh FYOO-kuh)

Kafue   (kuh-FOO-ee)
Kamchatka   (komm-CHUT-kuh)
Kariba   (kuh-REEBA)
Kaskawulsh   (KASS-kuh-WULSH)
Katabatic   (kat-uh-BATTIK)
Kauai   (KOW-eye)
Kazakhstan   (KUZZ-uck-stahn)
Kenya   (KEN-yuh)
Khamsin   (KAM-sin)
Khasi   (KAH-see)
Kibo   (KEE-boh)
Kilauea   (kill-uh-WAY-uh)
Kilimanjaro   (kil-uh-mun-JAH-roh)
Klamath   (KLAM-uth)
Koala   (koh-AH-luh)
Kodiak   (KOH-dee-ak)
Köppen, Wladimir   (KER-pin ["r" silent], VLAH-duh-meer)
Kosciusko   (kuh-SHOO-skoh)
Kotzebue   (kot-seh-BYOO)
Krakatau   (krak-uh-TAU)
Kunming   (koon-MING)
Kurashio   (koora-SHEE-oh)
Kurile   (CURE-reel)
Kuznetsk   (kooz-NETSK)
Kyushu   (kee-YOO-shoo)

La Niña   (lah-NEEN-yah)
Laccolith   (LACK-oh-lith)
Lahar   (luh-HAHR)
Lancashire   (LANKA-sheer)
Langarone   (lahn-gah-ROH-neh)
Lanier   (luh-NEAR)
Laredo   (luh-RAY-doh)
Larvae   (LAR-vee)
Lascaux   (lass-SKOH)
Laurasia   (law-RAY-zhuh)
Laurentian   (law-REN-shun)
Lead   [ice-surface channel] (LEED)
Lemur   (LEE-mer)
Levee   (LEH-vee)
Li   (LEE)
Liana   (lee-AHNA)
Lichen   (LYE-ken)
Linnaean   (LINNY-un)
Littoral   (LITT-uh-rull)
Loess   (LERSS)
Loihi   (loh-EE-hee)

Maasai   (muh-SYE)
Machu Picchu   (mah-CHOO PEEK-choo)
Madagascar   (madda-GAS-kuh)
Mafic   (MAFFIK)
Magellan   (muh-JELLUN)
Malachite   (MALLA-kite)
Malawi   (muh-LAH-wee)
Mali   (MAH-lee)

Manaus   (muh-NAUSS)
Manyara   (mun-YAH-rah)
Maquis   (mah-KEE)
Marsupial   (mar-SOOPY-ull)
Martinique   (mahr-tih-NEEK)
Massif   (mass-SEEF)
Maui   (MAU-ee)
Mauna Kea   (MAU-nuh KAY-uh)
Mauna Loa   (MAU-nuh LOH-uh)
Mauritania   (maw-ruh-TAY-nee-uh)
Mawensi   (mah-WENN-see)
Mawsim   (MAW-zim)
Maya   (MYE-uh)
Meander   (mee-AN-der)
Medieval   (meddy-EE-vull)
Melbourne   (MEL-bun)
Mercalli   (mair-KAHL-lee)
Mercator, Gerhardus   (mer-CATER, ghair-HAR-duss)
Meru   (MAY-roo)
Mesa   (MAY-suh)
Mescal   (MESS-kull)
Mesozoic   (meh-zoh-ZOH-ik)
Mica   (MYE-kuh)
Michoacán   (mitcho-ah-KAHN)
Mindel   (MINDLE)
Minot   (MYE-not)
Mirnyy   (MEAR-nyee)
Mistral   (miss-STRAHL)
Moho   (MOH-hoh)
Mohorovičić, Andrija   (moh-hoh-ROH-vih-chik, ahn-DREEA)
Mohs, Friedrich   (MOZE, FREED-rick)
Mojave   (moh-HAH-vee)
Monadnock   (muh-NAD-nok)
Mont Blanc   (mawn-BLAHNK)
Montreal   (mun-tree-AWL)
Moraine   (more-RAIN)
Moray Firth   (MAW-ray FERTH)
Morisawa   (mawry-SAH-wah)
Myanmar   (mee-ahn-MAH)

Nairobi   (nye-ROH-bee)
Namib   (nah-MEEB)
Namibia   (nuh-MIBBY-uh)
Naples   (NAYPLES)
Nazca   (NAHSS-kuh)
Neap   (NEEP)
Nefudh   (neh-FOOD)
Negev   (NEGGEV)
Nepal   (nuh-PAHL)
Nevado del Ruiz   (neh-VAH-doh del roo-EESE)
New Guinea   (noo-GHINNY)
New Madrid   (noo-MAD-rid)
Ngorongoro   (eng-gore-ong-GORE-oh)
Niche   (NITCH)
Niger [River]   (NYE-jer)
Niger [Country]   (nee-ZHAIR)
Nigeria   (nye-JEERY-uh)
Nkhata   (eng-KAH-tah)
Novaya Zemlya   (NOH-vuh-yuh zem-lee-AH)
Nuée ardente   (noo-AY ahr-DAHNT)
Nunatak   (NOON-uh-tak)

Oahu   (ah-WAH-hoo)
Obsidian   (ob-SIDDY-un)
Oja   (OH-jah)
Okhotsk   (oh-KAHTSK)
Orinoco   (aw-rih-NOH-koh)
Orogenic   (aw-ruh-JENNIK)
Orogeny   (aw-RODGE-uh-nee)
Orographic   (aw-roh-GRAFFIK)
Oscillation   (oss-uh-LAY-shunn)
Ouachita   (wah-CHEE-tuh)
Owasco   (oh-WAH-skoh)
Oxisol   (OCK-see-sol)
Oyashio   (oy-yuh-SHEE-oh)

Pahoehoe   (pah-HOH-hoh)
Paleontology   (pay-lee-un-TOLLO-jee)
Paleozoic   (pay-lee-oh-ZOH-ik)
Pamlico   (PAM-lih-koh)
Pampas   (PAHM-pahss)
Pangaea   (pan-GAY-uh)
Papua   (PAH-poo-uh)
Paricutín   (pah-ree-koo-TEEN)
Patagonia   (patta-GOH-nee-ah)
Pedon   (PED-on)
Pele   (PAY-lay)
Pelée   (peh-LAY)
Pelourinho   (peh-loo-REE-noh)
Penck, Walther   (PENK, VULL-tuh)
Peneplain   (PEEN-uh-plane)
Periglacial   (perry-GLAY-shull)
Petit Piton   (peh-tee-peet-TAW)
Philippines   (FILL-uh-peenz)
Phreatic   (free-ATTIK)
Physiography   (fizzy-OGG-ruh-fee)
Phytogeography   (FYE-toh-jee-OGGRA-fee)
Piedmont   (PEAD-mont)
Pierre (South Dakota)   (PEER)
Pinatubo   (pin-uh-TOO-boh)
Planar   (PLANE-ahr)
Platypus   (PLATT-uh-puss)
Pleistocene   (PLY-stoh-seen)
Pliny   (PLY-nee)
Pliocene   (PLY-oh-seen)
Podocarp   (POD-oh-carp)
Podsol   (POD-zol)
Pompeii   (pom-PAY)
Pontchartrain   (PON-shar-train)
Popocatépetl   (poh-puh-CAT-uh-petal)
Prague   (PRAHG)
Primeval   (pry-MEE-vull)
Puget   (PYOO-jet)
Pyrenees   (PEER-unease)

Quartic   (KWOAR-tik)
Quaternary   (kwah-TER-nuh-ree)
Qinghai-Xizang   (ching-HYE sheedz-AHNG)
Quartzite   (KWARTS-ite)
Quito   (KEE-toh)

Rainier   (ruh-NEER)
Raisz   (ROYCE)
Reykjavik   (RAKE-yah-veek)

Rhine   (RYNE)
Rhône   (ROAN)
Rhumb   (RUM)
Rhyolite   (RYE-oh-lyte)
Richter   (RICK-tuh)
Rio de Janeiro   (REE-oh day zhah-NAIR-roh)
Rio de la Plata   (REE-oh day lah PLAH-tah)
Roche Moutonnée   (ROSH moot-tonn-NAY)
Rondônia   (roh-DOAN-yuh)
Rotorua   (roh-tuh-ROO-uh)
Ruhr   (ROOR)
Ruwenzori   (roo-wen-ZOARY)

Sahel   (suh-HELL)
St. Lucia   (saint LOO-shuh)
St. Pierre   (sah PYAIR)
Saltation   (sawl-TAY-shunn)
Salvador   (SULL-vuh-dor)
San Andreas   (san an-DRAY-us)
San Gorgonio   (san gore-GOH-nee-oh)
San Joaquin   (san wah-KEEN)
San Juan   (sahn HWAHN)
Sangamon   (SANG-guh-mun)
Santorini   (santo-REE-nee)
São Gabriel de Cachoeira   (sau GAH-bree-ell day kah-choh-AY-rah)
Saragosa   (sarra-GOH-suh)
Schist   (SHIST)
Scotia   (SKOH-shuh)
Seine   (SENN)
Seismic   (SIZE-mik)
Seismograph   (SIZE-moh-graff)
Seneca   (SEH-neh-kuh)
Serengeti   (serren-GETTY)
Sesquioxide   (SESS-kwee-OXIDE)
Seychelles   (say-SHELLZ)
Sial   (SYE-al)
Siberia   (sye-BEERY-uh)
Sicily   (SIH-suh-lee)
Sierra Madre Oriental   (see-ERRA mah-dray orry-en-TAHL)
Sierra Nevada   (see-ERRA neh-VAH-dah)
Silicic   (sih-LISS-ik)
Silurian   (sih-LOORY-un)
Sima   (SYE-muh)
Slovenia   (sloh-VEE-nee-uh)
Soledad   (SOLE-uh-dad)
Solifluction   (sol-ih-FLUK-shun)
Solstice   (SOL-stiss)
Solum   (SOH-lum)
Somali   (suh-MAH-lee)
Soufrière   (soo-free-AIR)
Spatial   (SPAY-shull)
Spheroidal   (sfeer-ROY-dull)
Spodosol   (SPODDA-sol)
Sri Lanka   (sree-LAHN-kah)
Stalactite   (stuh-LAK-tite)
Stalagmite   (stuh-LAG-mite)
Steppe   (STEP)
Stratigraphy   (struh-TIG-gruh-fee)
Striation   (strye-AY-shunn)
Suffosion   (suh-FOH-zhunn)
Sulawesi   (soo-luh-WAY-see)

Sulfur   (SULL-fer)
Sumatera   (suh-MAH-truh)
Surficial   (ser-FISH-ull)
Surtsey   (SERT-see)
Susquehanna   (suss-kwuh-HANNA)
Svalbard   (SVAHL-bard)
Swakopmund   (SFAHK-awp-munt)
Swaziland   (SWAH-zee-land)
Synoptic   (sih-NOP-tik)
Syria   (SEARY-uh)

Tagus   (TAY-guss)
Taiga   (TYE-guh)
Taklimakan   (tahk-luh-muh-KAHN)
Talus   (TAY-luss)
Tana   (TAHN-nuh)
Tanganyika   (tan-gan-YEE-kuh)
Tangshan   (tung-SHAHN)
Tanzania   (tan-zuh-NEE-uh)
Tapir   (TAY-per)
Tephra   (TEFF-ruh)
Terrane   (teh-RAIN)
Tertiary   (TER-shuh-ree)
Teton   (TEE-tonn)
Thalweg   (THAWL-weg)
Thames   (TEMZ)
Thera   (THEERA)
Tianjin   (tyahn-JEEN)
Tiber   (TYE-ber)
Tibetan   (tuh-BETTEN)
Tierra del Fuego   (tee-ERRA dale FWAY-goh)
Titanium   (tye-TANEY-um)
Tombolo   (TOM-boh-loh)
Torricelli   (toar-ruh-CHELLY)
Travertine   (TRAVVER-teen)
Trivandrum   (truh-VAN-drum)
Tsavo   (TSAH-voh)
Tselinograd   (seh-LINN-uh-grahd)
Tsunami   (tsoo-NAH-mee)
Tucson   (TOO-sonn)
Tunguska   (toon-GOOSE-kuh)

Turkana   (ter-KANNA)
Tuscarora   (tuska-ROAR-ruh)

Uinta   (yoo-IN-tuh)
Ukraine   (yoo-CRANE)
Ustalf   (YOO-stalf)
Uvala   (oo-VAH-luh)

Vaiont   (vye-YAW)
Valdez   (val-DEEZE)
Venezuela   (veh-neh-SWAY-luh)
Verkhoyansk   (vair-koy-YUNSK)
Vesuvius   (veh-ZOO-vee-us)
Vicariance   (vye-CARRY-unss)
Vietnam   (vee-et-NAHM)
Viña del Mar   (VEEN-yah-del-MAHR)
Viscous   (VISS-kuss)
Viti Levu   (vee-tee-LEH-voo)

Waialeale   (wye-ahl-ay-AHL-ay)
Wasatch   (WAH-satch)
Weddell   (weh-DELL)
Wegener, Alfred   (VAY-ghenner, AHL-fret)
Whataroa   (wotta-ROH-uh)
Willamette   (wuh-LAMMET)
Wrangell   (RANG-gull)
Wrangellia   (rang-GHELLIA)
Wüste   (VISS-tuh)

Xerophyte   (ZERO-fite)
Xian   (shee-AHN)

Yosemite   (yoh-SEM-uh-tee)
Yucatán   (yoo-kuh-TAHN)
Yungay   (YOONG-GYE)
Yunnan   (yoon-NAHN)

Zagros   (ZAH-gross)
Zambezi   (zam-BEEZY)
Zimbabwe   (zim-BAHB-way)
Zoogeography   (ZOH-OH-jee-oggra-fee)

GLOSSARY

The number following an entry indicates the Unit in which it is discussed.

A horizon   Upper soil layer, which is often darkened by organic matter. (21)

Aa   Lava with a jagged, blocky surface formed from the hardening of not especially fluid lavas. (32)

Ablation (glacial)   Processes by which ice is lost from a glacier; melting is most important, but ice is also lost through sublimation (conversion directly to water vapor) and calving of icebergs into a lake or the ocean. (43)

Abrasion (glacial)   Glacial erosion process of scraping, produced by the impact of rock debris carried in the ice upon the bedrock surface below. (43)

Abrasion (stream)   Erosive action of boulders, pebbles, and smaller grains of sediment as they are carried along a river valley. These fragments dislodge other particles along the streambed and banks, thereby enhancing the deepening and widening process. (39)

Absolute zero   Lowest possible temperature at which all molecules cease their motion; occurs at –273°C (–459.4°F) or 0°K (Kelvin). (11)

Abyssal plain   Large zone of relatively low-relief seafloor, constituting one of the deepest areas of an ocean basin. (2)

Accretion   Process whereby bodies of rock derived from other lithospheric plates are attached to the margins of a continental landmass. These geologically consistent fragments can be regional in extent and are known as *terranes*. (31)

Accumulation (glacial)   Processes by which a glacier gains ice. The ice originates as snowfall, and as snow accumulates, lower layers are compressed and also undergo a certain amount of melting and refreezing; this first produces a denser material called *firn*, and eventually results in formation of solid glacier ice. (43)

Acid precipitation   Abnormally acidic rain, snow, fog, dust, smoke or other materials that fall out of the atmosphere. Most is produced by reactions of sulfur and nitrogen oxides that exist as industrial pollutants in the air. (17)

Acidity   For soils, the concentration of hydrogen ions dissolved in water within the soil, usually measured and expressed as soil pH (22)

Active dune   Unstable sand dune in a desert that is continually being shaped and moved by the wind; cannot support vegetation. (47)

Active layer   Soil above the permafrost table that is subject to annual thawing and freezing. (46)

Active volcano   Volcano that has erupted in recorded human history, or alternatively, one that has erupted within the past 10,000 years. (32)

Actual evapotranspiration (AE)   Amount of water that is lost to the atmosphere from the surface by the combination of evaporation and transpiration. (11)

Addition   Soil-layer formation process involving the addition of matter to the soil. A variety of materials can be added including organic matter from plants, dissolved ions in rainfall, dust particles from the air, and sediment eroded from soils farther uphill. (21)

Adiabatic lapse rate   When a given mass of air expands, its temperature decreases. If a parcel of air rises to a higher altitude, it expands and cools adiabatically. Its lapse rate is therefore referred to as an *adiabatic lapse rate*. (7)

Adiabatic process.   A process that does not involve the addition or removal of heat. Over a short period of time a rising or sinking air parcel gains little heat from its surroundings, and it follows an adiabatic process to a close approximation. (7)

Advection   Horizontal movement of material in the atmosphere. (7)

Advection fog   A fog that occurs when fairly moist air moves over a colder surface and loses enough heat to reach the condensation point. (11)

Aerosols   Tiny solid or liquid particles suspended in the atmosphere. (6)

Aggradation   Process in which a stream deposits more sediment than it erodes, raising the elevation of the stream's channel and floodplain and eventually filling the stream's valley. (39)

Air mass   Very large parcel of air (more than 1,600 km [1,000 mi] across) in the troposphere whose density, temperature, and humidity is relatively uniform in the horizontal dimension. Air masses can retain their properties for a week or more and routinely migrate for hundreds of kilometers as distinct entities. (12)

Air pollutants   Substances in the air that can negatively impact organisms or structures. (6)

Albedo   Proportion of incoming solar radiation that is reflected by a surface. Generally speaking the whiter the color of the surface the higher is its albedo (albedo derives from the Latin word *albus*, meaning white). (5)

Aleutian Low   Local name for the northernmost Pacific Ocean's Upper Midlatitude Low, which is centered approximately at 60 degrees N near the archipelago constituted by Alaska's Aleutian Islands. (9)

Alfisol   One of the 12 soil orders of the Soil Taxonomy, found predominantly under forest, but occasionally in savannas or grasslands; characterized by a thin or light-colored **A** horizon, often underlain by a light-colored **E** horizon, sizable clay accumulation in the **B** horizon, and moderate weathering. (23)

Alluvial apron   (*see* **bajada**)

Alluvial fan   Fan-shaped deposit consisting of alluvial material, and sometimes debris flow deposits, located where a mountain stream emerges onto a plain; primarily a desert landform. (41)

Alluvium   Sediment laid down by a stream on its valley floor. (39, 41)

Alpine glacier (*see* **mountain glacier**)

Alpine permafrost Patches of high-altitude permafrost at lower latitudes associated with major highland zones, such as the Tibetan plateau and the Canadian Rockies. (46)

Amensalism Biological interaction in which one species is inhibited by another. (24)

Amplification Increased shaking motion during an earthquake, in a local area where the underlying sediments are soft and water-saturated. Common settings for amplification include river valleys, former lake beds, and areas where a water body has been artificially filled. (33)

Andisol One of the 12 soil orders of the Soil Taxonomy; includes soils with high potential to retain nutrients and water, because they formed in volcanic ash or similar parent materials; found mainly in the Pacific Ring of Fire, Hawai'i, and the world's other volcanic zones. (23)

Angle of incidence (*see* **solar elevation**)

Angle of repose Maximum slope angle at which a material remains at rest and does not flow or slide downslope. Each material has its own angle of repose, which is closely related to the friction within that material that resists motion; greater resistance to motion leads to a greater angle of repose. (37)

Animal range (*see* **range, animal**)

Annual temperature cycle Pattern of temperature change during the course of a year. (7)

Annular drainage Concentric stream pattern that drains the interior of an excavated geologic dome (*see* Fig. 40.11B). (40)

Antarctic Circle Latitude (66.5°S) marking the northern boundary of the Southern Hemisphere portion of the Earth's surface that receives a 24-hour period of sunlight or darkness. (4)

Antarctic Ice Sheet Continental glacier that covers almost all of Antarctica; often considered to be made up of two adjoining ice sheets that behave somewhat independently. (44)

Antecedent stream River crossing a structural feature that would normally impede its flow because the river predates the structure and kept cutting downward as the structure was uplifted around it. (40)

Anthropocene Era of geologic time referring to the period when humanity's impact on global ecosystems has become significant. (19)

Anthropogenic climate change Climate change arising from human activity as opposed to natural processes. (19)**Anticline** Archlike upfold of rock strata with the limbs dipping away from its axis (*see* Fig. 34.12, left). (34)

Anticyclone Atmospheric high-pressure cell. Low-level anticyclones are characterized by sinking motion and spiraling airflow out of the central area of high pressure. In the Northern Hemisphere, winds flow clockwise around an anticyclone; in the Southern Hemisphere, winds flow counterclockwise around an anticyclone. (8)

Antipode Location on the exact opposite point of the near-spherical Earth. The North Pole is the antipode of the South Pole. (2)

Aphelion Point in the Earth's orbit that occurs every July 4, where the distance to the Sun is at its maximum (ca. 152.5 million km [94.8 million mi]). (4)

Aquiclude Impermeable rock layer that resists the flow of groundwater; consists of tightly packed or interlocking particles, such as those in shale. (38)

Aquifer Porous and permeable rock layer that can be at least partially saturated with groundwater. (38)

Aquifer, confined An aquifer that lies between two aquicludes and often obtains its groundwater from a distant area, where it eventually reaches the surface. (38)

Aquifer, unconfined An aquifer that is not separated from the ground surface by an aquiclude and obtains its groundwater from local infiltration; its upper boundary is the water table. (38)

Arctic Circle Latitude (66.5°N) marking the southern boundary of the Northern Hemisphere portion of the Earth's surface that receives a 24-hour period of sunlight or darkness. (4)

Arête Knifelike, jagged ridge that separates two adjacent glaciers or glacial valleys. (45)

Arid (B) climates Dry climates in which potential evapotranspiration exceeds the moisture supplied by precipitation; found in areas dominated by the subtropical high-pressure cells, in the interiors of continents far from oceanic moisture sources, and in conjunction with rainshadow zones downwind from certain mountain ranges. (15)

Aridisol One of the 12 soil orders of the soil taxonomy, and the most widespread on the world's landmasses; consists of dry soil (unless irrigated) associated with arid climates that is light in color and often contains horizons rich in calcium, clay, or salt minerals. (23)

Artesian well One that flows under its own natural pressure to the surface; usually associated with a confined aquifer that is recharged from a remote location where that aquifer reaches the surface (*see* Fig. 38.8). (38)

Artificial levees Ridges or berms constructed along most large rivers in settled areas to confine floods to the channel and protect farmland or towns from flooding. (41)

Ash (volcanic) Small particles that form when blobs of lava are exploded into the air, where they cool and solidify, during a volcanic eruption. Most fall to the ground around the erupting volcano. (32)

Aspect Directional orientation of a steep mountain slope. In the Northern Hemisphere, southerly aspects receive far more solar radiation than do northward-facing slopes, with all the environmental consequences such a differential implies. (5)

Asthenosphere Soft plastic layer of the upper mantle that underlies the lithosphere, which is able to move over it. (29)

Atmosphere Blanket of air that adheres to the Earth's surface and contains the mixture of gases essential to the survival of all terrestrial life-forms as well as liquid and solid particulates. (2)

Atolls Ringlike coral reefs surrounding empty lagoons; grew on the rims of eroded volcanic cones. (49)

Attenuation The weakening of the ground shaking produced by an earthquake with increasing distance from the focus. (33)

Aurora australis Name given to the *aurora borealis* phenomenon that occurs in the Southern Hemisphere's upper-middle and high latitudes (6).

Aurora borealis Vivid sheetlike displays of light in the nighttime sky of the upper-middle and high latitudes in the Northern Hemisphere; caused by the intermittent penetration of the thermosphere by ionized particles. (6)

Autotroph Organism that manufactures its own organic materials from inorganic chemicals. A good example is phytoplankton, the food-producing plants that manufacture carbohydrates. (24)

Avalanche Mass movement of snow and ice. (37)

Axis (Earth's) Imaginary line that extends from the North Pole to the South Pole through the center of the Earth. The planet's rotation occurs with respect to this axis. (4)

Azonal flow *Meridional* (north–south) flow of upper atmospheric winds (poleward of 15 degrees of latitude), particularly the subtropical and Polar Front jet streams. Periodic departures from the zonal (west-to-east) flow of these air currents are important because they help to correct the heat imbalance between polar and equatorial regions. (9)

Azores High Eastern segment of the North Atlantic Ocean's semipermanent subtropical high-pressure cell (often called the Bermuda High) that is usually centered above the Azores Islands; an important component of the North Atlantic Oscillation (NAO). (9)

B horizon Subsoil horizon, which is often marked by the accumulation of dissolved material and suspended particles from above. (21)

Backshore Beach zone that lies landward of the foreshore; extends from the high-water line to the dune line. (49)

Backwash Return flow to the sea of the thinning sheet of water that moved up the beach as swash. (48)

Bajada Coalesced assemblage of alluvial fans that lines a highland front (also known as *alluvial apron*); primarily a desert feature. (41)

Barchan Crescent-shaped sand dune with low arms or "horns" pointing downwind. Convex side of this dune is the windward side. (47)

Barometer Instrument that measures atmospheric pressure; invented by Torricelli in 1643. (8)

Barrier island Permanent offshore elongated ridge of sand, positioned parallel to the shoreline and separated from it by a lagoon. (49)

Basal ice Bottom ice layer of a glacier. (43)

Basalt Fine-grained, dark-colored igneous rock formed by magma reaching the surface before cooling to form an extrusive rock. (27)

Base level Elevational level below which a stream cannot erode its bed. (39)

Batholith Massive body of intrusive igneous rock that has destroyed and melted most of the existing geologic structures it has invaded. (27)

Bay Broad indentation into a coastline. (49)

Baymouth bar Sandspit that has grown across the mouth of a bay. (49)

Beach Coastal zone of sediment that is shaped by the action of waves; constructed of sand and other materials, derived from both local and distant sources. (49)

Beach drift Net movement of sediment laterally along a beach, as it is carried up the beach by swash and back down by backwash. One of the processes involved in longshore drift, which operates to move huge amounts of beach sand downshore in the direction of the longshore current. (48)

Bedload Particles that remain fairly close to the streambed as they are transported. Bedload usually consists of sand, gravel, or boulders, and is moved downstream because of force applied to it by the flowing water. (39)

Berm Flat or gently sloping part of the beach that lies in the backshore beach zone; deposited during storms and beyond the reach of normal daily wave action. (49)

Bermuda High Local name for the North Atlantic Ocean's subtropical high-pressure cell, which is generally centered over latitude 30 degrees N; also known as the Azores High. (9)

Biodiversity Shorthand for *biological diversity*; variety of the Earth's life-forms and the ecological roles they play. (26)

Biogeochemical cycles A complex web of pathways in which elements move between the lithosphere, hydrosphere, atmosphere, and biosphere. (20)

Biogeography Geography of plants (*phytogeography*) and animals (*zoogeography*). (1, 20)

Biological weathering Disintegration of rock minerals via biological means. Bacteria and lichens that allow faster chemical weathering, along with the physical breakage of rocks by plant roots and burrowing animals are all examples. (36)

Biomass Total living organic matter, encompassing all plants and animals, produced in a particular geographic area. (24)

Biome Broadest justifiable subdivision of the plant and animal world, an assemblage and association of plants and animals that forms a regional ecological unit of subcontinental dimensions. (25)

Biosphere Zone of terrestrial life, the habitat of all living things; includes the Earth's vegetation, animals, human beings, and the part of the soil layer below that hosts living organisms. (2)

Biota Total complement of living species found in a given area. (24)

Birdfoot delta Delta with long, narrow projections of land along major tributaries, such as the Mississippi Delta (*see* Fig. 41.13B). (41)

Blockfield A substantial area covered by boulders, without soil cover and often with little or no vegetation; common in Arctic and Alpine regions. Also known as a *rock sea* or *felsenmeer*. (36, 46)

Blocky soil structure Involves irregularly shaped peds with straight sides that fit against the flat surfaces of adjacent peds (*see* Fig. 22.5). (22)

Bluffs Low cliffs that border the outer edges of an alluvium-filled floodplain. (41)

Body wave Seismic wave that travels through the interior of the Earth; consists of two kinds—**P** waves and **S** waves. (29)

Boiling point (water) Key setting in the calibration of temperature scales. On the Celsius scale water boils at 100°C, whereas on the Fahrenheit scale water boils at 212°F. (7)

Bombs Just-solidified gobs of lava that rain down on a volcano during an explosive eruption. They originate as gas-rich magma whose gases explode when they reach the surface, hurling projectiles of solid lava into the air (which soon fall back to the ground). (32)

Bottomset beds   Finest deltaic deposits, laid down near the seaward edge of the delta, where fine particles settle out of deep water. (41)

Braided stream   One that has multiple intertwined channels, separating and then reuniting some distance downstream and giving a "braided" appearance. Braided stream channels tend to have relatively low sinuosity, that is, they do not wind back and forth a great deal as they flow down the stream valley. (41)

Breaker   Forward collapse of the wave crest as an incoming wave encounters the ever-shallower ocean bottom. (48)

Breccia   Occur in clastic sedimentary rocks when pebble-sized fragments in a conglomerate are not rounded but angular and jagged. (28)

Butte   Small, steep-sided, caprock-protected hill, usually found in dry environments; an erosional remnant of a plateau. (40)

C horizon   Bottom soil horizon which is the parent material, little-altered by soil forming processes. (21)

Calcite   A form of calcium carbonate ($CaCO_3$); this common mineral is the main constituent of limestone. (27)

Caldera   Steep-walled, circular volcanic basin usually formed by the collapse of a volcano whose magma chamber emptied rapidly. (32)

Calorie   One calorie is the amount of heat energy required to raise the temperature of 1 g (0.04 oz) of water by 1°C (1.8°F); not the same unit as the calories used to measure the energy value of food (which are 1,000 times larger). (11)

Calving   When an ice sheet enters the sea, the repeated breaking away of the leading edge of that glacier into huge, flat-topped, tabular icebergs. (44)

Canadian High   Local name for northern North America's polar high, centered over far northwestern Canada. (9)

Capillary action   Movement of water from wetter to drier soil, through the soil pores; similar to the way that water is drawn into a sponge. Capillary action often moves water upward through the soil above the water table. (38)

Carbon cycle   Pathways by which carbon (C) moves through the global environment and is stored in various pools such as rocks, the atmosphere, and the ocean. The carbon cycle is vitally important for understanding how fossil fuel use and human impacts on land cover are affecting atmospheric carbon dioxide levels and changing the global climate (20)

Carbon dioxide–weathering feedback   Negative feedback process in which warming conditions lead to lower CO_2 in the atmosphere, thereby causing cooling. (18)

Carbon sink   Parts of the global environment that are currently taking up more carbon than they release; for example, the ocean and biosphere currently act as sinks for carbon from the atmosphere (24)

Carbonate mineral   Mineral that contains the carbonate ion CO_3^{2-}. (27)

Carbonation   Process that occurs when carbon dioxide dissolves in water to form a weak acid; the acidity that is produced is often crucial for chemical weathering in natural environments. (36)

Carbonic acid   Weak acid (H_2CO_3), formed from water and carbon dioxide that is instrumental in chemical weathering processes. (36)

Caribbean karst   Rarest karst topography, associated with nearly flat-lying limestones; underground erosion dominated by the collapse of roofs of subsurface conduits produces the characteristic sinkhole terrain seen in Fig. 42.6. (42)

Carnivore   Animal that eats herbivores and other animals. (24)

Cartography   Science, art, and technology of mapmaking and map use. (3)

Catena   (*see* **soil catena**)

Cation   Positively charged ion, such as those of calcium. (22)

Cation exchange   Process by which one cation replaces another that is held on the surface of a soil colloid. Cation exchange allows storage and slow release of nutrients that are essential to plants. (22)

Cave   Any substantial opening in bedrock, large enough for an adult person to enter, that leads to an interior open space. (42)

Cave shaft   Vertical cave entrance. (42)

Celsius scale   Metric temperature scale most commonly used throughout the world (the United States is an exception). The boiling point of water is set at 100°C and its freezing point at 0°C. (7)

Cementation   During the lithification process, dissolved material in water within the pores of loose sediment is deposited to form a mineral cement that bonds the grains together. (28)

Cenozoic   Era of recent life on the geologic timescale (*see* Fig. 28.11), extending from 65 million years ago to the present; subdivided into the Tertiary and Quaternary periods. (28)

Centrifugal force   Outward force experienced by objects following a curved path. (2)

Chaparral   Name given to the dominant Mediterranean scrub vegetation in Southern California. (25)

Chemical compound   Created when at least two different elements combine to form a single molecule. (6)

Chemical weathering   Disintegration of rock minerals via chemical means. In any rock made up of a combination of minerals, the chemical breakdown of one set of mineral grains leads to the decomposition of the whole mass. (36)

Chinook   Name given to the *foehn* winds that affect the leeward areas of mountain zones in the western plateaus of North America. (8)

Chlorofluorocarbons (CFCs)   Collective term for a group of ozone-destroying gases. (6)

Chlorophyll   Green pigment found in plant surfaces that absorbs certain wavelengths of light and enables the process of photosynthesis to take place. (24)

Cinder cone   Volcanic landform consisting mainly of pyroclastics. Often formed during brief periods of explosive activity, they normally remain quite small. (32)

Cinders (volcanic)   Pyroclastic particles that are relatively coarse, the size of sand or gravel (32)

Circle of illumination   At any given moment on our constantly rotating planet, the boundary between the halves of the Earth that are in sunlight and darkness. (4)

Circum-Pacific Belt   Aspect of the Pacific Ring of Fire, the lengthy belt of subduction zones that girdles the Pacific Basin. Here the heaviest concentration of earthquake epicenters (as well as active volcanoes) occurs on the world map (*see* Fig. 33.8). (33)

Cirque Amphitheater-like basin, high up on a mountain, that now holds a mountain glacier, or once held one, and is formed by glacial erosion. (45)

Cirrus Cloud category that encompasses thin, wispy, streak-like clouds consisting of ice particles rather than water droplets; occur only at altitudes higher than 6,000 m (20,000 ft). (11)

Clastic sedimentary rocks Sedimentary rocks made primarily from particles of other rocks. (28)

Clay Smallest category of soil particles, smaller than 0.002 mm (0.00008 in), with the smallest in the colloidal range possessing diameters of less than one hundred-thousandth of a millimeter. (22)

Cleavage Tendency of minerals to break in certain directions along bright plane surfaces, revealing zones of weakness in the crystalline structure. (27)

Climate Long-term conditions of aggregate weather over a region, usually summarized by averages and measures of variability; a synthesis of the succession of weather events we have learned to expect at any given location. (6, 14)

Climate classification system A set of rules based on observed climatic variables that produces a relatively small number of distinct climatic types. (14)

Climatic controls Features of the Earth's surface—such as the distribution of land and water bodies, ocean currents, and highlands—that shape the climate of a locale by influencing its temperature and moisture regimes. (14)

Climatic normal The 30-year average of a location's temperature, precipitation, wind speed, or any other climatic element. Climatic normals are updated every 10 years; currently the averaging period is 1981-2010. (14)

Climatology Geographic study of climates. This includes not only climate classification and the analysis of climates' regional distribution, but broader environmental questions that concern climate change, interrelationships with soil and vegetation, and human–climate interaction. (1)

Climax community Achieved at the end of a plant succession. The vegetation and its ecosystem are in equilibrium with the soil, the climate, and other parts of the environment, or close to that state. (24)

Climograph Graph that displays the seasonal cycle of climatic average temperature and precipitation for a location. (15)

Closed system Self-contained system exhibiting no exchange of energy or matter across its boundaries (interfaces). (1)

Cloud Visible mass of suspended, minute water droplets and/or ice crystals. (11)

Coalescence process Raindrop-producing process that dominates in the tropical latitudes. (11)

Coast General reference to the strip of land and sea where the various coastal processes combine to create characteristic landscapes. The term *shore* has a more specific meaning. (48)

Cockpit karst In tropical karst areas, the sharply contrasted landscape of prominent karst towers and the irregular, steep-sided depressions lying between them. *Cockpit* refers to the depressions. (42)

Cohesion Tendency of slope materials to stick together and resist failure, which is not due to friction and does not depend on the weight of overlying materials. (37)

Cold-air drainage Category of local-scale wind systems governed by the downward oozing of heavy, dense, cold air along slopes under the influence of gravity; produces katabatic winds (such as southeastern France's *mistral*) that are fed by massive pools of icy air that accumulate over such major upland regions as the Alps and the Rocky Mountains. (8)

Cold conveyor belt Exists in the lower troposphere ahead of a warm front in a midlatitude cyclone. It flows toward the center of the low, parallel to the warm front. (13)

Cold currents Global-scale ocean currents that flow toward the equator (10)

Cold front Produced when an advancing cold air mass hugs the surface and displaces other air as it wedges itself beneath the preexisting warmer air mass. Cold fronts have much steeper slopes than warm fronts and thus produce more abrupt cooling and condensation (and more intense precipitation). (12)

Collapse sink Collapsed sinkhole in which the rock ceiling of the sinkhole collapses into the underground solution cavity. (42)

Collapse sinkhole In karst terrain, a surface hollow created by the collapse or failure of the roof or overlying material of a cave, cavern, or underground channel. (42)

Colloids Tiny soil particles made up of minerals or humus and less than 1 μm (0.000001 m [0.00004 in]) in diameter. (22)

Color (mineral) One of the most easily observable properties of a mineral; used in its identification. (27)

Column (cave) Coalescence of a stalactite and a stalagmite that forms a continuous column from the floor to the roof of a cave. (42)

Compaction Part of the process of lithification, whereby older sediments are compressed by the weight of newer, overlying sediments. This pressure squeezes the grains of older sediment more tightly together. Usually occurs in conjunction with cementation. (28)

Competence Refers to the largest particle size that a stream can set in motion. (39)

Composite volcano Volcano formed, usually above a subduction zone, by the eruption of a succession of lavas and pyroclastics that accumulate as a series of alternating layers; usually large, steep-sided and cone-shaped peaks, also called *stratovolcanoes*. (32)

Compressional stress Stress in which rock is squeezed together and forced to occupy less space, usually associated with the convergence of lithospheric plates. Rocks respond to this type of stress through folding and formation of reverse and thrust faults. (34)

Concave Refers to a surface that is rounded inward, like the inside surface of a sphere.

Condensation Process by which a substance is transformed from the gaseous to the liquid state. (7)

Condensation nuclei Small airborne particles around which liquid droplets can form when water vapor condenses; almost always present in the atmosphere in the form of dust or salt particles. (11)

Conduction Transport of heat energy from one molecule to the next. (5)

Conduits Pipe-like openings created when water dissolves rock surrounding pores and fractures, thereby enlarging them. (42)

Cone of depression   Drop (drawdown) in a local water table that immediately surrounds a well when water is withdrawn faster than it can be replaced by water flowing through the aquifer toward the well (*see* **drawdown**). (38)

Confined aquifer   (see **aquifer, confined**)

Conformal map projection   Map projection that preserves the true shape of the area being mapped. (3)

Conic map projection   One in which the transfer of the Earth grid is from a globe onto a cone, which is then cut and laid flat. (3)

Coniferous   Cone-bearing. (25)

Constant gases   Atmospheric gases always found in nearly the same proportions. Two of them constitute 99 percent of the air: nitrogen (78 percent) and oxygen (21 percent). (6)

Contact metamorphism   Metamorphic change in rocks induced by their local contact with molten magma or lava. (28)

Continental drift   Notion hypothesized by Alfred Wegener concerning the fragmentation of Pangaea and the slow movement of the modern continents away from that supercontinent. (30)

Continental effect   Lack of the moderating influence of an ocean on air temperature, which is characteristic of inland locations; this produces hotter summers and colder winters relative to coastal locations at similar latitudes. (7)

Continental glaciers   Huge masses of ice that bury whole large portions of a continent beneath them; also known as ice sheets. (43)

Continental rise   Transitional zone of gently sloping seafloor that begins at the foot of the continental slope and leads downward to the lowest (abyssal) zone of an ocean basin. (2)

Continental shelf   Gently sloping, relatively shallow submerged plain just off the coast of a continent, extending to a depth of ca. 180 m (600 ft/100 fathoms). (2)

Continental shield   Large, tectonically stable, relatively flat expanse of very old metamorphic and igneous rocks that forms the geologic core of a continental landmass. (29)

Continental slope   Steeply plunging slope that begins at the outer edge of the continental shelf (ca. 180 m [600 ft] below the sea surface) and ends in the depths of the ocean floor at the head of the continental rise. (2)

Continentality   Variation of the continental effect on air temperatures in the interior portions of the world's landmasses; the farther the distance from the moderating influence of an ocean (known as the maritime effect), the greater the extreme in summer and winter temperatures (northeastern Eurasia is the classic example of such extreme annual temperature cycles). (7)

Contouring   Representation of surface relief using isolines of elevation above sea level; an important basis of topographic mapping. (3)

Convection   Heat transfer accomplished by fluid flow, such as wind, ocean currents, or rising air. Within the atmosphere, the heat transfer can result from the relative warmth or moisture of the moving air or water vapor. These are known respectively as sensible and latent heat transfer. (5, 12)

Convection cell   Slow circulation of rock within Earth's mantle, rising below divergent plate boundaries, flowing from there toward subduction zones, sinking, and returning toward divergent boundaries at great depth within the mantle. One of the drivers of lithospheric plate motion. (31) The term also refers to plumes of rising air caused by atmospheric instability (12).

Convectional precipitation   Precipitation that occurs after condensation of air that rises because of instability. (12)

Convergence   A pattern of wind such that density increases in the downstream direction. It can occur because winds from different directions blow toward the same location (as in the center of a surface low pressure system) or because the air slows down along the direction of flow. (13)

Convergent evolution   Theory that holds that organisms in widely separated biogeographic realms, although descended from diverse ancestors, develop similar adaptations to measurably similar habitats. (26)

Convergent plate boundaries   Boundaries between lithospheric plates in which the relative motion of the plates is toward each other; that is, from a viewpoint on one plate it appears that the other plate is approaching. (30)

Convergent-lifting precipitation   Precipitation produced by the forced lifting of warm, moist air where low-level wind flows converge; most pronounced in the equatorial latitudes, where the Northeast and Southeast Trades come together in the Intertropical Convergence Zone (ITCZ), especially over the oceans. (12)

Convex   Refers to a surface that is rounded outward, like the outside surface of a sphere.

Conveyor belt   One of several broad air streams associated with a midlatitude cyclone. (13)

Coral reef   Aggradational reef formed from the skeletal remains of marine organisms. (49)

Core (drilling)   Tube-like sample of seafloor sediment that is captured and brought to the surface in a hollow drill pipe. The drill pipe is thrust perpendicularly into the ocean floor, and the sample it brings up shows the sequence of sedimentary accumulation. (18)

Core (Earth's)   (*see* **inner core; outer core**)

Coriolis force   Force that, owing to the rotation of the Earth, tends to deflect all objects moving over the surface of the Earth away from their original paths. In the absence of any other forces, the deflection is to the right in the Northern Hemisphere and to the left in the Southern Hemisphere; the higher the latitude, the stronger the deflection. (8)

Corrasion   Process in which waves move fragments of rock across eroding surfaces along the shoreline; those fragments greatly increase the erosive power of the waves. (48)

Corrosion   Process of stream erosion whereby certain rocks and minerals are dissolved by water; can also affect coastal bedrock that is susceptible to such chemical action. (39)

Counterradiation   Longwave radiation emitted by the Earth's atmosphere downward toward the surface (5)

Creep   Slowest form of mass movement; involves the slow, imperceptible motion of a soil layer downslope, as revealed in the slight downhill tilt of trees and other stationary objects. (37)

Crevasse   One of the huge vertical cracks that frequently cut the rigid, brittle upper layer of a glacier. (43)

Cross-bedding   Describes sedimentary structures in which thicker, nearly horizontal strata contain thinner beds inclined at angle; usually formed where sand is deposited on the sloping face of a dune or ripple. (28)

Crust   The outer, thin layer of Earth that forms the upper part of the lithosphere. (27)

Cryosphere   Collective name for the ice system of the Earth, which constitutes one of the five subsystems of the total Earth system. (2, 43)

Cryoturbation   Frost churning of mineral or organic soil materials in Gelisols. (23)

Cuesta   Long ridge that forms through erosion of gently dipping sedimentary rocks, with a steep escarpment cutting across rock strata on one side and a gently dipping slope underlain by an especially resistant rock layer on the other side. (40)

Cultural landscapes   Landscapes in which human intervention dominates to such an extent that physical processes have become subordinate. (2)

Cumulonimbus   Very tall cumulus clouds, extending from about 500 m (1,600 ft) at the base to over 10 km (6 mi) at the top, often associated with violent weather involving thunder, lightning, and heavy winds and rains. (11)

Cumulus   Cloud category that encompasses thick, puffy, billowing masses that often develop to great heights; subclassified according to height. (11)

Cycle of erosion   Cycle in which landscapes evolve through uplift followed by long-term erosion, as proposed by William Morris Davis (*see* Fig. 40.17). (40)

Cyclogenesis   Formation, evolution, and movement of midlatitude cyclones. (13)

Cyclone   Atmospheric low-pressure cell involving the convergence of air at low levels, which flows into and spirally rises at the center. The isobars around a cyclone are generally circular in shape, with their values decreasing toward the center. In the Northern Hemisphere, winds flow counterclockwise around a cyclone; in the Southern Hemisphere, winds flow clockwise around a cyclone. (8)

Cyclonic precipitation   Precipitation associated with the passage of the warm and cold fronts that are basic components of the cyclones that shape the weather patterns of the midlatitudes; often used as a synonym for *frontal precipitation* (12).

Cylindrical map projection   One in which the transfer of the Earth grid is from a globe onto a cylinder, which is then cut and laid flat. (3)

DALR   (*see* **dry adiabatic lapse rate**)

Daylight-saving time   By law, all clocks in a time zone are set one hour forward from standard time for at least part of the year. In the United States, localities (e.g., the state of Arizona) can exempt themselves from such federal regulations. (4)

Debris flow   Form of mass movement that involves a fluid, fast-moving slurry of sediment and water. (37)

Deciduous   Describes trees and other plants that drop their leaves seasonally. (24)

Deep ocean   Occurs at depths greater than 1,000 m and includes about 97 percent of the total ocean water volume. (10)

Deflation   Process whereby wind erodes the land surface and carries away the eroded material; may produce basins and troughs. (47)

Deflation hollow   Shallow desert basin created by the wind erosion process of deflation. (47)

Deforestation   (*see* **tropical deforestation**)

Deformation   Any change in the form and/or structure of a body of crustal rock caused by an Earth movement. (34)

Deglaciation   Melting and receding of glaciers that accompanies the climatic warm-up after the peak of a glacial period has been reached. (43)

Degree   A unit used to specify the size of an angle, which ranges from zero to 360 for a full circle. Latitude and longitude are almost always specified in degrees. (3)

Delta   Landform produced by sediment deposition around and beyond the mouth of a river where it empties into the sea or a lake; frequently assumes a triangular configuration—hence its naming after the Greek letter of that shape. (39, 41)

Deltaic plain   Flat, stable landward portion of a delta that is growing seaward. (39, 41)

Dendritic drainage   Tree limb—like stream pattern that is the most commonly observed of all such patterns (*see* Fig. 40.11E); indicates surface of relatively uniform hardness or one of flat-lying sedimentary rocks. (40)

Dendrochronology   Study of the width of annual growth rings in trees. (18)

Denitrification   Conversion of nitrate ions (NO_3^{-1}) to nitrogen gas (N_2) or in some cases, nitrous oxide (N_2O); carried out by bacteria in soil or water bodies. (20)

Density   Amount of mass per volume in an object or a portion of the atmosphere (6).

Denudation   Combined processes of weathering, mass movement, and erosion that over time act to lower the land surface. (39)

Depletion   Soil-layer formation process involving the loss of soil components as they are carried downward by water, plus the loss of other material in suspension as the water percolates through the soil from upper to lower layers. While the upper layers are depleted accordingly, the dissolved and suspended materials are redeposited lower down in the soil. (21)

Deposition (of water vapor).   Occurs when water moves from the vapor state to the solid state. (11)

Deposition   (of **sediment**) Occurs whenever solid material that has been transported some distance is dropped out of flowing water or wind, melts out of a glacier, or simply comes to rest at the foot of a slope. (35)

Desert biome   Characterized by sparse, xerophytic vegetation or even the complete absence of plant life, because of an arid climate. (25)

Desert climate (Bwh)   Most arid of climates, usually experiencing no more than 250 mm (10 in) of annual precipitation. Characteristics are displayed in Fig. 15.11. (15)

Desert pavement   Smoothly weathered, varnish-like surface of closely packed pebbles that has developed on the upper part of an alluvial fan or bajada, where the stable land surface is no longer subject to stream deposition. (47)

Desertification   Process of desert expansion into neighboring steppelands. (15)

Dew   Fine water droplets that condense on surfaces at and near the ground when saturated air is cooled. (11)

Dew-point   Temperature at which air becomes saturated and below which condensation occurs. (11)

Diffuse radiation Proportion of incoming solar energy (22 percent) that reaches the Earth's surface after first being scattered in the atmosphere by clouds, dust particles, and other airborne materials. (5)

Dike Intrusive igneous form in which magma has cut vertically across preexisting strata, forming a kind of barrier wall. (27)

Dip Angle at which a rock layer tilts from the horizontal. (34)

Direct radiation Proportion of incoming solar energy that travels directly to the Earth's surface. (5)

Disappearing climate Refers to locations whose present-day climate is so dissimilar from climates found anywhere in the late twenty-first century that their climate is said to "disappear." (19)

Discharge (stream) Volume of water passing a given location along a river channel within a given amount of time; measured as average water velocity multiplied by the cross-sectional area of the channel. (38)

Dispersal Movement of plants, animals, and other living organisms into new areas; animals may walk, swim, fly, while plant seeds can be carried by birds or on floating debris, for example. Numerous new species may evolve from members of a single species after they disperse into a new land mass or water body. (24)

Dissolution Chemical weathering process in which soluble rocks such as limestone are dissolved. (42)

Distributaries Several channels into which a river subdivides when it reaches its delta. (41)

Disturbance An event that disrupts an ecosystem, such as fire, an outbreak of tree-killing insects, or strong winds that blow down a forest. (24)

Diurnal temperature cycle Pattern of temperature change during the course of a day. (7)

Divergence A pattern of wind such that density decreases in the downstream direction. It can occur because winds blow away from a common location (such as outward from the center of a surface high pressure system) or because the air speeds up along the direction of flow. (13)

Divergent plate boundary Occurs when two tectonic plates move away from each other. (30)

Divide Topographic barrier, usually a mountain ridge, that separates two drainage basins. (39)

Doldrums Equatorial zone of periodic calm seas and unpredictable breezes where the Northeast and Southeast Trades converge. The crews of sailing ships dreaded these waters because their vessels risked becoming stranded in this becalmed marine environment. (9)

Dolomite A carbonate mineral ($CaMg[CO_3]_2$); term is also applied to a sedimentary rock made up mainly of that mineral. Karst topography may form in areas where there is dolomite, in addition to areas underlain by limestone. (28, 42)

Dormant volcano Volcano that has not erupted in recent geologic time, but an eruption is still considered possible. (32)

Downthrown block Block that moves downward with respect to adjacent blocks when vertical movement occurs during faulting. (34)

Drainage basin Region occupied by a complete stream system formed by the trunk river and all its tributaries; also known as a *watershed*. (39)

Drainage density Total length of the stream channels that exist in a unit area of a drainage basin. (40)

Drawdown Drop in a local water table that occurs when water is withdrawn through wells faster than it can be replaced by recharge of new groundwater to the aquifer. (38)

Drift (glacial) (*see* **glacial drift**)

Drift (ocean surface) Term often used as a synonym for ocean current when the rate of movement lags well behind the average speeds of surface winds blowing in the same direction. Currents are characterized by a slow and steady movement that very rarely exceeds 8 km/h (5 mph). (10)

Driftless Area Area in southwestern Wisconsin that was never covered by the continental glaciers that repeatedly buried adjacent areas of the U.S. Midwest. (44)

Drought Below-average availability of water in a given area over a period lasting at least several months (15).

Drumlin Smooth elliptical mound created under an ice sheet as it erodes or reshapes preexisting sediment. The long axis lies parallel to the direction of ice movement. (44)

Dry adiabatic lapse rate (DALR) Lapse rate of an air parcel not saturated with water vapor: $-1°C/100$ m ($-5.5°F/1,000$ ft). (7)

Dry conveyor belt Brings cold and dry air from the middle and upper troposphere into a cyclone. (13)

Dune Accumulation of sand that is shaped by wind action. (47)

Dust Bowl Environmental disaster that affected the U.S. Great Plains in the 1930s. During a prolonged drought when crops failed and there was little vegetation cover on agricultural fields, huge dust clouds formed, some of which drifted far east of the Great Plains. (47)

Dust dome Characteristic shape taken by the large quantities of dust and gaseous pollutants in a city's atmosphere (*see* Fig. 7.6). (7)

Dust (volcanic) Fine particles of volcanic ash that can be carried for substantial distances by prevailing wind flows if they are exploded high enough into the troposphere. (32)

Dynamic equilibrium State of a system in balance despite the continual flow of energy and/or mass through the system. A landscape that is in dynamic equilibrium has a stable form over time, although it continues to erode (1, 40)

E horizon Soil layer located between the **A** and **B** horizons, characterized by light color and relative low content of clay and humus; formed by eluviation, or loss of iron oxides, clay, and other materials to deeper horizons (**E** stands for eluviation). (22)

Early Anthropocene hypothesis Asserts that humans began modifying global climate about 8,000 years ago. (19)

Earth flow Form of mass movement in which soil or weathered bedrock moves downhill at a relatively slow rate; motion can continue for years (*see* Fig. 37.9). (37)

Earth system Shells or layers that make up the total Earth system range from those of the planet's deepest interior to those bordering outer space. This book focuses on the five key Earth layers: atmosphere, lithosphere, hydrosphere, cryosphere, and biosphere. (2)

Earthquake Shaking of Earth's crust; usually caused by fault motion that suddenly releases stresses that have been building up slowly. (29, 33)

Easterly wave   Wavelike perturbation in the prevailing easterly flow of the Northeast and Southeast Trades, which produces this type of distinctive weather system. Westward-moving air is forced to rise on the upwind side (producing often-heavy rainfall) and descend on the fair-weather downwind side of the low-pressure wave trough. (13)

Ecological niche   The environmental space within which an organism operates most efficiently. (26)

Ecological zoogeography   The study of animals' interactions with their environment or other animal or plant species (26)

Ecosystem   All of the living organisms in a particular area, their physical environment (soils, water, etc.) and the flows of energy and matter that link those living organisms to each other and to their environment. (24)

Edaphic factors   Factors concerned with the soil. (24)

Eddies   Localized loops of water circulation detached from the mainstream of a nearby ocean current, which move along with the general flow of that current. (12)

El Niño   Periodic, large-scale, abnormal warming of the sea surface in the low latitudes of the eastern Pacific Ocean that produces a (temporary) reversal of surface ocean currents and air flows throughout the equatorial Pacific. These regional events have global implications, disturbing normal weather patterns in many parts of the world. (9)

Electromagnetic spectrum   Continuum of electric and magnetic energy, as measured by wavelength, from the high-energy shortwave radiation of cosmic rays to the low-energy longwave radiation of radio and electric power (*see* Fig. 3.19).

Ellipsoid   A three-dimensional object that is elliptical in cross-section. (2)

El Niño–Southern Oscillation (ENSO)   Systematic changes in temperature, precipitation and the flow of ocean currents and prevailing winds in the equatorial Pacific Ocean. (10)

ELR   (*see* **environmental lapse rate**)

Eluviation   Refers to the soil process that involves downward transportation of dissolved material and microscopic, colloid-sized particles of organic matter, clay, and oxides of aluminum and iron, usually from A or E horizons into B horizons. (21)

Emergent coast   Coastal zone whose landforms have recently emerged from the sea, through either tectonic uplift or a drop in sea level, or both. (49)

Endemic species   Highly localized species, native to a particular location. (26)

Endemism   Biotic complex of an isolated (or once-isolated) area, many of whose species of plants, animals, and other life-forms exist nowhere else on Earth. The species in such places are particularly vulnerable to extinction through the environmental changes introduced by humans. (24)

Entisol   One of the 12 soil orders of the Soil Taxonomy, which contains all the soils that do not fit into the other 11 soil orders; these soils display only weak soil development, generally lack B horizons, and are found in many different environments. (23)

Environmental lapse rate (ELR)   The rate at which temperature declines with altitude in a column of air. The troposphere's average (or "normal") lapse rate is –0.64°C/100 m (–3.5°F/1,000 ft). (7)

Eolian   Pertaining to the action of the wind. (47)

Ephemeral plant   One that completes its life cycle within a single growing season. (25)

Ephemeral stream   Intermittently flowing stream in an arid environment. Precipitation (and subsequent stream flow) is periodic, and when the rains end, the stream soon runs dry again. (41)

Epicenter   Point on the Earth's surface directly above the *focus* (point of origin) of an earthquake. (33)

Epiphyte   Tropical rainforest plant that uses trees for support but is not parasitic. (25)

Equal-area map projection   One in which all the areas mapped are represented in correct proportion to one another. (3)

Equator   Parallel of latitude running around the exact middle of the globe, defined as 0° latitude. (3)

Equatorial jet stream   High-level wind belt blowing from east to west found at low latitudes. (9)

Equatorial low   Intertropical convergence zone (ITCZ) or thermal low-pressure belt of rising air that straddles the equatorial latitudinal zone; fed by the wind flows of the converging Northeast and Southeast Trades. (9)

Equilibrium   A system is in equilibrium with respect to some property if that property is unchanging over time. (1)

Equinox   One of the two days (around March 21 and September 23) in the year when the Sun's noontime rays strike the Earth vertically at the equator. In Northern Hemisphere terminology, the March 21 event is called the spring (vernal) equinox and the September 23 event is called the fall (autumnal) equinox. (4)

Erg   Sand sea; large expanse of sandy desert landscape. (47)

Erosion   Long-distance carrying away of weathered rock material and the associated processes whereby the Earth's surface is worn down. (35)

Erratics   Boulders carried far from their source area by glaciers. (44)

Escarpment   Steep slope or cliff; often extends for a considerable distance along the edge of a plateau or mountainous high-land. (40)

Esker   Glacial landform that is a steep-siding, sinuous (winding) ridge; formed by deposition of sand and gravel on the bed of a stream flowing through a tunnel in the glacier ice. When the glacier retreats, the esker is left behind (44)

Estuary   Drowned (submerged) mouth of a river valley that has become a branch of the sea. In the United States, the Chesapeake Bay is a classic example. (49)

Evaporation   Also known as vaporization, the process by which water changes from the liquid to the gaseous (water vapor) state. It takes about 2260 kilojoules of energy to change 1 kg (2.2 lb) of water from liquid to water vapor. (11)

Evaporites   Rock deposits resulting from the evaporation of the water in which these materials were once dissolved; include halite (salt) and gypsum. (28)

Evapotranspiration   Combined processes by which water (1) evaporates from the land surface and (2) passes into the atmosphere through the leaf pores of plants (transpiration). (11)

Evergreen   Of trees and other plants that keep their leaves year-round. (24)

Evolution   A change in the characteristics of a population that is heritable (passed on from one generation to the next). (26)

Exfoliation   Special kind of jointing that produces a joint pattern resembling a series of concentric shells, much like the layers of an onion. Following the release of confining pressure, the outer layers progressively peel away and expose the lower layers. (27)

Experiment   A general term involving a procedure or test that is designed to learn more about a hypothesis. (1)

Extinct volcano   Volcano that shows no sign of life and exhibits evidence of long-term weathering and erosion. (32)

Extreme events   Refers to temperatures above or below some threshold, such as 90°F for extreme heat or –20°F for cold snaps. (19)

Extrusive igneous rocks   Rocks formed from magma that cooled and solidified rapidly on the Earth's surface. (27)

Eye (hurricane)   Nearly cloud-free vertical tube that marks the center of a hurricane, often reaching an altitude of 16 km (10 mi). (13)

Eye wall (hurricane)   Cloud-filled rim of the eye at the center of a well-developed hurricane. The tropical cyclone's strongest winds and heaviest rainfall occur here. (13)

Fahrenheit scale   Temperature scale presently used in the United States. Water boils at 212°F (100°C) and freezes at 32°F (0°C). (7)

Fall (autumnal) equinox   In Northern Hemisphere terminology, equinox that occurs when the Sun's noontime rays strike the equator vertically around September 23. (4)

Fall (mass movement)   Form of mass movement that involves the free fall or downslope rolling of rock pieces loosened by weathering. These boulders form a talus cone or scree slope at the base of the cliff from which they broke away. (37)

Fault   Fracture in crustal rock involving the displacement of rock on one side of the fracture with respect to rock on the other side. (33)

Fault plane   Surface of contact along which blocks on either side of a fault move (*see* Fig. 33.12). (33)

Fault scarp   Exposed cliff-like face of a fault plane (*see* Fig. 33.12) created by geologic action without significant erosional change. (33)

Fault trace   Line on the surface where it is intersected by a fault (*see* Fig. 33.12). (33)

Fault-line scarp   Scarp that originated as a fault scarp but was modified, perhaps even displaced, by erosion (*see* Fig. 40.5B). (40)

Fauna   All of the animal species living in a particular region (20)

Feedback   Occurs when a change in one part of a system causes a change in another part of the system. (1)

Feldspars   Silicate minerals with silica tetrahedra joined in a three-dimensional structure, also containing aluminum and a variety of other elements. (27)

Felsenmeer   (*see* **blockfield**)

Felsic   Light-colored igneous rocks, with pink or light-gray feldspar and clear, almost-colorless quartz dominating. (27)

Ferrel cell   Large pattern within atmospheric general circulation found in the middle latitudes of both hemispheres. In cross-section the Ferrel cell has winds flowing poleward at low levels and equatorward at high levels. The prevailing westerlies are a feature of this cell. (9)

Fertigation   Contraction of "fertilize" and "irrigation," an Israeli-perfected technique for the large-scale raising of crops under dry environmental conditions. Via subterranean pipes, the roots of genetically engineered plants are directly fed a mixture of brackish underground water and chemical fertilizers that supplies just the right blend of moisture and nutrients for the crop to thrive.

Fetch   Uninterrupted distance across which the wind can blow; influences the height of waves on the ocean or lakes. (48)

Field capacity   Maximum amount of water that a soil can hold by capillary tension against the downward pull of gravity. (22, 38)

Fill terrace   (*see* **terrace, fill**)

Firn   Granular, compacted snow, a stage in the formation of glacier ice. (43)

Firn line   (*see* **snow line**)

Fissure   Linear opening in Earth's surface, from which lava erupts (32)

Fissure eruption   Volcanic eruption that comes not from a pipe-shaped vent but from a lengthy crack in the lithosphere. The lava that erupts does not form a mountain but extends in horizontal sheets across the countryside, sometimes forming a plateau. (32)

Fixed dune   Stable sand dune that supports vegetation, which slows or even halts the dune's wind-generated movement. (47)

Fjord   Narrow, steep-sided, elongated inlet of the ocean formed where a glacial trough has been inundated by seawater. (45)

Flood   A peak of stream discharge. In larger floods, water overflows from the stream channel and temporarily covers the floodplain. (39)

Flood basalts   Result from the accumulation of many lava flows that spread out over large areas, often from numerous vents. (32)

Floodplain   Flat, low-lying ground adjacent to a stream channel built and reshaped by successive floods as sediment is deposited as alluvium in some areas and eroded in others. (41)

Flora   All of the plant species living in a particular region. (20)

Flow (mass movement)   Gravity-driven downhill movement of sediment or weathered rock in which the material forms a fluid mass rather than discrete blocks (37)

Fluvial   Denotes running water; derived from Latin word for river, *fluvius*. (39)

Fluvial processes   The geomorphic processes associated with flowing water. (39)

Focus (earthquake)   Place of origin of an earthquake, which can be near the surface or deep below ground (33)

Foehn   Rapid movement of warm dry air (caused by a plunge in altitude), frequently experienced on the leeward or rain-shadow side of a mountain barrier, whose moisture has mostly been removed via the orographic precipitation process. The name is also used specifically to designate such local winds in the vicinity of central Europe's Alps; in the western plateaus of North America, these air flows are called *chinook* winds. (8)

Fog   Cloud layer in direct contact with the Earth's surface. (11)

Fold   Individual bend or warp in layered rock. (34)

Foliation   Banded appearance of certain metamorphic rocks, such as gneiss and schist; bands formed by minerals realigned into parallel strips during metamorphism. (28)

Food chain   Stages through which energy in the form of food moves within an ecosystem. (24)

Footwall   The rock surface that is below an inclined fault plane; opposite side from the hanging wall (see Figs. 34-2, 34-3, and 34-5). (34)

Forcings   In a general-circulation model, an external factor, such as Earth's topography, solar radiation, or atmospheric composition. (19)

Foreset beds   Sedimentary deposits that are built from the leading edge of the topset beds as a delta grows seaward; usually the most coarse-grained component of the delta. Later the foresets will be covered by the extension of the topset beds. (41)

Foreshore   Beach zone that is alternatively water-covered during high tide and exposed during low tide; zone of beach drift and related processes. (49)

Fossil fuels   Coal, oil, and natural gas, the dominant suppliers of energy in the world economy. (6)

Fracture (mineral)   When minerals do not break in a clean cleavage, they still break or fracture in a characteristic way. Obsidian, for example, fractures in an unusual shell-like manner that is a useful identifying quality. (27)

Freezing   Process by which a substance is transformed from the liquid to the solid state. (11)

Freezing nuclei   Small airborne particles (such as dust and salt) around which ice crystals can form when liquid water freezes. Freezing nuclei enhance freezing and allow it to occur at relatively warm temperatures just a few degrees below 0°C (32°F) (11)

Freezing point (water)   Key setting in the calibration of temperature scales. On the Celsius scale water ice melts at 0°C; on the Fahrenheit scale it melts at 32°F.

Freezing rain   Rain that reaches the surface as liquid but then freezes after reaching the ground to form a layer of ice where it lands. Occurs when the layer of below-freezing air is not deep enough to freeze the falling rain. (11)

Frictional force   Drag that slows the movement of air molecules in contact with, or close to, the Earth's surface; varies with the "roughness" of the surface. There is less friction with movement across a smooth water surface than across the ragged skyline of a city center. (8)

Frictional strength   Ability of geologic materials to resist slope failure through friction; depends on each material's properties but also increases with the component of the weight of overlying material pushing into the slope (37)

Front (weather)   Contact surface between two dissimilar air masses This narrow boundary zone usually marks relatively abrupt transition in air density, temperature, and humidity. A moving front is the leading edge of the air mass built up behind it. (12)

Frontal precipitation   Precipitation that results from the movement of fronts whereby warm air is lifted, cooled, and condensed; also frequently called *cyclonic precipitation*. (12)

Frost action   Form of mechanical weathering in which water penetrates the joints and cracks of rocks, expands and contracts through alternate freezing and thawing, and eventually shatters the rocks. (36, 46)

Frost creep   Downhill movement of particles within the active layer above permafrost under the influence of gravity, as they are heaved up when the soil freezes, and then settle somewhat farther downslope when the soil thaws. (46)

Frost heaving   Upward displacement of rocks and rock fragments within soils; triggered by the formation of ice in the ground that expands the total mass of rock materials. Heaving is especially common in cold regions where permafrost is present. (46)

Frost shattering   (*see* **frost wedging**)

Frost wedging   Forcing apart of a rock when the expansion stress created by the freezing of its internal water into ice exceeds the cohesive strength of that rock body. (36, 46)

Gelisol   One of the 12 soil orders of the Soil Taxonomy, found in cold environments where permafrost is present. Such soils show evidence of cryoturbation (frost churning) and/or ice segregation in the seasonally thawing active layer that lies above the permafrost table. (23)

General circulation   Global atmospheric circulation system of wind belts and semipermanent pressure cells. In each hemisphere, the former include the Trades, westerlies, and polar easterlies. The pressure belts include the equatorial low (ITCZ) and, in each hemisphere, the subtropical high, upper midlatitude low, and polar high. (9)

General circulation model   A computer model that can be run indefinitely, with constant or varying forcing values (external factors). Widely used to assess the role humans have played in climatic change and to project the effects of continued release of greenhouse gases. (19)

Geodesy   Precise study and measurement of the size and shape of the Earth. (3)

Geographic information system (GIS)   Assemblage of computer hardware and software that permits spatial data to be collected, recorded, stored, retrieved, manipulated, analyzed, and displayed to the user. (3)

Geography   Literally means *Earth description*. As a modern academic discipline, it is concerned with the explanation of the physical and human characteristics of the Earth's surface. "Why are things located where they are?" is the central question that geographical scholarship seeks to answer. (1)

Geologic structure   Folds, faults, and other rock structures originally formed by tectonic processes. (40)

Geologic timescale   Standard timetable or chronicle of Earth history used by scientists; sequential organization of geologic time units as displayed in Fig. 28.11. (28)

Geomorphology   Literally means *Earth shape* or *form*; geography of landscape and its evolution, a major subfield of physical geography. (1)

Geostrophic wind   Wind that results when the Coriolis and pressure-gradient forces balance each other out. Geostrophic wind has a constant speed and direction and blows parallel to straight isobars (*see* Fig. 8.7).

Geostationary orbit   Orbit in which a satellite above the Equator revolves at the planet's rotational speed. Therefore the satellite is "fixed" in a stationary position above the same point along the Equator. (3)

Geothermal energy   Energy whose source is underground heat. Where magma chambers lie close to the surface, groundwater is heated and emerges onto the surface as steam or hot water from a geyser or hot spring. (38)

Geyser   Hot spring that periodically expels jets of heated water and steam, an indication of magma not far below the surface. (32)

GIS   (*see* **Geographic information system**)

Glacial creep   One of the two mechanisms by which glaciers flow; involves the internal deformation of the ice, as a result of slippage on crystal planes and between ice crystals. (43)

Glacial drift   General term for sediment deposited by a glacier or by meltwater flowing from a glacier. (44)

Glacial erratic   Boulder that was transported by a glacier far from its source area. (44)

Glacial outwash   Meltwater-deposited sand and gravel, which are sorted by size and display distinct bedding. (44)

Glacial sliding   One of the two mechanisms by which glaciers flow; involves the sliding movement of the entire glacier over the rock or sediment below it; requires a thin film of water at the base of the glacier, so it does not occur in cold polar glaciers where the basal ice is below its melting point. (43)

Glacial striations   Scratches on the underlying rock surface made when boulders or pebbles (embedded in a moving glacier) were dragged across it. (43)

Glacial surge   Episode of unusually rapid movement by a glacier, as much as one meter per hour, which can last for a year or more. (43)

Glacial trough   Valley that has been eroded by a glacier; distinctively U-shaped in cross-sectional profile. (45)

Glacial/interglacial cycles   Refers to the glacial and interglacial periods throughout time, caused by variation in Earth's orbit. Periods with large ice sheets are known as glacial periods, and periods with reduced ice cover (such as the present) are known as interglacial periods. (18)

Glaciation   Period of global cooling during which continental ice sheets and mountain glaciers expand. (43)

Glacier   Body of ice, formed on land, which exhibits motion through flow of the ice. (43)

Global conveyor belt   Global system caused by thermohaline circulation, which moves masses of water throughout the world ocean. (10)

Global positioning system (GPS)   Constellation of more than two-dozen linked orbiting satellites that broadcast signals to portable receivers anywhere on the Earth's surface. The simultaneous detection of these signals enables the precise calculation of the receiver's latitude, longitude, and elevation. (3)

Global warming   A general term that refers to an observed temperature increase over the past 150 years. (19)

Global warming pause   A now-discredited theory that there has been a hiatus in global warming over the last fifteen years (19)

Gneiss   Metamorphic rock derived from granite that usually exhibits pronounced foliation. (28)

Gondwana   Southern portion of the primeval supercontinent Pangaea. (30)

GPS   (*see* **Global positioning system**)

Graben   Crustal block that has sunk down between two fairly parallel normal faults in response to tensional forces (*see* Fig. 34.7). (34)

Graded stream   Stream in which slope and channel characteristics are adjusted over time to provide just the velocity required for the transportation of the load supplied from the drainage basin, given the available discharge. (39)

Granite   Igneous rock consisting of feldspar, quartz, and other silicate minerals. Granite is felsic, meaning it formed from magma with relatively high silica content and is light-colored as a result; it is also intrusive since it cooled from magma deep below the surface and therefore has coarse grains. (27)

Granular soil structure   Involves peds that are usually very small and often fairly round in shape, so that the soil looks like a layer of breadcrumbs (*see* Fig. 22.6). (22)

Gravitational force   The primary force exerting an influence on air molecules. The atmosphere is "held" against Earth by this force, and our planet would have no atmosphere at all were it not for gravity. (8)

Gravitational potential energy   The potential energy an object possesses by virtue of its position above some reference level. When an object falls or moves downward potential energy is converted to kinetic energy, the energy of motion. (13)

Gravity   Force of attraction that acts among all physical objects as a result of their *mass* (quantity of material of which they are composed). (2)

Great circle   Circle formed along the edge of the cut when a sphere is cut in half. On the surface of the sphere, that circle is the shortest distance between any two points located on it. (3)

Greenhouse effect   Widely used but imperfect analogy describing the blanket-like effect of the atmosphere in the heating of the Earth's surface. Shortwave insolation passes through the "glass" of the atmospheric "greenhouse"; heats the surface; is converted to longwave radiation, which cannot penetrate the "glass"; and thereby results in the trapping of heat that raises the temperature inside the "greenhouse." (5)

Ground heat flow   Heat that is conducted into and out of the Earth's surface; also known as *soil heat flow*. (5)

Groundwater   Water contained within the lithosphere. This water, hidden below the ground, accounts for about 25 percent of the world's freshwater. (38)

Groundwater discharge   Flow of water out of an aquifer, often into streams; occurs where the water table intersects the ground surface. Locations of particularly noticeable groundwater discharge are referred to as springs (38)

Groundwater recharge   Replaces water lost from an aquifer through discharge. Most of that water has infiltrated and percolated downward from the land surface above the aquifer. (38)

Gyre   Cell-like circulation of surface currents that often encompasses an entire ocean basin. For example, the subtropical gyre of the North Atlantic Ocean consists of the huge loop formed by four individual, continuous legs—the North Equatorial, Gulf Stream, North Atlantic Drift, and Canaries currents. (10)

Habitat   Environment a species normally occupies within its geographical range. (26)

Hadley cell   Part of atmospheric general circulation found in the tropics and subtropics of both hemispheres. In cross-section the Hadley cells are like loops with air flowing toward the Equator at low levels, rising along the Equator, flowing poleward in the upper troposphere, and descending in the subtropics to complete the loop. (9)

Hail   Precipitation consisting of large ice pellets that do not melt before reaching the surface. (11)

Hanging valley   Valley formed by the intersection of a tributary glacier with a trunk glacier. When the ice melts away, the tributary valley floor usually is at a higher elevation and thus "hangs" above the main valley's floor. (45)

Hanging wall   The upper side of the fault plane, overhanging the other side, when a fault is tilted rather than perfectly vertical (see Figs. 34-2, 34-3, and 34-5). (34)

Hawai'ian High   Local name for the North Pacific Ocean's subtropical high-pressure cell, which is generally centered over latitude 30 degrees N; also known as the Pacific High. (9)

Headward erosion   Upslope extension, over time, of the "head" or source of a river valley, which lengthens the entire stream network. (39)

Hemisphere   Half-sphere; used precisely, as in Northern Hemisphere (everything north of 0 degrees latitude), or sometimes more generally, as in land hemisphere (the significant concentration of landmasses on roughly one side of the Earth). (2)

Herbivore   Animal that lives on plants, or, more generally, the first consumer level of a food chain. (24)

Heterosphere   Upper of the atmosphere's two vertical regions, which possesses a variable chemical composition and extends upward from an elevation of 80 to 100 km (50 to 63 mi) to the edge of outer space. (6)

Highland (H) climates   Climates of high-elevation areas that exhibit characteristics of climates located poleward of those found at the base of those highlands. The higher one climbs, the colder the climate becomes—even in the low latitudes. Thus, **H** climate areas are marked by the vertical zonation of climates, as shown for South America's Andes in Fig. 17.15. (17)

Histosol   One of the 12 soil orders of the Soil Taxonomy; soil that contains enough organic matter to dominate its physical properties; often form in wetlands where saturation of the soil limits organic matter decomposition. Common names for Histosols include peat and muck. (23)

Hogback   Prominent steep-sided ridge eroded from steeply dipping rock beds. (40)

Holocene   Current interglacial epoch, extending from 11,700 years ago to the present on the geologic timescale. (18)

Homosphere   The lower of the atmosphere's two vertical regions, which possesses a relatively uniform chemical composition and extends from the surface to an elevation of 80 to 100 km (50 to 63 mi). (6)

Horizontal datum   Unique geographic coordinate system created by examining a model ellipsoid and its related parameters. (3)

Horn   Sharp-pointed, Matterhorn-like mountain peak that remains when several cirques attain their maximum growth and intersect. (45)

Horst   Crustal block that has been raised between two reverse faults by compressional forces. (34)

Hot-house climate   A generally warm climate with little permanent ice like that in today's icecaps. Hot-house climates typically last hundreds of millions of years and have been the norm throughout Earth history. (18)

Hot spot   Location of persistent upward flow of hot rock from deep in the mantle, leading to repeated volcanic activity (the upward flowing rock is referred to as a mantle plume). Lithospheric plates move over hot spots, producing a linear series of shield volcanic centers, as happened in the case of the Hawai'ian island chain. (32)

Hot spring   Spring whose water temperature averages above 10°C (50°F); often emanates from a crustal zone that contains magma near the surface (see Fig. 38.10D). (38)

Humid continental climate (Dfa/Dwa, Dfb/Dwb)   Collective term for the milder subtypes of the microthermal climate, found on the equatorward side of the **D**-climate zone. Characteristics are displayed in Figs. 17.6 and 17.7. (17)

Humid subtropical climate (Cfa)   Warm, perpetually moist mesothermal climate type that is found in the southeastern portion of each of the five major continents. Characteristics are displayed in Figs. 16.3–16.5. (16)

Humus   Decomposed organic matter in the soil; forms a dark layer at the top of the soil and contributes importantly to a soil's fertility. (21)

Hurricane   Tropical cyclone capable of inflicting great damage. A tightly organized, moving low-pressure system, originating at sea in the warm, moist air of the tropical atmosphere, exhibiting wind speeds in excess of 33 m/s (74 mi/h). Like other tropical cyclones, it has a distinctly circular wind and pressure field. (13)

Hydraulic action   Erosional process in streams and on wave-eroded coastlines, in which blocks are broken loose from bedrock and carried away by flowing water or longshore drift. Fractures in the bedrock make this process much more effective; where waves strike a shoreline, air is trapped and compressed in fractures, helping to break boulders loose. (39, 48)

Hydrograph   Graph of a river's discharge over time (see Fig. 38.5). (38)

Hydrologic cycle   Complex system of exchange involving water in its various forms as it continually circulates among the atmosphere, lithosphere, hydrosphere, cryosphere, and biosphere (see Fig. 11.3). (11)

Hydrology   Systematic study of the Earth's water in all its states. (1)

Hydrolysis   General term for a group of chemical reactions that are important in silicate mineral weathering; all involve splitting of water (H_2O_3) molecules to release hydrogen ions (H^+), which then react with minerals. (36)

Hydrosphere   Sphere of the Earth system that contains all the water that exists on and within the solid surface of our planet and in the atmosphere above. (2)

Hydrothermal   Groundwater heated by magma that is relatively close to the surface (see Fig. 38.8D). (38)

Hygrophyte   Plant adapted to the moisture of wet environments. (24)

Hygroscopic water   Thin films of water that cling tenaciously to most soil particles but are unavailable to plant roots. (38)

Hypotheses   Tentative explanations of observations and measurements. (1)

Hypothetical continent model   Earth's landmasses generalized into a single, idealized, shield-shaped continent of uniform low elevation; elaborated in Fig. 14.2.

Ice-albedo feedback   Positive feedback process in which a change in snow cover and sea ice alters the albedo, which in turn affects snow and sea ice by modulating temperature. (18)

Ice age   Stretch of geologic time with large areas of permanent, year-round ice. Within an ice age ice cover expands and contracts in glacial/interglacial cycles. Earth is currently in an ice-age climate. The last glacial maximum was about 20,000 years ago; the present climate is an interglacial within the most recent ice age. (18, 43)

Icecap   Regional mass of glacier ice smaller than a continent-sized ice sheet (less than 50,000 km² [20,000 mi²] in size). While the Scandinavian Ice Sheet covered northern Europe, an icecap covered the Alps in central Europe. (44)

Icecap climate (EF)   World's coldest climate, the harsher of the two subtypes of the E-climate zone. Characteristics, insofar as incomplete data reveal, are displayed in Fig. 17.12. (17)

Ice-crystal process   Most common process whereby precipitation forms and falls to Earth; rainfall in the tropical latitudes, however, is produced by the coalescence process. (11)

Ice-house climate   Also known as glacial climates, relatively brief periods of colder temperatures on Earth, generally lasting only tens of millions of years having large areas of permanent (year-round) ice. (18)

Icelandic Low   Local term for the northernmost North Atlantic Ocean's upper midlatitude low, centered approximately at 60 degrees N near Iceland. (9)

Ice sheet   (*see* **continental glaciers**)

Ice shelf   Smaller ice sheet that is a floating, seaward extension of a continental glacier, such as Antarctica's Ross Ice Shelf. (44)

Ice streams   Stream-like zone of unusually fast-flowing ice within an ice sheet; the best examples are in Antarctica. (44)

Ice-wedge polygons   Polygonal networks of vertical, linear wedges of ice in permafrost. Formed when dropping temperature causes frozen ground to shrink and crack in winter; in summer, water percolates down from the thawed active layer into the still-open crack in the permafrost and freezes there, causing the ice wedge to grow over time (*see* Fig. 46.6). (46)

Igneous rocks   Rocks that formed directly from the cooling of molten magma; igneous is Latin for "formed from fire." (27)

Illuviation   Soil process in which downward-percolating water carries dissolved material and colloid-sized particles of organic matter and minerals into the **B** horizon, where these materials are deposited in pore spaces and against the surfaces of soil grains. (21)

Impermeable   Material that does not permit water to pass through it. (38)

Inceptisol   One of the 12 soil orders of the Soil Taxonomy; a weakly developed soil but has the beginnings of a **B** horizon, which differentiates it from an Entisol. (23)

Incision   Process in which a stream erodes downward into underlying bedrock or sediment, lowering the elevation of the channel and floodplain; an essential process in stream terrace formation. (39, 41)

Inert gas   Chemically inactive gas that does not combine with other compounds. Argon, an inert atmospheric gas, makes up almost 1 percent of dry air. (6)

Infiltration   Flow of water into the Earth's surface through the pores and larger openings in the soil. (38)

Infiltration capacity   Rate at which a soil is able to absorb water percolating downward from the surface. (38)

Infrared radiation   Radiation at wavelengths longer than visible wavelengths, including longwave radiation (see **terrestrial radiation** and **longwave radiation**). (5)

Inner core   Solid, most inner portion of the Earth, consisting mainly of nickel and iron (*see* Fig. 29.5). (29)

Insolation   *In*coming *sol*ar radi*ation*. (4)

Intensity (earthquake)   Size and damage of an earthquake as measured—on the Modified Mercalli Scale—by the impact on structures and human activities in the cultural landscape. (33)

Interactive mapping   In geographic information systems (GIS) methodology, the constant dialogue between the map user and the maping program. (3)

Interception   Blocking of rainwater through vegetation from reaching the ground. Raindrops land on leaves and other plant parts and evaporate before they can penetrate the soil below. (38)

Interface   Surface-like boundary of a system or one of its component subsystems. The transfer or exchange of matter and energy takes place here. (1)

Interglacial   Period of warmer global temperatures between the most recent deglaciation and the onset of the next glaciation. (43)

International date line   For the most part a line that is antipodal to the prime meridian and follows the 180th meridian. Crossing the line toward the west involves losing a day, whereas crossing the line toward the east means gaining a day. (4)

Intertropical Convergence Zone (ITCZ)   Thermal low-pressure belt of rising air whose average position is along the Equator; fed by the wind flows of the converging Northeast and Southeast Trades. (9)

Intraplate earthquakes   Earthquakes that occur in the interior of a tectonic plate. (33)

Intrusive igneous rocks   Rocks formed from magma that cooled and solidified below the Earth's surface. (27)

Inversion   (*see* **temperature inversion**)

Ion   Atom or group of atoms that has acquired electrical charge; often dissolved in water. (22)

Ionization   Process by which sunlight in the mesophere and beyond strips atoms and molecules of electrons, thereby creating electrically charged particles. (6)

Isarithmic mapping   Commonly used cartographic device to represent three-dimensional volumetric data on a two-dimensional map; involves the use of isolines to show the surfaces that are mapped. (3)

Island arc   Volcanic island chain produced in a zone where two oceanic plates are converging. One plate will be subducted below the other, forming deep trenches as well as spawning volcanoes that may protrude above sea level in an island-arc formation (Alaska's Aleutian archipelago is a classic example). (30)

Isobar   Line connecting all points having identical atmospheric pressure. (8)

Isohyet   Line connecting all points receiving an identical amount of precipitation (3).

Isoline   Line connecting all places possessing the same value of a given phenomenon, such as "height" above the flat base of the surface being mapped. (3)

Isostasy   The condition of vertical equilibrium between the lower-density crust and the higher-density mantle that the crust "floats" on. Crust with lower density and/or greater thickness adjusts to an equilibrium vertical position where the land surface is at higher elevation, while thinner and/or higher-density crust results in a lower surface elevation; this explains the high elevation of many mountain summits and the low-elevation of ocean basins. (31)

Isotherm   Line connecting all points experiencing identical temperature. (7)

Isotope   Related form of a chemical element; has the same number of protons and electrons but different numbers of neutrons. (6)

ITCZ   (*see* **Intertropical Convergence Zone**)

Jet stream   A concentrated, "river" of air. There are two high-altitude, west-to-east-flowing jets that are major features of the upper atmospheric circulation system poleward of latitude 15 degrees in both the Northern and Southern Hemispheres. Because of their general occurrence above the subtropical and subpolar latitudes, they are respectively known as the *subtropical jet stream* and the *polar front jet stream*. A third such corridor of high-altitude, concentrated wind flow is the *tropical easterly jet stream*, a major feature of the upper air circulation equatorward of latitude 15 degrees N. This third jet stream, however, flows in the opposite, east-to-west direction and occurs only above the tropics of the Northern Hemisphere. (9)

Joint   Fracture in a rock in which no displacement movement has taken place. (27)

Joint plane   Planes of weakness and separation in a rock; produced in an intrusive igneous rock by the contraction of cooling magma (sedimentary and metamorphic rocks also display forms of jointing). (27)

Jointing   Tendency of rocks to develop parallel sets of fractures without any obvious displacement of rock along the fractures. (27)

Kame   Rounded hill of sand and gravel deposited by meltwater within or at the base of the glacier. (44)

Karst   Distinctive landscape associated with the underground chemical erosion of particularly soluble bedrock, usually limestone or dolomite. (42)

Katabatic winds   Winds that result from cold-air drainage; especially prominent under clear conditions where the edges of highlands plunge sharply toward lower-lying terrain. (8)

Kelvin scale   *Absolute* temperature scale used by scientists, based on the temperature of absolute zero (–273°C [–459.4°F]). A kelvin is identical to a Celsius degree (°C), so that water boils at 373 K (100°C) and freezes at 273 K (0°C). (7)

Kettle   Steep-sided depression formed in glaciated landscapes that is the result of the melting of a buried block of ice. (44)

Kinetic energy   Energy of movement. (7, 13)

La Niña   A pattern in the tropical Pacific Ocean consisting of anomalously cold sea-surface temperatures, stronger than normal trade winds, and increased precipitation in the western Pacific. La Niña's are in contrast to El Niño events. (10)

Laccolith   Intrusive igneous form in which a magma pipe led to a subterranean chamber that grew, dome-like, pushing up the overlying strata into a gentle bulge without destroying them. (27)

Lagoon   Normally shallow body of water that lies between an offshore barrier island and the mainland shoreline (49)

Lahar   Mudflow largely composed of volcanic debris. Triggered high on a snowcapped volcano by an eruption, a lahar can advance downslope at high speed and destroy everything in its path; frequently solidifies where it comes to rest. (32)

Land breeze   Offshore air flow affecting a coastal zone, resulting from a nighttime pressure gradient that steers local winds from the cooler (higher-pressure) land surface to the warmer (lower-pressure) sea surface. (8)

Land hemisphere   The roughly one-half of the Earth that contains most of the landmasses (*see* Fig. 2.6); the opposite of the water (oceanic) hemisphere. (2)

Landform   Single and typical unit that forms parts of the overall shape of the Earth's surface; also refers to a discrete product of a set of geomorphic processes. (35)

Landscape   Aggregation of landforms, often of the same type; also refers to the spatial expression of the processes that shaped those landforms. (35)

Landslide   Form of mass movement involving the downslope movement of one or more discrete blocks of sediment or rock; may become a flow as it moves downslope. (37)

Lapse rate   Rate of decline in temperature as altitude increases. The term can apply to the air column (see **environmental lapse rate**) or a rising air parcel (see **dry adiabatic lapse rate** and **saturated adiabatic lapse rate**) (6)

Last glacial maximum   Last period on Earth in which the ice sheets were at their greatest extension, about 20,000 years ago. (18)

Late Cenozoic Ice Age   Last great ice age that spanned the entire Pleistocene epoch (1.8 million to 10,000 years ago) plus the latter portion of the preceding Pliocene epoch, possibly beginning as far back as 3.5 million years ago. (43)

Latent heat   "Hidden" heat involved in the processes of melting, freezing, evaporation, and condensation (*see* **latent heat of fusion**; **latent heat of vaporization**). (5)

Latent heat of fusion   Heat energy involved in melting, the transformation of a solid into a liquid. A similar amount of heat is given off when a liquid freezes into a solid. (11)

Latent heat of vaporization   Heat energy involved in the transformation of a liquid into a gas or vice versa. (11)

Latent heat transfer   Heat transfer associated with the flow of water vapor. (5)

Lateral moraine   Moraine formed along the sides of a mountain glacier. (45)

Latitude   Angular distance, measured in degrees north or south, of a point on Earth's surface. (3)

Laurasia   Northern portion of the primeval supercontinent Pangaea. (30)

Laurentide Ice Sheet Huge Late Cenozoic continental glacier that covered all of Canada east of the Rocky Mountains and expanded repeatedly to bury northern U.S. areas as far south as the Ohio and Missouri valleys. (44)

Lava Magma that reaches the Earth's surface. (27)

Leap year Occurs every fourth year, when a full day (February 29) is added to the calendar to allow for the quarter-day beyond the 365 days it takes the Earth to complete one revolution of the Sun. (4)

Leeward Protected side of a topographic barrier with respect to the winds that flow across it; often refers to the area downwind from the barrier as well, which is said to be in the "shadow" of that highland zone. (8)

Legend (map) Portion of a map where its point, line, area, and volume symbols are identified. (3)

Levee, artificial Artificially constructed ridge or berm built to limit the area covered by floods, most often along the channel of large rivers. (41)

Levee, natural River-lining ridge of alluvium deposited when a stream overflows its banks during a periodic flood; usually contains somewhat coarser sediment that is part of the floodplain more distant from the channel. (41)

Liana Tropical rainforest vine, rooted in the ground, with leaves and flowers in the tree canopy above. (25)

LiDAR *Light Detection and Ranging.* LiDAR has various uses, perhaps the most important of which is for highly accurate surveys of ground surface elevation. (3)

Limestone Nonclastic sedimentary rock mainly formed from calcite, often produced by marine organisms. It is susceptible to solution that can produce karst landscapes. (28, 42).

Lithification Rock formation; the process of compaction and cementation whereby a sediment is transformed into a sedimentary rock. (28)

Lithology Rock type of a local area, which greatly influences its landform and landscape development. (40)

Lithosphere Outermost shell of the solid Earth, lying immediately below the land surface and ocean floor (*lithos* means rock); composed of the Earth's thin crust together with the rigid uppermost portion of the mantle that lies just below. (2, 29)

Lithospheric plate One of the fragmented, rigid segments of the lithosphere (also called a *tectonic plate*, which denotes its active mobile character). These segments or plates move over the hot, plastic asthenosphere that lies just below the lithosphere. (29, 30)

Little Ice Age Period of decidedly cooler global temperatures that prevailed from about 1400 to 1850. During these four centuries, mountain glaciers in most parts of the world expanded considerably. (18)

Littoral zone Coastal zone. (48)

Loam Soil containing grains of all three texture-size categories—sand, silt, and clay—in similar proportions, as indicated in Figure 22.2. (22)

Local base level Lowest elevation to which a stream may erode, considered at the local scale; local base-level is often a larger river or lake that the stream flows into. (39)

Loess Deposit of dust laid down after having been blown some distance (perhaps tens to hundreds of kilometers) by the wind; predominantly made up of silt grains, and characterized by the productive soils that formed in it and its ability to stand in steep vertical walls. (47)

Longitude Angular distance of a location, measured in degrees east or west from the prime meridian. (3)

Longitudinal dune Long, ridge-like sand dune that migrates in a direction approximately parallel to the ridge crest, often formed as a result of strong winds from two predominant directions. (47)

Longshore bar Ridge of sand parallel to the shoreline that develops in the nearshore beach zone. (49)

Longshore current Shore-paralleling water current which is generated by the refracted, oblique-angled arrival of waves onshore (*see* Figs. 48.4 and 48.8); can also develop from tidal action and from coastal storms. (48)

Longshore drift Movement of sediment along the shoreline by *longshore currents* and *beach drift* (*see* Fig. 48.8). (48)

Longwave radiation Radiation at wavelengths longer than about 4 micrometers, emitted by the Earth and other objects at terrestrial temperatures. Stands in contrast to solar (shortwave) radiation emitted by the Sun, effectively all of which is at wavelengths below 4 micrometers (5)

Lower mantle Solid interior shell of the Earth, which encloses the liquid outer core (*see* Fig. 29.5). (29)

Luster (mineral) Surface sheen of a mineral that, along with color, can be a useful identifying quality. (28)

Mafic rocks Dark-colored igneous rocks, formed from magma poor in silica. (27)

Magma Fluid, mostly molten mass of rock. (27)

Magnetosphere Upper portion of the thermosphere, which constitutes the outermost of the atmosphere's layers. Here the Earth's magnetic field is often more influential than its gravitational field in the movement of particles. (6)

Magnitude (earthquake) Energy released by an earthquake, calculated from shaking as measured by seismographs. (33)

Mangrove Type of tree, exhibiting stilt-like roots and a leafy crown, that grows in seawater on low-lying, muddy coasts in tropical and subtropical environments. In certain places, such as tidal flats and river mouths, these trees congregate thickly. In the United States they are most evident along the southwesternmost shore of Florida where the Everglades meet the sea. (49)

Mantle (*see* **lower mantle** *and* **upper mantle**)

Map A model of the environment, distinguished by its graphics. (3)

Map projection A system for converting Earth surface coordinates to a plane. (3)

Marble Metamorphosed limestone. The hardness and density of this rock is preferred by sculptors for use in statues that can withstand exposure to the agents of erosion for millennia. (28)

Marine geography Physical side of this subfield treats coastlines and shores, beaches, and other landscape features associated with the oceanic margins of the continents. (1)

Marine terrace A rock platform originally eroded by waves but now standing above sea level, because of tectonic uplift of the land or lower global sea level (49)

Marine west coast climate (Cfb, Cfc) Perpetually moist mesothermal climate type associated with midlatitude coasts (and adjacent inland areas not blocked by mountain barriers) bathed by the prevailing Westerlies year-round, whose maritime influences produce temperature regimes devoid of extremes. Characteristics are displayed in Figures 16.7, 16.8, and 16.9. (16)

Maritime effect Moderating influence of the ocean on air temperature, which produces cooler summers and warmer winters relative to inland locations at similar latitudes. (7)

Mass balance Net gain or loss of a glacier's ice over the course of a year; can be measured for a single point on the glacier or for the glacier as a whole. (43)

Mass movement Spontaneous downslope movement of Earth materials under the force of gravity. Materials involved move *en masse*—in bulk. (35, 37)

Master horizons Soil horizons marked by capital letters: **O, A, E, B, C,** and **R** (*see* Fig. 21.6). (21)

Mathematical projections Maps that do not use a projection surface such as a cylinder or cone. (3)

Meander Smooth, rounded curve or bend of rivers that can become quite pronounced as their development proceeds; also characteristic of many ocean currents, which, after the passage of storms, can produce such extreme loops that many detach and form localized eddies. (41)

Meander belts On a wide floodplain, river-meander development zones that themselves form giant meanders (*see* Fig. 41.7). (41)

Mechanical weathering Involves the destruction of rocks by physical means through the imposition of certain stresses, such as freezing and thawing or the expansion of salt crystals; also known as *physical weathering.* (36)

Medial moraine Moraine—situated well away from a glacier's edges—formed by the intersection of two lateral moraines when a substantial tributary glacier meets and joins a trunk glacier. (45)

Mediterranean climate (Csa, Csb) West Coast mesothermal climate whose signature dry summers result from the temporary poleward shift of the subtropical high-pressure zone. Characteristics are displayed in Figures 16.11 and 16.14. (16)

Mediterranean scrub biome Consists of widely spaced evergreen or deciduous trees and often-dense, hard-leaf evergreen scrub; sometimes referred to as *chaparral* or *maquis.* Thick, waxy leaves are well adapted to the long, dry summers. (25)

Melting Change from the solid state to the liquid state. It takes about 336 kilojoules of energy to change 1 kg (2.2 lb) of water from a solid into a liquid. (11)

Mercalli Scale (Modified) Scale that measures earthquake intensity, the impact of a quake on the human landscape; ranges upward from intensity I to intensity XII. (33)

Mercator map projection Most famous of the cylindrical projections, one on which any straight line is a line of true and constant compass bearing. (3)

Meridian On the Earth grid, a north–south line of longitude. These range from 0 degrees (prime meridian) to 180 degrees E and W (180 degrees E and W, of course, are the same line—the international date line [written simply as 180°]). (3)

Meridional flow North–south (*azonal*) movement of air. (9)

Mesa Flat-topped, steep-sided upland capped by a resistant rock layer; normally found in dry environments. (40)

Mesopause Upper boundary of the mesosphere, lying approximately 80 km (50 mi) above the surface. (6)

Mesophyte Plant adapted to environments that are neither extremely moist nor extremely dry. (24)

Mesosphere Third temperature layer of the atmosphere, lying above the troposphere and stratosphere. Here temperatures again decline with increasing elevation as they do in the troposphere. (6)

Mesothermal Mild midlatitude climates exhibiting moderate seasonality in temperature; the **C** climates. (16)

Mesozoic Era of medieval life on the geologic timescale (*see* Fig. 28.11), extending from 248 million to 65 million years ago. (28)

Metamorphic rocks Rocks that were created from the transformation by heat or pressure of existing rocks in the solid state (without melting). (27)

Meteorology Systematic, interdisciplinary study of the short-term atmospheric phenomena that constitute weather.

Methane A variable gas, both more effective at absorbing longwave radiation and rarer than CO_2. Methane is produced naturally by bacteria found in wetlands and in the stomachs of many animals. (6)

Microclimate Climate on a localized scale. (14)

Microthermal (D) climates Weakly heated continental climates of the Northern Hemisphere's upper midlatitudes, where the seasonal rhythms swing from short, decidedly warm summers to long, often-harsh winters; mostly confined to the vast interior expanses of North America and Eurasia poleward of 45 degrees N. (17)

Mid-Atlantic Ridge Submarine mountain range that forms the midoceanic ridge extending through the entire North and South Atlantic oceans, from Iceland in the north to near-Antarctic latitudes in the south; marks a divergent boundary between lithospheric plates. (2, 30)

Midoceanic ridge High submarine mountain ranges, part of a global system of such ranges that mark the location of divergent plate boundaries, most often found in the central areas of the ocean basins. Here new crust is formed by upwelling magma, which continuously moves away toward the margins of the ocean basins. (2, 30)

Midstream bar Mid-channel ridge of sand or gravel formed by local deposition of bedload from a stream; plays an important role in the development of braided stream channels. (41)

Milankovitch cycles Refers to cyclical changes in the shape of Earth's orbit, the degree of tilt of Earth's axis and the direction of tilt over periods of 100,000 years, 40,000 years, and 20,000 years. The cycles lead to variations in the seasonal and spatial distribution of sunlight reaching the planet and thereby have the potential to influence climate. (18)

Mild midlatitude (C) climates Moderately heated climates that are found on the equatorward side of the middle latitudes, where they are generally aligned as interrupted east–west belts; transitional between the climates of the tropics and those of the upper midlatitudes where polar influences begin to produce harsh winters. Also called mesothermal climates. (17)

Millennium Period of 1,000 years. (18)

Mineral Naturally occurring inorganic element or compound having a definite chemical composition, physical properties, and a crystalline structure. (27)

Mistral **winds** Katabatic wind flow, affecting France's Rhône Valley each winter, which involves the cold-air drainage of the massive pool of icy air that accumulates over the nearby high-lying French and Swiss Alps. (8)

Mixed layer The surface layer of an ocean, extending to about 75 m (245 ft) in depth and containing about 3 percent of the total ocean water volume. (10)

Model An idealized representation of reality demonstrating specific aspects of interest to the researcher. (1)

Moho Abbreviation for the Mohorovičić discontinuity. (29)

Mohorovičić discontinuity (Moho) Contact plane between the Earth's crust and the mantle that lies directly below it. (29)

Mohs Hardness Scale Standard mineral-hardness measurement scale used in the Earth sciences; ranges from 10 (the hardest substance, diamond) down to 1 (talc, the softest naturally occurring mineral). (27)

Mollisol One of the 12 soil orders of the Soil Taxonomy, usually found in temperate grasslands; characterized by a thick, dark surface layer that is rich in organic matter and minimal acidity. (23)

Moment Magnitude Scale Most widely used measure of an earthquake's magnitude (energy release); evolved from the Richter Scale developed in the 1930s. (33)

Monadnock Prominent remnant of an upland on a peneplain that has not yet eroded. (40)

Monsoon Derived from the Arabic word for "season," a seasonal reversal of windflow and pressure found in various places around the planet (most spectacularly in Southeast Asia). The moist onshore winds of summer bring the *wet monsoon*, whereas the offshore winds of winter are associated with the *dry monsoon*. (9)

Monsoon rainforest climate (Am) Tropical monsoon climate, characterized by sharply distinct wet and dry seasons. Characteristics are displayed in Fig. 15.5. (15)

Moraine Ridge of debris deposited at the margin of a glacier. Forms as sediment is carried by flowing ice to the glacier margin, where it is melted out and accumulates over time (44)

Mountain breeze A breeze that flows downward; results from a greater radiative heat loss from the mountain slopes at night and the resulting increase in density. (8)

Mountain (alpine) glacier Glaciers that form in mountainous regions, including cirque and valley glaciers and small icecaps. (43)

Mudflow Rapid, gravity-driven flow of relatively fine-grained sediment; usually refers to a type of debris flow. (37)

Multispectral systems Remote-sensing platforms capable of capturing radiation intensity in multiple spectral bands. (3)

Muskeg In the northern coniferous forest biome, the particular assemblage of low-growing leathery bushes and stunted trees that concentrate in the waterlogged soil bogs and lake-filled depressions. (25)

Mutation A permanent change in the genetic information coded by DNA and passed on through reproduction; contributes to the genetic variation that is acted upon by natural selection. (26)

Mutualism Biological interaction in which there is a coexistence of two or more species because one or more is essential to the survival of the other(s); also called *symbiosis*. (24)

Natural landscapes Areas of the planet still essentially subject to natural physical processes. (2)

Natural levees (*see* **levee, natural**)

Natural selection Process whereby organisms better adapted to their environment have a greater tendency to survive and produce offspring. (26)

Neap tide Least extreme tide that occurs when the Moon's pull works at right angles against the Sun's attraction and the Earth's rotational bulge (*see* Fig. 48.10B). (48)

Nearshore Beach zone that is located seaward of the foreshore, submerged even during an average low tide. Longshore bars and troughs develop here in this zone of complex, ever-changing topography. (49)

Negative feedback mechanism A process that acts to resist or counteract change in a system. (1)

Net primary productivity The amount of biomass produced annually by the primary producers minus the amount of biomass that is lost when the primary producers themselves carry out respiration. (24)

Net radiation The difference between total absorbed radiation and emitted radiation, including both longwave and shortwave radiation. See Table 5.1 for global average values. (5)

Niche (*see* **ecological niche**)

Nitrogen fixation Process by which nitrogen gas (N_2) is taken from the atmosphere and converted to nitrogen compounds that are available for uptake and use by living organisms. (20)

Nonclastic sedimentary rocks Derived not from particles of other rocks, but from chemical solution by deposition and evaporation or from organic deposition. (28)

Nonrenewable resource One that with continued use will ultimately be exhausted (metallic ores and petroleum are good examples). (21)

Nonsilicates Minerals consisting of compounds that do not contain the silicon and oxygen of the silicates; include the carbonates, sulfates, sulfides, and halides. (27)

Normal fault Fault produced by tension, in which the hanging wall drops relative to the footwall, so that the rocks involved are extended, or stretched out (*see* Fig. 34.6). (34)

North Atlantic Oscillation (NAO) Variations in the pressure gradient between the North Atlantic Ocean's semipermanent upper midlatitude low-pressure cell (the Icelandic Low) and its semipermanent subtropical Azores high-pressure cell. Its fluctuations affect the warmth and moisture of wind flows in Europe, and the NAO's complicated workings also involve the North Atlantic Drift ocean current, the sinking of warm water near Greenland and Iceland, the upwelling of cold water off northwest Africa, and many lesser components. (10)

Northeast Trades   Surface wind belt that generally lies between the equator and 30 degrees N; the Coriolis force deflects equatorward-flowing winds to the right, thus recurving north winds into northeast winds. (9)

Northern coniferous forest biome   Upper midlatitude boreal forest (known in Russia as the snowforest or *taiga*); dominated by dense stands of slender, cone-bearing needleleaf trees. (25)

Nuée ardente   (*see* **pyroclastic flow**).

O horizon   In certain soils, the uppermost layer—lying above the **A** horizon—consisting mainly of organic material in various stages of decomposition. (21)

Ocean currents   Large-scale movements of water that form the oceanic counterpart to the atmospheric system of wind belts and semipermanent pressure cells. (10)

Oceanic trench   Prominent seafloor feature, often adjacent to an island arc, in a zone where two oceanic plates converge and one is subducted beneath the other. The greatest depths of the world ocean lie in such trenches in the Pacific east of the Philippines. (2)

Offshore bar   Sandbar that lies some distance from the beach and is not connected to land. A longshore bar is an example. (49)

Open system   System whose boundaries (interfaces) freely permit the transfer of energy and matter across them. (1)

Optimum (biogeographic) range   For each plant and animal species, the range of a particular environmental factor (temperature, for example), in which that species can maintain a large, healthy population. (24)

Orders of magnitude   Sizes of geographic entities. Figure 1.8 shows the entire range of magnitudes, including those in which geographers usually operate. (1)

Organic matter   One of the four soil components; material that forms from living matter. (20, 21)

Orogenic belt   Chain of linear mountain ranges. (29)

Orographic precipitation   Rainfall (and sometimes snowfall) produced by moist air parcels that are forced to rise over a mountain range or other highland zone. (12)

Outer core   Liquid shell that encloses the Earth's inner core, whose composition involves similar materials (*see* Fig. 29.5). (29)

Outwash plain   Plain formed just beyond the margin of an ice sheet, where sediment carried out of the glacier by meltwater is deposited; braided channels are a common feature. (44)

Oxbow lake   Lake formed when two adjacent meanders link up and one of the bends in the channel, shaped like a bow, is cut off (*see* Fig. 41.4). (41)

Oxidation   Refers to a large number of chemical reactions involving oxygen and other elements. Oxidation occurs when organic matter is burned or decomposed in the soil, and when animals and plants draw stored energy from organic matter through the process called *respiration*. Oxidation can also be a mineral weathering process. An important example is the oxidation of sulfide minerals in shales and salt-marsh sediments, which releases large amounts of acidity (6, 36)

Oxisol   One of the 12 soil orders of the Soil Taxonomy, found in tropical areas with high rainfall. Intense weathering has altered or destroyed many of the minerals originally present in the parent material and accumulation of iron oxides give many Oxisols a red or orange color. (23)

Oxygen cycle   Pathways by which oxygen moves through the global environment. (20)

Ozone hole   Seasonal depletion of ozone above the Antarctic region, primarily caused by the release of industrial gases containing chlorine (especially chlorofluorocarbons and halocarbons) (6)

Ozone layer   Also known as the ozonosphere, the ozone-rich layer of the stratosphere that extends between 15 and 50 km (9 and 31 mi) above the surface. The highest concentrations of ozone are usually found at the level between 20 and 25 km (12 and 15 mi). (6)

Ozonosphere   Synonym for ozone layer. (6)

P wave   Type of seismic body wave (moves through Earth's interior); a compressional wave that moves material back and forth parallel to the direction of wave movement and can travel through material in a liquid state. (29)

Pacific Ring of Fire   Circum-Pacific belt of high volcanic and seismic activity associated with subduction zones at convergent plate boundaries; stretches around the entire Pacific Basin counterclockwise through western South America, western North America, and Asia's island archipelagoes (from Japan to Indonesia) as far as New Zealand. (30)

Pack ice   Floating sea ice that forms from the freezing of ocean water. (44)

Pahoehoe   Ropy-patterned lava (*see* Fig. 32.11); forms where very fluid lavas develop a smooth "skin" upon hardening, which wrinkles as movement continues. (32)

Paleocene–Eocene Thermal Maximum (PETM)   A brief spike in Earth's warmth that occurred about 55 million years ago. (18)

Paleozoic   Era of ancient life on the geologic timescale (*see* Fig. 28.11) extending from 570 million to 248 million years ago. (28)

Pangaea   Primeval supercontinent, hypothesized by Alfred Wegener, that broke apart and formed the continents and oceans as we know them today; consisted of two parts—a northern Laurasia and a southern Gondwana. (30)

Parabolic dune   Crescent-shaped sand dune with its arms extending upwind; the arms are partially or completely stabilized by vegetation, which strongly influences this type of dune. The concave side of this dune is the windward side. (47)

Parallel   On the Earth grid, an east–west line of latitude. These parallels range from 0 degrees (equator) to 90 degrees N and S (the North and South Poles, respectively, where the east–west line shrinks to a point). (3)

Parallelism   Constant tilt of the Earth's axis (at 66.5 degrees to the plane of the ecliptic) that remains parallel to itself at every position in its annual revolution of the Sun. (4)

Parent material   Rocks of the Earth, and the deposits formed from them, from which a soil has formed. (21)

Particulates   Bits of solid or liquid matter suspended in the Earth's atmosphere. (6)

Patterned ground Periglacial rock and soil debris shaped or sorted in such a manner that it forms designs on the surface resembling rings, polygons, lines, and the like (*see* Fig. 46.8). (46)

Ped Naturally occurring aggregate or "clump" of soil and its properties. (22)

Pedology Soil science; study of soils. (20)

Pedon Column of soil drawn from a specific location, extending from the **O** horizon (if present) all the way down into the C-horizon, or parent material. (22)

Peneplain Concept of a "near plain" developed by William Morris Davis to describe the nearly flat landscape formed by extensive erosion over long periods of time. (40)

Perched water table Separate local water table that forms at a higher elevation than the nearby main water table; caused by the effects of a local aquiclude (*see* Fig. 38.9A). (38)

Percolation Downward movement of water through the pores and other spaces in the soil under the influence of gravity. (38)

Perennial plant One that persists throughout the year. (25)

Periglacial High-latitude or high-altitude environment with a cold climate that strongly influences soil and geomorphic processes; permafrost is generally present. (46)

Perihelion Point in the Earth's orbit, which occurs every January 3, where the distance to the Sun is minimized (about 147.5 million km [91.7 million mi]). (4)

Permafrost Permanently frozen layer of subsoil; generally overlain by a layer of soil that thaws in summer (17, 46)

Permafrost table Upper surface of the permafrost (analogous to the water table). The soil above is subject to annual thawing and freezing. (17, 46)

Permeable Material that permits water to pass through it. Most natural soils are able to absorb, in this manner, at least a portion of the water that falls on them, but rock or pavement may be almost completely *impermeable*. (38)

pH scale Used to measure acidity and alkalinity of substances on a scale ranging from 0 to 14, with 7 being neutral. Below 7, increasing acidity is observed as 0 is approached, whereas above 7, increasing alkalinity is observed as 14 is approached. (17)

Photochemical reactions Chemical reactions in which sunlight plays an important role. (6)

Photosynthesis Process in which plants convert carbon dioxide and water into organic matter and oxygen using energy from solar radiation. Photosynthesis is the main form of primary production, which captures and stores energy that is then used by all of the organisms in an ecosystem. (24)

Phreatic zone (*see* **zone of saturation**)

Physical geography Geography of the physical world. Figure 1.2 diagrams the subfields of physical geography. (1)

Phytogeography Geography of flora or plant life; where botany and physical geography overlap. (20)

Phytomass Total living organic plant matter produced in a given geographic area; often used synonymously with *biomass*, because biomass is measured by weight (plants overwhelmingly dominate animals in total weight per unit area). (24)

Phytoplankton Microscopic photosynthetic algae at the beginning of the aquatic food chain. (24)

Pingo Moundlike, elliptical hill in a periglacial zone whose core consists of ice rather than rock or soil (*see* Fig. 46.9). (46)

Planar map projection One in which the transfer of the Earth grid is from a globe onto a plane, involving a single point of tangency. (3)

Plane of the ecliptic Plane in which Earth travels during its annual revolution around the Sun. (4)

Plankton (*see* **phytoplankton**; **zooplankton**)

Plant nutrients In a soil, the elements essential for plant growth that circulate through the humus–vegetation system, including nitrogen, phosphorus, and numerous others. (21)

Plant succession Process in which one assemblage of plants is replaced by another in a predictable pattern that is repeatedly observed; succession follows disturbance of existing vegetation or an event that produces bare soil. (24)

Plate tectonics Study the processes by which lithospheric plates move over the asthenosphere, and the many consequences that follow from plate motion, including mountain uplift, faulting, and volcanism. (30)

Platy soil structure Involves peds that look like flakes stacked horizontally (*see* Fig. 22.3). (22)

Pleistocene Epoch that extended from 1.8 million to 10,000 years ago on the geologic timescale; includes the latter half of the last great ice age, the Late Cenozoic, which began about 3.5 million years ago, as well as the emergence of humankind. (43)

Plucking (quarrying) Glacial erosion process in which fragments of bedrock beneath the glacier are broken loose from the surface as the ice advances. (43)

Plunging fold Anticline or syncline whose axis dips from the horizontal. (34)

Plutonic igneous rocks Intrusive igneous rocks that form at a significant depth below the surface. (27)

Pluvial lake Lake that developed in a presently dry area during times of heavier precipitation associated with glaciations. The glacial Lake Bonneville, the (much larger) forerunner of Utah's Great Salt Lake, is a classic example. (44)

Polar cell Feature of the 3-cell model of atmospheric general circulation found at high latitudes in both hemispheres. Centered over the poles, they consist of high pressure at the surface and descending air. Air flowing out of the high gives rise to the polar easterlies. (9)

Polar (E) climates Climates in which the mean temperature of every month is less than 10°C (50°F). The tundra (**ET**) subtype has at least one month with average temperature between 0°C (32°F) and 10°C (50°F), whereas in the (coldest) icecap (**EF**) subtype the average temperature of every month is below 0°C (32°F). (17)

Polar easterlies High-latitude wind belt in each hemisphere, lying between 60 and 90 degrees of latitude. The Coriolis force is strongest in these polar latitudes, and the equatorward-moving air that emanates from the polar high is sharply deflected in each hemisphere to form the polar easterlies. (9)

Polar front Latitudinal zone, lying at approximately 60 degrees N and S, where the equatorward-flowing polar easterlies meet the poleward-flowing westerlies. The warmer westerlies are forced to rise above the colder easterlies, producing a semipermanent surface low-pressure belt known as the *subpolar low*. (9)

Polar front jet stream   Upper-atmosphere jet stream located above the subpolar latitudes, specifically the polar front; at its strongest during the winter season. (13)

Polar high   Large semipermanent high-pressure cell centered approximately over the pole in the uppermost latitudes of each hemisphere. (9)

Pollutant   Any substance that impacts an organism (or structures) negatively. (6)

Porosity (soil)   Proportion of the soil volume that is open space between mineral part, occupied by air or water. (22)

Positive feedback mechanism   A process that acts to amplify change in a system. (1)

Potential energy   Energy an object has by virtue of its position relative to another object. (13)

Potential evapotranspiration (PE)   The amount of water that would be lost to the atmosphere from an underlying well-watered vegetated surface. PE represents the atmosphere's demand for water. In dry climates PE is far in excess of actual evapotranspiration. (11)

Precambrian   Era that precedes the Paleozoic era of ancient life on the geologic timescale (*see* Fig. 28.11), named after the oldest period of the Paleozoic, the Cambrian; extends backward from 570 million years ago to the origin of the Earth, now estimated to be about 4.6 billion years ago. (28)

Precipitation   Any liquid water or ice that falls to the Earth's surface through the atmosphere (rain, snow, sleet, and hail). (11)

Predation   Biological interaction in which one species feeds on members of another species; in a broad sense it can apply only to animals feeding on other animals, but also to animals feeding on plants. (24)

Pressure (atmospheric)   Weight of a column of air per unit area at a given location, determined by the acceleration of gravity and the mass of atmosphere at that location. *Standard sea-level air pressure* produces a reading of 760 mm (29.92 in) on the mercury barometer. In terms of pressure, it is also given as 1,013.25 millibars (mb), or 14.7 lb/in^2. (8)

Pressure-gradient force   The difference in surface pressure divided by the distance between two locations is called the *pressure gradient*. Pressure gradient gives rise to a force that causes air to move (as wind) from the place of higher pressure to that of lower pressure. (8)

Prevailing Westerlies   (*see* **Westerlies**)

Primary circulation   (*see* **general circulation**)

Primary succession   Occurs when plants become established on bare ground or in a lake, where soils are not yet present (*see also* **plant succession**). (24)

Prime meridian   North–south line on the Earth grid, passing through the Royal Observatory at Greenwich, London, defined as having a longitude of 0 degrees. (3)

Prismatic soil structure   Involves peds arranged in vertical columns (*see* Fig. 22.4). (22)

Profile   (**stream**) Profile of elevation along a stream from its head in an interior upland to its mouth; often referred to as the *longitudinal profile* of a stream. (39)

Pyroclastic flow   A fluid mass of cooling bits of magma suspended in hot gases, which races downslope at speeds exceeding 100 km/h (65 mph). Also called a *nuée ardente*. (32)

Pyroclastics   Collective name for sediments and rocks made up of volcanic fragments, from fine ash particles to large cinders and bombs. Pyroclastics suspended in hot gases can flow down the slopes of a volcano and through adjacent river valleys; ash can also be blown high into the troposphere and settle in layers on the landscape far from the volcano. (32)

Quartz   One of the most common silicate minerals in the upper continental crust; compound of silicon and oxygen, which are arranged in tetrahedral structures. (27)

Quartzite   Very hard metamorphic rock that resists weathering; formed by the metamorphosis of sandstone (made of quartz grains and a silica cement). (28)

Quaternary   Second of the two periods of the Cenozoic era of recent life on the geologic timescale (*see* Fig. 28.11), extending from 1.8 million years ago to the present. (28)

R horizon   Soil horizon designation used for bedrock encountered at the base of a soil profile (21)

Radial drainage   Stream pattern that emanates outward in many directions from a central mountain (*see* Fig. 40.11A). (40)

Radiation   Transmission of heat energy in the form of electromagnetic waves. The energy of each wave is inversely proportional to its wavelength (i.e., shorter waves are more energetic). The totality of wavelengths comprises the electromagnetic spectrum (*see* Fig. 3.19). (5)

Radiation fog   Develops at night during clear skies, when the surface and lower atmosphere are cooled efficiently by radiation. (11)

Rain   Precipitation consisting of large liquid water droplets. (11)

Rain shadow effect   Dry conditions—often at a regional scale, as in the U.S. interior West—which occur on the leeward side of a mountain barrier that experiences orographic precipitation. The loss of moisture as air flows upward on the windward side together with adiabatic warming and falling relative humidity on descent greatly reduces the chances for precipitation in the rain shadow. (12)

Rainsplash erosion   (*see* **splash erosion**)

Range (animal)   Geographic area in which an animal species naturally occurs; often changes over time, and in some cases even seasonally. (26)

Recessional moraine   Morainal ridge marking a place where glacial retreat was temporarily halted, or the limit reached by a brief readvance of the ice during a period of overall retreat. (44)

Rectangular drainage   Stream pattern dominated by right-angle contacts between rivers and tributaries; these contacts are not as pronounced as in trellis drainage (*see* Fig. 40.11D). (40)

Recumbent fold   Highly compressed fold, which doubles back on itself with an axial plane that is nearly horizontal. (34)

Recurrence interval   The average time between floods that reach or exceed a given magnitude. (39)

Regime   (*see* **soil regime**)

Relative humidity   Proportion of water vapor present in a parcel of air relative to the maximum amount of water vapor that air could hold at the same temperature. (11)

Relief   Vertical distance between the highest and lowest elevations in a given area. (29, 35)

Remote sensing   Technique for imaging objects without the sensor being in immediate contact with the local scene. (3)

Renewable resource   One that can regenerate as it is exploited. (20)

Representative concentration pathways (**RCPs**) Greenhouse gas concentration scenarios for the future; used in general circulation models to project climatic responses to anthropogenic greenhouse gas emissions. (19)

Residence time   A measure of how fast a particular gas cycles in and out of the atmosphere. It is the length of time (usually expressed in years) a molecule remains in the atmosphere before being consumed in a chemical reaction or lost to the atmosphere in some other way. Residence time is equal to the amount a particular gas in the atmosphere divided by the rate of input or output of that gas. (6)

Residual parent material   Soil parent material that is formed in place though weathering of the local bedrock. (21)

Respiration   Process in which living organisms release energy stored in the chemical bonds of organic compounds; a form of oxidation in which oxygen reacts with the organic compounds, producing carbon dioxide. (24)

Reverse fault   Fault resulting from compression of the crust; the hanging wall rises relative to the footwall (*see* Fig. 34.2). (34)

Revolution   One complete circling of the Sun by a planet. It takes the Earth precisely one year to complete such an orbit. (4)

Rhumb line   A line of constant compass bearing. Rhumb lines appear as straight lines on the Mercator map projection, but as curving arcs on most others. (3)

Richter Scale   Open-ended numerical scale, first developed in the 1930s, that measures earthquake magnitude; ranges upward from 0 to 8+ and has evolved into the *Moment Magnitude Scale*. (33)

Rift   Opening of the crust, normally into a trough or trench, that occurs in a zone of plate divergence; *see* **rift valley**. (30)

Rift valley   Develops in a continental zone of plate divergence where tensional forces pull the crustally thinning surface apart. The rift valley is the trough that forms when the land sinks between parallel faults in strips. (30)

Rip current   Narrow, short-distance, stream-like current that moves seaward from the shoreline, cutting directly across the oncoming surf. (48)

Ripples   Small, regularly spaced ridges on the surface of a dune or streambed, usually less than 1 cm (0.4 in) high on a dune, but sometimes larger on streambeds. (27, 47)

Roche moutonnée   Landform created by glacial erosion of bedrock; asymmetrical mound that results from abrasion to one side (the side from which it advanced) and plucking on the leeward side. (43)

Rock   Any naturally formed, firm, and consolidated aggregate mass of mineral matter, of organic or inorganic origin, that constitutes part of the planetary crust; *see* **igneous**, **sedimentary**, and **metamorphic rocks**. (27)

Rock cycle   Pathways of transformation by which new rocks form, are weathered and eroded or undergo metamorphosis or melting, and thus become rocks of a different type; for example, an igneous rock weathered into particles that eventually become sedimentary rock, and that sedimentary rock can then be affected by tectonic processes that turn it into metamorphic rock (*see* Fig. 28.13). (28)

Rock flour   Very finely ground-up debris carried downslope by a mountain glacier; when deposited, often blown away by the wind to form *loess*. (45)

Rock sea   (see **blockfield**)

Rock steps   Step-like mountainside profile (in the postglacial landscape) often created as an eroding alpine glacier moves downslope. (45)

Rockslide   Type of mass movement consisting mainly of bedrock that breaks loose from a mountain side and slides or flows downslope. (37)

Rossby waves   Also called longwaves, airstreams that curve north and south as they move air generally eastward. They develop as a dynamic response to the equator-to-pole temperature gradient and are also affected by Earth topography and land/ocean contrasts. (9)

Rotation   Spinning of a planet on its axis, the imaginary line passing through its center and both poles. It takes the Earth one calendar day to complete one full rotation. (4)

Runoff   Removal of water that falls as precipitation, via the network of streams and rivers. Can refer to all of the water that is not evaporated or transpired, or just to the water that does not infiltrate into the soil but instead flows away over the land surface; in the latter case the term *overland flow* is also used. (11, 38)

S wave   Type of seismic body wave (moves through Earth's interior); causes motion at right angles to its direction of movement but (unlike **P** waves) cannot travel through material in a liquid state. (27)

Sahel   Derived from the Arabic word for shore, the name given to the east–west semiarid (**BSh**) belt that constitutes the "southern shore" of North Africa's Sahara; straddles latitude 15 degree N and stretches across the entire African continent (*see* Fig. 14.3). Suffered grievous famine in the 1970s that killed hundreds of thousands and since then has periodically been devastated by severe droughts. (14)

SALR   (*see* **saturated [wet] adiabatic lapse rate**)

Salt wedging   Form of mechanical weathering in arid regions. Salt crystals behave much like ice in the process of frost action, entering rock joints dissolved in water, staying behind after evaporation, growing and prying apart the surrounding rock. (36)

Saltation   Process in which sediment is moved by stream flow or wind that entails particles hopping along the streambed or across the ground surface; sand and gravel can saltate in a stream, while the wind generally moves only sand-sized particles in saltation. (39, 47)

Sand   Coarsest grains in a soil. Sand particles range in size from 2 to 0.05 mm (0.08 to 0.002 in). (22)

Sandspit   Elongated extension of a beach into open water where the shoreline reaches a bay or bend; built and maintained by longshore drift. (49)

Sandstone   Common sedimentary rock containing sand-sized grains. (28)

Santa Ana wind   Hot, dry, *foehn*-type wind that occasionally affects Southern California. Its unpleasantness is heightened by the downward funneling of this air flow from the high inland desert through narrow passes in the mountains that line the Pacific Coast. (8)

Saturated air   Air that is holding all the water vapor molecules it can possibly contain at a given temperature. (11)

Saturated (wet) adiabatic lapse rate (SALR) Lapse rate of an air parcel saturated with water vapor in which condensation is occurring; unlike the *dry adiabatic lapse rate* (DALR) the value of the SALR is variable, depending on the amount of water condensed and the corresponding latent heat released. A typical value for the SALR at 20°C (68°F) is –0.44°C/100 m (–2.4°F/1,000 ft). (7)

Saturated zone Belowground zone that is saturated, that is, all the pores are filled with water. The boundary between the saturated zone and the overlying *vadose zone* is the *water table*. (38)

Savanna biome Transitional environment between the tropical rainforest and the subtropical desert; consists of widely spaced trees with a grass understory. (25)

Savanna climate (Aw) Tropical wet-and-dry climate located in the transition, still-tropical latitudes between the subtropical high-pressure and equatorial low-pressure belts. Characteristics are displayed in Fig. 15.7. (15)

Scale Ratio of the size of an object on a map to the actual size of the object it represents. (3)

Schist Common metamorphic rock so altered that its previous form is impossible to determine; fine-grained, exhibits wavy bands, and breaks along parallel planes (but unevenly, unlike slate). (28)

Scientific laws Fundamental principles believed to hold without exception, for example, Newton's law. (1)

Scientific method A process consisting of certain elements that together constitute a convention for obtaining facts or learning the "truth" about phenomena of interest (*see* Fig. 1.3). (1)

Scree slope Steep accumulation of weathered rock fragments and loose boulders that rolled downslope in free fall; particularly common at the bases of cliffs in the drier climates of the western United States (also called a *talus slope*). (36)

Sea arch Small island penetrated by the sea at its base that originated as an especially resistant portion of a headland, now eroded away by waves. (49)

Sea breeze Onshore airflow affecting a coastal zone, resulting from a daytime pressure gradient that steers local winds from the cooler (higher-pressure) sea surface onto the warmer (lower-pressure) land surface. (8)

Sea cave Cave carved by undercutting waves that are eroding the base of a sea cliff. (49)

Sea cliff Especially steep coastal escarpment that develops when headlands are eroded by waves. (49)

Seafloor spreading Process wherein new crust is formed by upwelling magma at the midoceanic ridges and then continuously moves away from its source toward the margins of the ocean basin. (30)

Seamount Undersea abyssal-zone volcanic mountain reaching over 1,000 m (3,300 ft) above the ocean floor. (2)

Secondary succession Follows some form of vegetation disturbance, an event that disrupts an ecosystem, such as fire, an outbreak of tree-killing insects, or strong winds that blow down a forest (*see also* **plant succession**). (24)

Sediment yield Total mass of sediment leaving a drainage basin in a given period of time; often divided by basin area to allow comparison between basins of differing size. (39)

Sedimentary rocks Rocks that formed from the deposition and compression of rock and mineral fragments. (27)

Seismic reflection When a seismic wave traveling through a less dense material reaches a place where the density becomes much greater, it can be bounced back or reflected. (29)

Seismic refraction When a seismic wave traveling through a less dense material reaches a place where the density becomes greater, it can be bent or refracted. (29)

Seismic waves Pulses of energy generated by earthquakes that can pass through the entire planet. (29)

Seismograph Device that measures and records the seismic waves produced by earthquakes and Earth vibrations. (29)

Sensible heat flow Environmental heat we feel or sense on our skins. (5)

Shaft (solution feature) In karst terrain, a pipelike vertical conduit that leads from the bottom of a solution hollow. Surface water is funneled down the shaft to join a subsurface channel or the groundwater below the water table. (42)

Shale Finest grained of the soft sedimentary rocks; formed from compacted mud. (28)

Shear stress Stress that acts parallel to a surface. On a slope, shear stress is generated by the weight of soil or weathered, part of which acts as a force pushing material downslope; in this case the shear stress is resisted by the frictional strength of the material on the slope, plus cohesion. (37)

Sheet erosion Erosion produced by sheet flow as it removes fine-grained surface materials. (39)

Sheet flow Surface runoff of rainwater not absorbed by the soil (also called *sheet wash*); forms a thin layer of water that moves downslope without being confined to local stream channels. (39)

Shield volcano Formed from fluid basaltic lavas that flow in sheets that are built up gradually by successive eruptions. In profile their long horizontal dimensions peak in a gently rounded manner that resembles a shield (the main island of Hawai'i has some of the world's most active shield volcanoes). (32)

Shingle beach Beach consisting not of sand but of gravel and/or pebbles; associated with high-energy shorelines. (49)

Shoaling Nearshore impact of ever-shallower water on an advancing, incoming wave. (48)

Shore Has a more specific meaning than *coast*; denotes the narrower belt of land bordering a body of water, the most seaward portion of a coast (the *shoreline* is the actual contact border). (48)

Shoreline Within a *shore zone*, the actual contact border between land and water. (48)

Shortwave radiation Radiation coming from the Sun, which has much shorter wavelengths—and involves much higher energy—than the terrestrial (longwave, lower-energy) radiation emitted by the Earth. (5)

Siberian High Local name for northern Asia's polar high, which is centered over the vast north-central/northeastern region of Russia known as Siberia. (9)

Silicates Minerals consisting of silicon and oxygen and, mostly, other elements as well; basic building block is a tetrahedron in which a silicon atom is bonded to four oxygen atoms. (26)

Sill   Intrusive igneous form in which magma has inserted itself as a thin layer between strata of preexisting rocks without disturbing those layers to any great extent. (27)

Silt   Next-smallest category of soil particles after sand, the coarsest variety. Silt particles range in size from 0.5 to 0.002 mm (0.002 to 0.00008 in). (22)

Slate   Metamorphosed shale; a popular building material, it retains shale's quality of breaking along parallel planes. (28)

Sleet   Precipitation consisting of pellets of ice produced by the freezing of rain before it reaches the surface; if it freezes after hitting the surface, it is called *freezing rain*. (11)

Slide (*see* landslide).

Slip face   Leeward slope of a sand dune. (47)

Slope (river)   Drop in elevation divided by distance along the stream channel; the slope that drives downstream flow of water, erosion of the streambed, and transportation of sediment. (38)

Slope decline   Process in which slopes are worn *down*, becoming more convex in appearance, and then are flattened. (40)

Slope failure   Point at which slope materials begin to slide or flow away downslope. (37)

Slope retreat   Process in which slopes erode and shift horizontally without becoming much gentler over time, moving backward rather than downward. (40)

Slump   Type of landslide in which the failure surface is curved and the material in motion forms backward-rotating slump blocks (*see* Fig. 37.6). (37)

Small circles   Individual parallels splitting Earth in unequal pieces, each smaller than Earth's circumference. (3)

Smog   Poor-quality surface-level air lying beneath a temperature-inversion layer in the lower atmosphere. The word is derived from the contraction of "smoke" and "fog." (7)

Snow   Precipitation consisting of large ice crystals called snowflakes; formed by the ice-crystal process when crystals do not have time to melt before they reach the ground. (11)

Snowball Earth   Term used for the most extreme ice-house climates, called such because nearly the entire planet is covered with ice during these times. (18)

Snow line   High-altitude boundary above which snow remains on the ground throughout the year. (43)

Soil   Mixture of fragmented and weathered grains of minerals and rocks with organic matter and variable proportions of air and water. The mixture has a fairly distinct layering or horizonation, and its development is due to processes that are especially active just below a vegetation-covered land surface. (21)

Soil air   Air that fills the spaces among the mineral particles, water, and organic matter in the soil; often contains more carbon dioxide than atmospheric air does. (21)

Soil catena   Derived from a Latin word meaning chain or series, refers to a sequence of soil profiles appearing in regular succession on hillsides within a particular region (*see* Fig. 22.7). (22)

Soil components   These are four in number: minerals, organic matter, water, and air. (21)

Soil consistence   Subjective measure of a moist or wet soil's stickiness, plasticity, cementation, and hardness. This test is done in the field by rolling some damp soil in the hand and observing its behavior. (22)

Soil-formation factors   These are five in number: parent material, climate, biological agents, topography, and time. (21)

Soil geography   Systematic study of the spatial patterns of soils, their distribution, and the interrelationships with climate, vegetation, and humankind. (1)

Soil heat flow   Heat conducted into and out of the Earth's land surface; also known as *ground heat flow*. (5)

Soil horizon   Zone in a soil that is more or less parallel to the ground surface and differs from other horizons above and below it in terms of texture, color, structure, or other properties. The differentiation of soils into horizons is called *horizonation*. (21)

Soil moisture   Rainwater or snowmelt that has infiltrated into the soil; is present in the ground in either the liquid or frozen state. (11)

Soil order   In the Soil Taxonomy, the broadest possible classification of the Earth's soils into one of 12 major categories (*see* Table 23.1); a very general grouping of soils with broadly similar composition, the presence or absence of certain diagnostic horizons, and similar degrees of horizon development, weathering, and leaching. (23)

Soil profile   Entire array of soil horizons from top to bottom. (21)

Soil regime   A set of environmental conditions that occurs over substantial areas and favors certain processes of soil formation. (21)

Soil Taxonomy   Soil classification scheme used by contemporary pedologists and soil geographers; evolved from the Comprehensive Soil Classification System (CSCS) that was derived during the 1950s. (23)

Soil texture   Size of the particles in a soil; more specifically, the proportion of sand, silt, and clay in each soil horizon. (22)

Solar altitude   The angle of the Sun above the horizon. (4)

Solifluction   Slow downslope movement of soil that is common in periglacial environments; a combination of creep and slow flow of soil saturation by water during the thaw season. (37, 46)

Solstice   (*see* **summer solstice; winter solstice**)

Solum   The **A** and **B** horizons of a soil, that part of the soil in which plant roots are active and play a role in the soil's development. (22)

Solution sinkhole   In karst terrain, a funnel-shaped surface hollow (with the shaft draining the center) created by solution; ranges in size from a bathtub to a stadium. (42)

Source region   Extensive geographic area, possessing relatively uniform characteristics of temperature and moisture, where large air masses can form. (12)

Southeast Trades   Surface wind belt that generally lies between the equator and 30 degrees S. The Coriolis force deflects equatorward-flowing winds to the left, thus recurving south winds into southeast winds. (9)

Southern Oscillation   Periodic, change in pressure gradient in the atmosphere overlying the equatorial Pacific; associated with the occurrence of the El Niño and La Niña phenomena. (*see* Fig. 10.9). (10)

Species   Population of physically and chemically similar organisms within which free gene flow takes place. (26)

Specific humidity Ratio of the mass of water vapor in the air to the combined mass of the water vapor plus the air itself.

Spheroidal weathering Process by which rock flakes off in thin rounded layers in what looks like a small-scale version of *exfoliation* (*see* Fig. 36.5); common in certain rock types such as granite. (36)

Splash erosion Process involving the dislodging of soil particles by large heavy raindrops (*see* Fig. 39.1). (39)

Spodosol One of the 12 soil orders of the Soil Taxonomy, found in large areas of northern coniferous forest and in smaller patches in tropical forests; characterized by downward movement of iron, aluminum, and humus, to form dark reddish brown **B** horizons. A light gray **E** horizon often develops in the zone of eluviation where those materials are depleted (23)

Spring (season) Northern Hemisphere season that begins at the spring (vernal) equinox around March 21 and ends at the summer solstice on June 22. The Southern Hemisphere spring begins at the equinox that occurs around September 23 and ends at the solstice on December 22. (4)

Spring (water) Surface stream of flowing water that emerges from the ground; a location of groundwater discharge. (38)

Spring tide Unusually high tide that occurs when the Earth, Moon, and Sun are aligned, as shown in Fig. 48.10A. (48)

Spring (vernal) equinox In Northern Hemisphere terminology, the equinox that occurs when the Sun's noontime rays strike the equator vertically on or around March 21. (4)

Spur Ridge that thrusts prominently from the crest or side of a mountain. (45)

Stability (of air) The tendancy of an air parcel to rise or sink owing to buoyancy. The air is stable if air parcels return to their original positions after some upward movement. By contrast, if air parcels rise spontaneously (or after some initial force), the atmosphere is unstable. (7)

Stack Column-like island that is a remnant of a headland eroded away by waves. (49)

Stalactite Icicle-like rock formation hanging from the roof of a cave. (42)

Stalagmite Upward-tapering, pillar-like rock formation standing on the floor of a cave. (42)

Standard parallel Latitude of tangency between a globe and the surface onto which it is projected. (3)

Steady-state system System in which inputs and outputs are constant and equal.(1)

Steppe climate (BS) Semiarid climate, transitional between fully developed desert conditions (**BS**) and the subhumid margins of the bordering **A**, **C**, and **D** climate regions. Characteristics are displayed in Fig. 15.12. (15)

Stock Massive pluton that is smaller than a batholith. (27)

Storm Organized, moving atmospheric disturbance. (13)

Storm surge Wind-driven wall of water hurled ashore by the approaching center of a hurricane, which can surpass normal high-tide levels by more than 5 m (16 ft); often associated with a hurricane's greatest destruction. (13, 48)

Strata Layers of rock or sediment. (28)

Strath terrace (*see* **terrace, strath**)

Stratification Layering of rock or sediment. (28)

Stratigraphy Order and arrangement of rock strata. (28)

Stratopause Upper boundary of the stratosphere, lying approximately 52 km (32 mi) above the surface. (6)

Stratosphere Atmospheric layer lying above the troposphere. Here temperatures are either constant or increasing with altitude. (6)

Stratovolcano Alternative term for a composite volcano.

Stratus Cloud-type category encompassing layered and fairly thin clouds that cover an extensive geographic area; subclassified according to height. (11)

Streak (mineral) Color of a mineral in powdered form when rubbed against a porcelain plate; used in mineral identification. (27)

Stream piracy Capture of a segment of a stream by another river. (40)

Stream power A general indicator of a stream's ability to do work. The total stream power of a river increases with *discharge* and *slope* of the water surface. Therefore, the larger the discharge and the steeper the slope, the more sediment the river can transport downstream. (39)

Striations Scratches in rock that are made as a boulder or pebble embedded in a glacier is dragged along the glacier's rock bed. (43)

Strike Compass direction of the line of intersection between a rock layer and a horizontal plane. (33)

Strike-slip fault Transverse fault in which crustal blocks move horizontally, parallel to the fault trace. (34)

Subarctic climate Climate characterized by long, cold winters and short, cool summers. (17)

Subduction Process that takes place when two lithospheric plates converge and one plunges below the other one. The plate that plunges downward is oceanic and therefore is thinner, denser, and less buoyant than the overriding plate, which is often continental. The plunging plate descends into the mantle and is destroyed (30)

Subduction zone Area in which the process of subduction is taking place; common setting for earthquakes and volcanic activity. (30)

Subduction zone earthquakes Earthquakes that occur along the boundary of two lithospheric plates meet in a subduction; the plate boundary is, in effect, a large thrust fault. (33)

Sublimation Process whereby a solid can change directly into a gas. The reverse process is sometimes also called sublimation, but is more properly called vapor deposition. The heat required to produce these transformations is the sum of the latent heats of fusion and vaporization. (11)

Submarine canyon Submerged river valley on the continental shelf. Before the rise in sea level (of about 120 m [400 ft]) over the past 10,000 years, the shelf was mostly dry land across which the river flowed to its former outlet. (2)

Submergent coast Drowned coastal zone; submergence is caused in large part by the rise in sea level (about 120 m [400 ft]) of the past 10,000 years. (49)

Subpolar gyre Oceanic circulation loop found only in the Northern Hemisphere. Its southern limb is a warm current steered by prevailing westerly winds, but the complex, cold returning flows to the north are complicated by sea-ice blockages and the configuration of landmasses vis-à-vis outlets for the introduction of frigid Arctic waters. (10)

Subpolar low A belt of low pressure at the surface, produced by warm and moist air from a subtropical high rising over denser cold and dry polar air. (9)

Subsidence Vertical downflow of air toward the surface from higher in the troposphere.(7)

Subsystem Component of a larger system. It can act independently, but operates within, and is linked to, the larger system. (1)

Subtropical gyre Ocean circulation around the subtropical high that is located above the center of the ocean basin (*see* Fig. 10.4); dominates the oceanic circulation of both hemispheres, flowing clockwise in the Northern Hemisphere and counterclockwise in the Southern Hemisphere. (10)

Subtropical high Semipermanent belt of high pressure that is found at approximately 30 degrees of latitude in both the Northern and Southern Hemispheres. The subsiding air at its center flows outward toward both the lower and higher latitudes. (9)

Subtropical jet stream Jet stream of the upper atmosphere that is most commonly located above the subtropical latitudes; evident throughout the year. (9)

Subtropical monsoon climate Also known as the **Cw** climate, with an average annual rainfall similar to the tropical savanna climate, but with a distinct cool season. (16)

Suffosion sinkhole Collapse sinkhole created when an overlying layer of unconsolidated material is left unsupported. (42)

Summer In Northern Hemisphere terminology, the season that begins on the day of the summer solstice (June 22) and ends on the day of the fall (autumnal) equinox (around September 23). (4)

Summer solstice Each year, day of the poleward extreme in the latitude where the Sun's noontime rays strike the Earth's surface vertically. In the Northern Hemisphere that latitude is 23.5 degrees N (the Tropic of Cancer) and the date is June 22; in the Southern Hemisphere that latitude is 23.5 degrees S (the Tropic of Capricorn) and the date is December 22. (4)

Superimposed stream River that crosses a structural feature that would normally impede its flow because the feature was at some point buried beneath the surface on which the river developed. As the feature became exposed, the river kept cutting through it. (40)

Surf Water zone just offshore dominated by the development and forward collapse of breaking waves. (48)

Surface creep Movement of fairly large rock fragments by the wind actually pushing them along the ground, especially during windstorms. (47)

Surface detention Water that collects on the surface in pools and hollows, bordered by millions of tiny natural dams, during a rainstorm that deposits more precipitation than the soil can absorb. (38)

Surface layer Also known as the *friction layer*, the lower troposphere where frictional forces are important, generally the lowest 500 to 2000 m (1640 to 6562 ft) of the atmosphere. Above this layer frictional effects on wind flow are negligible. (8)

Surface winds Low-level winds within the friction layer. Such winds blow *across* the isobars instead of parallel to them, thus producing a flow of surface air out of high-pressure areas and into low-pressure areas at an oblique angle to the isobars. Surface wind speed is determined by the pressure gradient and the roughness of the surface. (8)

Surge (*see* **glacial surge**).

Suspect terrane Terrane consisting of rocks that are geologically mismatched to their current surroundings; apparently originated as part of another continent or island arc far from the location where the rocks they are adjacent to formed (*see* **terrane**). (31)

Suspension Process of sediment transportation by water or wind, in which the particles are too small to settle rapidly; therefore they are carried along with the flowing water or air for long distances without dropping to the streambed or ground surface. Clay, silt, and very fine sand grains are carried in suspension by both wind and flowing water, and even somewhat larger sand grains can be suspended in turbulent streams. Suspended sediment carried by the wind is often referred to as *dust*, and *becomes* loess when it is deposited (39, 47)

Swallow hole In karst terrain, the place where a surface stream "disappears" to flow into an underground channel. (42)

Swash Thinning sheet of water that runs up the beach after a wave reaches shore and has lost its form. (48)

Swells Long rolling waves that can travel thousands of kilometers across the ocean surface until they break against a shore. (48)

Symbiosis (*see* **mutualism**)

Syncline Trough-like downfold of rocks, with its limbs dipping toward its axial plane (*see* Fig. 34.12, right). (34)

System Any set of related objects or events and their interactions. (1)

Taiga Russian term for the northern coniferous forest biome, "snowforest" is also used. (25)

Taiga climate (Dfc/Dwc, Dfd/Dwd) Collective term for the harsher subtypes of the humid microthermal climate found on the poleward side of the **D**-climate zone. Characteristics are displayed in Fig. 17.3. (17)

Talus slope Steep accumulation of weathered rock fragments and loose boulders that rolled downslope in free fall; particularly common at the bases of cliffs in the drier climates of the western United States (also called *scree slope*). (36)

Tarn Small circular lake on the floor of a cirque basin. (45)

Tectonic plate (*see* **lithospheric plate**)

Tectonics Study of the movements and deformation of Earth's crust. (30)

Teleconnections Relationships involving long-distance linkages between weather patterns that occur in widely separated parts of the world; El Niño is a classic example. (12)

Temperate deciduous forest biome   Dominated by broadleaf trees; herbaceous plants are also abundant, especially in spring before the trees grow new leaves. (25)

Temperate evergreen forest biome   Dominated by needleleaf trees; especially common along western midlatitude coasts where precipitation is abundant. (25)

Temperate grassland biome   Occurs over large midlatitude areas of continental interiors; perennial and sod-forming grasses are dominant. (25)

Temperate karst   Marked by disappearing streams, jagged rock masses, solution depressions, and extensive cave networks; forms more slowly than tropical karst. (42)

Temperature   Index used to measure the thermal energy of matter. In gases most of this energy is in the form of kinetic energy possessed by the moving molecules; the more kinetic energy they have, the faster they move. In solids the energy is mainly vibrational, as atoms oscillate around their average position. (7)

Temperature gradient   Rate of temperature change per unit distance, most often taken in the horizontal direction. (7)

Temperature inversion   Condition in which temperature increases with altitude rather than decreases—a negative lapse rate. It inverts what we, on the surface, believe to be the "normal" behavior of declining temperature with increasing height. (6, 7)

Tensional stress   Stress that tends to pull apart rocks, causing faults to form; motion of rock on either side of those faults allows thinning of the crust and sometimes blocks of crustal rock sink down to form rift valleys. Tensional stress can be associated with the divergence of lithospheric plates, but is produced in some other settings as well. (34)

Terminal moraine   Ridge of debris formed at the outer margin of an ice sheet or the lower limit of a mountain glacier; built by sediment carried to the ice margin by the flowing glacier. These ridges are best developed when the ice margin remains at a particular location for a considerable period of time, and are important to Earth scientists because they mark the farthest extent of an ice advance. (44)

Terrace   A former floodplain that now forms a relatively flat bench standing above the active floodplain and channel of a stream. May be referred to as a *stream terrace* to differentiate it from *marine terraces*. When they lie at the same elevation on both sides of a river, terraces are said to be *paired*. (41)

Terrace, fill   Terrace that is underlain by thick deposits of the stream it is adjacent to; forms through *aggradation* that filled the stream valley, followed by *incision* that produced the terrace. (41)

Terrace, strath   Terrace that is underlain by a thin layer of stream sediment, resting on a bedrock bench (a *strath*); forms through *incision* following a period when the stream channel migrated back and forth across the valley floor at a relatively stable elevation. (41)

Terrane   A region with consistent ages and types of rock, which distinguish it from surrounding areas; usually used in reference to a fragment of crustal material that was accreted to a much larger continental plate through tectonic processes. (31)

Terrestrial radiation   Radiation emitted by the Earth or objects at terrestrial temperatures (*see* **longwave radiation**). (5)

Tertiary   First of the two periods of the Cenozoic era of recent life on the geologic timescale (*see* Fig. 28.11), extending from 65 million to 1.8 million years ago. (28)

Thalweg   Deepest part of a stream channel; line connecting the lowest points along a riverbed. (38)

Theory   A comprehensive body of explanatory knowledge that is both widely supported by experiment and generally accepted among scientists. (1)

Thermal energy   Heat. (5)

Thermocline   Ocean layer that extends beneath the mixed layer to a depth of about 1,000 m (3,300 ft) and is marked by a relatively sharp decrease in temperature with increasing depth. (10)

Thermohaline circulation   Describes the deep-sea system of oceanic circulation, which is controlled by differences in the temperature and salinity of subsurface water masses. (10)

Thermometer   Instrument for measuring temperature. Most commonly these measurements are made by observing the expansion and contraction of mercury inside a glass tube. (7)

Thermosphere   Fourth temperature layer of the atmosphere, lying respectively above the troposphere, stratosphere, and mesosphere. In this layer, temperatures increase as altitude increases. (6)

Thrust fault   Compressional fault in which the vertical angle of the fault plane is very low, essentially a low-angle reverse fault (*see* Fig. 34.3). (34)

Thunderstorm   Local storm dominated by thunder, lightning, heavy rain, and sometimes hail. (12)

Tidal bore   Rapidly rising high tide that creates a fast-moving wave front as it arrives in certain bays or rivers. (48)

Tidal current   Tidal movement of seawater into and out of bays and lagoons. (48)

Tidal range   Average vertical difference between sea levels at high tide and low tide. (48)

Tide   Cyclical rise and fall of sea level controlled by the Earth's rotation and the gravitational pull of the Moon and Sun. Two high tides and two low tides occur within a period slightly longer than 24 hours. (48)

Till   One of the two types of glacial drift; solid material (ranging in size from boulders to clay particles) carried near the base of a glacier that is deposited as an unsorted mass under the ice or at its outer margin. (44)

Time zone   Approximately 15-degree-wide longitudinal zone, extending from pole to pole, that shares the same local time. In the conterminous United States, from east to west, the time zones are known as Eastern, Central, Mountain, and Pacific. Figure 4.8 displays the global time zone pattern, which involves many liberties with the ideal 15-degree subdivision scheme. (4)

Tombolo   Sandspit that forms a link between the mainland and an offshore island. (49)

Topography   Natural and artificial features found at the surface of an area. (2)

Toponymy   Place names.

Topset beds   Horizontal layers of sedimentary deposits that underlie a deltaic plain. (41)

Tornado Small vortex of air, averaging 100 to 500 m (330 to 1,650 ft) in diameter, that descends to the ground from rotating clouds at the base of a severe thunderstorm, accompanied by winds whose speeds range from 50 to 130 m/s (110 to 300 mi/h). As tornadoes move across the land surface, they evince nature's most violent weather and can produce truly awesome destruction in the natural and cultural landscapes. (12)

Tornado Alley North–south corridor in the eastern Great Plains of the central United States, which experiences tornadoes with great frequency; extends from central Texas northward through Oklahoma and Kansas to eastern Nebraska. (12)

Tower (karst) In tropical karst landscapes, a cone-shaped, steep-sided hill that rises above a surface that may or may not be pocked with solution depressions. (42)

Trace gases Atmospheric gases with low concentration. (6)

Traction Transportation process that involves the sliding or rolling of particles along a riverbed. (39)

Trans-Eurasian earthquake belt Second in the world only to the Circum-Pacific Belt, a belt of high earthquake incidence that extends east-southeastward from the Mediterranean Sea across southwestern and southern Asia to join the Circum-Pacific Belt off Southeast Asia (*see* Fig. 33.8). (33)

Transform fault Special case of a strike-slip fault, one that marks the boundary between two lithospheric plates that are sliding past each other. California's San Andreas Fault is a classic example. (33)

Transform plate boundary Boundary at which two lithospheric plates are moving side-by-side, in a relative sense. (30)

Transformation Soil formation process involving the weathering of rocks and minerals' decomposition of organic material, or other changes in the materials already present in a particular soil horizon. (21)

Translocation Soil formation process involving the movement of dissolved and suspended particles from one horizon to another, usually downward. (21)

Transpiration Passage of water into the atmosphere through the leaf pores of plants. (11)

Transport capacity The maximum load of sediment that a stream can carry at a given discharge. Can be estimated for all types of sediment, or separately for suspended load and bedload (39)

Transported parent material Material deposited by wind, water, mass movement, or glacier ice. (21)

Transverse dune Ridge-like sand dune in which the ridge crest is oriented at a right angle to the prevailing direction of strong winds; usually straight or slightly curved. (47)

Transverse stress Lateral stress produced when two lithospheric plates slide horizontally past each other. (34)

Trellis drainage Stream pattern that resembles a garden trellis; channels have only two orientations, more or less at right angles to each other; often develops on parallel-folded sedimentary rocks (*see* Fig. 40.11C). (40)

Tributaries In a stream system, the smaller branch streams that connect with and feed the main artery (trunk river); terminology also applies to glacier systems. (39)

Tributary glacier Smaller glacier that feeds a trunk (main) glacier. (45)

Trophic level Each of the stages along the food chain in which food energy is passed through the ecosystem. (24)

Tropic of Cancer Most northerly latitude (23.5°N) where the Sun's noontime rays strike the Earth's surface vertically (on June 22, the day of the Northern Hemisphere summer solstice). (4)

Tropic of Capricorn Most southerly latitude (23.5°S) where the Sun's noontime rays strike the Earth's surface vertically (on December 22, the day of the Northern Hemisphere winter solstice). (4)

Tropical (A) climates Low-latitude climates with mean temperature of at least 180C (64.40F) every month. Tropical climates have a weak annual temperature cycle; some are moist throughout the year whereas others have a very large annual range in precipitation. (15)

Tropical cyclone (*see* **hurricane**)

Tropical deforestation Clearing and destruction of tropical rainforests typically for lumbering, livestock grazing, and to make way for expanding settlement. (15)

Tropical depression Tropical cyclone exhibiting wind speeds of less than 21 m/s (39 mi/h). Further intensification of the depression would next transform it into a tropical storm. (13)

Tropical karst Dominated by steep-sided, vegetation-covered hill terrain; solution features are larger than in slower-forming temperate karst landscapes. (42)

Tropical rainforest biome Vegetation dominated by tall, closely spaced evergreen trees; a teeming arena of life that is home to a greater number and diversity of plant and animal species than any other biome. (25)

Tropical rainforest climate (Af) Tropical climate without a dry season. Characteristics are displayed in Fig. 15.4. (25)

Tropical savanna climate (*see* **savanna climate**)

Tropical storm Intensified tropical depression, exhibiting a deep central low-pressure cell, rotating cyclonic organization, and wind speeds between 21 and 33 m/s (39 and 74 mi/h). Further intensification would transform the system into a hurricane. (13)

Tropopause Upper boundary of the troposphere along which temperatures stop decreasing with height. (6)

Troposphere Bottom layer of the atmosphere in which temperature usually decreases with altitude. (6)

Truncated spurs Spurs of hillsides that have been cut off by a glacier, thereby straightening the glacially eroded valley. (45)

Trunk glacier Main glacier, which is fed by tributary glaciers. (45)

Trunk river Main river in a drainage basin, which is fed by all the tributary streams. (39)

Tsunami Seismic sea wave, set off by a crustal disturbance, which can reach gigantic proportions; especially likely to be produced by subduction zone earthquakes (33)

Tundra biome   Plant assemblage of the coldest environments; dominated by shrubs, perennial mosses, lichens, and sedges. (25)

Tundra climate (ET)   Milder subtype of the **E**-climate zone, named after its distinct vegetation assemblage of mosses, lichens, and stunted trees. Characteristics are displayed in Fig. 17.11. (17)

Ultimate base level   Lowest elevation to which an entire stream network can erode; for stream systems reaching the ocean this is essentially sea level. (39)

Ultisol   One of the 12 soil orders of the Soil Taxonomy. Soils that have thin A horizons and B horizons marked by clay accumulation, like Alfisols, but more deeply weathered and acidic and often less productive when used for agriculture; located mainly in warm subtropical environments. (23)

Ultraviolet radiation   High-energy, solar radiation at wavelengths immediately below the visible portion of the electromagnetic spectrum.(6)

Unconfined aquifer   (*see* **aquifer, unconfined**)

Unconformity   The contact between eroded strata and the strata produced by resumed deposition of sedimentary rocks. (28)

Underfit stream   Small stream lying in a large river valley that seems incapable of having sculpted that valley. Stream piracy is often the explanation. (40)

Uplifted marine terrace   (*see* **marine terrace**)

Upper atmosphere   Atmospheric level above the surface friction layer. (8)

Upper mantle   Internal layer of Earth that is found below the crust. The uppermost part of this layer is relatively strong and rigid, and together with the crust it forms the *lithosphere*; with greater depth in the upper mantle, the rock becomes softer and more plastic (*see* Fig. 29.5). (29)

Upthrown block   Block that moves upward with respect to adjacent blocks when vertical movement occurs during faulting. (34)

Upwelling   Rising of cold water from the ocean depths to the surface; affects the local climatic environment because cold water lowers air temperatures and the rate of evaporation. (10)

Urban heat island   Form taken by an isotherm representation of the heat distribution within an urban region. The central city appears as a "highland" or "island" of higher temperatures on a surrounding "plain" of more uniform (lower) temperatures. (7)

Uvala   In karst terrain, a large surface depression created by the coalescence of two or more neighboring sinkholes. (42)

Vadose zone   Belowground zone that lies between the ground surface and the water table; normally unsaturated except after heavy rainfall or snowmelt (also called *zone of aeration*). (38)

Valley breeze   Occurs as mountain terrain heats up rapidly in the sun, causing lower pressure to develop over the mountain ridges, spawning a breeze that flows upslope during the day. (8)

Vapor pressure   Pressure exerted by the molecules of water vapor in air. (11)

Vaporization   Synonym for evaporation.

Variable gases   Atmospheric gases present in differing quantities at different times and places; three are essential to human well-being: carbon dioxide, water vapor, and ozone. (6)

Velocity (stream)   Rate of speed at which water moves in a river channel. This rate varies with distance above the streambed and away from the banks, as Figure 38.4 shows. (38)

Vent   Opening through the Earth's crust from which lava erupts. Most eruptions occur through pipe-shaped vents that build volcanic mountains, but fissure eruptions also occur through lengthy cracks that exude horizontal sheets of lava. (32)

Vertical datum   An agreed-upon standard for specifying elevation. The datum corresponds to "sea level" or zero elevation. Several vertical datums are in use, just as there are multiple horizontal datums (standards for longitude and latitude). (3)

Vertical zonation   Characteristic of **H** climates, the distinct arrangement of climate zones according to altitudinal position. The higher one climbs, the colder and harsher the climate becomes—as shown in Figure 17.15 for South America's Andes Mountains. (17)

Vertisol   One of the 12 soil orders of the Soil Taxonomy; found mainly in subtropical regions with pronounced wet and dry seasons. Vertisols are characterized by a tendency to shrink and swell greatly when drying and wetting, because of high clay content; this process causes large cracks to develop and can churn the soil as a whole. (23)

Vicariance   Dispersal of plant and animal species to new locations on Earth, and isolation of some species, because of lithospheric plate motion. (24)

Viscosity   Property of a fluid that resists flowing. (32)

Viscous   Flows relatively slowly. (32)

Volcanic arc   A chain of volcanoes. (30)

Volcanic ash   (*see* **ash [volcanic]**)

Volcanic bombs   (*see* **bombs [volcanic]**)

Volcanism   Eruption of molten rock at the Earth's surface, often accompanied by rock fragments and explosive gases. (30,32)

Volcano   Vent in the Earth's surface through which magma, solid rock, debris, and gases are erupted. This ejected material usually assumes the shape of a conical hill or mountain. (30, 32)

Wallace's Line   Zoogeographer Alfred Russel Wallace's proposed boundary line separating the unique faunal assemblage of Australia from the very different animal assemblage of neighboring Southeast Asia (mapped in Fig. 26.2). An example of boundaries between *zoogeographic realms* (26)

Warm currents   Global-scale ocean currents that travel from the tropics toward the poles, usually exhibiting temperatures that deviate by only a few degrees from those of the surrounding sea, although larger ranges in temperature can be found at the regional scale. (10)

Warm front   Produced when an advancing warm-air mass infringes on a preexisting cooler one. When they meet, the lighter warmer air overrides the cooler air mass, forming the gently upward-sloping warm front (usually producing more moderate precipitation than more steeply sloped cold fronts). (12)

Warm sector   In an open-wave, midlatitude cyclone, the wedge of warm air enclosed by the cold and warm fronts.

Warming hiatus   (*see* **global warming pause**)

Water balance   Analogous to an accountant's record (which tallies income, expenditures, and the bottom-line balance), the measurement of the inflow (precipitation), outflow (evapotranspiration), and net annual surplus or deficit of water at a given location. (11)

Water gap   Pass in a ridge or mountain range through which a stream flows. (40)

Water (oceanic) hemisphere   The roughly one-half of the Earth that contains most of the surface water; the opposite of the *land hemisphere* (*see* Fig. 2.6). (2)

Water resources   Subfield of physical geography involving its intersection with hydrology; systematic study of the surface and subsurface water supplies potentially available for human use. (1)

Water table   Top of the (phreatic) zone of saturation; generally has topography that is a subdued version of the land surface above, higher beneath hills and lower in valleys. (38)

Water vapor   Invisible gaseous form of water; a highly variable atmospheric gas found mainly in the troposphere. (11)

Water vapor feedback   The net effect of increasing water vapor in global warming. Consists of the stronger positive feedback due to increased specific humidity and the weaker negative feedback caused by decreasing saturated adiabatic lapse rate. (19)

Watershed   (*see* **drainage basin**)

Waterspout   Tornado that forms and moves over a water surface. (12)

Wave crest   Top of a wave. The wave's height is the vertical distance between the wave crest and the wave trough. (48)

Wave-cut platform   Platform that develops through wave erosion at the foot of a sea cliff, marking its recession. Its smooth bedrock surface slopes gently seaward. (49)

Wave height   Vertical distance between wave crest (top) and wave trough (bottom). (48)

Wave length   Horizontal distance between one wave crest (or wave trough) and the next. (48)

Wave of oscillation   A wind-driven water wave, offshore where the water depth is great. The water oscillates as the wave passes, but does not move horizontally with the wave (48)

Wave of translation   Swell nearing shore that has "felt" the rising ocean bottom and whose internal water motion (as a wave of oscillation) begins to be affected by it, signifying the beginning of the wave's erosional work. (48)

Wave period   Time interval between the passage of two successive wave crests past a fixed point. (48)

Wave refraction   Nearshore bending of waves coming in at an oblique angle to the shoreline. Shoaling slows part of the wave, which progressively bends as the faster end "catches up" (*see* Fig. 48.4). (48)

Wave trough   Bottom of a wave. The wave's height is the vertical distance between wave crest and wave trough. (48)

Weather   Immediate and short-term conditions of the atmosphere that impinge on daily human activities. (6)

Weathering   Chemical alteration and physical disintegration of Earth materials by the action of air, water, and organisms; affects bedrock and rock fragments in soils and other settings, and occurs in place. That is, it does not include the processes that remove the products of rock breakdown. (35)

Westerlies   Two broad midlatitude belts of prevailing westerly winds, lying between approximately 30 degrees and 60 degrees in both hemispheres; fed by the Coriolis force–deflected poleward wind flow emanating from the subtropical high on the equatorward margin of the Westerlies wind belt. (9)

Wilting point   Practical lower limit to moisture contained within a soil. Below this point, a crop dries out, suffering permanent injury. (38)

Wind   Movement of air relative to the Earth's surface. Winds are always named according to the direction from which they blow. (8)

Wind abrasion   Erosion of rock surfaces by windborne sand particles. (47)

Windward   Exposed, upwind side of a topographic barrier that faces the winds that flow across it. (8)

Winter   In Northern Hemisphere terminology, the season that begins on the day of the winter solstice (December 22) and ends on the day of the spring (vernal) equinox (around March 21).

Winter solstice   Day each year of the poleward extreme in latitude *in the opposite hemisphere* where the Sun's noontime rays strike the Earth's surface vertically. In the Northern Hemisphere, that date is December 22, when the Sun is directly above latitude 23.5 degrees S (the Tropic of Capricorn); in the Southern Hemisphere, that date is June 22, when the Sun is directly above latitude 23.5 degrees N (the Tropic of Cancer). (4)

Wrangellia   Assemblage of suspect terranes in northwestern North America, collectively named after the Alaskan mountain range where the phenomenon was first identified. (31)

Xerophyte   Plant adapted to withstand the aridity of dry environments. (24)

Yardang   Desert landform shaped by wind abrasion in the form of a narrow linear ridge lying parallel to the prevailing wind direction; most common in dry sandy areas underlain by soft bedrock. (47)

Younger Dryas   A brief return to full glacial conditions about 13,000 years ago that occurred early in the warming period following the last glacial maximum; thought to have been caused by changes in North Atlantic Ocean circulation . (18)

Zenith   Point in the sky directly overhead, 90 degrees above the horizon. (4)

Zonal flow   Wind blowing primarily along a parallel of latitude (east-west). (9)

Zone of ablation   Glacier's lower zone of net loss, where *ablation* (loss of ice) is greater than ice *accumulation*. (43)

Zone of accumulation Glacier's upper zone of net growth, where *accumulation* of ice (formed from snow) is greater than *ablation*, or ice loss. (43)

Zone of aeration   (see **vadose zone**)

Zone of intolerance Environmental conditions beyond the range that an animal or plant species can tolerate, except possibly for short, intermittent periods. (24)

Zone of physiological stress Environmental conditions that are marginal for a plant or animal species, but beyond its optimum range and likely to cause stress and limit population growth. (24)

Zone of saturation   (see **saturated zone**)

Zoogeographic realm Largest and most generalized regional unit for representing Earth's fauna; reflects evolutionary centers for animal life as well as the influence of barriers over time (*see* Fig. 26.7). (26)

Zoogeography Geography of animal life or fauna; where zoology and physical geography overlap. (20)

Zooplankton Microscopic animal life-forms that float in the ocean and freshwater bodies; eaten by small fish, which in turn are eaten by larger fish. Form a trophic level above *phytoplankton* in aquatic ecosystems. (24)

CREDITS

Credits for Line Art and Tables

Unit 2   Figure 2.7: Adapted from Sandwell et al., Science 3 October 2014: Vol. 346 no. 6205 pp. 65-67 DOI: 10.1126/science.1258213.

Unit 3   Figure 3.6B: National Geodetic Survey/NOAA. Figure 3.7: From Norman J. W. Thrower, *Maps and Man,* Prentice-Hall, 1972, P. 153. Figure 3.10: After Arthur Robinson et al., *Elements of Cartography,* 5 rev. ed., 1984, p. 99, published by John Wiley & Sons. Figure 3.12: After Edward A. Fernald and Donald J. Patton, *Water Resources Atlas of Florida,* 1984, pp. 26, 19, published by the Institute of Science and Public Affairs, Florida State University. Figures 3.13, 3.14: Adapted from Brian J. Skinner and Stephen C. Porter, *Physical Geography,* John Wiley & Sons, 1987, pp. 715, 711.

Unit 6   Figure 6.7B: Data Courtesy of NOAA. Figure 6.9: Courtesy of Environment Canada. Figure 6.10: Adapted from *Understanding Our Atmospheric Environment,* by Morris Neiburger et al., copyright © 1973 by W. H. Freeman and Company.

Unit 7   *Opener*: Image obtained using Climate Reanalyzer (http://cci-reanalyzer.org), Climate Change Institute, University of Maine, USA. Data courtesy of NOAA National Climatic Data Center. Figure 7.5: From *Geography: A Modern Synthesis,* 3/ed., by Peter Haggett, copyright © 1983 by Peter Haggett, published by HarperCollins Publishers. Figure 7.6: From *Urbanization and Environment* by Thomas R. Detwyler and Melvin G. Marcus, © 1972 by Wadsworth Publishing Company, Inc. Figure 7.9: After AVISO. Figure 7.13: Data from K. Fennig, A. Andersson, S. Bakan, C. Klepp, M. Schroeder, 2012: Hamburg Ocean Atmosphere Parameters and Fluxes from Satellite Data—HOAPS 3.2—Monthly Means/6-Hourly Composites. Satellite Application Facility on Climate Monitoring. doi:10.5676/EUM_SAF_CM/HOAPS/V001.

Unit 8   Figure 8.2: Adapted from Roger G. Barry and Richard J. Chorley, *Atmosphere, Weather and Climate, Methuen* (Routledge), 2 rev. ed., pp. 8, 43, © 1987 by Roger G. Barry and Richard J. Chorley. Figure 8.3: Adapted from Roger G. Barry and Richard J. Chorley, *Atmosphere, Weather and Climate, Methuen* (Routledge), 2 rev. ed., pp. 8, 43, © 1987 by Roger G. Barry and Richard J. Chorley. Additional data after Robin McIlveen, *Fundamentals of Weather & Climate,* 2nd ed., © 2010. New York: Oxford University Press, p. 298, and Dennis L. Hartmann (1994), *Global Physical Climatology,* San Diego: Academic Press, p. 199. Figure 8.5: Adapted from Aguado, Edward; Burt, James E., *Understanding Weather and Climate,* 6th Edition, © 2013. Reprinted by permission of Pearson Education, Inc., Upper Saddle River, NJ. Figure 8.6: From Howard J. Critchfield, *General Climatology,* 4/E, © 1983, p. 85, published by Prentice-Hall, Inc., Englewood Cliffs, N.J. Figure 8.7: Adapted from Lutgens, Frederick K.; Tarbuck, Edward J., *The Atmosphere: An Introduction to Meteorology,* 11th ed. © 2010. Pearson Education, Inc. Figure 8.10: Adapted from Steven A. Ackerman and John A. Knox, *Meteorology: Understanding the Atmosphere,* 3rd Edition, © 2012. Sudbury, MA: Jones & Bartlett Learning, LLC, p. 205, 207.

Unit 9   Figure 9.2: Adapted from Steven A. Ackerman and John A. Knox, *Meteorology: Understanding the Atmosphere,* 3rd Edition, © (2012). Jones & Bartlett Learning, LLC, Sudbury, MA, p. 216. Figure 9.4: After Dennis L. Hartmann (1994) *Global Physical Climatology.* Academic Press, San Diego, CA, p. 16. Figure 9.6: Adapted from Robin McIlveen, *Fundamentals of Weather & Climate.* 2nd ed.,

© 2010. Oxford University Press, New York, NY, p. 111. Figures 9.7, 9.8: Adapted from Robinson, Peter J.; Henderson-Sellers, Ann, *Contemporary Climatology,* 2nd Edition, © 1999. Pearson Education, pp. 130, 136. Figure 9.10: Data courtesy of NASA/Goddard Earth Sciences Data and Information Services Center. Figure 9.11: Adapted from Robin McIlveen, *Fundamentals of Weather & Climate,* 2nd ed., © 2010. Oxford University Press, New York, NY, p. 108. Figure 9.12: After Dennis L. Hartmann (1994) *Global Physical Climatology.* Academic Press, San Diego, CA, p. 145, 156.

Unit 10   Figure 10.1: Adapted from Steven A. Ackerman and John A. Knox, *Meteorology: Understanding the Atmosphere,* 3rd Edition, © 2012. Jones & Bartlett Learning, LLC, Sudbury, MA, p. 237. Figure 10.2: O. Brown, R. Evans, and M. Carle, University of Miami Rosenstiel School of Marine and Atmospheric Science, Miami, Florida. Figure 10.4: From *Oceanography: An Introduction to the Marine Sciences* by Jerome Williams, copyright © 1962 by Little, Brown and Company. Figure 10.5: (top) José P. Peixoto and Abraham H. Oort (1992). *The Physics of Climate,* American Institute of Physics, New York, NY, p. 180, (bottom) Steven A. Ackerman and John A. Knox (2012). *Meteorology: Understanding the Atmosphere,* 3rd ed., Jones & Bartlett Learning, LLC, Sudbury, MA, p. 240. Figure 10.6: Jesse Allen/NASA Earth Observatory. http://earthobservatory.nasa.gov/Newsroom/NewImages/images.php3?img_id=17405. Figure 10.8: Same source as Fig 3.13, p. 382. Figure 10.9: Adapted from Alastair G. Dawson and Greg O'Hare (2000) Ocean-Atmosphere Circulation and Global Climate: The El Niño-Southern Oscillation. *Geography,* 85(3): 193–208. Figure 10.11: Data courtesy of NASA/Goddard Space Flight Center/Tropical Rainfall Measuring Mission. Figure 10.12: Adapted from Bruce T. Anderson and Alan Strahler (2008) *Visualizing Weather and Climate,* John Wiley & Sons Inc., Hoboken, NJ.

Unit 11   Figure 11.2: Adapted from Morris Neiberger, James G. Edinger, and William D. Bonner (1982) *Understanding our Atmospheric Environment.* San Francisco, CA: W. H. Freeman and Company, p. 117. Figure 11.3: Same source as Fig 3.13, p. 241. Figure 11.5: Adapted from Kenny, J. F., Barber, N. L., Hutson, S. S., Linsey, K. S., Lovelace, J. K., and Maupin, M. A. (2009) Estimated use of water in the United States in 2005. *U.S. Geological Survey Circular 1344.* Figure 11.8: Adapted from Brian J. Skinner and Stephen C. Porter, *The Blue Planet: An Introduction to Earth System Science,* John Wiley & Sons, 1995, p. 329. Figure 11.11: Adapted from Robert V. Rohli and Anthony J. Vega (2008) *Climatology.* Sudbury, MA: Jones and Bartlett Publishers, pp. 211, 217, 223. Figure 11.12: After W. D. Sellers, *Physical Climatology,* 1965, p. 84, published by The University of Chicago Press, © The University of Chicago Press. Figures 11.13, 11.14: Same source as Fig. 7.12.

Unit 12   Figure 12.2: Hal Pierce/NASA Earth Observatory. Figure 12.3: Adapted from Steven A. Ackerman and John A. Knox (2012) *Meteorology: Understanding the Atmosphere.* 3rd ed., Jones & Bartlett Learning, LLC, Sudbury, MA, pp. 291 (top), 294 (bottom). Figure 12.4: Adapted from Robinson, Peter J.; Henderson-Sellers, Ann *Contemporary Climatology,* 2nd Edition, © 1999. Pearson Education, p. 87. Figure 12.6 (bottom): After K. Hindley, "Learning to Live With Twisters," Vol. 70, 1977, p. 281, published by *New Scientist,* and Robinson, Peter J.; Henderson-Sellers, Ann *Contemporary Climatology,* 2nd Edition, © 1999. Pearson Education, p. 88. Figure 12.7: Marit Jentoft-Nilsen/NASA Earth Observatory. Figure 12.8C: Data courtesy of NOAA/National Weather Service's Storm Prediction

Center. Figure 12.10: Data from *Climate and Man: The 1941 Yearbook of Agriculture,* U. S. Department of Agriculture (USDA), 1941, p. 1085.

Unit 13 Figure 13.2: Adapted from Morris Neiberger, James G. Edinger, and William D. Bonner (1982) *Understanding our Atmospheric Environment.* San Francisco, CA: W. H. Freeman and Company, p. 280–281. Figure 13.4: From the *Cape Cod Times,* August 19, 2001, p. 15. Figure 13.5: Robert A. Rohde/NASA Earth Observatory. Figure 13.8: Adapted from Morris Neiberger, James G. Edinger, and William D. Bonner (1982), *Understanding our Atmospheric Environment.* San Francisco, CA: W. H. Freeman and Company, p. 264. Figure 13.11B: Adapted from Aguado, Edward; Burt, James E., *Understanding Weather And Climate,* 6th Edition, © 2013. Reprinted by permission of Pearson Education, Inc., Upper Saddle River, NJ.

Unit 15 Figure 15.4: Adapted from press release map, © 1985 National Geographic Society.

Unit 16 Figure 16.12: Data courtesy of Government of Australia, Bureau of Meteorology.

Unit 17 Figure 17.8: From Anne La Bastille, "Acid Rain: How Great a Menace?," *National Geographic,* November 1981, p. 667, © 1981 National Geographic Society.

Unit 18 Figure 18.2B: After Fig. 3, Gill, J. L., J. W. Williams, S. T. Jackson, J. P. Donnelly, and G. C. Schellinger. 2012. Climatic and megaherbivory controls on late-glacial vegetation dyanmics: A new, high-resolution, multi-proxy record from Silver Lake, Ohio. *Quaternary Science Reviews* 34:66–80. Figure 18.6: Adapted from Aguado, Edward; Burt, James E., *Understanding Weather And Climate,* 6th Edition, © 2013. Pearson Education, Inc., Upper Saddle River, NJ. Figure 18.7: Adapted from Fig 3 from David Archer, *The Global Carbon Cycle.* © 2012. Princeton University. Press. Figure 18.8: Adapted from Aguado, Edward; Burt, James E., *Understanding Weather and Climate,* 6th Edition, © 2013. Pearson Education, Inc., Upper Saddle River, NJ. Figure 18.9: Adapted from Jürgen Ehlers & Philip Gibbard, *Quaternary Glaciations—Extent and Chronology,* © 2004. Elsevier. Figure 18.10: Adapted from Alley, R.B., "The Younger Dryas cold interval as Viewed From Central Greenland," *Quaternary Science Reviews,* Volume 19, Issues 1–5, 1 January 2000, pp. 213–226. Figure 18.11: Adapted from Robert A. Rohde/Global Warming Art, globalwarmingart.com.

Unit 19 Figure 19.1: Adapted from "Can natural or anthropogenic explanations of late-Holocene C02 and CH4 increases be falsified?," W. F. Ruddiman, J. E. Kutzbach, and S. J. Vavrus, *The Holocene* 2011 21: 865. Figure 19.3B: Adapted from Fig 2.14, State of the Climate in 2011, special supplement to *Bulletin of American Meteorological Society,* Vol. 93, No. 7, July, 2012. Figure 19.4: Data courtesy of Climatic Research Unit, University of East Anglia. Figure 19.5: FAQ Fig 3.1 in Trenberth, K. E., P. D. Jones, P. Ambenje, R. Bojariu, D. Easterling, A. Klein Tank, D. Parker, F. Rahimzadeh, J. A. Renwick, M. Rusticucci, B. Soden and P. Zhai, 2007: Observations: Surface and Atmospheric Climate Change. In: *Climate Change 2007: The Physical Science Basis.* Contribution of Working Group I to the Fourth Assessment Report of the Intergovernmental Panel on Climate Change [Solomon, S., D. Qin, M. Manning, Z. Chen, M. Marquis, K.B. Averyt, M. Tignor and H.L. Miller (eds.)]. Cambridge University Press, Cambridge, United Kingdom and New York, NY, USA. Figure 19.6: After National Climatic Center/NESDIS/NOAA, http://www.ncdc.noaa.gov/sotc/service/national/Statewidetrank/201201–201206.gif. Figure 19.8: Data courtesy of U.S. Carbon Dioxide Information Center. Figure 19.9: After Fig. 2, NOAA Annual Greenhouse Gas Index/NOAA Earth System Research Laboratory, http://www.esrl.noaa.gov/gmd/aggi/. Figure 19.10: Adapted from Intergovernmental Panel on Climate Change, Fifth Assessment Report, 2013, Figure TS 9. Figures 19.11, 19.12: Same source as Fig. 19.10, Fig TS.15. Figure 19.13: Same source as Fig. 19.10, Figures TS.15 and TS.16. Figure 19.14: Same source as Fig. 19.10, Figures TFE.1, and

3n (pp. 45). Figure 19.15: Data courtesy of Williams, J. W., Jackson, S. T., Kutzbach, J. E. (2007) Projected distributions of novel and disappearing climates by 2100AD. *Proceedings of the National Academy of Sciences* 104: 5738–5742.

Unit 21 Figure 21.3: From Léo F. Laporte, *Encounter With the Earth,* Canfield Press, p. 300, © 1975 by Léo F. Laporte.

Unit 22 Figure 22.1: Figure from *Fundamentals of Soil Science,* 8th rev. edition, by Henry D. Foth, reprinted by permission of John Wiley & Sons, Inc. Figure 22.2: After USDA, Soil Conservation Service, n.d.

Unit 23 Figure 23.1: Adapted from *Environmental Science,* Third Edition, by Jonathan Turk, Amos Turk, and Karen Arms, copyright © 1984 by Saunders College Publishing. Figures 23.13, 23.14: Data from Natural Resource Conservation Service (USDA).

Unit 24 Figure 24.1: From Léo F. Laporte, *Encounter With the Earth,* Canfield Press, p. 120, © 1975 by Léo F. Laporte. Figure 24.3: After H. Lieth and E. Box, *Publications in Climatology,* C. W. Thornthwaite Laboratory of Climatology, Vol. 25, No. 3, 1972, p. 42. Figure 24.4: After Natural Resource Conservation Service (USDA). Figures 24.6, 24.7: From *Geography: A Modern Synthesis,* 3/ed., by Peter Haggett, copyright © 1983 by Peter Haggett, published by Harper-Collins Publishers. Figure 24.8 Adapted from *Fundamentals of Ecology,* Third Edition, by Eugene P. Odum, copyright © 1971 by Saunders College Publishing. Figure 24.9: © *Biol. Bull.,* Vol. XXI, No. 3, pp. 127–151, and Vol. XXII, No. 1, pp. 1–38. Figure 24.10: After C. B. Cox et al., *Biogeography,* 1973, p. 61, © Blackwell Scientific Publications Ltd.

Unit 25 Figures 25.2, 25.4: Figures adapted from *Environmental Science,* Third Edition by Jonathan Turk, Amos Turk, and Karen Arms, copyright © 1984 by Saunders College Publishing. Figure 25.3: Adapted from *Plants and People,* Resource Publication of the Association of American Geographers, Thomas R. Vale, 1982.

Unit 26 Figure 26.3: After *Guide to the Mammals of Pennsylvania,* by Joseph F. Merritt, published by the University of Pittsburgh Press, © 1987 by the University of Pittsburgh Press. Figure 26.4B: After "Atlas of Relations Between Climatic Parameters and Distributions of Important Trees and Shrubs in North America" by Robert S. Thompson, Katherine H. Anderson, and Patrick J. Bartlein, U.S. Geological Survey Professional Paper 1650 A&B, http://pubs.usgs.gov/pp/p1650-a/ Figure 26.8: Adapted from MacArthur, R. H., and Wilson, E. O. 1967. *The Theory of Island Biogeography.* Princeton University Press, Princeton, New Jersey.

Unit 28 Figure 28.4: Adapted from Brian J. Skinner and Stephen C. Porter, *The Dynamic Earth: An Introduction to Physical Geology,* John Wiley & Sons, 3 rev. ed., 1995, p. 427.

Unit 29 Figure 29.1: After USGS.

Unit 30 Figure 30.3: National Geophysical Data Center/ NOAA. Figure 30.6: Data from the U.S. Coast and Geodetic Survey, and NOAA. Figure 30.9: Adapted from USGS. Figure 30.11: Same source as Fig. 3.13, p 490.

Unit 32 Figure 32.14: From Stephen L. Harris, *Fire Mountains of the West: The Cascade and Mono Lake Volcanoes,* © 1991 Mountain Press Publishing Company.

Unit 33 Figure 33.8: Data from the US Coast and Geodetic Data Survey. Figure 33.9: USGS.

Unit 34 Figure 34.2: Same source as Fig. 3.13, pp. 416. Figure 34.5: Same source as Fig. 3.13, pp. 416. Figure 34.10: Same source as Fig. 3.13, p. 424. Figures 34.12, 34.13: Same source as Fig. 3.13, pp. 424, 426.

Unit 35 Figure 35.5: Same source as Fig. 3.13, p. 293.

Unit 37 Figure 37.5: Same source as Fig. 28.4, p. 234. Figures 37.6, 37.7A: Same source as Fig. 3.13, p. 293. Figure 37.9: Same source as

Fig. 3.13, p. 293. Figure 37.12: After U. S. Geological Survey, Professional Paper 950, n.d.

Unit 38 Figure 38.2: After *Physical Geography: Earth Systems,* Scott Foresman & Co., Glenview, Illinois, 1974, © John J. Hidore. Figure 38.4C: Adapted from Brian J. Skinner and Stephen C. Porter, *The Dynamic Earth: An Introduction to Physical Geology,* John Wiley & Sons, 1989, p. 221. Same source as 38.7, p. 221. Figure 38.5: After J. P. Bruce and R. H. Clark, *Introduction to Hydrometeorology,* Pergamon Press, Elmsford, New York, 1966, © J. P. Bruce and R. H. Clark. Figure 38.7: Same source as Figure 38.4C, pp. 199. Figure 38.8: From Léo F. Laporte, *Encounter With the Earth,* Canfield Press, p. 249, © 1975 by Léo F. Laporte. Figure 38.9: Adapted from *Introduction to Physical Geography,* Second edition, by H. M. Kendall, R. M. Glendinning, C. H. MacFadden, and R. F. Logan, copyright © 1974 by Harcourt Brace Jovanovich, Inc. Figure 38.12: Same source as Fig. 11.8, p. 245.

Unit 39 Fig. 39.9: USGS.

Unit 41 Figure 41.5A: Same source as Fig. 3.13, pps. 274, 287, 290. Figures 41.11, 41.12: Same source as Fig. 3.13, pps. 274, 287, 290. Figure 41.15: Same source as Fig. 3.13, pps. 274, 287, 290.

Unit 42 Figure 42.3: Same source as Fig. 3.13, p. 256. Figure 42.4: From a portion of Glasgow North (Kentucky) Quadrangle, 7.5 minute series, U. S. Geological Survey, n.d.

Unit 43 Figures 43.3A, 43.3B: Same source as Fig. 3.13, both p. 345. Figure 43.7: Same source as Fig. 38.7, p. 270.

Unit 44 Figure 44.3: From J. F. Lovering and J. R. V. Prescott, *Last of Lands . . . Antarctica,* Melbourne University Press, 1979, data from Gordon and Goldberg, "Circumpolar Characteristics of Antarctic Waters," American Geographical Society, Antarctic Map Folio Series 1970, and Kort, "The Antarctic Ocean," *Scientific American,* Vol. 207, 1962. Figures 44.5, 44.8: Adapted from Richard Foster Flint, *Glacial and Pleistocene Geology,* John Wiley & Sons, 1957, pp. 227 ff., copyright © Harrison L. Flint. Figure 44.7: Same source as Fig. 3.13, p. 365.

Unit 45 Figure 45.7: After Armin K. Lobeck, *Geomorphology,* McGraw-Hill Book Company, 1932. Figure 45.9: Same source as Fig. 38.7, p. 268.

Unit 46 Figure 46.3: Same source as Fig. 3.13, p. 361.

Unit 47 Figures 47.3, 47.5, 47.6: Same source as Fig. 3.13, pp. 316, 322, 324. Figure 47.8: After U. S. Bureau of Reclamation, 1960.

Unit 48 Figures 48.1, 48.4, 48.8: After Keith Stowe, *Essentials of Ocean Science,* 1987, pp. 84, 87, and 124, published by John Wiley & Sons, copyright © 1987 by John Wiley & Sons, Inc. Figures 48.2, 48.3A, 48.5: Same source as Fig. 3.13, pp. 385, 386, and 387.

Unit 49 Figures 49.1, 49.4: Same source as Fig. 3.13, pp. 393, 393. Figure 49.11: Same source as Fig. 38.7, p. 297. Figure 49.12: Same source as Fig 11.8, p. 302.

Credits for Photographs

Title Page: © Ian Phelps Photography.

Unit 1 *Opener:* Goddard Space Flight Center/NASA. Figure 1.1: © David Liu/iStockPhoto. Figure 1.5: H. J. de Blij. Figure 1.7: © aolr/iStockPhoto. Figure 1.9: © Ablestock.com.

Part I Opening Photo: © kan_khampanya/Shutterstock.

Unit 2 *Opener:* © kavram/iStockPhoto. Figure 2.8: Tibor G. Toth/National Geographic Image Collection.

Unit 3 *Opener:* H. J. de Blij. Figure 3.1A: © Print Collector/Getty Images. Figure 3.16: Tom D'Avello, USDA-NRCS. Figure 3.18: © Stephanie Konfal/USGS/Handout/Reuters/Corbis. Figure 3.20: Terranova/Science Source.

Unit 4 *Opener:* © AP Photo/NTB Scanpix, Terje Bendiksby. Figure 4.5: © Kennan Ward/Corbis.

Unit 5 *Opener:* James E. Burt. Figure 5.6: Courtesy of SOHO (ESA & NASA).

Part II Opening Photo: © Martin Harvey/Corbis.

Unit 6 *Opener:* Goddard Space Flight Center/NASA. Figure 6.1: NASA. Figure 6.2: © Dmitry Kalinovsky/Shutterstock. Figure 6.4: H. J. de Blij. Figure 6.8: NASA.

Figure 6.11: © Сергей Сидоров/iStockPhoto.

Unit 7 *Opener:* NASA/ Goddard Institute for Space Studies. Figure 7.1: Time Life Pictures/Getty Images. Figure 7.4: NASA/Science Source. Figure 7.7 (left): Izzy Schwartz/Getty Images, (right): © Chad Ehlers/Getty Images.

Unit 8 *Opener:* Dzarek/Shutterstock. Figure 8.9A: Publio Furbino/Shutterstock. Figure 8.9B: © 2011 Alex S. MacLean/Landslides Aerial Photography.

Unit 9 *Opener:* Goddard Space Flight Center/NASA. Figure 9.3: AirPhoto. Figure 9.9: GEOSPACE/Science Photo Library.

Unit 10 *Opener:* H. J. de Blij. Figure 10.3: NASA/"Perpetual Ocean," http://www.youtube.com/watch?v=CCmTY0PKGDs.

Unit 11 *Opener:* H. J. de Blij. Figure 11.6: H. J. de Blij. Figure 11.7: Novarc Images/Alamy.

Figure 11.8A: © Tom Bean and Susan Bean, Inc./DRK PHOTO. Figure 11.8B: © Johnny Johnson/DRK PHOTO. Figure 11.8C: © Darrell G. Gulin/DRK PHOTO. Figure 11.8D: © Warren E. Faidley/DRK PHOTO. Figure 11.15: University of Wisconsin-Madison.

Unit 12 *Opener:* Jack Fields/Science Source. Figure 12.5: JSC/NASA. Figure 12.6 (top): © Mike Hollingshead/Corbis. Figure 12.8A: Benjamin Krain/Getty Images. Figure 12.8B: Associated Press.

Unit 13 *Opener:* University of Dundee/Science Source. Figure 13.1: NOAA-NASA GOES Project. Figure 13.3: © Warren E. Faidley/DRK PHOTO. Figure 13.6: © Steve Starr/Corbis. Figure 13.7A: NOAA GOES Project. Figure 13.7B: AP Photo/John Minchillo. Figure 13.11A: NASA Earth Observatory.

Part III Opening Photo: NASA's Scientific Visualization Studio.

Unit 14 *Opener:* Steve McCurry/Magnum Photos.

Unit 15 *Opener:* H. J. de Blij. Figures 15.3, 15.6: H. J. de Blij. Figure 15.8: © Wolfgang Kaehler/Corbis. Figure 15.10: John Moss/Science Source. Figure 15.14: Steve McCurry/Magnus Photos.

Unit 16 *Opener:* tagstiles.com - S.Gruene/Shutterstock. Figure 16.6A: © Kevin Fleming/Corbis. Figure 16.6B: Frank Carter/Getty Images. Figure 16.7: H. J. de Blij. Figure 16.13: © Ed Darack/Science Faction/Corbis.

Unit 17 *Opener:* Nigel Pavitt/AWL Images. Figure 17.4: © Doug Wilson/Corbis. Figure 17.5: John Warden/Alaskastock.com. Figure 17.10: © Robert Glusic/Getty Images. Figure 17.13: © Ocean/Corbis. Figure 17.16: H. J. de Blij.

Unit 18 *Opener:* © Andres Ello/Shutterstock. Figure 18.2A: Louisa Howard/Science Source. Figure 18.12: © Wolfgang Kaehler/Corbis.

Unit 19 *Opener:* Chris Graythen/Getty Images. Figure 19.3A: Dr. Juerg Alean.

Part IV Opening Photo: Shinelu/Shutterstock.

Unit 20 *Opener:* © Joe Mason. Figures 20.1, 20.2, 20.6: © Joe Mason.

Unit 21 *Opener:* © Joe Mason . Figure 21.1: © Joe Mason. Figures 21.2, 21.4: H. J. de Blij. Figures 21.8, 21.9: Randall Schaetzl.

Unit 22 *Opener:* H. J. de Blij. Figures: 22.3B, 22.4B, 22.5B, 22.6B: Randall Schaetzl.

Unit 23   *Opener:* © Joe Mason. Figure 23.2: © Joe Mason. Figures 23.3, 23.4, 23.5: Randall Schaetzl. Figure 23.6: H. J. de Blij. Figures 23.7, 23.8, 23.9, 23.10, 23.11, 23.12: Randall Schaetzl.

Unit 24   *Opener:* H. J. de Blij. Figure 24.2: © Morley Reed/iStockPhoto. Figures 24.5, 24.11: H. J. de Blij. Figure 24.12: Mitsuaki Iwago/Minden Pictures /National Geographic Stock.

Unit 25   *Opener:* zstock/Shutterstock. Figures 25.5, 25.6: H. J. de Blij. Figure 25.7: © Jim Steinberg/Animals Animals. Figure 25.8: © John Eastcott/Yva Momtiuk/DRK PHOTO. Figure 25.9: © brytta/ iStockphoto. Figure 25.10: H. J. de Blij. Figure 25.11: © John Cancalosi/DRK PHOTO.

Unit 26   *Opener:* H. J. de Blij. Figures 26.1, 26.5, 26.9: H. J. de Blij. Figure 26.4A: © Steve Gettle/Minden Pictures/Corbis.

Part V Opening Photo: Michael Collier.

Unit 27   *Opener:* Bruce M. Herman/Science Source. Figure 27.2: © Steve Kaufman/DRK PHOTO. Figure 27.3: © Gibson, Mickey/ Animals Animals. Figure 27.4B: © Degginger, Phil/Animals Animals. Figure 27.6: H. J. de Blij. Figure 27.7: Copyright Larry Fellows/Image Courtesy AGI Image Bank, www.earthscienceworld.org/images. Figure 27.8: © Joe Mason.

Unit 28   *Opener:* © Fotofeeling/Westend61/Corbis. Figures 28.1, 28.3: © Joe Mason. Figure 28.5: N. R. Rowan/Science Source. Figure 28.6: © Tom Bean/Corbis. Figure 28.7: © Thomas McGuire. Figure 28.8: © Braud, Dominique/Animals Animals. Figures 28.10, 28.12: H. J. de Blij.

Unit 29   *Opener:* Juancat/Shutterstock. Figures 29.7A, 29.7B, 29.8: H. J. de Blij. Figure 29.9: © Joe Mason. Figure 29.11: © Tom Bean and Susan Bean, Inc./DRK PHOTO. Figure 29.12: Richard Bergmann/ Science Source.

Unit 30   *Opener:* H. J. de Blij. Figure 30.2: Tibor G. Toth/National Geographic Image Collection. Figure 30.10: © John S. Shelton. Figure 30.12: © David Butow/Corbis.

Unit 31   *Opener:* kavram/Shutterstock. Figure 31.2: © OSF/C. Monteath. Figure 31.4: © Carson Baldwin, Jr./Animals Animals. Figure 31.5: NASA. Figure 31.6: H. J. de Blij. Figure 31.7: AGI/USGS.

Unit 32   *Opener:* Rigucci/Shutterstock. Figure 32.1: Images & Voclans/Science Source. Figure 32.3: © Doug Wechsler/Animals Animals. Figure 32.4A: Jeffrey B. Banke/Shutterstock. Figure 32.6: © Steve Raymer/National Geographic, Inc. Figure 32.7: Dave Harlow, USGS. Figure 32.8: H. J. de Blij. Figure 32.10: © Degginger, Phil/ Animals Animals. Figure 32.11: H. J. de Blij. Figure 32.12A: Tibor G. Toth/National Geographic Image Collection. Figures 32.13A, 32.13B: H. J. de Blij.

Unit 33   *Opener:* Peter Menzel/Science Source. Figure 33.3: © Corbis/Corbis. Figure 33.4: © Bettmann/Corbis. Figure 33.5: © Kapoor Baldev/Sygma/Corbis. Figure 33.6: © Corbis. Figure 33.7: © AP Photo/La Prensa Grafica. Figure 33.10: © Corbis/Corbis.

Unit 34   *Opener:* Sarah Fields Photography/Shutterstock. Figure 34.6B: © Joe Mason. Figure 34.9: © Georg Gerster/Comstock. Figure 34.11: H. J. de Blij. Figure 34.15: USGS.

Part VI Opening Photo: © Larry Lee Photography/Corbis

Unit 35   *Opener:* Meg Wallace Photography/Shutterstock. Figures 35.1, 35.2: © Joe Mason. Figure 35.3: © Wilburn, Jack/Animals Animals. Figure 35.4: Walter Bibikow/Getty Images. Figure 35.6: © Jeremy Woodhouse/DRK PHOTO. Figure 35.7: H. J. de Blij. Figure 35.8: © Joe Mason.

Unit 36   *Opener:* © Joe Mason. Figures 36.1, 36.3, 36.6: H. J. de Blij. Figures 36.2, 36.7: © Joe Mason. Figure 36.4: Dave Rock /

Shutterstock. Figure 36.5: Courtesy Bruce Molnia, USGS/AGI Image Bank, www.earthscienceworld.org/images.

Unit 37   *Opener:* H. J. de Blij. Figure 37.1: Dr. George Plafker/USGS. Figure 37.3: H. J. de Blij. Figure 37.4: William W. Bacon III/Science Source. Figure 37.7B: © Robert Holmes/Corbis. Figure 37.8: © Bill Young/San Francisco Chronicle/ San Francisco Chronicle/Corbis. Figure 37.10: © Imaginechina/Corbis. Figure 37.13: © Lowell Georgia/Corbis.

Unit 38   *Opener:* © Mableen/iStockPhoto. Figure 38.1: © Kim Heacock/DRK PHOTO. Figure 38.3: © Dennis, David M./Animals Animals. Figure 38.10: © Sid & Shirley Rucker/DRK PHOTO.

Unit 39   *Opener:* Nagel Photography/Shutterstock. Figure 39.1: slavik65/Shutterstock. Figures 39.2, 39.4: © Joe Mason. Figure 39.5: © Corbis. Figures 39.6, 39.9: USGS. Figure 39.8: © CC Lockwood/ DRK PHOTO. Figure 39.10: Scott Prokop /Shutterstock. Figure 39.13: G. K. Gilbert/USGS.

Unit 40   *Opener:* H. J. de Blij. Figure 40.1: © Grant Heilman/Grant Heilman Photography, Inc. Figure 40.7: H. J. de Blij. Figure 40.9: © Kent, Breck P./Animals Animals. Figure 40.10: H. J. de Blij. Figure 40.15: From the American Geographical Society Library, University of Wisconsin-Milwaukee Libraries.

Unit 41   *Opener:* H. J. de Blij. Figure 41.1: © Jeff Foott/DRK PHOTO. Figure 41.2: © Tom and Susan Bean, Inc./DRK Photo. Figure 41.5B: © Joe Mason. Figure 41.9: © Nati Harnik/AP/Corbis. Figure 41.10: © Clay Cartwright/iStockPhoto. Figure 41.14: © Earth Satellite Corporation/Science Source.

Unit 42   *Opener:* © Marie Hickman/Corbis. Figure 42.1: © Tony Waltham/Robert Harding/Getty Images. Figure 42.2: H. J. de Blij. Figure 42.6: © MDA Information Systems/Science Photo Library. Figure 42.7: Phil Eschbach/Winter Park History & Archives Collection, Winter Park Public Library, Winter Park, FL. Figure 42.8: Robert Semeniuk. Figure 42.9: © Kent, Breck P./Animals Animals .

Part VII Opening Photo: Jim Richardson/National Geographic Stock.

Unit 43   *Opener:* Bruce Molnia, USGS. Figures 43.1A, 43.1B: H. J. de Blij. Figure 43.2: © Marli Miller, University of Oregon. Figures 43.5, 43.8: H. J. de Blij.

Unit 44   *Opener:* H. J. de Blij. Figure 44.1: USGS/Science Photo Library. Figure 44.4A, 44.4B, 44.4C: H. J. de Blij. Figure 44.6: © Stephen J. Krasemann/DRK Photo. Figures 44.10A, 44.10B, 44.13: © Joe Mason. Figures 44.11, 44.14: USGS.

Unit 45   *Opener:* H. J. de Blij. Figure 45.1: Brian Moorman. Figures 45.2, 45.3, 45.8: H. J. de Blij. Figure 45.4: Stocktrek Images/Getty Images. Figure 45.5: © NASA/Science Source.

Unit 46   *Opener:* © Steven Kazlowski/Science Faction/Corbis. Figure 46.2A, 46.2B: USGS. Figure 46.4: © Murray, Patti/Animals Animals. Figure 46.7: © Stephen J. Krasemann/DRK Photo. Figure 46.8: M.A. Kessler. Figure 46.9: Paolo Koch/Science Source.

Unit 47   *Opener:* © Joe Mason. Figures 47.1, 47.9: © Joe Mason. Figure 47.2: © Warren Faidley/DRK PHOTO. Figure 47.4: © Georg Gerster/Science Source.

Unit 48   *Opener:* H. J. de Blij. Figure 48.3B: Jim Corwin/Science Source. Figure 48.6, 48.11A, 48.11B: H. J. de Blij. Figure 48.7B: Peter Chadwick/Science Source. Figure 48.9: Georg Gerster/Science Source.

Unit 49   *Opener:* H. J. de Blij. Figure 49.3: © Don and Pat Valenti/ DRK PHOTO. Figure Figure 49.5: Photo by G.M. Ashley, Rutgers University, New Brunswick, NJ. Figure 49.6: Science Source/Getty Images. Figure 49.8: © Bettmann/Corbis. Figure 49.10: H. J. de Blij.

INDEX

A. *See* tropical climates
ablation, ice, 524–525, *525*
above mean sea level (AMSL), 30
abrasion, 476, 527, 528, *529*, 563–564
abyssal plains, 22, *23*
accretion, 394
accumulation, ice, 524–525, *525*
accuracy, mapping, *31*
acidity, 285–286
acid precipitation, 220
acid rain, 221–222, *222*
actual evapotranspiration (AE), 141, 145
addition, 276, *276*
adiabatic, 87
adiabatic lapse rates, 87–88
advection, 94, 141
AE. *See* actual evapotranspiration
aerial photography, 40
Af. *See* tropical rainforest zone
Africa, 21, 340–341, *341*, 545
aggradation, 479, *481*, 481–482, *482*,
    503–504
air, 274, *274*
    molecules, movement, and, 99–102,
        100, *101*, *102*
    pollution, 79, *89*, 89–91, *90*, *91*
    stability, 87, *87*, *88*, 88–89, *89*
air masses
    in atmosphere, 151
    classifying, 151
    key terms for, 162
    movements of, 151
    objectives regarding, 150
    review questions on, 163
    source regions and, 151, *152*
    thunderstorms of, *155*, 155–156
air pressure
    air molecules forces and, 99–102, *100*,
        101, *102*
    concept of, 98, *98*
    Coriolis force in, 101, *101*, *102*
    in daily life, *104*, 104–105
    frictional force in, 101–102
    gravitational force in, 98
    key terms, 107
    large-scale wind systems and, *102*,
        102–103
    objectives for, 97
    pressure-gradient force in, 99–101, *100*
    review questions, 107
    small-scale wind systems and, 103, *105*,
        105–107, *106*
    winds and, 97–107
Airy, George, 389–390
albedo, 59

Aleutian subpolar lows, 115
Alfisols, 293t, 295–296, *296*, *301–302*
alkalinity, 285–286
alluvial fans, *497*, 497–498, *498*
alluvium, 473, 498, 500
altitude
    atmospheric pressure and, 98–99, *99*
    climates of high, 216, 223–225, *225*, *226*,
        227–228
    latitude, vegetation, and, 321, *321*, 323
    solar, 52
Am. *See* monsoon rainforest climate
amensalism, 315–316
amplification, earthquake, 413
AMSL. *See* above mean sea level
Andisols, 295, 299–300, *301–303*
angle of repose, 450–451
animals, *334*, 334–335, *335*, 336,
    340–342, *341*
annual cycles, 92, *92*
annular drainage, 489, *491*
Antarctic
    Circle, 47
    Ice Sheet, 532–535, *533*
Antarctica, 21–22
antecedent streams, 491, *492*
anthropocene era, 243
anthropogenic
    climate change, 243
    emissions, 252–254, 252t, *253*, *254*
anticlines, *427*, 427–428, *428*
anticyclone, 103, *103*
aphelion, 46
aquicludes, 465–466, *466*
aquifers, 465–466, *466*
Ar. *See* argon
Arctic Circle, 47
areal coverage, 193t, 201t, 205t, 218t, 220t
argon (Ar), 74–75
arid climates (B)
    areal coverage in, 201t
    BS among, 198–201
    BW in, 198
    in climate type regional distribution,
        186–188
    key terms for, *203*
    by latitude, *197*
    major, 197–202, *200*, *202*
    objectives for, 191
    review questions for, 203
    Sahara in, 198, *198*
Aridisols, 293, *294*, 295, *301–303*
artificial levees, 501, *501*
Asia, 547, *549*
asthenosphere, 372

Atlantic ocean currents, 128
atmosphere
    air masses in, 151
    air movement in, 99–102, *100*, *101*, *102*
    circulation systems in, 67
    climate and, 69
    in Earth system spheres, 16, 17–18, *18*
    lower, 85–96
    potential energy in, 175–177, *176*
    solar radiation in, 59, *59*
    system of ocean and, *122*, 122–123, *123*
    upper, 117–119, *118*, *119*
    See also circulation patterns,
        atmospheric
atmospheric
    circulation, 99
    constituents, 69, 69–70, 72, 72–73
    content, 73–79, *74*, *75*, *76*, *78*
    moisture, 134–149
    pressure, 98, 98–99, *99*
    rivers, 146–147, *148*
atmospheric structure
    atmospheric content, arrangement,
        and, 73–79, *74*, *75*, *76*, *78*
    constituents of, 69, 69–70
    evolution, origin, maintenance, and, 70,
        72, 72–73
    key terms, 83–84
    as layered, 79–81, 83
    mesosphere in, 80, 81
    objectives for, 68
    particulates in, *71*, 71–72, 71t, *73*, 79
    review questions, 84
    stratosphere in, 79–81
    temperature inversions in, 80, *80*
    thermosphere in, 80, 81, 83
    troposphere in, 79, 80
    water vapor in, 70, 72, 76–77
    weather and, 69
atolls, 591, *591*
attenuation, earthquake, 413
aurora borealis, 81, *81*, 83
Australia, 21t, 22, *212*, 212–213
avalanche, 453, *453*
    See also mass movement
Aw. *See* savanna climate
axis, 45–47, *47*
axis tilt, 46–47, *47*

B. *See* arid climates
backshore, 584, *584*
backwash, 574
bajada, 497–498
barchan dunes, 566, *568*
barometer, 98, *98*

as soil component, 273–274, *274*
types of, 349–350
models, 5, 9–10, *10*, 36
Modified Mercalli Scale, 413, *413*
Mohorovičić discontinuity, 370
Mollisols, 293, *294*, 295, *301–303*
monadnock, 485
monsoon rainforest climate (Am), *195*, 195–196
monsoons, 115–117, *116*, *117*, *195*, 195–196
moraines, 539, *539*, *552*, 552–553
mountain breeze, 105
mountain glaciers, 522, *522*
in Africa, 545
in Asia, *547*, *549*
depositional landforms of, *552*, 552–553
erosion and, *549*, 549–552, *550*
in Europe, 546–547, *548*
fjords in, 551–552
glacial troughs in, *549*, 549–550, *550*
global distribution of, 545–547, *546*, *547*, *548–549*
global warming industrial temperature and, *246*
high-mountain, 550–551
key terms for, 553
lakes in, 551
in New Zealand, 545–546, *548*, *551*
in North America, 545, *547*
overview of, 545
review questions on, 553
in South America, 545
from space, 546, *546*
mountains
glaciers in, 522, *522*
in high-altitude H climates, 223–225, *226*
isostasy involving, 388–391, *389*, *391*
mountain/valley breeze systems, 105, *106*
mP. *See* maritime polar
mT. *See* maritime tropical
mutations, 336
mutualism, 316

N$_2$. *See* nitrogen
NAO. *See* North Atlantic Oscillation
natural
landscapes, *24*
levees, 501, *501*
processes, *24–25*
selection, 332, 336
nearshore, 584, *584*
negative feedbacks, 8, 8–9
net primary productivity, 308, *308*
net radiation, 63, 63–64, *64*
New Hampshire, *39*
New Zealand, 545–546, *548*, *551*
nitrogen (N$_2$)
as constant gas, 74
denitrification in cycle of, 266, *266*, *267*
fertilizer and cycle of, 268–269, *269*
fixation, 265–266
in organic compounds, 262

reactive, 268
soil, vegetation, and global cycle of, 265–266, *266*, *267*, 268
nitrous oxide, *251*, 262
nonclastic sedimentary rocks, 357–359
nonrenewable resources, 272–273
nonspherical Earth, 15–16, *16*, *17*
normal faults, 424–425, *425*
North America
continental glaciation in, 536–537, *537*, 540
landmass of, 21
mountain glaciers in, 545, *547*
vegetation regions of, 321–323, *322*
weather systems of, *171*
North Atlantic Oscillation (NAO), 132, *132*
northern coniferous forest biomes, 329, *329*
Northridge earthquake, 385, *385*, 386

O$_2$. *See* oxygen
O$_3$. *See* ozone
obsidian, *348*
ocean-atmosphere system, *122*, 122–123, *123*
ocean circulation
climate and, 129
ENSO in, 129–131, *130*, *131*
key terms of, 133
NAO in, 132, *132*
objectives for, 121
ocean-atmosphere system in, *122*, 122–123, *123*
ocean current generation in, 123–124, *124*
ocean current geography in, 126, *127*, 128, *129*
ocean currents as, 122
review questions, 133
wind-driven flow in, 124–126, *126*
ocean hemisphere. *See* water
oceanic-continental plate convergence, 383, *383*
Oceanic Niño Index (ONI), 129–131, *130*, *131*
oceanic-oceanic plate convergence, 383, *384*
oceans
basins in, 22, 22–24, *23*
circulation systems in, 67
currents of, 122–124, *124*, 126–128, *127*, *129*
in Earth system, 20, 22, 22–24, *23*
lithosphere, *378–379*, 378–380
temperature zones of, 122, *122*, *123*
offshore bar, 586
oilfields, 358, *358*
ONI. *See* Oceanic Niño Index
open systems, 7
orders of magnitude, *10*, 11
Oregon, United States, *161*, *162*
organic
compounds, 261–262, *262*
matter, 274, *274*

organisms, living, 261, 274–275, *275*
orogenic belts, 375
orographic precipitation, 151, 159, 161, *161*, *162*
outer core, of Earth's interior, 369, *369*
outwash, 539, 541
outwash plain, 541
oxbow lake, 498, *499*
oxidation, 445, 445t
Oxisols, 295, 296–298, *297*, *301–303*
oxygen (O$_2$), 74, *74*, 262, 445, 445t
ozone (O$_3$), 77–78, *78*, 80–81, *82–83*
ozone layer, 80–81, *82–83*

P. *See* phosphorus
Pacific
Northwest, 418–419, *419*
ocean currents, 126, 128, 129–131, *130*, *131*
Ring of Fire, 381
pacific highs, 114–115
pack ice, 535, *535*
pahoehoe, 402
Panama Canal, *193*
Pangaea, 377, *377*, 392–393
parabolic dunes, 568, *568*
parallelism, 46
parallels, 29, *29*, 33
parent material, 274
particulates, *71*, 71–72, 71t, *73*, 79
patterned ground, 559, *560*
PE. *See* potential energy; potential evapotranspiration
pedology, 261
pedon, 282, *282*
penaplain, 493
periglacial environments
geomorphic processes in, 556–558, *557*, *558*
humans in, 556, *556*
ice wedges in, 558–559, *559*
key terms for, 561
landforms of, 558–560, *559*, *560*
objectives for, 554
patterned ground in, 559, *560*
permafrost and, 555, *555*, *557*, *558*
pingos in, 559–560, *560*
review questions on, 561
understanding, 560–561
perihelion, 46
permafrost, 218, 555, *555*, *557*, *558*
permeability, 461
perpetually moist climates (Cf)
areal coverage of, 205t
Cfa in, 205–206, *206*, *207*, *208*, 301
Cfb, Cfc, and, 206–207, *209*, *210*, 210–211
global distribution of, *206*
as mild midlatitude C climate, 205–207, *207*, *208–209*, *210*, 210–211
temperate forest biomes in, 327, *327*
Phoenix, Arizona, *3*

closed, 7
equilibrium in, 8
feedbacks in, 8
geographic coordinate, 30
glacial, 523–525, *524, 525*
global positioning, 38–40, *39*
interfaces in, 7–8
large-scale wind, 102–103, *103*
nature of, 6
ocean-atmosphere, *122,* 122–123, *123*
open, 7
in physical geography, 5–9, *6, 7, 8, 9*
small-scale wind, 103, *105,* 105–106, *106*
soil development, 287–288, *288*
South Florida Atlantic coast in, *7*
streams as, *480,* 480–482, *481*
subsystems within, 6–7
See also circulation systems; Earth
system; fluvial system landforms;
weather systems

tambora, 406
tectonic activity, 489
temperate forest biomes, 327, *327*
temperate grassland biomes, 326
temperate karst landform, 511
temperature
advection in differences of, 94
annual cycles in, *92, 92*
climate change and extremes in, *247*
dew-point, 136
diyurnal cycles in, 92
in geographic dispersal, 313
glacial movement and, 527
global variations in, 94–95, *95*
gradient, 95
greenhouse gas emissions and impact
on, 250–251, *251*
heat compared to, 86–87
horizontal distribution of, 91–95, *92,
93, 94, 95*
inversions, *80, 80, 89,* 89–91, *90, 91*
key terms, 95–96
kinetic energy measured by, 86
land, water, and differences in, 92–94
latitude and differences in, 92–94, *93, 94*
of lower atmosphere, 85–96
nature of, 86–87
objectives for, 85
ocean zones of, *122, 122, 123*
overview of, 86, *86*
review questions, 96
of severe midlatitude D climates,
217–218
thermometers for measuring, 86
in troposphere, *87,* 87–91, *88, 89, 91*
tropospheric, *87,* 87–91, *88, 89, 91*
vapor pressure and, 136, *136*
See also climate change, human
impacted; climate change, natural;
climograph; global warming
temperature change

from anthropogenic emissions,
252–254, *252t, 253, 254*
global warming and industrial period,
244–245, *245, 246–247,* 247–249, *248*
tensional faults, 424–426, *425, 426*
terminal moraines, 539, *539*
terraces, 501, *503,* 503–504, *504*
terracing, 273, *273*
terranes, 393–394, *395*
terrestrial
geography, 2
radiation, 57, 59–61, *61,* 63
terrestrial biomes
Af, 324
BW, 325–326, *326*
in global plant distribution, 324–327,
325, 326, 327, 328, 329, 329
major, *323*
Mediterranean scrub, 327, *328, 329*
northern coniferous forest, 329, *329*
polar E, EF, and, 329
principal, 324–327, *325, 326, 327, 328,
329, 329*
temperate forest, 327, *327*
temperate grassland, 326
tropical Aw, 324, *325*
theory, 5
thermal energy, 56
thermocline, 122
thermohaline circulation, 128
thermometers, 86
thermosphere, 80, 81, 83
thrust fault, 423–424, *424*
thunder, 158–159
thunderstorm-related weather
phenomena, 158–159
tides, 578–580, *579, 580*
till, 537–538
time measurement, 50–51
timescale, *306, 362,* 362–363
tombolo, 586, *586*
topography, 15–16, *16, 37,* 275–276,
372, 373
tornadoes, 158–160, *160*
trace gases, 70
trade winds, 110
transformation, 276, *276*
transform plate boundaries, 383–384,
384, 386
translocation, 276, *276*
transpiration, 137, *137*
transportation, 477, *477, 564,* 575–578,
576, 577
transport capacity, stream, 477, 478
transverse
dunes, 567–568, *568*
faults, 426, *426*
trellis drainage, 489, *491*
TRMM. *See* Tropical Rainfall Measuring
Mission satellite
trophic level, 308
Tropic

of Cancer, 46–47
of Capricorn, 47
tropical
cyclones, *166,* 166–171, *167, 168, 168t,
169, 170*
deforestation, *195,* 195–196
depressions, 165
gyre, 128
jet, 111
karst landform, 511, 513, *515*
savanna biomes, 324, *325*
storms, *166,* 166–171, *167, 168, 168t,
169, 170*
tropical climates (A)
Af, *192,* 192–195
Am, 195–196
areal coverage of, *193t*
in climate type regional distribution,
185–186
key terms for, 203
by latitude, *193*
major, *192,* 192–197
objectives for, 191
review questions for, 203
water balance, heat balance, and, *194*
Tropical Rainfall Measuring Mission
Satellite (TRMM), *153*
tropical rainforest zone (Af), *192,* 192–195,
194, 324
tropopause, 79
troposphere, 79–80, *87,* 87–91, *88, 89, 91*
truncated spurs, 549
tsunamis, 413–414, *414, 417*
tundra (EF), 220, *220t,* 223, *224, 226,* 329

ultimate base level, 479
Ultisols, 295, 296, *296, 301–303*
unconformity, 360
United States (U.S.)
acid rain in, *222*
air pollution in, 91, *91*
animal species conservation in, 341
climate change and temperature
extremes in, *247*
hurricanes in, 168–169, *169, 170,*
170–171
loess deposits in, 569, *569*
mapping of, 37–38, 300–301, *301*
mass movement risk in, *458*
midlatitude cyclones in, *173*
Soil Taxonomy of, 300–301, *301*
tornadoes in, 159–160, *160*
water usage, 139–140, *140*
upwelling, 125–126, *126*
U.S. *See* United States

vadose zone, 465, *465*
valley breeze, 105
See also mountain/valley breeze systems
valleys, 286, 486, *487*
See also river valley landforms
valley soil, 286

vapor pressure, 136, *136*
variable gases, *75*, 75–79, *76*, *78*
vegetation
   carbon global cycle and, 263–265, *264*
   in infiltration, 462
   in karst landforms, 511
   latitude, altitude, and, 321, *321*, 323
   lithospheric surface water for, *461*
   N_2 global cycle and, 265–266, *266*,
      267, 268
   North American regional, 321–323, *322*
   shoreline evolution impacted by,
      591–592
   in soil formation, 275, *275*
   soil in relation to, 262–266, *263*, *264*,
      266, *267*, 268–269, *269*
velocity, stream water flow, 463, *464*
vents, 399
vertical
   datum, 30
   position, 30
   zonation, 223–224
Vertisols, 295, 299, *300*, *301–303*
vicariance, 316
volcanic activity, 398–399, 405–407, *407*
volcanic landforms, 399, *488*
   basalt plateaus, basalt plains, and,
      402, 404
   calderas as, *400*, 400–401
   cinder cones as, 402, *403*
   composite volcanoes as, *401*, 401–402
   shield volcanoes as, 402, *403*, *404*
volcanism
   atolls, coral reefs, and, 591, *591*
   in Hawai'i, 402, *404*, 404–405, *405*, *406*
   hot spots of, 404–405, *406*
   in human history, 405–407, *407*
   key terms for, 408
   landforms of, 399–404, *400*, *401*, *403*,
      404, *488*
   landscapes of, 407–408, *408*
   magma properties involved in, 399, *399*,
      400
   objectives for, 397
   overview of, 398, *398*
   review questions on, 409
   volcanic activity distribution in,
      398–399

Wallace's Line, 332, *333*, 334
warm
   currents, 125
   fronts, 154
warming hiatus, 244
water (H_2O)
   atmospheric moisture and balance of,
      134–149
   carbonation involving, 444
   chemical weathering involving rock
      and, 443–444
   cloud condensation, 141–142

evaporation of, 138–141
gap, 491
geographic dispersal and availability of,
   313–314
heat differences for land and, 92–94
hemisphere, 20, *20*
hydrologic cycle for, 136–138, *137*, *138*
hydrolysis involving, 445, *445t*
in karst landform development, 510–511
key terms for, 148
landscape, landforms, and running,
   434, *434–435*
in lithosphere, 460–467, *461*, *462*, *463*,
   464, *465*, *466*
objectives regarding, 134
in photosynthesis, 306
physical properties of, 135, *135*
as precipitation, 142–145, *143*, *144*
resources, *3*, *4*
review questions, 149
in slope erosion, 472, *473*
slope failure involving, 453
as soil component, 274, *274*
surface, 145–148, *146*, *147*, *148*, *461*,
   461–464, *462*, *463*, *464*
table, 465–467
usage, U.S., 139–140, *140*
waves in open, 573, *573*
water balance
   atmospheric rivers and, 146–147, *148*
   BW, *199*
   Cfa, *207*
   condition range, 145, *146*
   Cs, 211, *211*
   global warming industrial temperature
      change and, *246*
   human population in relation to,
      147–148
   latitude and variations in, 145–146, *146*
   surface, 145–148, *146*, *147*, *148*
   in tropical A climate, *194*
water vapor (H_2O)
   atmospheric river of, *148*
   in atmospheric structure, 70, 72, 76–77
   dew in, 136
   GCM projections and feedback of, *255*,
      255–256, *256t*
   humidity and, 136
   measuring, 135–136, *136*
   saturated air and, 136
   vapor pressure in, 136, *136*
   as variable gas, 76–77
   as water property, 135, *135*
wave-cut platform, 589, *589*, *591*
waves
   coastal landforms impacted by,
      584–585, *584–585*, 587, 589,
      589, 591
   in coastal processes, *573*, 573–574, *574*,
      575, 575–578, *576*, *577*
   corrasion of, 575–576

erosion, sediment transportation, and
   action of, 575–578, *576*, *577*
height of, 573, *573*, *574*
length of, 573, *573*, *574*
in longshore drifts, 578, *579*
in open water, 573, *573*
of oscillation, 573
refraction of, 574
against shore, 573–574, *575*
in shore-zone currents, 577–578, *578*
weather, 69, *173*, 173–174, *247*, 247–249, *248*
weathering
   biological, *442*, 446
   chemical, *442*, *443–444*, 443–446, *445*,
      445t, *446*
   erosion in, 433
   feedback, *233*, 233–234, *234*
   geography of, 446–448, *447*
   key terms for, 448
   landscape, landforms, and, 433, *433*
   mechanical, 441–442, *442*
   objectives concerning, 440
   overview of, 441
   processes, 440–448
   review questions, 448
   spheroidal, 443–444, *445*, *446*
weather systems
   energy and moisture within, 174–177, *176*
   key terms for, 177
   low-latitude, 165–171, *166*, *167*, *168*,
      168t, *169*, *170*
   middle latitudes, higher latitudes, and,
      171, 171–174, *172*, *173*
   of North America, *171*
   objectives concerning, 164
   review questions for, 177
   storms as, 165
Wegener, Alfred, 377–379, 392–393
wells, 467–468, *468*
westerlies, 110–111
Western hemisphere, 20
wetness, 38, *39*
wind
   erosion, 563–564, *565*, 565
   flow, 125
   transportation, 564, *564*
wind-driven flow, 124–126, *126*
winds
   air pressure and, 97–107
   in atmospheric pressure, 98
   chinook, 107, 161–162
   in daily life, *104*, 104–105
   foehn, 151
   geomorphic, 562–571, *563*, *564–565*,
      566, *569*, *570*
   geostrophic, 102–103
   katabatic, 105–106
   key terms, 107
   landscape, landforms, and, 435, *435*
   large-scale systems of, *102*, 102–103, *103*
   in leeward locations, 104